Microsoft 365 Mastery Bible

The Complete Illustrated Guide to Word, Excel, PowerPoint, Outlook & the Full Office Suite

Gally Renders

Disclaimer and Terms of Use

The author and publisher of this book and the accompanying materials have used their best efforts in preparing this book. The author and publisher make no representation or warranties with respect to the accuracy, applicability, fitness, or completeness of the contents of this book. The information contained in this book is strictly for informational purposes. Therefore, if you wish to apply the ideas contained in this book, you are taking full responsibility for your actions.

Neither the author nor the publisher shall be held liable for any loss or damages arising directly or indirectly from the use of the information contained herein.

Non-Affiliation Disclaimer

This publication is **not affiliated, associated, authorized, endorsed by, or in any way officially connected with Microsoft Corporation**, or any of its subsidiaries or affiliates.

The names *Microsoft*, *Office 365*, *Word*, *Excel*, *PowerPoint*, *Outlook*, *Access*, *Publisher*, *OneNote*, and *OneDrive* are registered trademarks of **Microsoft Corporation**. All product names, logos, and brands are property of their respective owners and are used in this book **for identification purposes only**.

Printed in the United States of America

CONTENTS

OVERVIEW OF OFFICE 365

Office 365 Launched in 2001 is a cloud-based, subscription version of Microsoft's popular suite Microsoft Office. Office 365 comprises the same essential applications as traditional versions of Office, including Word, Excel, Access, PowerPoint, OneNote, Outlook and others (depending on the plan purchased). Other inclusive apps are Publisher, OneDrive, SharePoint, Skype, Microsoft Teams, etc.

Features of Office 365

1. Explore and Integrate: Office 365 Integrates flawlessly with Microsoft Office and other Microsoft programs. It takes just a few minutes to create a trial account that users can test and integrate with the cloud.

2. Collaborate Anywhere and at Anytime: With Office 365, users get web-enabled access to documents, emails, contacts, and calendars irrespective of whichever device or browser they use. This accessibility enables integrated collaboration among employees.

3. Stay Up and Running AT ALL TIMES: By running your applications in the cloud and storing your data thereon, this will ensure all your business documents and applications, such as emails and documents, will always be available – even if your company experiences any disaster.

4. Flexibility: With several options in the Office 365 suite, enterprises buy only the office solutions they need and convert CAPEX into OPEX.

5. Single Sign-on (SSO) – Users can store their password in the system tray of their machine to avoid having to repetitively sign in.

6. Support: Users receive business-class service including IT admin phone support, a financially-backed 99.9% uptime guarantee, geo-redundant data centers, robust security, disaster recovery, privacy controls, and standards and lots more.

7. Security: Office 365 has built-in security and constant compliance. You always know where your data is and who can access it. You can also remotely wipe all data from your mobile device if it's stolen or lost.

8. No Licensing Issues: Office 365 includes all licensing and it can be deployed company-wide. This means everyone has the same version of Microsoft Office, and all users receive upgrades in unison.

9. Organization: When you use Office 365, email, calendar and contacts are all synchronized. For example, if you update a contact on your mobile phone, it will be automatically updated across all your devices. You can also access the same data from any device or save a document on your PC and continue working on it, on your smartphone or tablet.

10. Real-time Collaboration: Teams can collaborate with Microsoft Word, Excel and PowerPoint. Each user can see instantaneous edits or additions their colleagues are making to a document. It also saves each version robotically, so you can return to older versions easily. This ensures a quicker and simpler collaboration between employees and departments, hence, resulting in improved productivity.

Apps Integrated into Office 365 for Business

What's included in Office 365 for Business depends on which subscription level the user chooses, but the essential apps and services available through Office 365 are:

- Word
- Excel
- PowerPoint
- OneNote

- Outlook
- Exchange
- Publisher
- SharePoint
- Microsoft Teams
- Skype for Business
- OneDrive for Business

Some Office 365 applications also have mobile versions to help business do more on the go, including Word, Excel, PowerPoint, OneDrive for Business, Skype for Business, Yammer, and OneNote. In addition, Office 365 subscriptions also consist of 1 TB of OneDrive for Business cloud storage for individual user.

Office 365 for Business pricing plans for Small and Medium Businesses (SMBs)

	Office 365 Business Essentials	**Office 365 Business**	**Office 365 Business Premium**
Apps included	*Web versions* of Word, Excel, and PowerPoint	*Desktop versions* of Outlook, Word, Excel, PowerPoint, OneNote, Access, and Publisher *Web versions* of Word, Excel, and PowerPoint	*Desktop versions* of Outlook, Word, Excel, PowerPoint, OneNote, Access, and Publisher *Web versions* of Word, Excel, and PowerPoint
Services included	Exchange SharePoint Skype for Business Microsoft Teams 1TB of OneDrive file storage and sharing Microsoft Planner 24/7 phone and web support	1TB of OneDrive file storage and sharing 24/7 phone and web support	Exchange SharePoint Skype for Business Microsoft Teams Email Hosting 1TB of OneDrive file storage and sharing Microsoft Planner Microsoft StaffHub Microsoft Connections Microsoft Listings Outlook Customer Manager Microsoft Invoicing Microsoft Bookings MileIQ Business center 24/7 phone and web support
Email hosting	50 GB mailbox and custom email domain address	None	50 GB mailbox and custom email domain address
Device licensing	Services for use on 5 phones and 5 tablets only	Office apps and services on 5 phones, 5 tablets, and 5 PCs or Macs per user	Office apps and services on 5 phones, 5 tablets, and 5 PCs or Macs per user

Office 365 for Business Pricing Plans for Enterprise

	Office 365: ProPlus	**Office 365: Enterprise E1**	**Office 365: Enterprise E3**	**Office 365: Enterprise E5**
Apps included	*Desktop versions* of Outlook, Word, Excel, PowerPoint, OneNote, Access and Publisher *Web versions* of Word, Excel, and PowerPoint	*Web versions* of Word, Excel, and PowerPoint	*Desktop versions* of Outlook, Word, Excel, PowerPoint, OneNote, Access and Publisher *Web versions* of Word, Excel, and PowerPoint	*Desktop versions* of Outlook, Word, Excel, PowerPoint, OneNote, Access and Publisher *Web versions* of Word, Excel, and PowerPoint
Services included	1TB of OneDrive file storage and sharing 24/7 phone and web support	Exchange SharePoint Skype for Business Microsoft Teams Yammer Microsoft StaffHub Microsoft Planner 1TB of OneDrive file storage and sharing 24/7 phone and web support	Exchange SharePoint Skype for Business Microsoft Teams Yammer Microsoft StaffHub Microsoft Planner Unlimited personal cloud storage 24/7 phone and web support	Exchange SharePoint Skype for Business Microsoft Teams Yammer Power BI Pro Microsoft StaffHub Microsoft Planner Office 365 Cloud App Security Advanced Data Governance Customer Lockbox Exchange Online Advanced Threat Protection Unlimited personal cloud storage 24/7 phone and web support
Email hosting	None	50 GB mailbox and custom email domain address	100 GB mailbox and custom email domain address	100 GB mailbox and custom email domain address
Device licensing	Office apps and services on 5 phones, 5 tablets, and 5 PCs or Macs per user	Office apps and services on 5 phones, 5 tablets, and 5 PCs or Macs per user	Office apps and services on 5 phones, 5 tablets, and 5 PCs or Macs per user	Office apps and services on 5 phones, 5 tablets, and 5 PCs or Macs per user

Office 365 for Business Cost

Plan	Cost per user/per month
Business Plans	
Office 365 Business Essentials	$5.00
Office 365 Business	$8.25
Office 365 Business Premium	$12.50
Enterprise Plans	
Office 365 ProPlus	$12.00
Office 365 Enterprise E1	$8.00
Office 365 Enterprise E3	$20.00
Office 365 Enterprise E5	$35.00

System Requirements for Office 365 for Business

Computer and processor	• PC: 1 gigahertz (GHz) or faster x86-bit or x64-bit processor with SSE2 instruction set • Mac: Intel processor
Memory	• PC: 2 GB RAM • Mac: 4 GB RAM
Hard disk	• PC: 3.0 GB of available disk space. • Mac: 6 GB of available disk space. HFS+ hard disk format (also known as Mac OS Extended or HFS Plus).
Display	• PC: 1024 x 768 1024 by 768 screen resolution. • Mac: 1280 x 800 1280 by 800 screen resolution.
Graphics	Graphics hardware acceleration requires a DirectX 10 graphics card.
Operating system	• PC: Windows 10, Windows 8.1, Windows 8, Windows 7 Service Pack 1, Windows Server 2016, Windows Server 2012 R2, Windows Server 2012, or Windows Server 2008 R2.

	• Mac: Mac OS X 10.10 or later.
Browser	Up-to-date version of Internet Explorer, Microsoft Edge, Safari, Chrome, or Firefox.
.NET version	.NET 3.5 required. Some features may require .NET 4.0, 4.5, or 4.6 CLR also to be installed.

How to Create Office 365 Account

Follow the steps below to set up a Microsoft Office 365 account.

1. Open your web browser and log on to the Microsoft Office website **www.office.com**
2. At the top-middle of the screen, click the **Buy Office 365** button.
3. Here, there are three available options: **Office 365 Home**, **Office 365 Personal**, or **Office Home and Student**. Choose the option that best suits your needs. Click **Buy now** for the option that you want.
4. A sign in window will appear. If you currently do not have an account, click on **Create one**.
5. Create an account by entering an e-mail address. If you would rather use a phone number, click **Use a phone number instead**, and then enter your phone number. Click **Next**.
6. Enter a password. *I recommend that you click the **Show password option**, so you can be sure of what you are typing*. Kindly note that the system does not ask you to type in your password twice to confirm it, so make sure you type it correctly. Click **Next**.
7. Enter your first and last name, as requested. Click **Next**.
8. Depending on the method you used to set up this account, Microsoft sends you a verification code to either the **e-mail** or **phone number** that you entered. You will need to access your e-mail or phone to get this verification code. Enter the security code in the Microsoft set up screen.
9. Select or deselect the box, asking if you want information, tips, and offers (the choice is yours). Click **Next.**
10. Enter the characters in the reCAPTCHA.
11. Select payment method. The available options are **credit card** or **debit card**, **bank account**, or PayPal. Enter your correct information accordingly and then click **Save**.
12. Select or deselect the box, asking if you want promotional e-mails from Microsoft. Click **Subscribe**.

13. Once the payment is processed successfully, you will be directed to your Office 365 dashboard. At this stage now, you can install Office on your computer. Click **Install Office** and follow the prompts on the screen to install it. Depending on the speed of your computer and Internet connection, this may take up to 10 or 15 minutes.
14. If you set this up on a computer, Microsoft then asks if you want the free apps for your phone, as well. If you select yes, Microsoft texts you a link to download the apps. Click **Next**.
15. You now have access to all programs in the Office 365 suite. Access them by clicking the Start menu.

BOOK 1

MICROSOFT WORD

CHAPTER ONE

MICROSOFT WORD AT A GLANCE

In this chapter, we'll begin by looking at the Microsoft screen, gaining an overview of everything we'll be looking at on the screen so you can begin to recognize some of the options available to you. You will get an overview of some of the features inside of Microsoft Word and grasp some of the languages we'll be using throughout the book.

The First Step

Before you even think of getting any job done in Microsoft Word, you should know this very important part which is how you can get it up and running on your computer. Without further ado, let's see how you can get Microsoft Word entirely for free. First, you'll see how you can get Word on the web, and secondly, you'll know how to install Word on your pc also entirely for free. Well, let's jump on the PC.

From The Web

You can get Microsoft Word for free on the web by going to the website **office.com**. Once you land on office.com there are three different options. If you already have a Microsoft account, let's say you use skype or you've used office before, chances are you already have an account and you could simply **sign in**.

If you don't have a Microsoft account, don't worry, you can also **sign up** for the free version of Office by clicking on the **sign up for the free version of office** link as shown below and this will prompt you to set up a new Microsoft account.

There's also a third option over here to "**Get Office**". This brings you to a marketing website where you can purchase Office.

So, you go ahead and sign in. Once you click on sign in, **office.com** gives you free access to Microsoft word; not only do you get Microsoft Word, you can also get Excel, PowerPoint, and a whole bunch of other very powerful apps but for this book, we're focused on Word.

Over on the left-hand side to kick off a new Word document you can click on the "**Create**" icon and right up at the top you can start a new Word document.

Also, if you go down here you can click on the Word icon and this drops you to the Word start page. Here too you can also start a new document or you could jump into one of the available templates or browse for more templates.

Once you've been using Word on the web, you'll be able to get back to all of your recent documents down below. Back on office.com, let's say that you have a Word document that you had started on your computer or you started working on it somewhere else and now you want to work on that document on the web. Well, you can easily get a document from your computer or maybe a flash drive and send this to the web. How do you do this? All you need to do is simply get your document and then drag and drop it onto office.com.

You'll see a blue highlight appear and once you release it, here you see that it automatically uploads it for you and if you wait a moment this will automatically open it in Word.

This has now opened this document in word and although you're working on a document using Word online that doesn't mean you can't take that document and get it back onto your computer. Up in the top left-hand corner under the File menu, there's the option to "**Save as**" and when you click on that, you can save a copy online under any name you want.

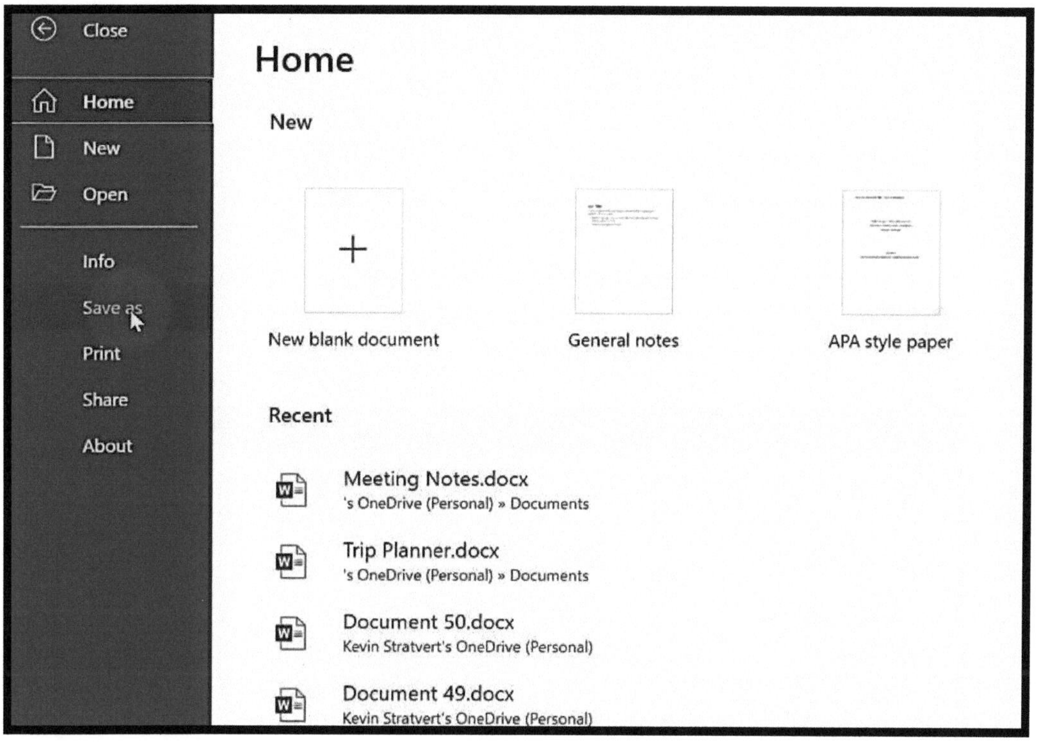

Also, right down below there's the option to download a copy to your computer; you could take a document, upload it, work on it using Word online, and then you can download a copy again to your computer, and not only can you **download a Word document** you could also **get a PDF** and an **ODT**.

You might be wondering why you would ever want to install Office when you could use Word on the web. Well, with Word on the web you have to be online. Also, it's missing some of the functionality that you find in the desktop app but it has the functionality that you're likely going to use most often. For Example, the Table of contents is now available on the web.

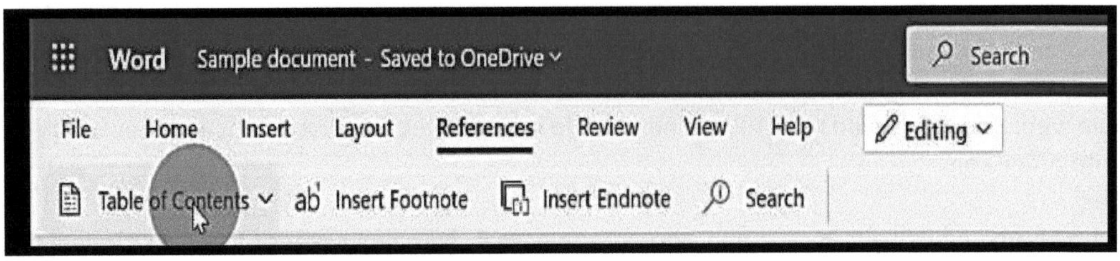

Also, one thing that might surprise you is that the web has some functionality that the desktop version of Word does not have. Here for example, under the home tab if you go over to "**Dictate**", you can take an audio file and have Word online transcribe that for you; that's not something you could do on the desktop.

An even easier way that you can access Word on the web if you're on a Windows 10 or 11 computer is if you have Windows 10 or 11 and it includes an app called **Office**. To access this app simply go down to the taskbar, click on "**Search**" and then type in "**Office**".

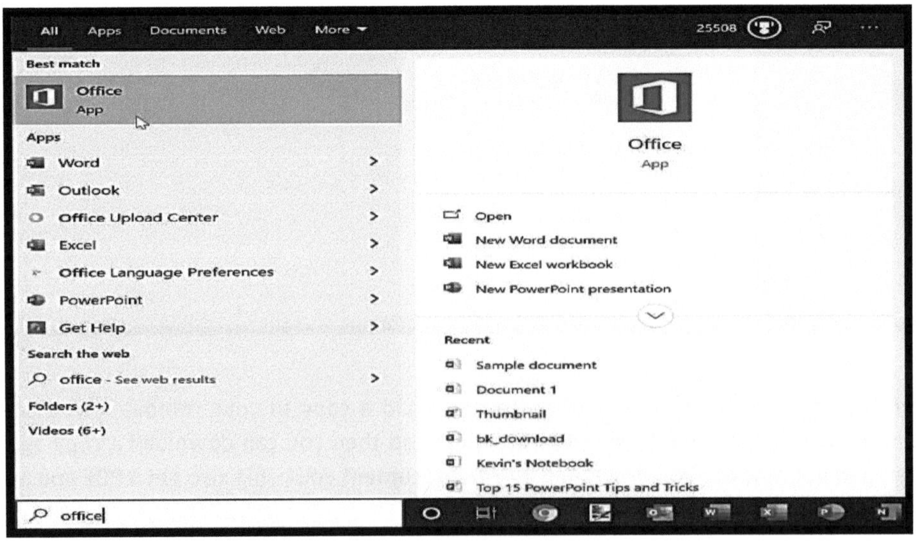

You'll see the best match appear for the Office app, you go ahead and click on that. This opens up the Office app and it looks very familiar to what you just saw on Office.com.

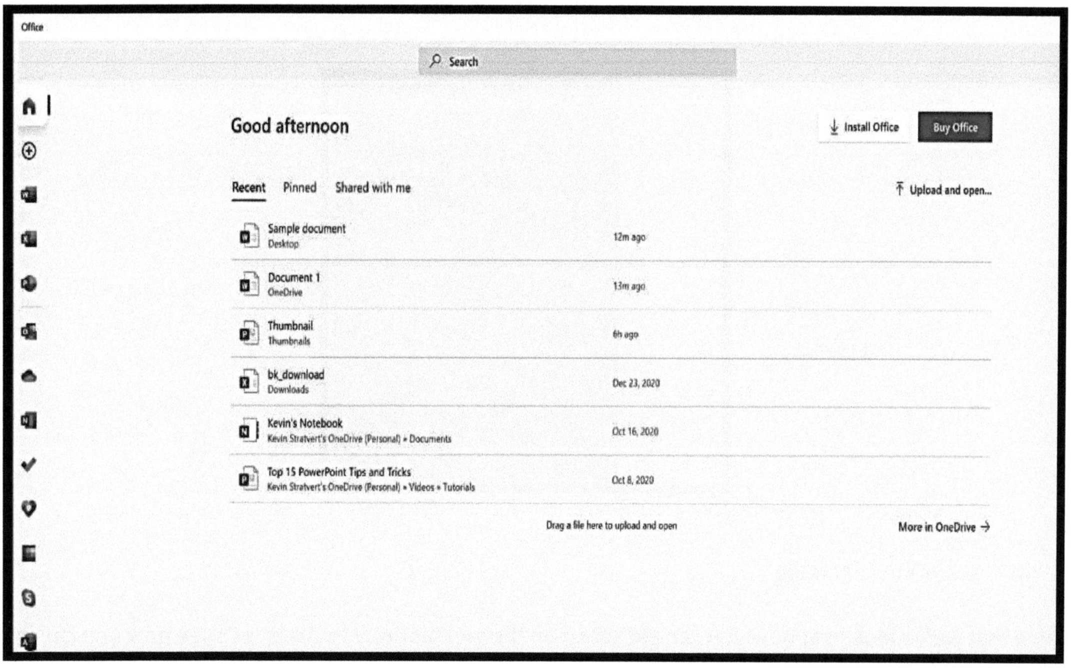

Here too you can access Word the same way. In the top left-hand corner once again, you can click on the "**Create**" icon and here you could kick off a blank new Word document. Also, if you click on the Word icon on the left-hand side this will bring you to the Word start page, and just like before you could kick off a new document, jump into one of these templates and you can also get back to recent documents.

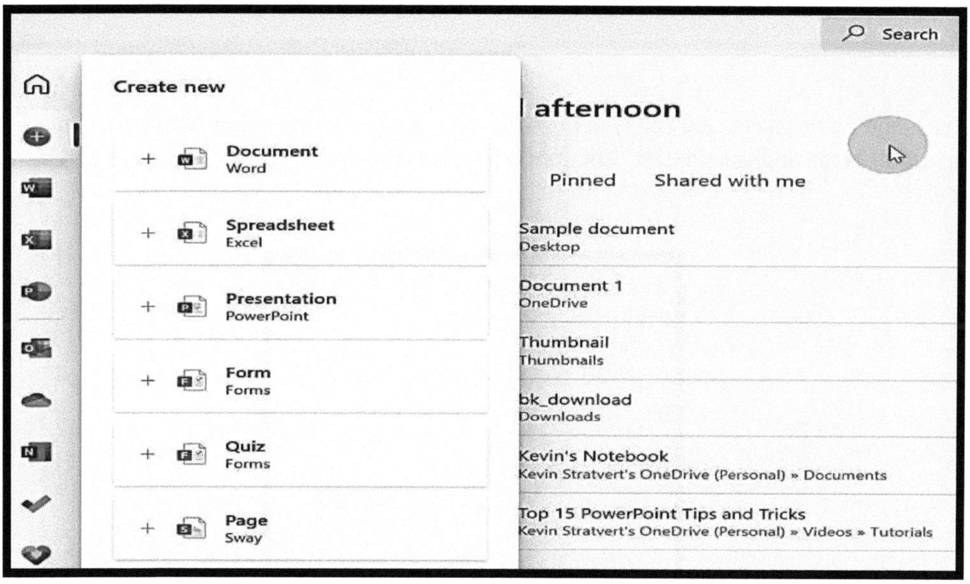

This is one more way to access Word on the web. One of the very nice features is that down here on your taskbar you'll see the Office icon, if you right-click on this you can get back to recent documents, PowerPoints, Excel files, and you can also kick off a new document directly from here so you don't even have to open the app to kick off a new word document.

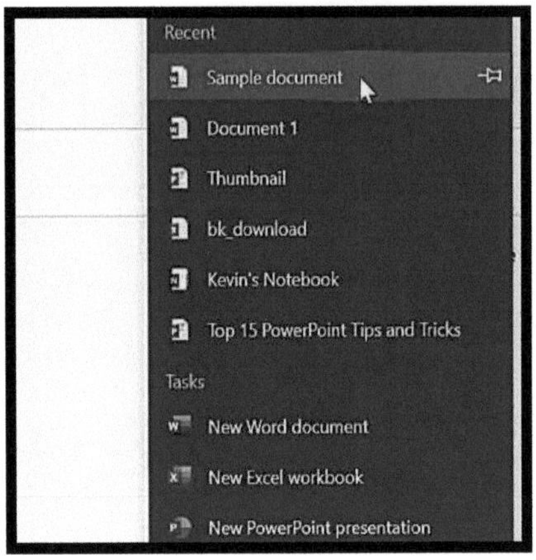

Switching On Your Desktop

Now that we've looked at how you can get Word on the web entirely for free, let's see how you can **install Office** on your desktop. First, let's look at the approach that you can use if you're a student, a teacher, or an administrator at an educational institution. Head over to the URL below.

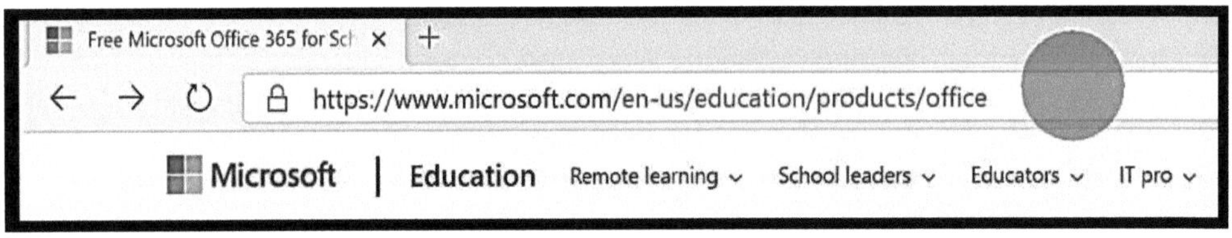

Once you land on this page, you can type in your school email address below, and once you finish typing in your email address click on "**Get started**" to see if you're eligible to download and install Office for free.

After you type in your email address, it'll ask you if you're a student or a teacher, so select the one that you are and when you finish signing up for your account, once again you'll land on the website that we saw earlier and that's office.com.

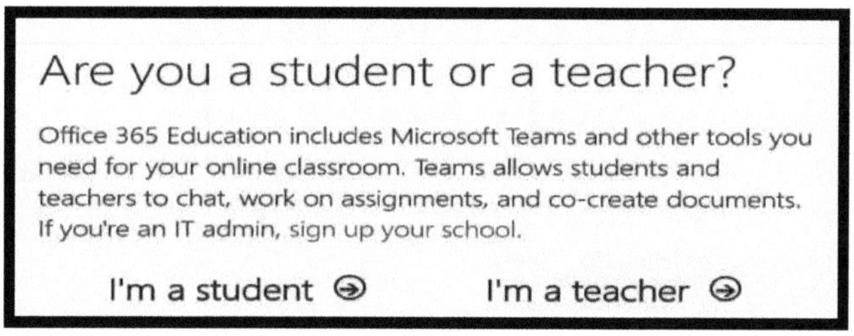

If you're eligible on office.com, in the top right-hand corner you'll see an option to **install Office**. When you click on this you can then install all of the Office 365 apps; it includes things like Outlook, and OneDrive for business, and here you get Microsoft Word.

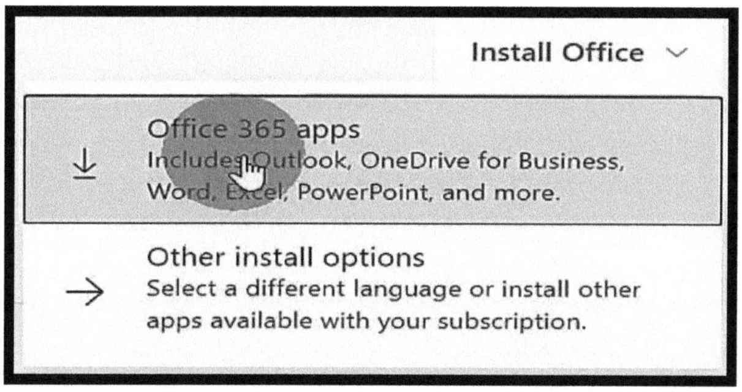

Now if you don't like using Word on the web and you're not a student or a teacher (so you don't qualify for the education deal), you can still get Word on your computer entirely for free and you're going to use something called **Microsoft Rewards** to get Microsoft Word for free.

You might be wondering what Microsoft Rewards are and how this will help you get Microsoft Word.

Well, Microsoft Rewards is a program that Microsoft created to reward you for using Microsoft products, which means the more you use Microsoft products, the more points you get, and then you can redeem different things with those points. For example, you could get Microsoft Word but you could also get all sorts of other things like gift cards from many different retailers.

To get to Microsoft Rewards head to the URL below.

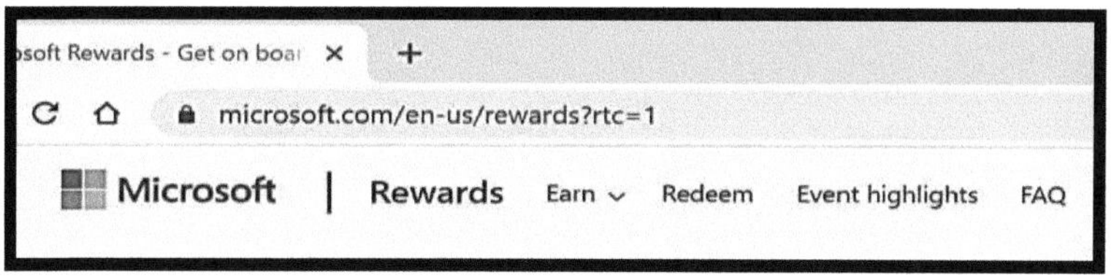

Here you could either sign up for free (if you don't yet have a Microsoft account), but if you already have a Microsoft account you can go ahead and click on sign in.

After you've signed in you might be wondering how you earn points. There are three different ways that you can earn points. The first way to earn points is to search on **Microsoft Bing**; simply head to **bing.com** and then you can type in your search query. The second way to earn points is to search using Bing on your phone; here you're going to type in a word, click on search, and once again you just earned even more points. The third way to earn points is to use the **Edge** browser when you're searching on bing.com. Those are the three different ways that you could earn points and if you want a quick reminder of how you could earn points on the Bing rewards homepage, over on the right-hand side you can click on the option that says Points breakdown.

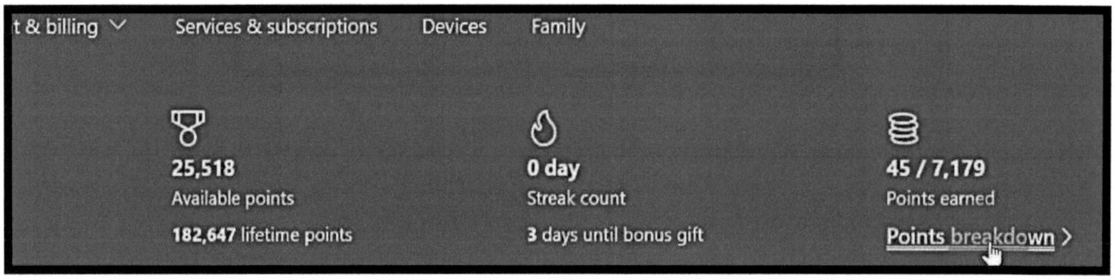

Here you'll see how many points you've earned daily and how many potential points you can earn. You can earn up to 270 points per day or about 120 per year and that's more than enough to buy Microsoft Word.

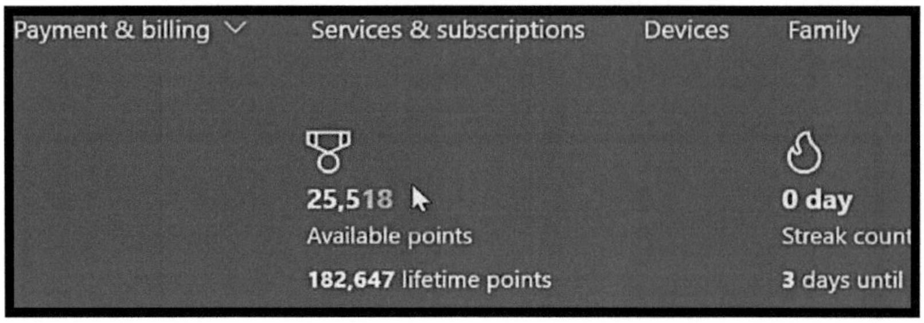

If it feels painful to have to earn all of these points, think of it this way: you likely already search on the web anyway and so if you just make Bing part of your daily habit of searching on your pc and searching on your phone, you're going to start accruing points very quickly and you're going to get word and lots of other things entirely for free.

After accumulating your points over some time, you can redeem your points. On the left-hand side, you'll see a tab that says "**Redeem**".

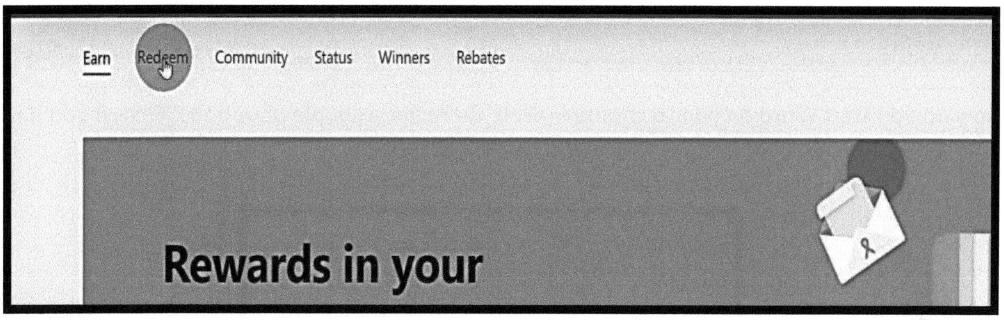

Click on that and within this page, on the left-hand side, click on **Shop**.

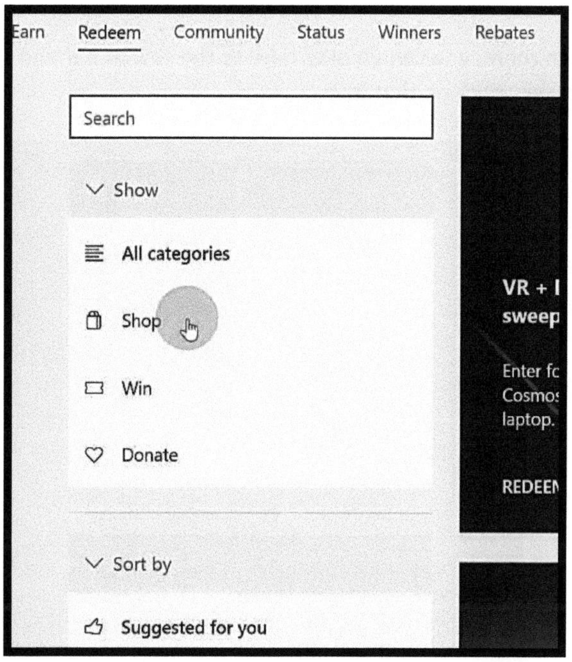

Within Shop you'll see all different types of gift cards that you can redeem with your Microsoft points. Here, for example, you can get an Xbox gift card, you can get Skype credit, and below that, you can shop with different retailers. To get Microsoft word for free you're going to redeem either a gift card at Target or Walmart.

If you're in a market where there is no Walmart and there is no target, there should be local retailers in your market where you can get a gift card, and then you can use that gift card to purchase Office. Once you pick up your gift card with either Walmart, Target, or whoever the retailer is in your local market, you can search for Microsoft 365, and here you can get a personal version of Microsoft 365. This gives you access to Word along with all of the other Microsoft 365 applications. The only downside with Microsoft Rewards is that it's currently not available in all markets. If you're in a market that doesn't support Microsoft Rewards, well, unfortunately, this technique won't work for you but hopefully, some of the other techniques will work and you'll be able to use Word entirely for free.

Starting Word

So how do you start Word on your computer? Well, there are a couple of options. First, if you have the icon already pinned to your taskbar, you can click on that button.

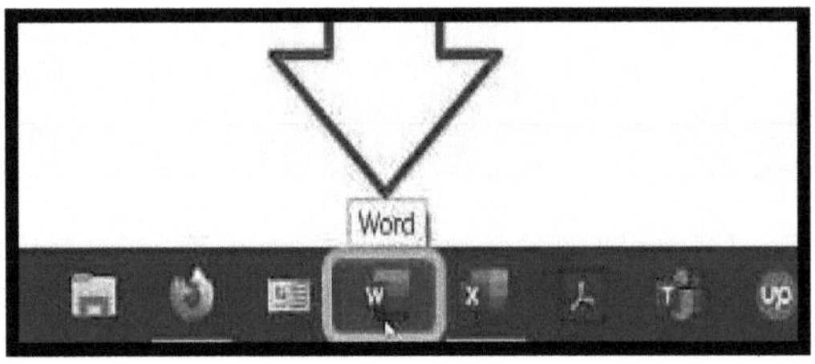

If you don't see this down there, you can go over here to the search bar and type "**Word**", it should appear up there, and you can click on that icon.

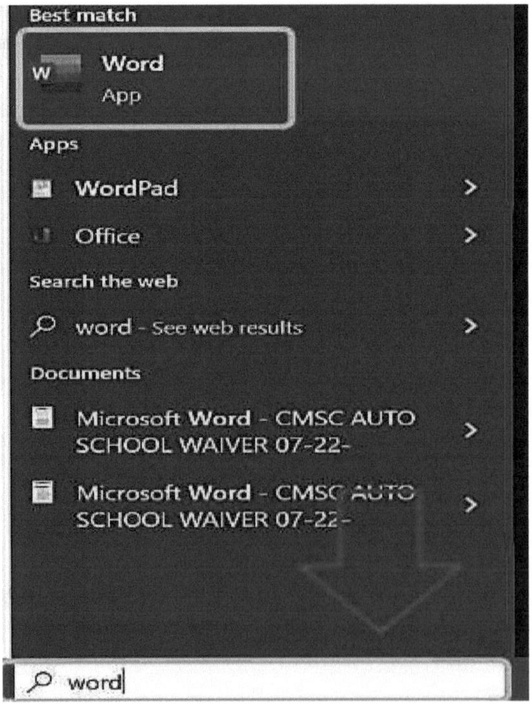

A Blank Document

When you first launch Word, you'll be confronted with the home page, which displays a list of recent documents in the main body of the screen. They're all displayed under here, you can just double-click on any of them to open that document.

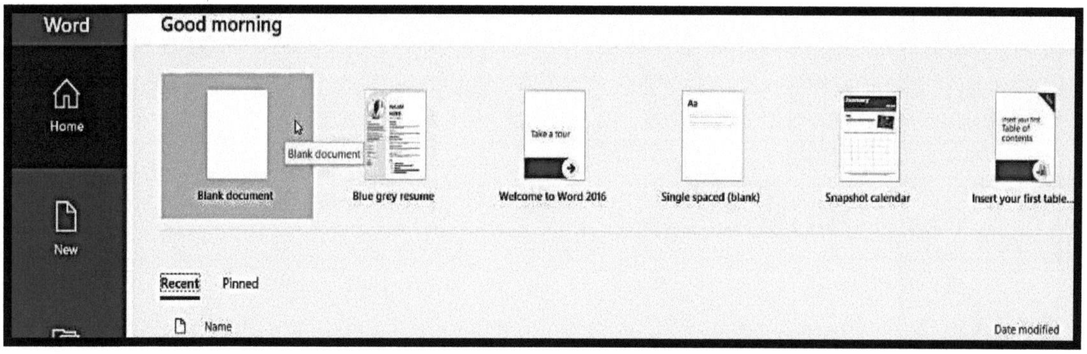

This is a convenient way to quickly access files you've recently worked on. So that's a little bit of information on how to access recent documents, but what we want to focus on here is how to create a blank document, and if you look just above that Recent list, the first thing you'll see is a Blank document. It's quite simple; just choose "**Blank document**" to make a new one. But before we do that, let's have a look at some of the other options you have in this area for creating new documents. If you don't want a blank document, you may create one based on a template, and Word provides a variety of different templates to choose from.

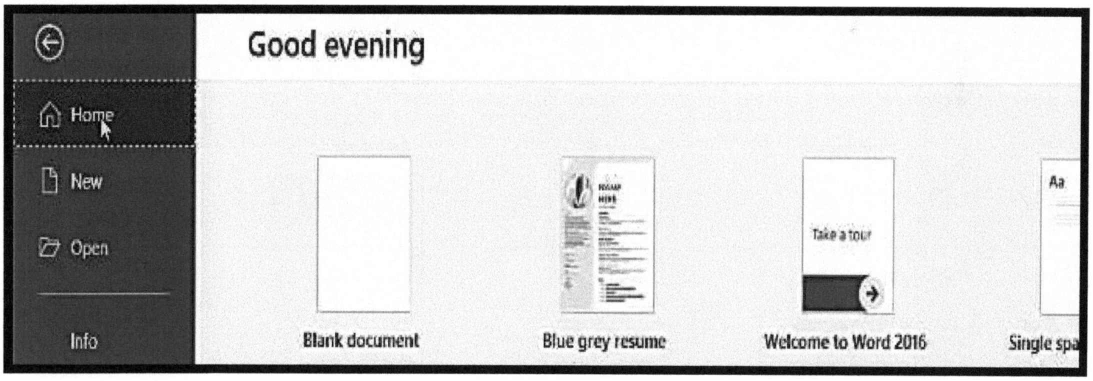

Now, if you've never used a template before and aren't sure what they are, it's just a good starting point; for example, if you know you need to create a resume or some meeting notes, you can search for a specific template related to a resume or meeting notes, and it will already have a lot of the information or layout that you need for that particular document. You can see some templates at the top here, and if you click on the "**More templates**" link, you can go through and take a look at some of the templates that are available for you to use.

There are a lot of them in here, and they're worth checking out if you've never been into this section before.

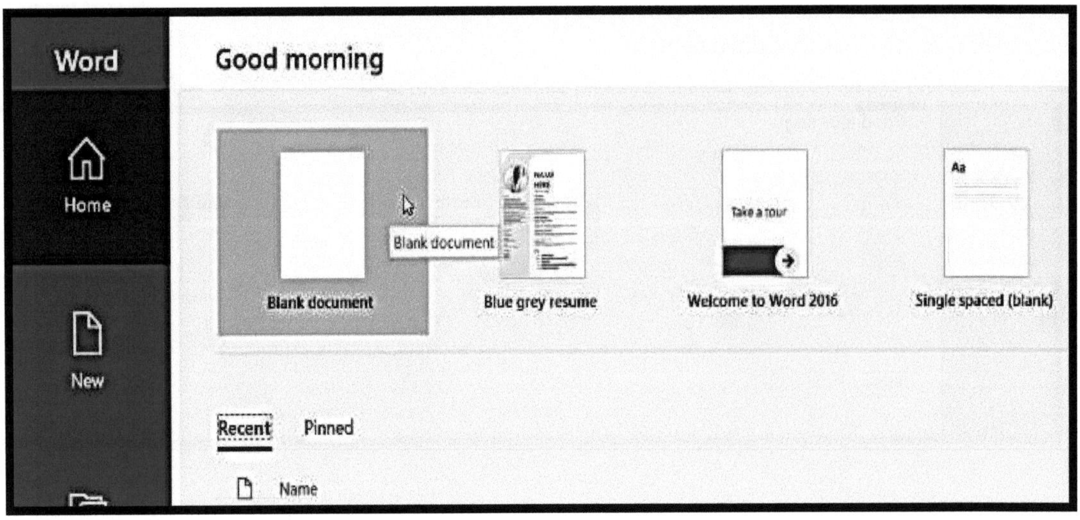

If you're looking for something extremely particular, such as a resume, you can enter it in, click the magnifying glass, and it will search the templates, bringing up a list of all of those Resume templates. If you find one you like, you can just click on it and begin working on it.

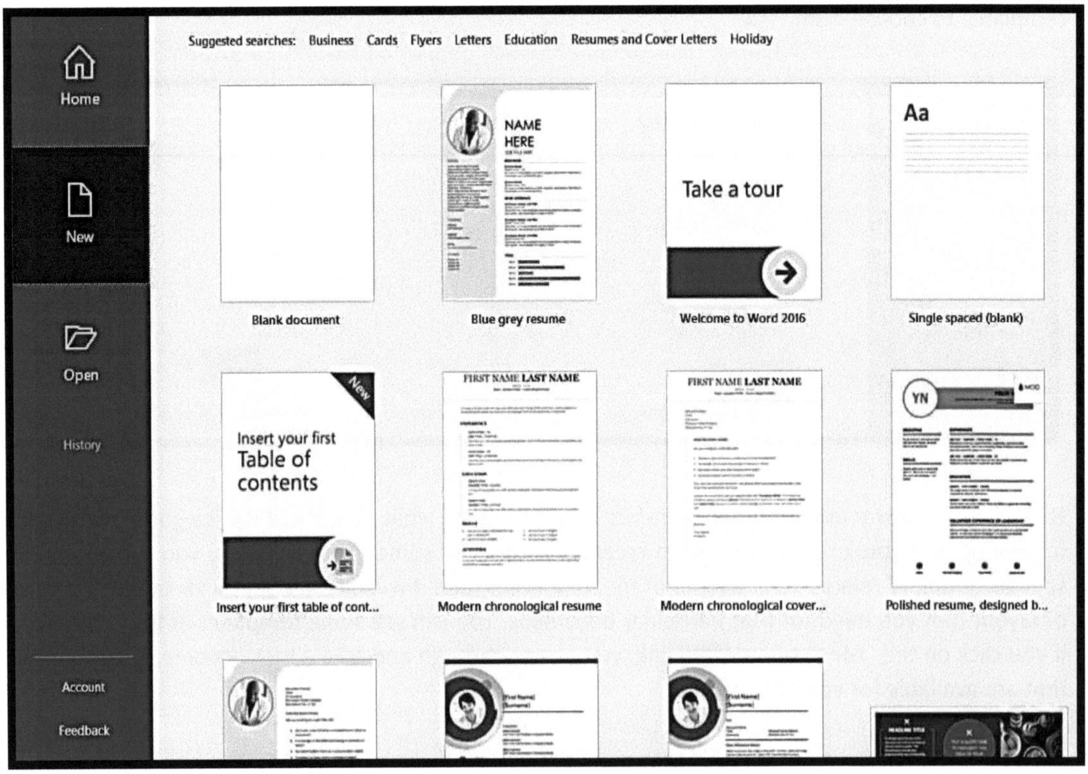

You can just pick a Blank document if you don't want to utilize a template. So now we've generated a blank document, the first thing you should notice is that if you look up to the top of the screen, you'll see what we refer to as the **Title bar**, which currently states Document 1, indicating that this document hasn't been saved yet.

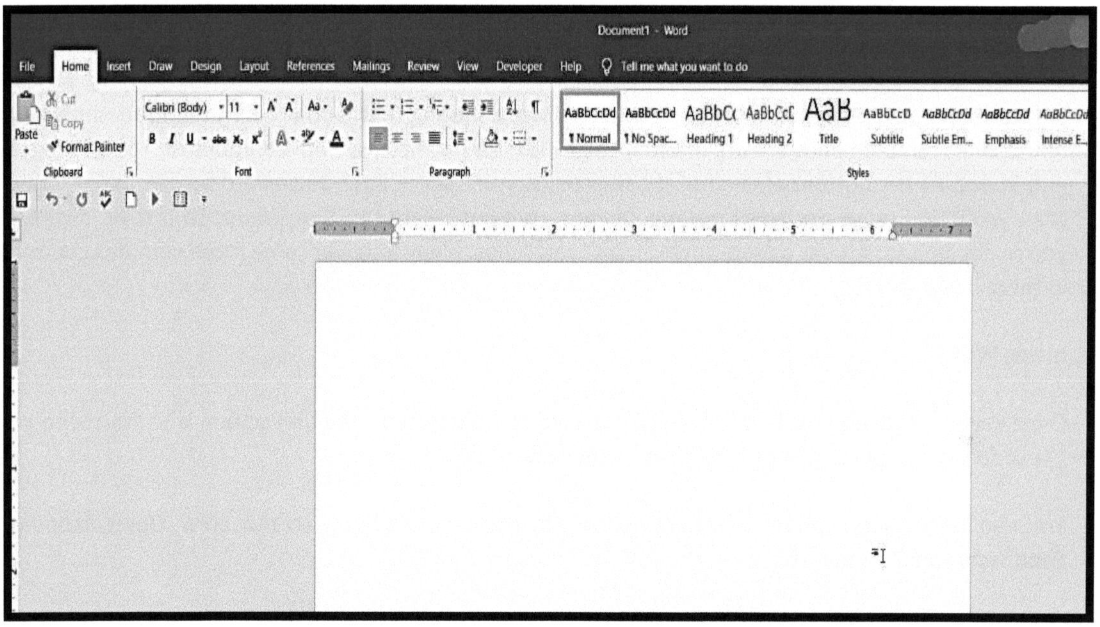

You'll see that when you create a new blank document, the default name is document 1, 2, or 3, and so on, just be aware of that since you'll notice this change when we save this document later.

Ribbons and Tabs

Home, **Insert**, **Layout**, **View**, **Draw**, **Mailings**, **Review**, **Developer**, and **Help** are some of the tabs on this screen.

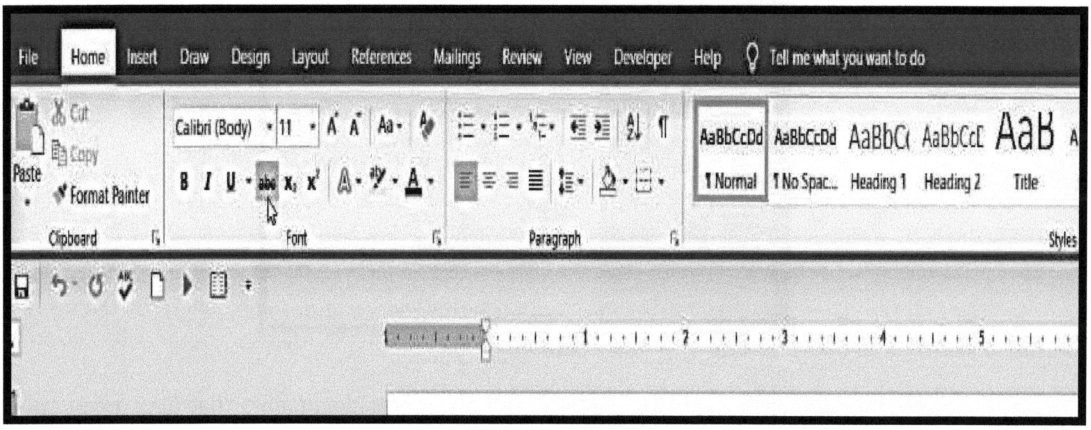

What you see on your computer may alter somewhat from what we have here, depending on which of these tabs you have selected, but in general, you will see these tabs running across the top, and what these tabs contain or what we refer to as **Ribbons,** which contain all of the commands that you require. If you click on the home tab, for example, you'll see the home ribbon, which has all of the commands that run across the top of the screen, and these commands are generally organized into their appropriate ribbons. The commands that you use the most are found on the home ribbon, and we'll go over some of them later in this chapter, but just know that they're ordered logically with their respective commands.

Before we go any further with these ribbons, it's worth noting that if you're on the home ribbon, you'll notice that the commands are organized into groups; for example, we have **Clipboard**, **Font**, **Paragraph**, and **Styles**. At the bottom of each of those groups, you'll see a little downward arrow, and if you click that, you'll find more advanced options or more options related to that group, so just be aware that you're not limited. If it has a drop-down arrow, you'll most likely discover some more complex commands in there if you click on it.

Inside Word

Once you're inside Microsoft Word, you have a lot of options here. The first option is to start with a blank document, which is most likely what you want to do.

You also have several options over here on the left. You have the **home screen**, **New**, **Open**, **Account**, **Feedback**, and **Options**.

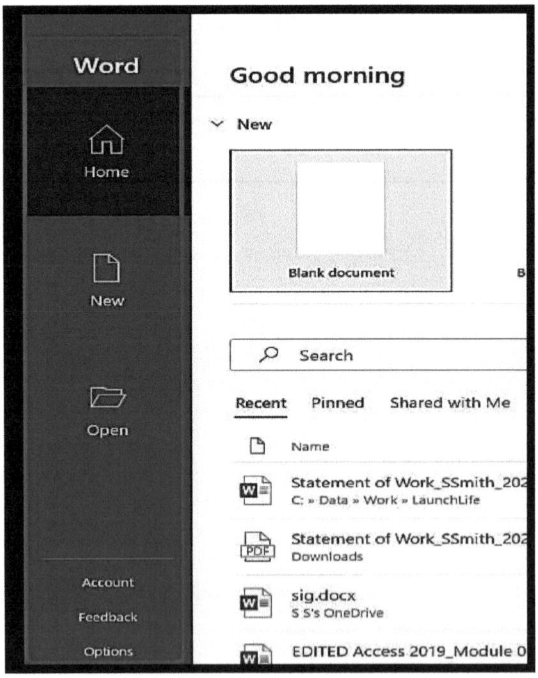

Home is selected by default. Under the "**Recent**" tab are recent documents that have been opened. Under the "**Pinned**" tab is any document that you want to open often, these will be very frequently used files, and the way you get them in the pinned list is by going to "**Recent**" and clicking on the pin icon right there, then when you go back to the home button, go back to the Pinned tab and you will see that document now resides under the Pinned list. "**Shared with me**" will be any documents that have been shared with you over OneDrive.

If you go to "**New**", you can either select a blank document or you can look through all of the templates down here that Microsoft Word has provided for you. These templates allow you to quickly begin a resume, a calendar, or a lot of other frequently used documents such as brochures and reports.

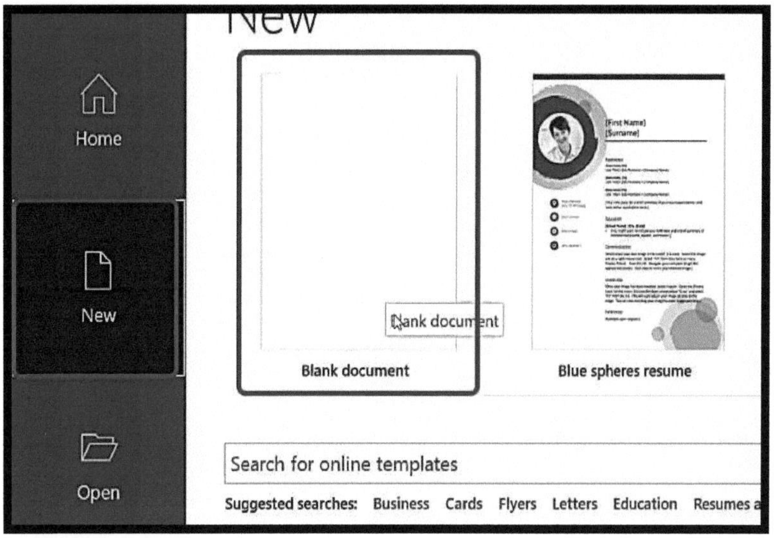

If you go to the **Open** menu, you can see a lot of files that you have opened recently, more so than you can see on the home screen. If you go to **Browse**, you'll be able to find a document on your computer. And if you go to **OneDrive**, you'll be able to see any kind of documents that you've shared with everyone else, or people have shared with you.

The **Account** menu tells you information about your Microsoft Office subscription and gives you different options for setting defaults for Microsoft Word

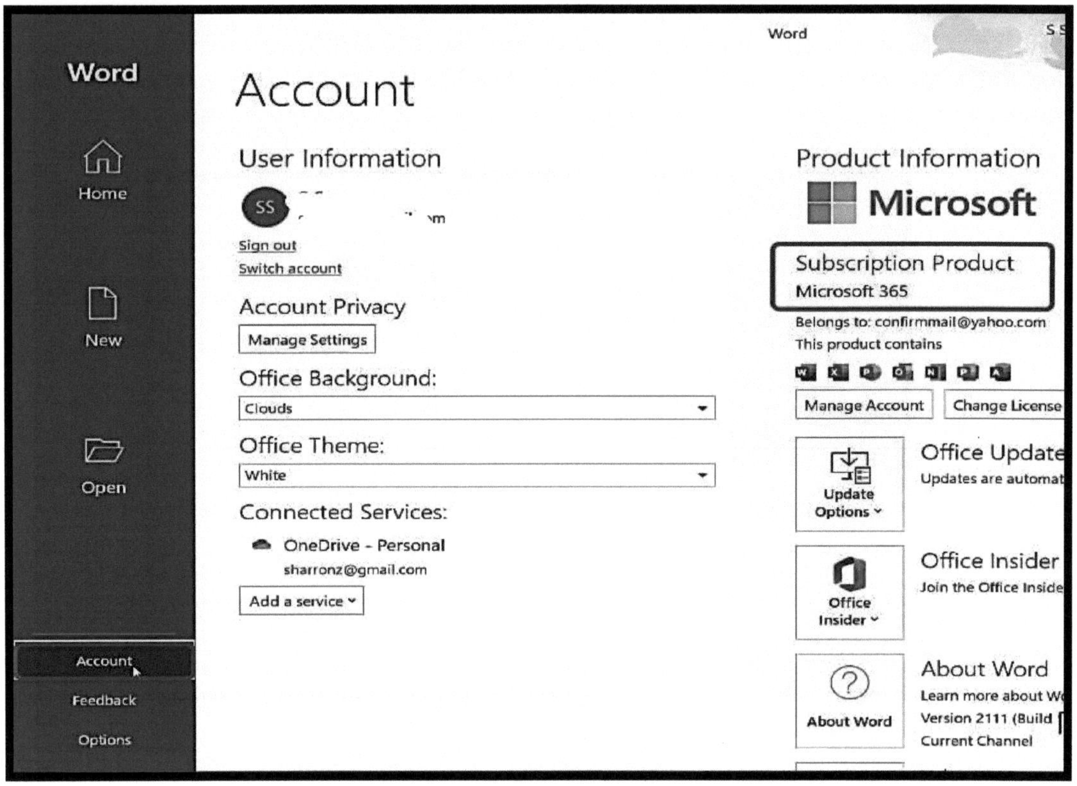

The Backstage View

It's worth mentioning that the **backstage view** is available in all Microsoft Office programs; for example, if you're using PowerPoint, Excel, or even Outlook, you'll see a backstage view. You might be curious as to where it is and how to get to it. To put it simply, it's the **File** tab at the top.

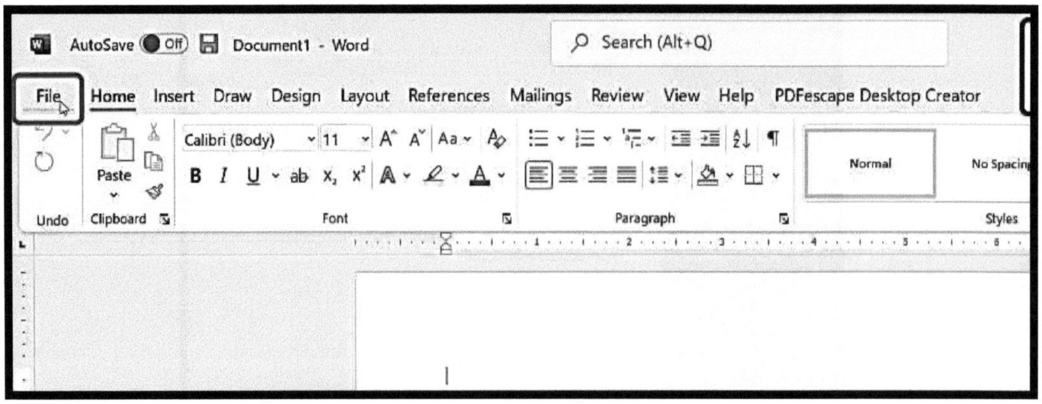

If you go to **File**, you'll see the backstage view, which is where you'll see all of your document's mini-tasks. So, starting with the menu that runs along the left-hand side, we can see that it's divided into a few divisions. We have three icons in the top section: Home, New, and Open. If you click on "**Home**," you'll be sent to the home screen, which we'll discuss in more detail later.

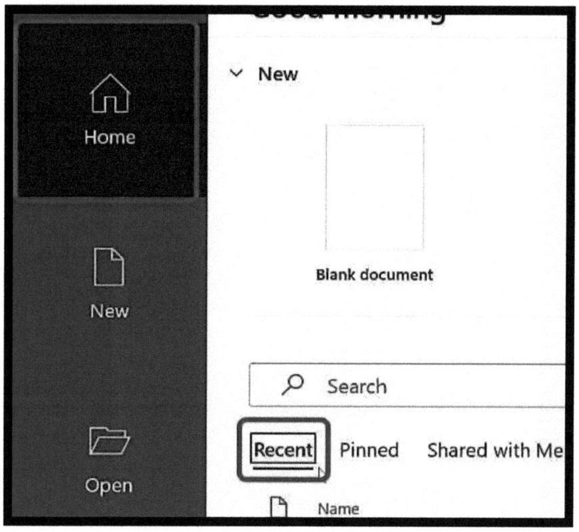

Then there's "**New**," which is where you'd go if you wanted to make a new document, either a blank one or one based on an existing template.

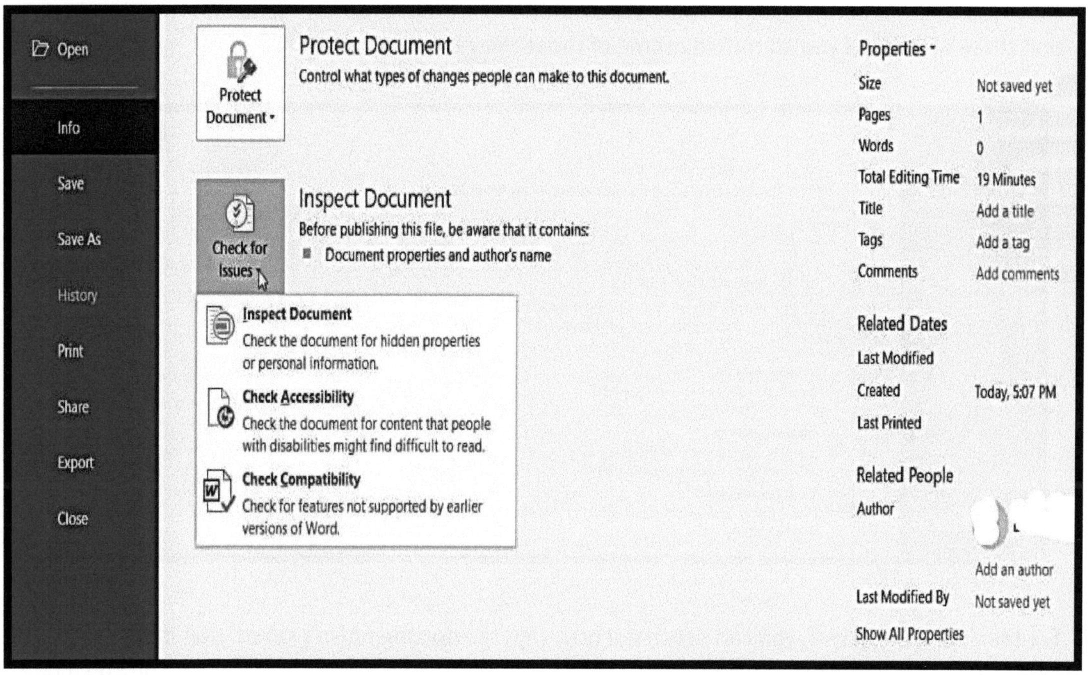

There's also "**Open**," which is where you'd go if you wanted to open a fresh file or one, you'd saved somewhere.

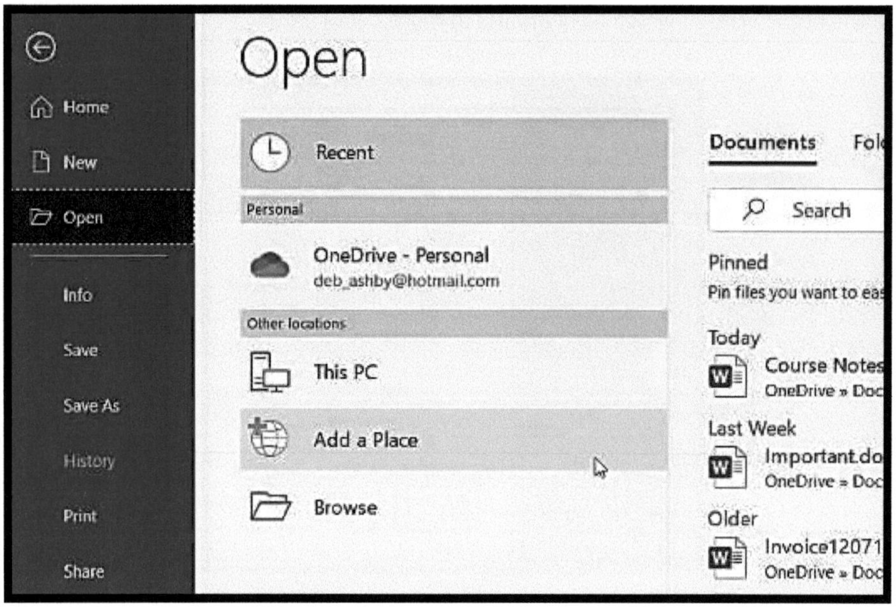

In the "**Info**" section, you'll find all of the details about the document you're now working on. If you haven't saved your document or it doesn't have any content in it yet, there isn't much information to show you about it in this section. You'll discover all the attributes on the right-hand side, and you'll see that these will alter if you start doing some of those things.

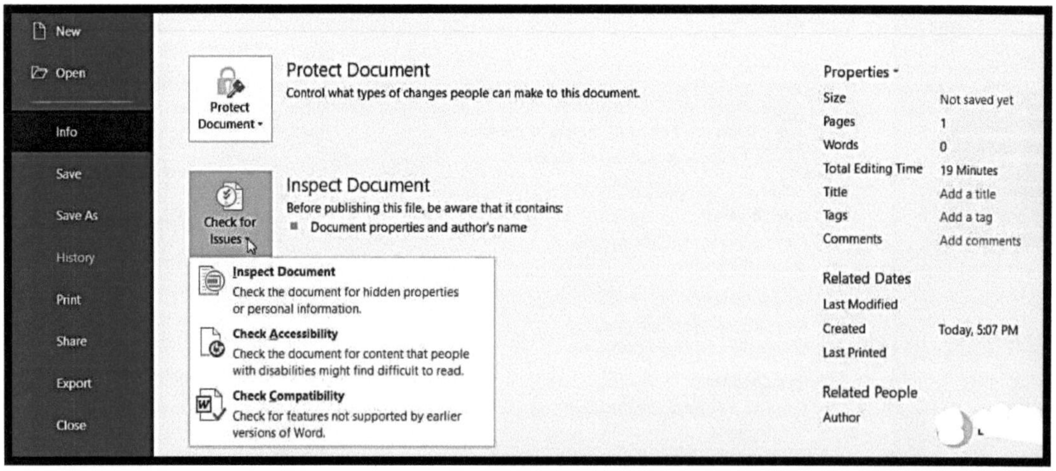

For the document above, you can see that it now says the document isn't saved, that it has one page and zero words, and that it tells us how long we've been in that file. We could go through and add a title and tags to help with searching. So just keep in mind that if you're seeking information regarding the file you're currently working on, this is a good place to start.

If we return to the left-hand side, we'll find some more items that we may utilize in our document. The first option is "**Protect Document**," which allows you to limit the types of changes that others can make to your document. So, if you're sharing this document with others, say, if you've sent it to a client, a coworker, or someone else on your team, you can have some control over what they can do with it; for example, you can apply some editing restrictions if you want, and that's where you'll find all of those kinds of options, which we'll go over in more detail later.

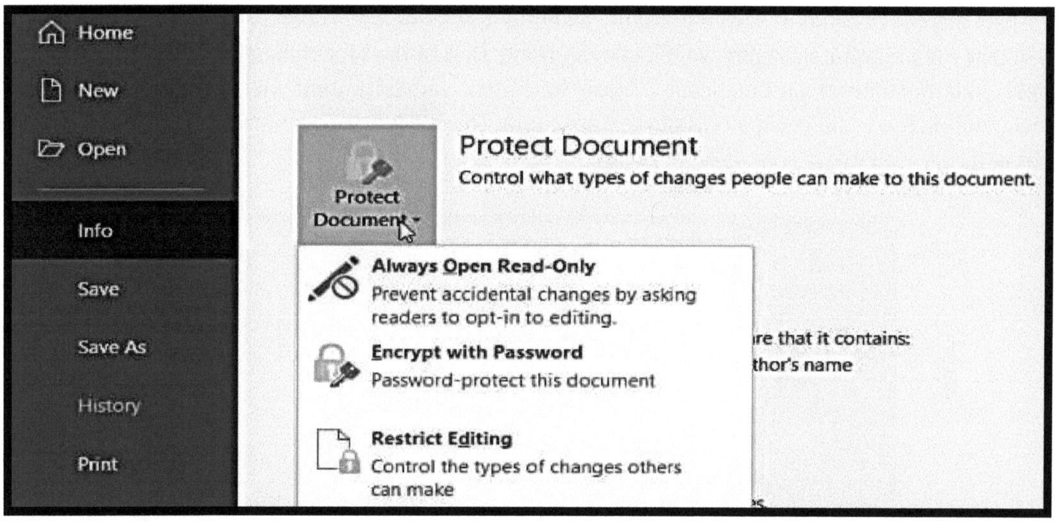

We then have an "**Inspect Document**" option, which allows you to inspect or conduct a quick check on your document once you've done it to ensure that it has some features that make it available to others and that there are no incompatibility issues. This means that if you're using Word 2025, you should be aware that someone you're sharing the document with might be using an older version of Word, so it's worth pointing out or highlighting in your document or figuring out which elements they won't be able to see because some functionality in 2025 isn't available in older versions.

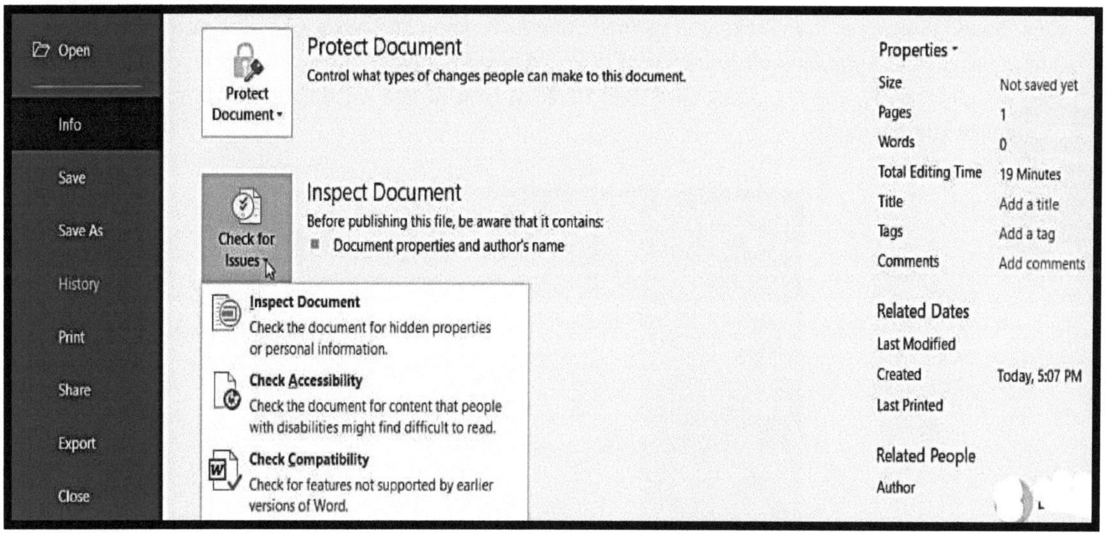

Let's imagine a new typeface was released in 2025, and you used it in your document; if you shared it with someone who doesn't have that version of Word, they might not be able to see the font you chose, and the same is true for other things. It's always a good idea to check if your document is compliant and make any necessary adjustments before sending it out, and you can do that by clicking the "**Inspect Document**" option.

We also have a section called "**Manage Document**," where you may go back and recover unsaved versions of your document. We've probably all done it at some point: begun typing forgot to save, and then quit the program, believing we'd lost everything. One of the nice things about Word 2025 is that it saves your documents automatically. If you had prior versions, they will appear on this "Manage Document" screen, and you may click to restore them. This is a lifesaver, so make a mental note of where that button is located.

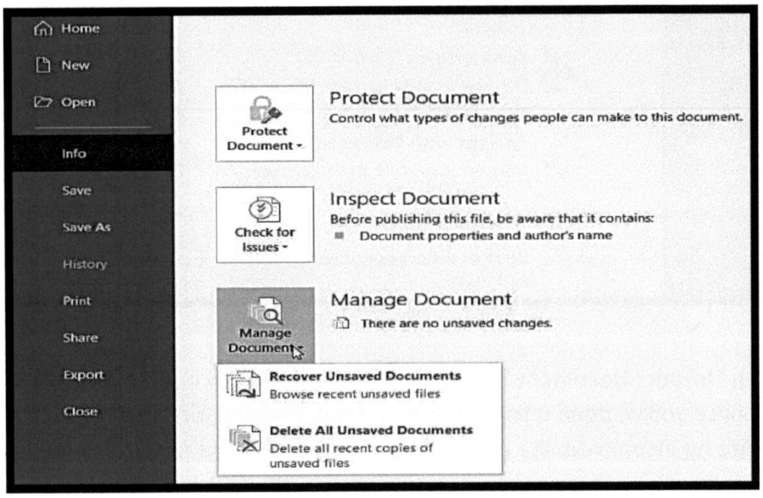

Then there's "**Save**" and "**Save As**" beneath that. These two operate in a somewhat different manner. You'll notice that when you click "**Save**," you must first save your document before you can use it. If you click "**Save**" right now, it will take you to the "**Save As**" region and ask you where you want to save your document because you haven't done so yet. If you'd already saved your work and merely wanted to save some modifications, you could just click the "**Save**" button, which will appear to do nothing but save the changes to the file name.

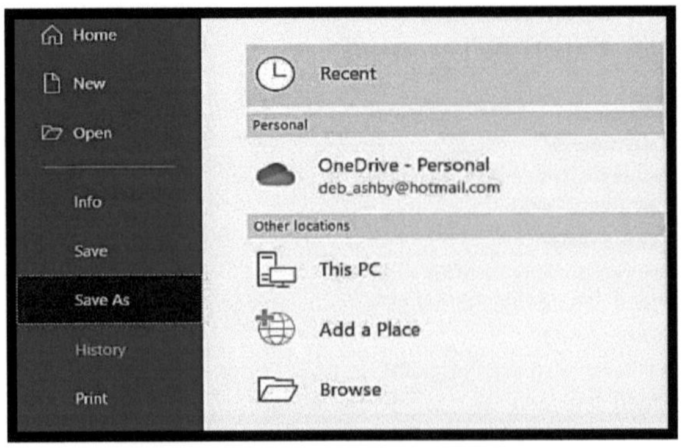

The difference between the two is that one is used to save a document that has previously been saved as a name, while the other is used to pick a folder and save a document for the first time.

You'll now be able to find past revisions of your work in the "**History**" section.

Then there's "**Print**," which does exactly what it says on the tin. If you've written a beautiful document and want to print it, you can come in here, choose your printer, and see all of your printing options.

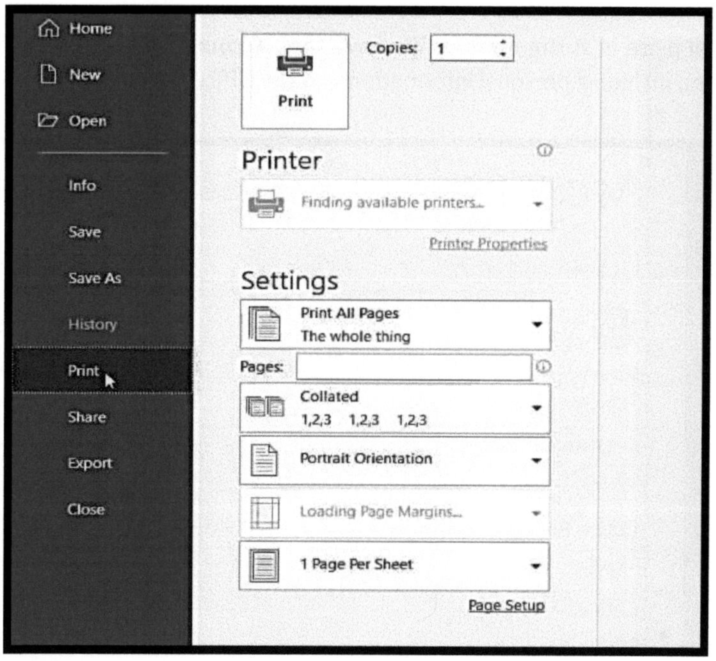

There is a "**Share**" option in this section that will allow you to share your document in a variety of ways. You have the option of emailing it, saving it to the cloud and sending a link to individuals, presenting it online, or even posting it to a blog. When it comes to sharing, there are numerous alternatives.

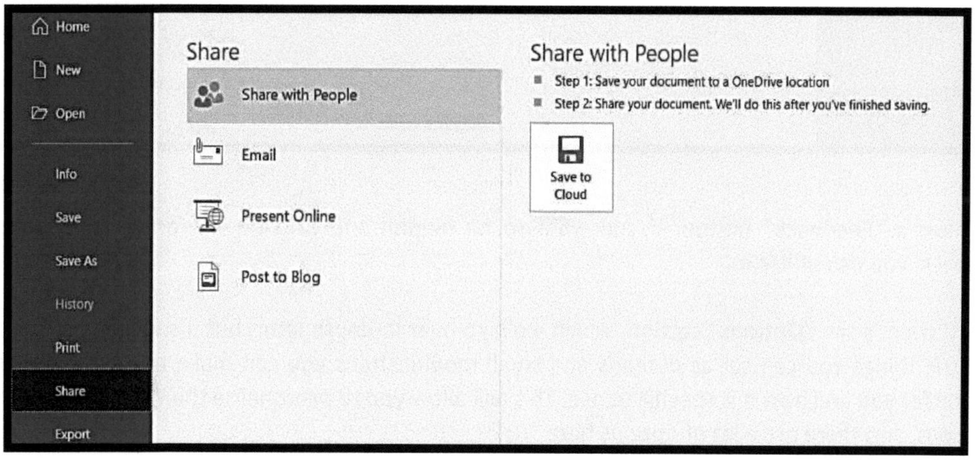

You should use "**Export**" primarily if you want to **create a PDF file**. If you've never used a PDF before, it's essentially a version of your document that's very difficult to modify, so if you want to secure your document or don't want people to be able to simply go in and alter it, it's always worth saving or producing a PDF of it before sending it. There is a lot of software available these days that allows users to alter PDF files, but in general, if you want to add an extra layer of security, producing a PDF is a good option, and this is where you would go.

Finally, at the bottom of the page, there is a "**Close**" button that will close your document.

Our third and final piece is at the bottom. We have the "**Account**," which simply displays information about your account, including personal information and the Office version you're using.

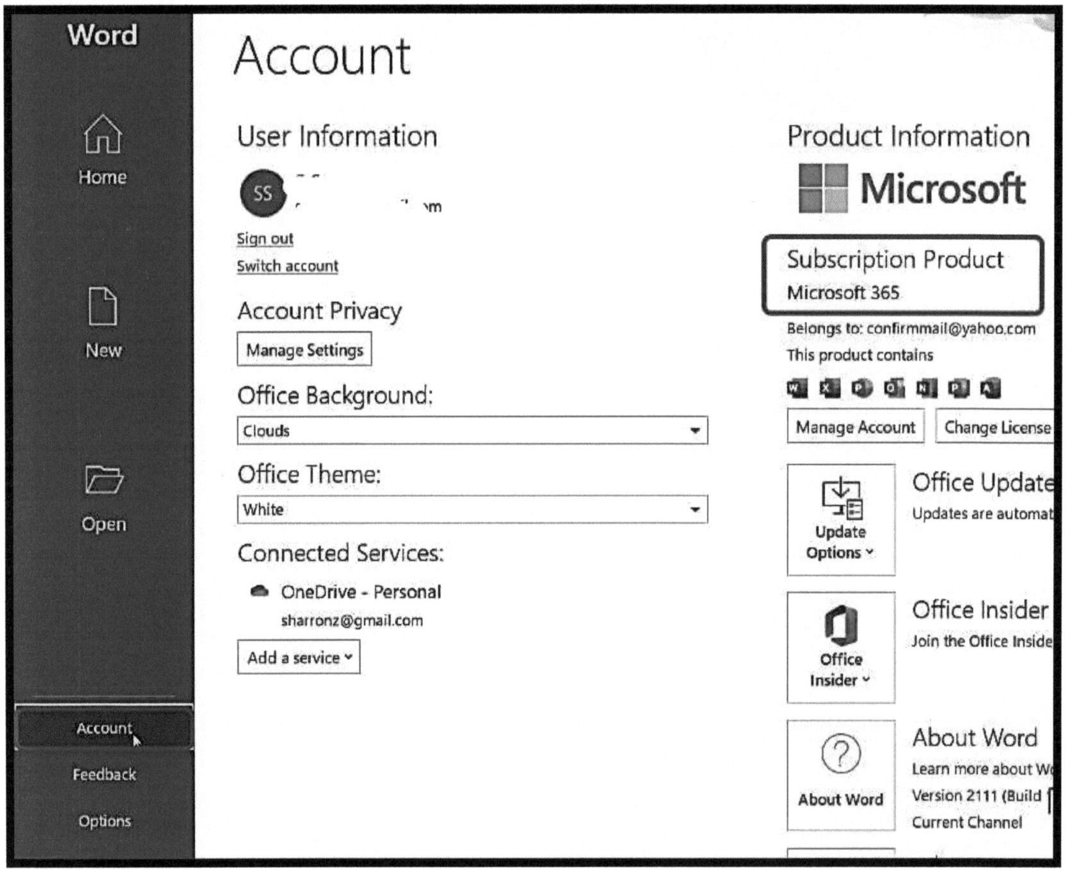

We have a "**Feedback**" button, if you want to be helpful and provide Microsoft with some useful feedback, you can utilize it.

Finally, there's the "**Options**" section, which we'll go over in-depth later, but it's where you'll find all of the little things you can set as defaults and small modifications you can make to how Word works in general for you and how it is specific to use. This will allow you to personalize the way your Word version functions, and there are a lot of options here.

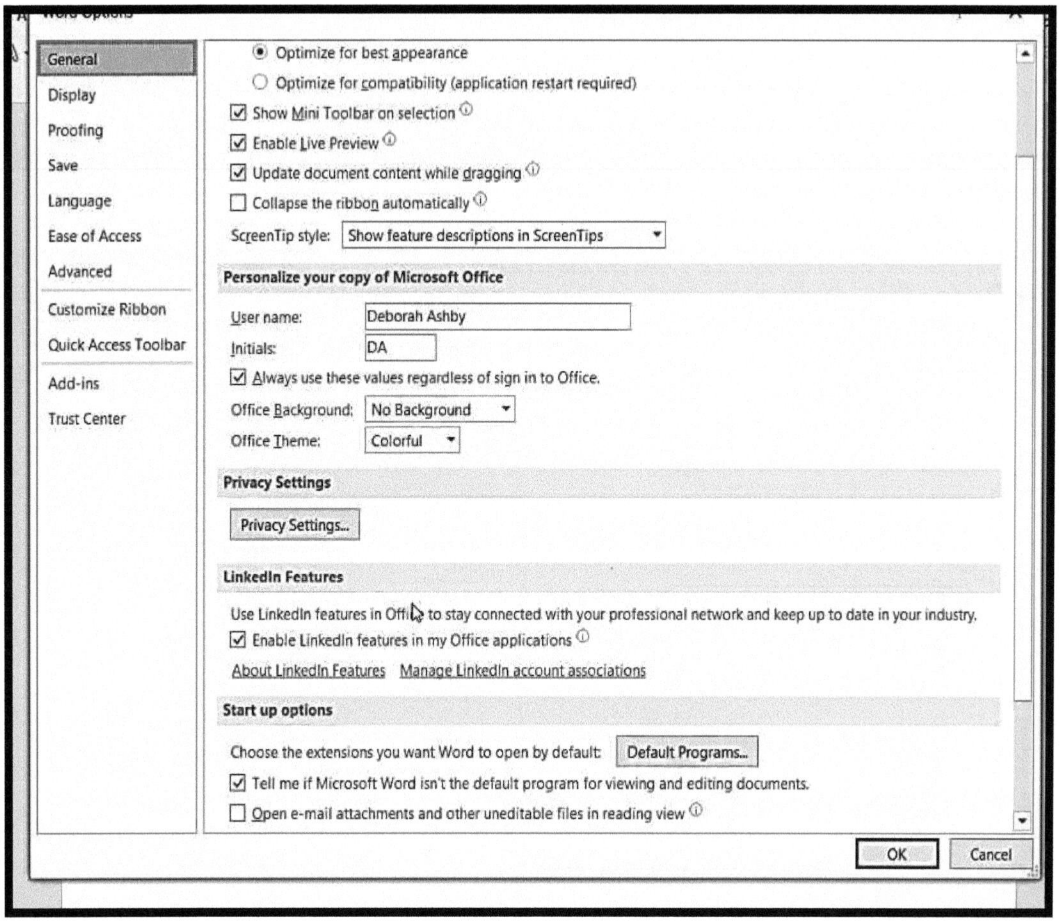

One other thing you'll notice in this backstage view is the **back arrow** at the top, which does exactly what you'd expect it to do: it returns you to the document you were in.

Now, let's go back to Microsoft Word.

The Home Tab

If you click the home tab, you will see that the **home ribbon** comes up, and you have a lot of options here. Now if you have some text in your Word document and you would like to be able to see this closer up, you can go to **View**, **Zoom**, and change this so that you can see closer up what you are doing with all of these options on the home tab.

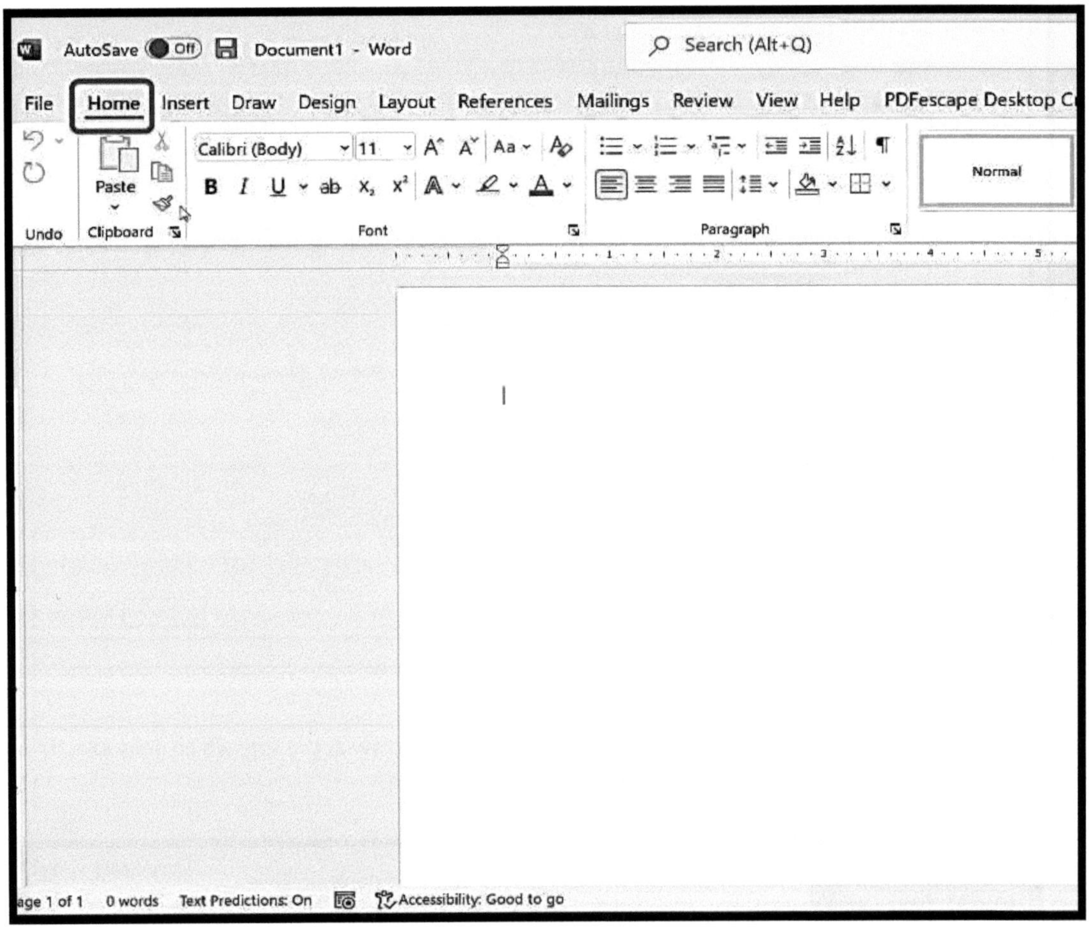

Back to the Home tab, if you double click with your mouse on any word, you will select the entire word. Here you can **bold**, **italicize**, or **underline** your text.

You can also use the **strike-through** option. If you're editing someone else's paper but you want to leave what they had originally put in the text, you can use the strike-through. You can also **change the color** of your font, you can **highlight** all of your text and **center** it, **right-justify** or **left-justify** it. You can also **change the spacing** when you have **paragraphs** right here. You can **add bulleted points**, and **number points** and you can even make this an **outline** if you have multiple lines on here. With the **Indent** button, you can indent or you can out-dent. You can also **highlight** your text. Over here on the right side, you can **dictate** a new sentence if you would like. These are some of the options in the home ribbon.

The Insert Ribbon

In this tab, there are a couple of options we should review. First of all, in Word, you can **insert a table** of information so that all of your rows and columns are lined up and are formatted well.

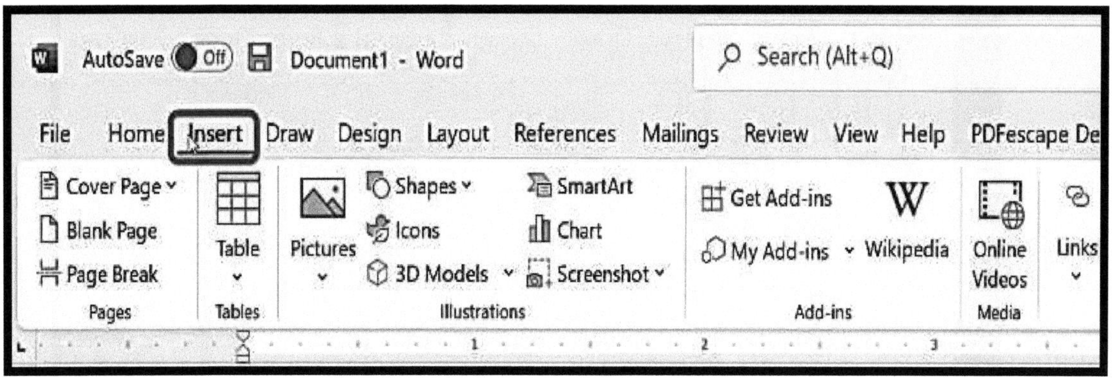

Clicking on the **Pictures** tab allows you to **insert a picture**. You can insert pictures from your device, you can use stock images, or even get pictures online and into your document.

There are a couple of other things you can insert including **Shapes**, **Icons**, **Screenshots**, **Online videos**, **Links**, **Symbols**, and **Equations**, among others.

You can also **insert the page number** on your document by going to the button on top of the page, whatever your preferences are for page number, and then you can make your page number be on the left, center, or right.

The Layout Ribbon

On the Layout, you can **change the margins of your page**. The normal setting is 1inch all around but you can also customize those margins and make them whatever size you would like by hitting "**Customize**".

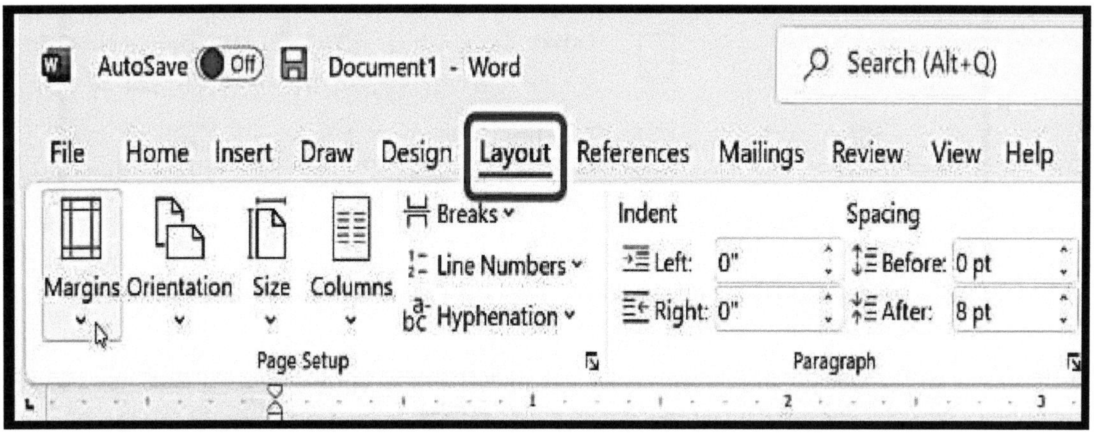

You can **change the orientation of your page** from Portrait to landscape.

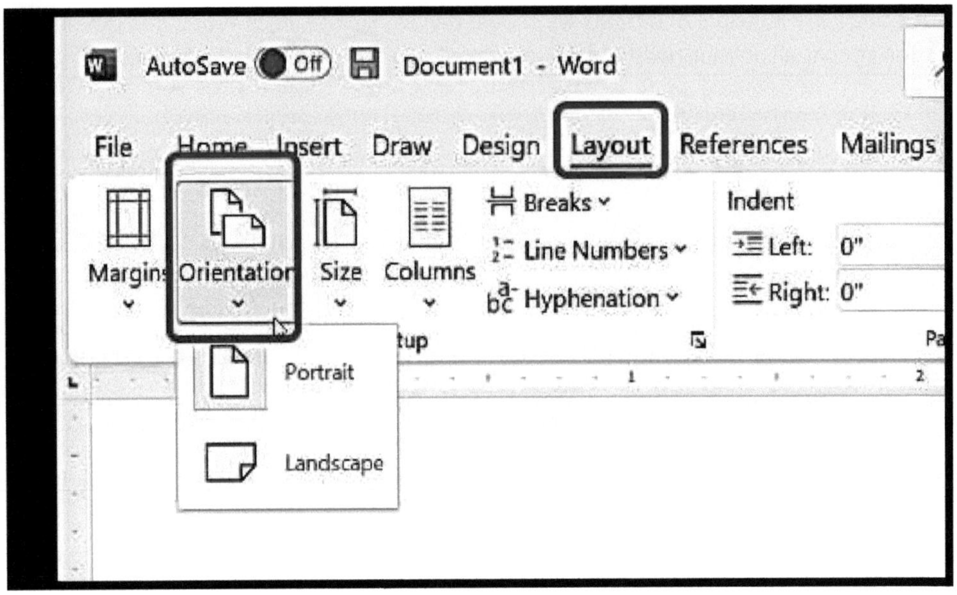

You can also **change the size of your paper** here.

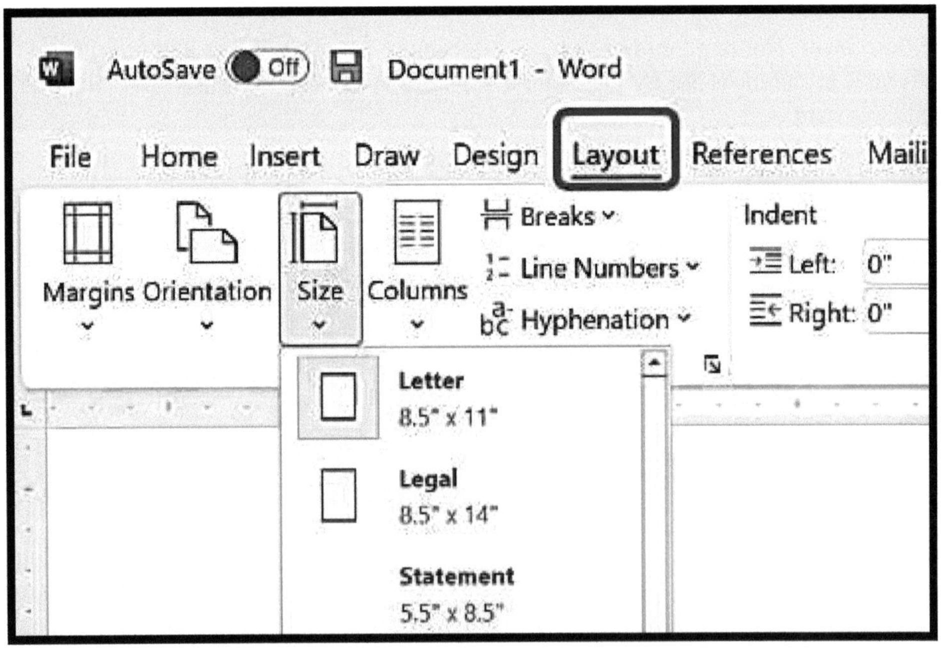

Here's one that is used often; sometimes you may want to **insert a page break** and have information print on a different page. What you would do is go to "**Breaks**" and hit “**Page**”.

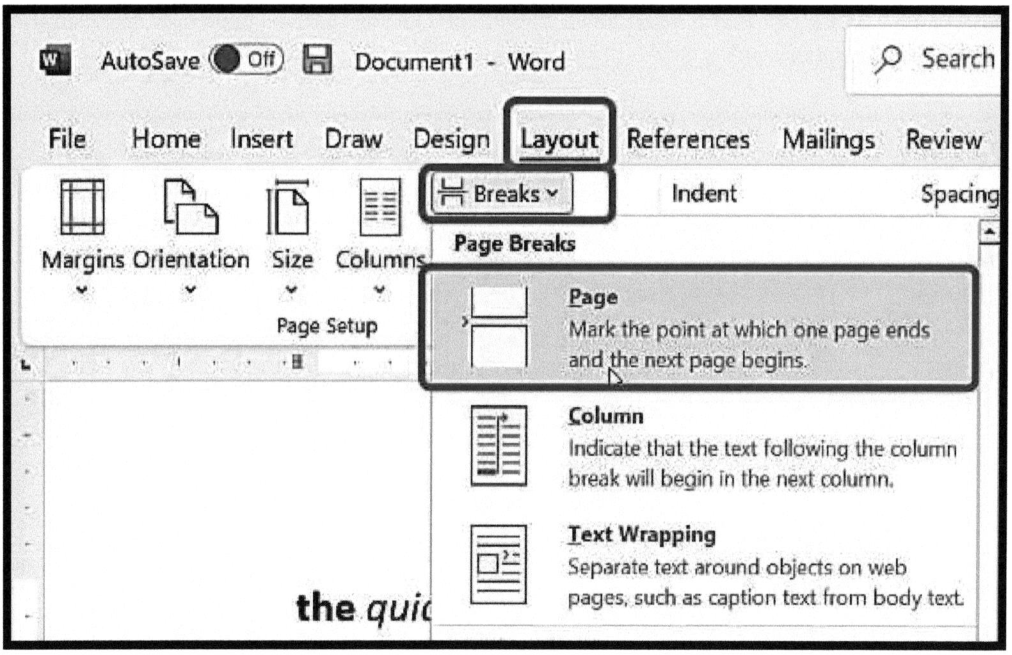

If you scroll down, you will see that the page break is right there. Sometimes that's very useful when you have one sentence appearing at the bottom of a document that you would prefer for it to go onto the next page. But be careful with that because, as you add words to your page, the page break will remain in place, and your text may spill over onto another page, yet leave the page break and it might not look right.

The Review Ribbon

There are some options here that are very useful when you are working with Microsoft Word. Let's say you want to make your paper a little more exciting and you have a word in your document that you don't like and you would like to use something more emphatic. if you double-click on the word and go up to "**Thesaurus**", you can now see different words that could be used to replace that word.

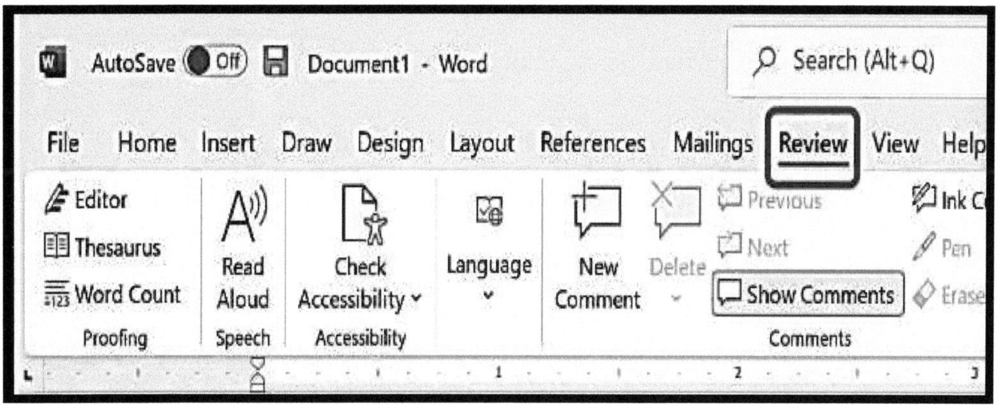

If you click on the new words and insert any of them, you can see that you have now replaced the synonym with the initial word.

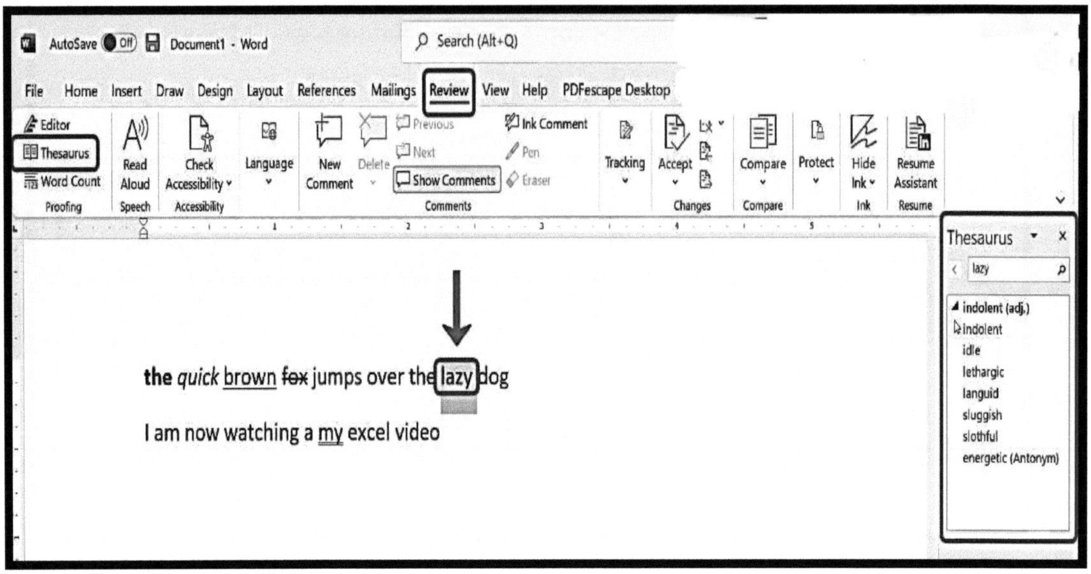

You can also click on "**Word count**" and see how many words you have on this page. In the Status bar, you can see how many words you have on your page as well.

The View tabs

On the View tab, you can **zoom** in and out of the document that you are working with.

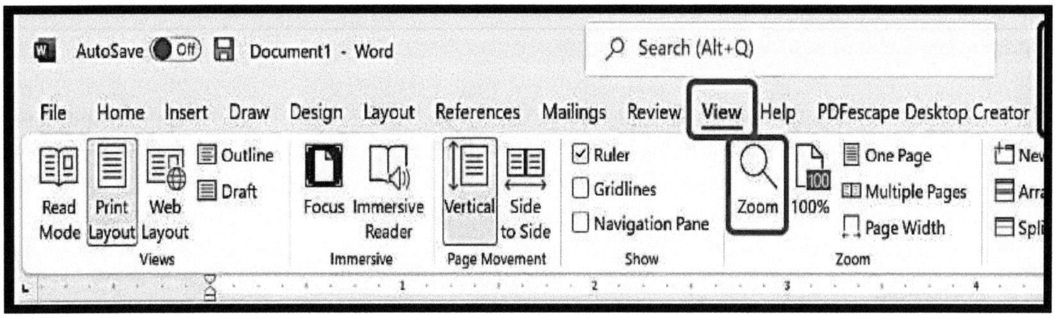

You can also pick "**Multiple**" to see how many pages you have within your document and get a rough idea of what it's going to look like when you print it out.

You can also **switch windows** to any other document you may have opened at this time, or for a shortcut on that you can press **Alt + Tab** on your keyboard and switch between word documents as well. To go back to the regular view that you've been working with, click "**One Page**", click on **zoom**, and go back to your regular view.

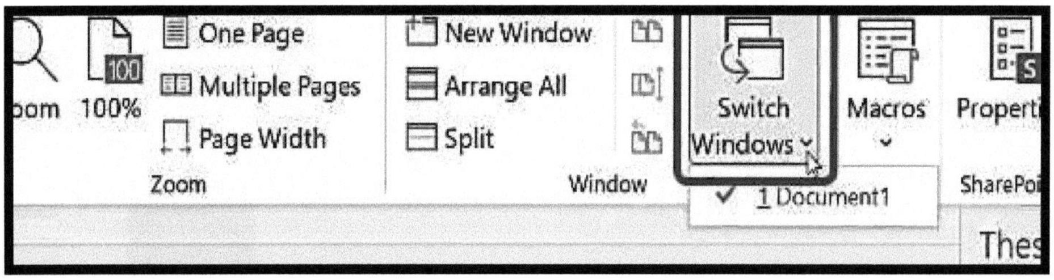

What if You Can't Remember All of These?

One more thing you should note is up there in the "**Search**" **bar**. This is true for all Microsoft Office applications.

If you can't remember one of the commands we went over and you click in here and type a word, you will see it right up here. With that, instead of having to search for the ribbon, you can type in the command of what you would want to do until you get a little more familiar with the environment inside of Microsoft word

Changing the View of your Document

This is where we'll find our views, all the way over on the right-hand side. This is just a quick way of switching between views, and these buttons correspond to how you're now looking at your document. As you can see, the first one is Read Mode, and switching to it provides you with a fresh perspective on your paper. It removes almost everything on the screen except your document, allowing you to see more clearly what you're looking at. This is useful if you're trying to read a document. The Print Layout view is the next one we have; it's the one you'll be in by default, and it's the one you'll be in most of the time when you're working on your papers. The Web Layout view is the next one. So, if you're creating something for the web, or if you want to incorporate things like HTML code, you'd go into this Web Layout view and create your document that way.

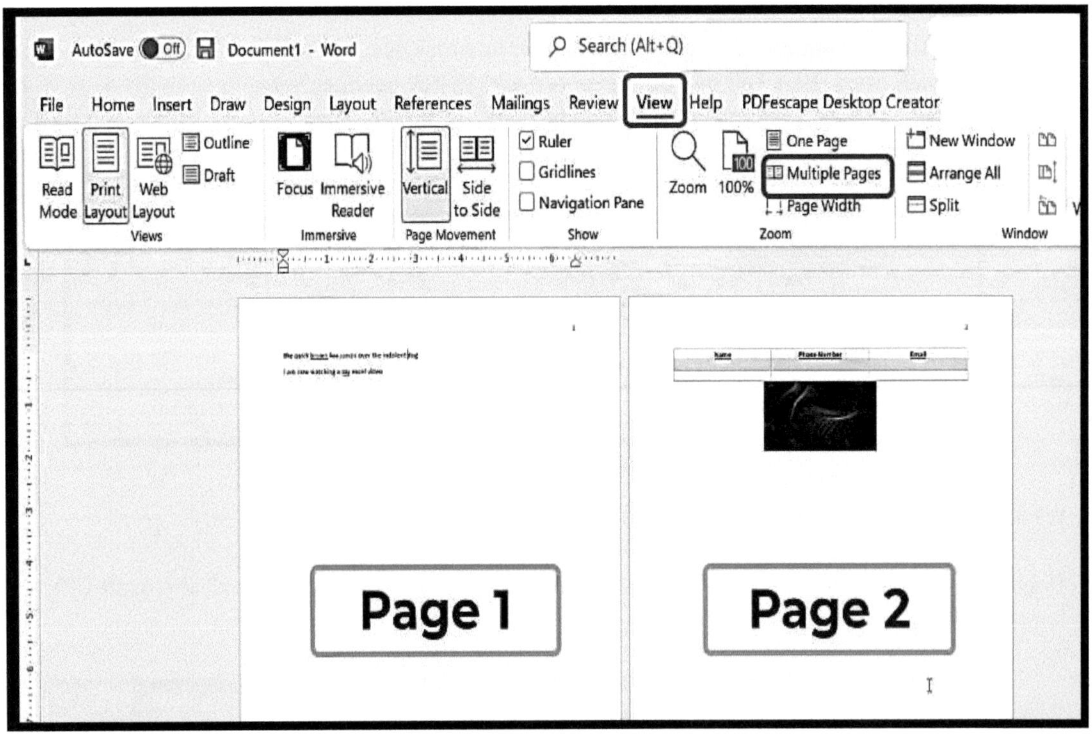

That's Not All

We have what we term the **Quick Access Toolbar** found directly beneath the ribbons, and this toolbar contains a few different commands that you can configure. Its entire purpose is to provide you with quick access to the commands that you use the most frequently, which will differ widely from person to person based on what you do in your everyday work.

If you click the drop-down at the bottom, you'll notice that you have a lot of alternatives for adding to that fast access toolbar. The ones that are currently checked are already in the toolbar, but you can add any of the ones listed here if you want. Alternatively, if you want a command that isn't mentioned here, you can browse for **more commands** and choose something to add to the Quick Access Toolbar from there.

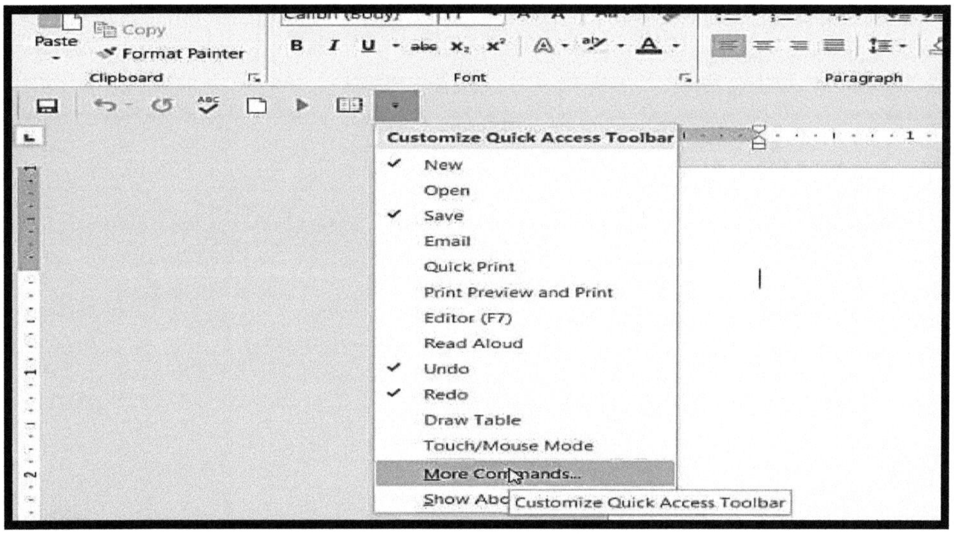

Moving to the top of your document, you'll notice a **ruler** running across the top and a ruler running down the side. On this top ruler, you might notice some small triangles in the little block at the bottom. It says Left indent, Hanging Indent, and First Line Indent when you hover over it, and this helps you align the text in your document; so be aware of the ruler and the Indent functionality in there.

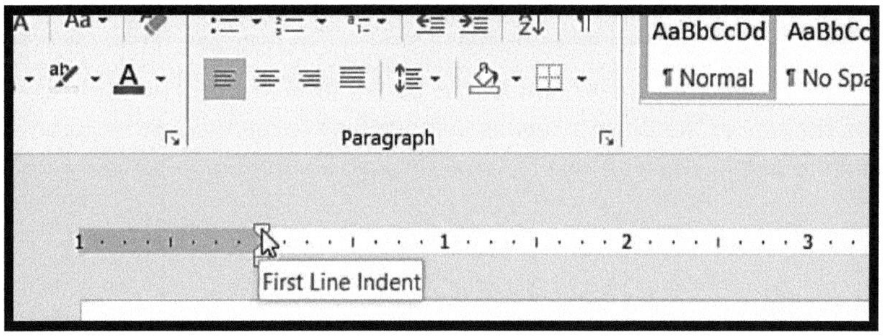

The **Status bar** is located at the bottom of the screen, and on the left-hand side of the status bar is some general information about our documents. You can see from the image above that we're in section 1 and on page 1 of 1. We presently have no words in the document, and it also indicates that our language is set to English (United States), but these will undoubtedly alter as we begin to add content.

It's also worth mentioning that the View ribbon at the top can also be used to access various options, such as **views** and **zoom**. You can see in the first group that we have our layouts, as well as a couple of extra views and some zoom possibilities.

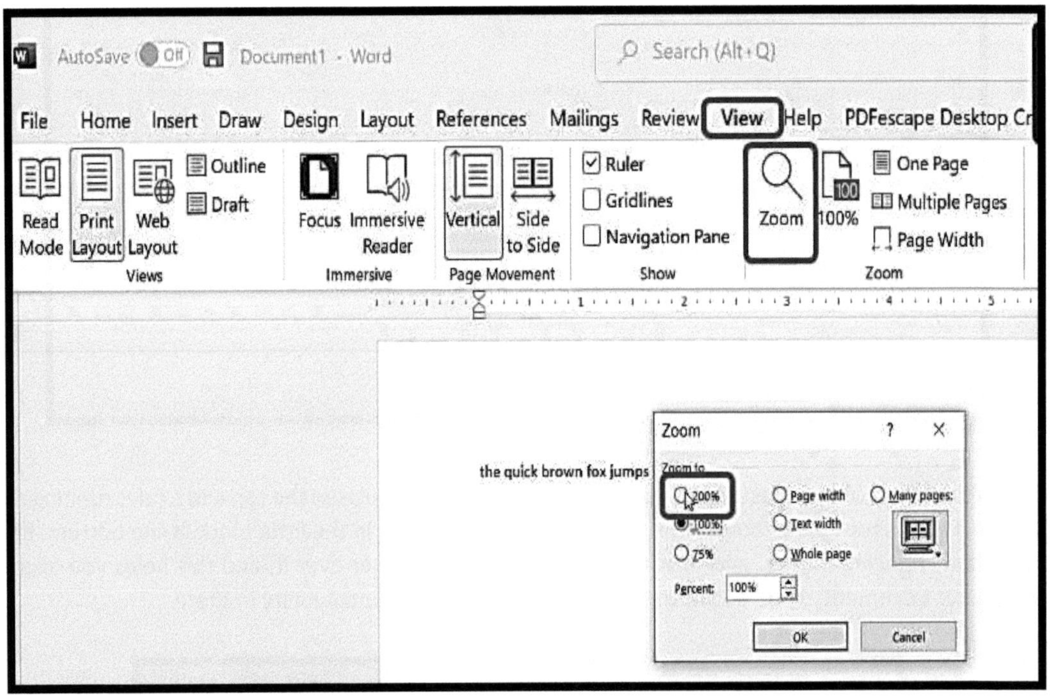

The **Scroll bar** on the side of the Word screen is the last thing to mention. So, you can go up and down, and if you have more pages, you'll be able to scroll through all of them.

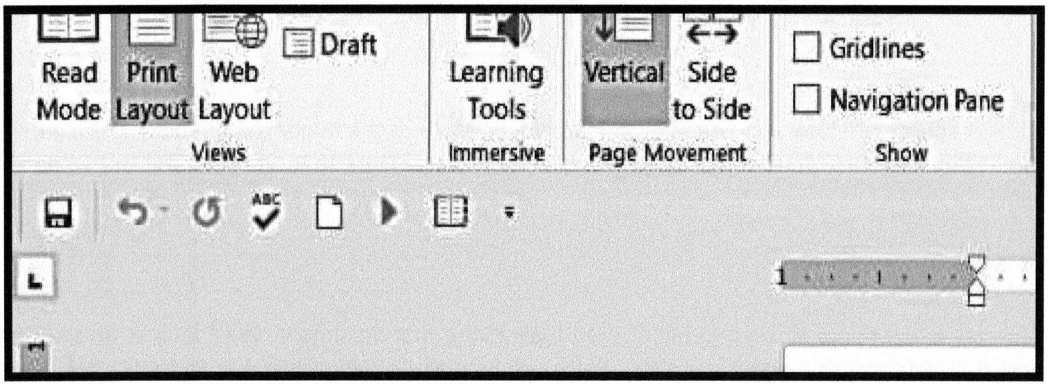

So that's a summary of what you're looking at on your Word screen right now, and hopefully, that's gotten you a little bit more familiar with the terminology we'll be using throughout this book, and it's a nice foundation for us to leap off into our following chapters.

CHAPTER TWO

TYPING MASTERY

In this chapter, we'll look at how to get started typing, some of the minor things you'll notice when you start typing your first documents, and several keyboard shortcuts that will help you out while you're typing.

What Happens When You Type?

There are a few things to keep in mind when typing your first document. We'll look at these things to familiarize you with the typing interface, which will help you prevent mistakes and make typing easier and more enjoyable.

The Margins

You'll notice that the cursor is flashing, but it isn't flashing directly at the page's left-hand edge. This is because there is a left margin of approximately an inch in there, and you can see that as you start typing, all of your text will start coming out there.

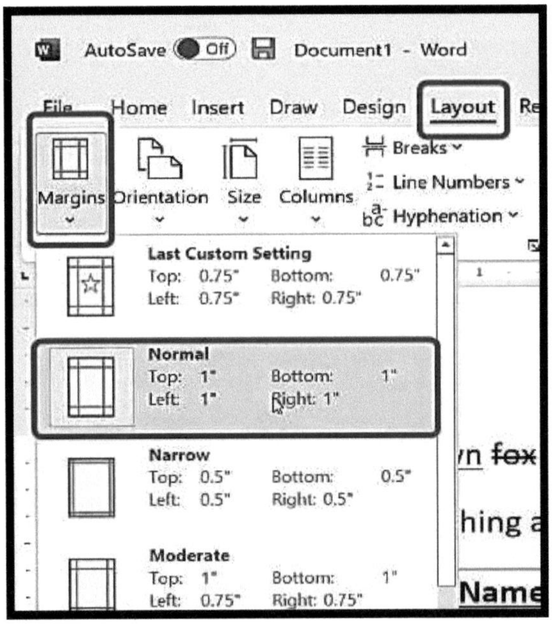

So, after typing a basic line, if you continue typing when you reach the end of a row, you won't need to hit the enter key to move on to the next line; it will automatically wrap as you type, you can just keep typing and it will wrap itself around, eliminating the need to hit enter at the end of a line.

If you want to force a line break, for example, if you were on a line and wanted to do anything on the following line, you can hit the enter key, which will take you down to the next line and allow you to continue typing.

Capitalizing the First Letter

Another thing to keep in mind is that if you start typing a sentence at the beginning of a line and don't capitalize the initial letter, Word will capitalize it for you. So, if you just start typing, you'll notice that it has capitalized it for you without you having to go back and change it. This is because of Word's autocorrect feature, which recognizes certain words and changes them for you, making your life much easier. In the following chapters, we'll go over a lot of the autocorrect settings.

Colorful Underlines Explained

There are a few more things to keep in mind.

If you misspell a word, you'll notice a red squiggly underline, which indicates that you've made a mistake with that word. So, you can hover over the word and right-click your mouse to bring up a menu of possibilities, and more often than not, the term you're looking for will be among them. If it isn't, you'll have to go in and manually make the change.

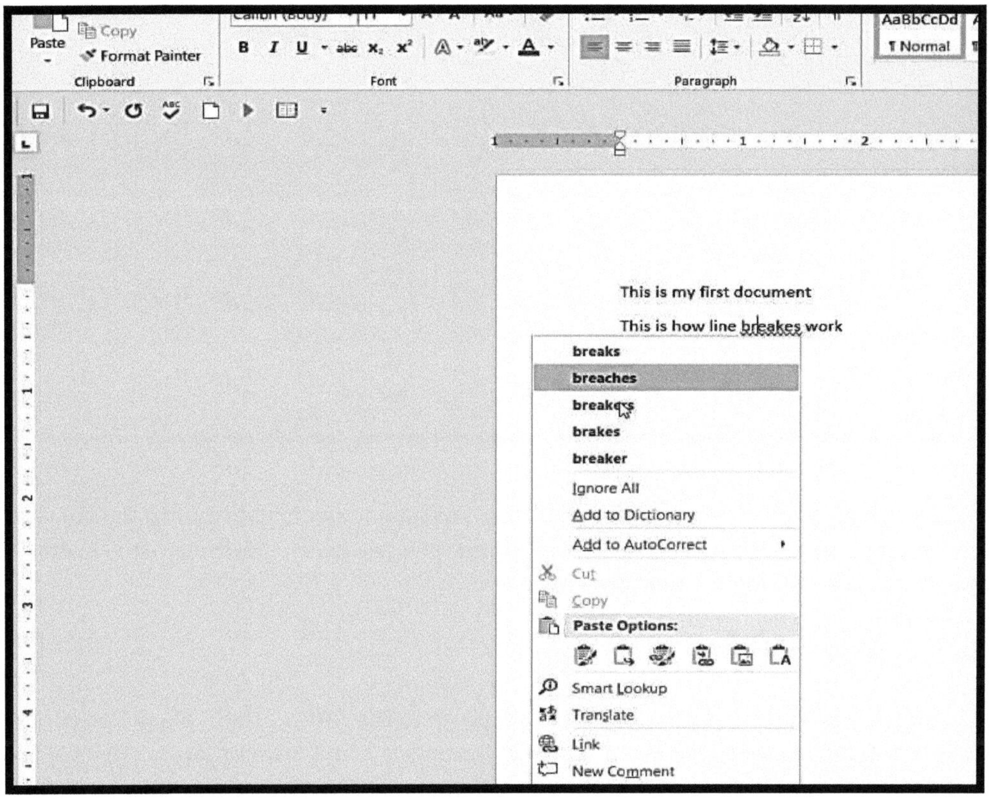

If you add a comma and/or a double space to a sentence, Words will detect a problem. There's a grammatical issue, and this type of error is fairly common, especially when you're working quickly. You might accidentally type a comma, or hit the spacebar twice, but you can see now with the double underline that if you right-click it tells you how to rectify it.

Unknown Words?

When you misspell a word, you can right-click and add the word to the dictionary. There will be times when your sort of word is the name of someone or a place. Word doesn't always recognize things like that and will tell you it's spelled incorrectly when it isn't; it's just not recognized by Word.

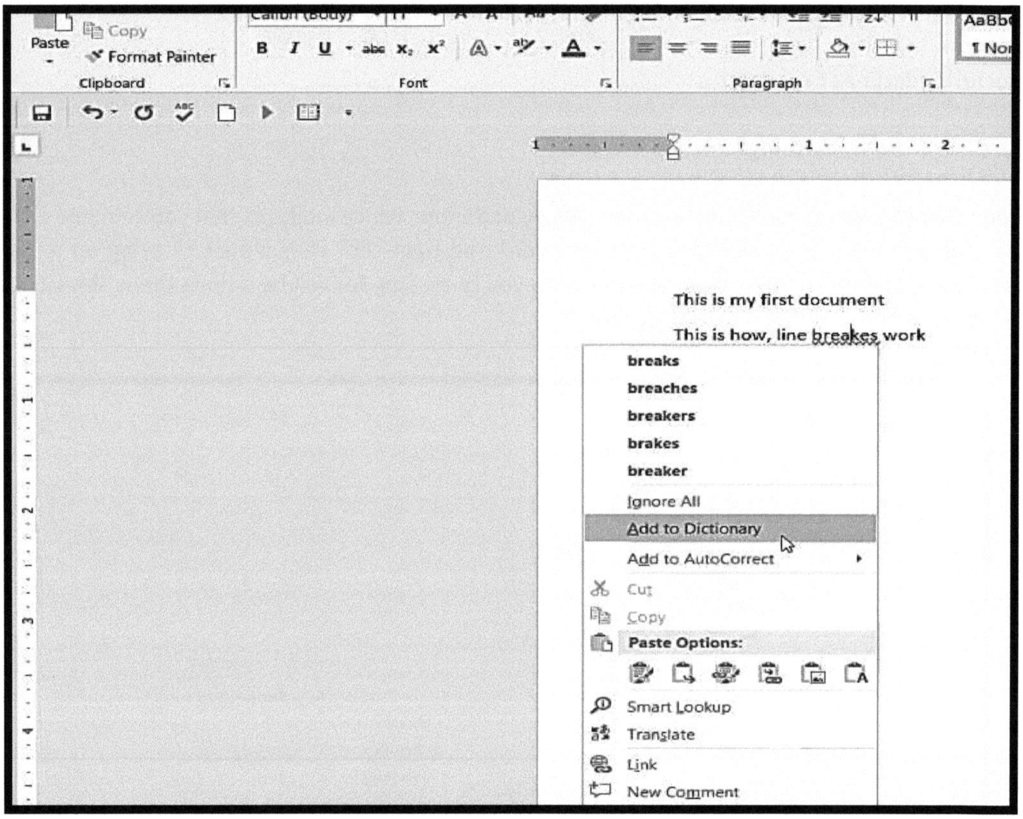

If you, have it, you can pick "**Add to Dictionary**," and it will add the word to the Word dictionary, ensuring that it does not register as a spelling problem the next time you type it. So, keep that in mind, and all of these options are available from the right-click menu.

Switching Between Documents

If you want to create another blank document, simply go to **File**, "**New**," then "**Blank Document**." If you check up in the title bar, you'll see that it now says Document 2. This means that your original document hasn't been closed; it's just been moved beneath the one you're currently working on.

There are a couple of options for switching between the two documents, or in this example, switching back to the original. You can access your Word icon by hovering your mouse over it in the bottom status bar. If you have the Word icon pinned to your toolbar, you can see both of your documents and choose which one you wish to work on. As a result, you can return to Document 1 at any time.

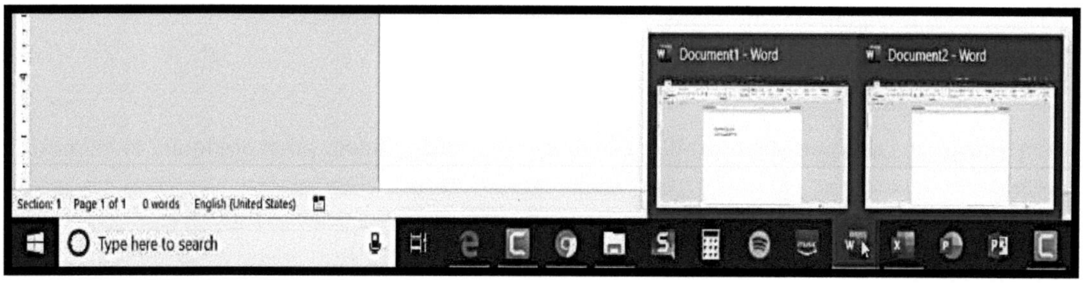

You may also go to the View ribbon and select "**Switch Windows**," which will display a list of all of your open documents. You can go back to Document 2 once more. Being able to flip between any or all of the documents that you have open is quite convenient.

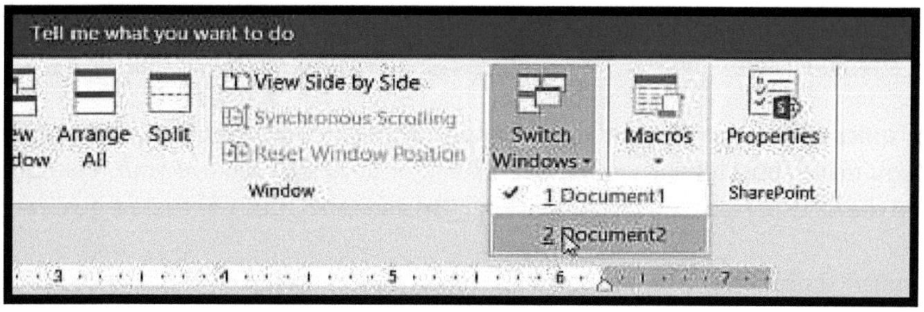

Closing Your Document Without Closing Word

If you want to close your document, you can do it quickly and conveniently. A word of caution: if you look right up in the upper right-hand corner, you'll see a large cross icon that states "**Close**," which will close Word completely, so be wary of that.

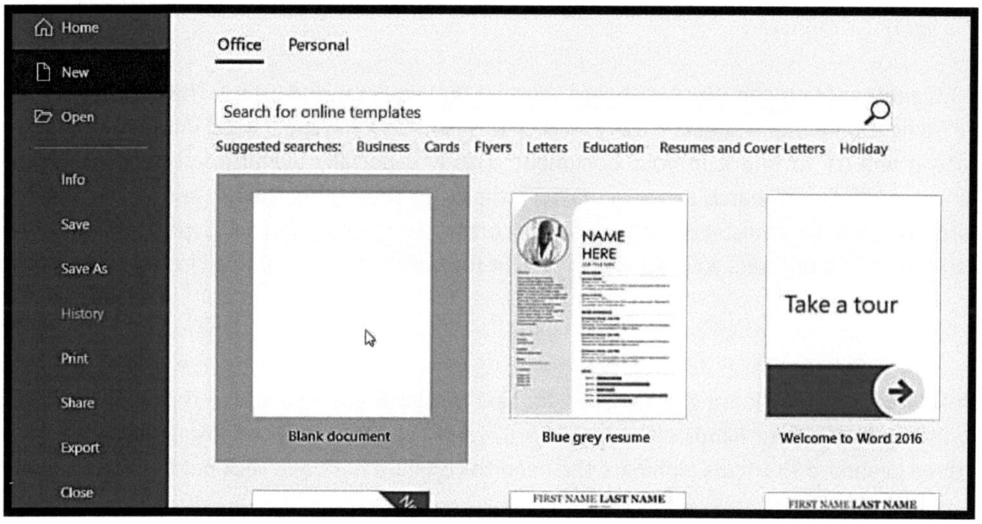

Note: You can just go to File and select "**Close**" if you only want to close one file.

Some Useful Keyboard Shortcuts

Here, you get to see some Microsoft Word keyboard shortcuts; the ones that are found to be most useful.

Your Familiar Commands

To begin, let's review some of the most commonly used keyboard shortcuts, which are **Ctrl + C**, **X**, and **V** for **Copy**, **Cut**, and **Paste**, respectively, as well as **Ctrl + Z** to **undo** a mistake and then **Ctrl + Y** to **redo** whatever you've undone. Those are some basic keyboard shortcuts that everyone should be familiar with; they can be used in nearly any program, but let's look at the most useful Microsoft Word keyboard shortcuts.

Quick Save and Print Commands

The first is "**quick save**," and we all know the importance of saving frequently. It's a good idea to save whenever you make substantial changes, but instead of having to remove your hand off the keyboard and go up to the Save button, you can simply hold **Ctrl** and tap **S** (**Ctrl + S**) to do a quick save.

There's a similar shortcut for printing quickly. Holding **Ctrl** and tapping **P** (**Ctrl + P**) brings up the print choices, which you can adjust if desired, before clicking Print. So, **Ctrl + S** and **Ctrl + P**, are some good easy time savings.

Spelling and Grammar Check

The **F7** key is yet another simple keyboard shortcut. The **spelling** and **grammar check** can be accessed by pressing F7.

The Navigation Command

Ctrl + F is another Microsoft Word keyboard shortcut that you may find useful. The **Find panel**, also known as the **Navigation panel**, appears when you click this. So, you can type a word into the navigation search box and it will try to find it in your document. This is especially useful if you're working on a large document, such as a research paper, a master's thesis, or something similar, and you've got thousands of words. If you make a mistake, such as using incorrect terminology in your paper, and you want to find all instances of the mistake, you can simply type in the word and it will find all instances of that mistake.

Text Formatting

The following three keyboard shortcuts are for text formatting. When you're typing in Microsoft Word, every time you take your hands off the keyboard, you're wasting time and losing efficiency. To aid with this, three keyboard shortcuts eliminate the need to come up here and click bold, italics and underline.

To underline or italicize text, press the Ctrl key and tap U (Ctrl + U). As soon as you begin typing, you'll see that the content is underlined. Hold Ctrl and hit U again to stop the highlighting; nothing will be underlined now. **Ctrl + B** creates bold text, and **Ctrl + I** similarly create italicized text. Of course, if you don't undo "bold," your italics will be bolded and italicized, so remember to turn off the italics; the same goes for underlining and bold.

Automatic Page Break

Automatic page break is another useful Microsoft Word keyboard shortcut. Everyone has gone into Word, placed the cursor where they want a page to stop, and then tapped enter numerous times to transfer the rest of the text down to the next page at some point. That is completely unnecessary; you don't need to press enter fifteen times to get your text to the correct location. Instead, if you wish to finish a page, break it, and start a new page below, simply hold the **Ctrl** key and touch **enter**; the page will be broken, and the new page will begin below.

The Repeat Command

F4 is another helpful Microsoft Word keyboard shortcut. What is the purpose of F4? Whatever you've done recently will be **repeated**. If you touch the F4 key after you've tapped "delete" or "backspace" a few times, it will continue to delete. If you press F4, anything you were doing at the time will be continued. Another example: if you type the phrase "Hello" followed by a space, then press F4, it just repeats the word you just wrote.

The Return Command

Shift + F5 is the next keyboard shortcut. Holding down the Shift key and pressing F5 returns you to the last location where your cursor was. So, let's say you want to copy a paragraph from the bottom of a page and paste it at the top of it. You'll highlight the text or sentence, copy it with Ctrl + C, and then hold **Shift** and hit **F5** to return to where you last had your cursor, where you can now paste in that sentence or word with Ctrl + V.

Extending Your Highlighted Texts

The "**Shift + Left arrow**" and "**Shift + Right arrow**" keys are also handy. Remember that a double-click on a word highlights the word, and a triple-click on a word highlights the phrase or paragraph, depending on the application you're using, which is quite beneficial if you want to copy the entire paragraph or word. But what if you're dragging or double-clicking and want to enlarge the highlighted text before copying it? If you hold down the Shift key and then press the right arrow key, the highlighted text will extend to the point where you want it to stop, and you can now copy and paste it. You won't have to click and drag to get exactly what you want; simply double-click the initial word, hold Shift, then press the right arrow to stretch it as far as you like. The left arrow returns you to the previous position and expands to the left, however in Microsoft Word, this is a highly helpful keyboard shortcut.

Duplicating your Format

Then there's **Ctrl + Shift + C**, which performs what? It duplicates the formatting rather than the text. So, let's assume you've set up your document with the layout you want, and you'd like to duplicate it elsewhere. You select the formatting you wish to copy, hold Ctrl, Shift, then hit C, and the formatting is copied. Now you go down and highlight the area where you want to paste the formatting, and if you hold **Ctrl**, **Shift**, and hit **V**, it copies the formatting rather than the words.

There are many more Microsoft Word keyboard shortcuts available, but these are the most commonly used.

Now that you know how to get started with Word, we've gone over all of the fundamentals that will serve as a wonderful springboard and a solid basis for moving on to some of the more advanced options, which we'll begin to explore in the next part.

CHAPTER THREE

EDITING YOUR DOCUMENT

In this chapter, we'll explore how you can make your text look presentable and easy to read. We'll be focusing on non-printing characters and what we mean by non-printing characters is that there are some characters that you can't see, such as space and enter tab, and we'll show you how to make them visible so that you can work with your document a little bit better. We'll also look at line spacing and how to modify it while typing in your manuscript. As we proceed, we'll look at paragraphs and other components that you will come across as you type in Word.

To edit text, you have to **select** it first. Just **click** and **drag** on it as you need or hold down the Shift key and move the **insertion point** character per character with the arrow keys. You can also double-click on any word to select it quickly or click once from the left to select entire lines.

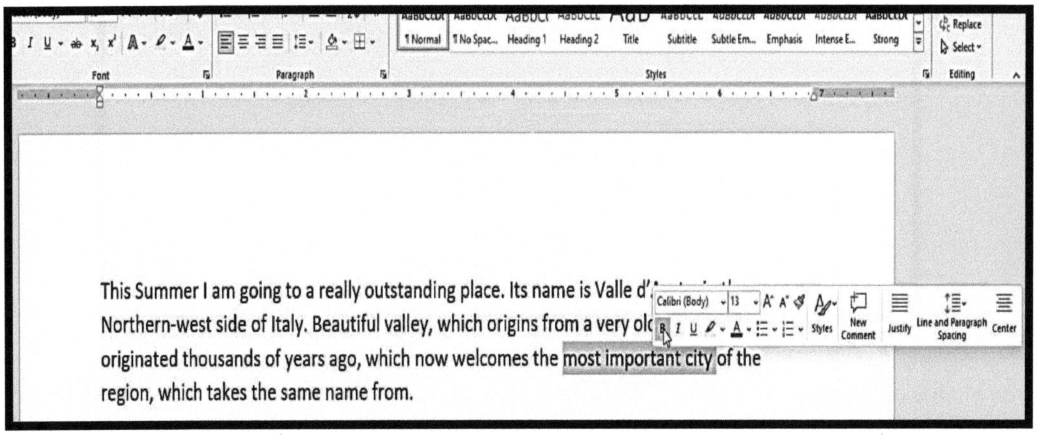

At this point, you can edit text with the dialog box showing up or the home tab above. You can **adjust font family**, **size**, **style**, **change the font color**, **add highlighting** or apply amazing effects, including **outline**, **shadow**, or **reflection**. Hover over these to get a quick preview before applying any.

Working with characters that aren't printable

If you start with a blank document, the first thing you'll do is set it up. So, you'll start by typing a word or a sentence, then pressing the Enter key, which will move you down a line.

You might come across the **tab key** on your keyboard when working on your document, which will offer you a little bit of space in there, and what the tab does is that it pauses at the half-inch point on the ruler, so you can press tab many times and it will travel half an inch across the page each time. The advantage of utilizing tabs rather than putting in a lot of space to get your cursor where you want it is that everything will be lined up, so if you go beneath and start typing something, pressing the tab key will jump you precisely underneath where the tab was in the above line. So instead of tapping the spacebar over and over to move your text along, it's a lot easier to line things up. So just keep in mind that you have those tabs while you're working on a document, and you'll see how to alter them and create your custom tabs later so things line up where you want them to line up on the page.

A couple of things to keep in mind while typing: you'll notice that as you get to the end of these sentences, it just wraps around; you didn't have to hit Enter or insert a line break to get it to wrap around to the next line, which is perfect; you'll also notice that you may have a little squiggly line under some texts; if you right-click and look at your options, the top line is probably what you want. After typing, you'll realize that you've used some tabs and that there are some spaces in there that you can't see. "Well, I don't need to see the tabs and spaces," you might say, but it's sometimes quite important, and you'll see some examples of why in a moment, plus there's a feature in Word called the **Show/hide** button that will display all of those non-printing characters. When you go to the paragraph group on the home ribbon, there's a little object that looks like a paragraph marker that says show/hide when you hover over it, and the dialog box says display paragraph markings and other hidden formatting symbols when you click it.

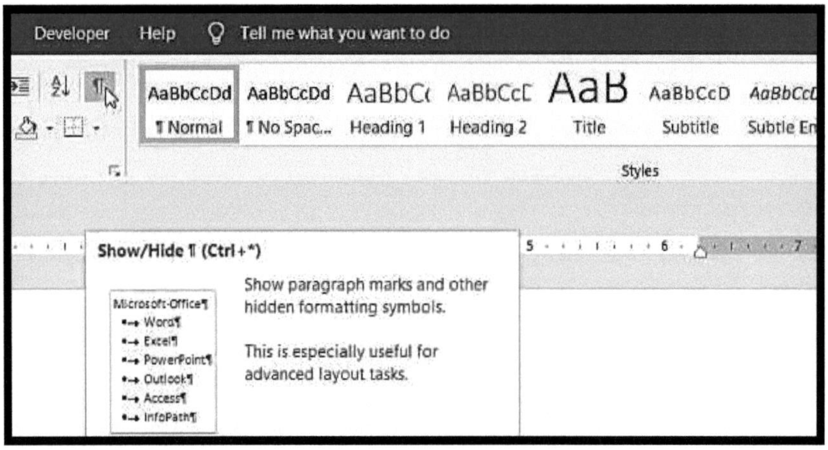

If you enable this, you will see that the little paragraph mark appears everywhere there is an Enter, as well as the little arrows that indicate the tab.

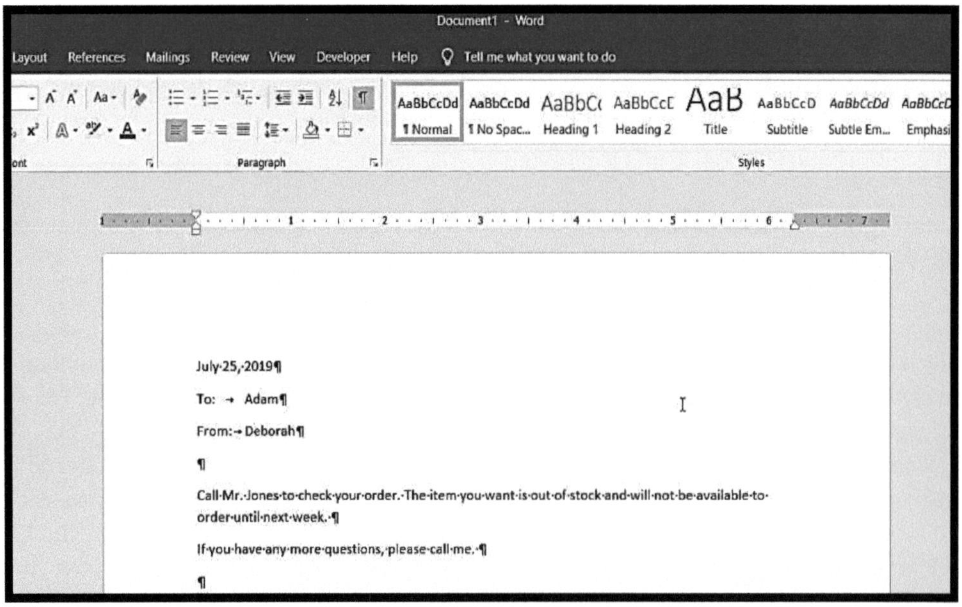

Have you ever printed a document and received a blank sheet of paper at the end, which you couldn't figure out why because you didn't have anything on that piece of paper? If you turn on your paragraph marks, you'll usually find that you have a rogue or a few rogue paragraph markers or extra Enter keys on the next page, which is why that page is printing. All you have to do now is go in and delete those extra lines, which you wouldn't be able to see if you didn't turn on that show/hide feature. So having these non-printing characters visible is beneficial, and you can turn it off by clicking the same icon again.

Spacing between lines

When you press the Enter key in Word, you'll notice that there's a large space between the lines that don't appear to be single line spacing; its double line spacing by default. If you want to clean this up a little bit, you can select all of your document or all of the text in your document using the shortcut key **Ctrl + A**, which highlights everything. Then go to the Paragraph group in your home ribbon and pick the little dropdown for line and paragraph spacing. You may browse through the options and as you hover over them, more or less space will appear between the lines in your text, so it's really up to you which one you choose.

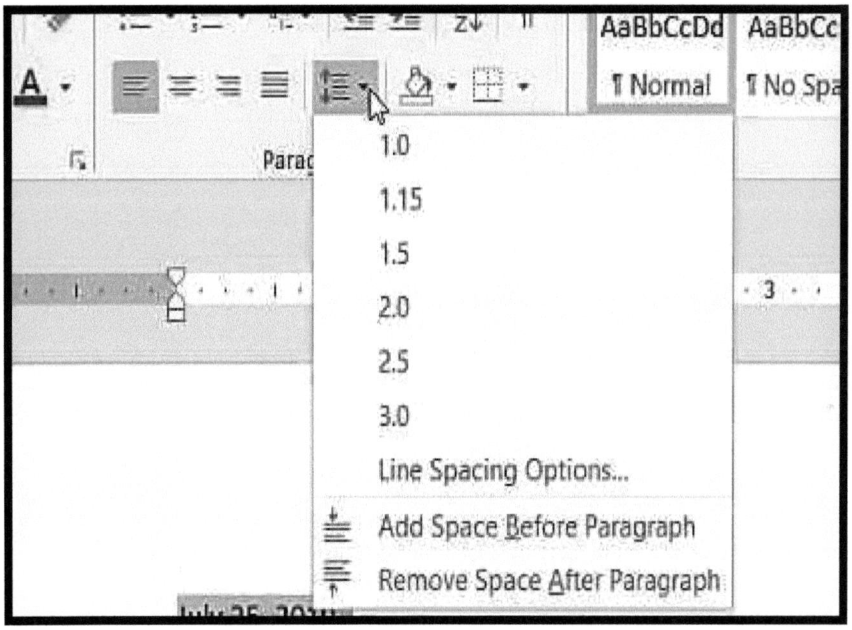

If you select your entire document, it will adjust the line spacing based on what was already there. So, if you only wanted to adjust the line spacing in one area, you could simply highlight that area, go up and make it smaller or bigger, or you can go into **line spacing options** and play around with how much spacing you want, and if you don't like the space in between, you can reduce it to zero, click **OK**, and you'll have no space in between.

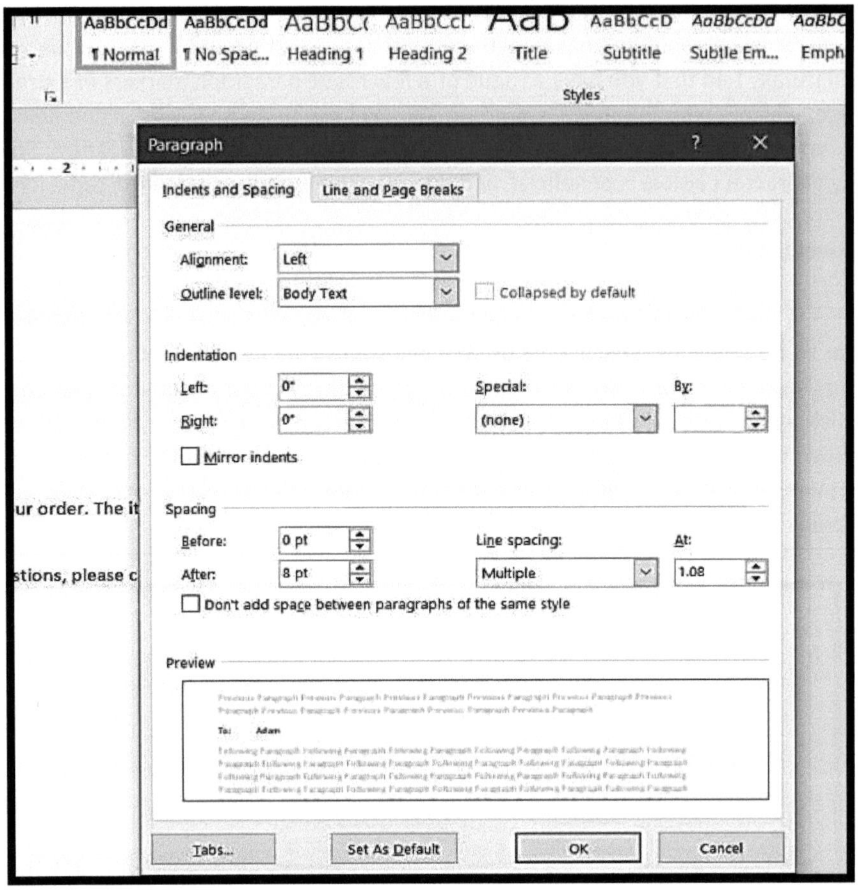

It's certainly worthwhile to experiment with some of the line spacing possibilities. You also have a couple at the bottom where you may **add space before or after paragraphs**, depending on what you want to do, but everything in this area is quite configurable.

So far, we've learned how to turn off and on paragraph markers, as well as how to show and hide non-printing characters and change line spacing. Let's move on to paragraphs now.

Paragraphs

With the paragraphs, you can easily manage the text distribution. Every time you make a new line with the Enter key you create a text paragraph, such as a collection of sentences that share the same topic and properties. You can show all of these by clicking on Show-Hide under the Home tab. Each paragraph has its alignment and distribution options. Click within the paragraph and go to **Home**, then **Paragraph** to set its alignment, choosing Justify to spread the words to cover the whole page width.

With Line Spacing adjust the distance between the paragraph lines as you need. Use Shading to apply a custom background color to the whole paragraph and Borders to add borders around it. Switch to the Layout tab to manage the paragraph indentation and spacing, concerning the previous or the following paragraph.

Before we move on to the next section, you'll see how to make some basic modifications to a document. These are just some fundamental editing functions that you should be familiar with. We've already spoken about how to edit a document in a few different ways, and now I just want to show you how to **delete**, **backspace**, **undo**, and **redo** items in your documents.

Getting Rid of Unwanted Text

When it comes to deleting, the Delete key on your keyboard always deletes everything to the right of it, so if you start pressing it, you'll notice that it deletes the word on the right.

Backspace, on the other hand, deletes everything to the left of it, so if you start typing, you'll see it deletes the word to the left. Just keep in mind that they function in slightly different ways.

Undoing and Redoing

Another thing to remember is that the **Undo** and **Redo** buttons are really useful.

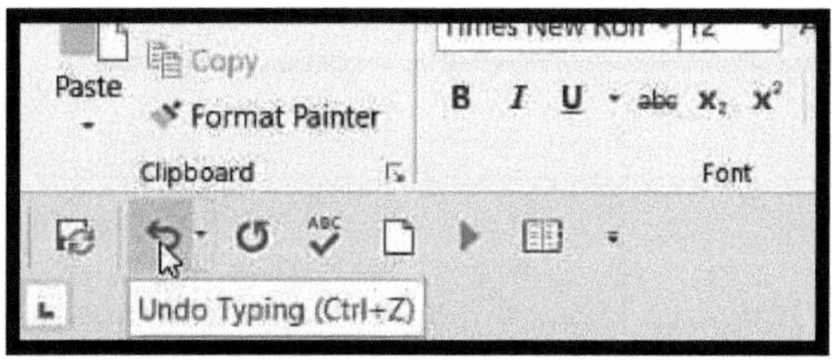

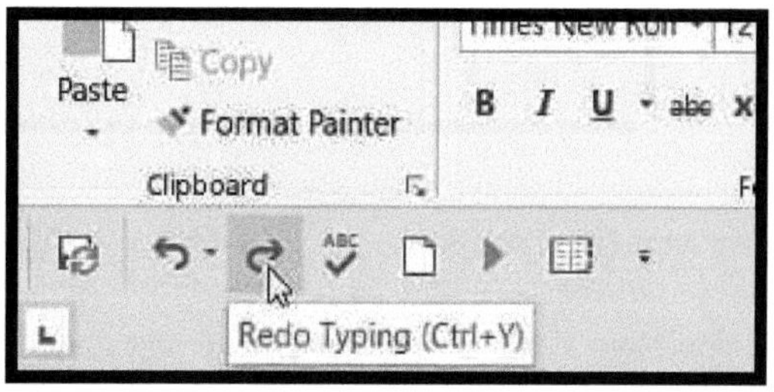

It's amusing how many people can't function without their undo button. These two buttons will always be there by default on the quick access toolbar to make them incredibly easy to find. If you look up there, you'll notice the Undo button, which allows you to undo what you've just done. If you've just backspaced the word "Property," for example, clicking undo will give you the P back, the R back, and so on, and you may keep undoing to reverse your last action.

The same thing happens when you redo something. If you decide you don't want that word there, you can redo and the deletion will be undone. It's also worth mentioning that beneath or adjacent to your undo icon is a small drop-down menu that allows you to go back and select how far you want to undo; it also allows you to undo many times in one go, so keep that in mind.

These have a shortcut key as well. The shortcut key for Undo is Ctrl + Z, so if you accidentally do anything and then think, "Oops, I want to backtrack out of that," a simple Ctrl + Z is your greatest buddy. The shortcut key for Redo is Ctrl + Y.

CHAPTER FOUR

SAVING YOUR DOCUMENT

Updating and Saving a Document

If you had previously saved this document and then made modifications, you could just click **Save** and nothing would happen; it would simply save your changes to the file name.

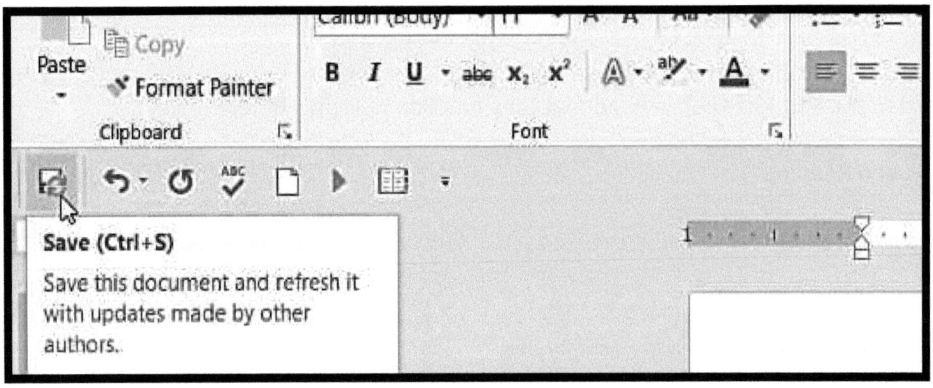

The First Time You Save

If you have a document that you've just generated, it's tough to know if it's been saved or not at first glance, but there are a few indications. The greatest clue is that if you look up the title bar at the top of the document, it only says Document 1 at the moment. You haven't saved this document since it doesn't have a name; recall that Document 1, 2, 3 is the default name you get when you create a new blank document. So now you know that this document hasn't been saved, and the first thing you should do is save it to a location of your choice.

There are several different ways to save a document in Microsoft, as there always are. You can go to your **Quick Access toolbar** and find the **Save** icon.

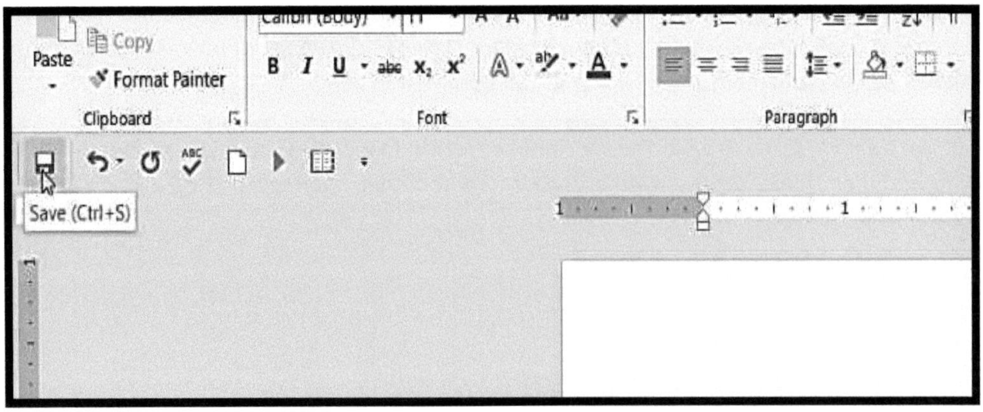

Because you haven't saved this document yet, clicking **Save** will take you to the "**Save as**" area.

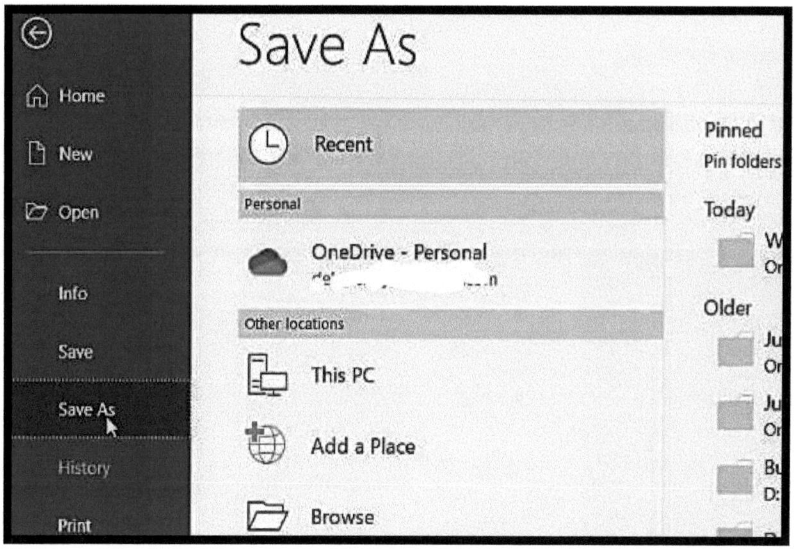

So that's the first option; if you have the symbol on your **Quick Access toolbar**, you may click on it.

The second option, which essentially takes you to the same location as we just left, is to enter the backstage area. You navigate to **File** and then to **"Save as.**" If you're not used to it, this "**Save as**" page may appear strange; nonetheless, it's the same in all Microsoft products. On the left-hand side, you'll see a list of recent folders, which will show you all of the folders that you've lately saved.

Putting files on OneDrive

Alternatively, you have other options for saving your document; for example, you could save it to OneDrive, a cloud storage service, which would be a wonderful option if you wanted to collaborate on it.

So, if you wanted to share it with others and also make sure that you could access it from anywhere, you would save it to OneDrive; to the Cloud.

Any Other Location for Saving

You can save it to any other area; for example, you can save it to your PC's local disks. You can also opt to browse to a certain location, and if you click on browse, you can now select a destination from your local files.

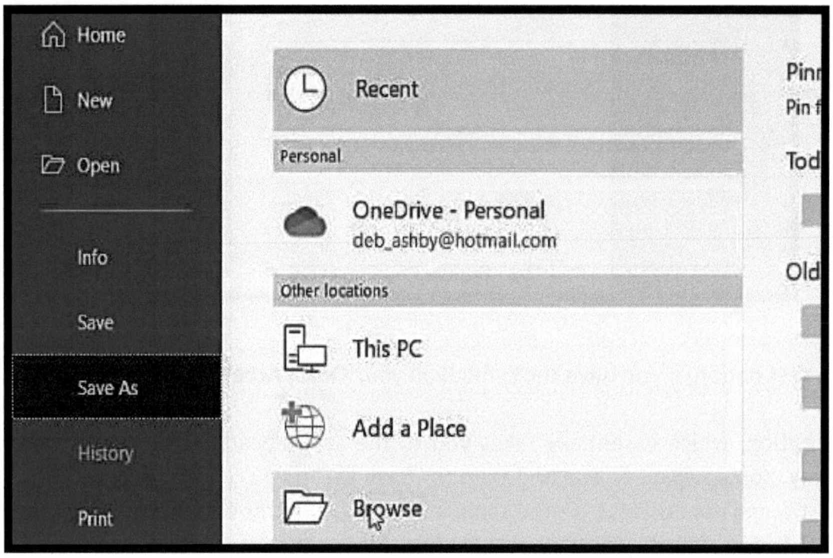

So, you choose a folder and drop your file into it, after which you can give it a name.

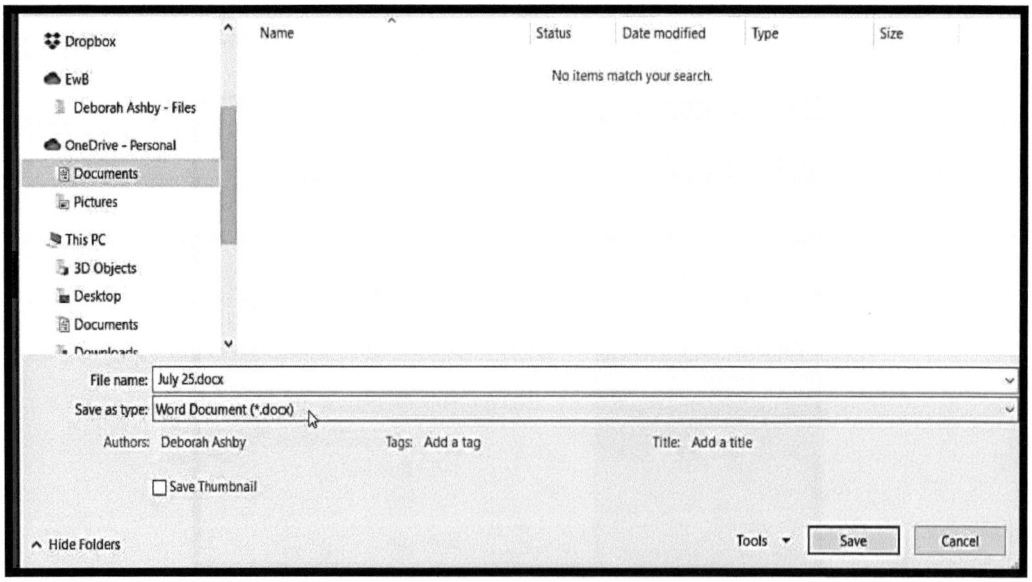

As you can see in the File name field, it took the first line of the document and used that as the default file name; it will always use the first line of your document as the default file name, which is usually not what you want to call your files. So, you're going to swoop in and make a shift. Also, if you click the dropdown under Document type, you can save your document as a variety of alternative file kinds.

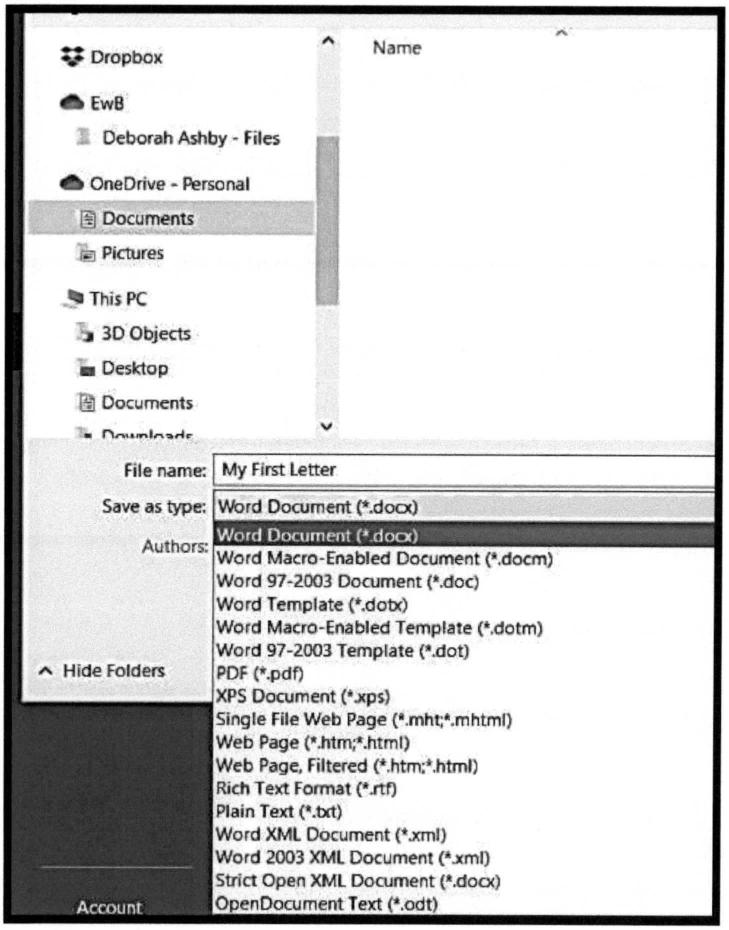

You might want to use some of these, and that will store them in that format. You can also save your work in the old Word format; but, why would you want to save your document in an older version?

Well, as mentioned briefly in one of the earlier sections, if you plan to send this to someone who you know doesn't have the same version of Word as you or has a much older version, you can help them out a little by saving your document as a docx file. This ensures that when they open it, it will open in compatibility mode and they will be able to read it, so just be careful who you're sending it to.

You can also save it as a template, save it as a PDF if you want to reuse it, and we'll go over some of these options as we go through this book, but for now, keep it as a docx file extension, then click Save, and that's all there is to it.

If you look up to the top of your document, you should notice that it now has a name.

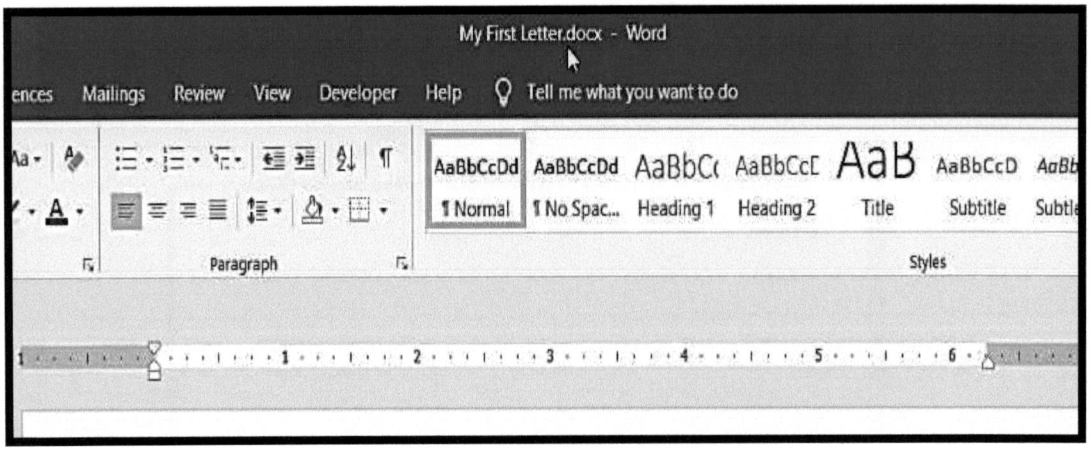

What this means in terms of saving is that if you make a change to this and want to save it, you can go to your quick access toolbar and click on **Save**; you don't need to go back into "**Save as**," and if your Save button has those little arrows over the top, it's because you've saved it into a cloud location, so when you save it, it's essentially doing a refresh with the cloud version, so if you click Save, it just saves. Another option is to use **Ctrl + S** on your keyboard, which is a shortcut for saving as well.

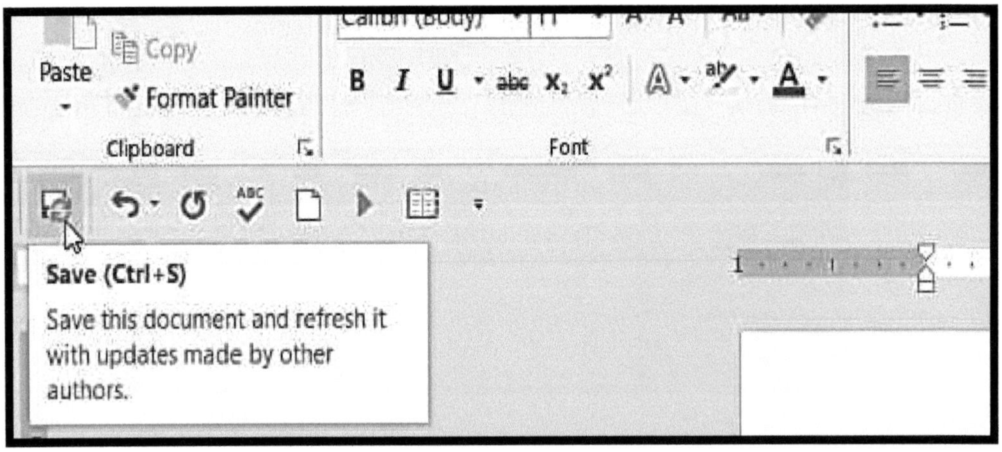

When working on a document, if you've saved it and know the filename, it's a good idea to press **Ctrl + S** now and then to ensure that your changes are saved. Now, if you make one more change and don't save it, but instead choose to dismiss your file by clicking the cross at the top, you'll notice that it prompts you to save your changes. As you exit Word, it will detect any unsaved changes and prompt you to save them, allowing you to choose whether or not to save them.

CHAPTER FIVE

DOCUMENT NAVIGATION

We've arrived at the chapter where we will talk more about Documents. **In this chapter,** we will focus on opening a document, we'll have a look at a couple of different methods to examine your document once it's been opened.

Opening A Document

It's quite straightforward because it's essentially the inverse of saving. If you've already closed a document and wish to reopen it, head to the File tab, which, as you may recall, takes us to the backstage area. If you have nothing open, it may take you to the open area automatically; Word is expecting that you will want to open a file. That could be the case, or it could be that you wish to start over.

The "**Open**" page appears to be identical to the "**Save as**" screen we were in earlier; you'll notice that "**Recent**" is highlighted, and you also have documents underlined, indicating that it's showing you underneath all of your recent documents, which you should be able to see right at the top. Now, if you want to open a document, all you have to do is select it from the list and it will open.

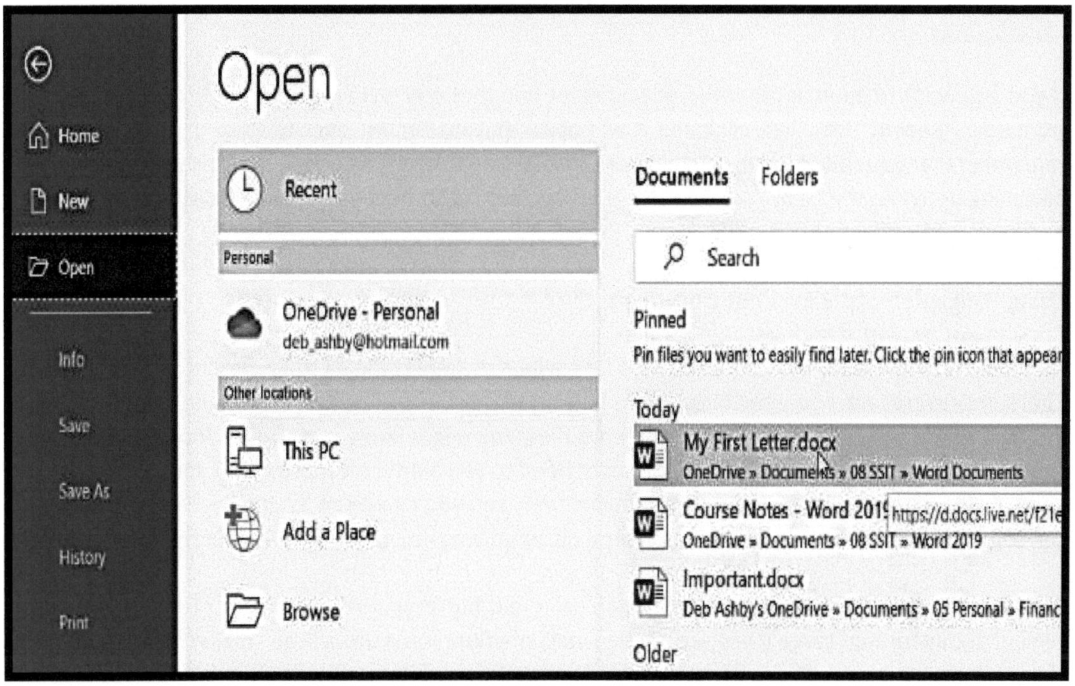

You can also open a Word document by browsing your computer and selecting the file you wish to open, then clicking open.

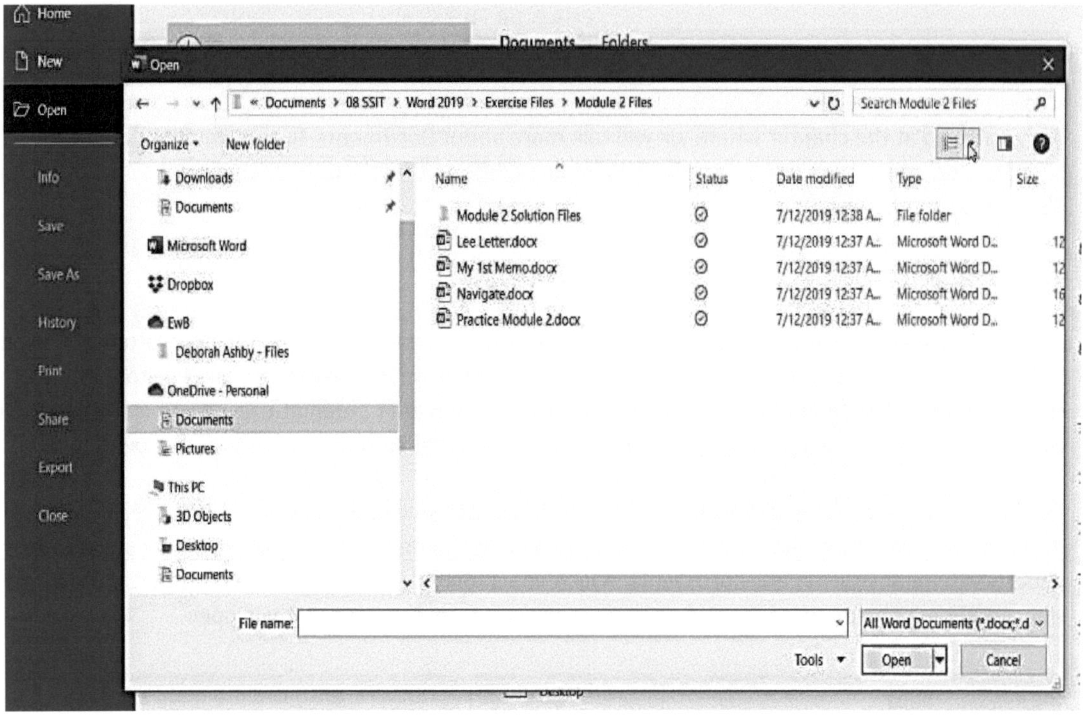

If you also want to open another file, say a recent one that you just saved, go back to "**File**" and select it from your "**Recent**" list, then click and it will open that document over the top of the other one, and remember that you can switch between them by going up to View, selecting "**Switch windows**," and then switching between the two. You may also use your mouse to hover over your word icon at the bottom and navigate between the two of them.

Using File Explorer to View Your Document

Also, concerning the Explorer window, let's take a look at something a little different but very significant. There are several ways to view files before you open them if you go to browse and then move up to a location where you have several different word files. You might know what you're looking for most of the time, but sometimes you won't, and you might need to see a little bit more detail about a file. There are several different ways to view the file list in Explorer, and you can do so by clicking on the little option at the top right-hand side. It will give you several different ways to view the file list in the Explorer window.

You may currently have the "**Details**" option selected, but if you switch to Extra Large icons, they will appear in this format. Large icons are a little small, medium icons are a little smaller, and small icons are the smallest. We have a popular feature called Lists. Most people prefer the Details option because it not only provides you the list, but also offers you a little bit more information about each file, and you can see in the status column that they're all saved to the cloud and synchronized. You can also view when these files were last edited, as well as their kind (whether they're Word documents, PDFs, or Excel files) and size. It's useful to be able to view the file size sometimes, especially when working with very huge files.

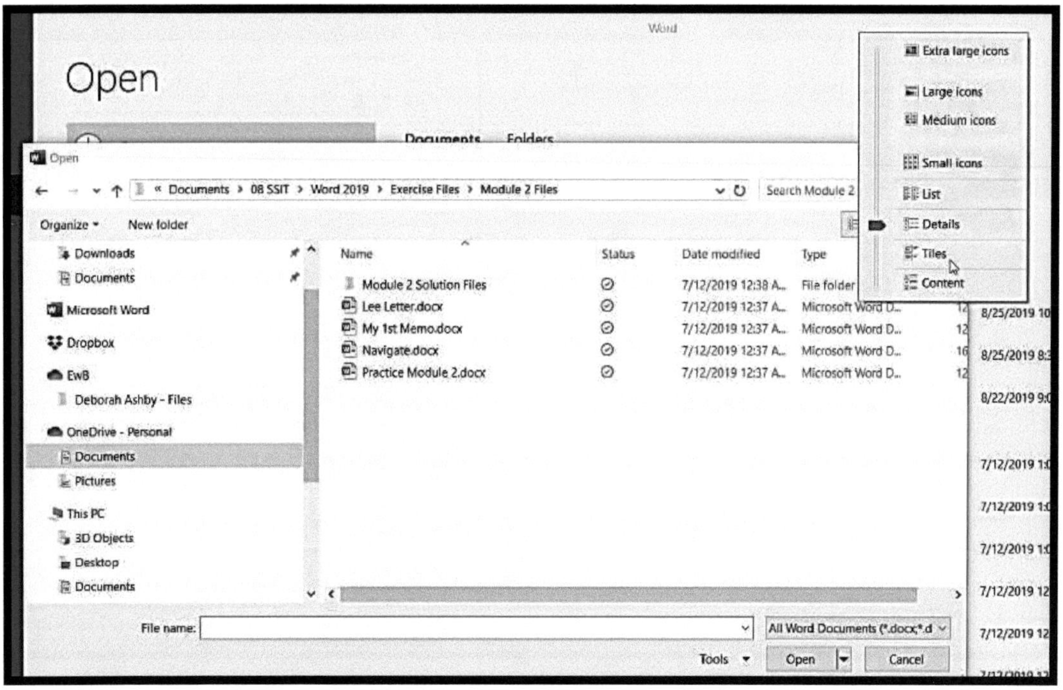

There are a couple of others in here, such as Tiles and Contents, which provide a little bit more information such as the author, date modified, and other details; just be aware that you have different ways to view your files so you can see a little bit more information and decide which one you want to open.

Side-by-Side Document Viewing

Another feature that comes in handy from time to time is the ability to compare two papers side by side. If you're comparing two papers or cutting something from one and putting it into another, this is useful because you don't want to continually switch between them. You can put them side by side and compare them. You'll see a "**View Side by Side**" option on the View ribbon; click it and voilà, both of your papers are now taking up exactly half of the screen.

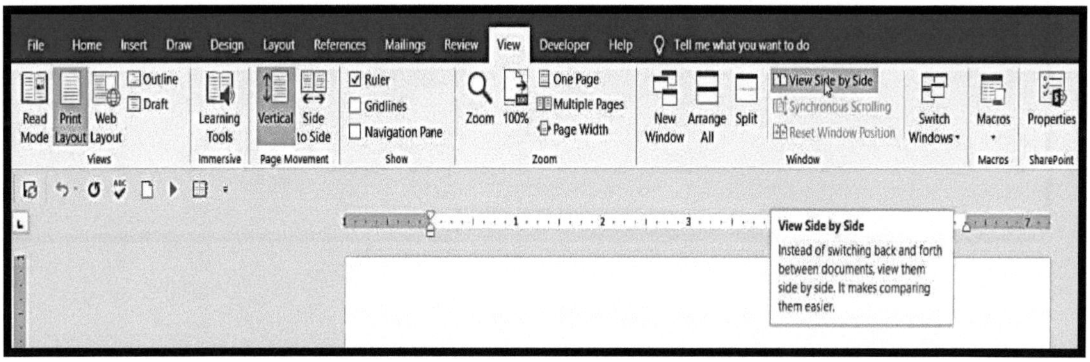

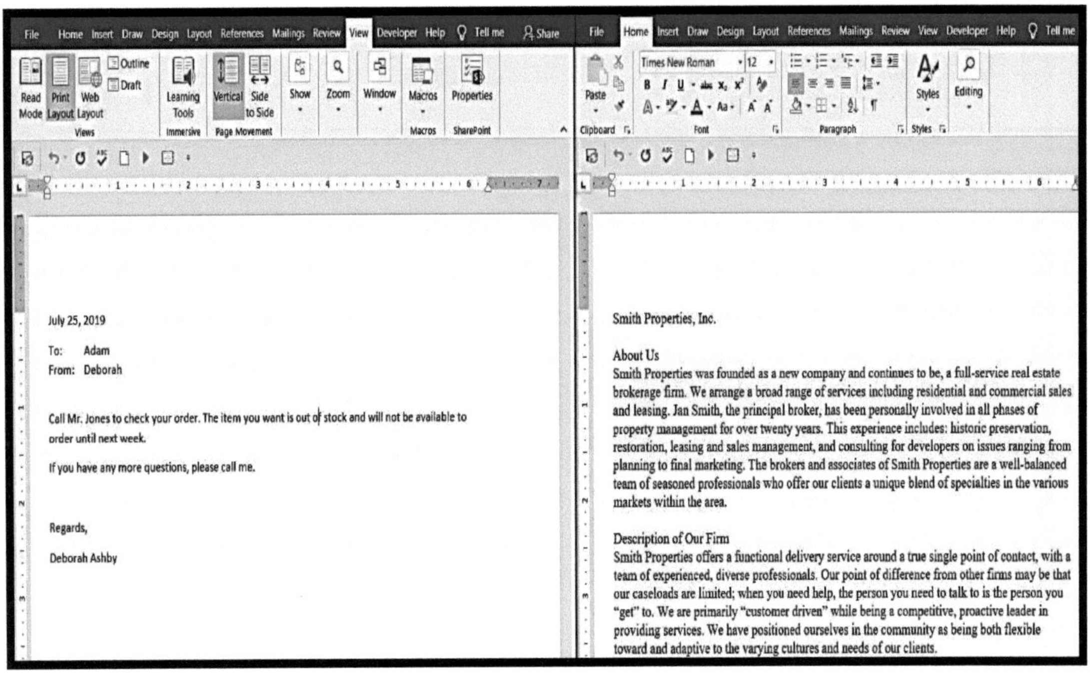

If you simply want to restore one of the documents to its original size, simply click on the "**Maximize**" button in the corner.

Creating A Document Split

On that same View ribbon, you can also use the "**Split**" option, which is quite fascinating. Let's see how that works out. When you click "**Split**," it will split the same document into two parts of the screen, allowing you to scroll each document independently.

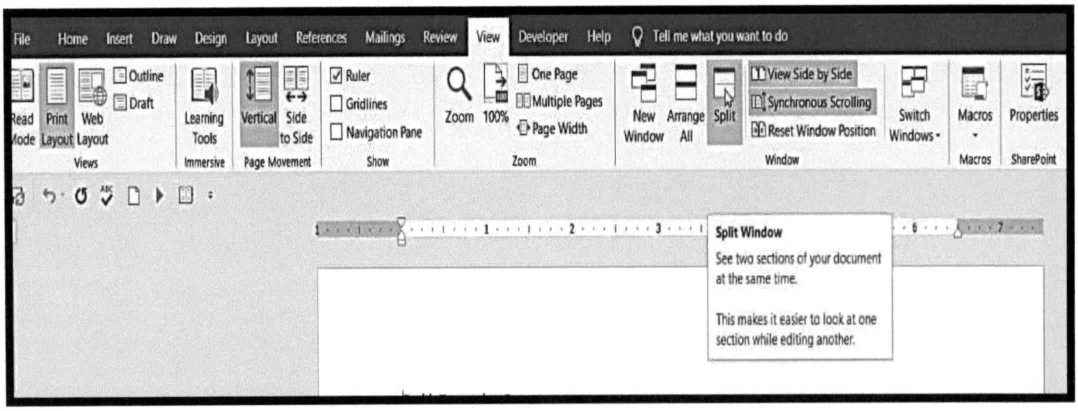

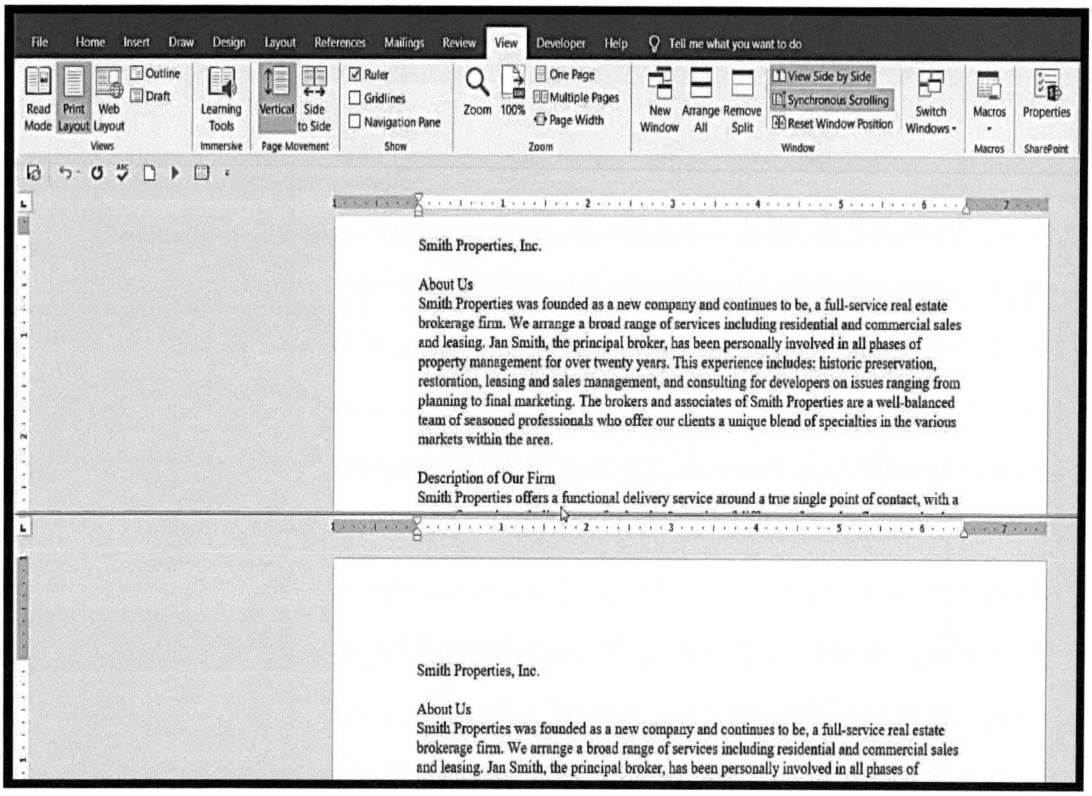

This is useful if you're doing some comparisons or copying and pasting; you can perform a variety of things with the Split option if you want to see two separate portions of the same page at the same time. If you wish to return to it being just one document, select "**Remove Split**" and you're done.

Arranging Your documents

The "**Arrange All**" button is a final option that you might choose to use. This is found on the View tab, and when you click it, it tiles the documents over each other. If you have a lot of documents open, definitely more than two, being able to see them all arranged next to each other can be beneficial.

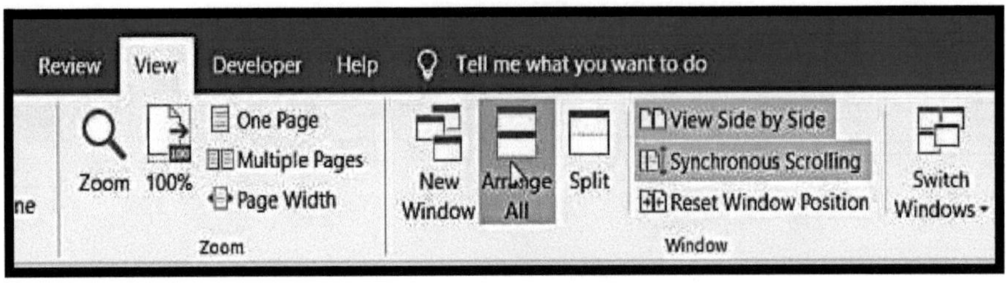

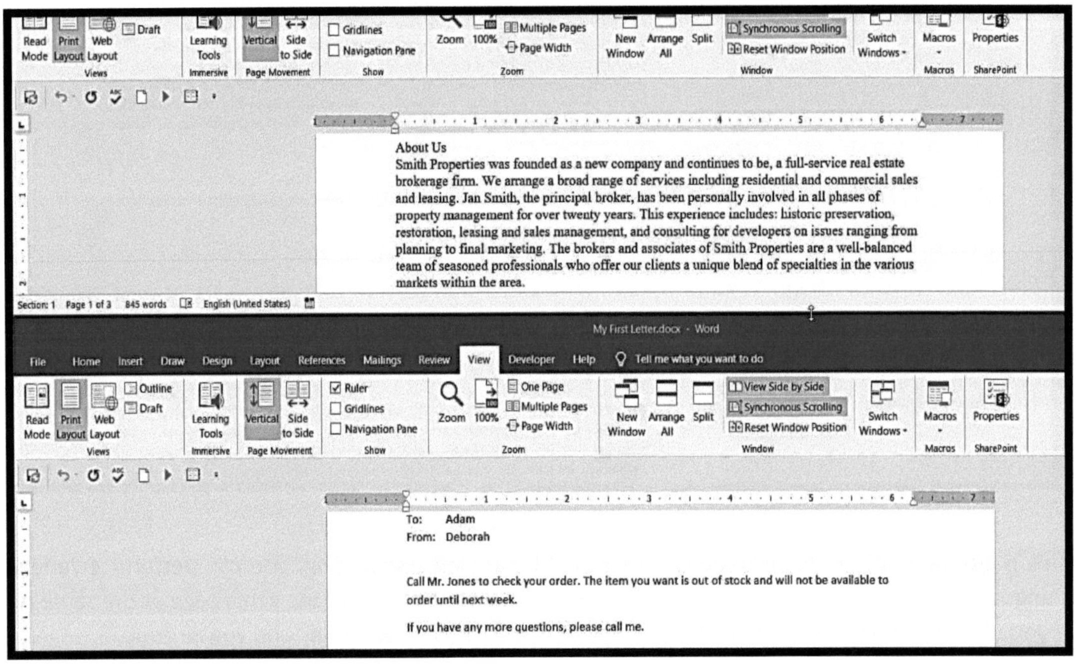

Simply click the "**Maximize**" button on one of the papers to exit.

Getting Around in Documents

In this part, we'll look at how to navigate your document.

Let's imagine you have a document that has been created; the first thing you should do is look down to the bottom left hand of the screen and look at Section 1, Page 1 of 3, 845 words, as shown in the document below. We're obtaining a little bit of information about this document, and you can see from our status bar that it has three pages, the number of words and that we're presently in section 1. We'll touch on how to section up documents a little bit later, so doesn't worry about that for now. What we want to talk about today is how you can rapidly navigate your document using some of the capabilities in Word, which is especially beneficial if you have a big document.

Although the document above is only three pages long, many of the documents you write will be hundreds of pages long, so being able to hop around fast to get what you need is critical and will improve your efficiency. Let's start with the most obvious point.

Scrolling through Your Document

The right or left-hand Scroll bar is a simple method to navigate through your page. If your document only has a few pages, you can simply utilize it to scroll up and down.

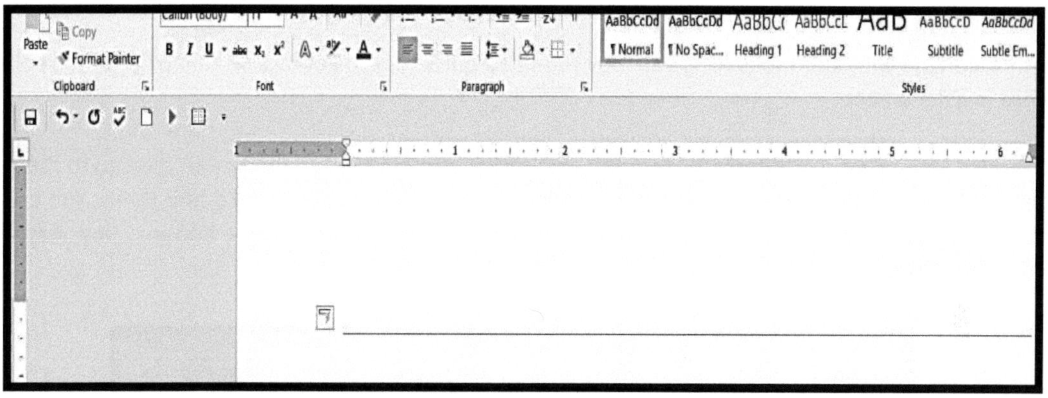

Other Alternatives

Let's look at how you can use various utilities to make a lengthier document. If you had a manuscript with hundreds of pages and wanted to jump to the bottom of it, using the scroll bar and scrolling down would take a long time; by the time you finished, you'd probably be taking a risk. A quick way to do it is to use your keyboard to hop straight down to the bottom; in Word, use **Ctrl + End** to leap down to the bottom. You may confirm this by going down on the left-hand side of your screen and checking your status bar to see if you're on the last line of that document's last page.

If you want to go back to the top, use **Ctrl + Home**, which will do the opposite.

A couple of other shortcuts: holding down **Ctrl** and **pressing your arrow keys,** such as the **right arrow key**, jumps you around that document per word, and **using the left arrow key**, you may go the other way. **Arrow keys + Ctrl** Up and Down will jump you up one paragraph at a time, and if you want to select everything in your document, press **Ctrl + A**, which will select all of the text in your document. We'll be using **Ctrl + A** a lot as we go through this book because it's a really useful feature when you want to make mass changes to your document.

This chapter covered a lot of ground on documents, and hopefully, it gave you a better understanding of how to open one or more files and the many methods to view them while working on them. You also saw a few simple techniques to navigate about your page using your keyboard, and you can use these to make Word more efficient.

CHAPTER SIX

GO TO, FIND, AND REPLACE FUNCTION

We started looking at some of the basic ways to use your keyboard to navigate around your document in the previous chapters, and in this chapter, you'll be introduced to a few more utilities that you can use in Word to navigate around your document and increase your efficiency when working with your documents. In this chapter, we'll concentrate on the Go to, Find, and Replace options.

Looking for a Word

When working with enormous files, you may require a quick way to get to a certain page or to replace a word in a document.

To access your "**Find**" and "**Replace**" options, first click at the top of your document, then go to the **home** ribbon. Across on the right-hand side is an editing group, which is where you'll find them. You can see that "**Find**" has a little drop-down arrow next to it, indicating that it offers three options: **Find, Advanced Find**, and "**Go to.**"

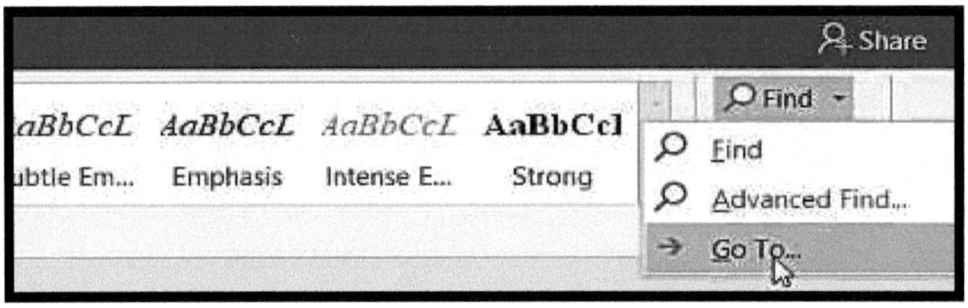

In this example, we'll start on "**Find**", which opens up a little navigation window on the left-hand side where you can put in exactly what you're looking for. It brings up results once you type in the term you're looking for, and you can see them listed underneath in the results section, as well as highlighted in your document, making it incredibly easy to find them. By clicking on the cross, you can close this navigation panel. That's one approach to searching your document for a specific word.

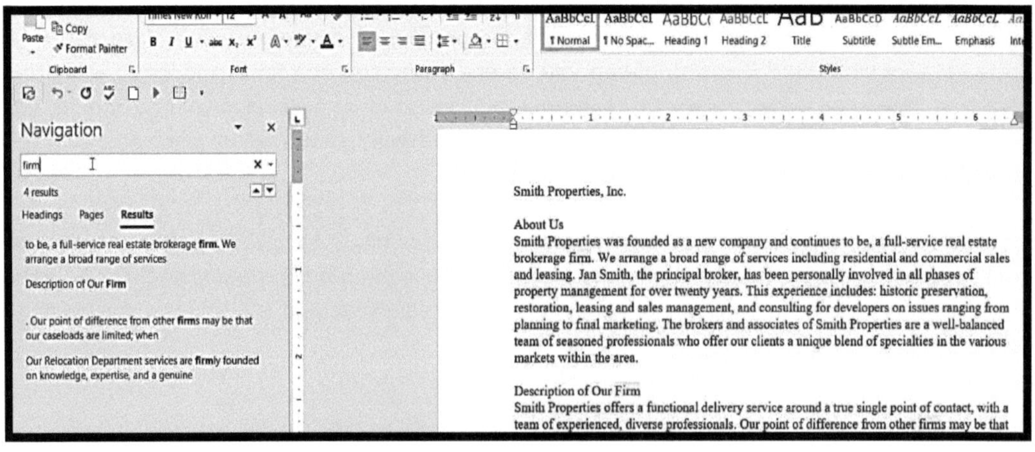

Using advanced Find to Search Your Document

If you select Advanced Find, you'll be sent to a little dialog box that asks you what you're looking for and displays the last item you looked for. If you want to skip this one by one, you can say "**Find Next**," and it will highlight the word in the document the first time it finds it. You can then say "**Find Next**" again to step through each one in your document one at a time, and when you get to the end, it will tell you it's finished searching, and you can click OK.

Getting specify with Find

The "**More**" option in this Find and Replace box is also something to be mindful of. When you click it, you'll see that you have a lot of options. For example, you can tell it to **match the case**, which means that it will only find the word "firm" if it matches that case. If the word "firm" is in this document with an uppercase F, it won't find it because you told it to match the case.

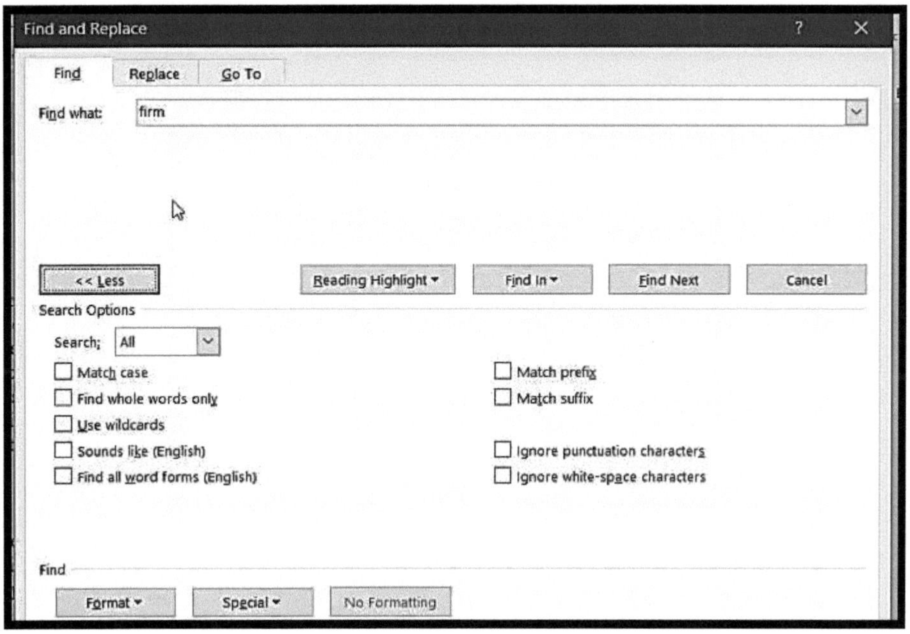

You can choose to "**Find whole words only**," which means that if what you're looking for is a part of another word, it won't be found.

You can also say "**Use Wildcards**" and place a wildcard in front of the word, midway in the middle, or at the end. If you type in A*, for example, it will discover everything in that document that begins with an A, regardless of what comes after it. You can also include a wildcard at the start, such as *S, which implies it will look for anything that ends in S, regardless of what follows before it. It's worth noting that wildcards come in handy when looking for certain items. Another thing you can do up here is to type "A" and then two question marks (A??) to indicate that the word must begin with an A and contain no more than three characters in total. It doesn't matter what those three characters are, but they must be three.

You can also use the "**Sounds Like**" option, which will find any words that sound similar to the one you're looking for. It might pick up words like turn, burn, study, or anything along those lines for a word like "firm", for example.

"**Find all word forms**" will find any form of that word, and you can then choose between "**Match prefix**," "**Match suffix**," "**Ignore punctuation**," and "Ignore whitespace characters," among other possibilities. Just keep in mind that beneath the "More" drop-down, you have a variety of different options for customizing what you're looking for in your document, and it can become very detailed.

Replace One Word with Another

Make sure you're at the top of my document and select the "**Replace**" option this time. We have "**Replace**" under "**Find**" in the editing group, which allows you to **replace one word with another**. For instance, if you want to replace the name "Smith" in your document with "Ashby," you may say find "Smith" and replace it with "Ashby," and you'll have all of the prior possibilities to choose from. If you merely want to replace Smith with Ashby, pick "**Replace all.**" You'll get a little window telling you that the replacements have been made, and if you look at the document behind you, Smith has been replaced. This is a quick and easy technique to replace several words in a document.

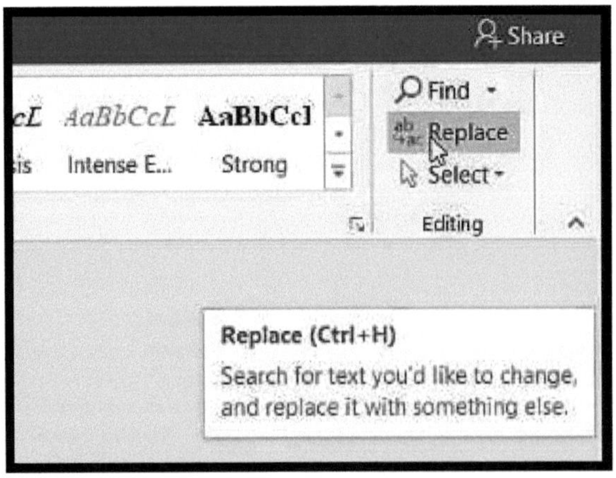

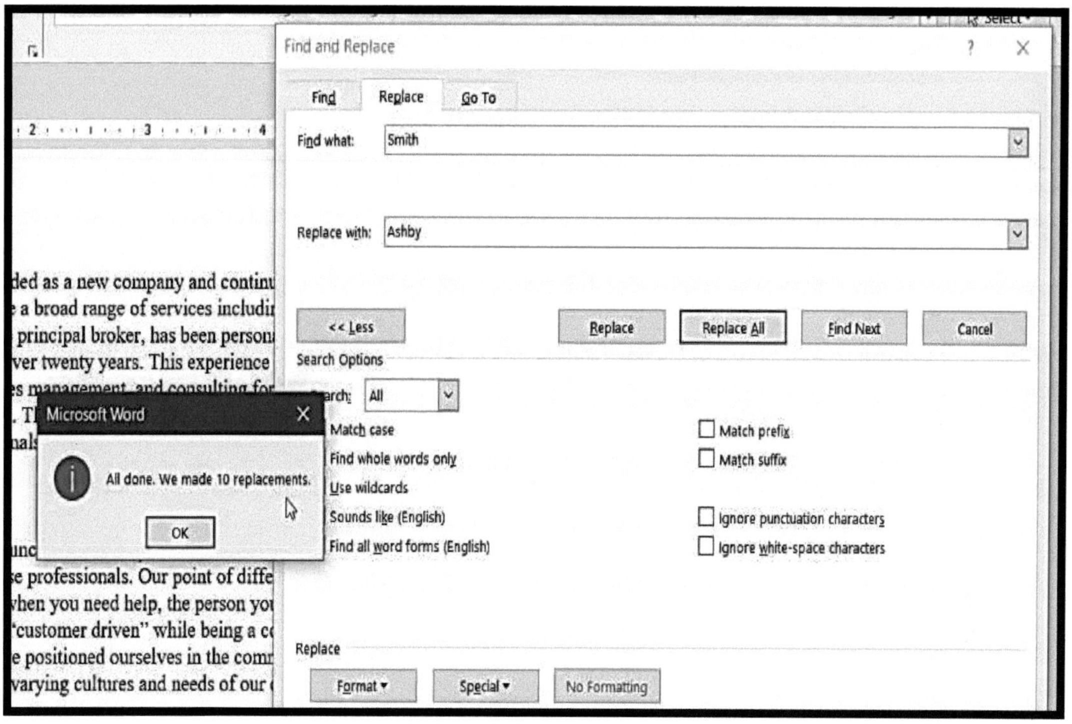

Go to any point in your document

The "**Go to**" option, which lets you perform different things in your document, is the last item you should know in this chapter. When you go to Find and select the "**Go To**" option, you'll be able to browse around your document. You can accomplish this by entering a page number and saying "**Go To**"; for example, if you input page 2 and say "Go To," it will take you to page 2.

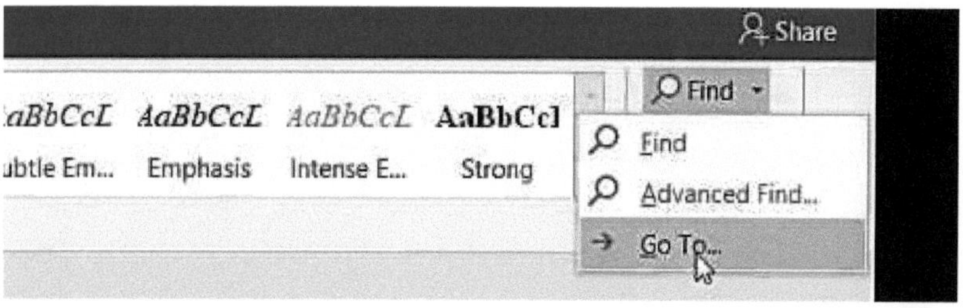

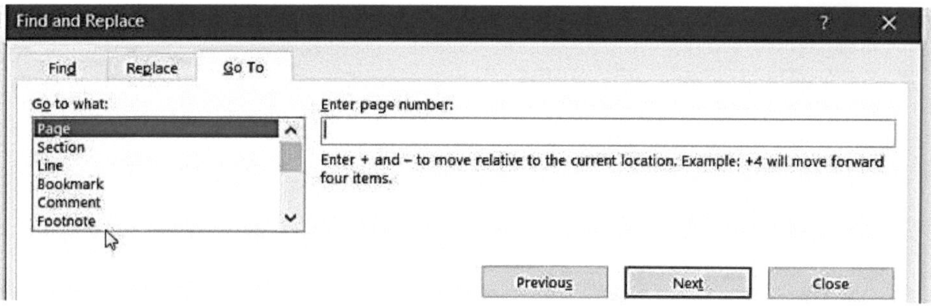

You can go by **Section**; you'll learn how to break your text into sections in later chapters, and then you can go to any area you want.

By going to a specific line number, you can navigate by **Line**.

You can navigate via **Bookmark**; bookmarks are something we'll talk about later, but they're a way of placing a small bookmark in a certain spot on a page so you can jump to it fast.

You can also use any **Comments** you have in the text to browse.

We also have **Footnotes**, **Endnotes**, **Fields**, **Tables**, and a variety of other features. Which one you use depends on what you have in your document at the time. But keep in mind that those Go to choices are still available.

Let's move on to the next chapter now that we've covered the basics of Find, Replace, and Go to.

CHAPTER SEVEN

SPELLING CHECK ON WORD

In this chapter, we are going to discuss how to check your work for typing and spelling mistakes. We will go further to discuss AutoCorrect settings, how to adjust, add and undo AutoCorrect correction. Also, we will look at how to get rid of grammatical mistakes to make sure that everything is correct and finally on document proofing.

Introduction

It is said that one of the most important parts of writing well is spelling, which is the art of putting words together correctly from their letters and helps people read better. Hence, it is very important in Microsoft Word.

Checking Your Spelling

To start a check of your document spelling and grammar, use the shortcut F7, or follow these steps:

- Click on the **Review tab** is on the ribbon.
- Then Click on **Spelling or Spelling and Grammar** as shown in figure 1 below.
- The program finds spelling mistakes and a dialogue box shows all the words that are wrongly spelt.
- You can use Word's Spell Check right away. The red zigzag line on any word indicates wrongly spelt words and mistakes as shown in figure 2. In figure 2, the word 'Content' was wrongly spelt as 'contrent' hence a red zigzag line is displayed on the word. The spell-check feature also tells you if you've used the same words over and over again by underlining them with a red zigzag. You can either delete the word that is repeated or you can choose not to pay attention to it.

Note: You can choose to correct the word, delete the word, leave it if it's correct and add it to the dictionary if it doesn't recognize that word.

Figure 1: The Spelling and Grammar icon on Microsoft Word

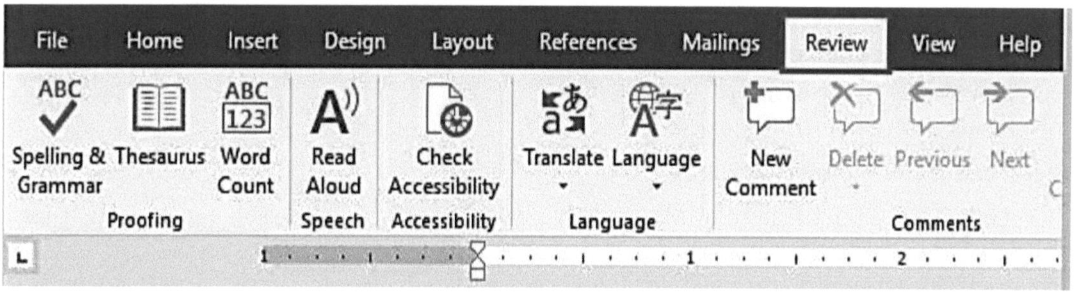

Figure 2: Microsoft Word showing a Red Zigzag line

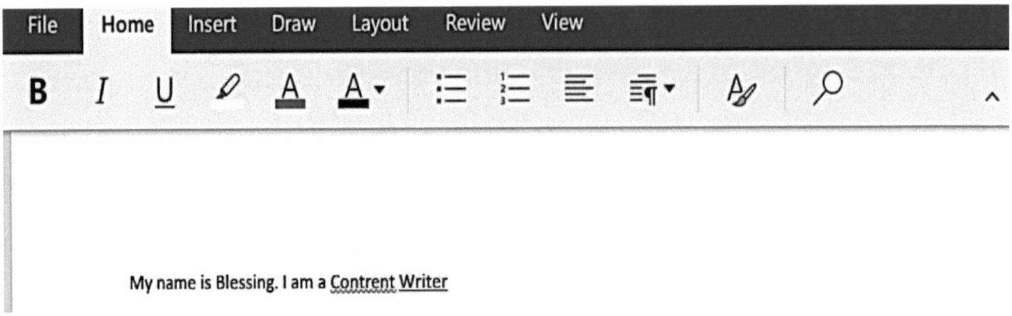

Turning On or Off Spell Check on Microsoft Word

- Click on **File**
- Click on **Options**
- Next, click on **Proofing**
- Clear the Check spelling as you type box, and then click **OK.**

To Turn Spell Check Back on, all you have to do is repeat the same process and select the Check spelling as you type box.

Fixing Words That Aren't Spelled Right

- Choose Check Spelling from the Edit menu to make sure your words are correct.
- Select the misspelt words that you want to find, then press the Enter key
- The spelling submenu is displayed
- From the submenu, type the correct and click Ok

Words That Have Been Mistakenly Flagged.

You might have typed a name that Microsoft Word doesn't know. To correct a word that has been falsely accused, you can do any of the following:

- Right-click it and choose the correct word from the list.
- **Add to the Dictionary:** click on add the word to the dictionary. The word will no longer be marked as spelled incorrectly in this or any other document because it is now correct.
- **Ignore** This command tells Word to ignore the word and accept it as spelt correctly through the whole document.

AutoCorrect in Microsoft Word 2025

AutoCorrect is a useful feature in Microsoft Word that checks your spelling and corrects it for you automatically. Word quickly fixes a lot of common typos and spelling mistakes on the spot, so you may not even notice the red zigzag when the AutoCorrect feature is on. AutoCorrect fixes that typo as soon as you hit the spacebar or punctuation marks at the end of a word. It also turns common text shortcuts into the single characters they should be. It corrects common punctuation mistakes, when you start a sentence, it capitalizes the first letter of it. When you forget to capitalize your name, AutoCorrect corrects it, it fixes the inverse caps lock problem and other common typos, as well as a lot of other mistakes that can be made when typing.

How To Turn On and Off AutoCorrect on Microsoft Word

- Go to **File** and click on **Options**
- Click on **Proofing** and choose AutoCorrect Options from the drop-down menu on the right.
- Click on the **AutoCorrect button.** This is shown in figure 3
- On the AutoCorrect tab, you can choose or choose not to replace text as you type.

For AutoCorrect to work, make sure that the Replace text as you type box is checked or unchecked.

Figure 3: The AutoCorrect Feature on Microsoft Word

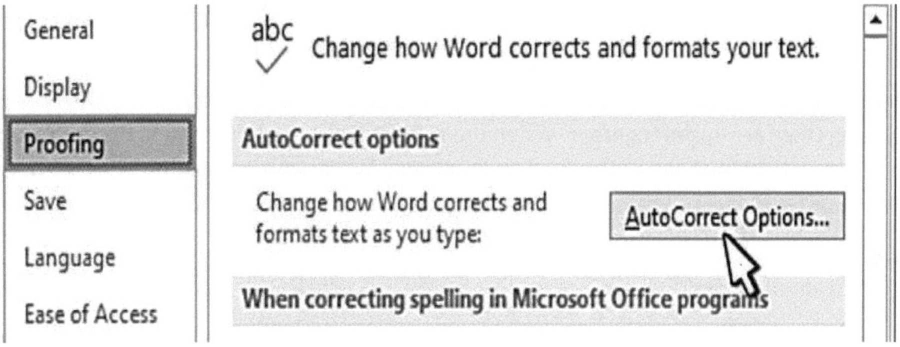

Undoing a Correction Made by AutoCorrect

You can easily undo a correction made by AutoCorrect. To do this, click on Ctrl+Z (the Undo command).

Note: You need to press it right after AutoCorrect makes its correction.

Adding a New AutoCorrect Entry

To do this follow the steps below:

- Right-click a word that you want to add to your AutoCorrect list.
- Type a word or phrase that you often misspell in the Replace column, and then click Replace.
- In that column, type the correct word.
- Then, click **Add**.

Note: If you want more information about the AutoCorrect tab, check out the AutoCorrect options page and you can add as many entries as you want. Add a new entry for each new thing that you do and click add.

Adjusting AutoCorrect Settings

To change how AutoCorrect works and the words it corrects, follow these steps:

- Click on the **File** tab.
- Then, choose Options and the Word Options dialogue box appears
- Choose the **Proofing** category on the left side of the window.

- Then click on the **AutoCorrect** Options button on the right. When you open the AutoCorrect tab, you can see all of the things that AutoCorrect can do, like capitalize the first letter of a sentence, change words that don't follow the rules, click the Exceptions button and so on

To remove an entry from the AutoCorrect list, scroll down until you find the item you want to remove, click on it, and then click the **Delete** button on the right.

Grammar Check

Microsoft word grammar checker works the same way as to spell check. The main difference is that offenses are marked with a frigid blue zigzag. When you right-click the blue-underlined text the pop-up menu shows to find out what's wrong. You can also choose to ignore it if you wish.

The grammar proofing tool looks at the text and thinks about how the actions and the people or things that are doing them are linked together. When the subject and verb don't match, this is the most common cause of grammatical mishaps in the English language. The grammar checker is very good at spotting when there are two spaces between words when there should only be one space between the words.

Turning on Grammar Check

- First, put on Microsoft Word
- Click on the Tools menu, then click on **Options**.
- When you're on the Spelling and Grammar tab, change the Writing style box to Grammar and Style.
- Then click **OK**.

An All-in-One Document Proofing

There is an all in one process for spelling and grammar checker if you don't want to look at the red zigzag underline or if you don't have that feature turned on, you can just simply do a final document proof with these steps:

- Click the **"Review"** button.
- The "**Spelling & Grammar button**" is on the Proofing group. To get to it, click it.
- The **Editor** pane shows up. Errors are shown one at a time as they happen in your text. You must check each of them one after the other.
- And then correct each of them.

Note: The Spelling pane or the Grammar pane shows up depending on what the person did. You can choose to ignore, ignore all, however, if you make the same spelling mistake or similar grammatical mistakes again, you'll be reminded of them. You can choose to delete repeated words, add to dictionary etc. You can choose to select the correct answer from the list that's shown and then click the Change button to change the text. You can also click the correct word, then click Change All to change all of the places where you made that spelling mistake.

- Afterwards, click the **OK** button.

Note: Another way to go through spelling and grammar mistakes one by one is to use the Spelling and Grammar Check button on the status bar. You can use that button to jump from one spelling or grammar mistake to the next.

Checking a Document Again

The Ignore or Ignore All button has consequences. If you've used the ignore all options too many times, you can tell Word to recheck your document, which brings forgotten misspellings and grammatical mistakes back into the word document proofing, so you can fix them again, too. Follow these steps:

- Click the **File** button.
- Choose **Options** to open the Word Options window.
- Then choose the Proofing option on the list.
- To check the document again, click the Recheck Document button on the toolbar. It's under the title "**When Correcting Spelling and Grammar,"** and it says "When."
- Click the Yes button to say that you want to un-ignore things you've chosen to ignore in the past.
- Click the OK button to close the Word Options window.
- Then the document shows itself again, highlighting all the words and things you've chosen to not read before.

Note: Rechecking a document doesn't erase any changes you made to Word's dictionary. If spelling and grammar errors don't show up when you recheck them, it might be because the option to hide them has been turned on, that is "Hiding all proofing errors in a document" is turned on.

Settings for Document Proofing

There are a lot of settings and options you can use to control Word's document-proofing tools.

i. Making Changes to the Custom Dictionary

You build the custom dictionary by adding words that are correctly spelt but have been marked as misspelled to the list of words. If you want, you can also manually add words, remove words, or just look at the dictionary to see if you're making any mistakes. Take these steps:

- Click the **File** button.
- Choose Options to open the Word Options window.
- Then, choose **proofing.**
- Click the button that says "**Custom Dictionaries**" to make your own.
- The Custom Dictionaries dialogue box comes so you can add words to it.
- Make sure you put the words you want in a custom dictionary in the text box that says Word(s). When you're done, click "**Add**."

To remove a word from the custom dictionary: click on the word in the scrolling list and then click Remove from the custom dictionary. To delete, click the Delete button.

Note: Click OK when you're done with your dictionary. Then, click the OK buttons on the open dialogue boxes to close them and go back to your text.

ii. **Putting an end to automatic proofing**

To get rid of the red zigzag underlines and blue underlines from your document, while typing follow these steps:

- Click on the File tab and then choose Options. The Word Options dialogue box comes up
- On the left side of the dialogue box, click on Proofing and then click on the button.
- You can uncheck the box that says, "Check Spelling as You Type."
- Remove the checkmark from the item while you're at it. Mark Grammar Errors as You Type, then Check Your Work.
- Choose "OK."

iii. Cutting back on grammar-checking

To change grammar settings, do the following:

- Click the **File** button.
- Choose **Options** to open the Word Options window.
- Then, choose proofing.

People who use Word to correct their spelling and grammar should look for the Writing Style item below the title. Here you can choose what kind of grammar to use, and how you want things to look. To the right, there is a button that says "Setting."

Click the Settings button to go to the next step.

The Grammar Settings dialogue box comes up.

In Word, you can uncheck the boxes for things that you no longer want to be marked as offensive.

iv. All proofing mistakes in a document can be hidden.

The r**ed zigzag** and the chilling **blue underlines** can be hidden if you tell Word to keep checking your spelling, but you can also tell it to hide them. Words are still marked as misspelled, but they don't show up in the text. Make sure you follow these steps:

- When you open the Word Options dialogue box, look for the Proofing area.
- There are options in Word called "**Options**" that let you hide spelling and grammar mistakes only in the document you're working on.
- To turn them off, check the boxes next to them and uncheck them.
- The proofing flags disappear all over your document.

This is how to hide spelling and grammar mistakes in a paper.

How to Hide a Word Document That Has No Spelling or Grammar Errors.

If you only want to hide spelling and grammar mistakes in one Word document and not all of your other documents, follow these steps:

- Click on the "**File**" tab.
- When you get to the bottom of the left-hand pane, click on "**Options"**

- The "**Word Options"** window will show up, so you can make changes.
- Click on the **"Proofing" tab. T**he proofing tab is where you go to make sure everything is
- Make sure the "Hide spelling errors in this document only" and "Hide grammar errors in this document only" boxes are checked at the bottom of the window. Click "Yes." This is shown in figure 4 below.

Figure 4: The 'Hide All Error in a Document only' icon

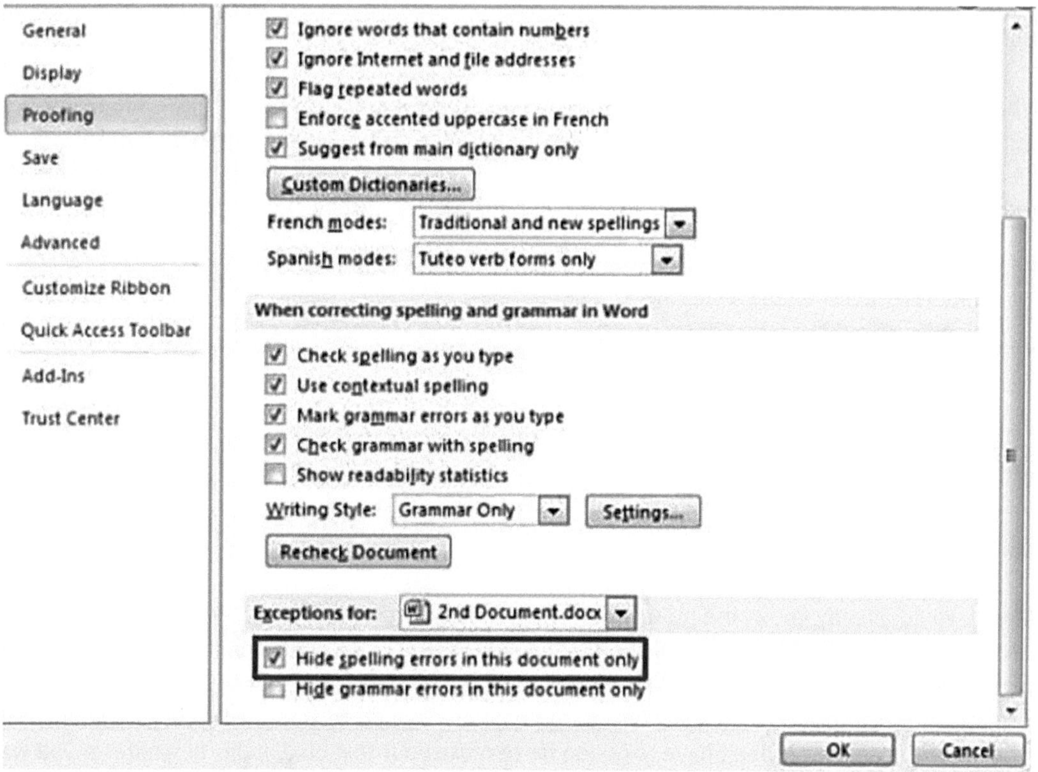

Finally, In Word, when you check your document for spelling and grammar, it keeps track of which mistakes you choose not to fix. And with all mistakes corrected, the document is ready.

CHAPTER EIGHT

WORKING WITH TEXT BLOCKS

You'll learn how to use Microsoft Word's Blocks in this chapter. We'll begin by discussing what text blocks are, and then we'll go over the many options for picking text blocks, which may seem simple at first, but you must grasp the numerous alternatives available. If you master the fundamentals first, you will be more efficient.

It's Time to Get Acquainted

What is Microsoft Word blocks and how do you use them? A block is nothing more than a grouping of text. It could be a single word, a group of words, a sentence, a line, or an entire paragraph.

You can move, copy, and edit text blocks once they've been selected in your project.

Selecting a Block

There are numerous methods for selecting a text, and we will go over all of them.

Using the Mouse and Keyboard

As a general rule, the **text** is **highlighted** or **selected** when you **drag your mouse** across it, so if you want to change the color, size, font, or anything else related to it, make sure it's selected first. Otherwise, Word won't know which text to apply the changes to.

As you would already know, the most frequent method is to **click** and **drag**. When you click and drag, a tiny **toolbar** appears, which you can see off to the right. This is just a brief toolbar that appears with some formatting options. Word has presumptively assumed that you'll want to do something with the formatting of this text, so instead of having to go all the way up to the ribbon, it's just popped up this little mini toolbar, so you don't have to go too far to make your text bold, italic, or whatever you want to do with it; Just keep in mind that it will appear.

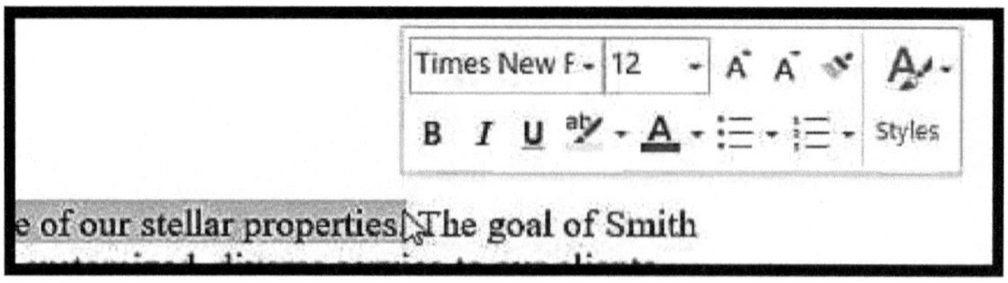

If that bothers you, you can turn it off in the backstage area, which I'll demonstrate later. Simply **click anywhere else on the screen** to **deselect** the text. If you want to pick just one word, double-click on it; if you want to select the complete sentence, **hold** down the **Ctrl key** and **click anywhere** in it; it will select the text or highlight only that sentence up until the first full stop. If you hover your mouse over the white space along any line and click, it will choose the entire line, regardless of whether it's a sentence or not. If you stay clicked in the margin with that arrow and drag down, you will select all of the sentences or all of the lines.

If you wish to select from one point to another, you can do so in another way. If you **click** your mouse just before the text, **hold** down your **Shift key**, then **click** after the word you want to end, and it will pick that section. If you want to **add** to that section, **hold** down your **Shift key** again and use your **arrow keys** to make additional additions to that selection.

If you want to **choose multiple sections** that aren't next to each other, **click** the first area and then **drag** and **select** that sentence while holding down the **Ctrl key**. To make numerous selections, remember to hold down the Ctrl key.

The final option, as we've suggested a few times before, is to use **Ctrl + A** to select everything in your document, and then use the formatting choices in the toolbar to make whatever adjustments you need.

Using Ribbon

Ribbons are another item on your list. We've mostly used the keyboard and mouse, but if you go back to the **home ribbon**, all the way over on the right-hand side, there's a select option that lets you choose everything, similar to **Ctrl + A**.

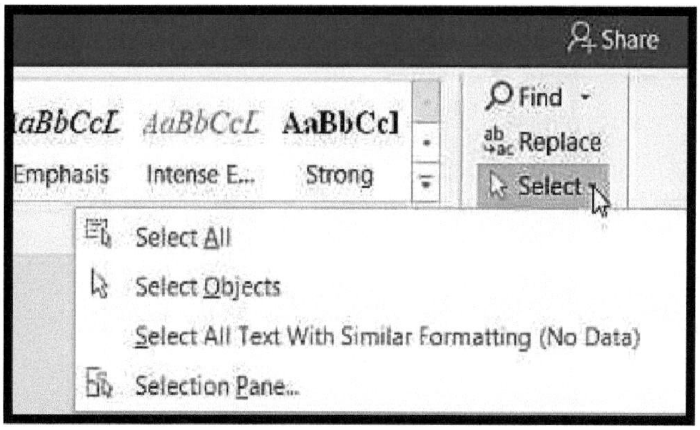

You also have the option to "**Choose Objects**," which is useful if you have things like shapes on your screen and want to select them all; it will allow you to create a spotlight around them and it will select all of those objects, which we will use later in this book. Always keep in mind that such selections are also available on the home ribbon.

A Unique Selection

Remember that if you **click the left mouse button** and **drag it down** to choose a block of text, it will select all of the lines you cover, but what if you don't want that? What if you simply want to pick a portion of a text block? To do so, hit the **left mouse button** and **drag** it down while **holding the Alt key** on the keyboard. It will now just choose the region you want to format.

The Good Old Cut, Copy and Paste

Cut, **Copy**, and **Paste** are the backbone of Microsoft applications, and we'll start exploring them in this section. These are probably instructions you've used before, and they're the same in all Microsoft apps;

whether you're working in Word, Excel, or PowerPoint, the functionality is identical, including the keyboard shortcuts. What we want to accomplish in this section is to make sure you understand what Cut, Copy, and Paste can do, and if you already know how to use them, you might pick up a couple of new tricks.

Each of them has four steps, the first of which is making a selection. You might wish to choose a sentence and cut it. It's crucial to understand the difference between Cut and Copy. When you **cut** something in your document, you're effectively **deleting** or **removing** it and pasting it somewhere else; it's similar to moving in many respects, thus cutting the sentence and inserting it somewhere else will move it. You can highlight the same line, copy it, and paste it elsewhere on the page using "**Copy**," and it will **generate an exact copy** without moving it.

Let's take a look at the "**Cut**" option first. There are a few different ways to cut a line of text in Microsoft, as there are always.

If you right-click on your selection, you'll notice a "**Cut**" option in the right-click menu.

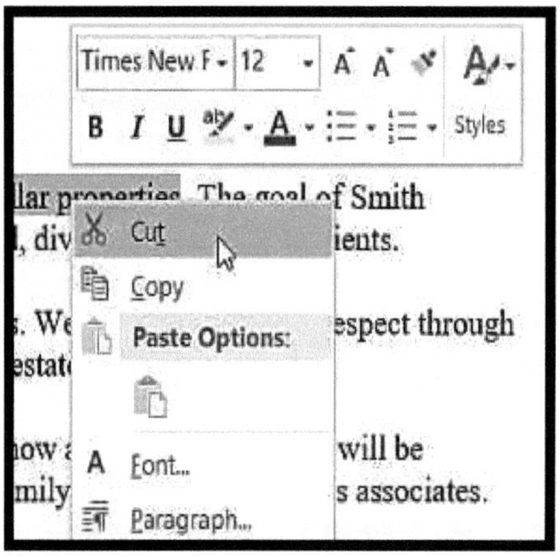

Alternatively, you can navigate to the home ribbon and find a "**Cut**" option in the first group under the Clipboard group.

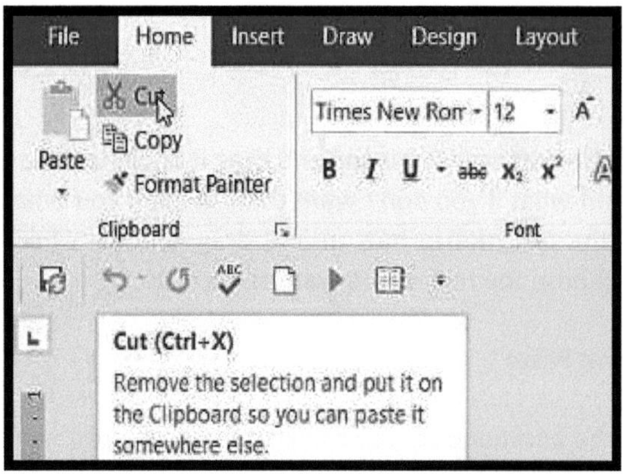

A third option is to utilize the **Ctrl + X** keyboard shortcut, which you can see when you hover your mouse over the cut icon on the home ribbon.

After clicking "**Cut**," go to the area where you wish to move the text and paste the sentence you just cut. You can paste text in a few different ways. You can select one of the pasting choices by **right-clicking your mouse**. You can also use the keyboard shortcut **Ctrl + V or jump up to the home ribbon and click the paste button there.**

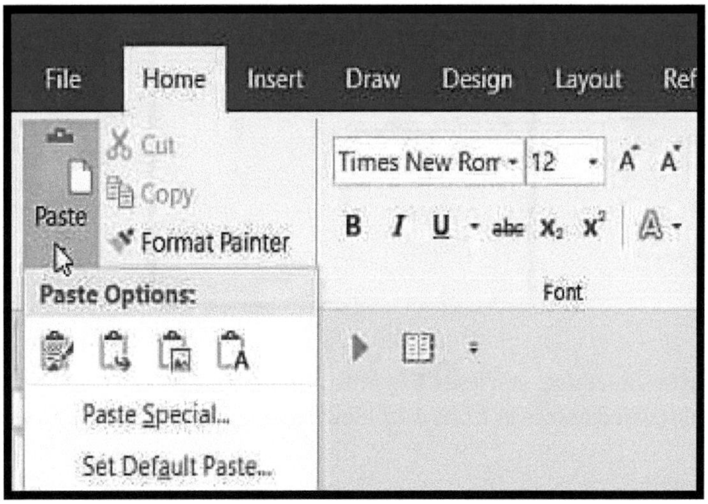

The Paste button is separated into two pieces; the top half is marked in gray, and if you click the lower half, it will give you a couple of different options to paste that line.

- You can opt to paste and keep the source formatting, which means that the formatting will be carried over to wherever you've cut the sentence from.
- You can select to merge formatting as you paste, so if you're pasting this line into a document with completely different typefaces, you can choose to merge the formatting to match the font style or color of the page you're pasting into.
- You could paste it as a photo, which turns it into a picture object rather than text.
- You can choose to keep text only if you just want simple text without any styling; you have a few alternatives there.
- If you just want to perform a direct paste, click the top half of the paste button, your cut-out sentence will be pasted, and you'll notice that it's no longer in its original spot.

There are many more possibilities under "**Paste special**," but we'll go over those in greater detail later.

Let's go on to "**Copy**." **Ctrl + C** is the keyboard shortcut to remember. You can alternatively utilize the "**Copy**" button on the home ribbon, browse to the location where you want this line or word to appear, then paste by clicking your Paste button. This time, it will appear as a copy rather than a move.

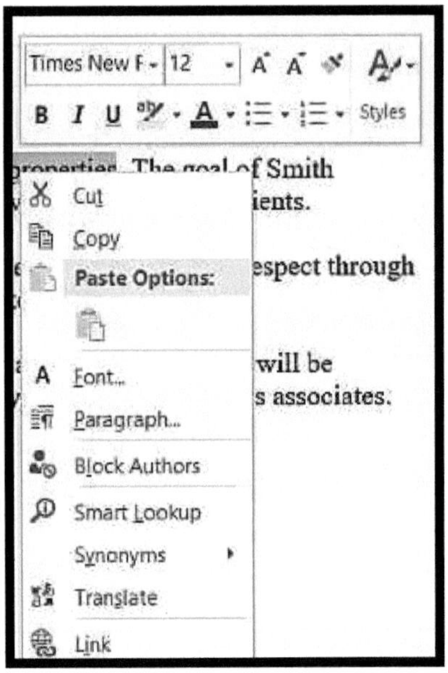

Also, when you paste something, you'll see this little pop-up appear, and if you click it, it'll bring you the paste options we discussed before in a fast drop-down menu in case you wish to use them.

Viewing the Clipboard

You should also be aware of the Clipboard. It's not immediately obvious how to bring up the clipboard, but if you go up to the **home** ribbon, in the first group where we have **Clipboard** written at the bottom, click on the little drop-down arrow, you'll get this little pane open up on the side that says "**Clipboard**," and you can see it has the last item that you copied or cut on it.

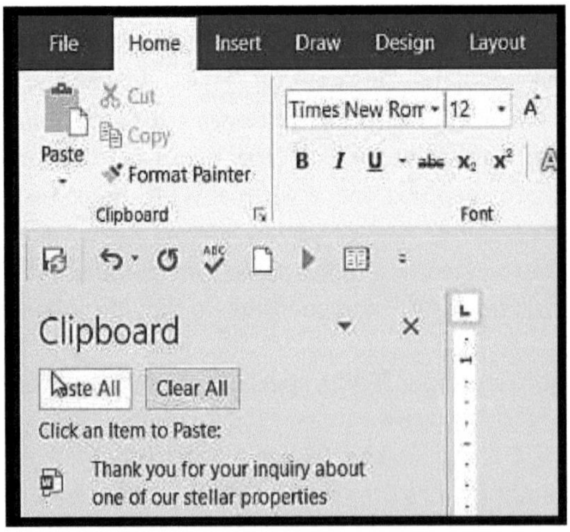

Every time you cut or copy something, it is temporarily saved to your clipboard, and as you continue to copy, a list of all of them appears on your clipboard. "Why is that useful?" you may ask. It's great if you want to go through and cut and copy a lot of things, then paste them in different orders that aren't necessarily the same as the ones you copied. You may now go to your Clipboard and select to paste this bit of text into the document by selecting "**Paste**" from the drop-down menu.

This Clipboard can hold up to **24 items**, allowing you to copy a large number of items and paste them wherever you like on the page.

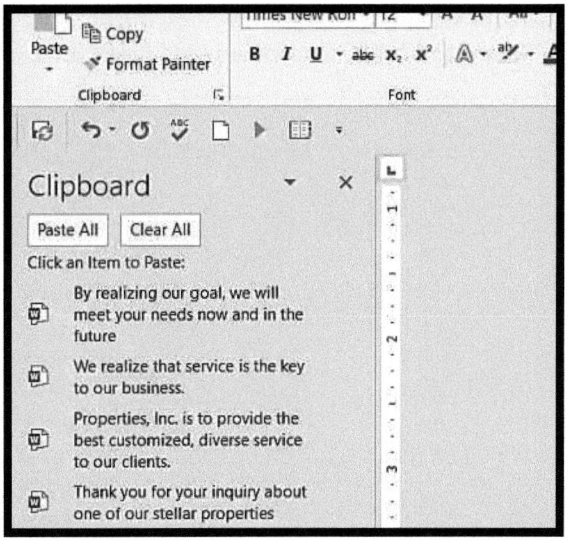

If you wish to **empty your clipboard**, simply click "**Clear All**," and everything on that clipboard will be removed. You can also dismiss your clipboard by clicking the cross icon. If you want to use it, just be aware that it's lurking in the background.

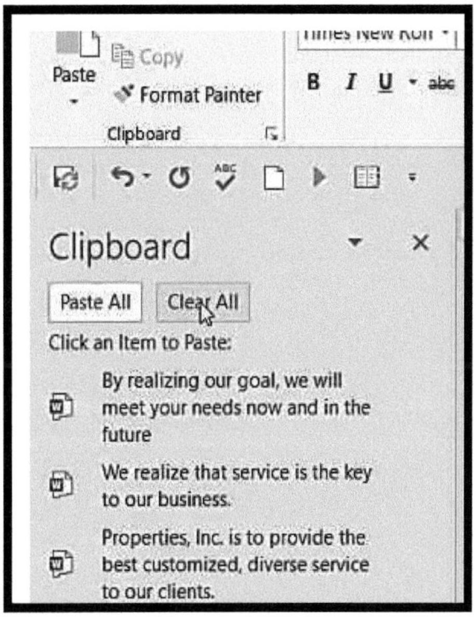

Changing the Position of Your Text Block

There are a few additional options for moving items about in your text. You can use your keyboard and mouse to execute a Cut or a move if you highlight a word, sentence, or paragraph. You can drop it wherever you want and move it that way if you hold down the **Ctrl** key and then click and drag with your mouse.

Hopefully, you now have a better knowledge of the Text block, the **Cut**, **Copy** and **Paste** commands, how to move things around with your mouse and keyboard, and how to copy and paste items from the clipboard into your project.

CHAPTER NINE

CHARACTER FORMATTING

In this chapter, we'll go deeper into character formatting possibilities. After you've finished working on your documents, you might notice that nothing sticks out; all of the text is very uniform, it's all the same size, it's all the same font, there are no real headings, and nothing shines out. You can use some formatting at this point to emphasize particular points and make it more interesting and simpler to read. When you create a new document in Word, you'll notice that everything stays the same until you change it, thus whatever font size you choose will stay the same until you change it, and the font color will always be black unless you go in and change it. The term "**formatting**" refers to the process of altering the appearance of text.

Text in a Basic Style

You can use some formatting to spice up your document and make it more interesting than it is now.

If it relates to your document, the most apparent place to start would be the Title. You can highlight your title here, and the first thing you might want to do is modify the font to make it stand out from the rest of the text.

Choosing a Font

Your current font may be set to **Times New Roman** on the **home** ribbon's Font group, but if you click that drop-down, you can choose from a variety of fonts to apply to your selected text.

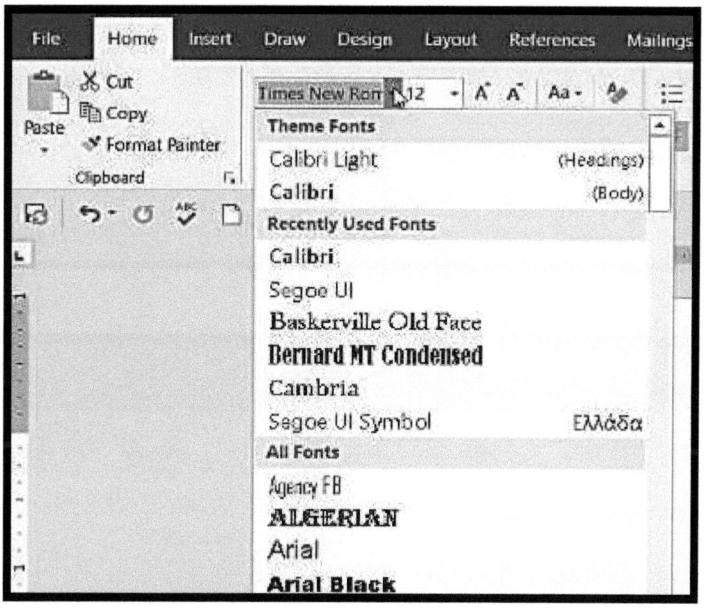

It's worth noting that the default font in later versions of Word is Calibri, which is a lovely font to use, but you can change it to something else if you want to make it stand out.

Text Size Selection

You can make your text a little bigger than the rest of the text after choosing your font. Moving up to the **Font** group, you'll notice that next to the font style is the font size, and if you hover over any of these, a live sample of what it will look like appears.

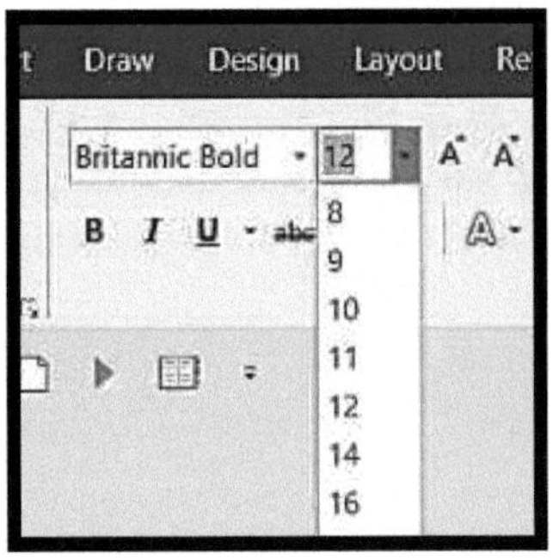

It's also worth mentioning that these font sizes are listed in increments, so if you want font size 15, for example, you won't find it in that list, but you can manually type it into the box above to acquire that font size.

Next to your font size, you'll see two buttons: **Increase font size** and **Reduce font size**.

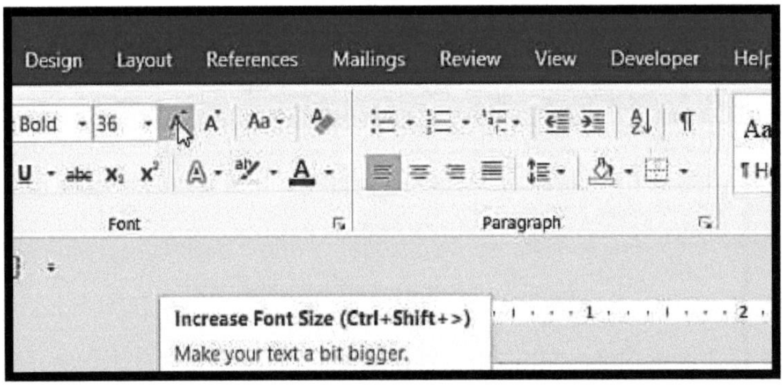

These are useful if you only want to modify the size of your font in increments without having to go inside the drop-down menu. You may either go in and increase the font size by one each time, or you can select to lower it back down.

Changing Text Case

Change the case with the option next to "**Reduce font size.**" You can choose uppercase, sentence case, lower case, capitalize each word, or toggle case.

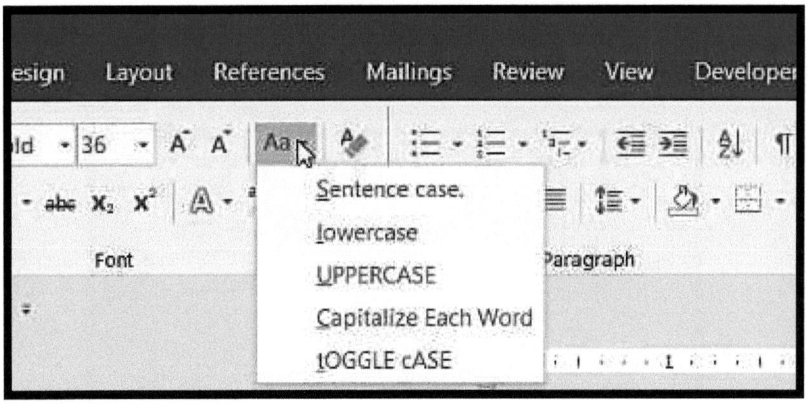

Clear Formatting

The "**Clear All Formatting**" option is next in this small group. If you decide you don't want all of the formatting you've applied, simply click the "**Clear All Formatting**" option to remove it all.

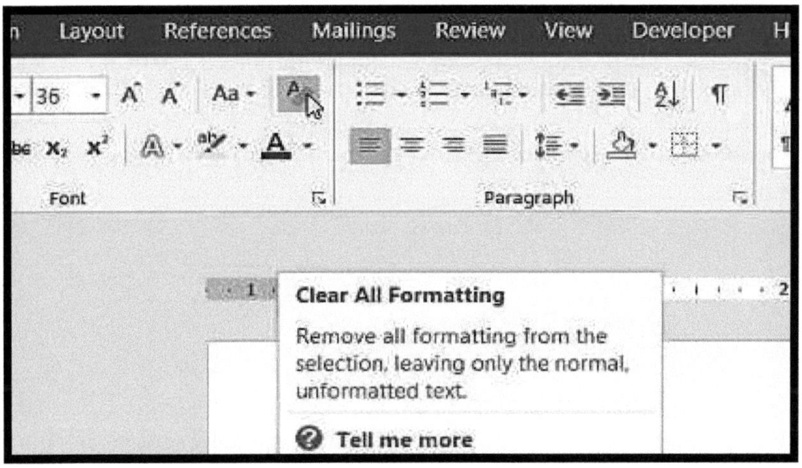

Other Formatting Options

In the bottom row of this small group, we have "**Bold**," which will make your text bold, "**Italics**," which will make your text italic if you like, and "**Underline**," which will underline your text.

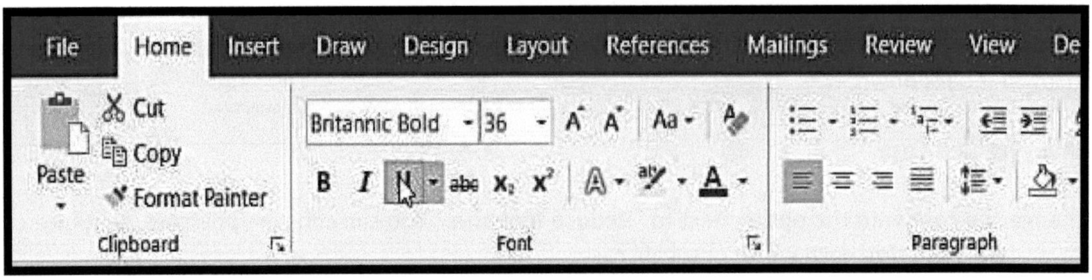

You'll note that the underlining has a small drop-down menu that allows you to select an underline style, such as **solid line**, double line, dash line, and so on; so, you do have some more options down there.

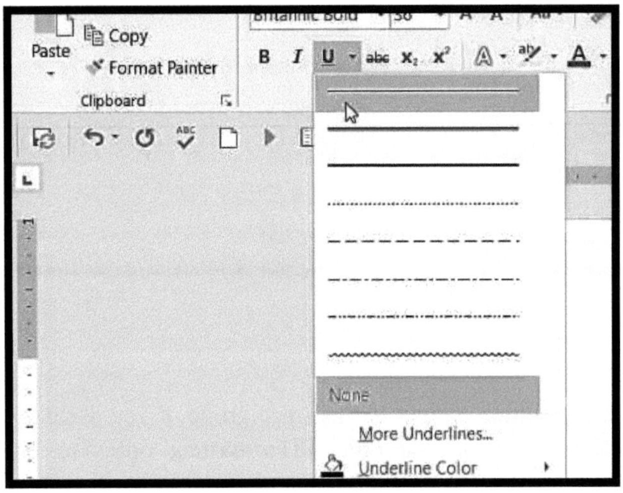

You also have a "**Strikethrough**" option, which you could find useful if you deal with a lot of contracts; it allows you to strikethrough a section of selected text.

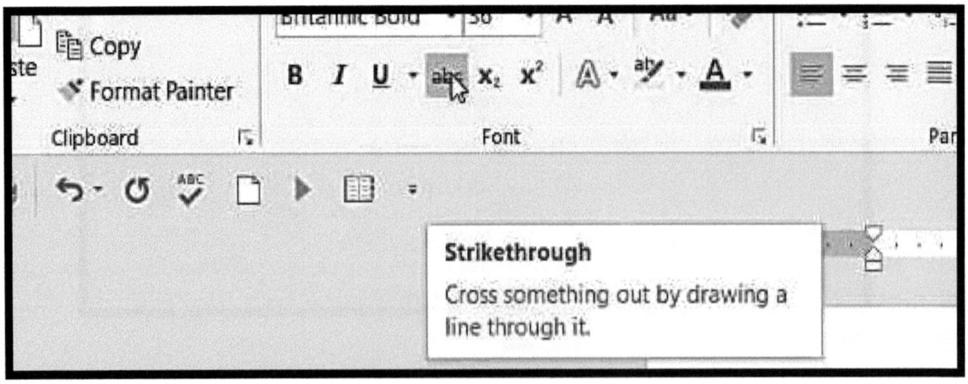

Then there are **Subscript** and **Superscript** options, which are useful for words like H20 and other related terms.

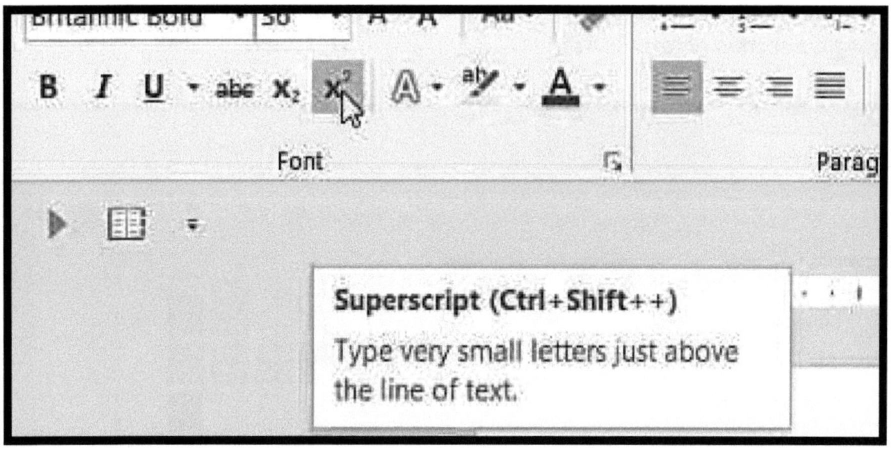

Aside from that, there are some text effects. This changes the color and appearance of the selected text, and there are several alternatives available, some of which are much nicer than others, but depending on the sort of document you have, some of these can seem very effective.

Always keep in mind who your audience is when applying effects, colors, and other elements to your documents; you don't want to design anything that looks too jokey, too humorous, or too unprofessional if it's for work purposes, so keep that in mind. If you're making a flyer, a newsletter, or something similar, go ahead and fill in the blanks, but keep in mind who your target audience will be if you choose this option. You have more options in there for changing things like the **Outline color**, applying a **Shadow**, **Reflection**, or **Glow**, and so on. There are a few alternatives for you to experiment with.

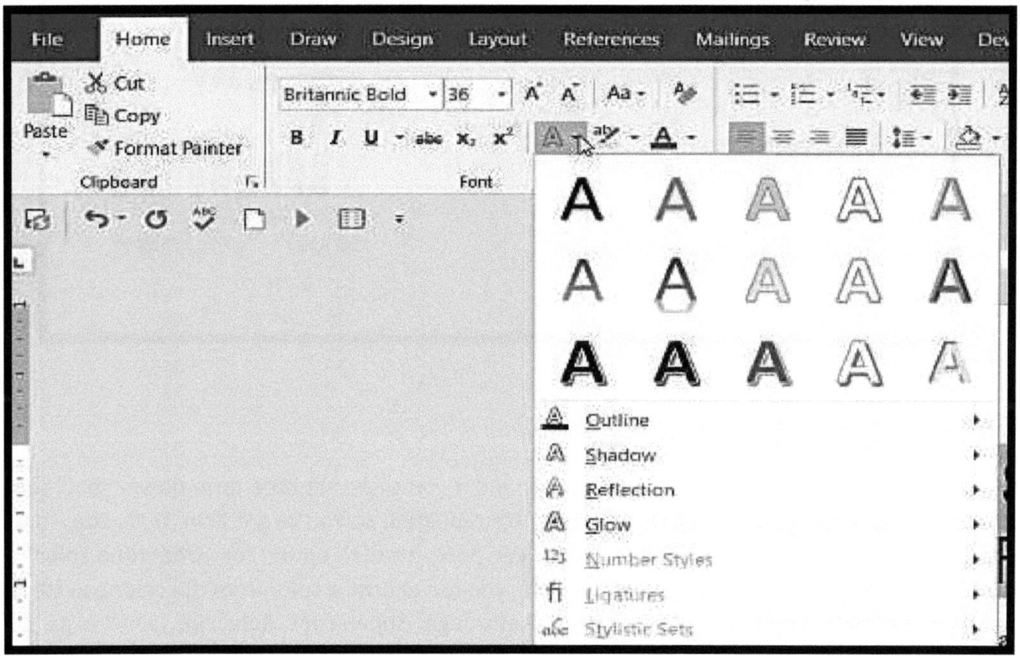

The next button is a **Highlight color**, which is perhaps best found farther down on this page and allows you to highlight a section of text.

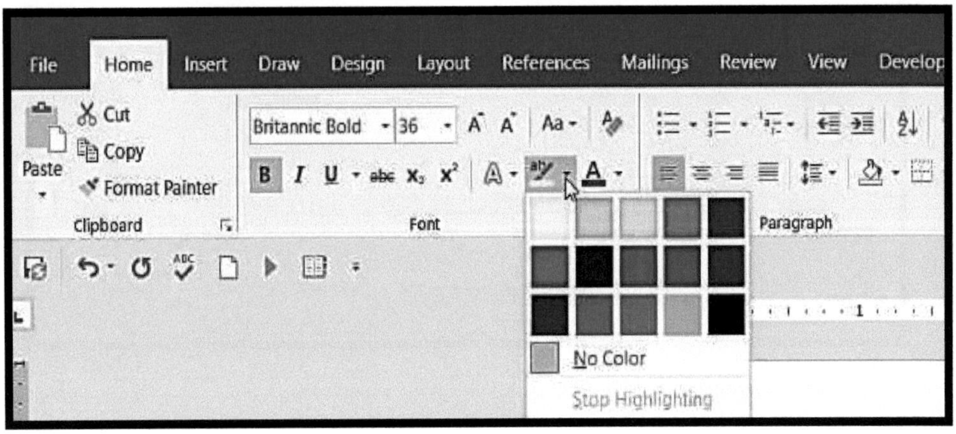

This is equivalent to simply striking through a piece of paper with a highlighter pen.

We also have **Font Color** options, so if you want to alter the font color, simply choose it and then choose from the color palette.

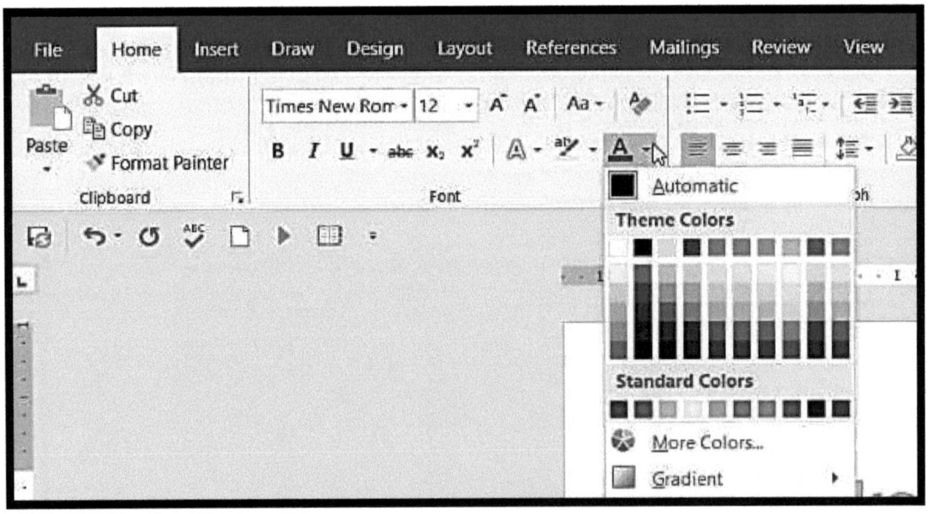

The Dialogue Box for Fonts

There are several possibilities in that font group, and if you click that little drop-down, you'll see a few more options down here as well; many of them are repeated, so you've got Font style, size, font color, Underline style, and all of those on the ribbon. We don't have an option for "**Underline color**" on the ribbon, therefore if you choose an Underline style, you can choose a color from the color palette for that particular Underline. Strikethrough, Double strikethrough, Superscript, Subscript, Small caps, all caps, plus a few other things that aren't on the ribbon are all present.

You also have an **Advanced** tab, which allows you to change things like Character spacing, which is the amount of space between each letter. If you want to broaden it out, this will assist, and there are many additional options here as well in the Advanced section.

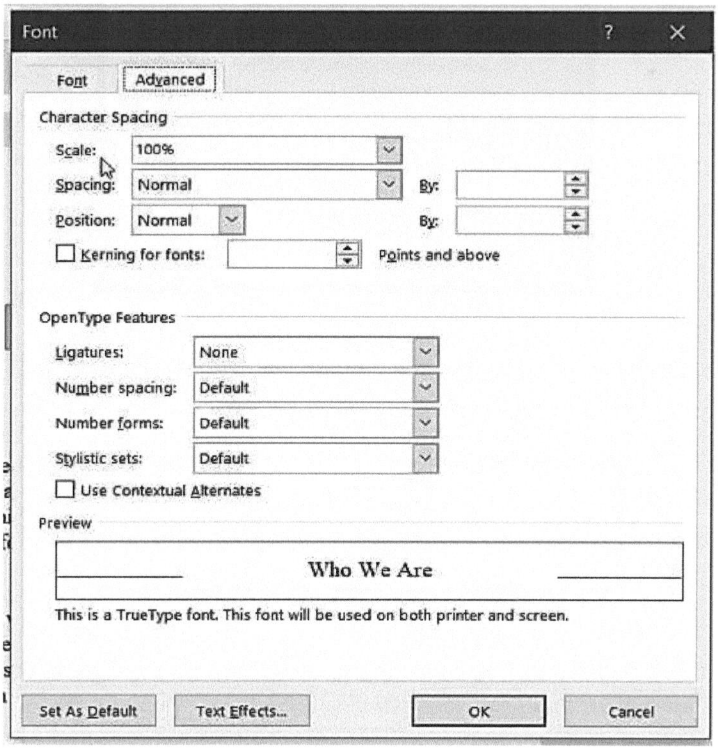

Using a Formatting Copy and Paste

We'll use the Format Painter here, which is a very effective means of copying formatting from one piece of text and putting it into another. We looked at some of the formatting options for the text in your document in the previous section of this chapter. If you have some formatting applied to a text or sentence and want to apply it to another sentence or paragraph further down in the document, you could do so by highlighting the text, going into your font group and applying those different attributes, selecting the text effects, changing the font size, font color, and so on, but that's not a practical solution.

A much better approach to do it is to use the **Format Painter** to **copy the formatting** from the text and paste it over the top of the other text. Let's look at how you might go about doing that.

The first step is to pick the text with the formatting you wish to copy, then go to the **home** ribbon and select **Format Painter** from the first group, which is the **Clipboard** group.

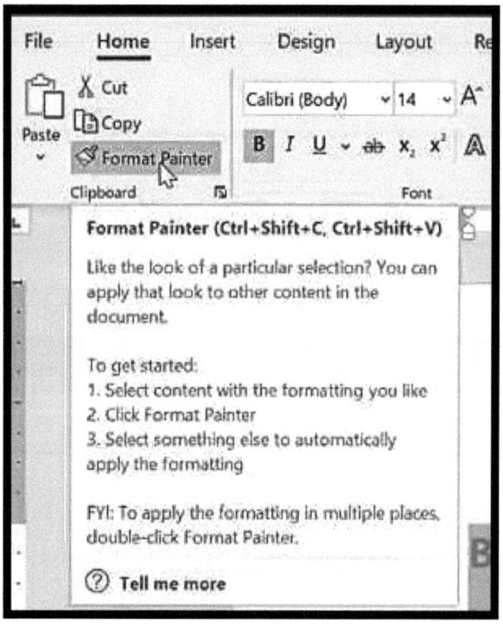

You'll click it once, and when you return your mouse to your text, you'll notice that your cursor has changed to a miniature paintbrush, indicating that your Format Painter is now active.

All you have to do now is click and drag it over the text where you want to paste the formatting, and there you have it; it's applied that same formatting.

What if you wanted to accomplish the same thing with a different text? Once you've copied it once, the Format Painter will be off, and your cursor will return to its normal appearance. The trick is to double-click on the format painter if you want to do a lot of Format Painting in a document. After selecting your text, double-click on Format Painter and swipe it over your text. You will see that the Format Painter is still active because you double-clicked, and you can keep swiping throughout your page. You can exit Format Painter by pressing the **ESC** key on your keyboard or just clicking on format painter again to return to a standard cursor. Instead of manually applying those individual font properties, this is a much more efficient approach to copying formatting over.

In this chapter, you studied the fundamentals of applying Character Formatting to selected text and you also learned how to copy formatting to other areas of your document.

CHAPTER TEN

STYLE FORMATTING

Styles are a very nice way to format your document if you've never used them before. It's simple to update, and you can even include an automatically updated table of contents. They provide numerous advantages, and if you've never used them before, this chapter will completely transform the way you work with Word documents. Hopefully, this has piqued your interest; why don't you fire up your computer; it's time to show you how to use Styles.

A Quick Overview

"**What is the advantage of using a Style**?" you might be wondering. Let's take a look at an example of how Styles can help.

You may already have a Microsoft Word document open that you've created or are working on, and you'll need to refer to this sheet from time to time. Let's say you're working on a document and have a lot of varied text on this page, as well as a lot of headers in your sheet or pages. Simply put, it's a basic Word document that hasn't been formatted in any way.

In this situation, you might want to make the headers stand out a little more; simply highlight and bold them, increase the font size or even alter the font color, and you'll have a new header. The only issue is that if you go down to other headers, you'll see that they're still in your original design.

You could redo all of the formatting by using the format painter to copy the format, paste it there, apply it, and then go through your entire sheet and do that, but there are a lot of different headers in a sheet, so that will be a lot of work. This is where Styles come into play, and they can help you save a lot of time.

Finding and Applying a Style

You can identify your current style by highlighting all of the different headings in your document and looking under the **home** tab for a section named "**Styles.**"

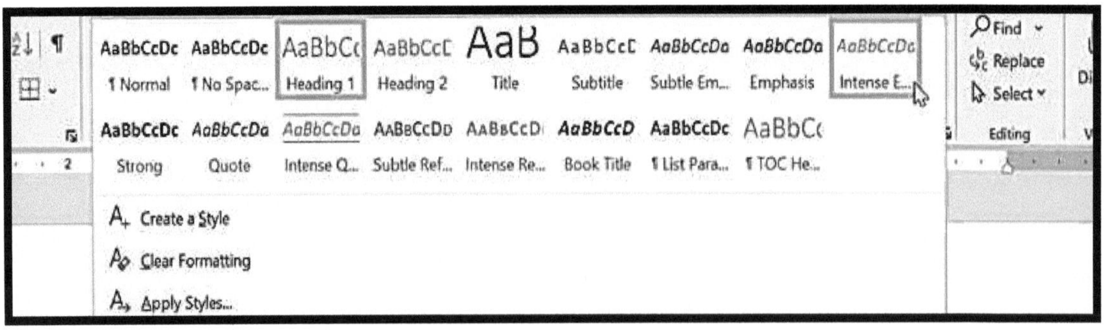

You can check the current style of the headers here, and you can also **apply a style** to a specific header by clicking on it. You can also scroll through them or see them all at once if you click the small drop-down next to it. This way, you'll be able to see all of the Styles you have.

Modifying a Style

After you've applied a Style, the cool thing is that any changes you make to it now will affect all of the other locations where it's applied. This implies that if you **right-click** on "**Heading 1**" for example, you can choose "**Modify**" from the drop-down menu. You can change the fonts, size, bolding, and color, as well as the paragraph spacing and borders around it, when you click Modify. There are a lot of different things you can tweak for the style of this header when you click Modify.

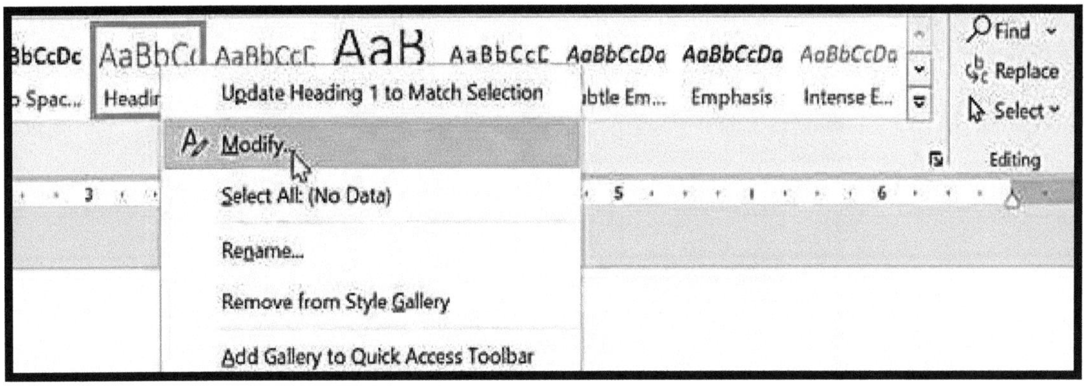

After you've made your changes, click "**OK**," and you'll notice that all of the headers in your document have been automatically modified, and any place that references that style will automatically update as you work on it.

As you can see, the benefit is that as you make changes to a style, it will rapidly apply those changes to all areas where that style is used, providing you a lot of flexibility and making it very easy to update a document's appearance and feel.

In Microsoft Word, there's a **design** pivot up top, and if you click there, you'll see all of the available numerous themes, all of which take advantage of the styles that you've set.

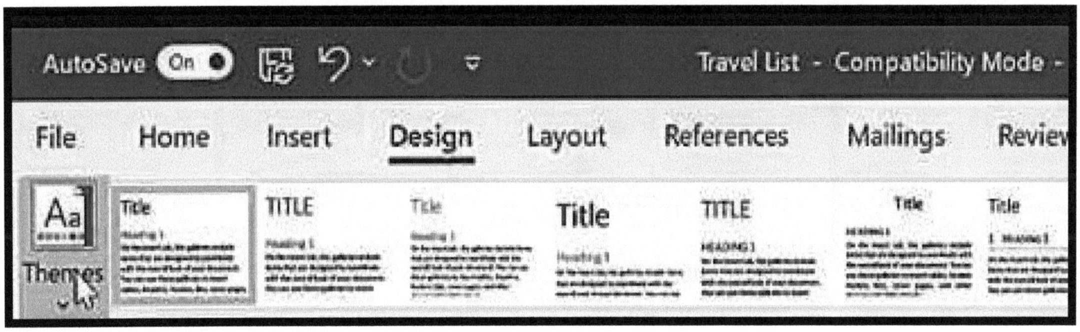

If you select a theme, it will now search for all of the headers that have a specific style that you are using and apply that theme. You can choose from a variety of themes here, and it takes advantage of the fact that you can define multiple styles for different elements inside your document, making it incredibly simple to change the appearance and feel of your document. You can even use different colors and font types, which will swiftly scan and update throughout your text. One of the advantages of styles is the ability to update them fast.

Adding a Table of Contents

Another reason to use Styles is that you can not only adjust the document's appearance and feel, but you can also add some functionality. If you want to provide a table of contents for all of your content, you'll usually have to fill in headings first, then add each of them as a part to your table of contents. It's difficult to maintain that up to date manually, so you'll go to the "**References**" page and select the option to **insert a table of contents**.

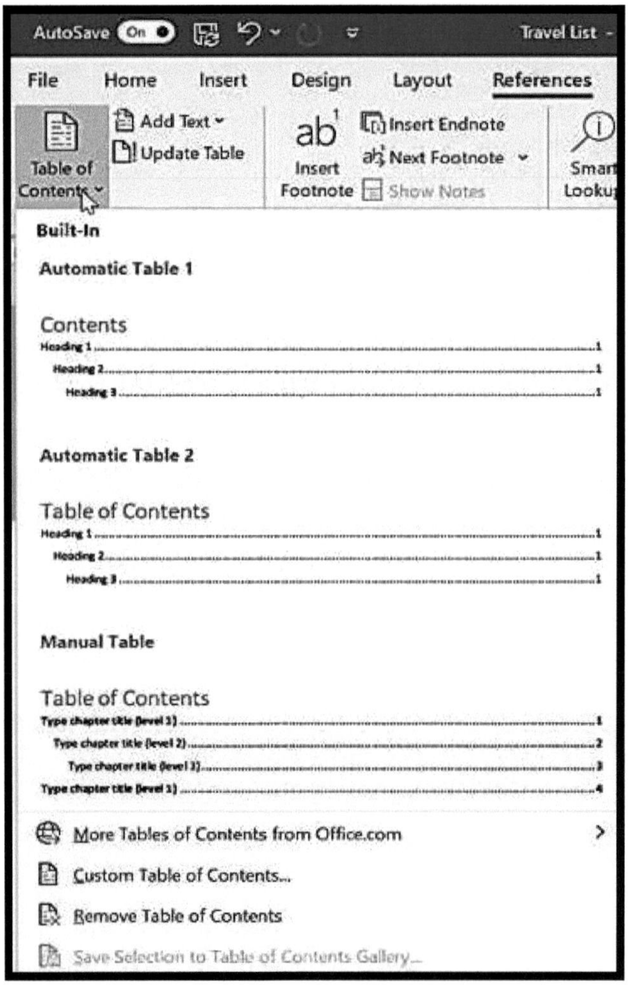

You can either **insert the contents of the automatic table of contents** here or **create a custom table of contents**. When you use the automatic table of contents, you'll notice that all of the page numbers correspond to the page where the headings appear, indicating that the table of contents is taking advantage of the fact that you've included all of these headings in your document. You can delete the table of contents if you want to get rid of it.

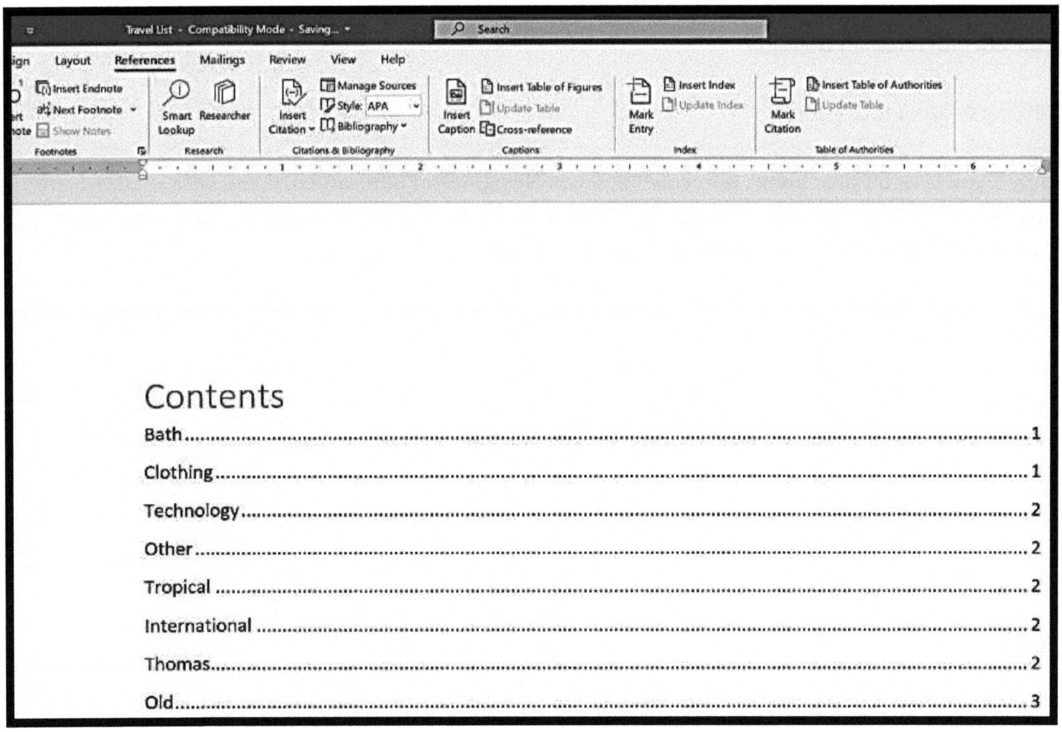

If you wish to create a custom table of contents instead, one thing you'll notice is that you can have the table of contents take advantage of the styles (Heading 1, Heading 2), so as you define additional styles in your document and more heading Styles, the table of contents will benefit.

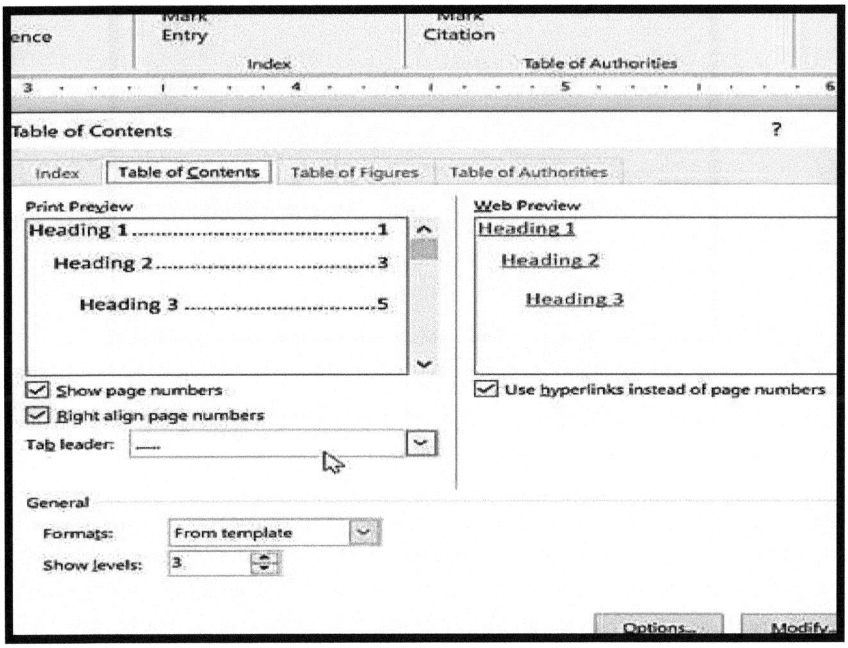

This is a useful feature if you wish to use a table of contents, and it's also important to employ styles in your document when doing so.

Getting Around in Style

Also, if you click on the "**View**" tab, you'll find the **Navigation Pane**, which, if you click on it, will offer you a quick view of your document headings. If you want to jump down to a specific heading, simply click on it, and it will take you to that location in the document.

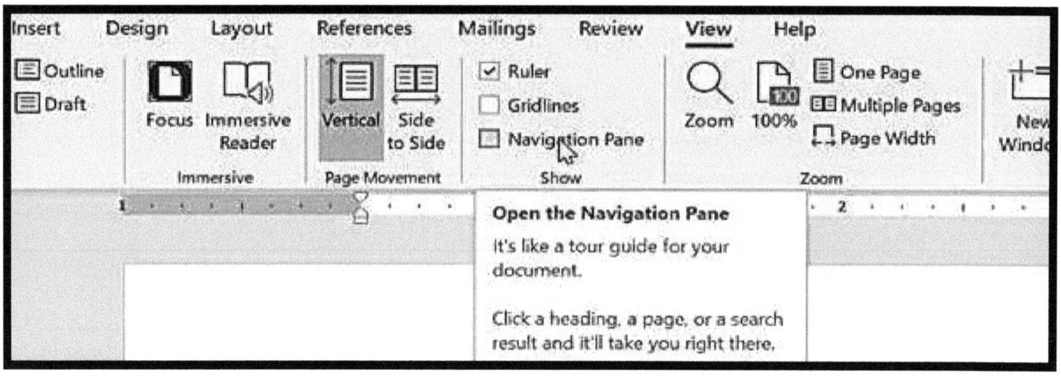

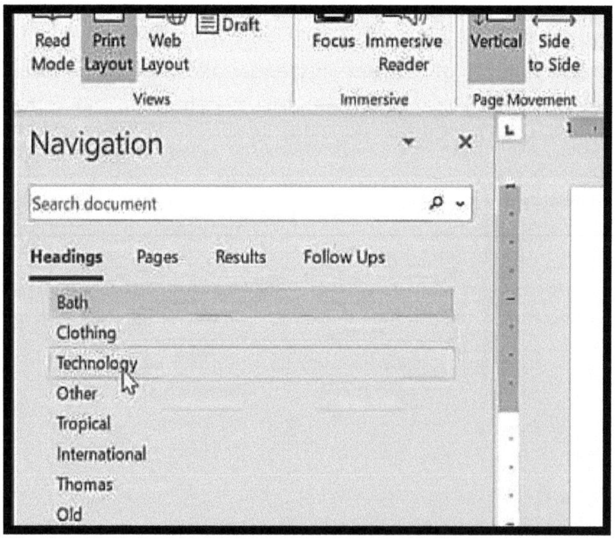

This is a great way to quickly navigate your page using headings you've set, and you may do it by using multiple Styles. This is just another advantage of using Styles.

In this chapter, you learned why you should use Styles in your document and what benefits you should expect. One of them is that you can quickly update the look and feel of your document without much effort; you can also use styles if you want to insert a table of contents and have it automatically generated; you want to keep the page numbers current, and you can also use styles to navigate throughout your document and jump to different sections within your document.

CHAPTER ELEVEN

PARAGRAPH FORMATTING

We're down to alignment alternatives in this chapter. We looked at how to change the formatting of individual characters in the last chapter, and now we're going to look at how to format full paragraphs.

If you have a document with no formatting and everything is left-aligned, you'll notice that the text on your page that is left-aligned wraps around, so there is never a broken word at the end of the line. It will just put it on the following line, and there will be no hyphenations or anything like that.

The Formatting Commands for Paragraphs

These are a set of commands that can be used to format a paragraph

The most obvious component of formatting is to center the title on the page. To do so, first, choose your title by hovering your mouse over the margin and clicking once, and then go to your **Paragraph** group on the **home** ribbon, where you'll see your alignment options right here.

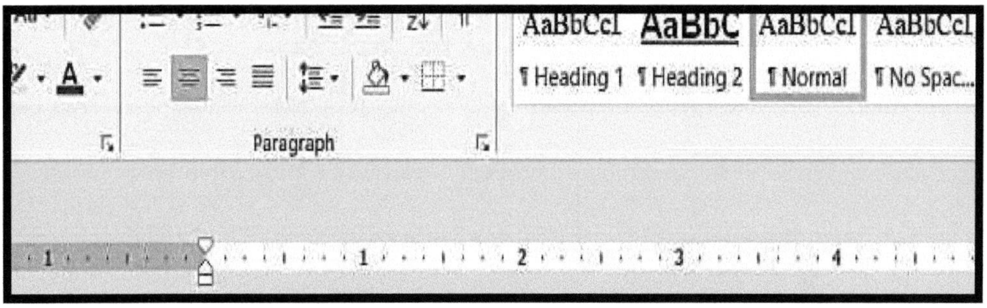

You may have **left alignment** enabled; you may also have **Center** or **Right alignment** enabled. There's also one here called "**Justify**," which will center everything equally on the page, ensuring that the text goes up to the end of the left and right margins, respectively. This is common in newspaper printing and books, and it implies that if Word needs to space out a word a little bit more, it will do so to align your text on both borders. The paragraph group on the home ribbon contains only those four buttons.

Justification and alignment

Everything in the middle

Click the "**Center**" align button (there's also a keyboard shortcut of **Ctrl + E**) to move a title in your document to the center of the page.

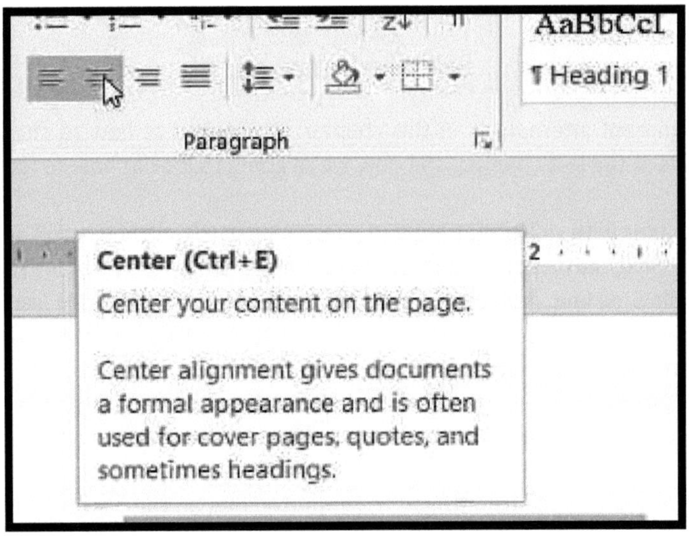

Left or Right

You can use the left align button to simply realign that back to the left.

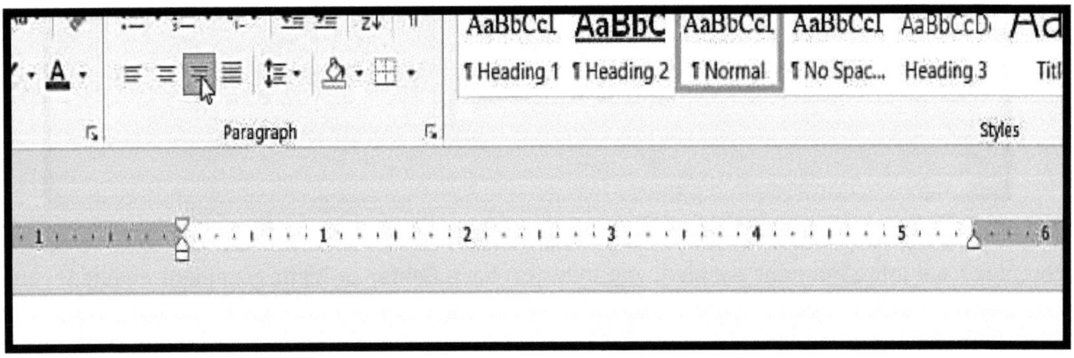

Let's see what occurs if you Right-align your text. To do this, use the **Ctrl + A** keyboard shortcut to select all of the text in the document, then click "**Right-align**", and as you will now see, everything is aligned on the right side, but not on the left.

A document that entices

Now select all of our text once more, and look at the "Justify" option that was stated earlier. If you click "**Justify**" and look at your sentences and paragraphs, especially where the text wraps around, you'll notice that they're aligned both on the right and left sides, so Words goes ahead and adds some extra spacing to make that work.

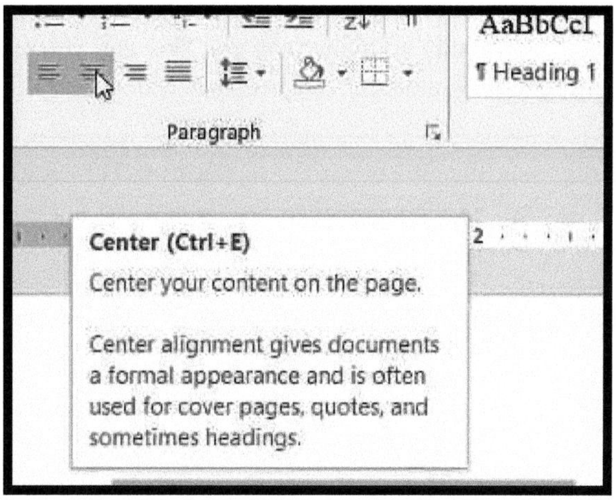

Paragraphs and line spacing

Let's look at how line spacing works in paragraphs now. It's self-evident that when we talk about line spacing, we're referring to the amount of space between each line of text.

If you input a word and tap **Enter**, you'll notice that the cursor doesn't appear right beneath the text you typed, but rather a bit further down, leaving a small gap.

This was created mostly because many users did not want to type right beneath the text above them. You generally want a little bit of space after a title, so you'll find that the default in Word nowadays is to have a little bit of a gap in there, and of course, you can alter it, which is what we'll discuss in this section.

How can you identify what line spacing you're using right now?

Finding out what your current line spacing is

If you go back to that Paragraph group, you'll notice that there's a button right in the middle labeled "**Line and Paragraph Spacing**" with a drop-down arrow that, when clicked, will show you where you have your line spacing set to, as indicated by the tick next to it.

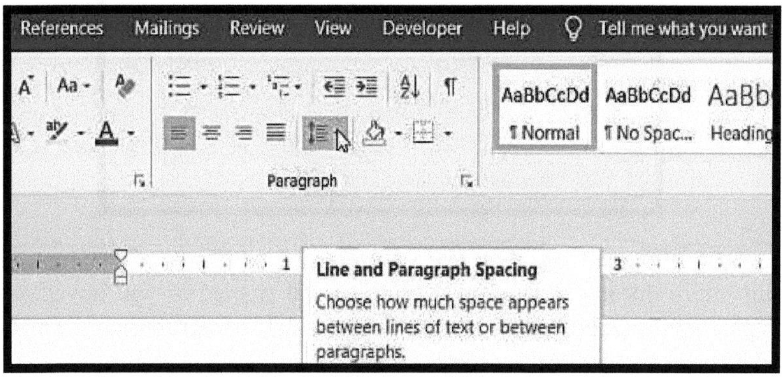

However, if you don't use any, it won't display any line spacing to you. If you're not happy with the amount of line spacing, you can modify it to anything you like, and you can see how the line spacing changes as you hover over them. As you descend, it becomes wider.

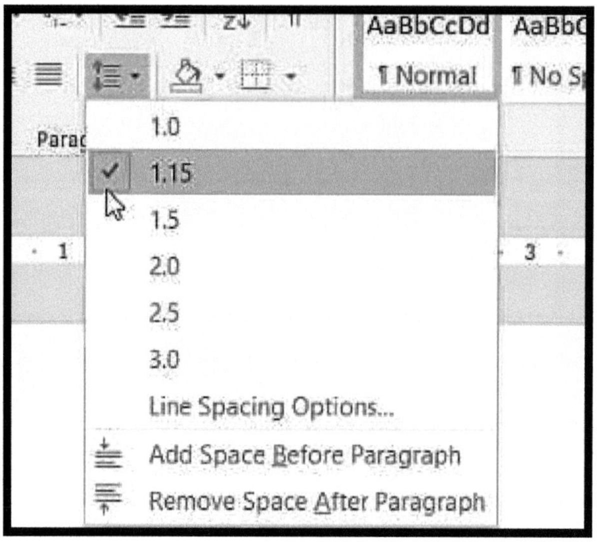

In general, 1.15 is easier to read than 1.0, which is a touch too close together; hence, if your document is pretty long and others will be reading it, 1.15 is a good line spacing, but you can raise it if you like; it's entirely up to you.

You have the option to "**Remove Space After Paragraph**" at the bottom, which is another sign that line spacing is there.

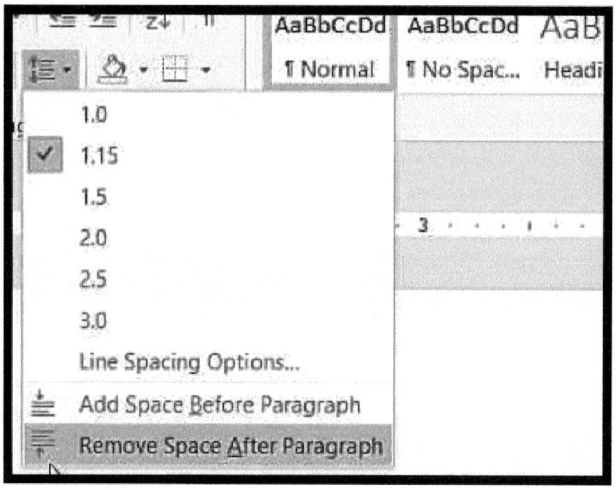

If you have your entire document chosen, everything will change as you hover over these options; however, if you only want to modify the line spacing on a single paragraph, simply make the appropriate selection before going in and changing the line spacing.

Choosing a Line Spacing

Another thing to keep in mind is that if you go to Line spacing options from the Line spacing drop-down menu, a small dialog box will appear. The spacing group is the third small group here, and this is where you can get quite specific about how much space you want before or after the text. You may already have these parameters set, but you can modify them to **single line spacing**, **1.5 lines**, and see what it looks like in the preview at the bottom. You can pick "**At least**" and then adjust how much space you want in there with double line spacing. As a result, you can make it seem any way you like.

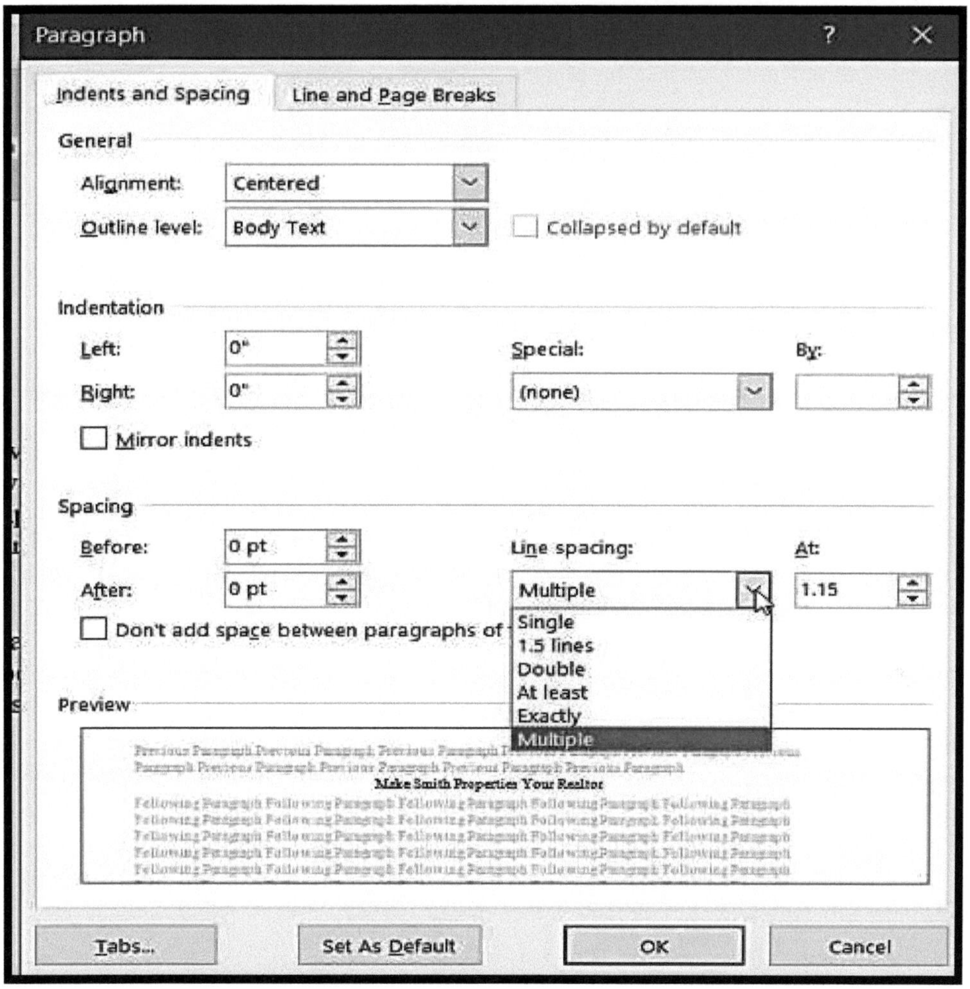

You can also opt to "**Set as Default**" if you've already set your spacing and how you want it to be for all of your documents.

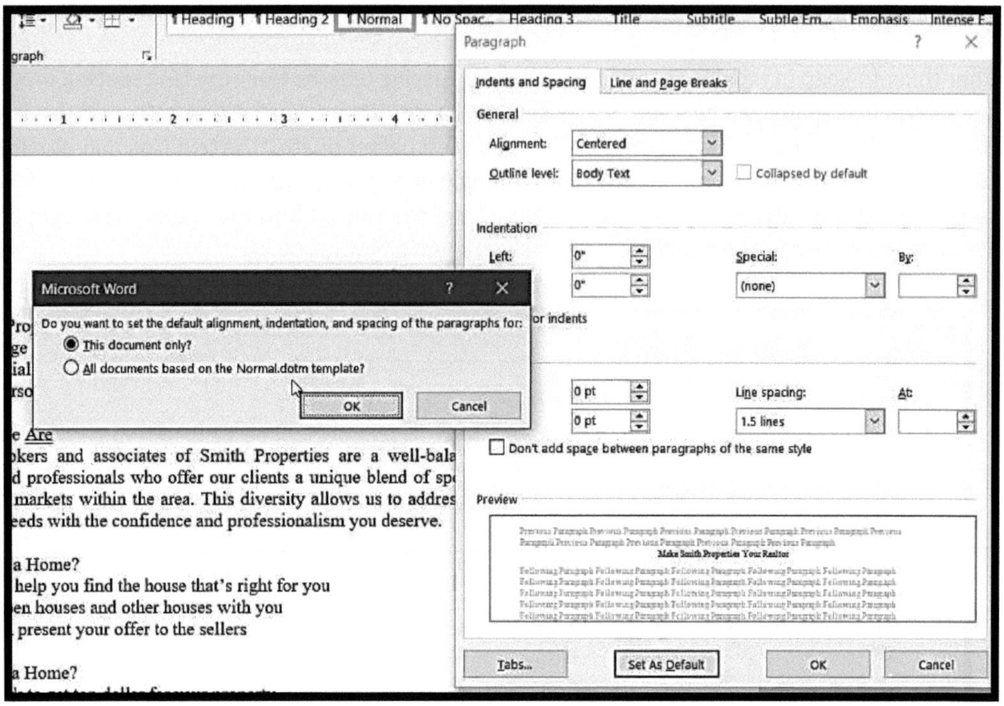

It will then ask you if you want it to be the default spacing for "**This document only**" or "**All documents based on the normal.dotm template**." It's all up to you. Just keep in mind that if you set it for all papers, you'll get whatever line spacing you specify in this box every time you create a new document. That's all there is to line spacing, so have fun with it. If you want to use the same line spacing throughout all of your documents, come in here and set it as default.

Working with Indents

We'll now discuss how to use indents in your documents. We've talked a lot about formatting options for paragraphs in the previous sections, and you may want to indent the left side of a paragraph when you have a document. You could indent the first line with the tab key, but it will only do so for the first line. Indenting the entire paragraph is what we're talking about here, and you can do it with the small indent marks that we have on the ruler. You might notice that a line has been changed half an inch while trying to alter things with your indent marker, as shown below.

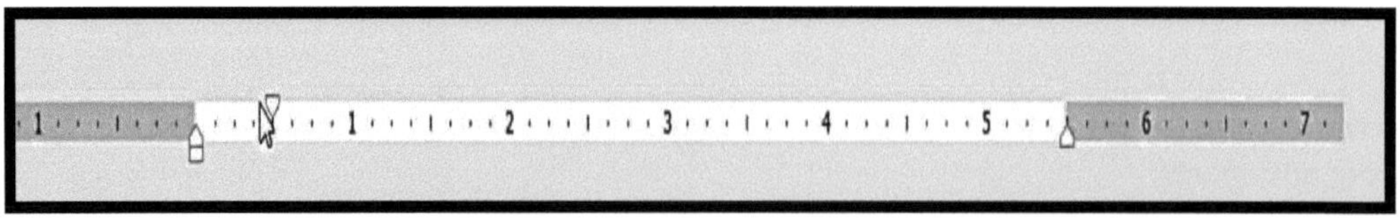

How would you indent the entire paragraph if you didn't want it this way or if you couldn't use the tab key? You can use the small markers on the ruler at the top to establish the spacing between your paragraphs or indents exactly how you want it.

Indenting the First Line of a Paragraph

Let's have a look at the functions of these tiny buttons. When you hover your mouse over them, starting with the **downward-facing triangle** at the top, you'll notice that it says "**First Line Indent**," which means that this small button only controls the first line of your paragraph.

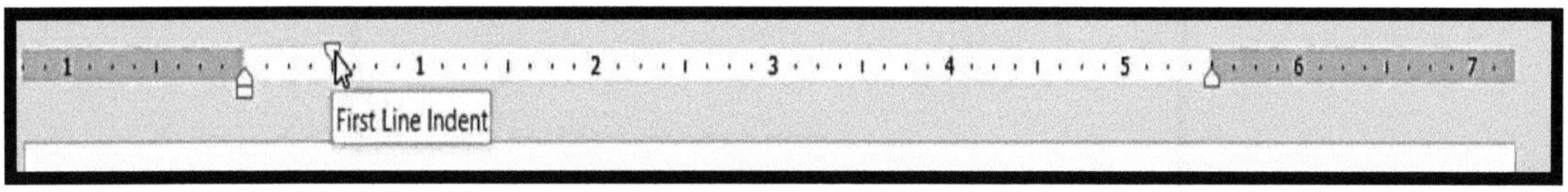

You can see that because it's the First Line Indent, it only moves the first line if you **highlight a paragraph**, **go up to your ruler**, then **click on your First Line Indent button** and **drag** it to roughly three-quarters of an inch and let go.

It works in the same way as the tab, but this time you can specify the size of the indent; keep in mind that the tab will only allow you to advance half an inch by default.

The hanging Indent

Let's look at the little **upwards-pointing triangle** below that on the ruler. This is the "**Hanging Indent**," and dragging it across the page will show you that it moves everything but the first line, almost like the First Line Indent.

Can you think of any examples of a hanging indent at this point? Let's say you have a list that you've numbered. You can see what's going on here if you glance at your ruler. The First Line Indent, as well as the Hanging Indent, have both moved out.

Using Indents to Adjust a List

You can use them to change the appearance of your numbered list. If the items in the list appear to be too near to the numbers, click and drag the Hanging Indent out to move them further away.

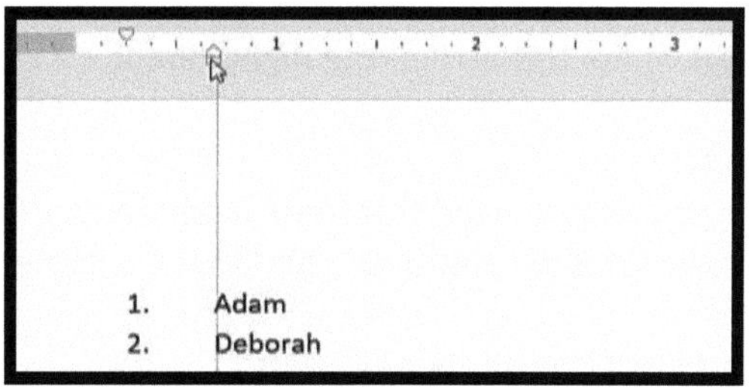

The numbers, which are set to the First Line Indent, are the same way. You can change those by dragging that in or out.

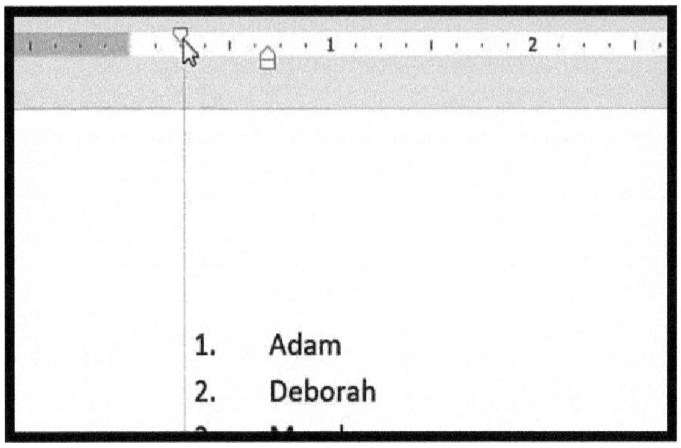

That's an example of indents in your document, which Word does for you when you apply Numbering.

Indenting a Paragraph

The **Left Indent**, which is positioned beneath the Hanging Indent, is the other object we have here.

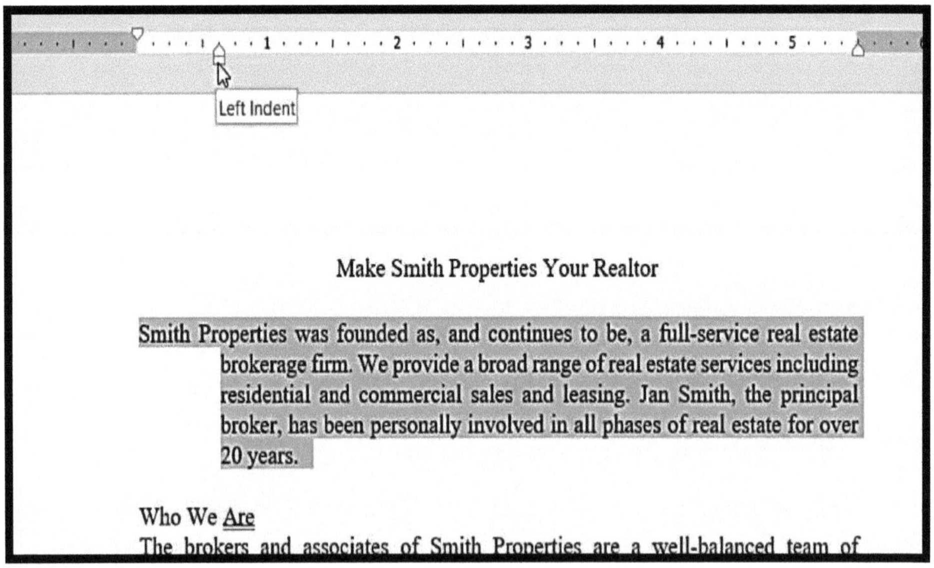

If you drag that, it will drag the entire text across. The **Right Indent** is a corresponding one on the right-hand side of the document.

You can grab it and drag it in, but keep in mind that these modifications only apply to the paragraph you've chosen; if you want to do it for the entire text, hit **Ctrl + A** and then drag your indents in and out.

That's a quick rundown on how to use your ruler's indents.

CHAPTER TWELVE

TAB FORMATTING

Welcome to the Tabs section of the book. This is where we'll discuss how to use tabs in your documents, which is a valuable feature in Microsoft Word.

This means that pressing the Tab key will allow you to override the half-inch tab spacing that is set by default. As we've seen before, clicking your mouse on the top of a document and pressing the Tab key will move you half an inch each time. You can modify the space between tabs to anything you wish, and you can even change the type of tab you're using in your project.

If you look at the document below, you'll see that everything is neatly organized in columns and everything is in order.

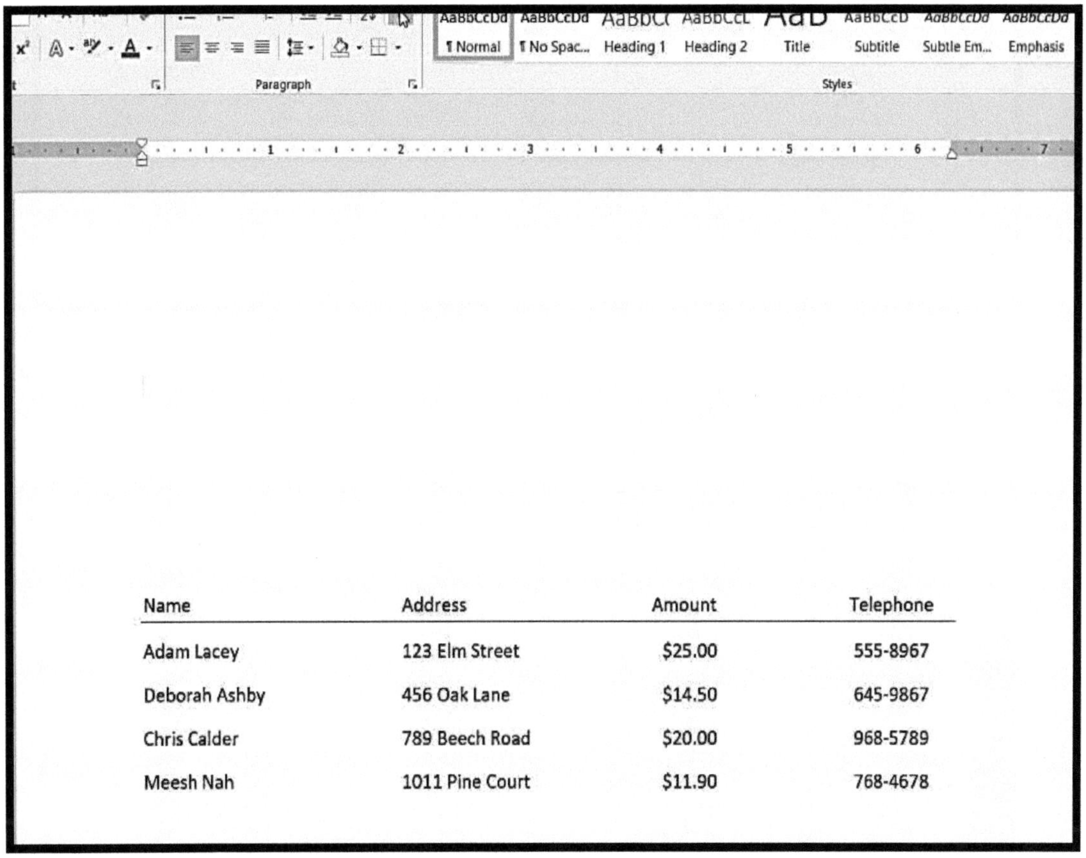

This may appear to be a table, but it is not. It was made with tabs, and if we turn on our **Show/ Hide paragraph markers** (which we looked at in a previous chapter), you can see all of the tabs in there, so everywhere there's a tiny arrow, there are tabs.

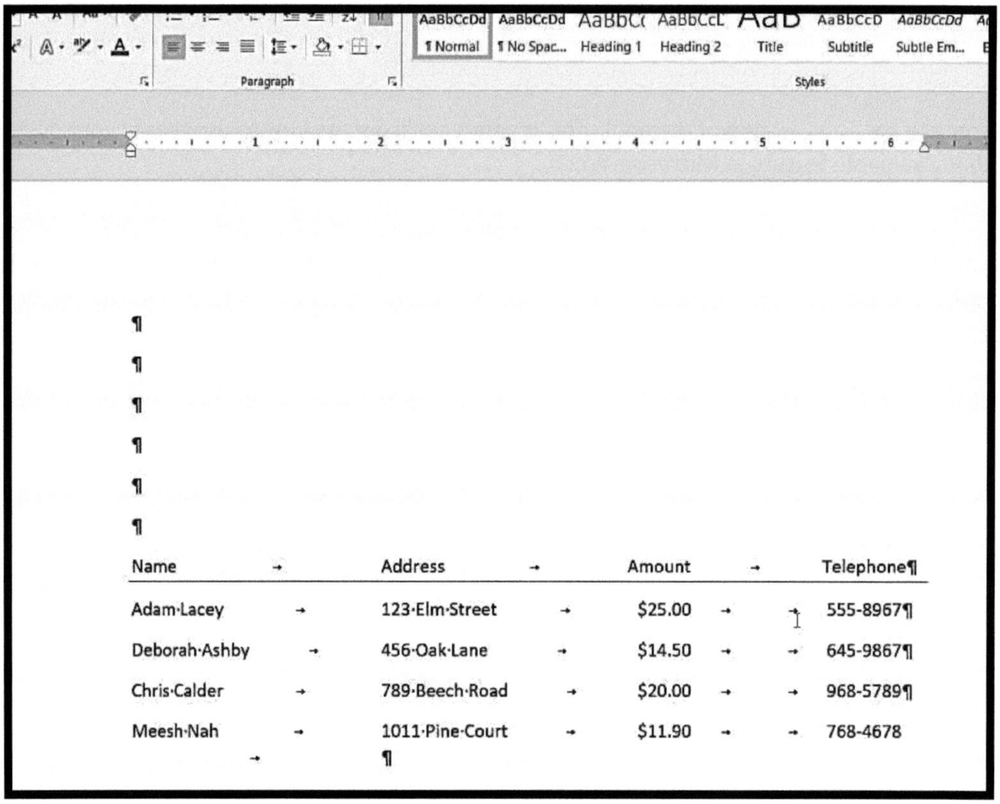

Identifying Your Tab Stops

Let's have a look at how to create tabs in a document now. If you're wondering where your tabs are, they're fairly difficult to find if you don't know where they are. If you glance up at the ruler and cast your gaze to the left, you'll notice a very small **back-facing** "**L**" **sign** just there, and if you hover your mouse over it, you'll see the "**Right tab.**"

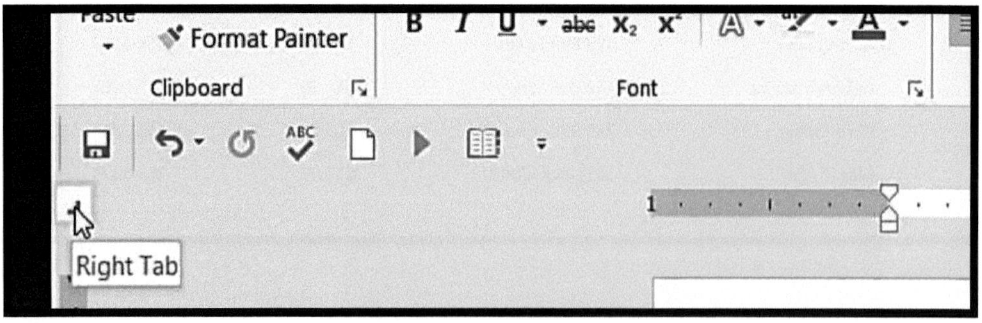

If you click it, you'll be taken to the "**Decimal tab,**" which is the following tab.

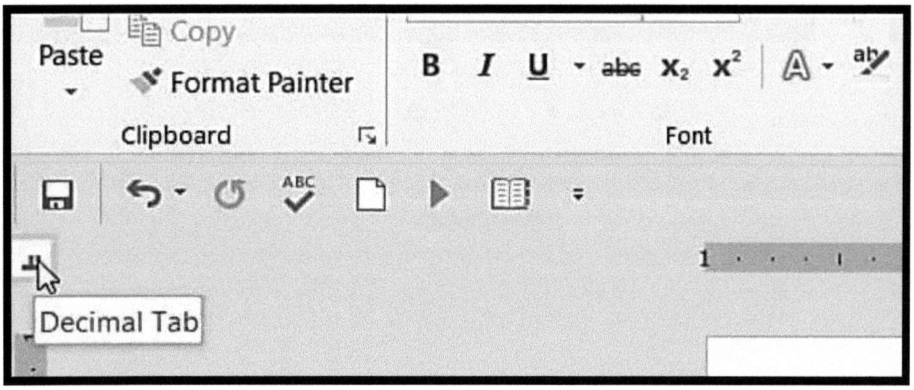

Let's have a look at what they are.

The Left and Right Tab Stops

The Left and Right tabs allow you to align your text to the left or right, respectively.

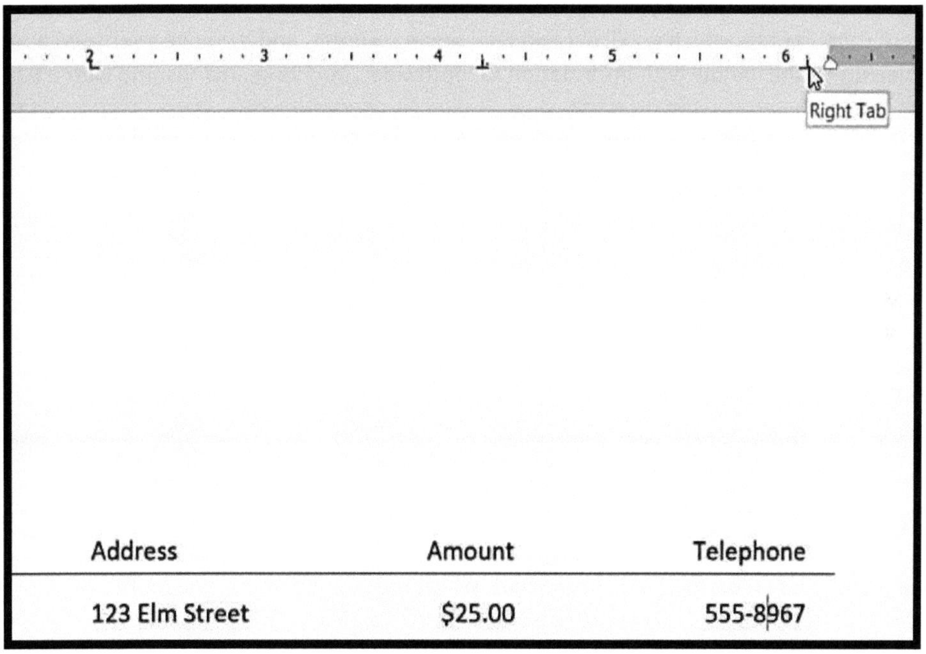

The Decimal Tab

The Decimal tab allows you to align your text using decimal points, so if you have numbers in your document, you can use that option to align all of the decimal points.

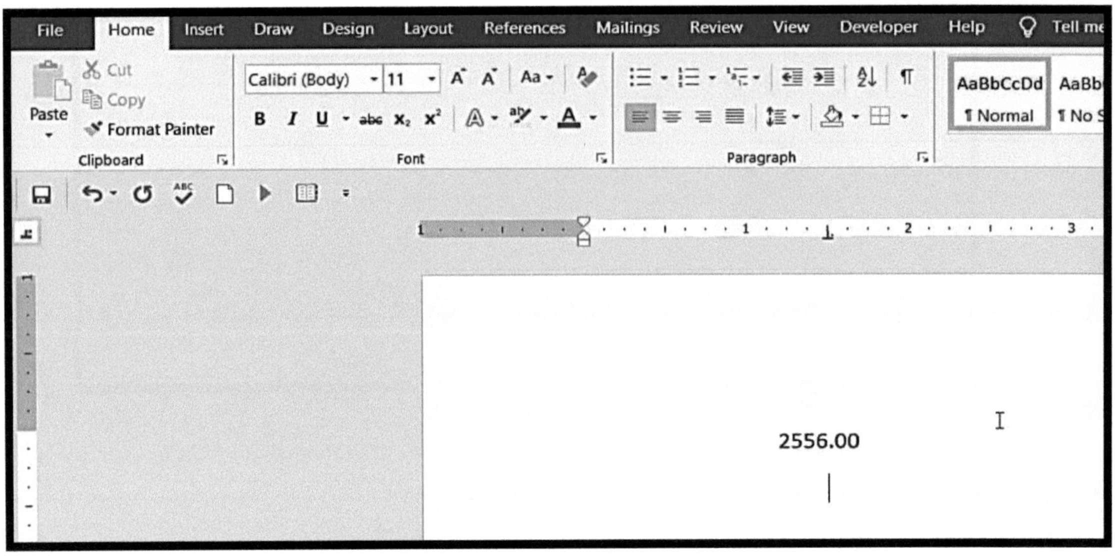

The Bar Tab

Then there's the Bar tab, which you'll occasionally see in resumes, and if you've ever seen a resume with a line going down the middle with information on both sides, you've seen a resume with a Bar tab.

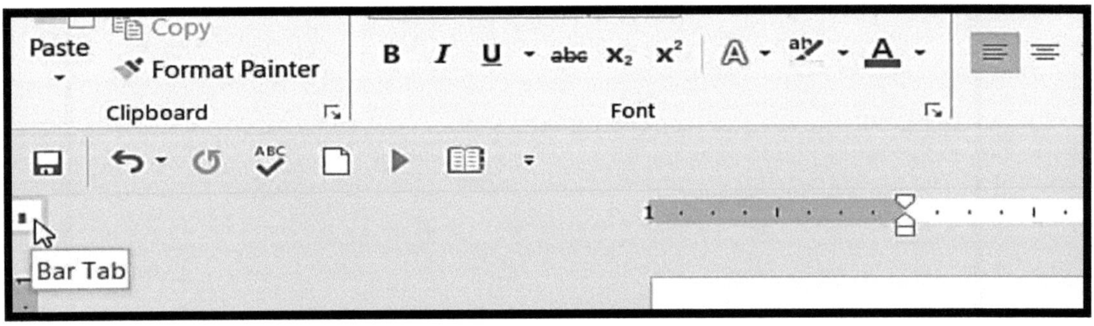

Working With Tabs

Then there are indents like the **First Line indent**, which is the same as the one on the ruler.

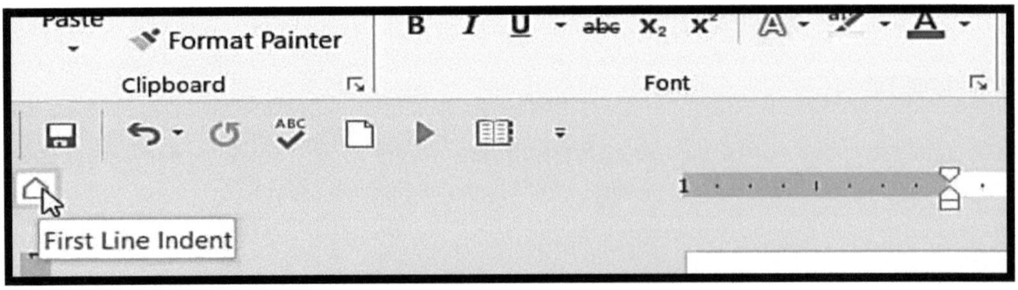

If you click again, you'll see the **Hanging Indent**, and then you'll be back on the left tab.

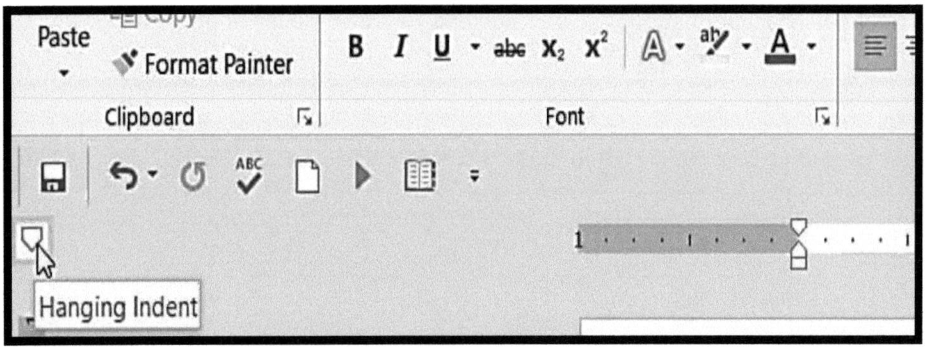

You may just click on them to rotate between them and then select the tab you want to use in your project. Let's have a look at how they function. You can see the small "**L**" if you select the Left tab and then go over to the ruler and click where you want to drop it.

When you press the tab key, you'll be taken exactly underneath the tab you've set, where you may input your text. The same thing happens if you press Enter, then tab to the following line, and you'll be immediately underneath the line above.

To remove a tab, simply click and drag it off the ruler.

- Let's have a look at some of the other tabs and see what they can achieve. Returning to your small tab indication, if you select the **Decimal tab**, go to your ruler and, by clicking on the ruler, drop this at roughly one-and-a-half inches. When you type a number with a decimal point, you'll notice that the decimal point shows just beneath the decimal tab.

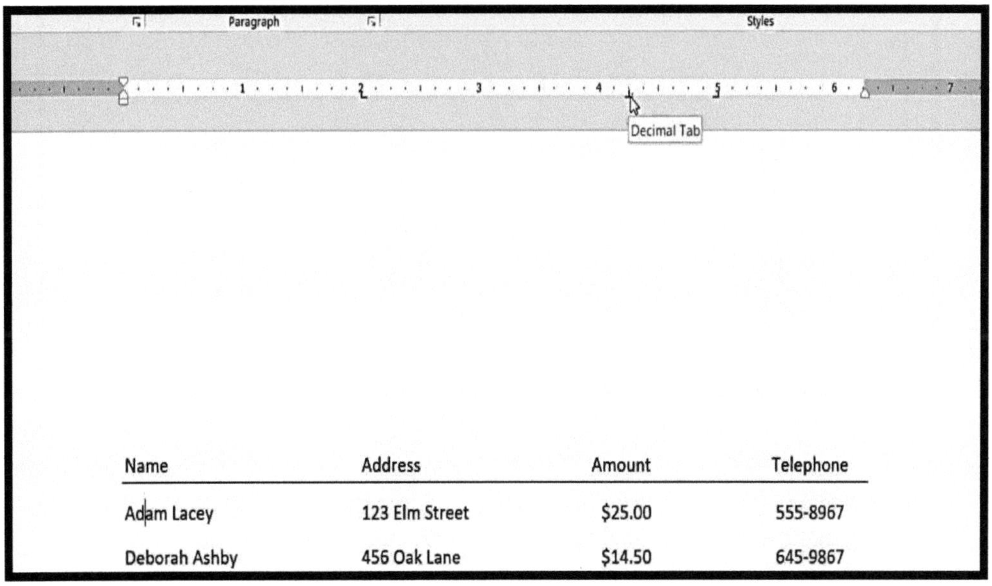

- Moving on to the **Bar tab**, if you place it anywhere on your ruler, it creates a small line in there where you can input some text, hit the tab, and go to the other side, as seen on resumes where your personal information is on one side and you're schooling or prior career is on the other.

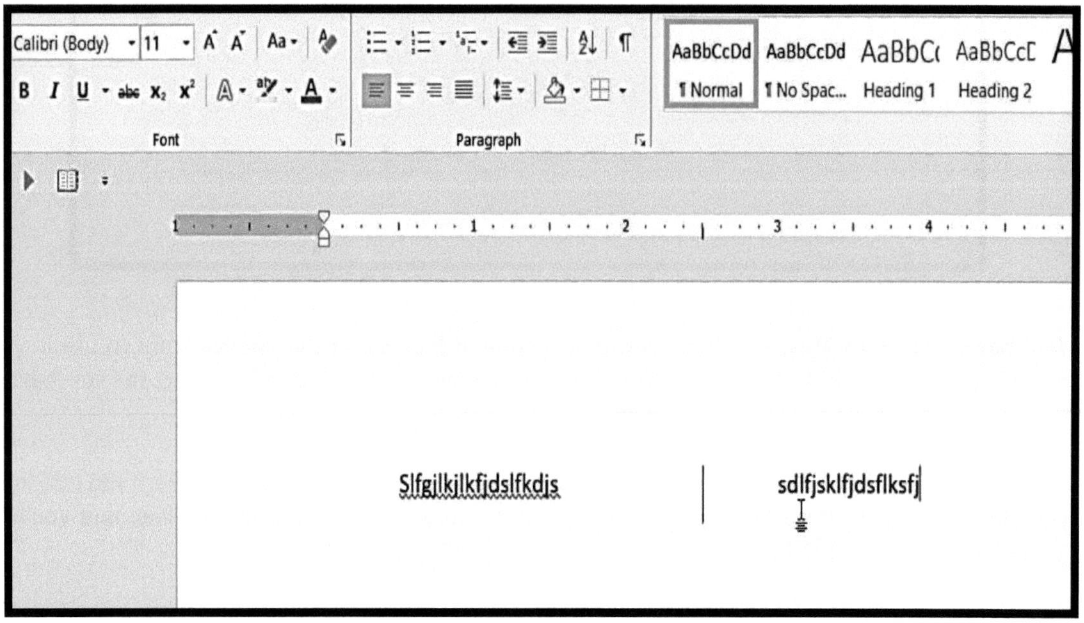

- Moving on to the following tab, **Indents**, we've already looked at those in the last chapter, so we'll skip over them and concentrate on the **Center tab**. You can see that it's exactly centered underneath that tab if you set it to roughly 2.5 inches, click on the ruler, then type your words.

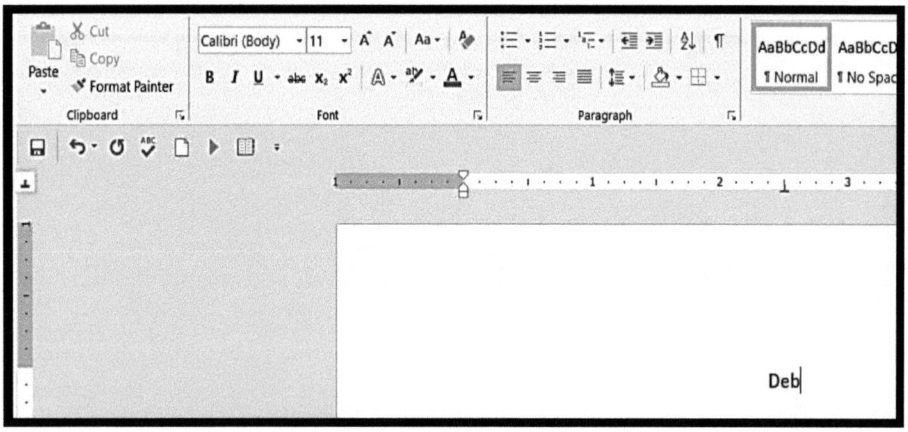

- The **Right-align tab** is the last one we'll have a look at. If you click on the end of the ruler, your cursor will jump to the right-hand side, and as you type, your characters will print to the left, ensuring that everything is lined up beautifully on the right-hand side.

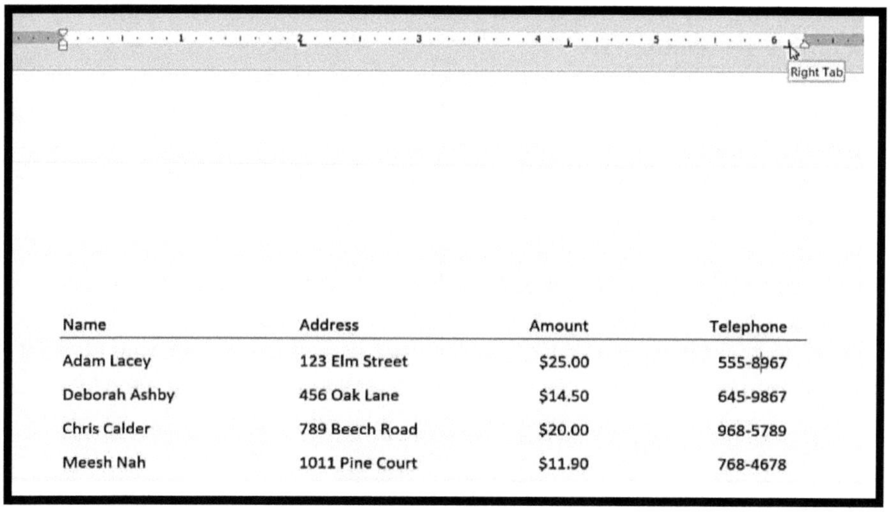

These are the many tabs available in your document, and which ones you select will depend on what you're trying to align.

Creating a Tabbed Table

If you look at this table, you can see the tabs that have been set up for these columns down here if you click within it.

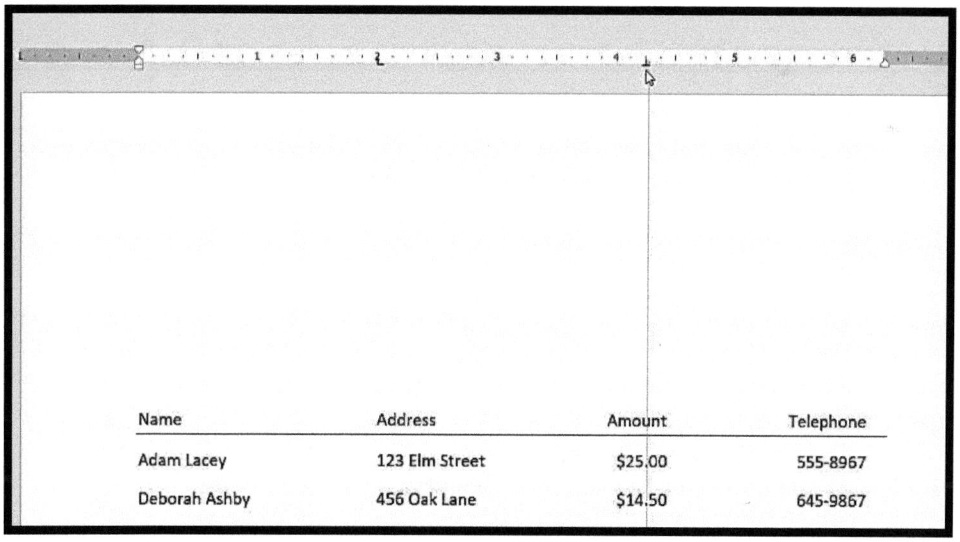

If you look at the ruler, you can see that we have a left tab set up there, which aligns all of the addresses to the left, a decimal tab set up for the "Amount" column, which aligns everything to the decimal points, and finally, a right tab set up for the last column, which aligns everything to the right.

You can reproduce this tiny table that we have above in your document for mastering, and let's do it together for further assistance.

- First and foremost, you'll set up your tabs. All you have to do is grab the Left tab and drag it to the desired location.
- After that, you'll take your Decimal tab and place it where you want it.
- Finally, you'll take your Right tab and place it at the very end of the document.

After you've set up your tab on your ruler, you can begin recreating the information from the previous section. Remember to press your tab key to go across to the tabs as you text, then hit Enter to continue to the next line.

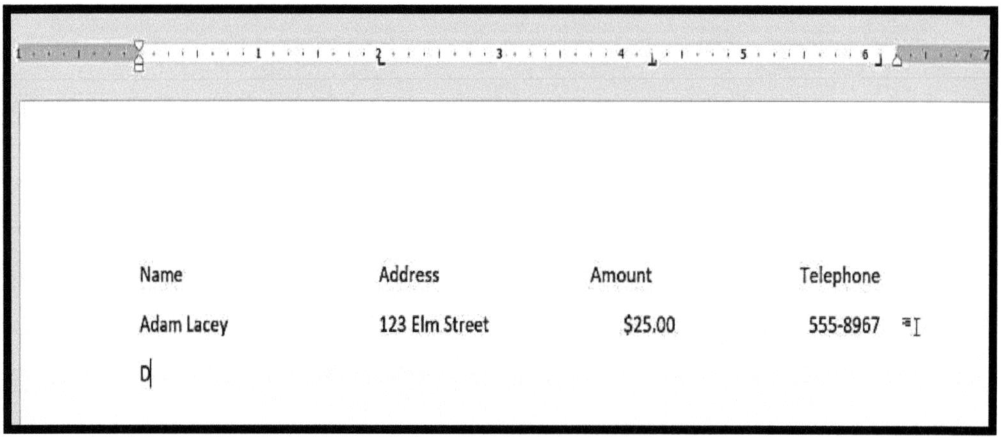

You can see how we made that table above, and if you don't like where the tabs are now or how the document looks, simply drag and drop these tabs left or right.

Leaders of Tabs

The last point we'll talk about is something called **Tab Leaders**. This may not be a term you're familiar with, but if you've ever looked at a table of contents in a book, document, or manual, you'll notice the page number, a few dots, and the title of the chapter or topic are frequently included. A tab leader is essentially that (those dots).

You'll see how to add those on when you set up your tabs in a few seconds.

- The first step is to highlight all of the text where you want tab leaders, and there are a couple of different ways to get to tabs: you can **double-click** on any tab stop on the ruler to go to your tab box, or you can click the little drop-down arrow at the bottom of this box to go to the same place.

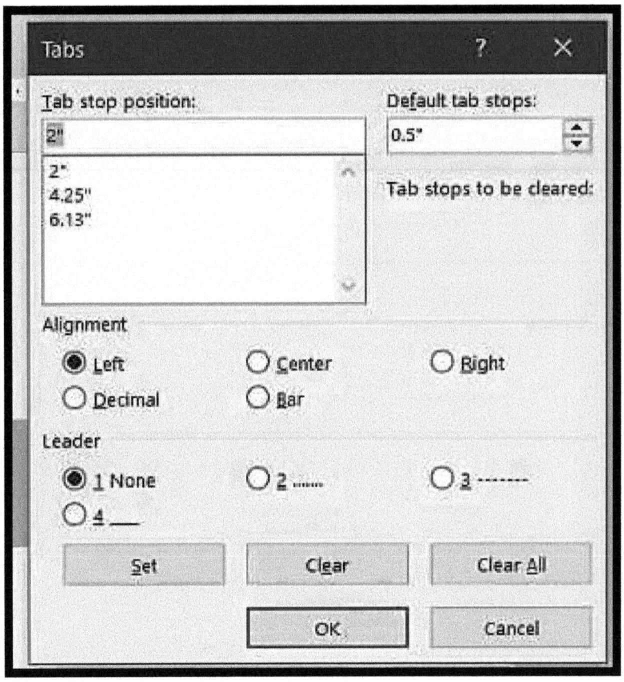

- All of the tabs you've set up for the selected region are displayed here. We have tabs at 2 inches, 4.25 inches, and 6.13 inches, as you can see above. We have "**Leader**" at the bottom here, which is now set to none. We'll now set a leader, and all you have to do is choose your initial tab stop and a leader style (dots, dashes, or a solid horizontal line), then click "**Set**" (very important and very easy to forget).

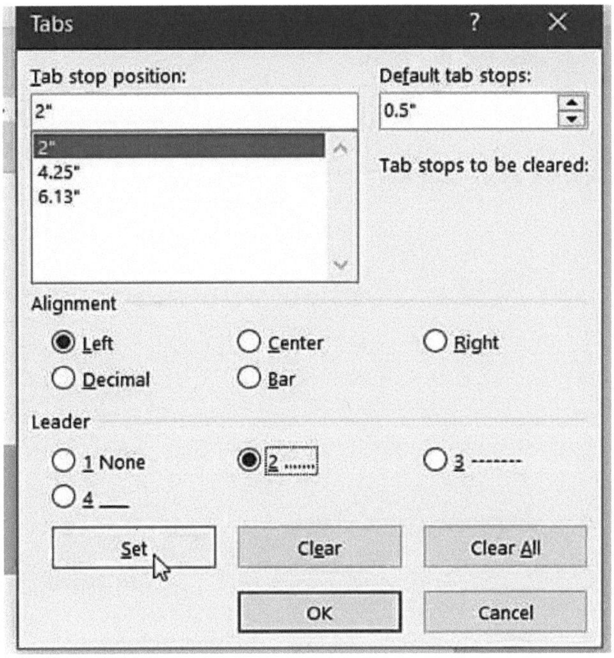

- You must repeat this process for each of the tabs, so go to the next tab (which is 4.25 in this example), select the **Leader style**, click "**Set**," and then go to the last tab (6.13 in this case), click "**Set**," and then click "**OK**."

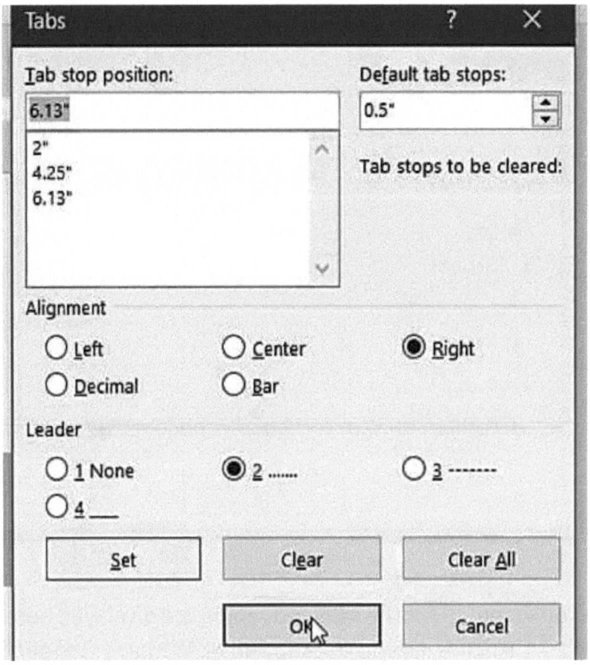

In between your columns, you'll see your leaders. If you're doing anything like a table of contents, that's great; it just adds a professional touch. Hopefully, it gives you a better sense of how to build up different sorts of tab stops on your ruler, as well as how to incorporate those leaders.

CHAPTER THIRTEEN

PAGE FORMATTING

Welcome to this chapter. Here, you'll learn how to set up your website so that you can start working on a simple document right away.

Page Setup

When you initially start a document, it comes with only the default pages, which are all portraits. The first thing you should look at is your **margins**. You may be okay with them by default, but some individuals like to adjust them just to vary the look of their work or to fit more content on their page. You'll be able to fit more content on the page if the margins are smaller.

Setting Up your Margins

You have two rulers in your document: one at the top and the other on the side. If you can't see your rulers, click up to **View** and check **Ruler**. You'll be able to view your margin line after that.

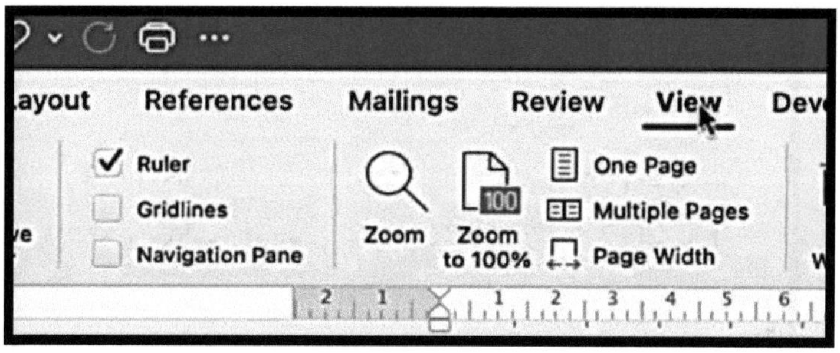

Remember that **your margin is the gray area**, while **your document is the white area** where your words will appear.

If you want to adjust them, go up to your ruler, place your mouse between the gray and white sections until it changes into a **double-headed arrow**, click, and you'll see a faint line come up a vertical line, and all you have to do now is pull your margin to the right.

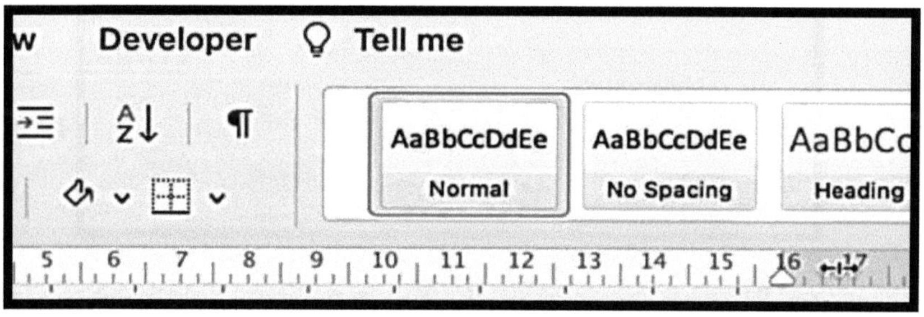

As you can see, this has reduced your margin and increased the amount of text on your page. You can repeat the process on the left. These arrows can get in the way, so sometimes you just have to move them out of the way; this will shift part of your text, but don't worry, just grab the margin, in the same manner, we did the other side, make sure your cursor is a double-headed arrow, and drag it over to the left once more. Because your arrows up here have moved, you may notice that some paragraphs have been indented.

All you have to do now is move those back into place, and everything will fall into place. The left-hand half of this method is a little tricky. Moving your margins up and down to the top and bottom of your page, on the other hand, is significantly easier; you simply click and drag to shift those margins up or down, and you can do the same with the bottom. By decreasing your margins, you can fit a lot more text on your website.

If none of it appeals to you, click the Layout tab and then "**Margins**." When you click the drop-down menu, you'll find a variety of options.

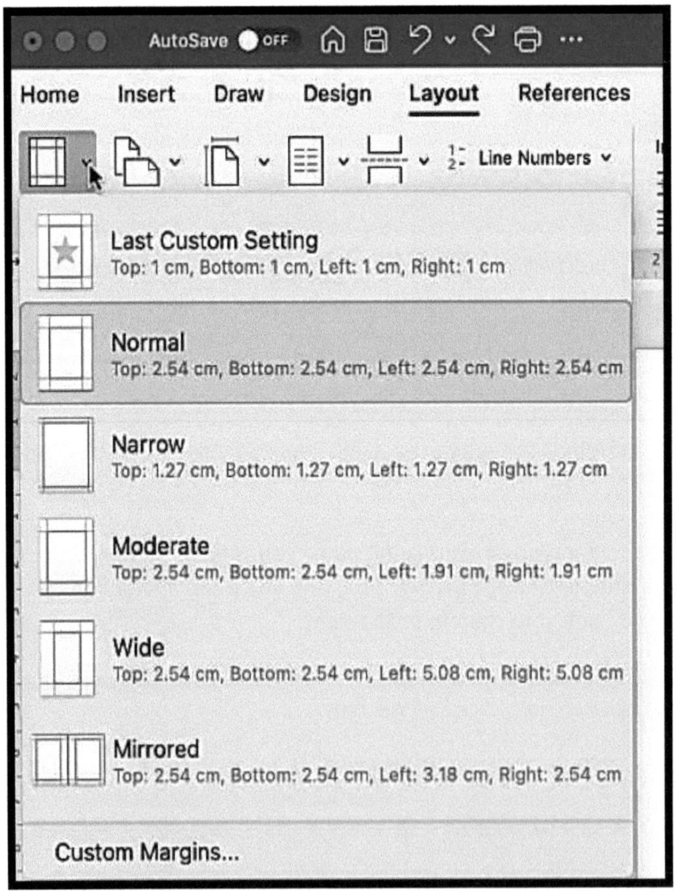

If you merely want to decrease the size of your margins, "**Narrow**" is an excellent option, and if you click on it, it will automatically modify your margins.

Creating Your Margins

If you wish to choose your margins, navigate to "**Custom Margins.**"

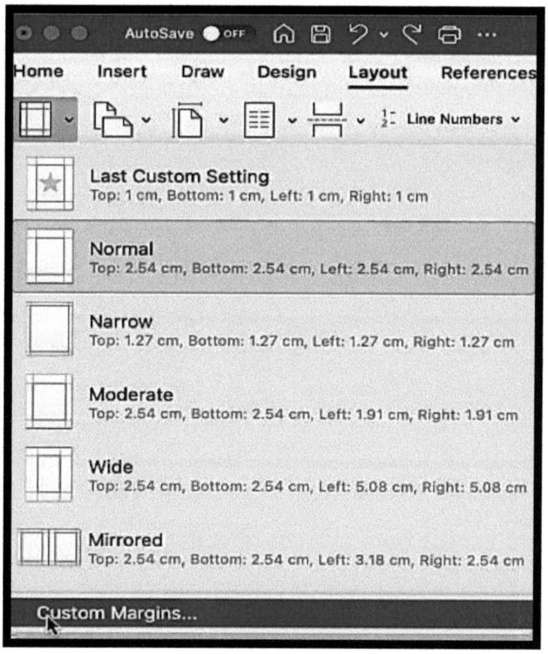

When you click that, a dialog box will emerge, allowing you to modify the top, bottom, left, and right margins. So, you can either use the up or down arrows and as you do, a tiny little preview will appear here, showing you precisely what you're doing.

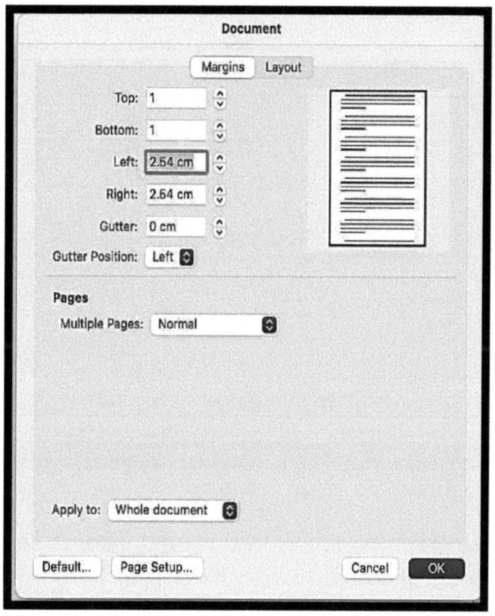

Alternatively, you can just insert a figure for each of these and click "**OK**" to see how the margins have changed. Once you're satisfied with all of your margins, you can go on to your headers and footers.

Headers and Footers

Page numbers, document titles, report titles, and other data will appear in the headers and footers at the top and bottom of your page. They'll be in the margin all through your document.

Accessing Your Headers and Footers

You can get to your headers and footers in two ways. The first option is to go to the "**Insert**" ribbon and then to "**Header and Footer**."

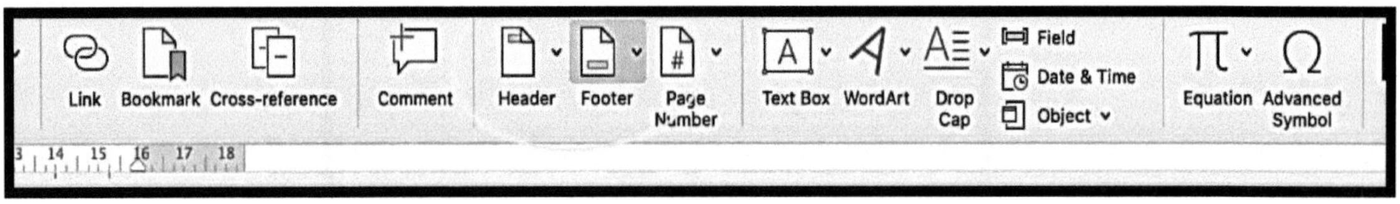

Simply click on "**Header**" and choose from a variety of options and graphics; similarly, you can do the same with the Footers.

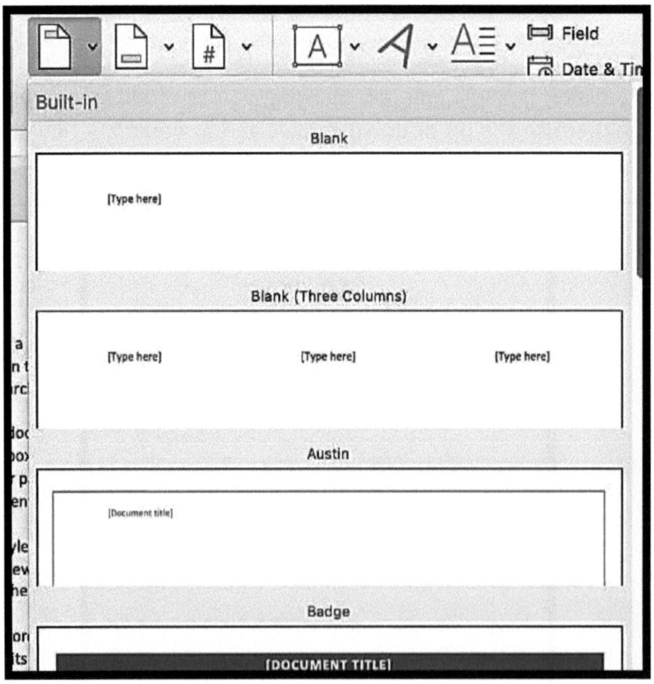

If you don't want any of these graphics, simply move your cursor to the top of your page and double-click. You can go to your headers and footers from here. Your header will appear on each page as you type.

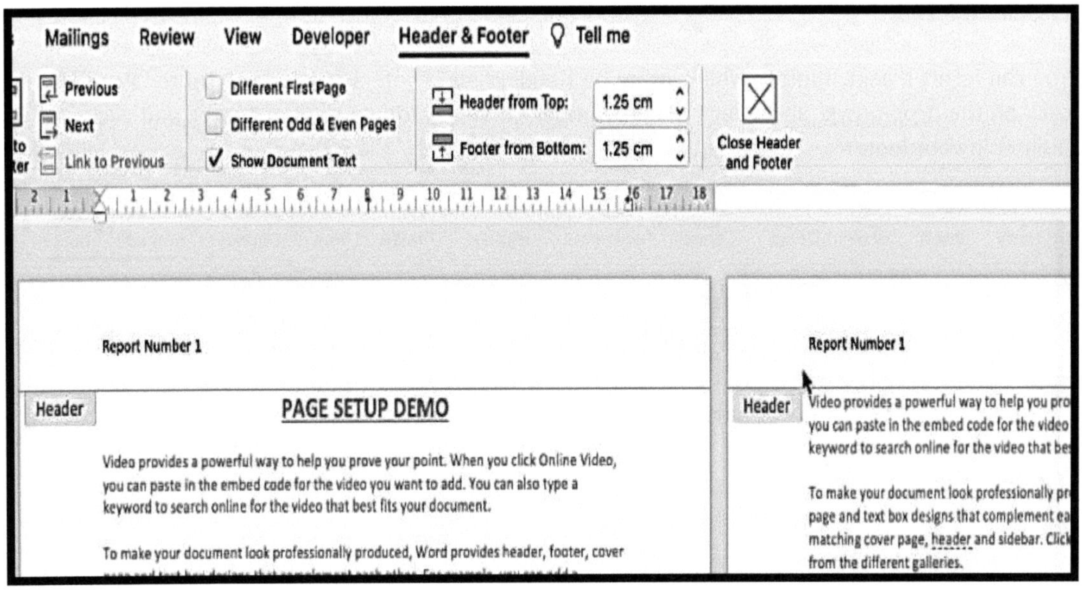

If you don't want this header on your first page, go to the header and footer menu and choose "**Different First Page**." If you do so and then double-click on your main document, you'll see that a header appears only on the second page.

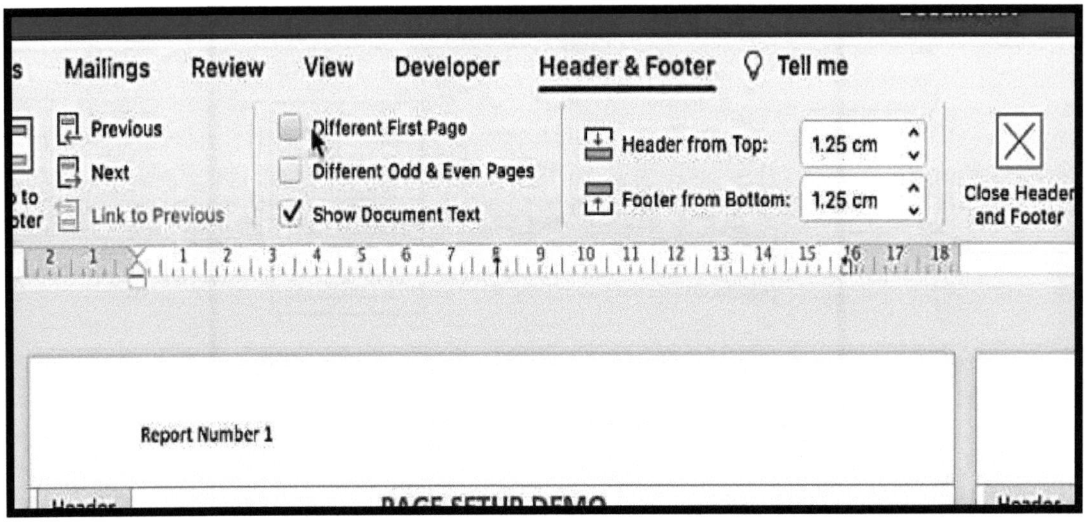

Return to "**Headers and Footers**," uncheck it, and it should display again.

Removing Headers and Footers

To exit your Headers and Footers, simply **place your cursor** within your main document and **double-click**. The Headers and Footers are grayed out, which means you can't alter them by clicking on them unless you double-click back into the Headers and Footers.

Page Numbering

You can insert a page number while you're on Headers and Footers. You can go up to "**Page Number**," click on the drop-down, and select "**Page Number**" if you go down to your footer and click with your pointer in your footer.

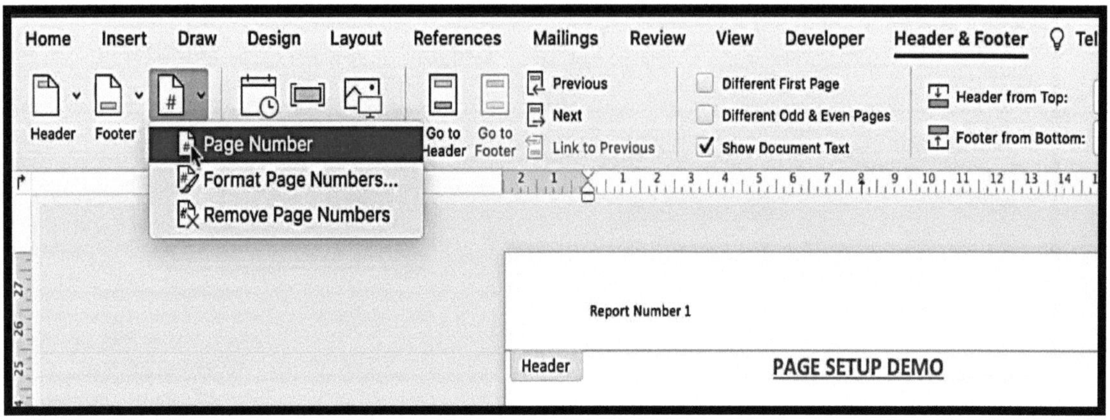

You'll get a dialog box once more, this time with the choice to position your page number on the left, middle, or right of your page.

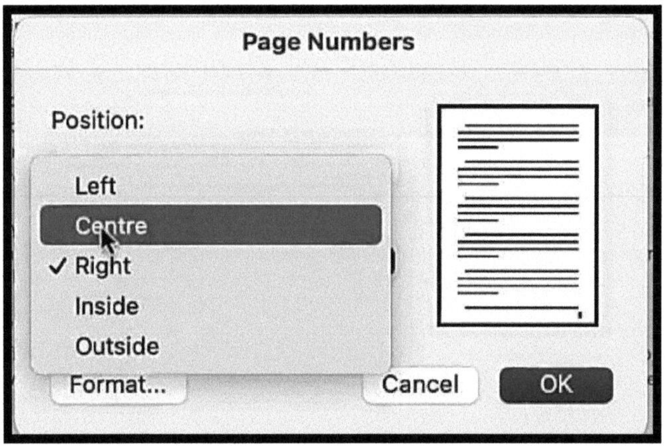

Choose your preferred option, and you'll be given the option of whether or not that page number displays on the first page. If you check or uncheck the box, then click "**OK**," your page number will appear on your pages. You can also add text here if you want.

Changing the Orientation of Your Page

If you wish to change the orientation of your page, for example, if you've completed your work and believe it should be in landscape format, go to the "**Layout**" tab, go to "**Orientation**," click on the drop-down menu and select "**Landscape**."

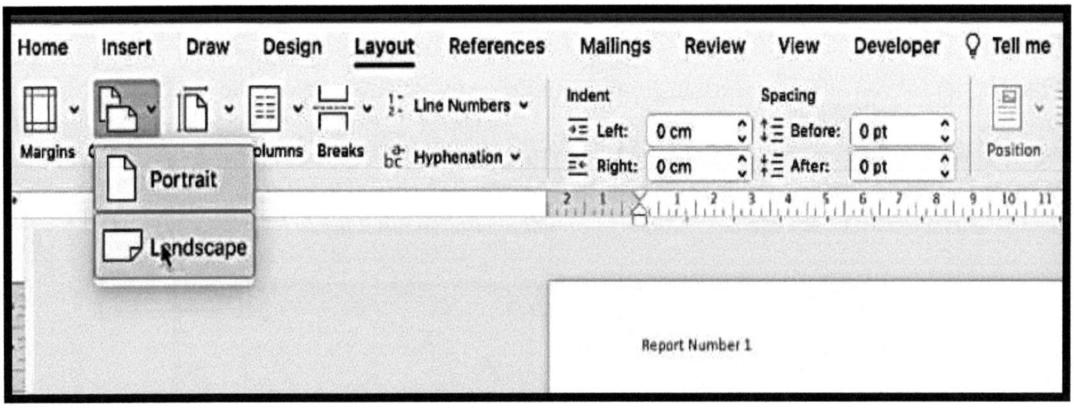

All of your margins will remain the same once you select "**Landscape**," but if they don't meet your needs, return to "**Margins**," select a different margin, and your document will adjust accordingly.

A Fresh Start

You're working on a document and you've reached the end of one page; how do you insert a new page inside or between some previous pages? This section explains how to add a new page to any document.

Text on a New Page

If you're typing and you reach the end of the page, the obvious solution is to place your cursor there and press Enter. A new page will be automatically created and added.

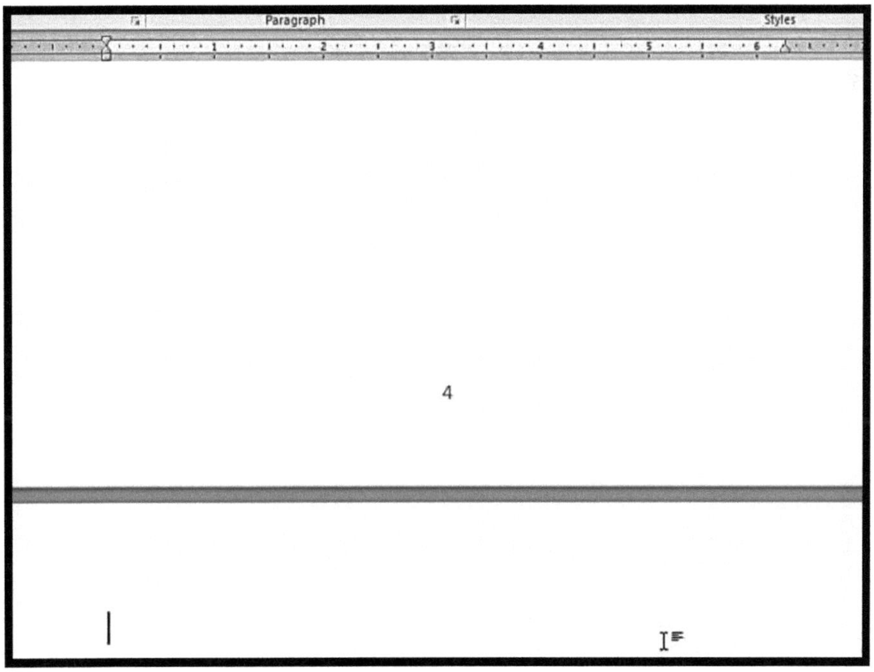

Inserting a Blank Page

What happens if you need to insert an empty page inside the one, you're working on now?

Simply move your cursor to the end of the current page.

To create a new blank page, go to the **Insert** tab and select "**Blank Page**."

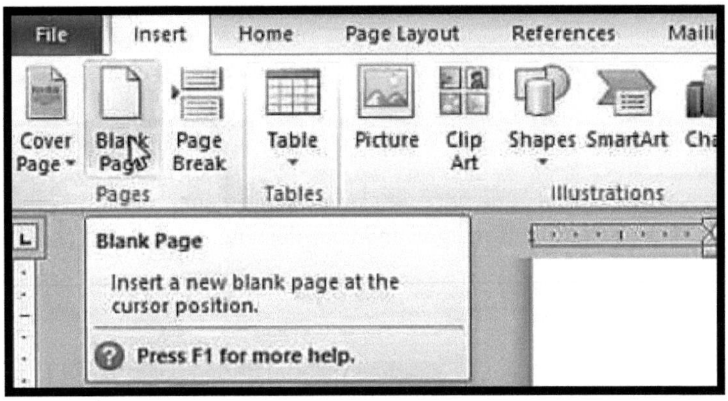

Please keep in mind that if you position your cursor at the start of a page before inserting a blank page, the blank page will appear first.

A unique Page Background

In this section, you'll learn how to change your word document's background color, add textures to the background if desired, and insert images. We'll also go through how to print coloring pages in a Microsoft Word document, as well as how to add a watermark.

Changing the Color of Your Pages

The word document's background color is set to white by default, but you can alter it to any other color you choose. You can do it in an already-created document or a blank document before you begin typing.

To do so, go to the **Design** tab, select "**Design,**" and then pick the second-to-last option, "**Page Color,**" on the right-hand side.

Hover your mouse over these several colors and select one by clicking on the small drop-down arrow. The default is restored if you click where it reads "**No Color**." You can see a preview by hovering your mouse over the color. If you'd want to make a custom color, go to "**More Colors**" and make one.

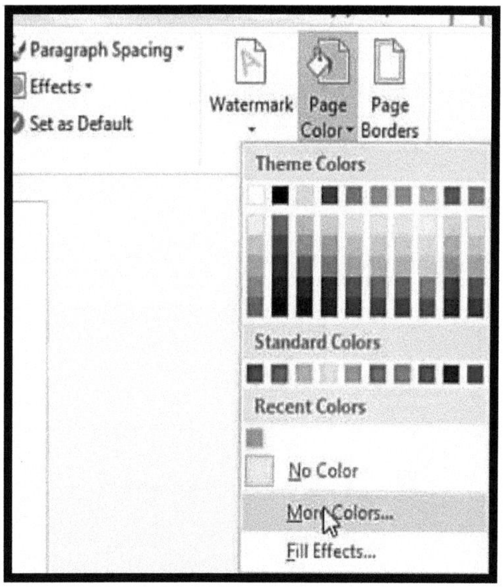

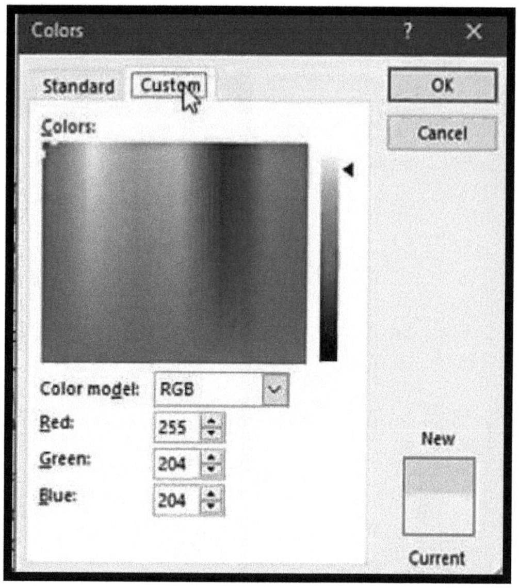

When you're satisfied with the color you've chosen, click "**OK**," and the color will appear throughout the page.

Adding Texture, Patterns, and Images to Your Page

To apply texture, go back to the page color option and select "**Fill Effects**."

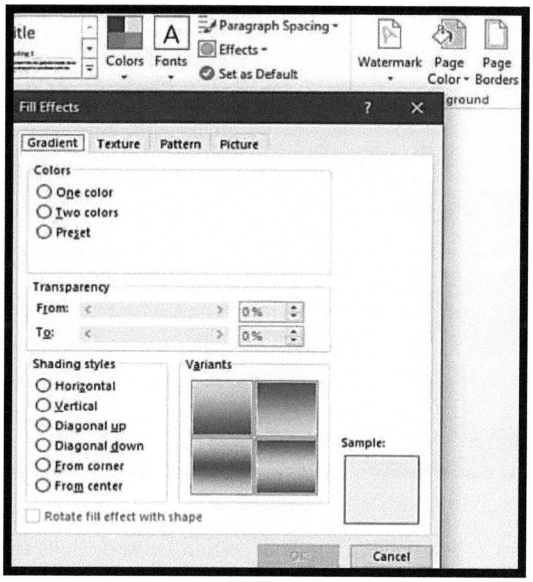

This opens a window with four tabs: **Gradient**, **Texture**, **Pattern**, and **Picture**.

Let's begin with the **Gradient**. You can have two colors with Gradient; one will be darker, while the other will be lighter. A preview of how an option will appear in your document appears while you are selecting it.

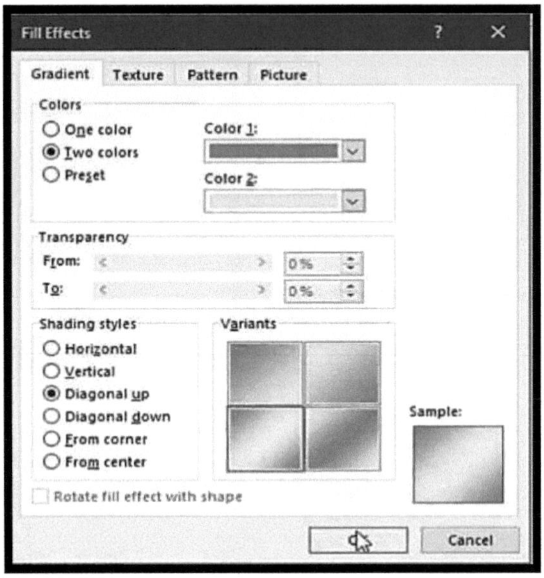

If you go to the Color area and choose "**Two colors**," you can then choose your first color and your second color. The Shading Styles allow you to customize the appearance of the gradient. After you've made your selections, click "**OK**," and the document's entire background will be transformed into the gradient you chose.

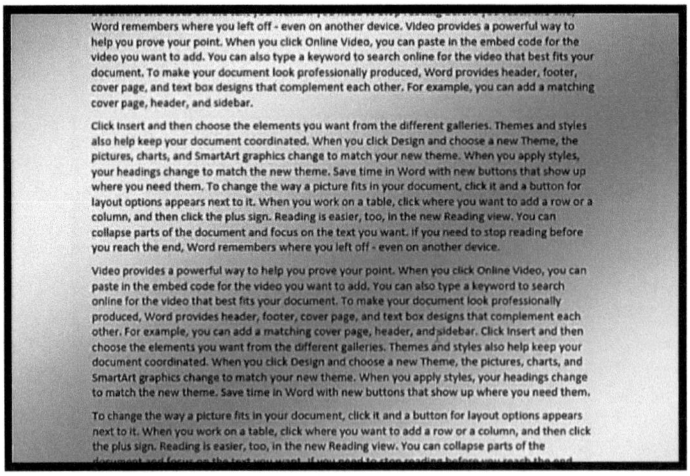

Word remembers where you left off - even on another device. Video provides a powerful way to help you prove your point. When you click Online Video, you can paste in the embed code for the video you want to add. You can also type a keyword to search online for the video that best fits your document. To make your document look professionally produced, Word provides header, footer, cover page, and text box designs that complement each other. For example, you can add a matching cover page, header, and sidebar.

Click Insert and then choose the elements you want from the different galleries. Themes and styles also help keep your document coordinated. When you click Design and choose a new Theme, the pictures, charts, and SmartArt graphics change to match your new theme. When you apply styles, your headings change to match the new theme. Save time in Word with new buttons that show up where you need them. To change the way a picture fits in your document, click it and a button for layout options appears next to it. When you work on a table, click where you want to add a row or a column, and then click the plus sign. Reading is easier, too, in the new Reading view. You can collapse parts of the document and focus on the text you want. If you need to stop reading before you reach the end, Word remembers where you left off - even on another device.

Video provides a powerful way to help you prove your point. When you click Online Video, you can paste in the embed code for the video you want to add. You can also type a keyword to search online for the video that best fits your document. To make your document look professionally produced, Word provides header, footer, cover page, and text box designs that complement each other. For example, you can add a matching cover page, header, and sidebar. Click Insert and then choose the elements you want from the different galleries. Themes and styles also help keep your document coordinated. When you click Design and choose a new Theme, the pictures, charts, and SmartArt graphics change to match your new theme. When you apply styles, your headings change to match the new theme. Save time in Word with new buttons that show up where you need them.

To change the way a picture fits in your document, click it and a button for layout options appears next to it. When you work on a table, click where you want to add a row or a column, and then click the plus sign. Reading is easier, too, in the new Reading view. You can collapse parts of the

Under "**Colors**," there's also a "**Preset**" option that, when clicked, pulls up a selection with a variety of options including Early Sunset, Late Sunset, and others.

Moving on to **Texture**, a large number of textures are available in Microsoft Word if you click on it. By **selecting a texture** and pressing "**OK**," it is added to the background.

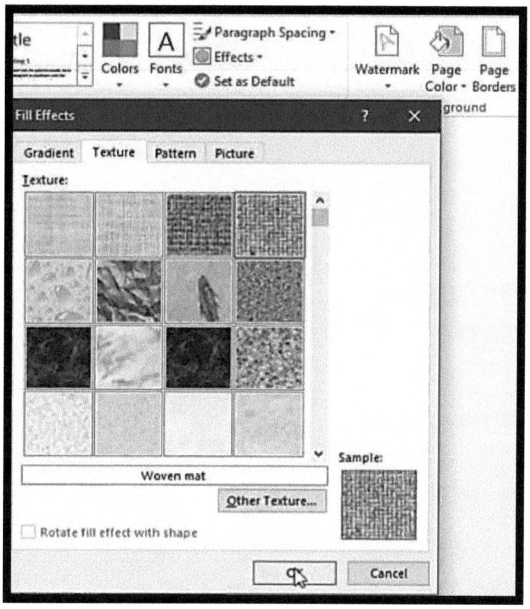

You can alter the color of your text if you don't like how the texture looks with it. Simply pick the text, navigate to the **home** tab, and then to the Font color option to alter the font color.

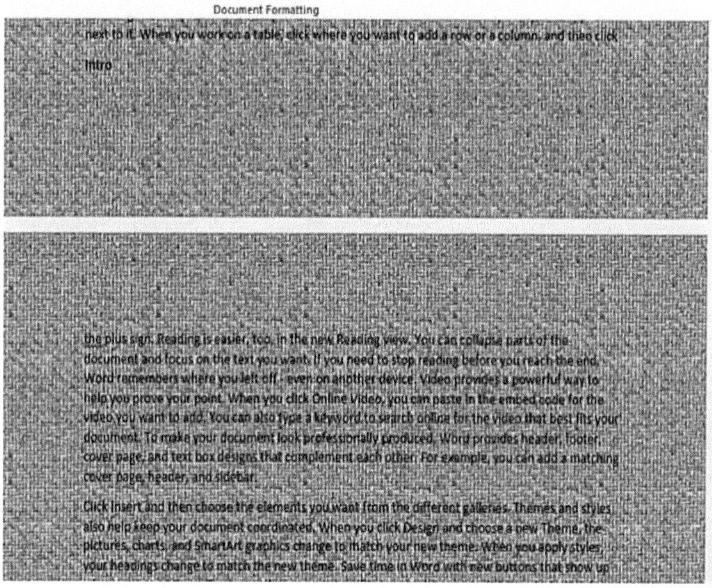
Document Formatting

next to it. When you work on a table, click where you want to add a row or a column, and then click

Intro

the plus sign. Reading is easier, too, in the new Reading view. You can collapse parts of the document and focus on the text you want. If you need to stop reading before you reach the end, Word remembers where you left off - even on another device. Video provides a powerful way to help you prove your point. When you click Online Video, you can paste in the embed code for the video you want to add. You can also type a keyword to search online for the video that best fits your document. To make your document look professionally produced, Word provides header, footer, cover page, and text box designs that complement each other. For example, you can add a matching cover page, header, and sidebar.

Click Insert and then choose the elements you want from the different galleries. Themes and styles also help keep your document coordinated. When you click Design and choose a new Theme, the pictures, charts, and SmartArt graphics change to match your new theme. When you apply styles, your headings change to match the new theme. Save time in Word with new buttons that show up

Let's have a look at the next effect, **Patterns**. A large number of patterns (in black and white) are displayed when you click on "**Pattern**". It's worth noting that if you've previously chosen a gradient, the pattern in the gradient colors will be displayed.

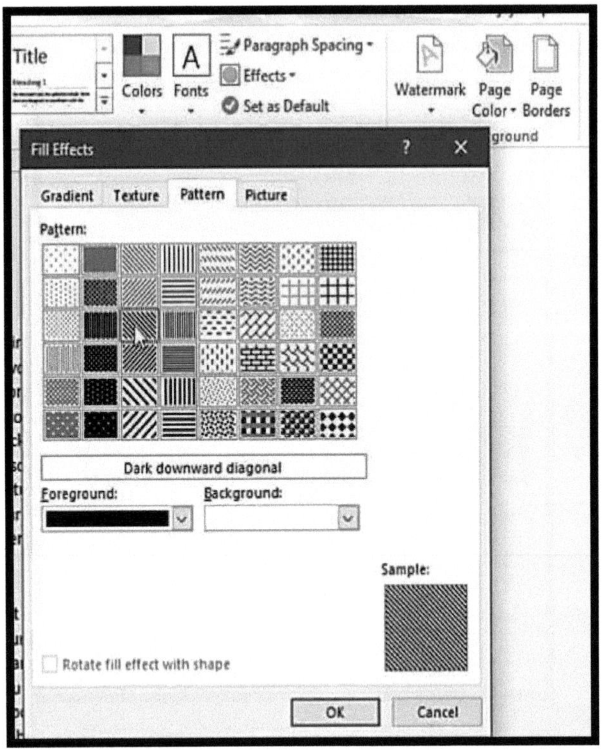

After you've chosen a pattern, click "**OK**," and the pattern will be applied to the entire document.

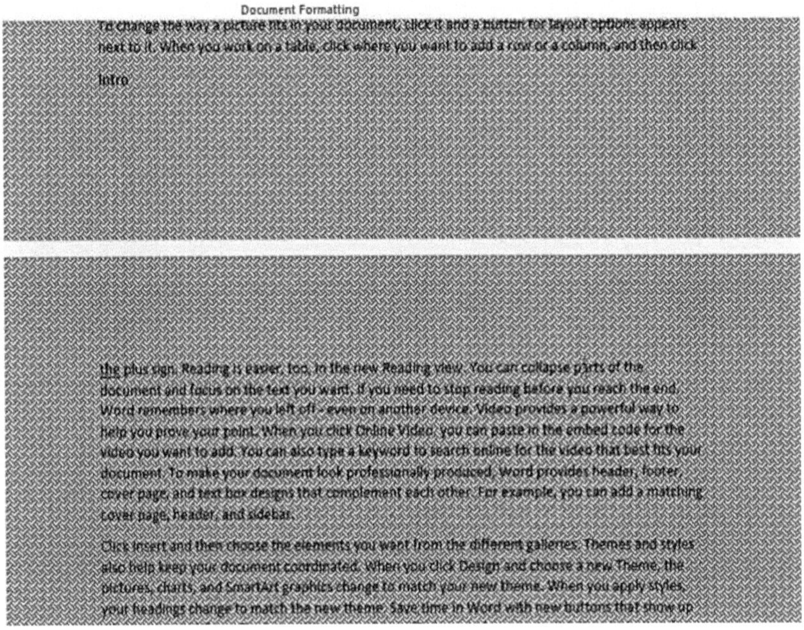

Finally, let's take a look at the last option in this window: uploading a background image. Go to "**Picture**" and click the "**Select Picture**" button.

Then go to the folder with the background image you wish to insert, click on it, and then click "**Insert**."

As you can see, it will display a sample before allowing you to click "**OK**."

The image has now been added to the document in Word.

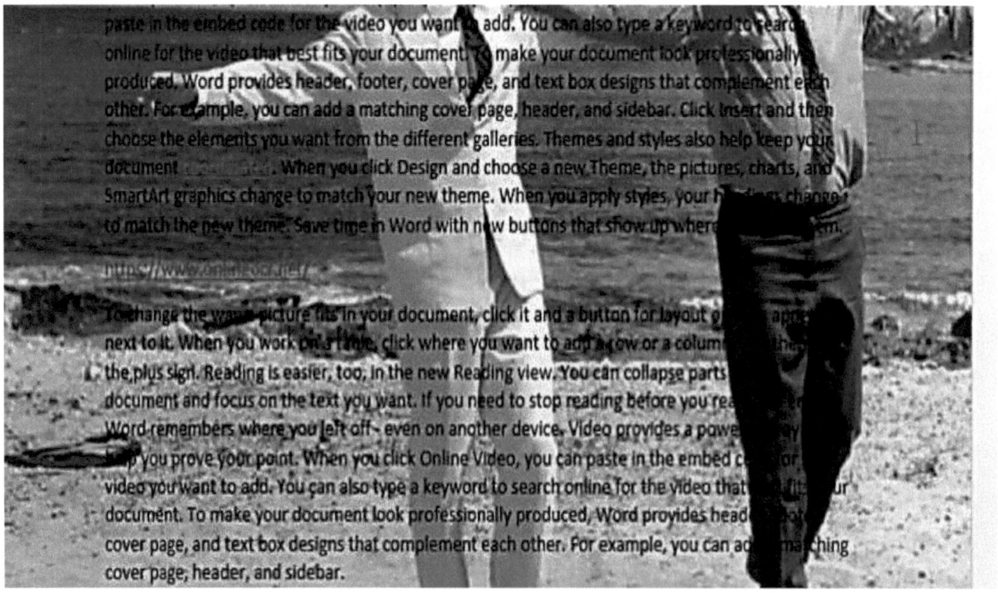

This is how you can make your page look more interesting by changing the background color and adding textures, patterns, and photos.

Inserting A Watermark

Here, you will see how you can insert a Watermark into a Word document. Let's say you have a secret recipe or a logo and you want to make sure that no one shares; you can add a watermark to keep your document a lot safer.

First, we will look at how you can insert a preset watermark and we will also look at how you can customize it.

To insert a watermark, on the top ribbon, click on the **Design** tab. Within this tab, over on the right-hand side and you will see an option that says "**Watermark**".

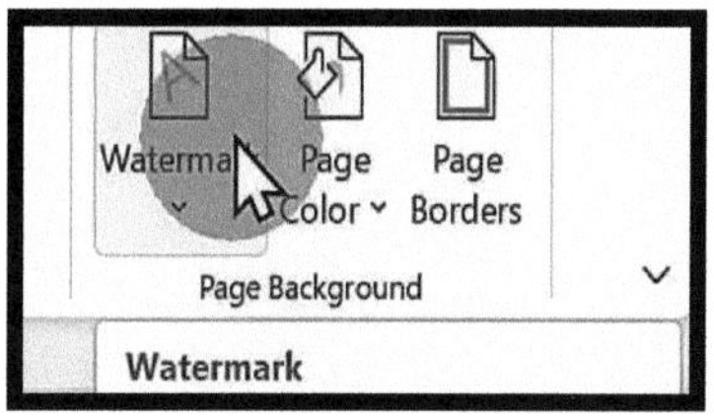

Clicking on the drop-down menu brings up a lot of the preset categories. There is a set of confidential watermarks, Disclaimers, and also Urgent.

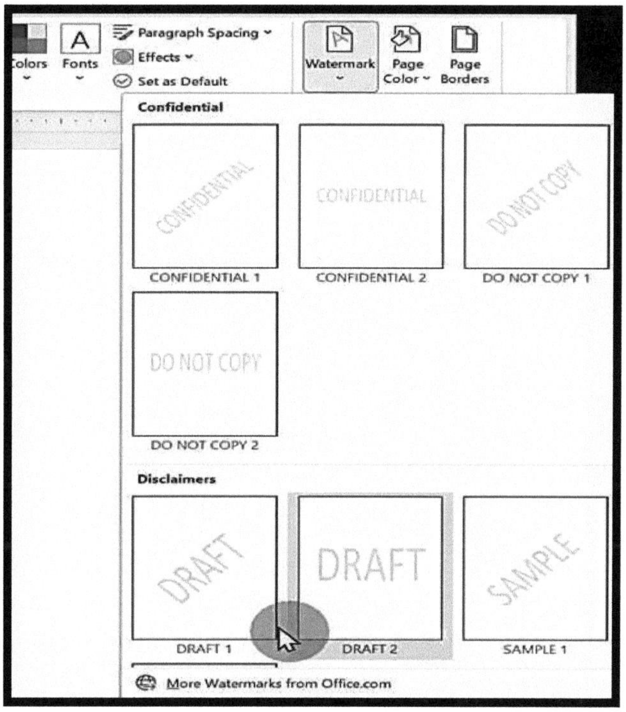

If you select any under the confidential category, that inserts a confidential watermark into your document.

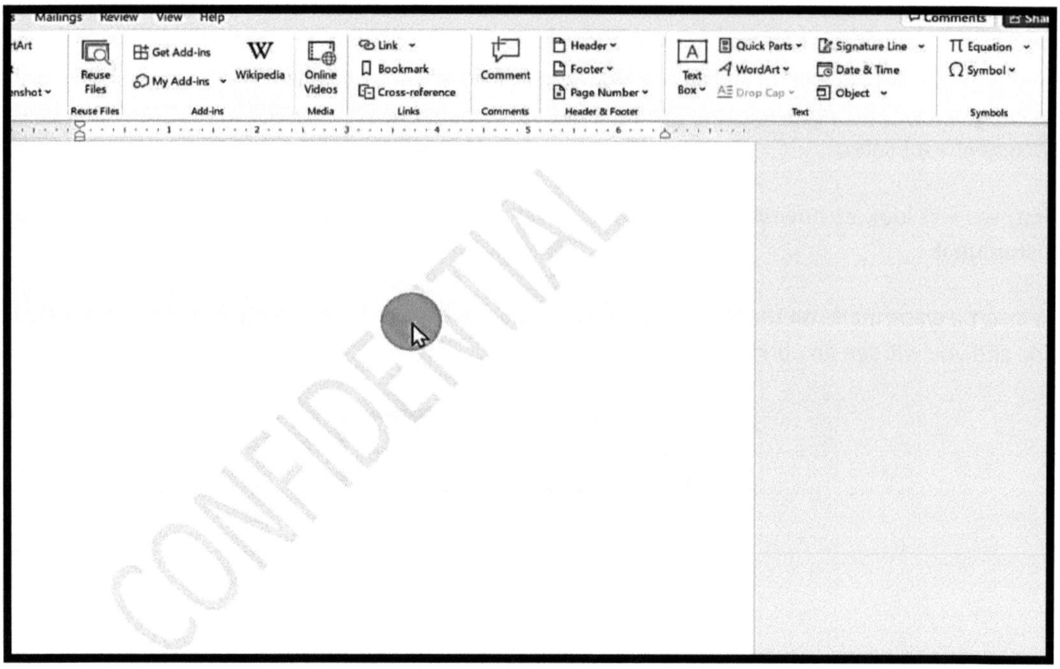

If you insert another page into your document, you can see that the watermark carries over to the new page; that's one thing to keep in mind when you are working on a document.

Customizing a Watermark

Now if you want to customize your watermark to make it more personal, go back to the **Design** tab and once again, click on "**Watermark**", and at the bottom, there is the option to customize the Watermark by simply clicking on "**Custom Watermark**".

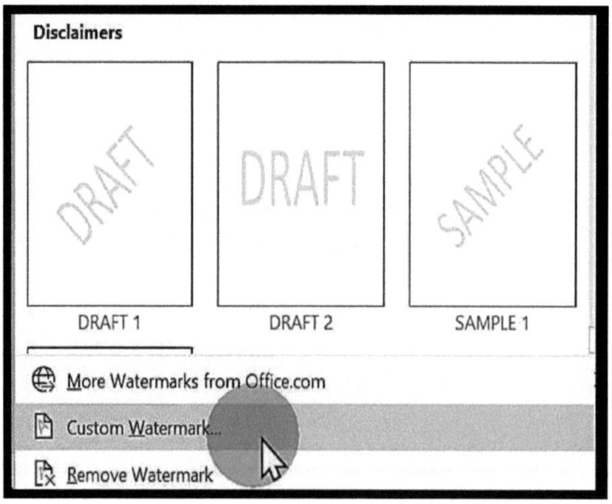

This opens up a dialogue where you can now customize your watermark.

Text-based

If you want to do a text-based watermark, go over to "**Text Watermark**" and start by choosing a language.

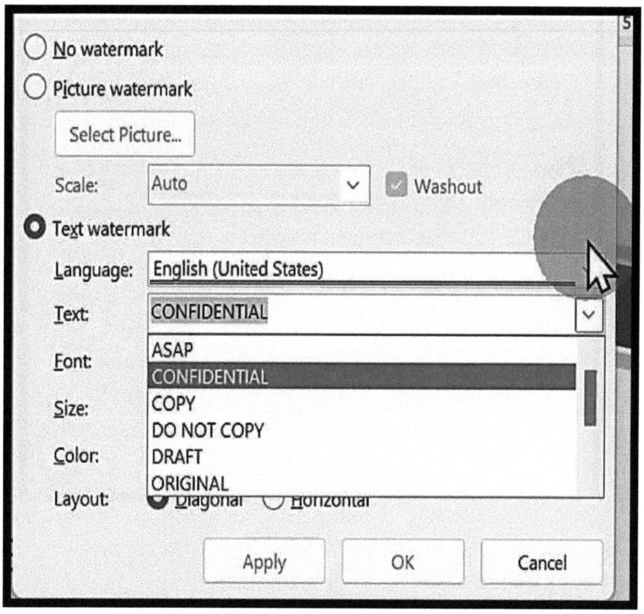

You can view all the preset watermark options but you can also type in a text.

You can also set the font, size, and color, and right next to that, there is an option called **"Semitransparent"**. This would wash out the watermark just a little bit more. So, if your watermark is a little too dark and interferes with your text, you can set it to Semitransparent.

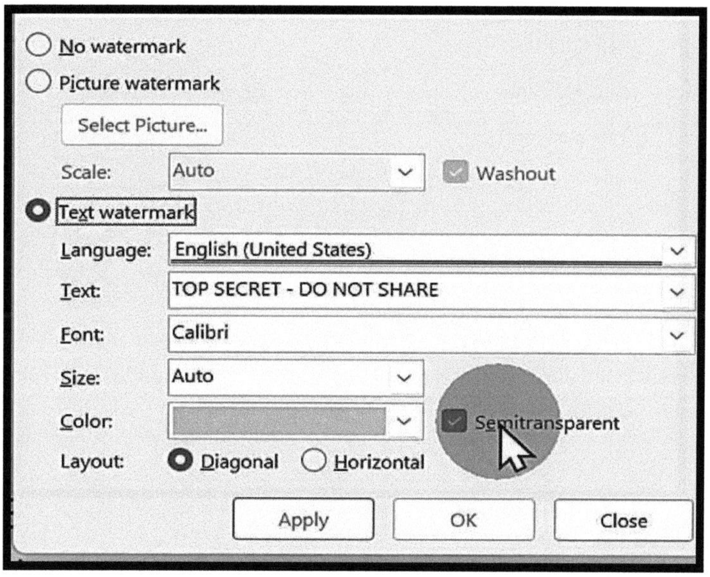

Down at the bottom, you can also choose the layout. You can have the text go diagonally across your document, or you can have it go across horizontally.

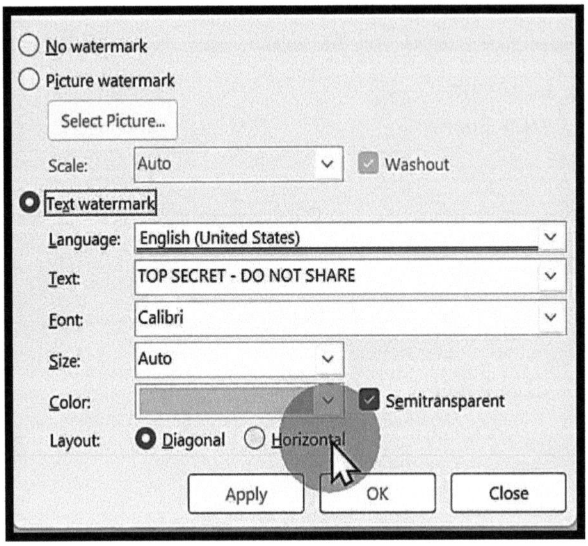

Picture-based

The text watermark works fine, but if you need something more serious, you can try inserting a **Picture Watermark**. Right here, you select a Picture Watermark and select a file on your PC.

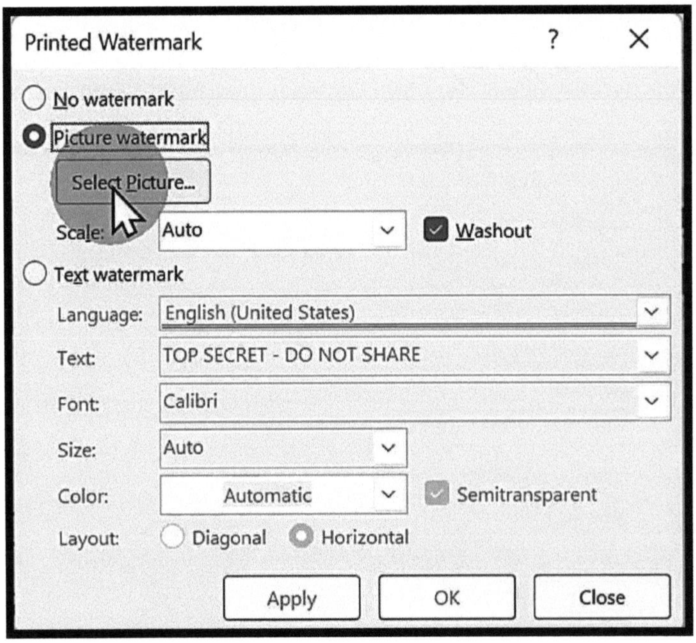

Below that, you can adjust the scale but by default, it is set to Auto and that will ensure the best fit.

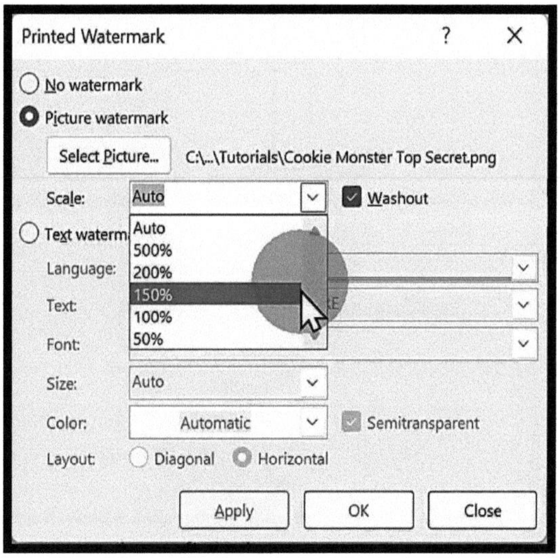

To the right of that, there is a check-box called "**Washout**" and this is similar to the Semitransparent option we talked about earlier. When you click on it, it washes out the image a little bit more and when you uncheck it, you can see the image a little bit better.

Removing A watermark

To remove a watermark, within the dialogue we talked about, at the very top, you can simply click on "**No Watermark**".

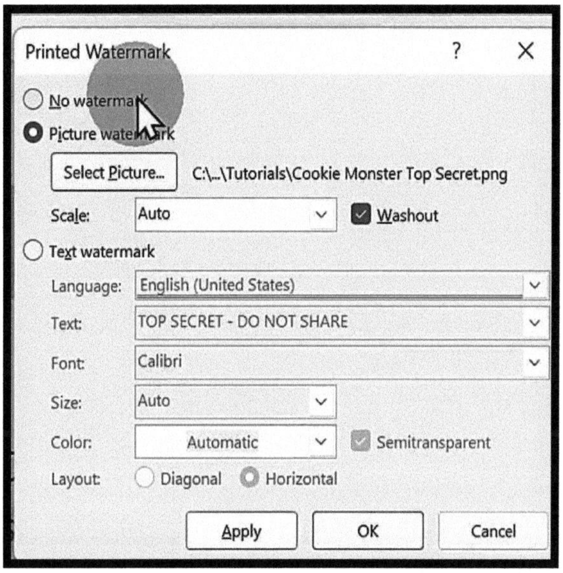

Back on the main screen, you can also remove the watermark by going back up to the **Design** Tab, the "**Watermark**" **drop-down**, and down towards the bottom there is the option to "**Remove Watermark**".

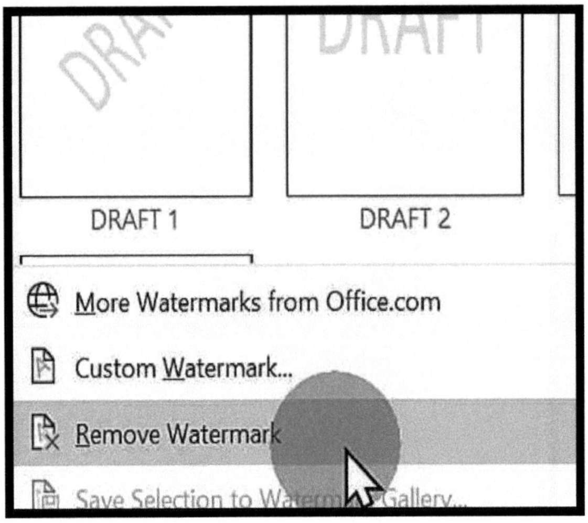

That is how easy it is to insert a watermark and keep your document safe.

Printing Colored pages

To print this document with the entire background color after applying a background color to your page, simply select "**File**" and then "**Print**."

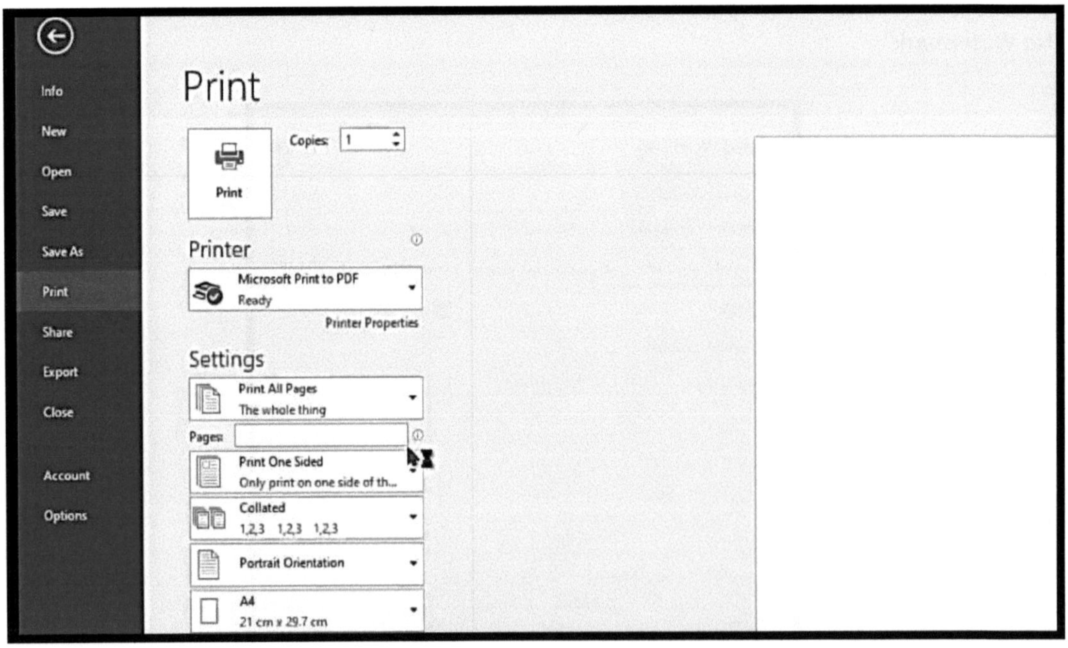

Please note that if you try to print a document with a colored background, you will notice that there is no color applied and instead, you have a white background.

To print the page with the background color you choose, go to "**Options**".

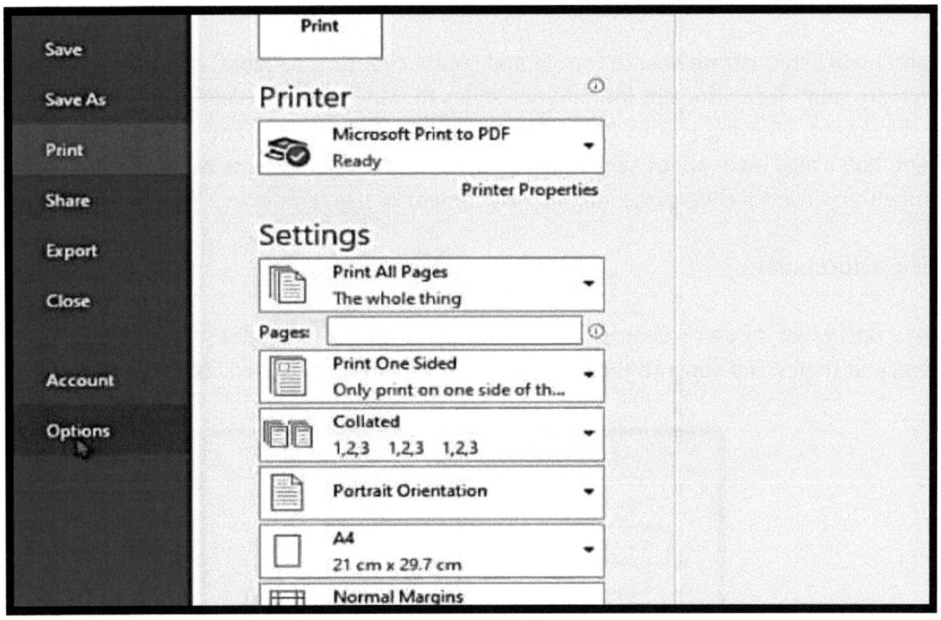

Then select "**Display**" from the menu. Here, there is an option called "**Print background color and images**", simply check the box to enable printing with background color and click "**OK**".

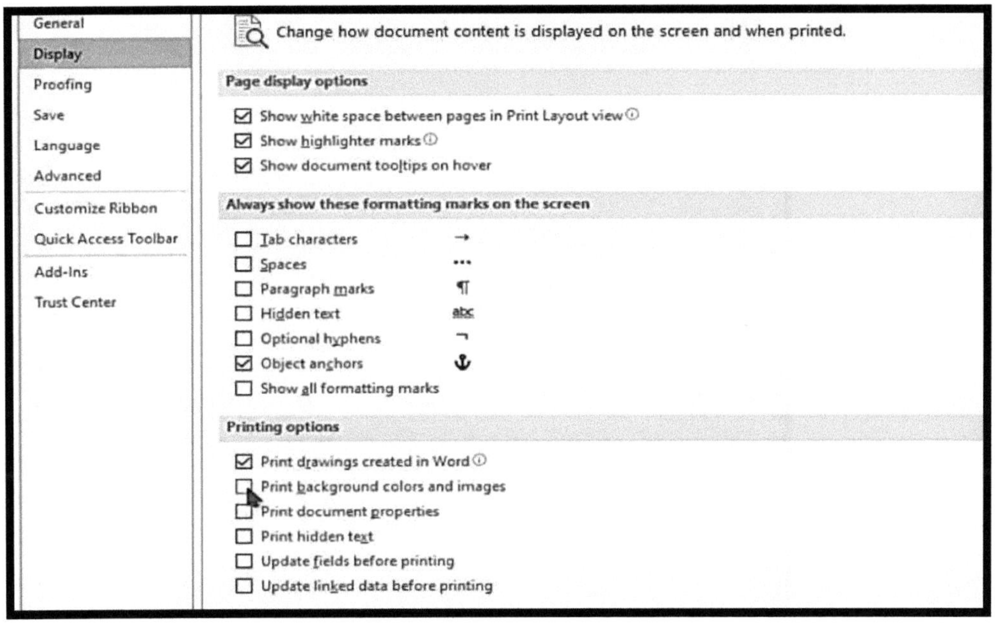

When you return to print, the background color is now visible in the Print Preview, indicating that you can print your Word documents with the full background color.

CHAPTER FOURTEEN

SECTION FORMATTING

This chapter will demonstrate how to format and create chapters, sections, and subsections. You'll learn how to create multi-level lists and format your styles to match the different styles, headings, and levels in your document, allowing you to easily format all of your headings and sections and organize your document. You'll also learn about section breaks and how to use them as we go. Finally, you'll discover how to manually insert a cover page into your document or use templates.

Sectioning a document

While your document is open in Word, go to the **home** tab, under the **Styles** section, click the small button right here, and then scroll down to the third button over, which is labeled "**Manage Styles.**"

After that, go to the "**Recommend**" tab.

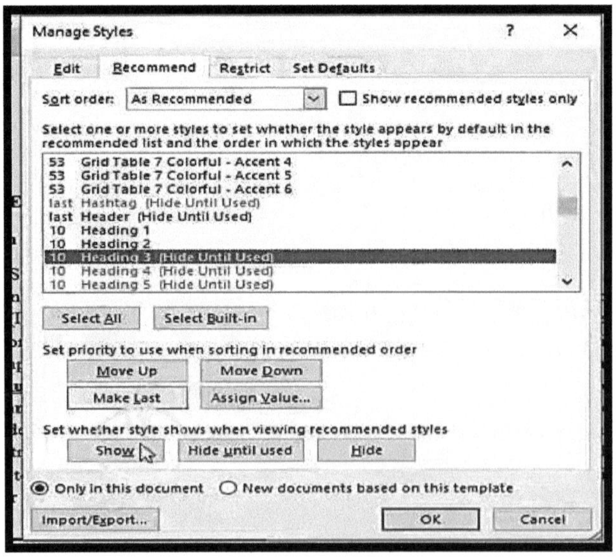

You're turning on the various headings because you want three levels: Chapters, Sections, and Subsections, and you want each of them to correspond to a distinct heading in your Styles section.

Heading 1 may already be visible, while the other headings are grayed out. To make them visible, click on a heading like Heading 2 and select "**Show**," do the same for Heading 3 then select "**OK**." That way, all three of those headings will be shown.

You can now close the Styles box because you don't need it any longer, and you can start creating and formatting your levels.

Creating Chapters, Sections, and Subsections

While staying in the home tab, go to the Paragraph section and look for the multi-level list.

Select "**Define New Multilevel List**" at the bottom of the page.

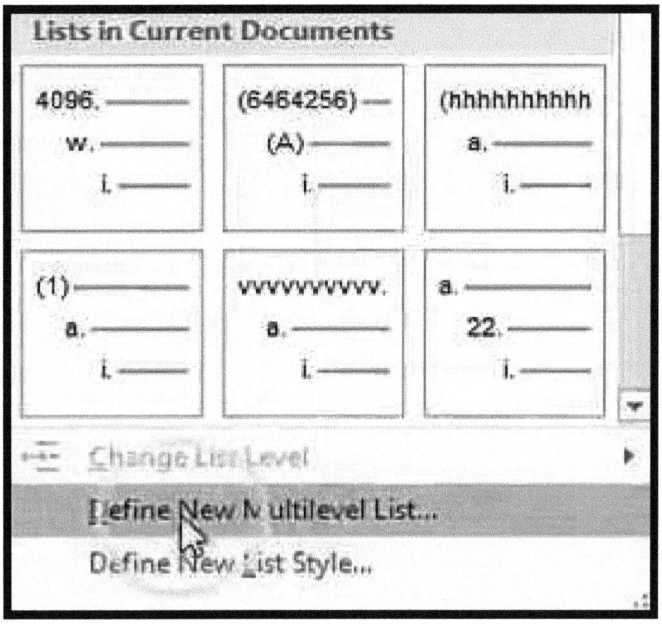

From here, you'll create three levels: Level 1 for chapters, Level 2 for sections, and Level 3 for subsections, with a variety of tasks to complete for each.

To begin, go to Level 1 and remove the "**Enter formatting for number**" box and type in "Chapter," then select the "**Number style**," and last, go to the bottom right corner and "**Follow number with**" a space. That will prevent the first chapter's title from colliding with the chapter's real title; you must include that space.

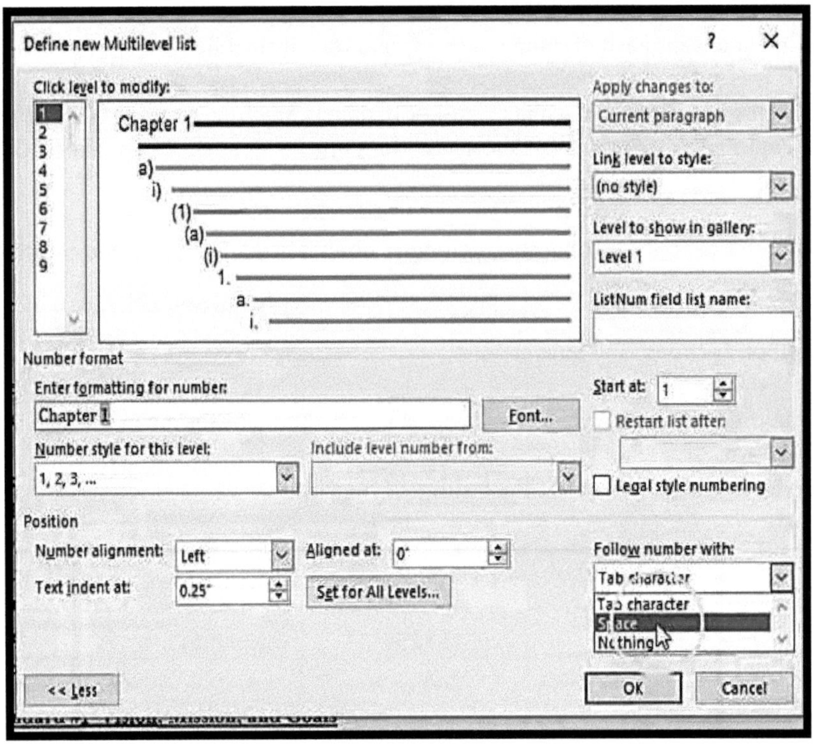

You may also make it so there are no indents by clicking the "**Set for all levels"** button, adjusting the numbers to zero, and then clicking "**OK.**"

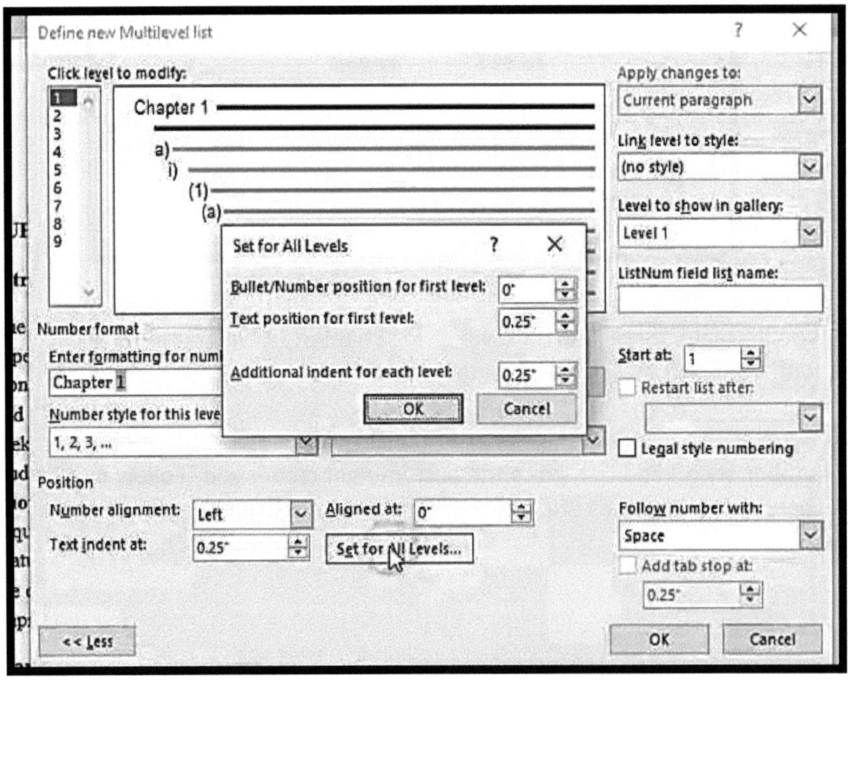

The next step is to format the font, which you can accomplish by selecting "**Font**" and making the desired changes.

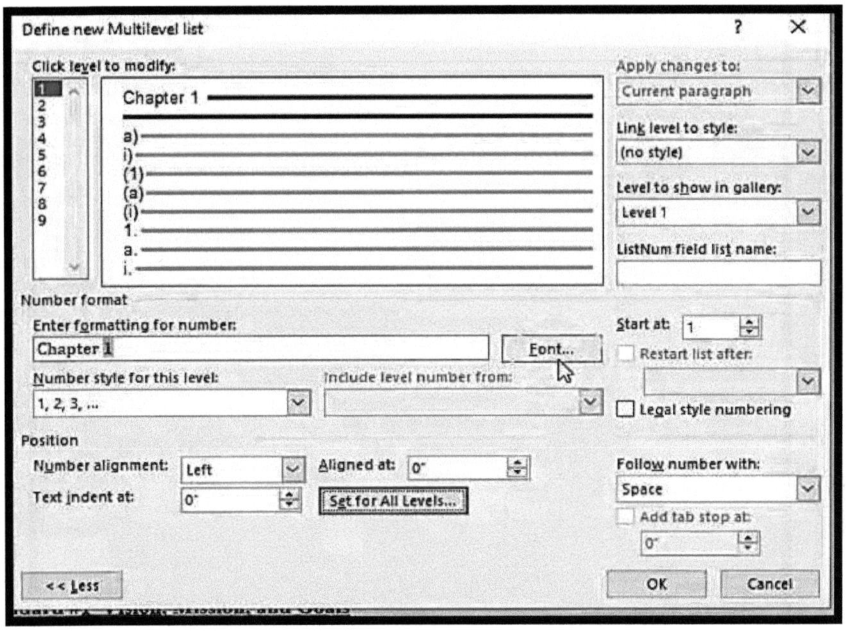

Before moving on to Level 2, you must attach this level to the style you desire, in this case, heading 1 because that is the style you will use for your chapters.

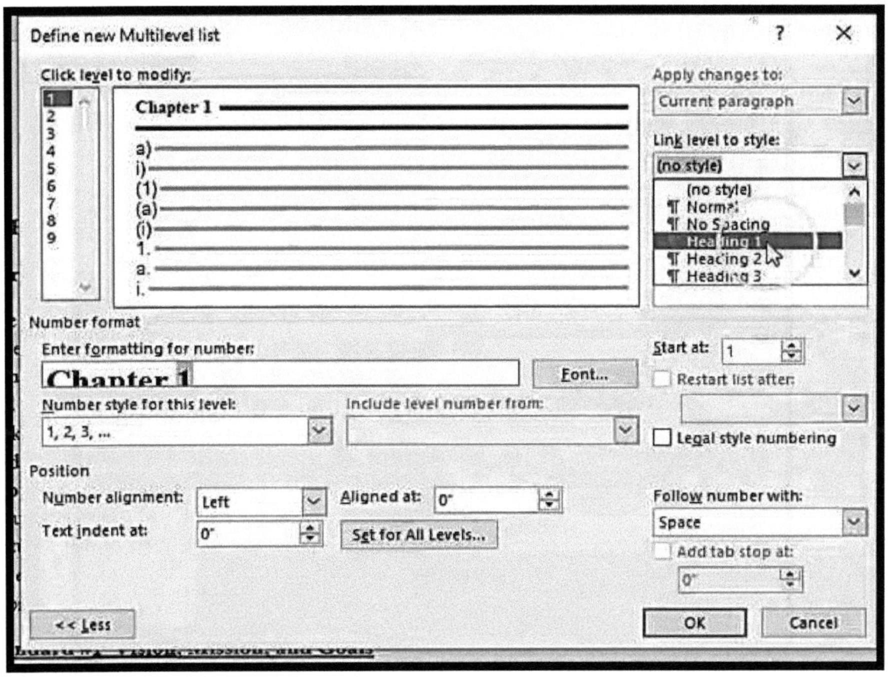

Once you've completed that, you're ready to move on to Level 2, which is your section. You'll start with "**Enter formatting for number**," then continue to "**Include level number from**" and choose level 1, which will insert the chapter number.

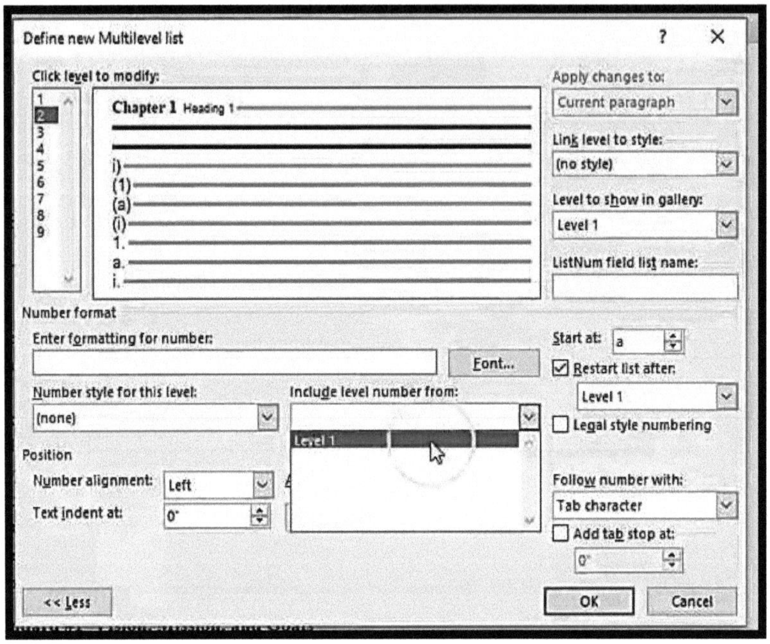

You can start with a period or a stop, then choose a number style, and finally finish with a space. Because you already set your indent for all levels, you don't need to do it again. You can also proceed to format the font and size before clicking "**OK.**"

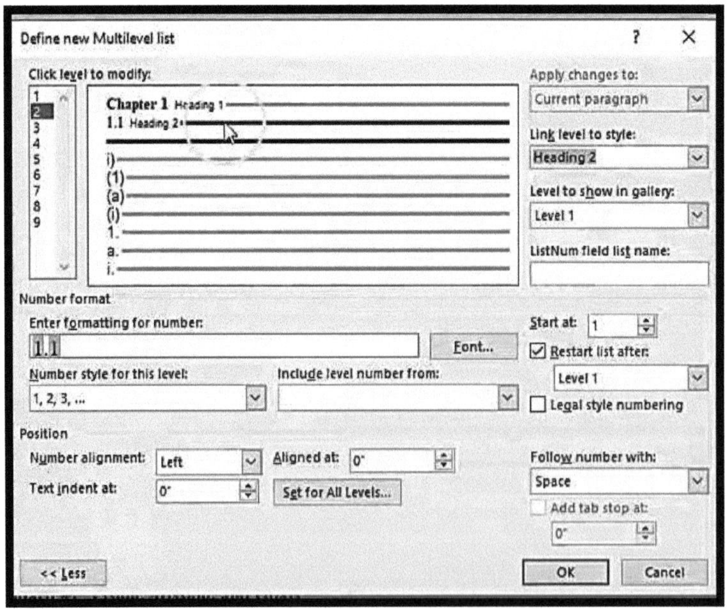

Before moving on, you must link level 2 to the desired style, which will be heading 2. You will see an example of it here once you've done that.

The next step is to go to Level 3, which is where your subsections are located. You'll start with your **chapter number from Level 1** and the full stop, then your **section number from Level 2** and the full stop, and finally, you'll **select a Numbers style for this level** and place a space after the number. You won't need to set your indent; simply pick a **Font**, select a **size**, and click "**OK.**" The next step is to **attach this level to your Heading 3** style, and once you've done so, you've specified all of your multi-level lists; simply click "**OK.**"

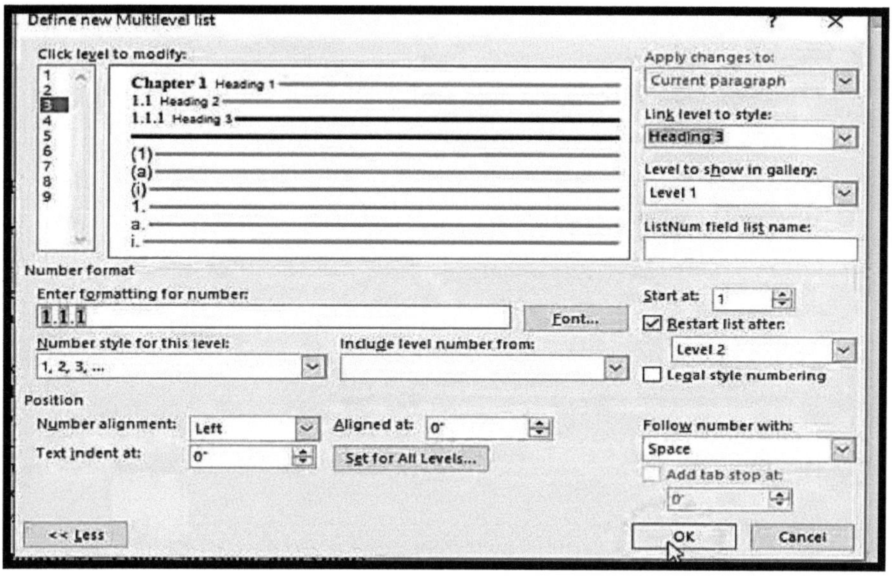

You can now see that chapter 1 for your heading 1 is right there.

Before continuing, go up to Styles, heading 1, right-click, and select "**Update Heading 1 to Match Selection.**" This will ensure that everything is properly formatted, including the chapter, number, and title.

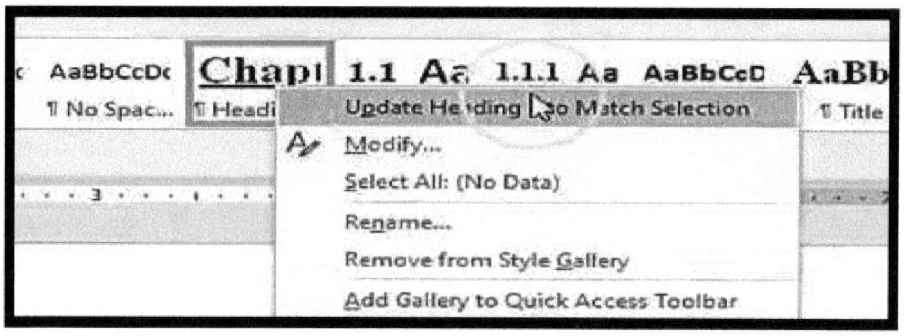

Now that you've done that, go down the page and look for another title. When you've done that, click there and then on Heading 1, which will add that to Chapter 2. It'll put in your chapter, the number, and then the real title of the chapter once you've got everything set up and they've matched your levels and heading styles.

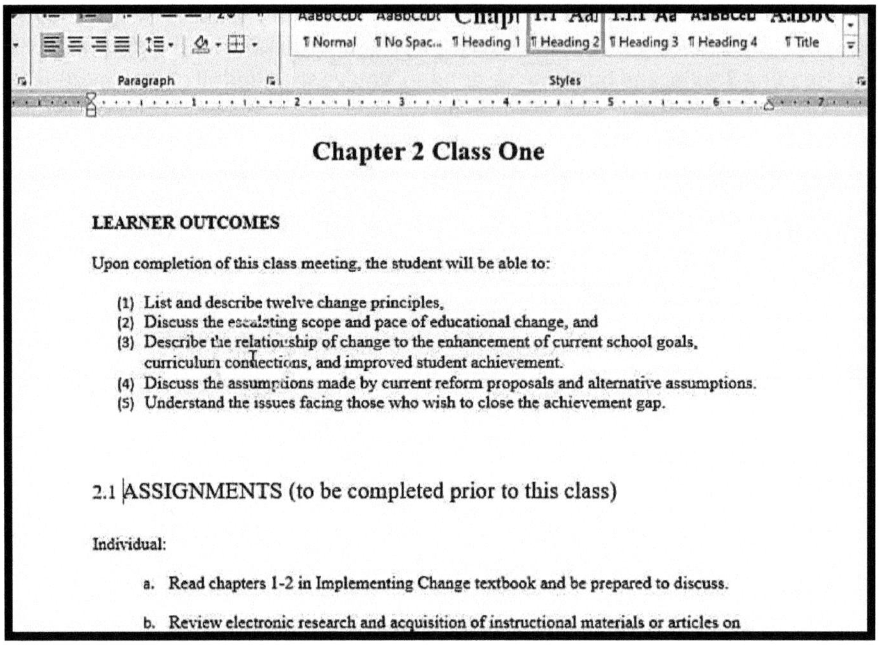

Now you can look at your level 2 or sections by scrolling down. Simply click a phrase or sentence to make it your first section, then go to heading 2 and do the same thing as before (match to your actual level formatting), then go to Heading 2, right-click, and update Heading 2 to reflect your selection.

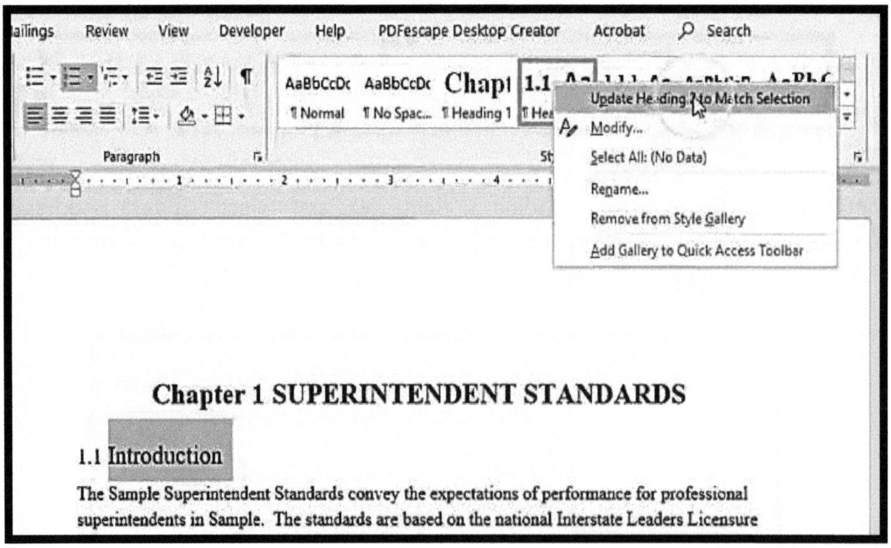

As you do this for your sections, you'll see that it puts in 2.1 for a certain chapter, such as chapter 2, because you're presently in section 1 of chapter 2.

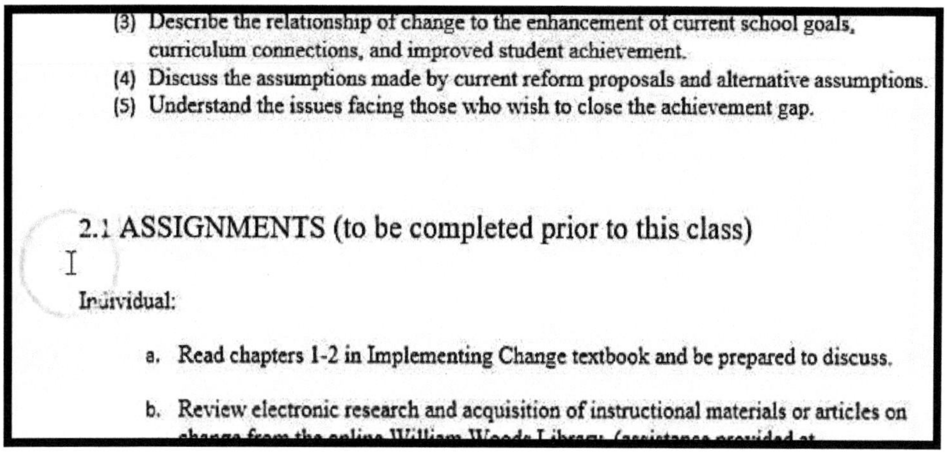

You can now add your subsections by picking your text and going to Heading 3, updating heading 3 to reflect the selection, and you're done.

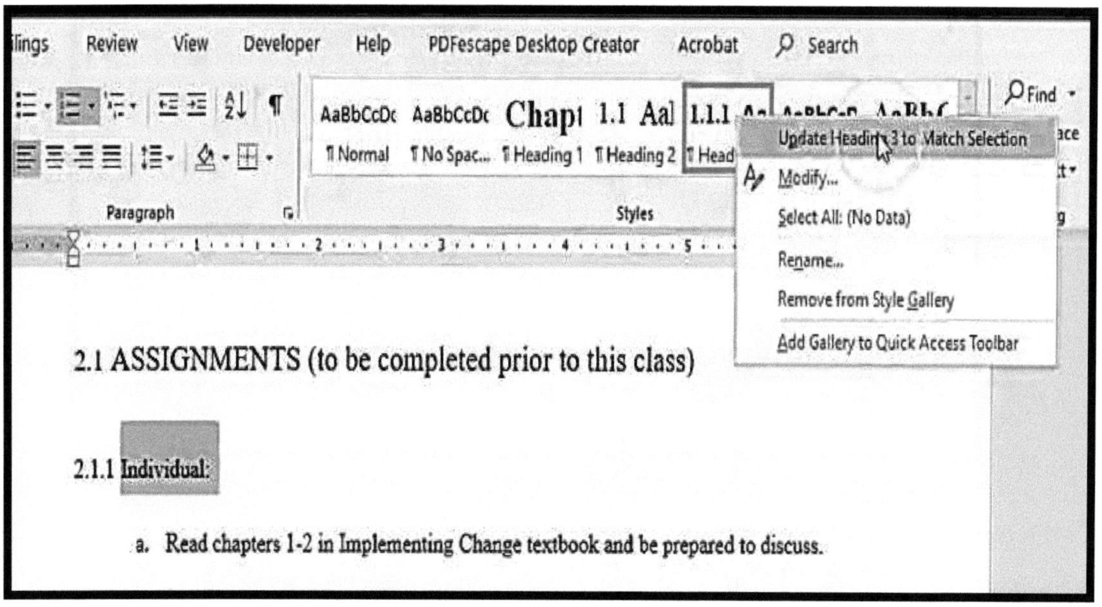

Section Breaks

Let's imagine you want your new chapters to always start on a new page, regardless of what content you add to the prior chapter because the heading just goes down the entire time when you add information. That's what we're going to talk about here, so let's get started.

Adding a Section Break to a Document

To do so, go to the page just before a new chapter, select the **Layout** tab, and then "**Breaks.**" Go to "**Section Breaks**" and select "**Next Page.**"

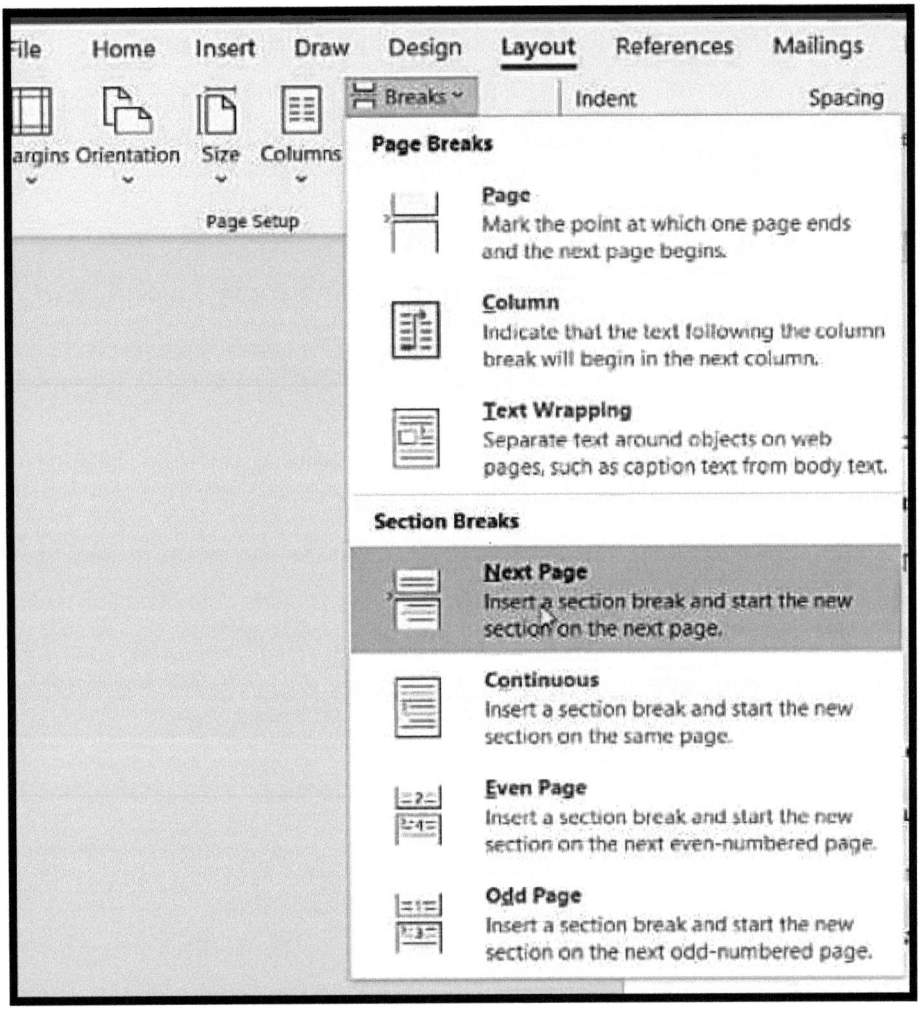

This creates a section break and moves on to the next page to begin the next section.

If your page numbers aren't aligning, for example, if you have a page 2 and instead of seeing a page 3 (which is the following page), it goes back to page 1, that's not what you want and it's simple to change. Simply select a particular page number, **right-click**, and select "**Format Page Numbers.**"

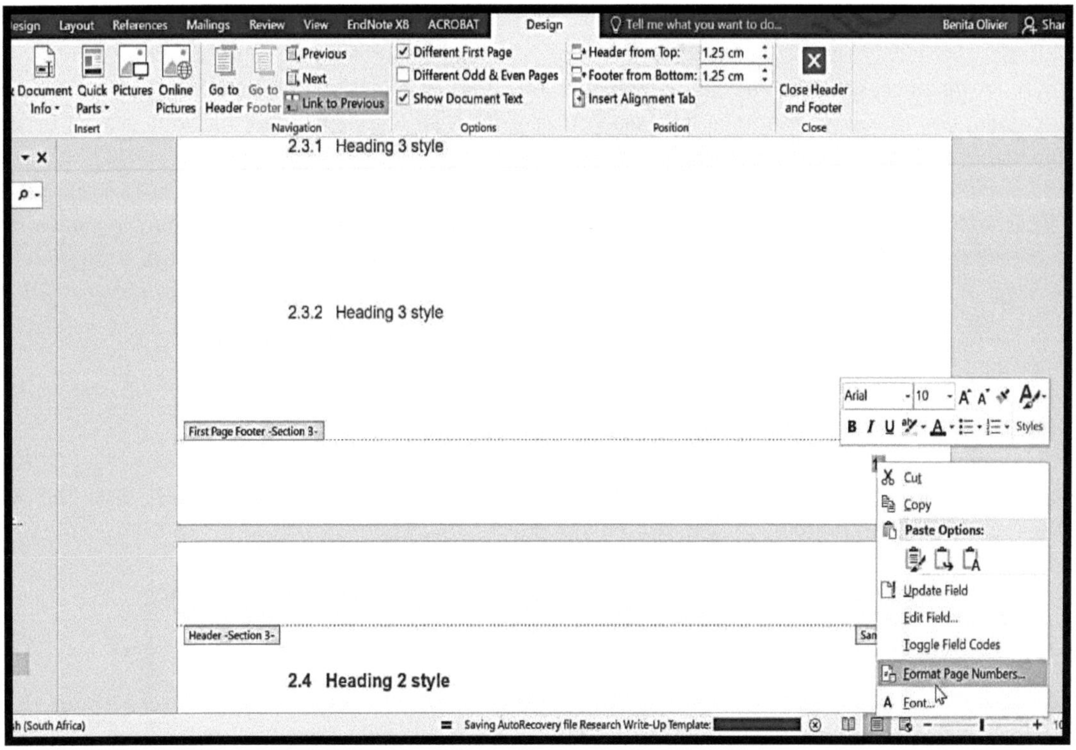

Click the "**Continue from the previous section**" button, and it will remember to continue from the previous section on the next page. That is something you can do for each of your chapters.

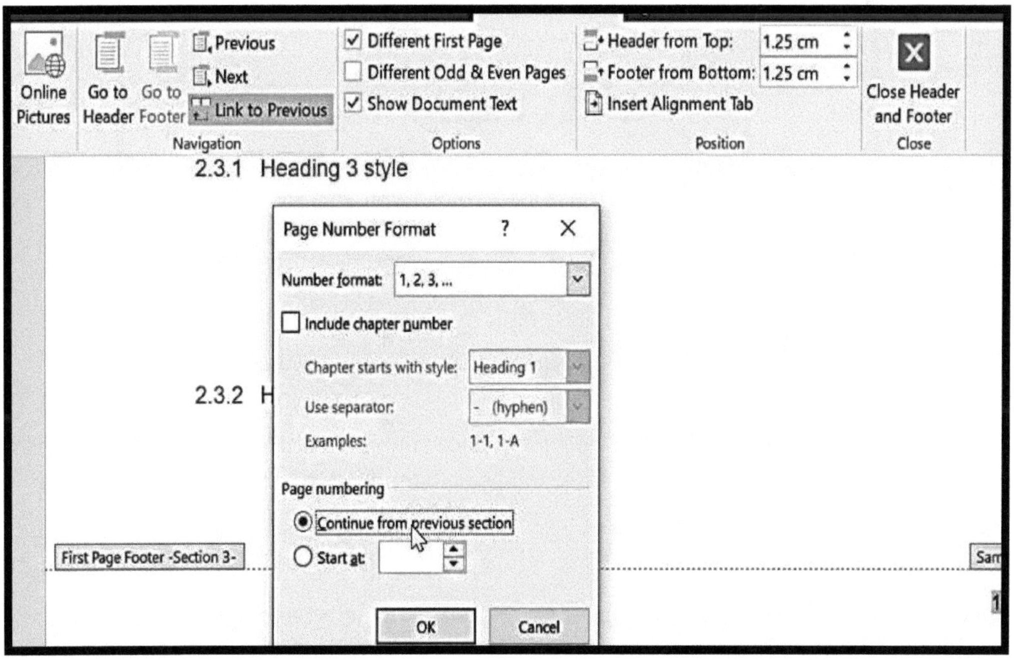

The Very First Page

There are instances when you'll want to include a cover page in an assignment, presentation, or any other document you make; for example, let's say you've written an excellent paper for your project or work and want to include a good cover page.

This is a straightforward process. There are some really attractive templates that you can use for this, and we'll show you how to add a cover page or insert a cover page in a Microsoft Word document step by step.

Adding a Cover Page

The first step is to learn how to include a cover page in a document.

Place your mouse at the start of the document, then go to the Layout tab, select "**Breaks**," and then "**Next Page**" under **Section Breaks**.

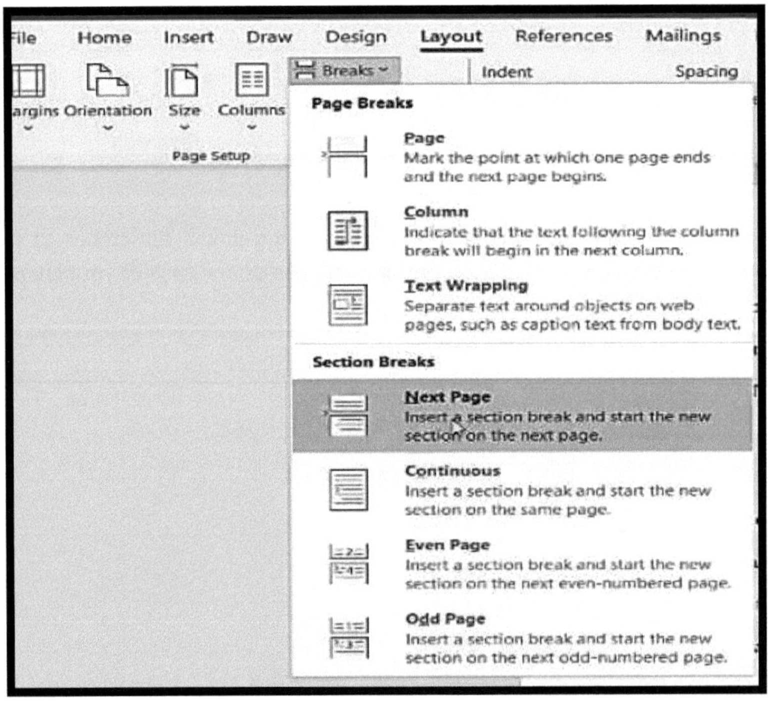

You'll have a distinct cover page for your document once you've done this. You can fill in details for the cover page here, such as your name, the date, and any other information you like.

Inserting a Cover Page Using template

The "**Cover Page**" button can be found under the **Insert** tab. By selecting the drop-down menu, you will be presented with all of the available options. If you don't find one that you like here, scroll down to see a variety of other options.

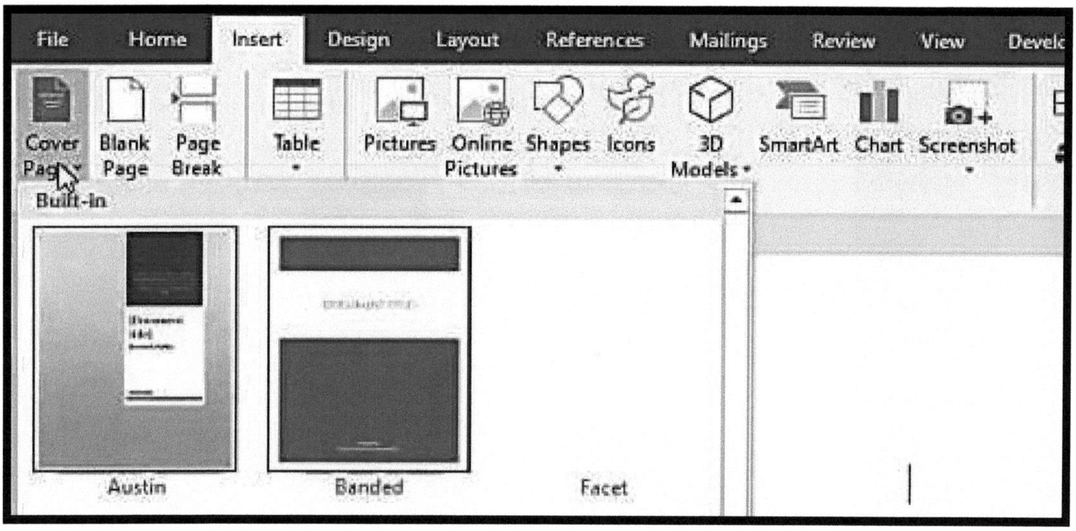

You can go to the section that says "**More Cover Pages**" and click on there to see a few more alternatives.

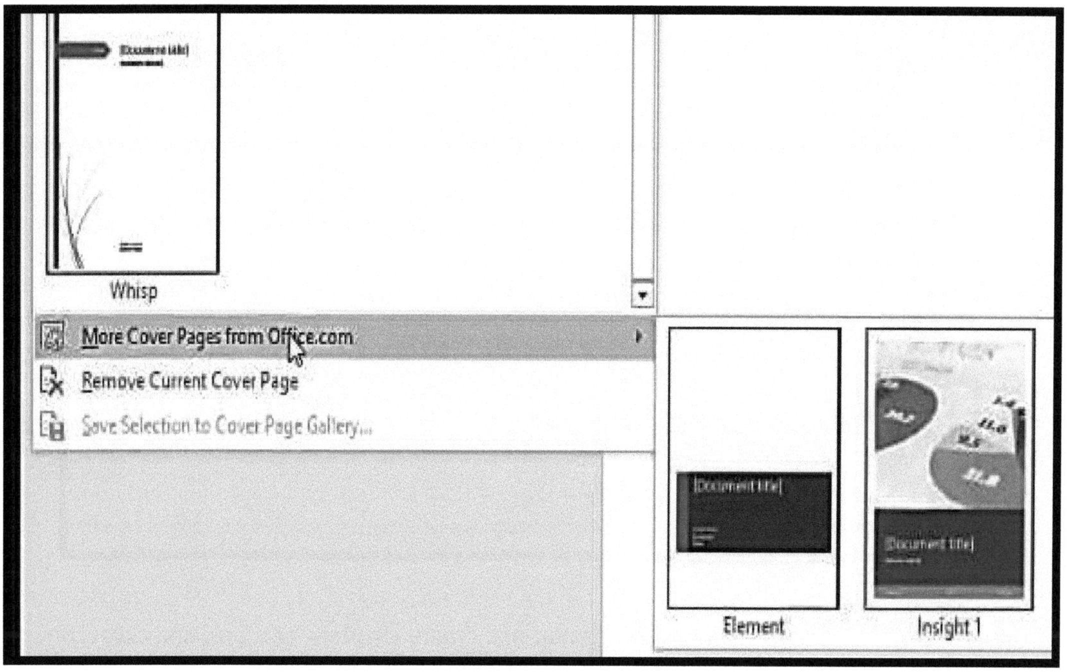

Because there are only a few, go to the **File** tab, select "**New**," then type "**cover page**" into the search bar. This opens up lots of new alternatives you weren't aware of previously. When you click "create" after selecting a template, the template is downloaded and inserted into your document. You can now type in your title, subtitle, and other relevant information.

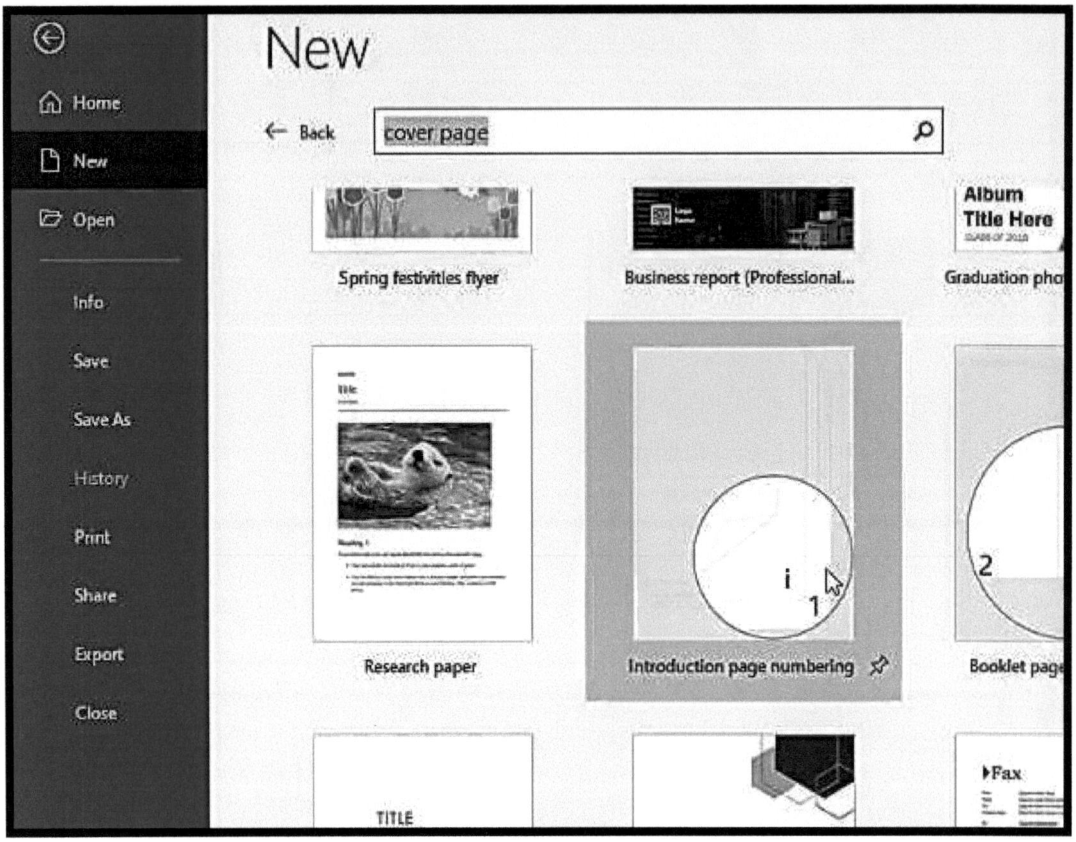

If you decide you no longer want the cover page, simply select "**Remove the cover page**" and the cover pages will be removed.

In this chapter, we demonstrated how to format and create chapters, sections, and subsections. You learned how to create multi-level lists and format your styles to match the different styles, headings, and levels in your document, which is helpful and allows you to easily format all of your headings and sections and organize your document. You also learned about section breaks and how to use them. Finally, you saw how to manually insert a cover page into your document and with templates.

CHAPTER FIFTEEN

TEMPLATE AND THEMES FORMATTING

We've arrived at the chapter where we'll delve into the amazing world of Word templates and themes. You've seen how to get access to Microsoft's templates and how to load them up in the previous chapters.

In this chapter, we'll look at how to create custom templates from scratch or an existing document, as well as how to save a document as a template to reuse it later. We'll also go through how to make themes in Word, how to use them, and how to save a current theme as the default.

Making your templates

The great thing about templates is that you can put whatever you want on them, whether it's text, images, styling, or formatting, and then it'll be ready to go the way you want it every time you open it. The nice thing is that when you open it and click "**Save**," it'll want to save it as a new document. This is handy because the majority of users will open an existing document (such as an invoice), enter in the data, and then "**Save as**." As you surely know, it's easy to hit save by accident, and autosave has the potential to overwrite an existing document, which you don't want to happen. This is a fail-safe function, as it will produce a new document each time you open one. You can have as many templates as you need to complete your assignment.

If you go to "**File**," then "**New**," then "**Blank Document**," it will appear at the top of the page as a new document. We'll use an invoice template as an example, but this process of building a template can be used for any document you want to make. When you click on your document template (in this case, an invoice), it opens as a new document, preventing you from losing it. You can now type in some information, and when you hit the save button at the top or go to File and save, it will save it to the location you specify.

Using an Existing Document to Create a Template

Now we'll look at how to make one from an existing document. You can notice in the example invoice sheet below that the date and invoice number have been cleaned away, as well as the address and some other information.

The good news is that we're going to save this as a template, and when you save something as a template, it always goes into a specific folder for templates. You'll go into File, choose "**Save as**," and over here, where it says **Word document**, you'll click the drop-down and select "**Word Template**"; this is the one with "***.dotx**" at the end of it.

There's also a "***.dotm**" if you've put macros in there, which is something different entirely, but you'd have to save it that way. As you can see, there is an earlier format; this should not be a concern. You simply go to the current version of the Word template here, click "**Save**," and it will save it to the appropriate folder.

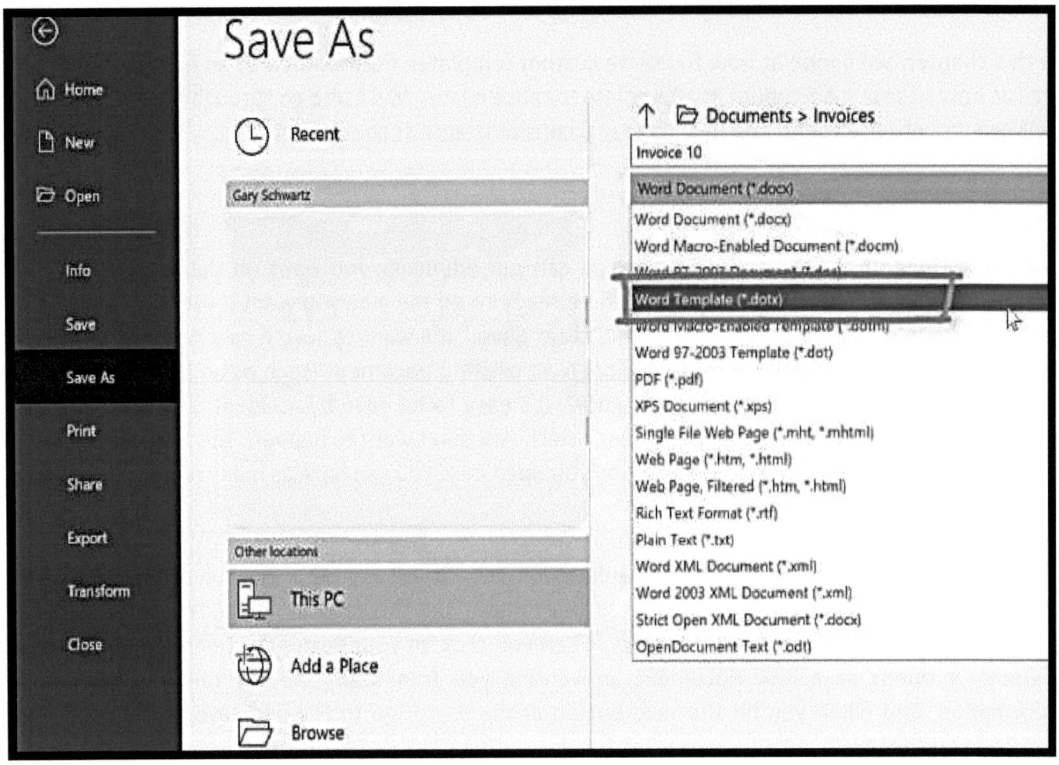

If you go to **File**, then "**New**," your newly created template will be found under the "**Personal**" menu.

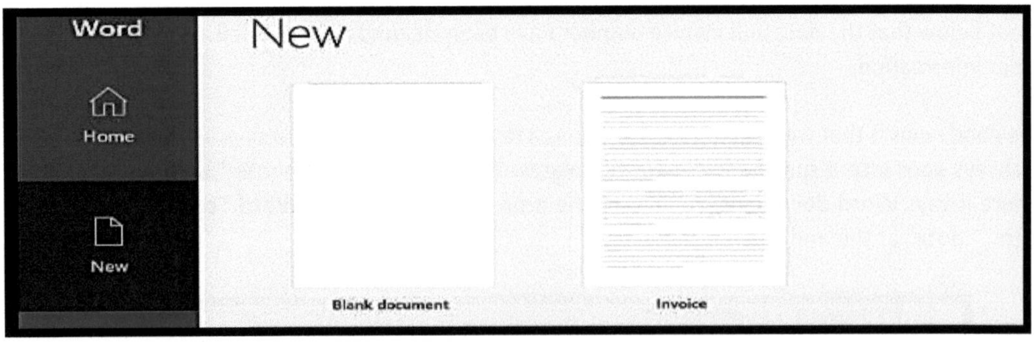

It's as simple as clicking on it to open it, and it's ready to use. The good news is that if you hit "**Save**," it will prompt you to save it in, so you won't mistakenly overwrite it.

Making Modifications to a Template

So, you've got a template and want to be able to customize it. It is suggested that you make any necessary modifications here, then "**Save as**" and rename the file. The beautiful part is that you can go into "**File**," then "**New**," and it will be under the "**Personal**" tab, with the other one intact, exactly like before.

Using the Mac

When using a Mac, this is in a slightly different location. You'll notice that it looks a little bit different on your Mac; all you have to do is go to "**File**," select "**New from templates**," and you'll find a variety of templates to choose from.

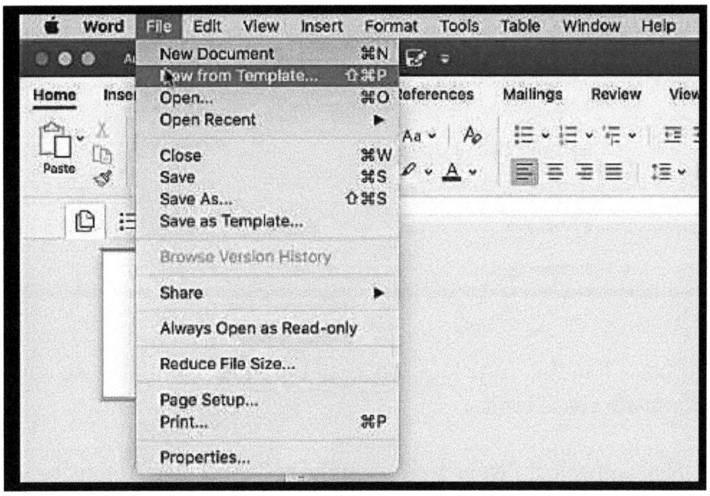

These are all Microsoft-provided, and if you looked at the ones on Windows, you would have seen ones that were similar, if not identical. The difference is up top where it says "**Featured**," and next to it is "**Personal**," which is where your templates are located. Simply double-click on it, and it will open as a document.

If you've just finished creating a new template and want to save it, go to "**File**" and then "**Save as Template.**"

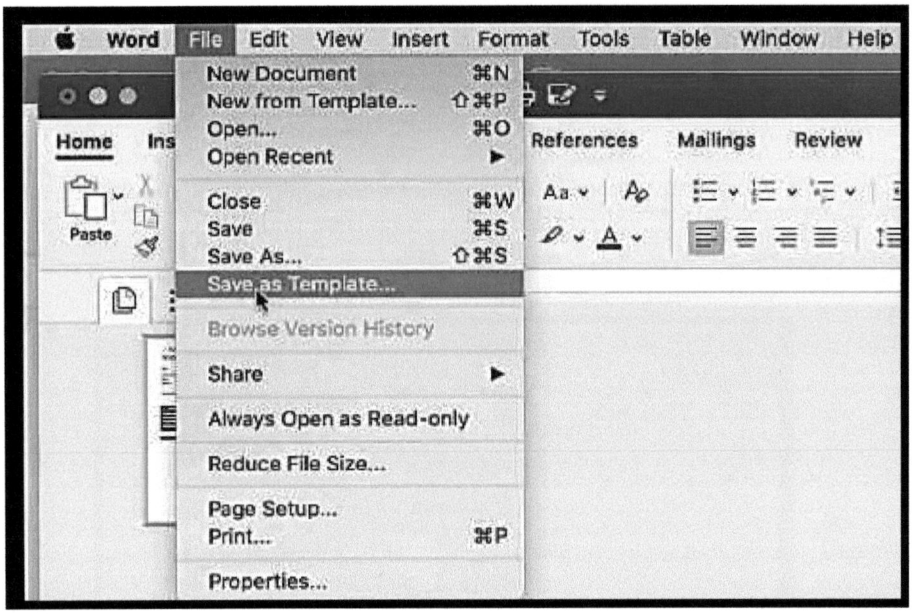

It will automatically place it in the same location as all the other templates, so you don't have to bother about choosing a template space for it.

If you go back to **File** and select "**New from templates**," your new template will appear under the "**Personal**" option. One intriguing feature is that if you have a template on a Mac or Windows machine, you can copy it to another machine's template folder and it will operate just as well.

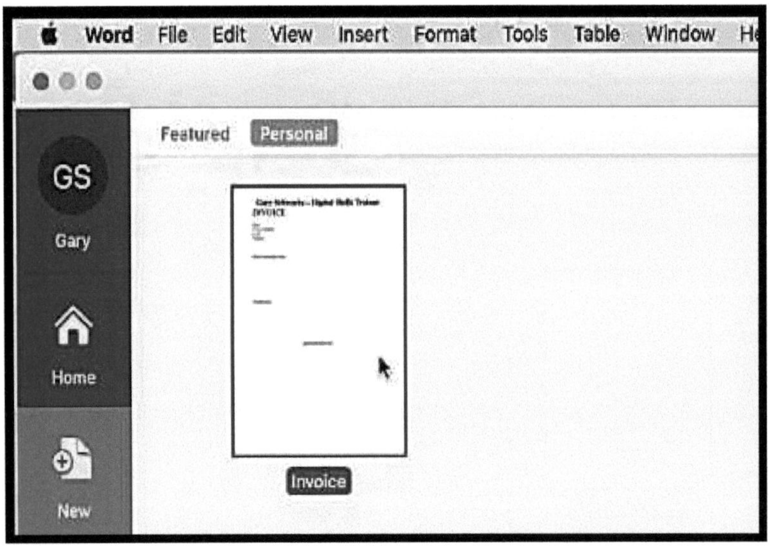

Creating a Template from Scratch

We'll take a look at the stages involved in building a template from scratch in Word.

The first step would be to **open a basic Word document**, then **format** it and type any content you want to add each time you utilize the template.

Finish any formatting modifications, such as bolding the title and headings, then **save your work as a template** with the extension "***.dotx.**" Word will save the template to your custom templates folder by default, but you can specify another place.

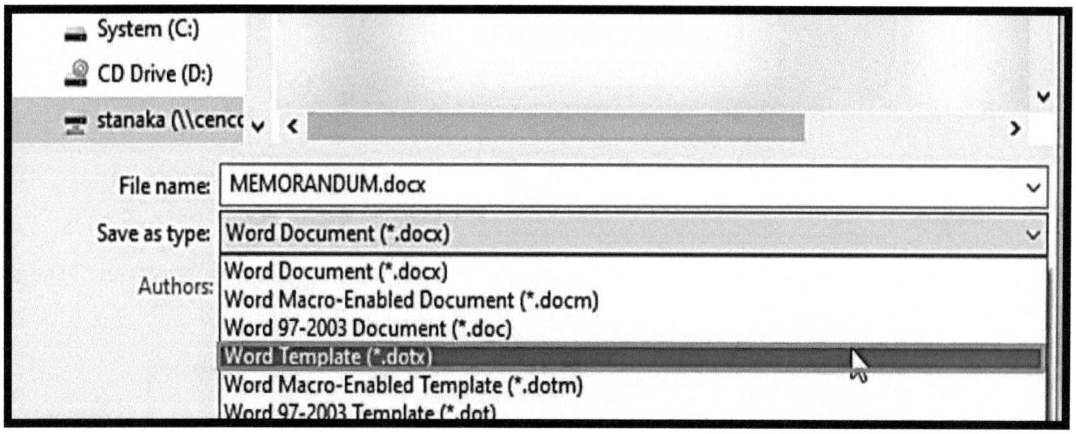

When you return to the usual interface, where it previously said Document 1 as a placeholder, you'll see the name of your template at the top of the screen.

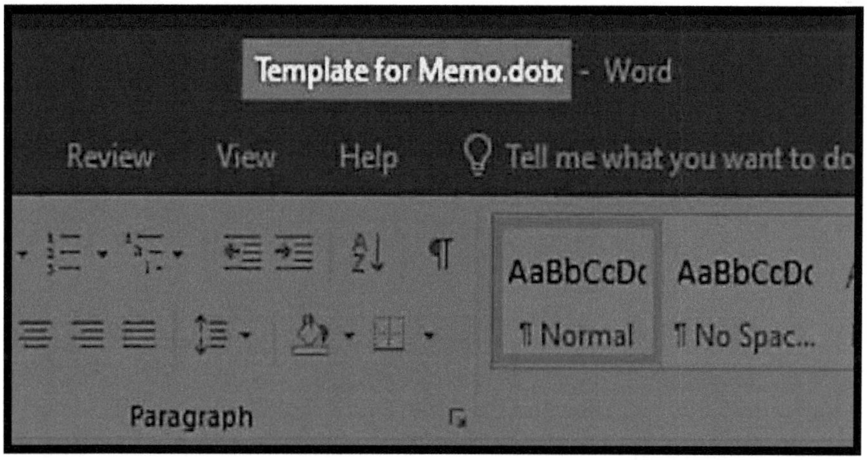

Double-click on the template file's icon, which looks like a notepad, to open a new Word document based on the template you just made. This is a new file that you can modify just like any other Word document; the only difference is that you've given yourself a good start.

Applying a Theme to a Document

In Word, a theme is a set of styles, colors, fonts, effects, and paragraph spacing that you can use to customize your document.

To apply a theme, go to the **Design** tab, then to the **Document Formatting group**, and select "**Themes**" from the drop-down menu.

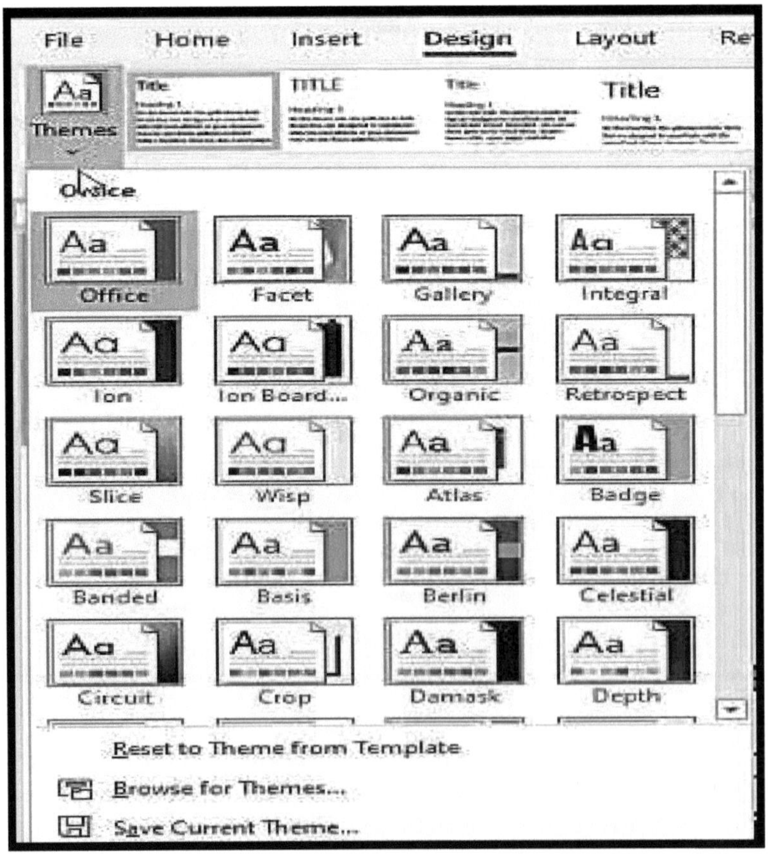

Hold your mouse over a theme in the drop-down menu to preview it in your document, then click it to **apply** it to your work.

To delete an applied theme from your document, go to the **Design** tab in the ribbon and select the "**Reset to theme from template**" option from the themes drop-down button in the document formatting button group.

Saving and Customizing a Theme

Individual aspects of a theme can be customized by changing the style, color, font, and paragraph spacing, and after making changes to a theme, you can save it as a new theme to use in subsequent Word documents.

- To do so, simply **change a theme** to your liking.
- Then, under the ribbon, select the **Design** tab, then the themes drop-down button in the document formatting button group, after which "**Save current theme**."

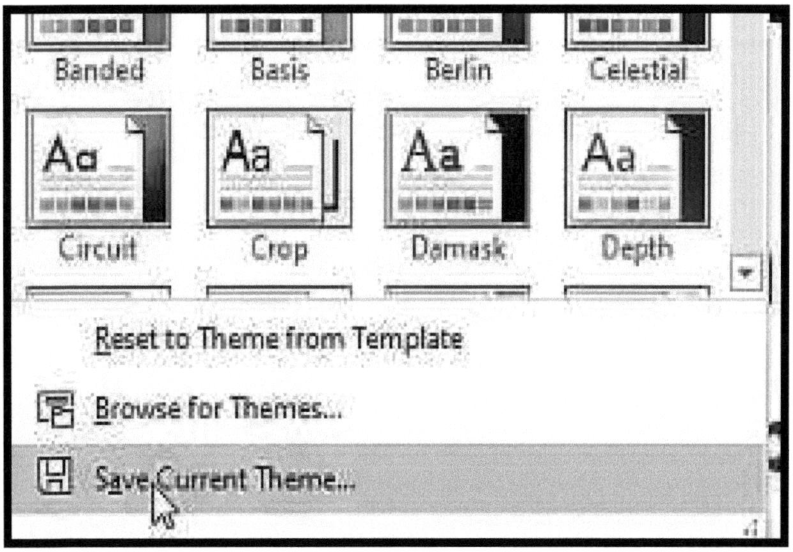

This opens the theme dialog box, which saves the theme to the document themes folder by default. To make it easier to apply custom themes in the future, leave the folder location unchanged, then type the name of your custom theme in the file name field, and click the Save button in the dialog box to save it as a theme file with a "***.thmx**" file extension in the selected directory.

Your custom theme will now display as a theme option in the theme buttons drop-down menu of options, which you can select from the list.

Click the **Design** tab in the ribbon, then click the "**Set as Default**" button in the document formatting button group to make the currently applied theme the default theme for any new blank documents you create.

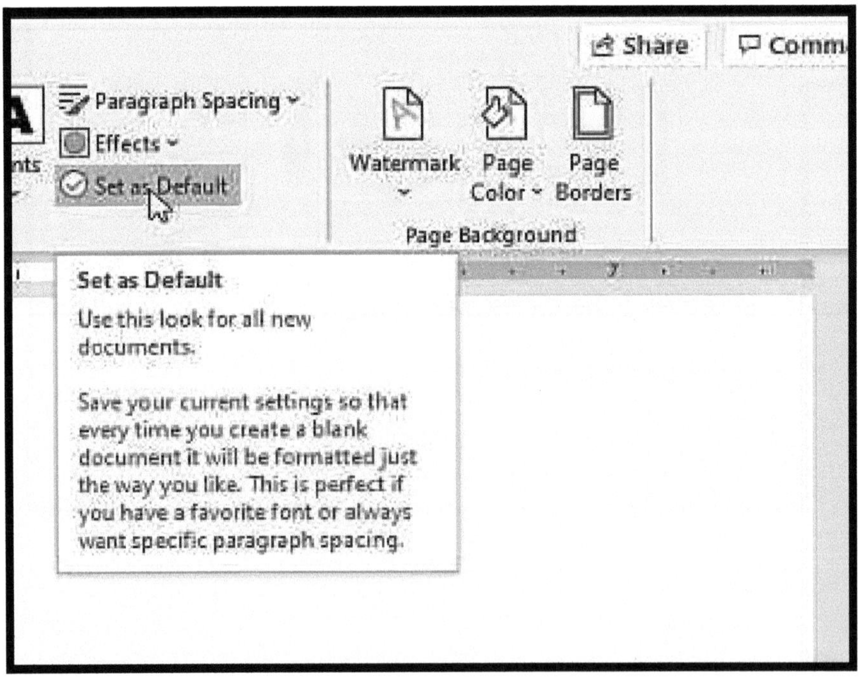

We talked about templates a lot in this chapter, and you learned how they can save you time and prevent you from accidentally overwriting existing files if you open up a file that appears similar to the one you already have. You learned how to make a template from scratch as well as how to apply a template created from an existing document. Finally, we discussed themes and how to create and use them in Word documents.

CHAPTER SIXTEEN

LISTS AND NUMBERINGS

You may have a list in your Word document that you'd like to number, and by numbering, we mean using numbers, letters, Roman numerals, or something similar. You can also number paragraphs, which is what this chapter will go over. We'll also look at using bullets and outlines as an alternative to making lists stand out.

Making Numbered Lists for Your Document

We'll start by making a list. You have the option of working with an existing document or starting from scratch. Now you'll make a simple list and number each item. The next step, as with most things in Word, is to choose what you want to number: thus, you'll use your mouse to highlight all of those names, then go to the **home** ribbon, where you'll find your **Numbering** drop-down in the **Paragraph group**.

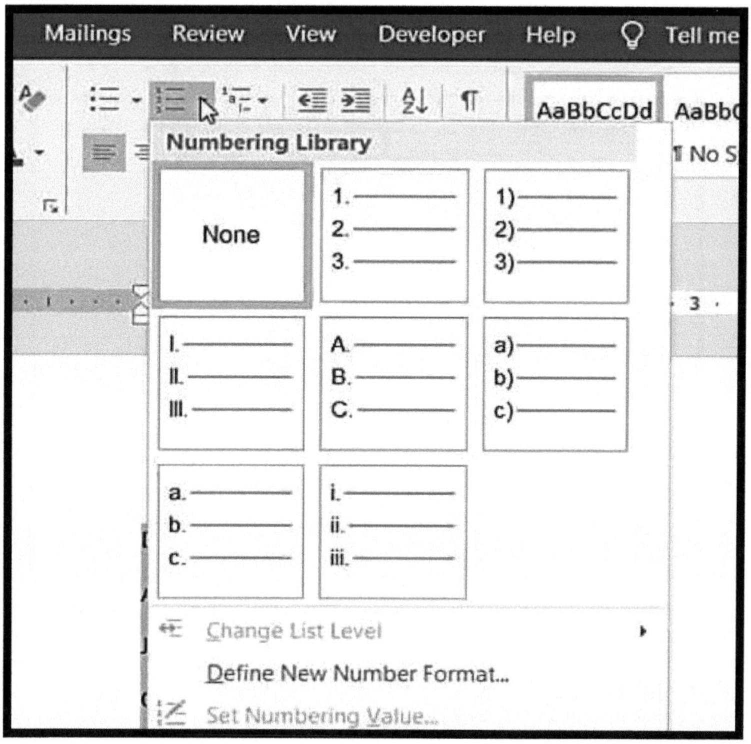

You will notice the tour advice their states "**Create a numbered list**" as you hover over it. You can also click the arrow for more numbering formats, and you can see if you have any selected, but if you don't, you'll notice it has the gray box around the outside and also gives you several options for your list, so it depends on whether you want it numbered (1,2,3), with Roman numerals, or with letters instead (A, B, C).

You can also see that we have capital and lowercase letters, as well as many variations on those. You can start with a simple list by selecting one of these options, and your list will be numbered. The advantage of utilizing numbering in this manner is that if you add a name to the bottom of the list and tap **Enter**, the following number is immediately entered. If you decide at this time that you don't want that number there, you can either hit backspace or opt to turn off numbering by clicking the numbering button again, which will return you to the margin where you may continue putting paragraphs or anything you want into this page.

Also, if you want to change the way this is numbered, all you have to do is highlight it, go back up to your numbering options, and choose the option you want.

When you go to the numbering drop-down, you'll notice that the formats you've recently used are placed at the top, making them much easier to find.

Making Your List More Personal

If you go down to the bottom, you'll see "**Change List Level**," which we'll look at when we look at outline numbering. Our focus here is on the "**Define New Number Format**" option, which allows you to change the style of your numbering.

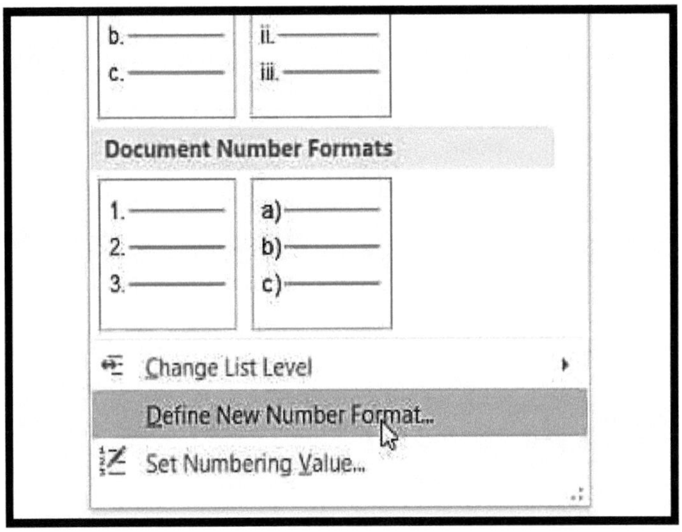

If your number style is set to A, B, C, you can change it to something else if you want, and you can also change the font of your numbering; so, if you want it to look slightly different, you can change the font style, color, and size, and you can see what that looks like in the preview at the bottom.

You have the option of selecting your alignment as well. If yours is set to left alignment, you can select to center it or move it to the right, which will change the positioning of the numbering.

Set Numbering Value

We'll also take a look at another numbering option called "**Set Numbering Value.**" If you want to make a list underneath, this is beneficial. If you start a new list and add some numbering, you'll notice that it follows the numbering from the previous list. This is where you would define the new number if this is a distinct list and you wish to restart it.

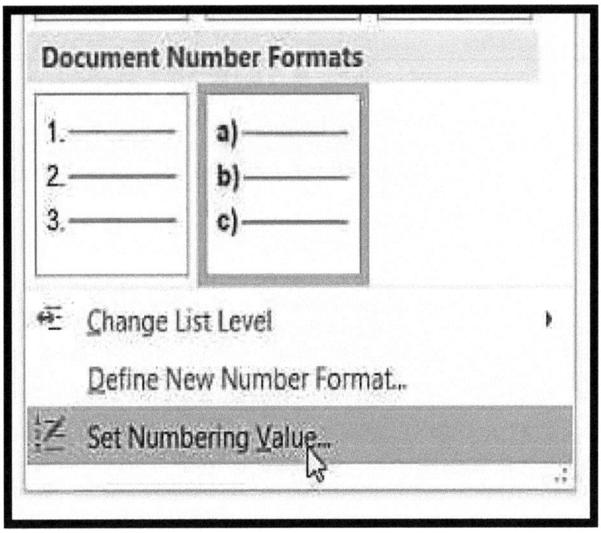

Return to Numbering, select "**Set numbering value**," select "**Start New List**," and set your value to the first letter (a) or number one, then click "**OK**" to create your distinct list. If you don't specify that, it will continue from the previous list, so keep that in mind while you work with your numbering.

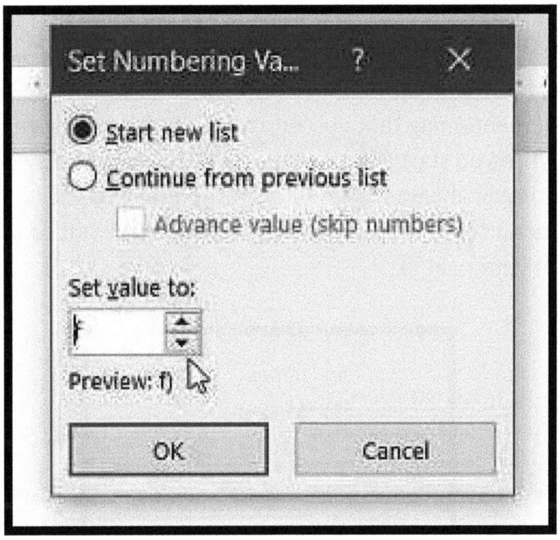

Numbering in Paragraphs

Let's have a look at how Numbering works in paragraphs. If you click at the beginning of the first paragraph in your document and then on the Numbering button, you'll see that it's numbered the entire paragraph, and if you do the same for the second, it continues.

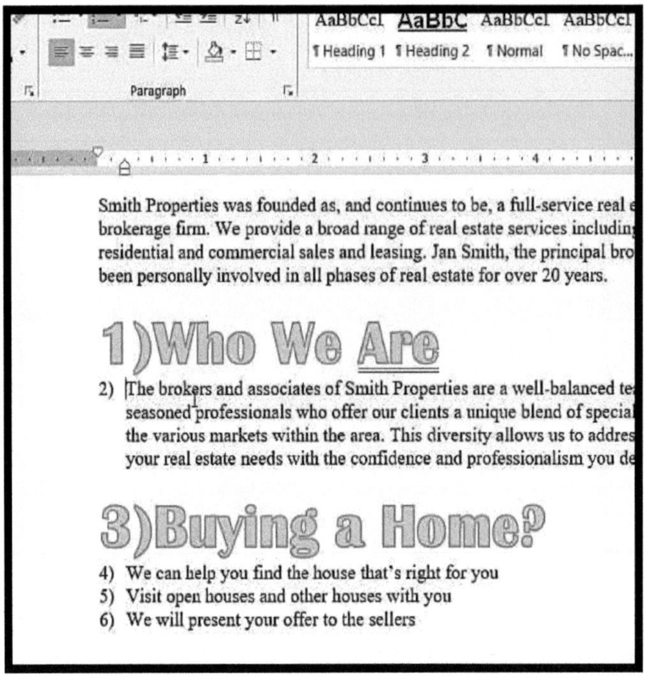

When you click at the beginning of a paragraph and apply the numbering, you'll see that only the first line is numbered at first. The reason for this is that this paragraph has a line break in it, so if you hit Enter and move to the second line, Word considers those to be whole new paragraphs, whereas the other paragraphs, which do not have line breaks, are continuous sentences that just keep going.

This is to demonstrate that numbering is not limited to the lists you build; it can also be applied to paragraphs. It's also worth mentioning that you don't have to have a list ready to use Numbering. If you wish to use Numbering before you start creating your list, you can do so by simply clicking the numbering button. Keep in mind that it will always continue from the previous list unless you instruct it to restart. You can go ahead and click on the drop-down menu, where you'll notice a useful "**Restart numbering**" button, and then construct your next list.

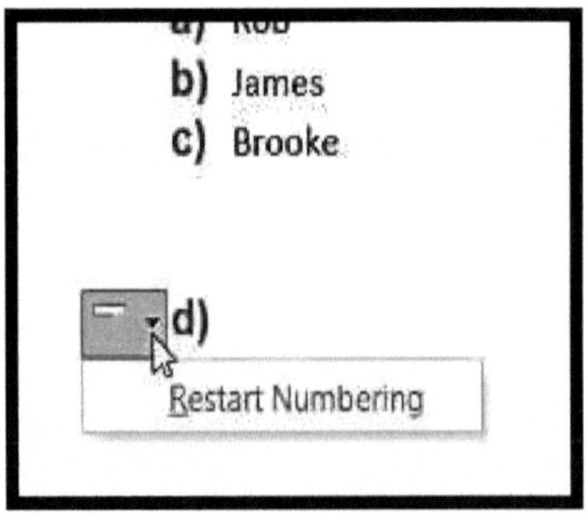

That is how you work with fundamental numbering in a document, both in terms of lists and paragraphs.

Making Use of Bullets

We've just finished working with numbered lists in a document, and now we'd like to focus on working with bullets, which is significantly different from numbering but similar in concept.

Let's suppose you have a basic list in your document that is currently numbered, and you want to replace the numbers with **bullets**. If you're not sure what a bullet is, it's just a symbol, and there are many various symbols that you can use as bullets; most of the time, you'll see some Round solid black bullets in documents, but as we'll see as we go through this part, you can use a variety of different things as bullets.

- The first thing you'll do is do your selection, which will **highlight your list**.
- Next, head back up to the **paragraph group**, and next to numbering, there's a small button that says "**Bullets.**" "**Create a bulleted list,**" the tooltip says, and "**Click the arrow to customize the look of the bullets.**"

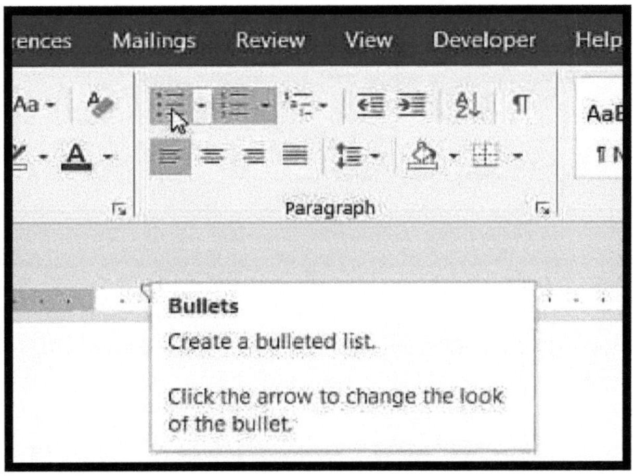

When you click the arrow, you'll be provided with a bullet library, which will show you some of the most popular bullets that people might use, as well as a handful of ones that have been used previously. Still, you can select any of these and see what they look like by hovering over them.

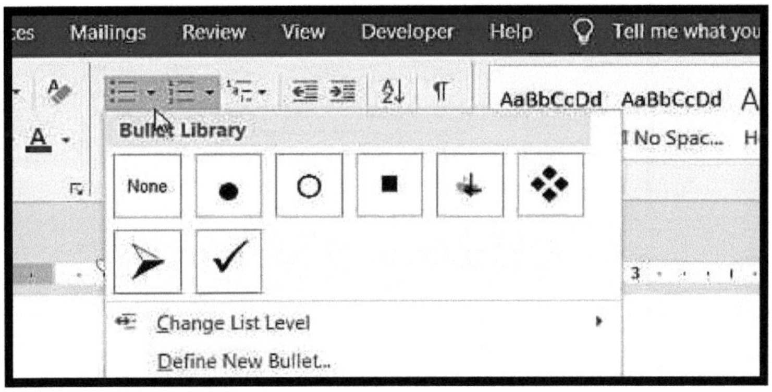

The trick is ideal for to-do lists, and you can select the type of bullet that best suits the document you're working on.

Defining New Bullets

Alternatively, you can go to the "**Define New Bullet**" option and pick a symbol that will work for your document.

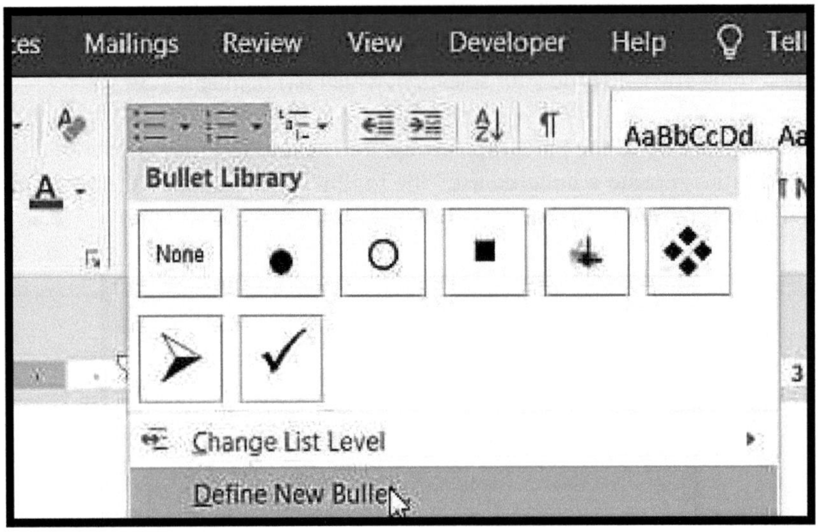

As you can see, bullet characters are available, as well as the options of Symbol, Picture, or Font.

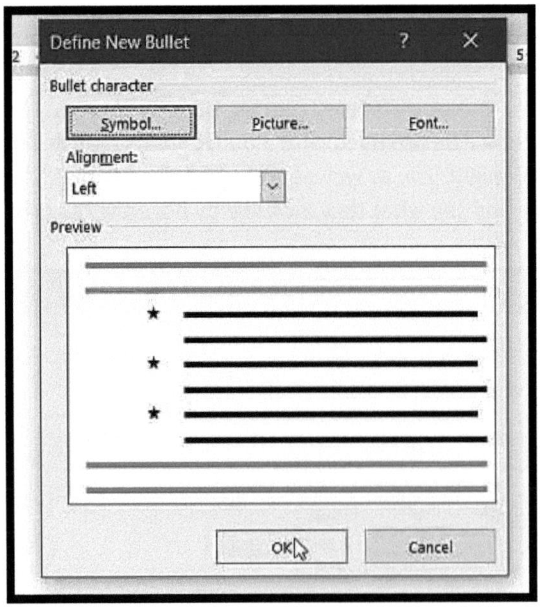

Let's use Symbols

First, we'll take a look at "**Symbol.**" This will open your symbol dialog box, and symbols are essentially just fonts divided into fonts, so there are a lot of different symbols. Because there are so many, it would take a long time to go through them all looking for something appropriate, but a lot of the best ones that you'll use or some of the most popular you'll find under "**Webdings,**" where we have lots of interesting things that you can add into a document.

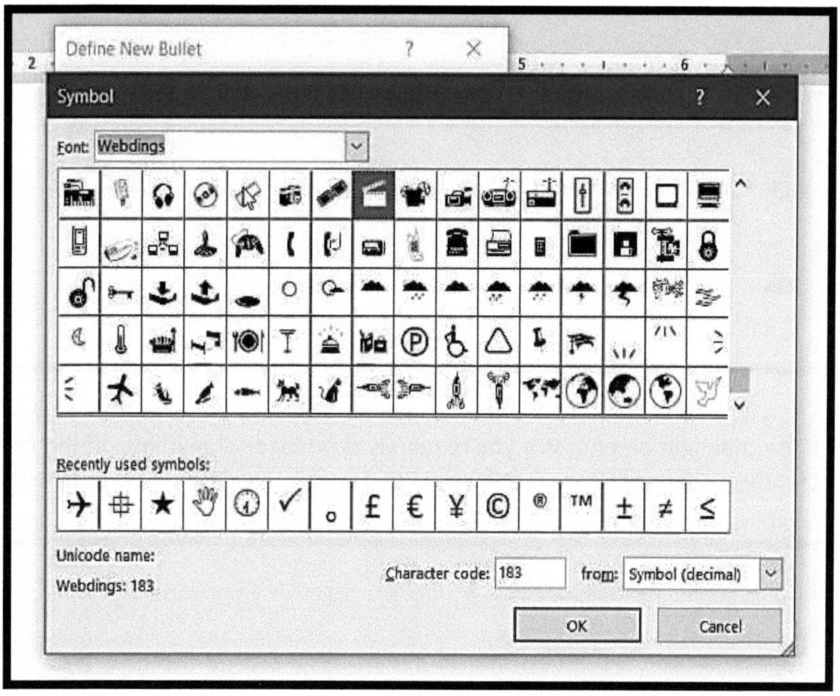

You can **scroll through this selection of symbols** until you find one you like, then click **OK**. You'll receive a preview of what that symbol will look like in your document, then click **OK** again, and your numbering will be changed to that symbol.

How about some pictures?

Pictures that are saved or that you can obtain on the internet. Keep your list highlighted and move back up to the bullets drop-down, then back down into "**Define New Bullets**" and select the **picture** option in the middle. You have a few options here: you may choose a picture from a **file**, so if you have one saved to your desktop in "**My Documents**" or a **local Drive**, you can select it from there; you can also choose to search the **internet** for an image, or you can select one from your **OneDrive cloud storage**.

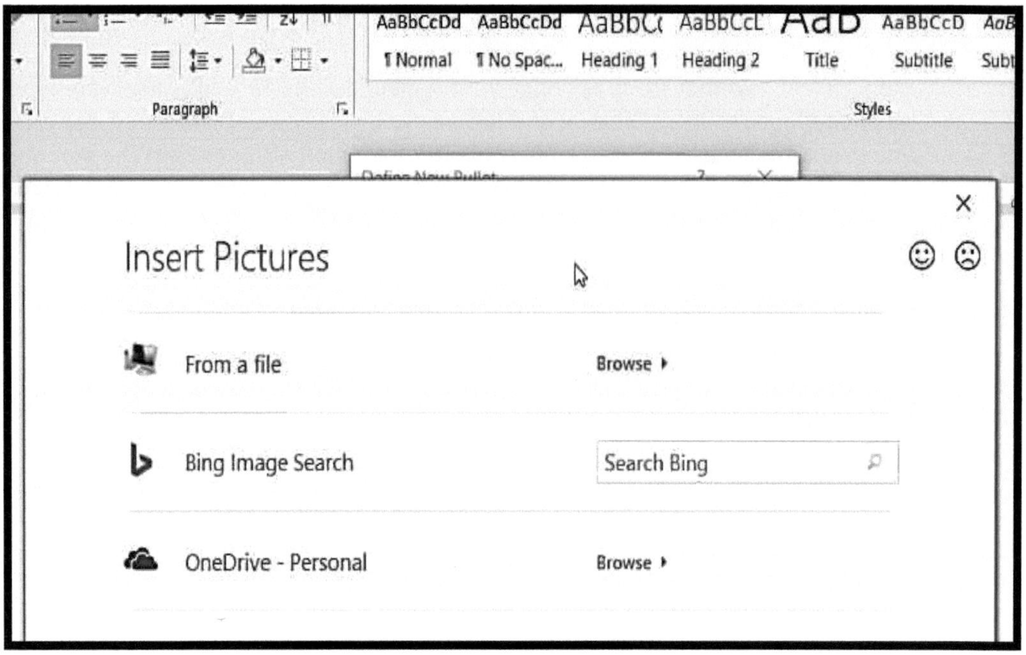

Always check the copyright on photos if you're merely downloading anything off the internet to make sure it has a **Creative Commons License** so you can use it in your work if you're searching for a Bing image.

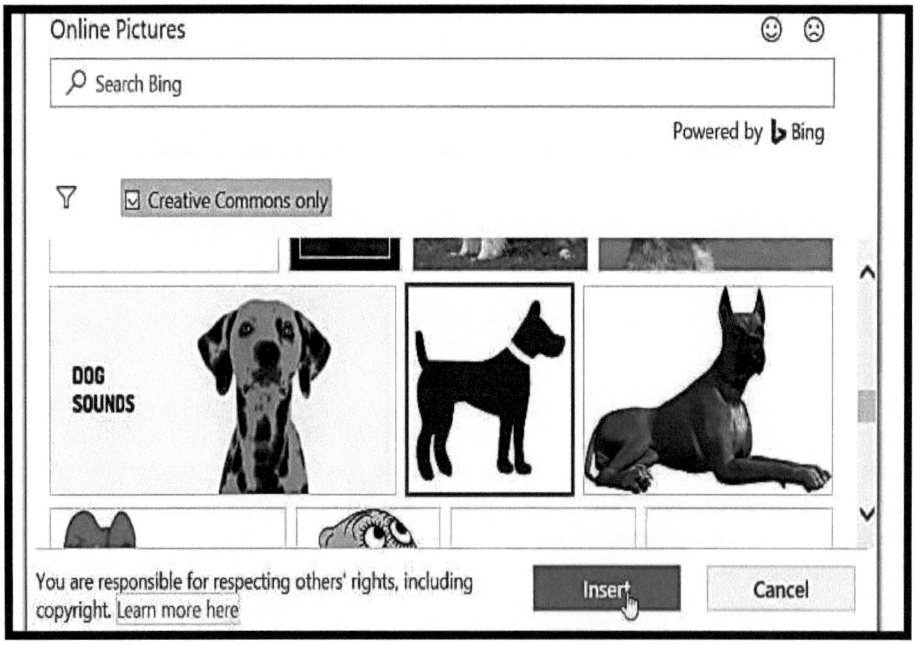

After selecting an image, click "**Insert**" and "**OK**," and you're done; your document now has a cute little image that adds a little more interest.

An Example in Action

Another example of when you might want to include bullets, graphics, or even symbols is if you're typing something about a phone number and want to include the phone symbol in your phrase. To begin, place your cursor where you want it, then select the "**Insert**" ribbon and scroll across to the "**Symbols**" group on the right-hand side.

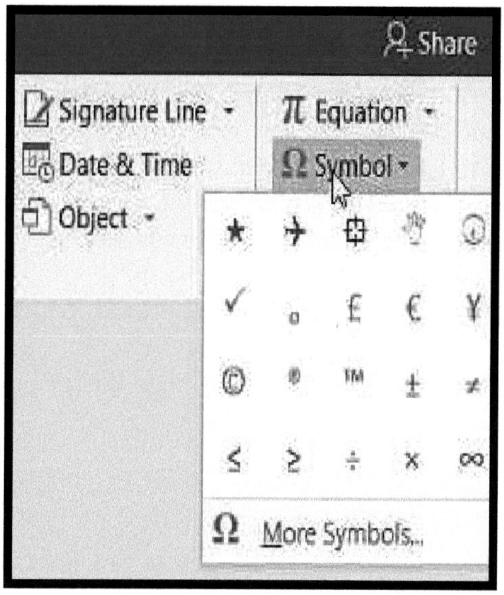

Click on "**Symbol**" and it will show you a list of all the ones that have been used recently, but you'll go to "**Add additional symbols**" and select the telephone icon under the "**Wingdings**" font, then click on "**Insert**" and **close**.

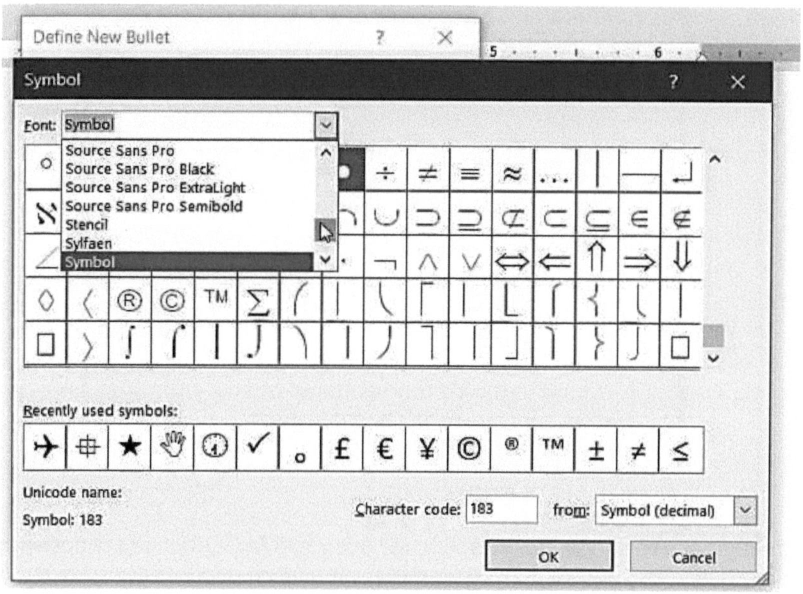

You can see how you've used those symbols not only in your bullets but also within a single line of text. That's another nice thing you can do with symbols in your documents, and it's worth remembering that they're regarded like fonts, so you can do things like alter their color by highlighting them, and use your font settings because they're essentially fonts, not photos. Hopefully, that provides you an idea of how to use basic bullets, as well as some more complicated bullets, in your documents utilizing symbols and images you've stored.

Creating an Outline

We're going to look at how to outline from scratch in this section. If you click your cursor after the first item on your numbered list and hit the Enter key now, it will just resume the numbering. What if you wish to add a second level to this number? You can simply use the tab key on your keyboard to add a second level of numbering to your list. You must be familiar with this terminology; for example, using the **tab key** to **demote a level** and the **Shift + tab** to **advance a level**.

If you enter something in here and press **Enter**, the numbers will continue at that level.

You can exit by using the **Shift + tab**, which will return you to the previous screen. If you want to get rid of it completely, you can **backspace**.

A More Organized Paragraph

Now let's look at how you can number paragraphs at the outline level. To begin, you must first identify and choose the paragraphs to which you want to apply this.

- To get started, go to the home ribbon and then to the "**Multilevel List**" button.

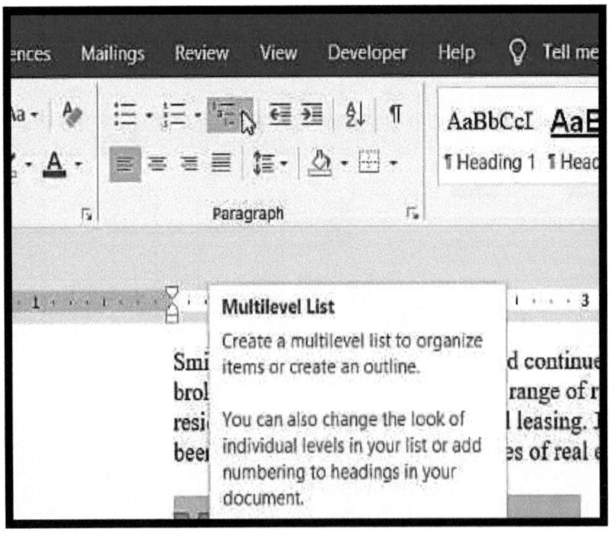

- Then click the **drop-down** menu, which will give you a few options to choose from. It all depends on the style you want for your outline-level numbering.

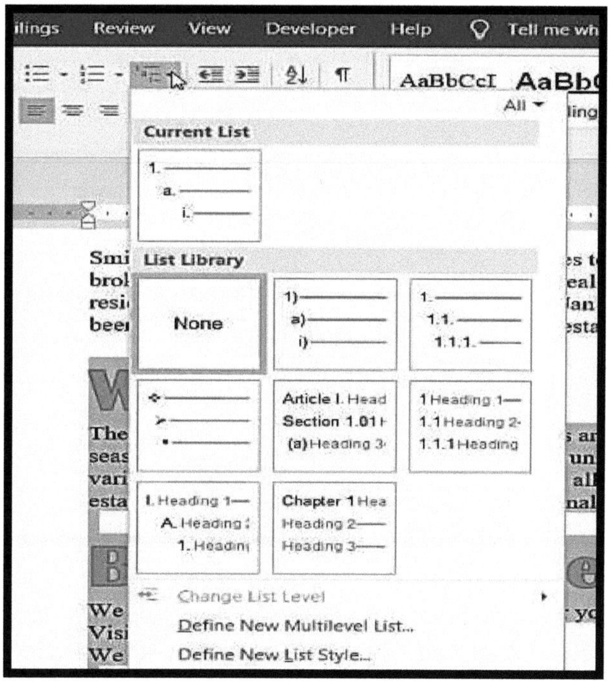

After selecting an option, you'll see that it has been applied to the paragraphs you've chosen, but you'll notice that it hasn't recognized what should be on the second level.

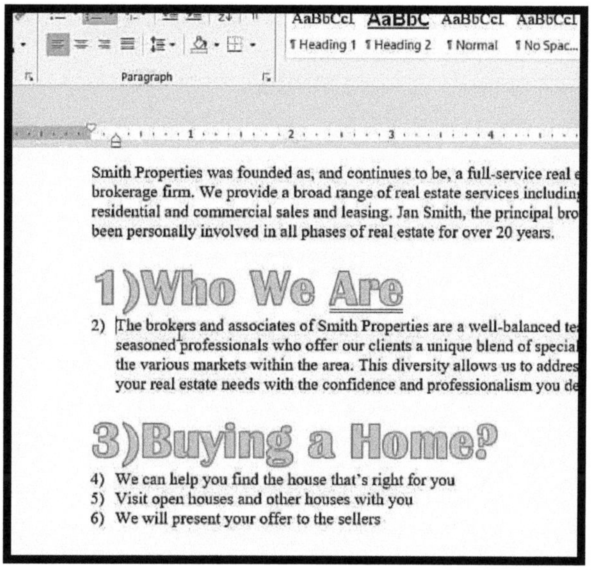

- Simply click in front of the first sentence in the first selected paragraph and hit the tab key to demote it to the second level, and you'll see that everything follows through. You can do the same for the ones down here: just hit the tab and keep going until you get everything looking just right.

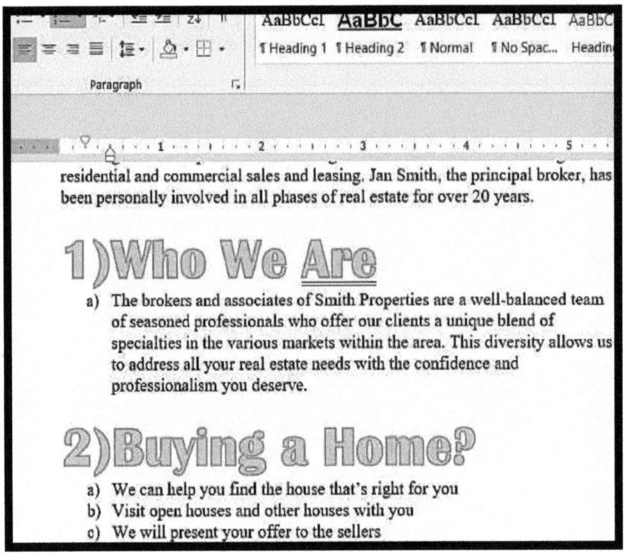

It's fairly simple to outline, and you can even add a third level if you wish. If you press the tab key again after completing the second level, you'll be taken to the third level, where you can make those little tweaks.

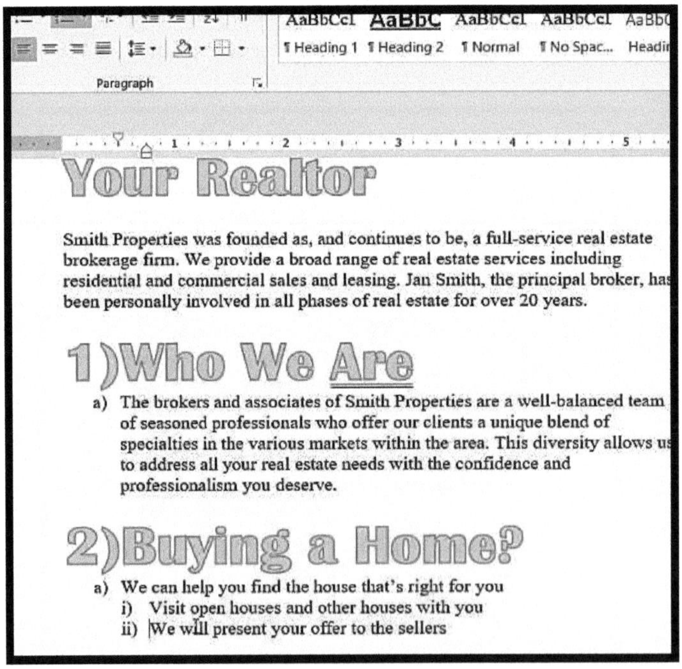

A Simple Modification

What if you wanted to redo the entire outline? Make another selection, then move up to the outline level numbering and "**Define New Multilevel List**."

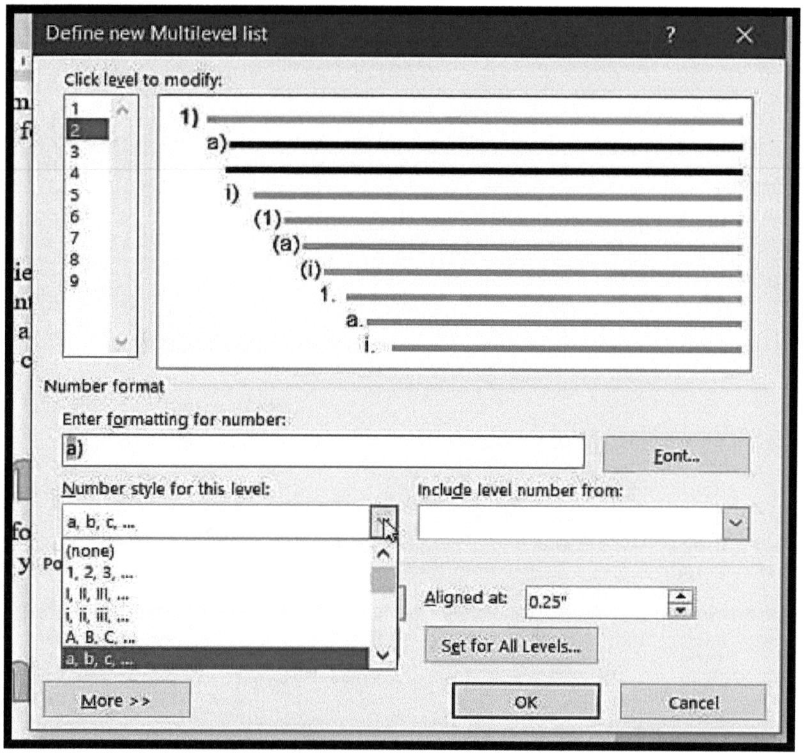

You'll be able to go more specific and personalize the appearance of your Multilevel List as a result of this. There are nine levels listed here, as well as those that are listed elsewhere. Number one will be the first level, followed by "a," "I," and so on. You can go in, choose a level, and tweak the formatting to your heart's content.

You can alter the font for your level 1s by going into the options and changing it to something completely different. You can modify the number styles for this level, the number alignment, the text-indent, and a lot of other things on level two, so it's worth getting in there and customizing so you can quickly apply your newly customized Multilevel numbering to your document. Just keep in mind that you can define that new Multilevel list in this section.

If you already have a document written out, you can go in, highlight everything, select your multi-level list, and then promote and demote as needed, or even get detailed and edit each level of your list.

In this chapter, we looked at how you can use bullets, lists, and an outline to add numberings to your documents to make them stand out and be more exciting to read.

CHAPTER SEVENTEEN

TABLES

This chapter covers a lot of ground in Microsoft Word when it comes to tables. When it comes to document structure, tables are extremely important. They enable you to present data in columns and rows, making data more ordered and easier to understand. Tables have their contextual ribbon, which includes a variety of choices for restructuring, modifying, and formatting the table to match the overall appearance and feel of any page.

We'll walk over the process of adding a table into a document in this chapter, as well as some formatting tips and tricks.

Adding a Table in Your Document

There are a few different approaches you can take, and we'll go over a number of them. The first method is to go to the **Insert** ribbon and look for the **Tables group**.

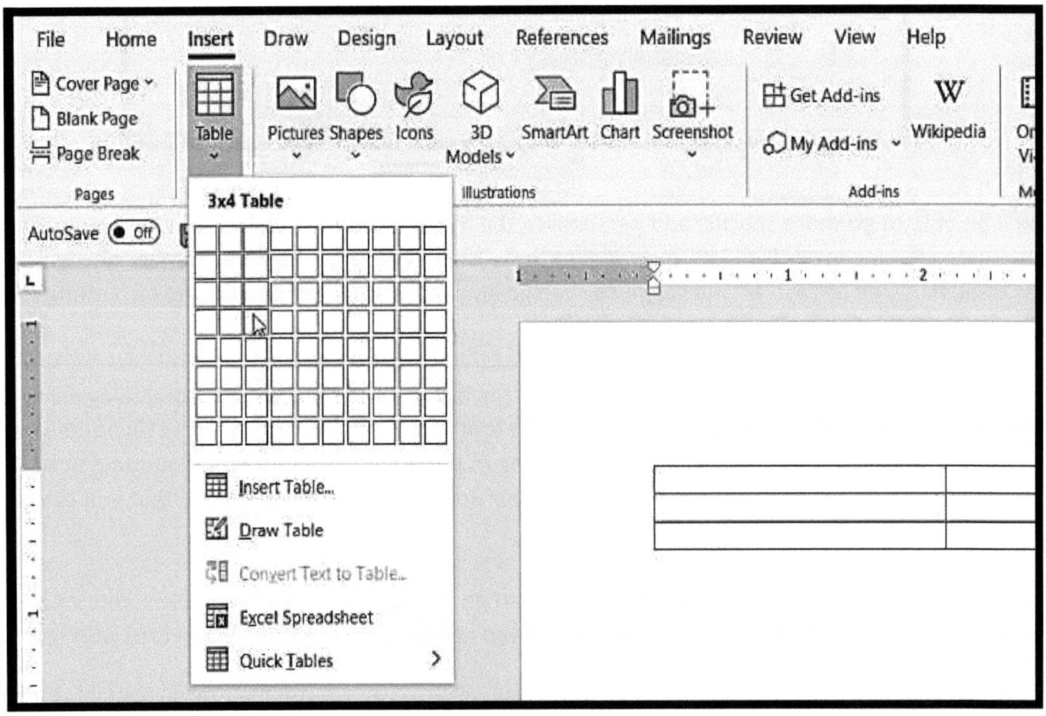

There's only one option in there, which is "**Table**", and when you click on it, you're taken to a small grid where you can choose how many rows and columns you want in your table. When you move your cursor over this grid, it creates the table instantaneously in your document; for example, if you want a table with three columns and four rows, all you have to do is select that many squares in the small grid and click, and your table is ready.

Inserting a table

When it comes to inserting tables, you can go back to the **Insert** tab, select the **Table drop-down**, and choose "**Insert Table**" instead of utilizing the grid.

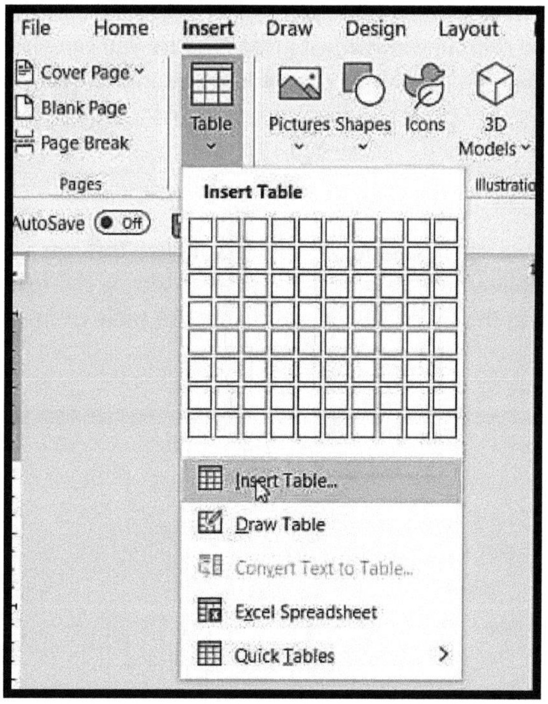

This is where you can choose the number of columns and rows you would like in your table.

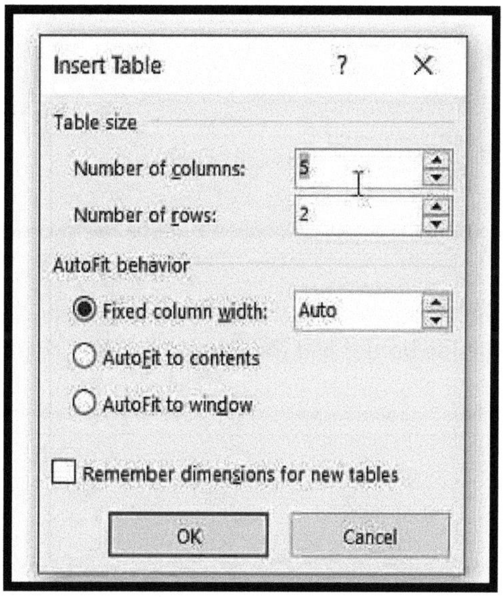

You also have the option of selecting "**AutoFit behavior**." "**Fixed column width**" in this case indicates that each column in the table will be the same width as determined by Word, and you have this set to automatic so it will do the best fit. "**AutoFit to contents**" implies that each column will extend to be consistent with the table's left and right margins, and "**AutoFit to window**" means that the columns will expand to be even with the table's contents.

Most of the time, the **Fixed column width** is just great because you can resize or change the width of the columns once you've created the table. After you've made your selections, click "**OK**," and your table will appear.

Make a table for yourself

You can also draw your table when it comes to inserting tables. This can also be useful since it gives you greater control over the appearance of your table, such as setting the row, height, column, width, and overall table size. Return to the **Insert** ribbon and select the table drop-down menu; there is a "**Draw Table**" option there.

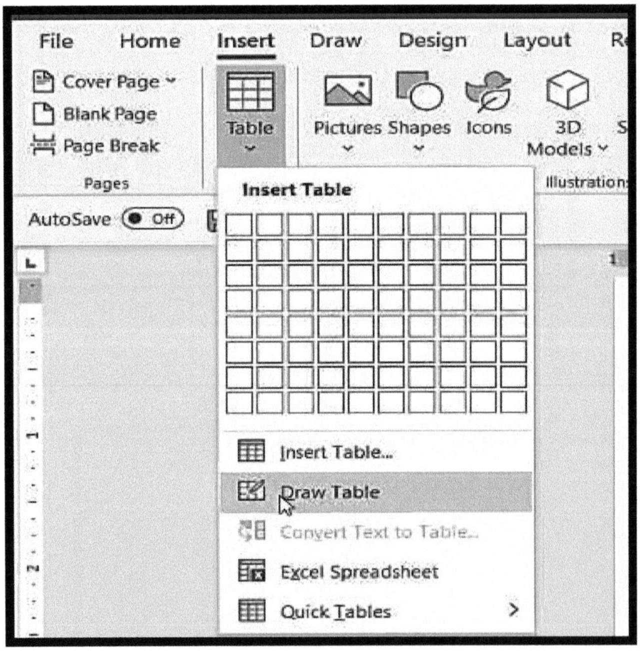

When you click on it, your mouse pointer will change to a pencil icon. This will allow you to draw your table; you'll begin with the outside border and then draw rows and columns using your pencil.

This option is much more flexible and allows you a little more control than the others. If you see that you still have the pencil icon after you've completed designing your table, simply go to your ribbon and deselect the "**Draw Table**" option.

After you've drawn or inserted your table, you'll see the table tools' contextual ribbon. You must first choose the complete table. If you look up at your ribbons, you will notice that because you have a table selected, you have access to the **Table Design** ribbon and the **Table Layout** ribbon, and you can pretty much manage every part of your table with these two ribbons.

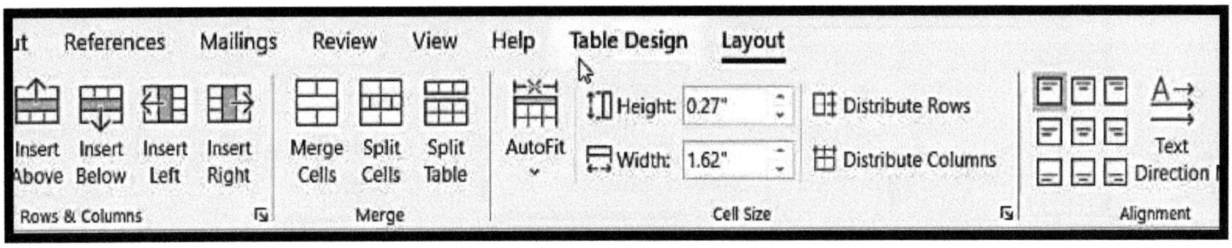

Deleting a table

You can also delete tables. You'll notice a small cross in the top left-hand corner, which when clicked will instantly select the entire table.

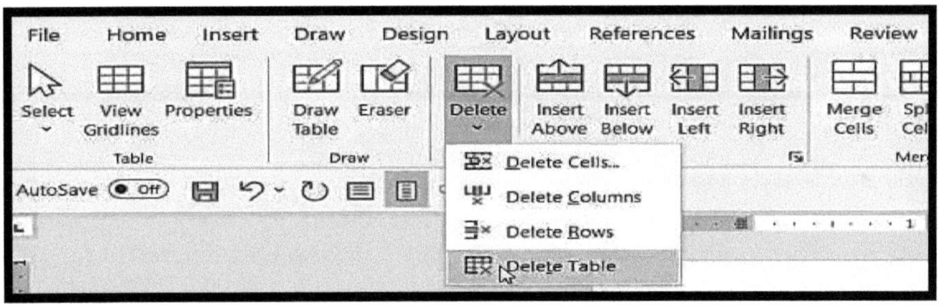

You can delete this by right-clicking and selecting "**Delete Table**" from the right-click menu. You can also use the "**Delete Table**" option on the **Layout** ribbon.

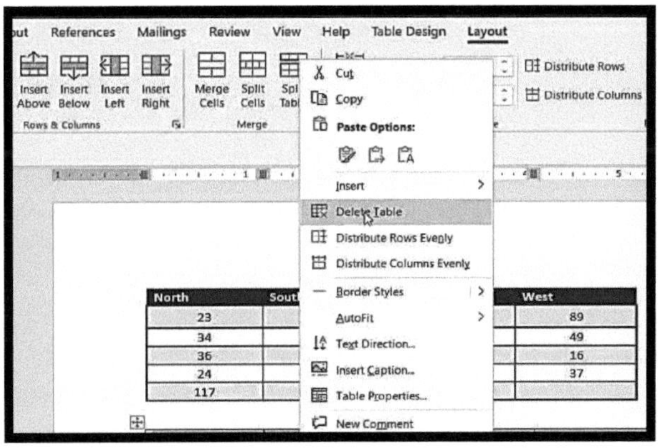

Designing your table

Let's start with the design of the table. This is where you may make changes to your table's formatting. You have different table styles in the middle here that you can apply, and as you mouse over them, you get a live preview so you can see which one you want to use before you apply it; it's a nice simple method to add formatting.

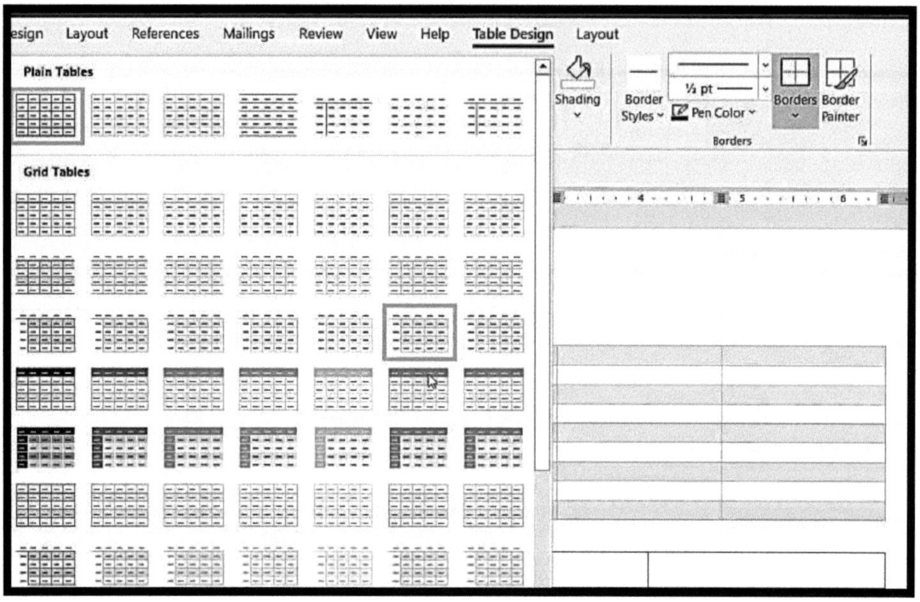

You can apply a table style and then change various formatting elements, so you're not limited to your current color scheme. If you like the table design you've chosen but want to change the background shading of the header row, simply click it, go to "**Shading**," and choose a different color.

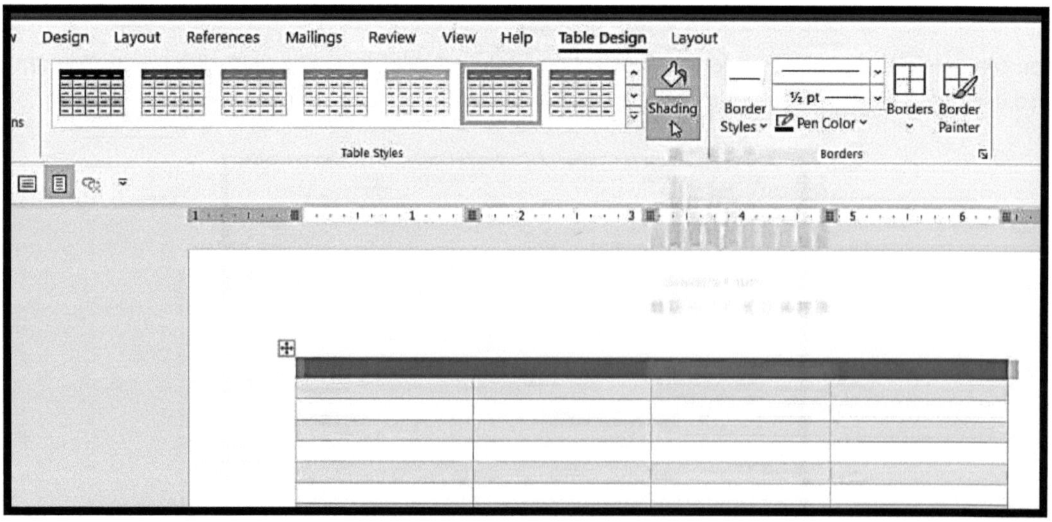

Tables with text

You can add text to your table by clicking in the first cell, typing the word, and then pressing the tab key to proceed to the next cell. It's worth noting that hitting the tab key will allow you to travel between your various cells and enter data.

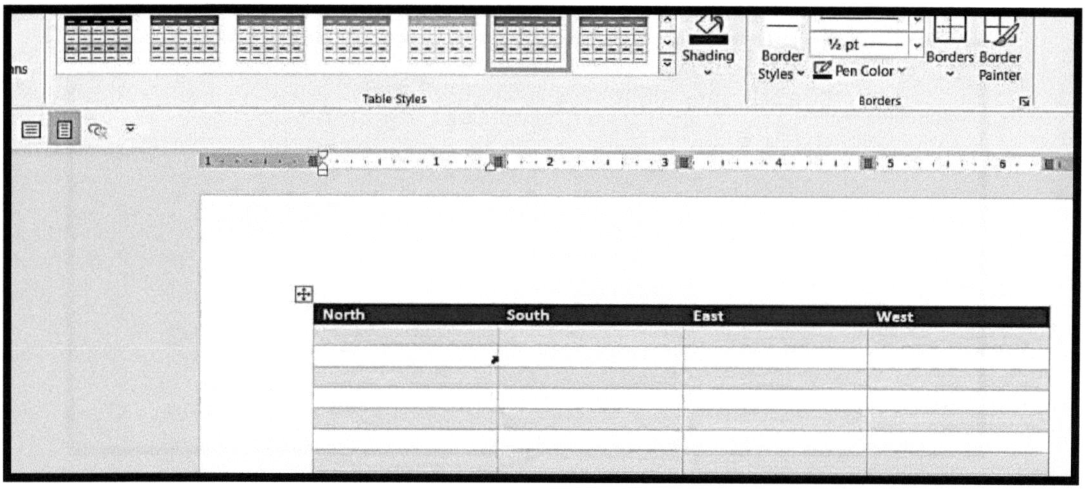

It's also worth mentioning that clicking Enter while clicked in a cell will just make that cell taller, rather than moving you down to the next row. You can either click on the cell below with your mouse or use your down arrow to move to the next row.

Aligning text in a cell

Another thing you might do to improve the appearance of your table is to align all of your text. You'll first highlight everything in this table, and then under the alignment group, you'll find a variety of options for aligning your items within the cell. If you choose "**Align Center**," all of those will be aligned in the center.

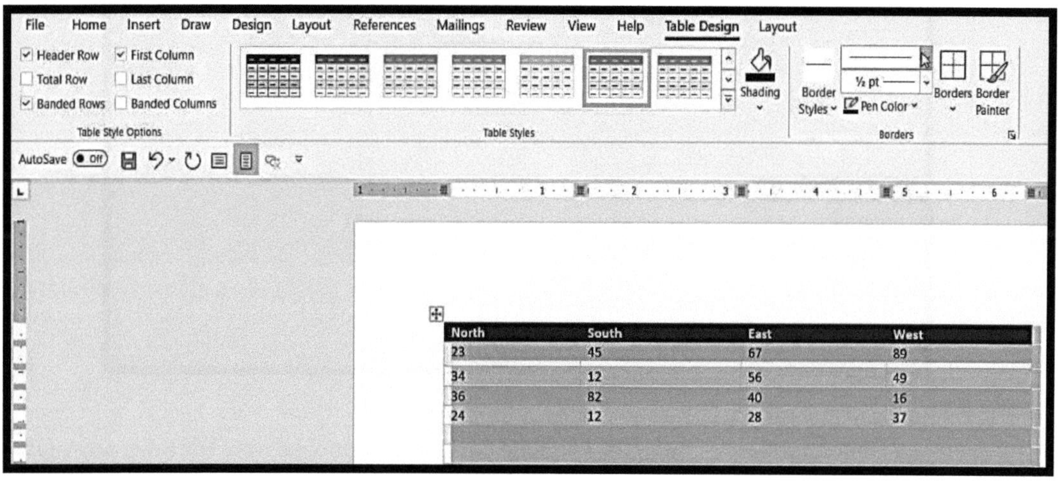

Converting tables to text

Converting tables to ordinary text is something else you might wish to do. If you wish to delete the table's borders and headings and just have them as plain text on your page, Word has a convenient little button for you. Select your table first, and then click the "**Convert to Text**" button on the **Layout** Ribbon's Data group at the bottom.

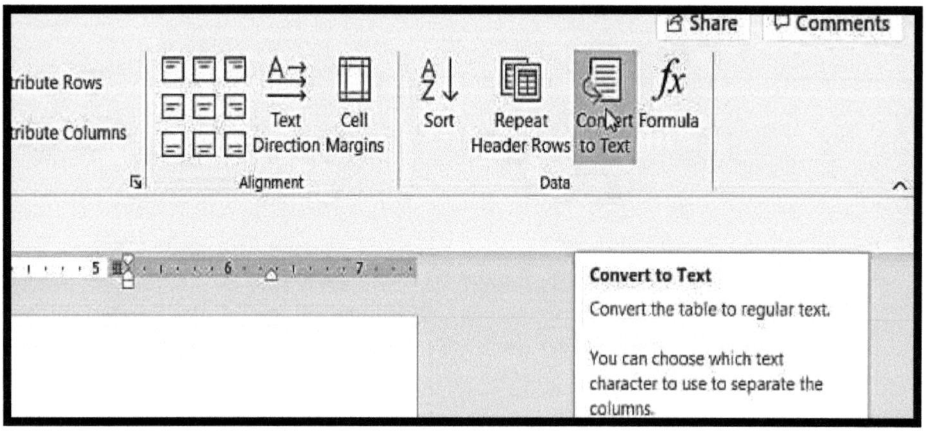

When you click that, you'll be asked how you'd like these fields separated. Because there will be no columns or rows to separate these values once it is removed from the table, Word will ask you how you want to divide the text: with paragraph marks, tabs, commas, or anything else.

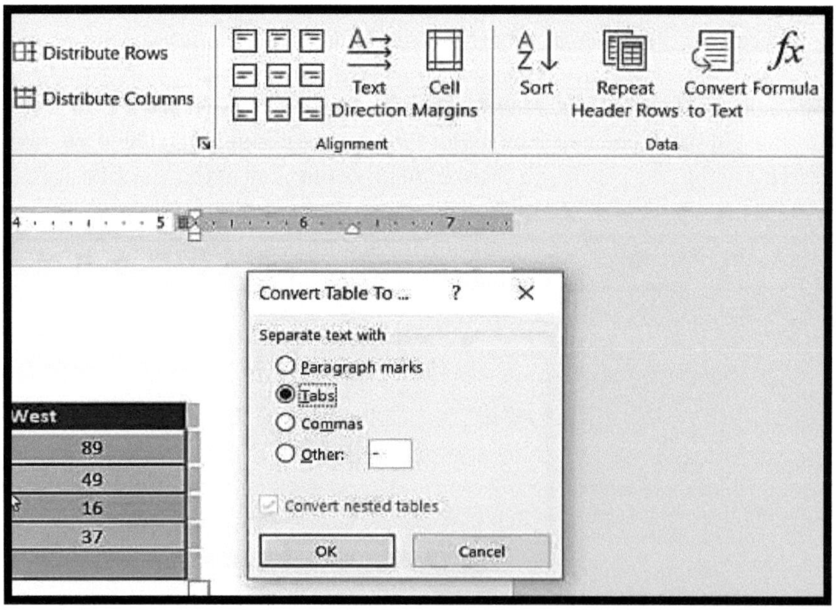

If you opt to use tabs, for example, click "**OK**," and the data will be removed from the table and separated by a tab.

You can put this data back in a table by selecting it, going to the Insert ribbon, navigating to Tables, and selecting "**Convert Text to Table**."

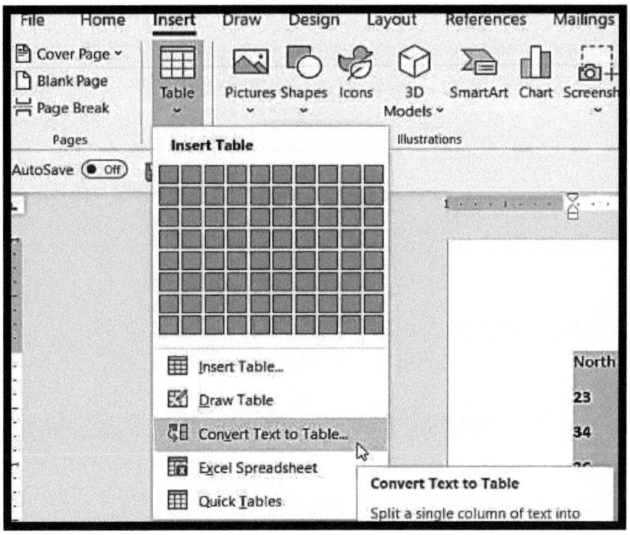

When you click "**OK**," Word will recognize the number of columns and rows in your table.

Modifying a Table

On the Table Design ribbon, you can choose from a variety of styling options. The Layout ribbon is the other ribbon we have here, and it allows you to do things like insert rows, and columns, merge cells, split cells, change the text direction, and align your text within your cells. Let's take a look at a few of these fundamentals.

Adjusting rows and columns

You can also change the width of these columns by hovering your mouse over the column border and dragging it in or out. You can do the same thing with the rows by dragging them up and down.

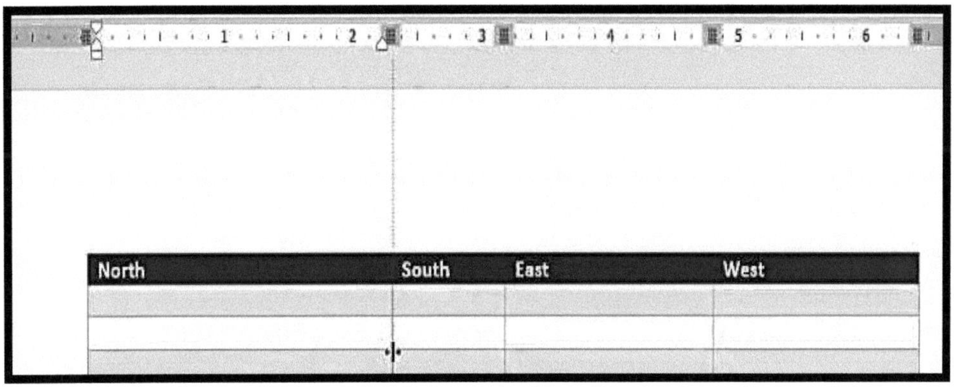

After you've entered some data into the top part of your table, there are a few formatting options available. You have several table designs options, so if your table has some alternate colors, it's because you've enabled the "**Banded Rows**" option. If you turn them off, they will be gone.

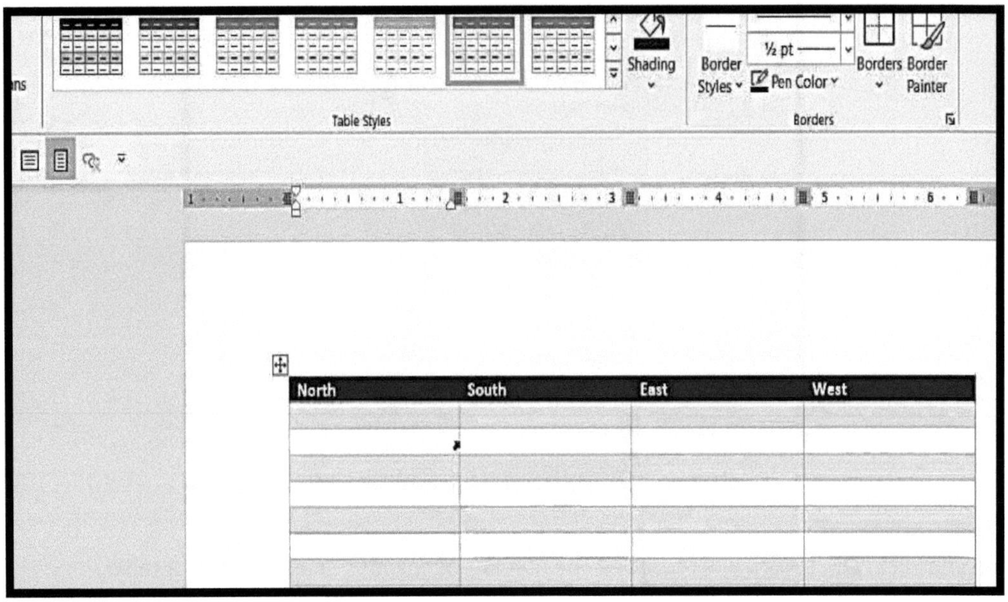

Instead, you can use "**Banded Columns.**"

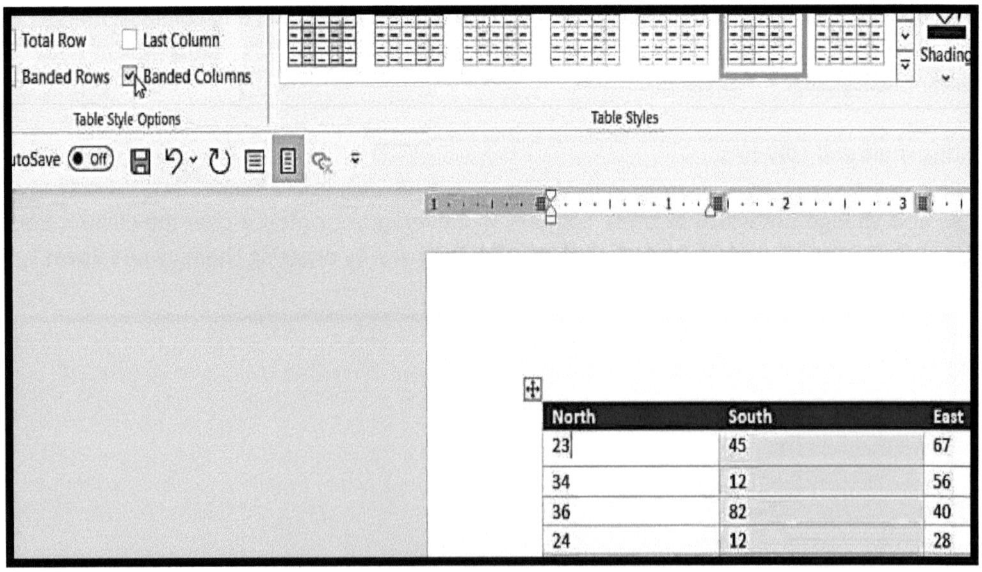

You can format the "**Last Column**," and you can see how it has bolded all of the numbers.

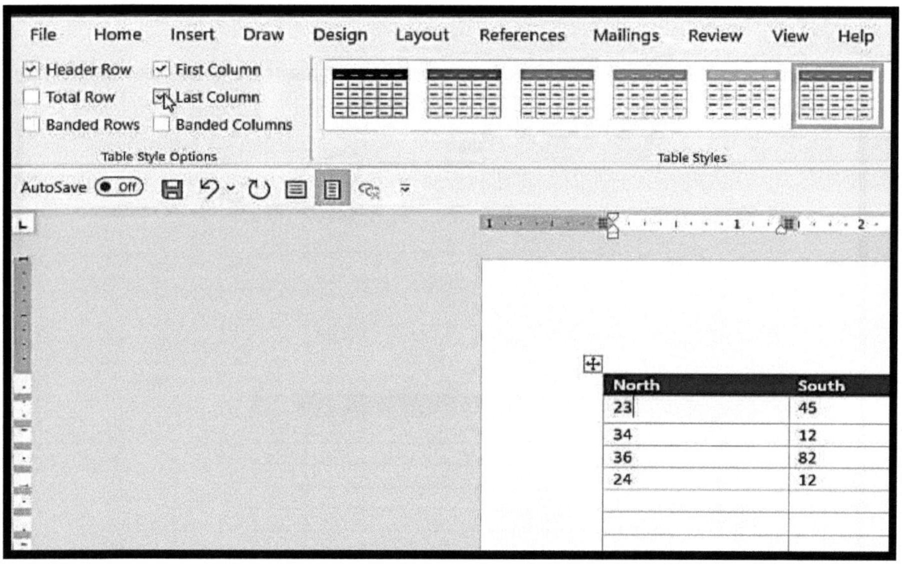

A **Total Row**, a **Header Row**, or a **First Column** could all be added.

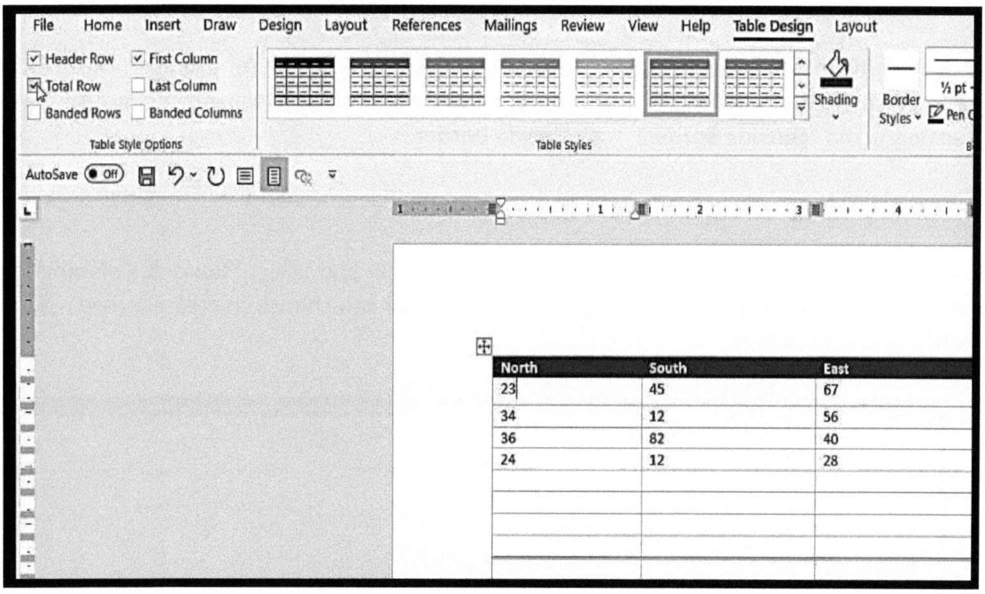

Adding Borders

You may also use borders to add some spice to your table. You may currently have extremely faint borders going through your table; however, if you want to change this, choose your table and look up the borders group.

By picking a solid line, you can give it a pretty thick boundary. You can even change the color of the border entirely.

Then, using the "**Borders**" drop-down, you can select the type of border you desire.

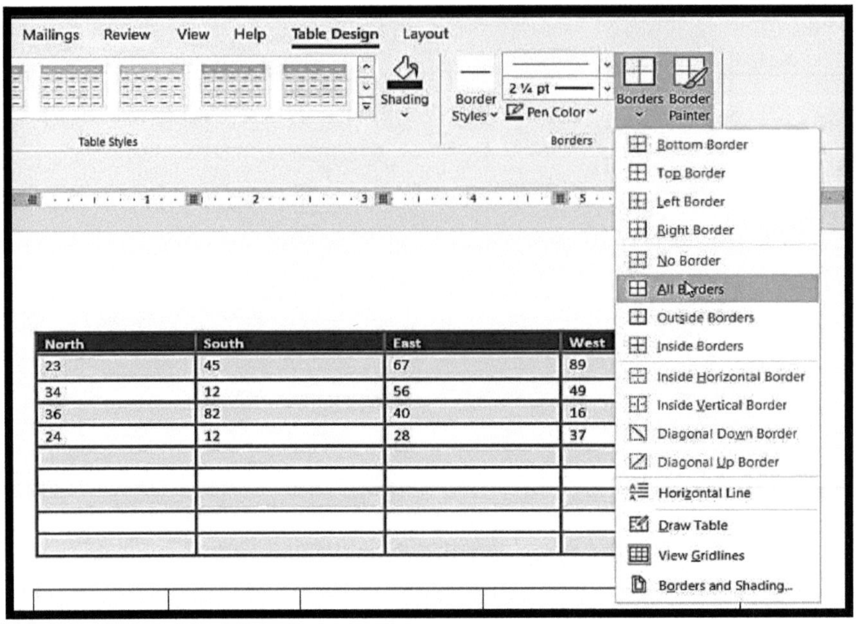

You can see how these different options apply if you hover over them; for example, the "**Left Border**" option will give you a left border. You can create a border around everything by doing "**All Borders**," or you can only do an "**Outside Border**" or an "**Inside Border**."

Adding and removing rows and columns

To insert a new row between cells, simply highlight the row and select "**Rows & Columns**" from the **Layout** ribbon's "**Rows & Columns" group**. From there, you can choose to create a new row above or below where you've clicked.

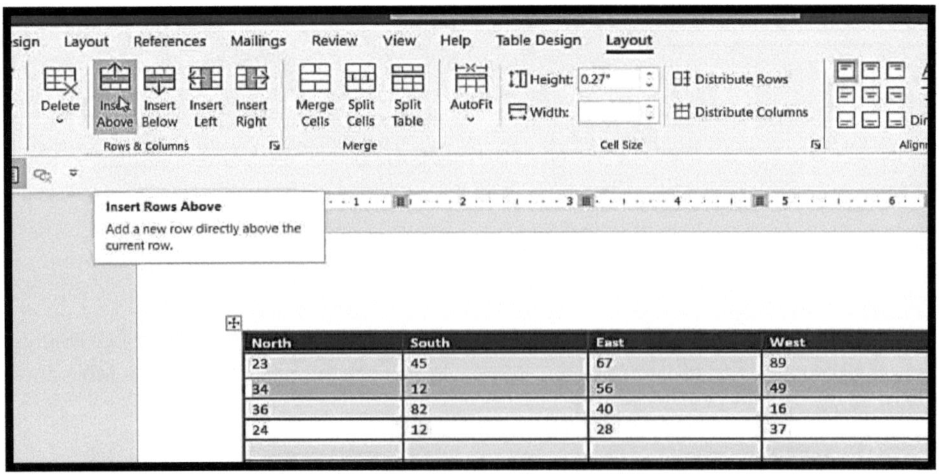

Columns work the same way, so you can highlight one and choose whether to put a blank column to the left or right.

Another easy way to add a row is to click slightly outside the table and then press the enter key. This will add a new row to the table. Similarly, if you want to remove a row or column that you've added, you can select it and then click the "**Delete**" option in the Rows and Columns group, where you can say "**Delete rows**" or "**Delete columns.**"

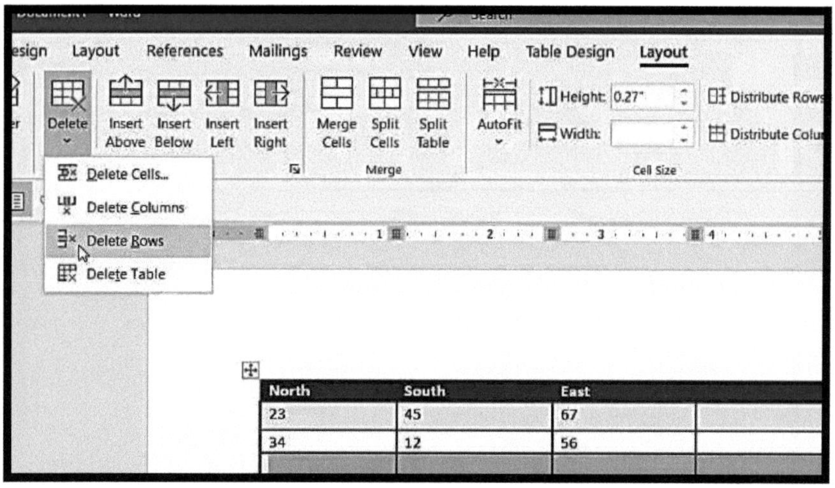

You can also delete rows or columns by right-clicking your mouse and selecting Delete Rows or Columns from the right-click menu.

Merging and Splitting Cells

Another important aspect of tables is the ability to split and merge cells. If you want to split a cell into two, you can do it by clicking within the cell, going to the Layout ribbon, and selecting "**Split Cells**" from the **Merge group**.

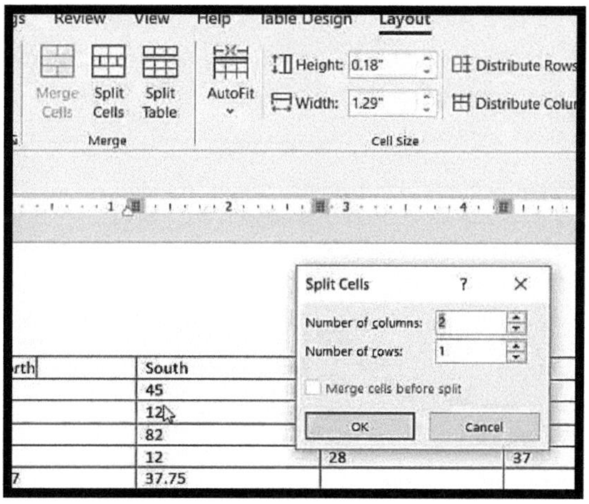

It will ask you how many columns and rows you want this cell to be divided into.

You can split it into two, three, or four cells as long as there is enough room; click "**OK**," and you'll notice that the cell has been split.

Merging columns is the inverse of that. If you want to merge these two cells back into one, click both cells, go to Layout, and pick "**Merge Cells**" from the drop-down menu.

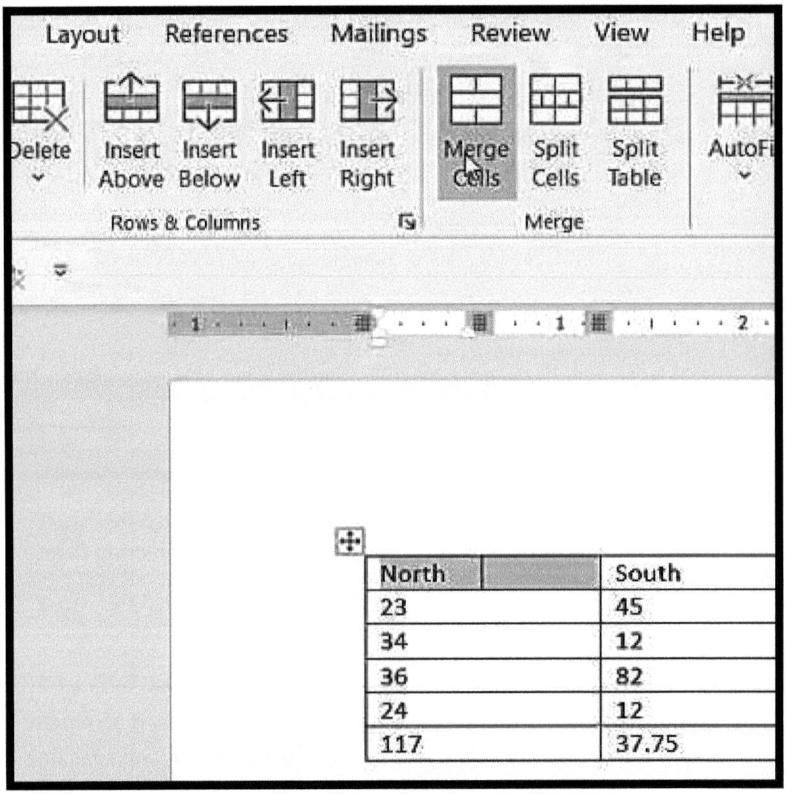

As you can see, there are a lot of options when it comes to table formatting, and there's a lot more than what you've seen in this chapter, which should pique your interest. You can add a table to a document and play about with it to see what you can do with it.

CHAPTER EIGHTEEN

BORDERS

In this chapter, we will discuss the borders and shading of a page in Windows Word.

Let's say you have a fairly simple document, such as a certificate, a notice, a poster, or something similar, to which you'd like to add a border.

Accessing Borders

To access and insert borders, go to the Design tab and then to the icon here that reads "**Page Borders**" along with the ribbon.

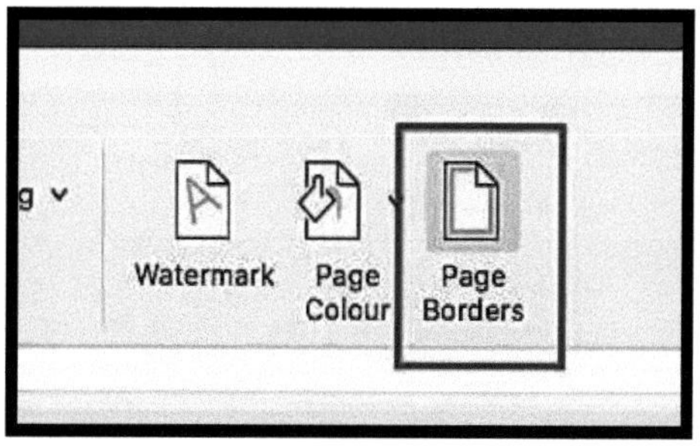

When you click on that, a dialog window will pop up. You can choose from several options on the left-hand side, including **No Borders**, a **Box border** (which is just a basic box), a **Shadowbox**, a **3D box**, or a **Custom box**.

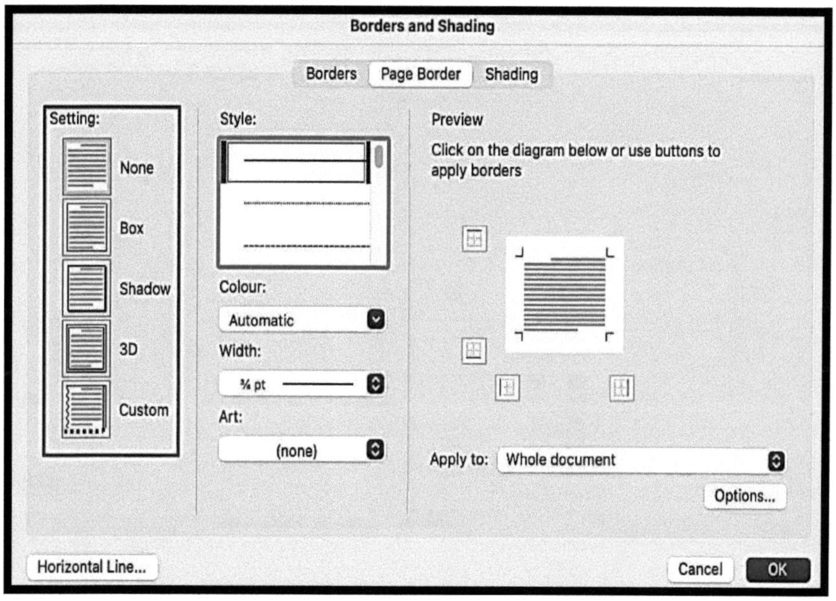

Applying a Page Border

While they are all fairly self-explanatory, let's have a look at the options to the right of the Box border. If you want to place a Box border around your page, you have a variety of styles to choose from, and if you grab the slider and drag it down, you'll see all of the different lines that you can use for your Box border.

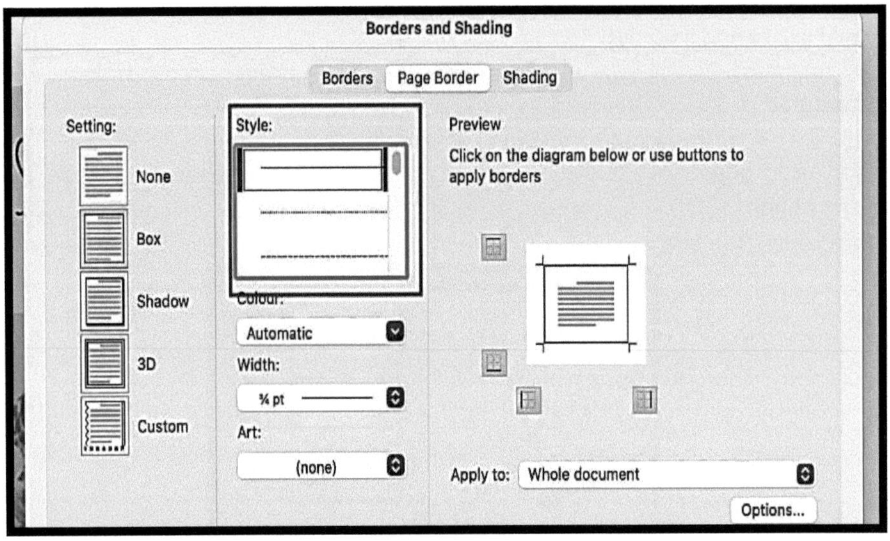

You can get that border by selecting one of those lines and clicking "**OK.**"

In addition to choosing a style, you can also choose a **color**. You can select any color from this drop-down menu by clicking on it.

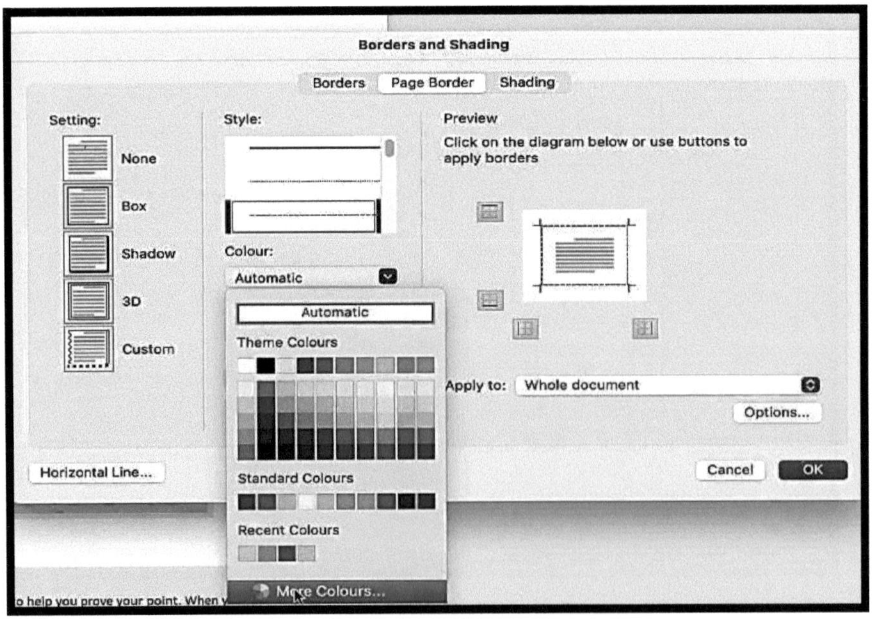

You can also go down to "**More colors**" and look at your color wheel, place your cursor anywhere on the color wheel, and move this slider to the left or right. The color of your choice will appear in this box; simply select "**OK.**"

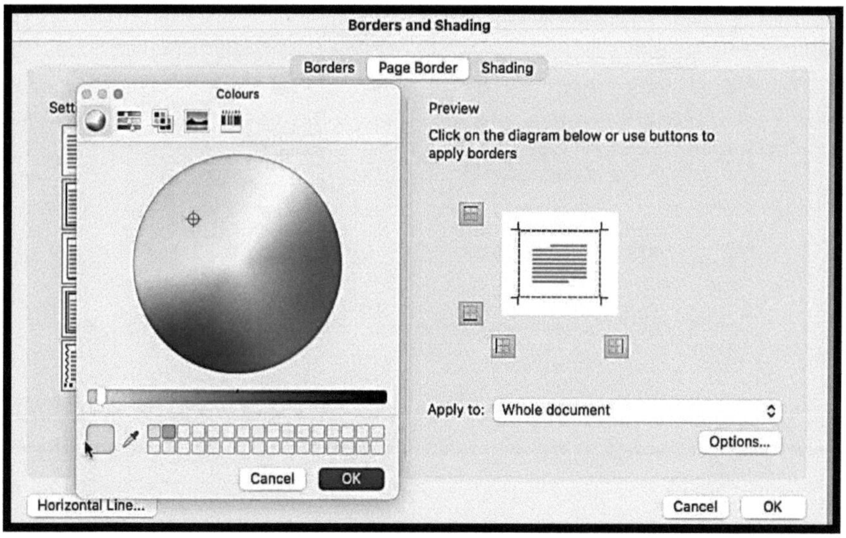

If you choose a color, let's say red, you'll notice that all of your preview boxes have turned red as well.

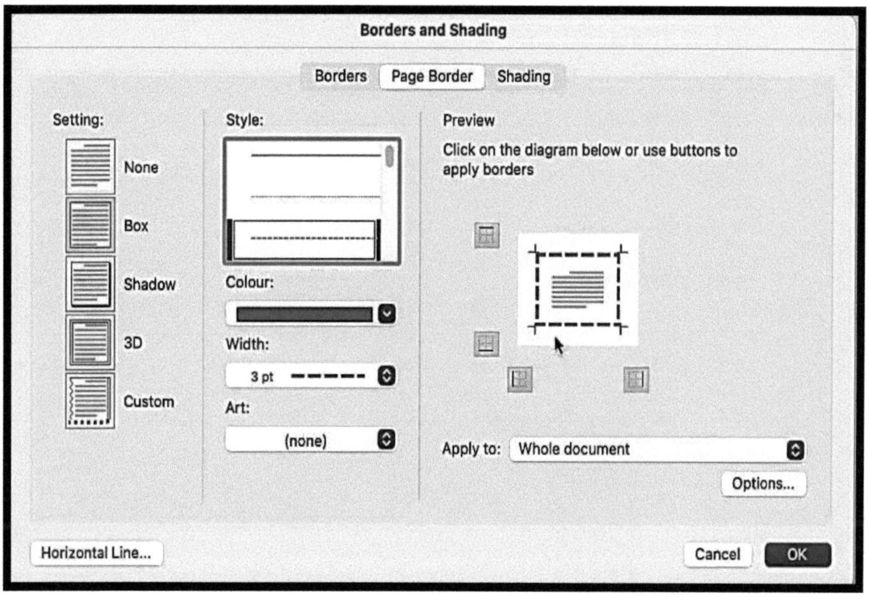

You can also customize the border's weight and width. If you click on the drop-down menu, you can choose a heavier weighted line if you wish, and you can see how it will look in the preview. Simply click "**OK**," and your thicker border will appear.

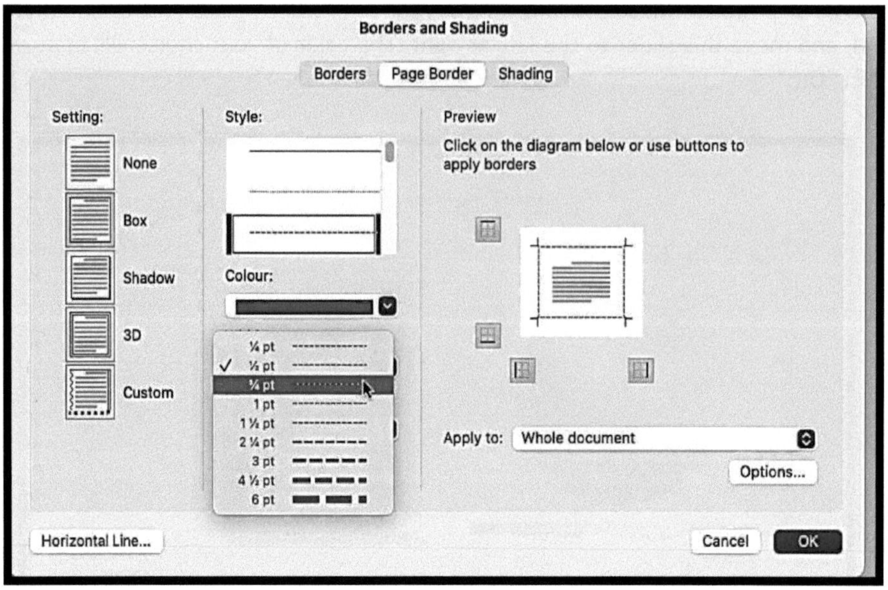

You can choose some more creative options right here at the bottom. You can scroll down the drop-down menu, and there's an arrow at the bottom that allows you to browse down and look at all the numerous options available.

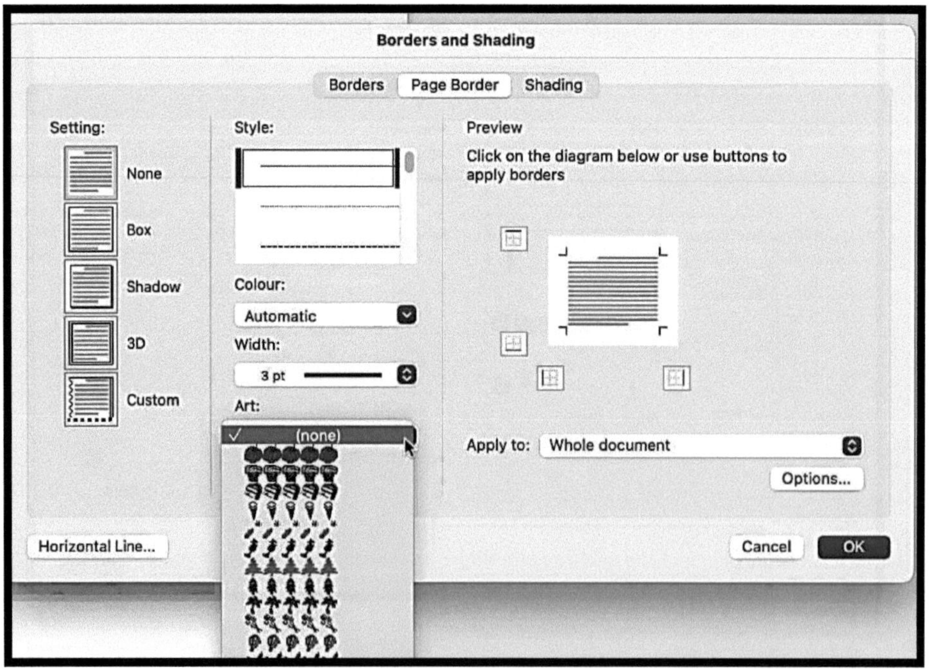

When you have colored options, such as globes, these colors are set, and the color option is grayed out, indicating that you can't change the color of this option but can modify the size.

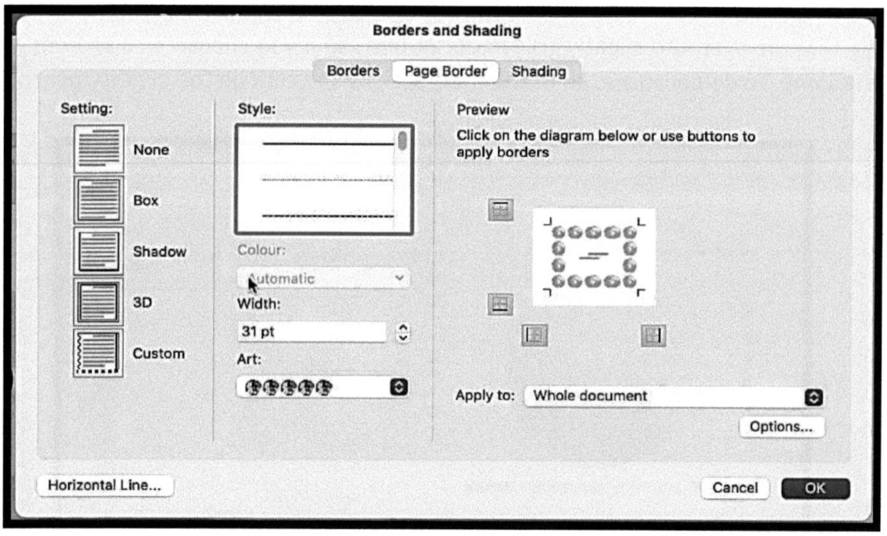

You can see from the preview that the globes change the size and become smaller if you make them smaller by clicking on the Width drop-down, and if you click "**OK**," you can see that you now have slightly smaller globes for this particular border.

You'll find some black and white options if you return to the art area and scroll down. These options are a little more flexible, and you can see that the color option is no longer grayed out when you select a design. You can now select any color by clicking and selecting it, so let's select red once more and click "**OK**." You can see that you now have the same design as before but in red.

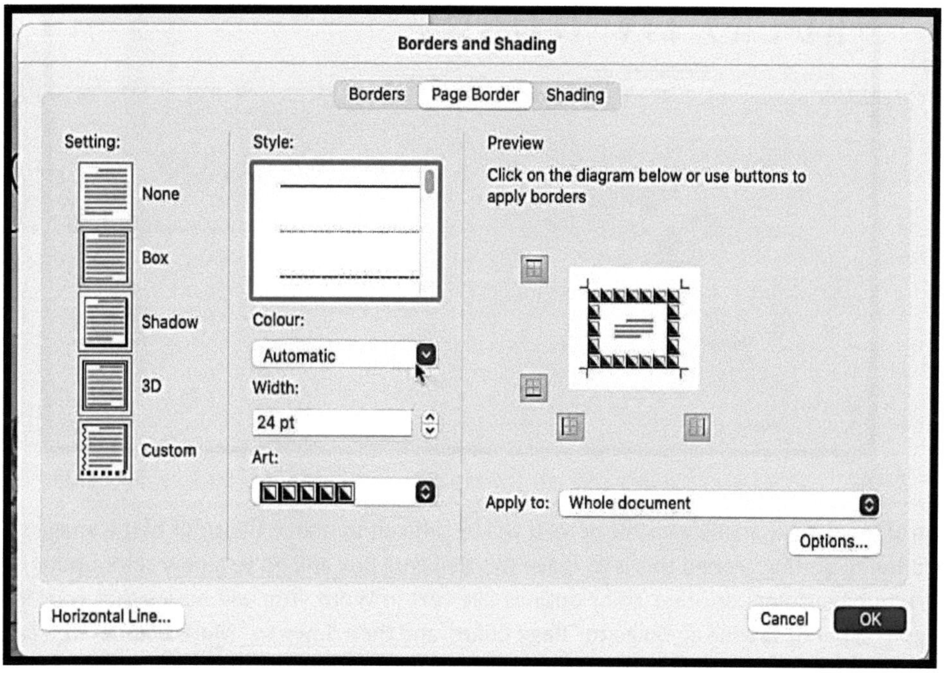

You may want your border to match a color in your text or an image. When you click on Page Border, you'll notice that the overlay is slightly grayed out or blue, so if you choose a color from your image, it won't be the same. To demonstrate, go to the color options and click on the eyedropper tool here.

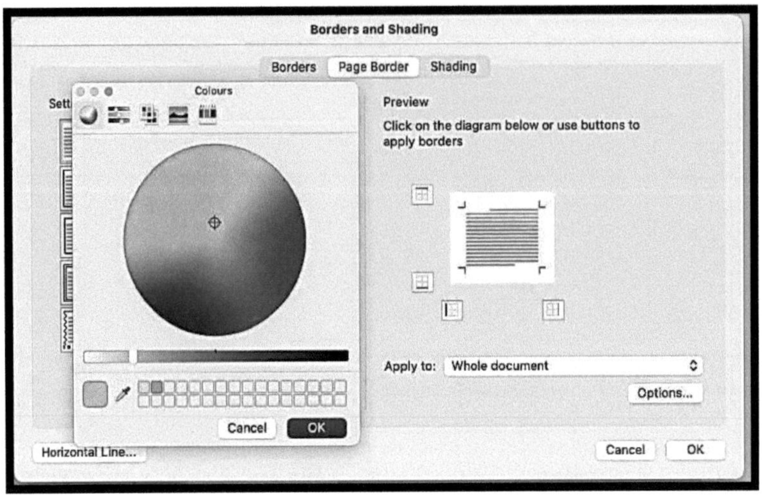

It allows you to navigate through your document and identify whatever color you want by magnifying it, and then using the middle square, you can just click on it and it will match the color in your color wheel.

Because of the overlay shading visible here, it will be difficult to match the color of the image's backdrop. The solution to getting around this is to leave this dialogue box and go to a new color option, which you can do with page colors or other color options like text in Word. You will notice that your image is no longer grayed out or shaded if you go to "**Page Color**" and then down to "**More Colors**." You can now use the eyedropper tool to select, then go to the backdrop and select that color.

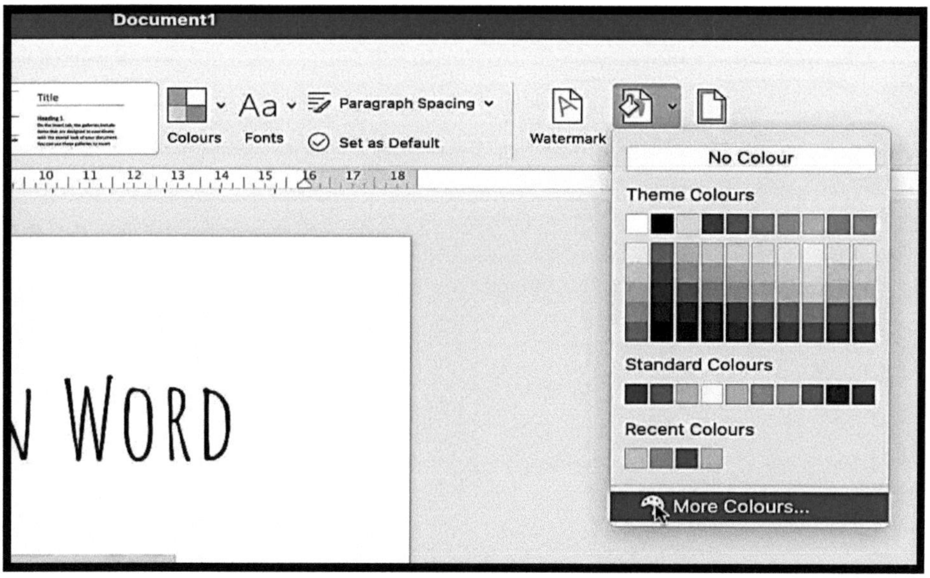

You can see that it's been identified in this window, and then simply click "**OK.**"

Although the page background has changed, you can simply return to "**Page Color**" and choose "**No color**." When you go to Page Borders and then to "**Color**," you'll notice that your color has shown in recent colors.

All you have to do now is select the desired option, and then scroll down to the artwork you've chosen; you can see how it appears in the preview. You can now apply it to the "**Whole document**," but if you don't want borders around your page, for example, if you only want to remove the right-hand border, you can use the icon shown below.

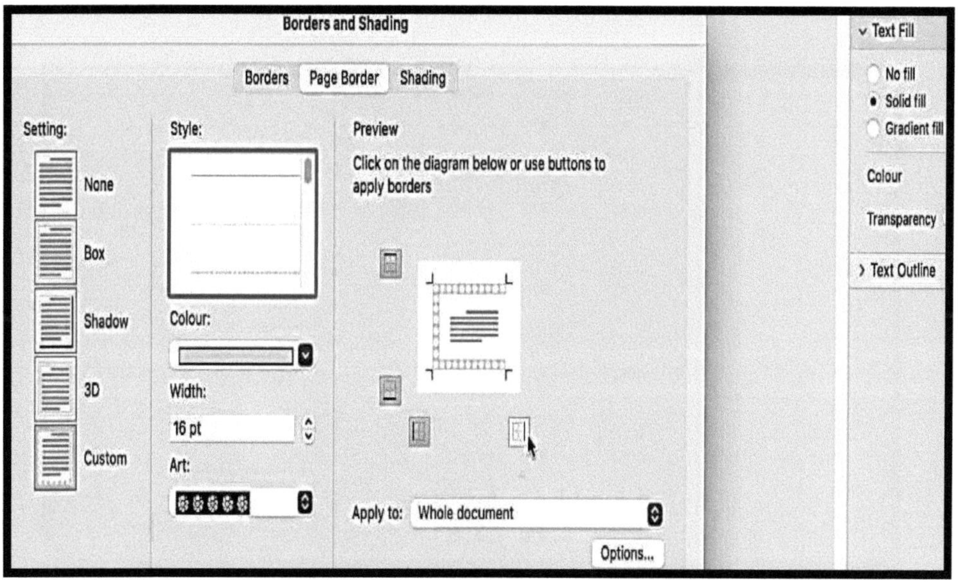

It will send the border across your page, not just to where it would connect up with the next border, and if you click "**OK**," you can see how it goes all the way to the page's edge, not just to where it would link up with the vertical border there.

If you don't want that, return to Page Borders, select the option you want, then click "OK," and your page borders will appear.

An Advanced Technique

Also, if you want to improve or extend the borders in some way and have a few more options, there is a more complex procedure. Go to the "**Shapes**" on the **Insert** tab, and then to the square here.

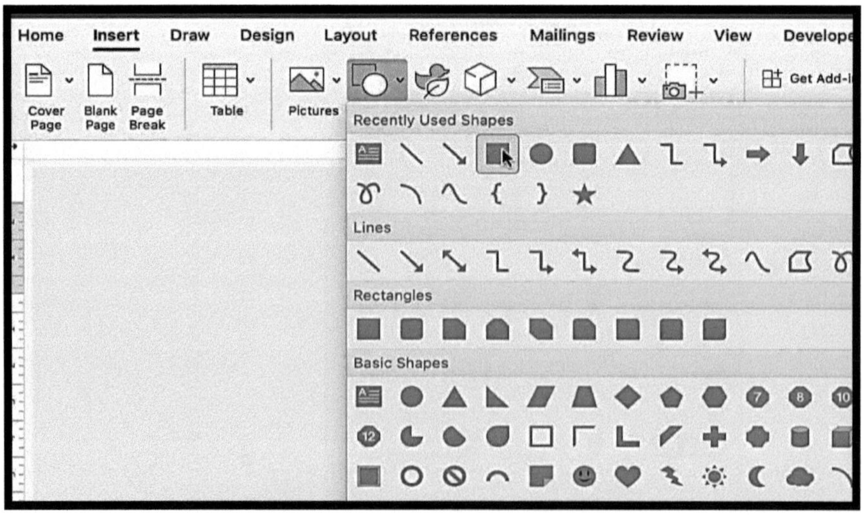

Click and drag it after you've clicked on it. Attempt to make this form slightly larger than your borders. You can do this with your cursor, but if it gets a little awkward and doesn't quite fit, say the bottom bits are a little too broad in comparison to the top, you can select it, go to "**Shape Format**," and then to the **Height and Width section**.

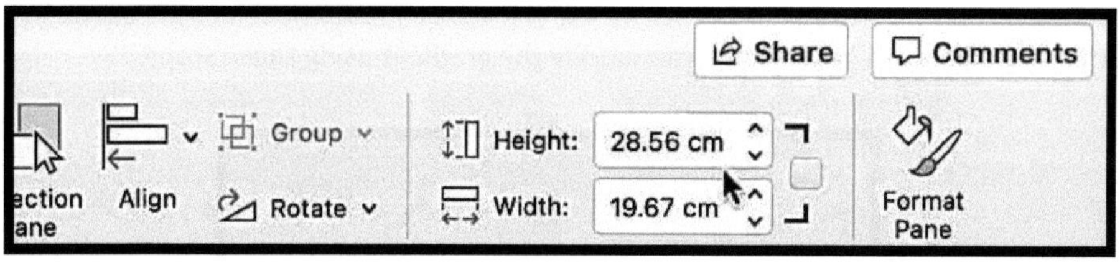

Because you want to reduce the height of it slightly, you can move over and use the arrow keys to select the down arrow to reduce it, and you can see how it will be equal to the top with one click. Once you've done that, you'll notice that all of your text and everything is hidden, and all you have to do now is make sure it's selected, go to Shape Format, and select the "**Send Backwards**" option.

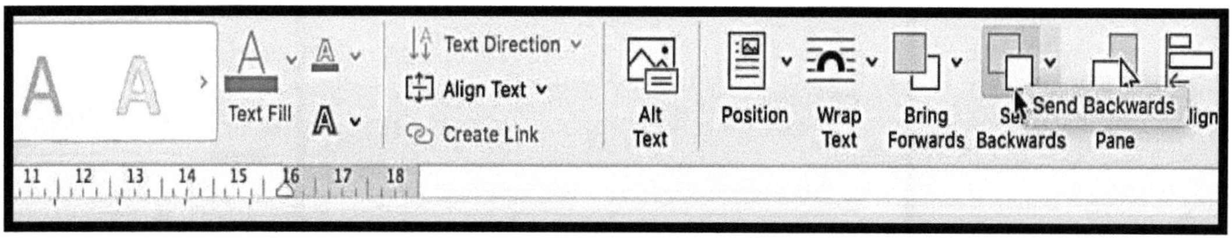

You can click once to see your picture, then again to see the text, which is gradually moving this shape backward while bringing all of the other information forward.

You can double-click on the shape or move up to **Shape Format** and along to the Format Pane button here if you merely want to apply it as a border.

This will open the Format Shape menu, where you can choose from a variety of alternatives under "**Fill**": You can either fill the existing color or remove it and just fill it with a transparent fill.

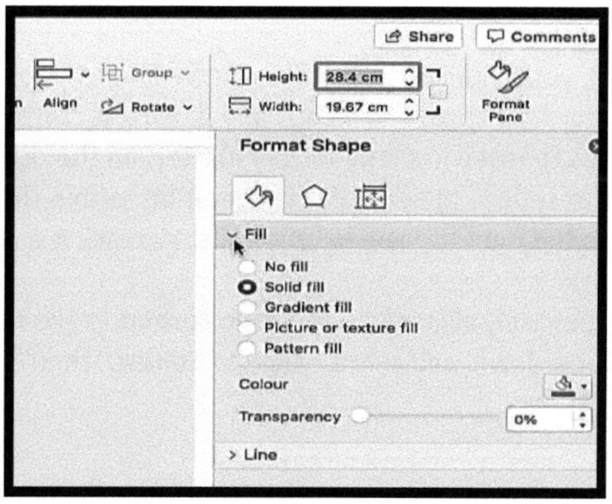

You may then make the line thicker or alter the color by going down to "**Line**" and clicking on "**Solid line**."

If you wish to change the color, go to the drop-down menu and select Recent Colors, then select the desired color. As you can see, you now have an additional border.

You can also use this to extend your border by filling it with a gradient. Select the shape again, then go to "**Fill**," then "**Gradient fill**," where you can modify the gradient using these stops.

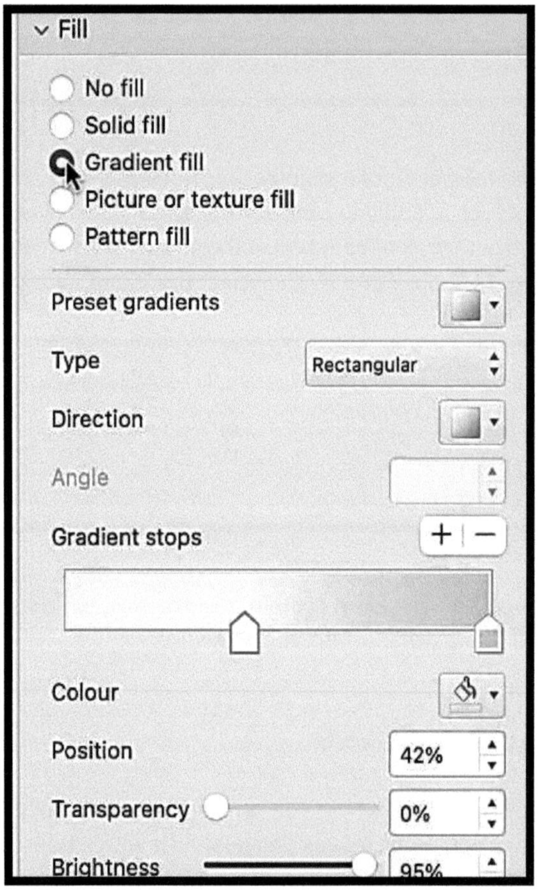

Once you've completed that, you'll notice that it looks wonderful on the outside, but the border patterns here are fairly close to the border's edge. If you copy and paste this current shape that you've placed in, you can see it's copied across to make it a little thicker like it is on the right-hand side. You must send it backward, and then you must reduce the size of it so that it fits within the border arts, and if it's a little clunky, you can go up and to the right to the height and just increase it one stop.

To make sure everything is precisely aligned, go to **Shape Format**, make sure the shape is selected, then go to "**Align**," click on the drop-down, and select "**Align to Center**," then "**Align to Middle**."

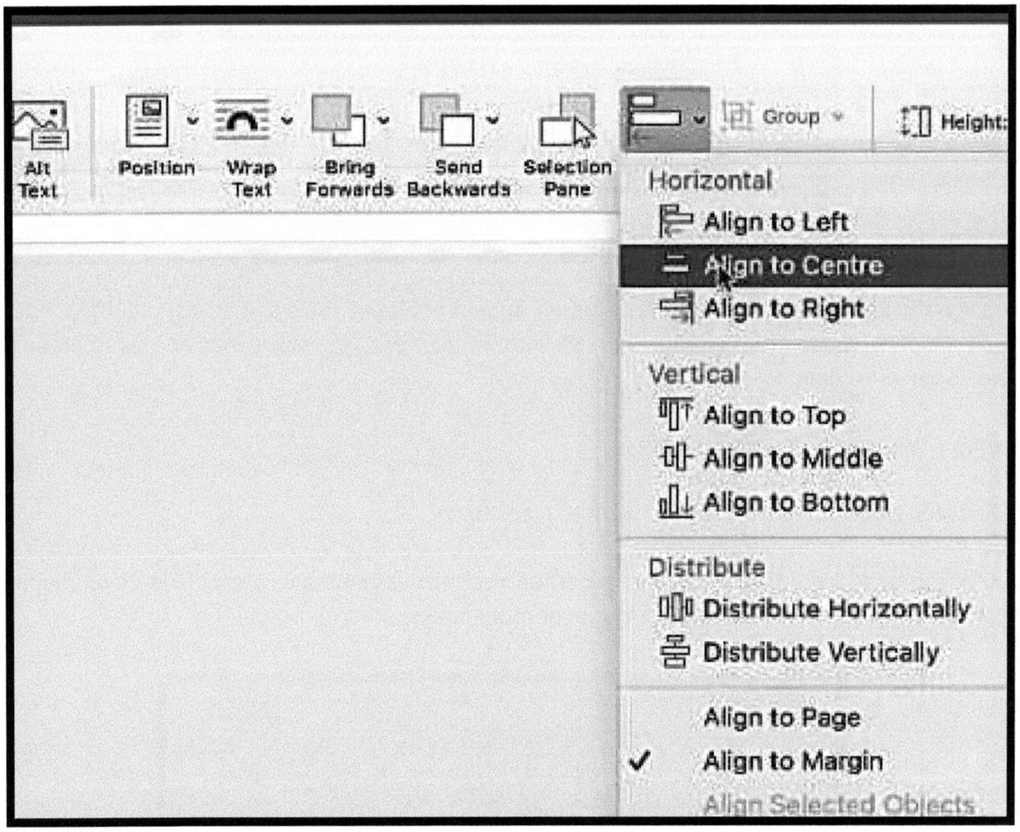

This will place this shape in the center of your document, and you can do the same thing with the outside border: click on the outside border, go up to "**Align**," **align to the center**, and **align to the middle**. That means all of your borders are now precisely aligned, and you can modify the gradients if you wish.

So, if you don't want the gradient, go across and select "**No fill**," and you'll just have a thicker line on the side of the border arts, or you can go back and select "**Gradient fill**," and then adjust the sliders accordingly.

Remember, in your Gradient fill, you can modify all of these colors. You don't have to use white for your Gradient fill; you can use whatever color you like, and you can add markers if you want. Simply select a different hue by clicking on the plus symbol. You can also experiment with all of these colors to achieve the precise effect you want.

Also keep in mind that the Border can be applied to a paragraph or a text, while the Page Border can be applied to the whole document, a section, the first page only, or all except the first page. If you are writing any documentation and you want the page border to be there from the second page onwards, or after the contents page, you can go with the last option.

CHAPTER NINETEEN

COLUMN OF TEXTS

Welcome to this chapter, where we'll look at word processors in Microsoft Word and discuss Columns and how you may make your text look like columns.

What you should know

When you think of a newspaper, the text doesn't appear like what you'd see on a standard Word page; instead, it's divided into little columns that run across the page. Let's have a look at how to make your content appear in columns.

Creating a Two- or Three-Column Text

The first step is to choose the text you want to transform.

Columns can be found under the **Layout** tab. Then you can click on the columns drop-down menu, specifying whether you want one, two, three, or more columns.

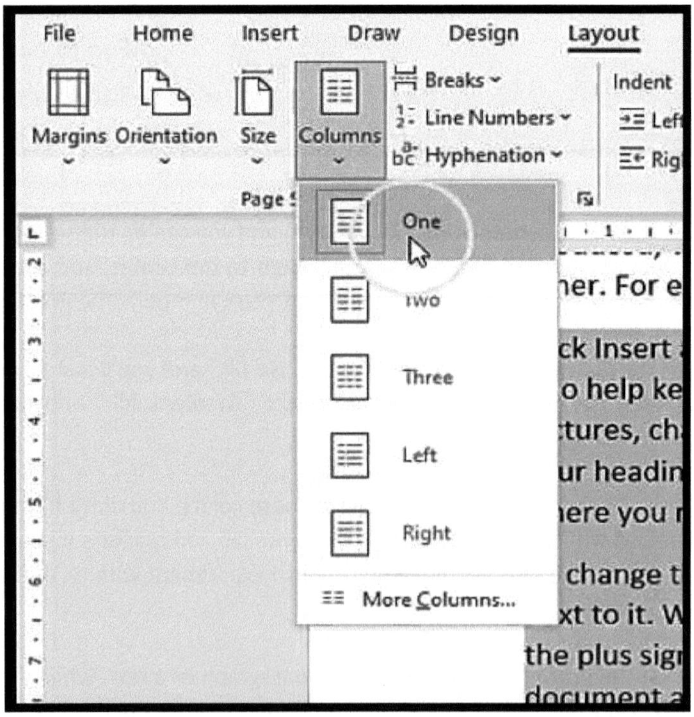

Let's use the three-column option for this illustration, and when you click on it, it will appear in your document.

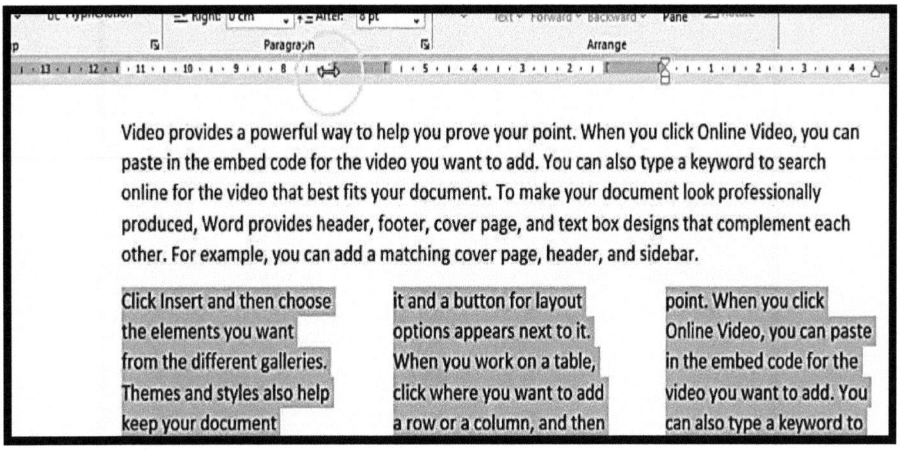

You'll note on the ruler that you can change the width of each one or the margins between them all if you wish.

That is one method to go about it. The other option is to go to "**More Columns**" and see what else is there.

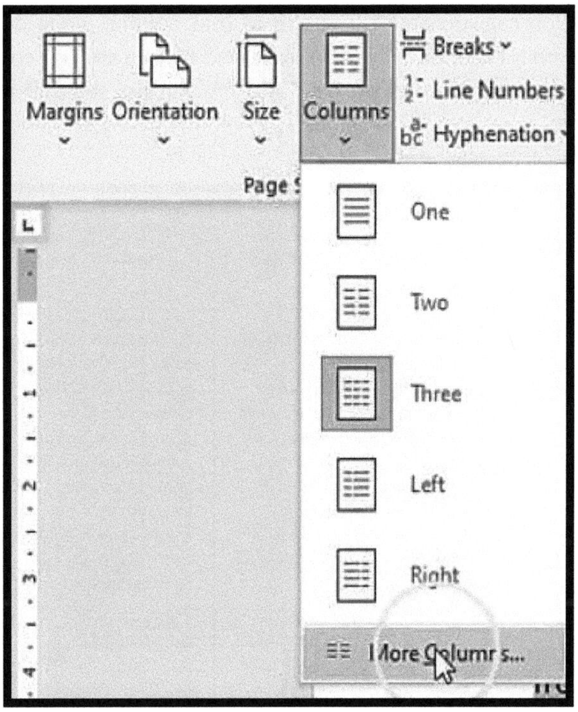

You can specify the number of columns you want and even draw a line between them, so if you only want to draw a line between them to separate them, you can do so by selecting that option. If you like, you can do it with just the selected text or the entire document.

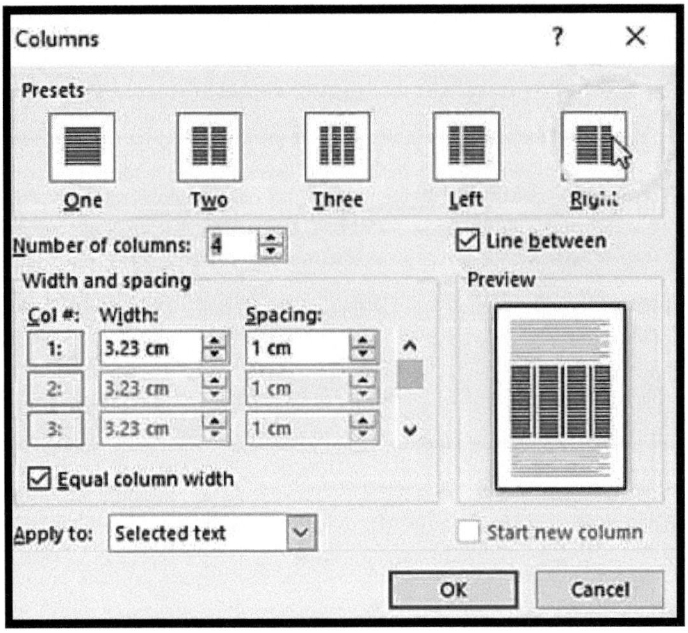

You may also make changes to these columns from here. Let's say you want the first column to be a little broader than the rest. You can increase the width of that column while decreasing the width of the others. So, the first column is exceptionally wide, followed by two smaller columns. You will see the line in the middle of the two columns if you click "**OK**," and you will also see that the first one is broader than the other two.

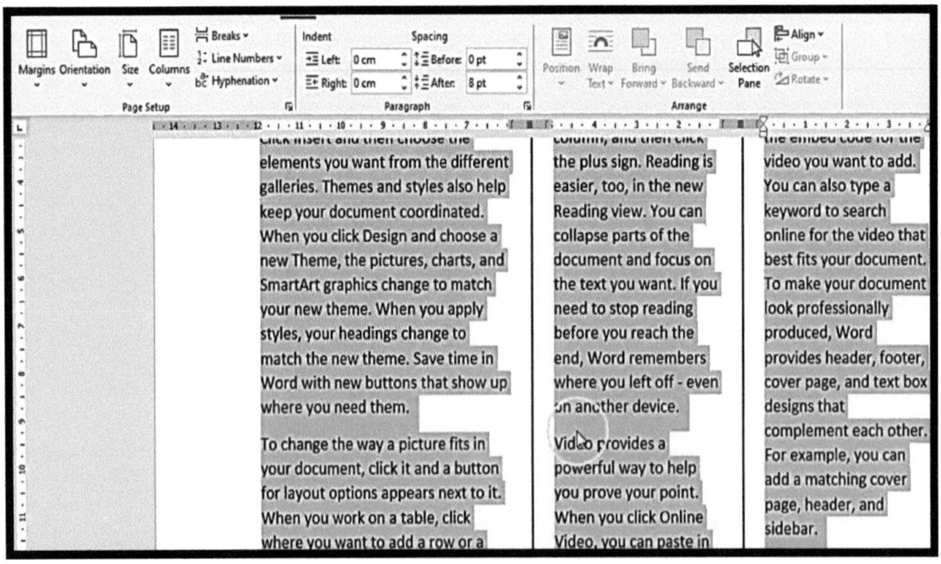

If you want to make it equal again, go back to "**More Columns**," make them all the **same width and spacing**, and it'll do the math for you automatically. You can also divide them into two columns if you choose.

Creating a four-column document

What if you want it to be divided into four columns? I'm going to demonstrate one small thing you can do with columns. You go to "More Columns" and change the "**Number of columns**" option to 4 to make it into four columns.

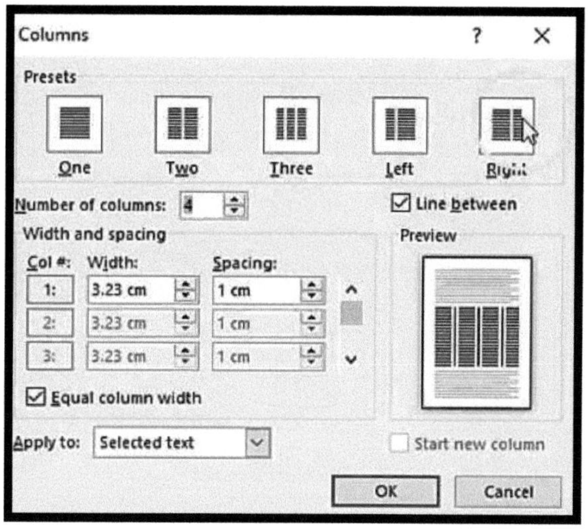

You'll see that there's also a "**Left**" and "**Right**" option, with a very thin first one on the left and a very thick first one on the right; you won't be able to use these options if you're doing 4-column.

A column on the left or right

Finally, let's take a short look at the Left and Right alternatives. If you go to the Columns drop-down and select "**Left**," you'll see that there's only a left column, but the rest of the text is still there.

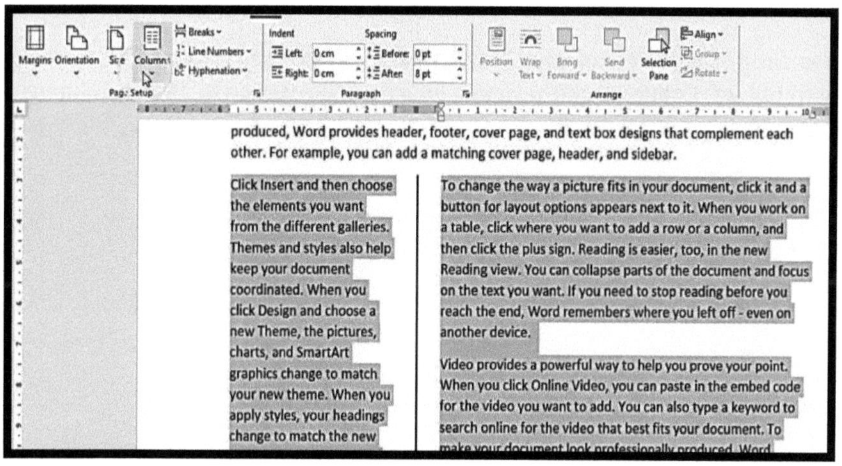

When you do it as a "**Right**," you'll get a large first column and everything on the right side.

Column Trick

Suppose you want a specific text to appear at the top of a column rather than in the middle. **Breaks** will be used to accomplish this, and they will be inserted at that specific point. Select the text, then go to "**Breaks**," under Page Breaks, and select "**Column**."

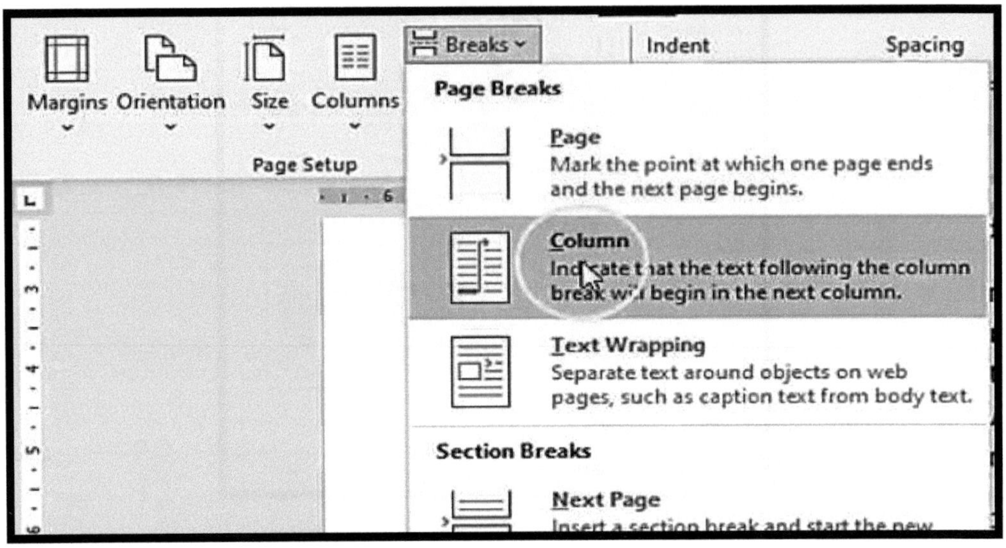

This shifts the text to the top of the page and moves everything down. You'll need to use a column break if you want to add additional text to the top of each column. That way, you can describe exactly how the columns should look, what should be at the top of each column, and how they should be brought up so that if you add additional text over here, it will not move up to the top since you already have what you want at the top.

A Quick Rundown

This concludes our discussion of columns. Remember to select the text you wish to convert to columns before going to **Layout**, **Columns**, and selecting your option or going to "**More Columns**" and using the options provided to you.

CHAPTER TWENTY

DOCUMENT GRAPHICS

In this chapter, we'll talk about inserting images from files or the internet into existing documents, as well as making the text flow properly around those pictures. We'll also go over the many options for modifying and editing any images in your document.

Adding Images

You can add pictures to a document in a few different ways. You can insert a picture saved to your hard drive or local drives, your desktop, or your pictures folder, into a document in one of two ways. The first is if you have a picture saved to your hard drive or local drives, and you've saved it to your desktop or your pictures folder, and the second is if you want to search for a picture available online and insert it directly into your word document.

From Your Local Drive

Let's take a look at the first option, which is inserting a locally saved image. It's worth mentioning that the photo will insert where you presently have your cursor in your document, so keep that in mind. Go to the **Insert** ribbon by clicking at the top of your page or the spot where you want the picture to be inserted, and you'll see "**Pictures**" and "**Online Pictures**" options in the Illustrations group.

- If you have an image stored locally or in a file, select the "**Pictures**" option. If you click on it, your local drives will open up, and you can navigate through them to find the location where your images are saved.

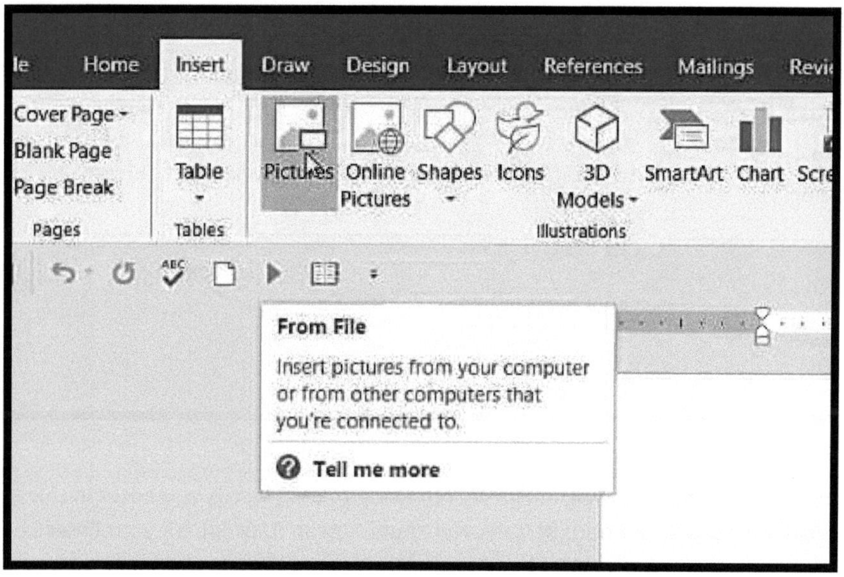

- You can now select it and then click "**Insert**," and your image will appear in your document.

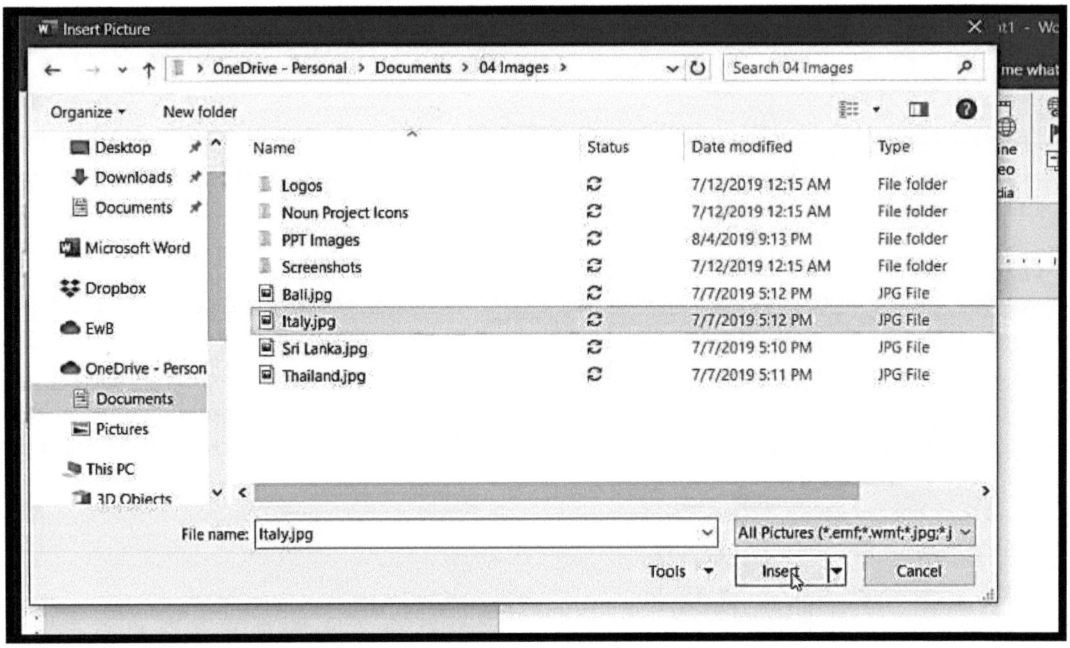

Inserting an Image from the Internet

Let's have a look at the second option, which is to insert an image from the internet. Return to your "**Insert**" ribbon and select the second option here, "**Online Pictures**."

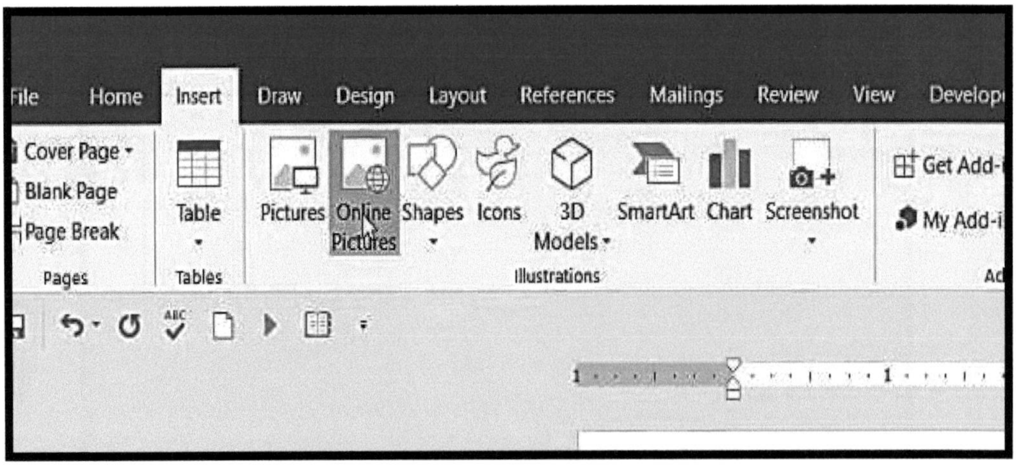

This will allow you to conduct a Bing image search using the keywords you enter in this field at the top. So, if you wanted to look for a photo of Italy, you could type in Italy and hit your Enter key, and it would look for pictures of Italy.

One thing to note is the "**Creative Commons only**" checkbox, which has a small tick in it. You have to be careful with photos because of copyright issues; you can't just go to Google and look for an image to use in your document, presentation, or whatever else you want to use it for. You should double-check that you have the authorization to use that image. So, what you're looking for in your photos is images with solely a Creative Commons License, which means you can use them without worrying about copyright issues. You can now select a photo you like and click "**Insert**," and your image will appear in your document.

You can simply drag those handles in if you want to, and you'll notice that with these photos from the internet, you get a photo credit under it, but it also has a license, so you're okay to go.

There's a lot more you can do with these images, which we'll get into later in this chapter.

Inserting Images into Existing documents

Now let's look at how to add images to a document that already has text in it. You saw how to insert a few images earlier, but it was only into a blank document. What's the difference if the document already has text? Let's see what happens.

You already have your document and want to include an image anywhere within it. The first thing to remember is to place your cursor where you want to put the image and to select a photo from your local storage. So you'll go to the **Insert** menu, select "**Pictures**," and then navigate to your image. Select your image, then click "**Insert**" to place it in your document.

You can **resize** this. If you want it to fit on the page above, make it small but not too small.

Editing Images

When working with images, you'll notice a "**Picture Tools**" contextual ribbon appear, and you should be aware of what these contextual ribbons are for.

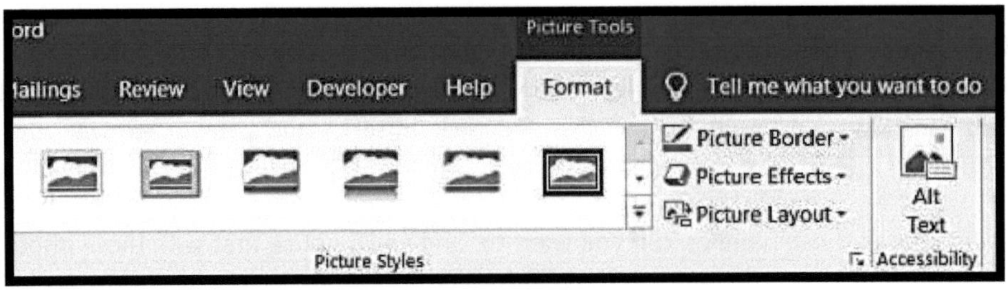

They only appear when you need them, so after you've clicked on your image, you'll see this Picture Tools ribbon with the Format sub ribbon, which contains all of the options you'll need to edit this image in your document. We'll go over some of these options right now.

Resizing an Image

If you wish to resize your image, look around the outside in the corners and in the middle for the resize controls or resize knobs. To resize the image, simply grab one of these controls and drag your image in. Now, if you hover over any of these at the top or bottom, you'll notice that they become a little diagonal double-headed arrow, which signifies that dragging them in diagonally will make them smaller. You can also utilize the side ones, although dragging them in would distort your image.

Image Alignment

If you wish to center this image on the page, go to your **home** ribbon and use the **Paragraph** group's alignment options. Simply choose "**Center**" to place the image in the center of the page.

Changing the position of an Image

You can also move this image from its present location to another in the document. Simply **click on your image**, **drag it up or down**, and **drop** it right **where your cursor is**, then let go of the mouse.

Changing the appearance of an image

For this, we'll start with the large group in the middle that reads "**Picture Styles**," and if you click the drop-down menu, you'll find that there are a lot of various styles you can use to change the look and feel of your image. You can get an idea of what they look like if you hover over them.

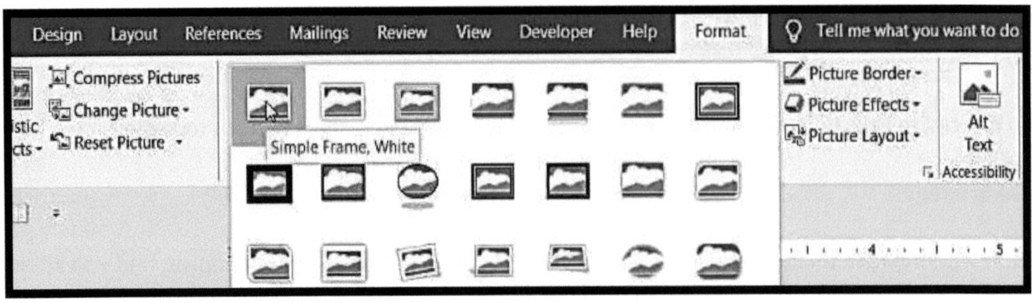

We have some **simple white frames** at the top, which create a great look.

There's the **rounded rectangle**, which also has a reflection, creating a pretty wonderful effect.

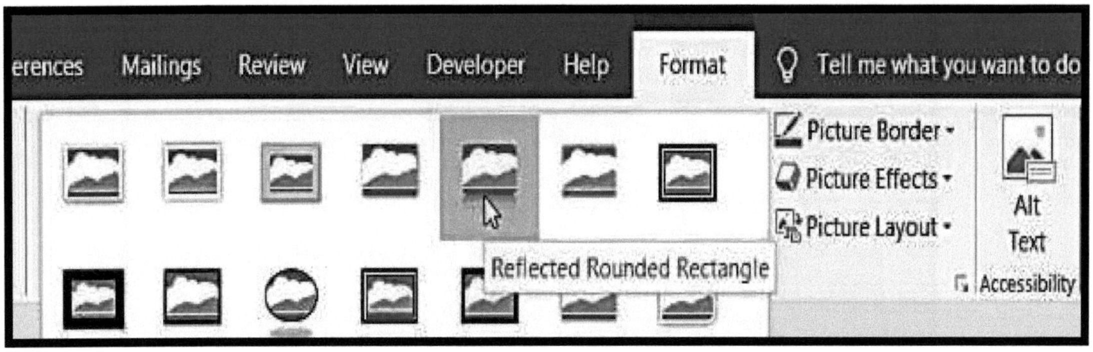

We have some **soft edges** and **Shadows**, and you can adjust things like the **Perspective** as well, tilt the image back, and modify the shape; there are a lot of things you can do here. Just keep in mind that you have a variety of picture styles to choose from.

Modifying an Image

Next to that Picture style, we have three options: "**Picture Border**, **Picture Effects**, and **Picture Layout**.

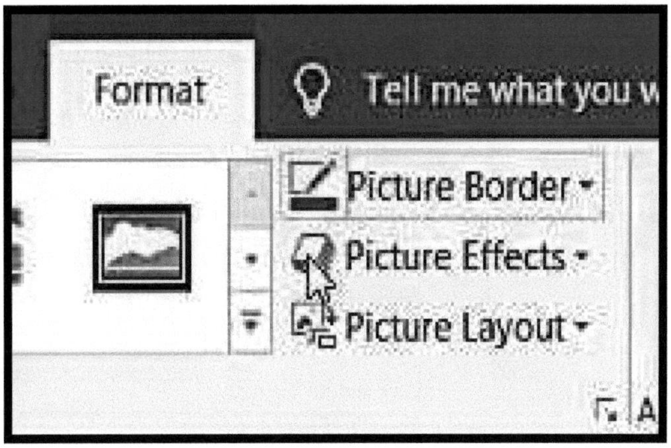

Adding and Removing an Image Border

If you simply want a plain border, **Picture Border** is the way to go. You can choose "black" from the palette and modify the border's weight to make it easier to see. As you'll see, the outside of the page now has a great plain black border.

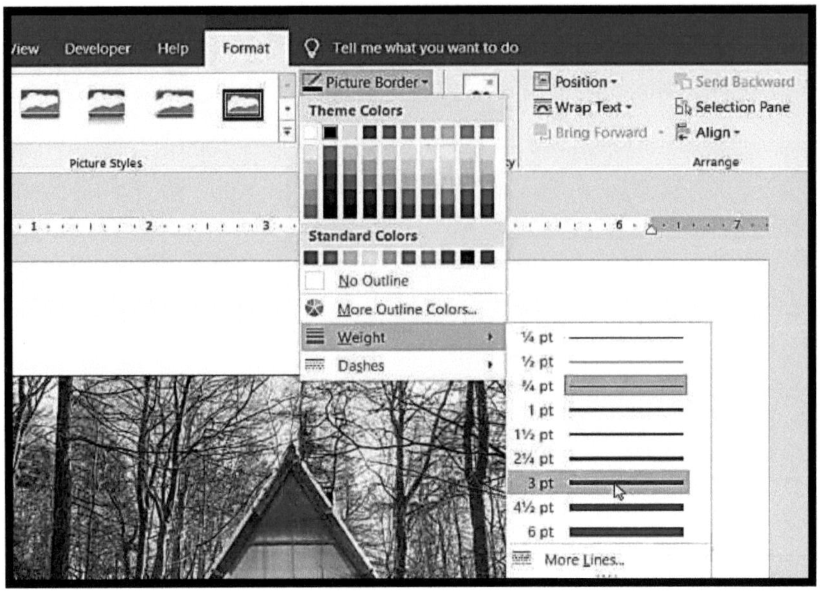

To eliminate your border, click "**Picture Border**" then "**No Outline.**"

Changing the Image's Effects

You can use Picture Effects to add a **shadow**. Some of them are faint and difficult to notice, but you should now notice a small shadow across the bottom of the image. Inner shadows and Perspective shadows are also available, and some of these can be pretty dramatic. Just be mindful that you have shadow options.

Reflections come in a variety of forms. You may have noticed it in one of the image styles; they can be pretty effective, and there are a variety of methods to add a reflection.

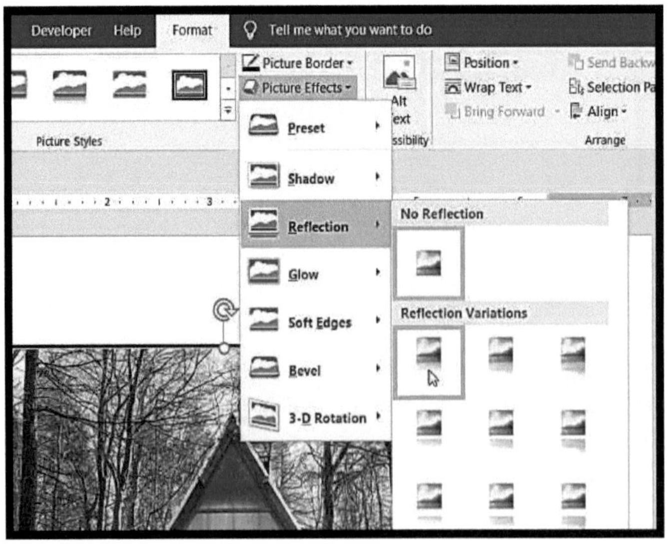

You can add a **Glow** to the outside if you wish, and you can choose from a variety of colors for your glow down here.

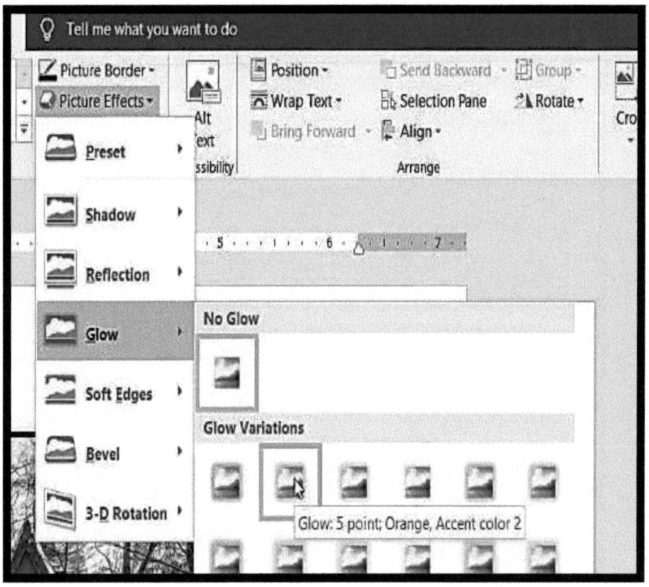

You could wish to add some **soft borders** to your image; however, some are more dramatic than others.

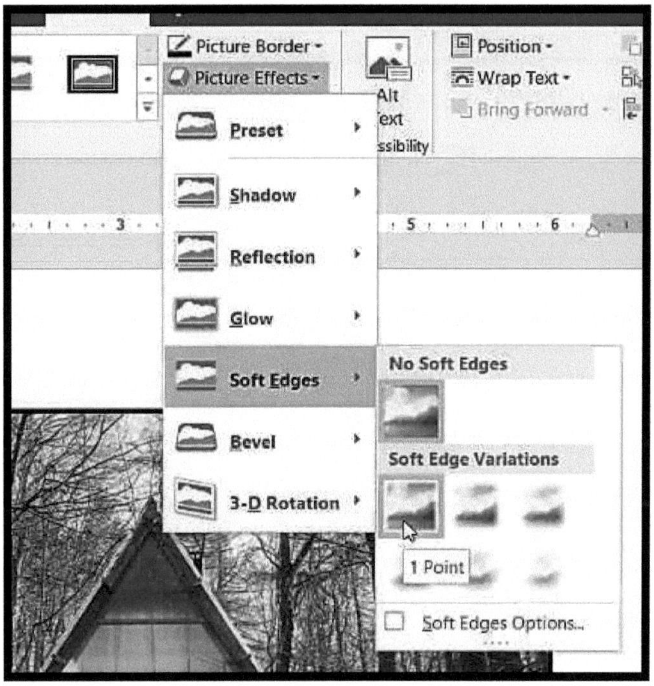

You can also give your image a **Bevel** effect.

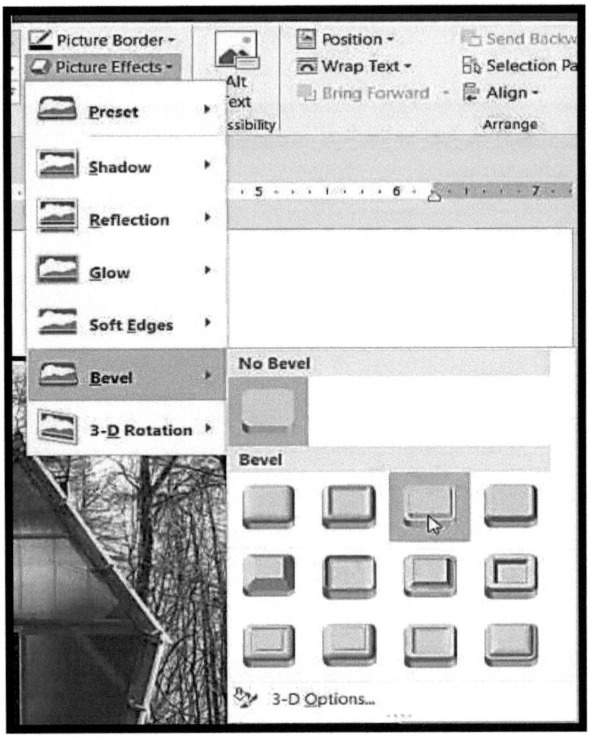

Finally, we have a **three-dimensional rotation**.

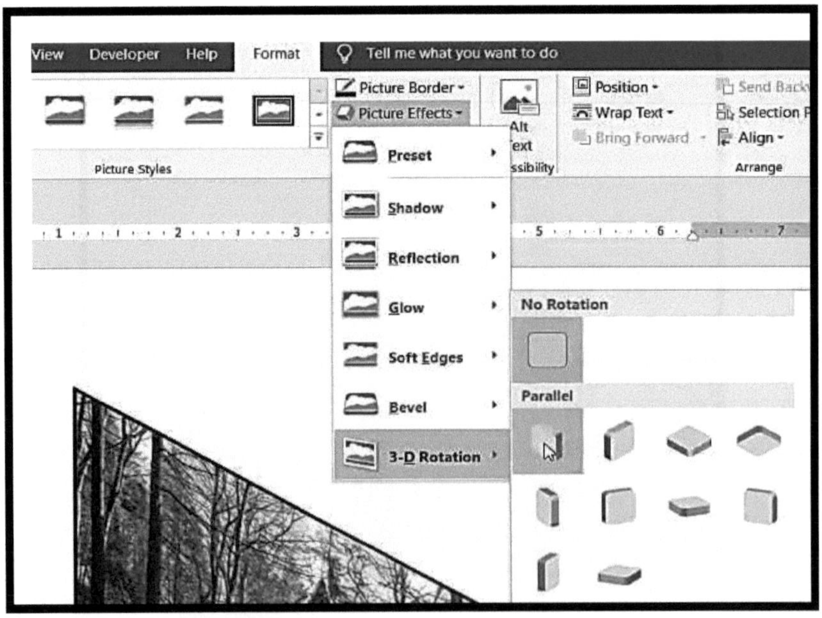

You can rotate your picture in a variety of ways if you want to, and it can be pretty attractive in some cases, so it's worth going into Picture Effects and playing around with those options.

Image layout

Then there's the **Picture Layout**. This could be useful if you want to make something that looks great with some writing over the top, and we have a variety of layouts here that you can hover over to see the many ways you can show a photo with text. That's a good one to use if you want the transparency to go over the top of a document.

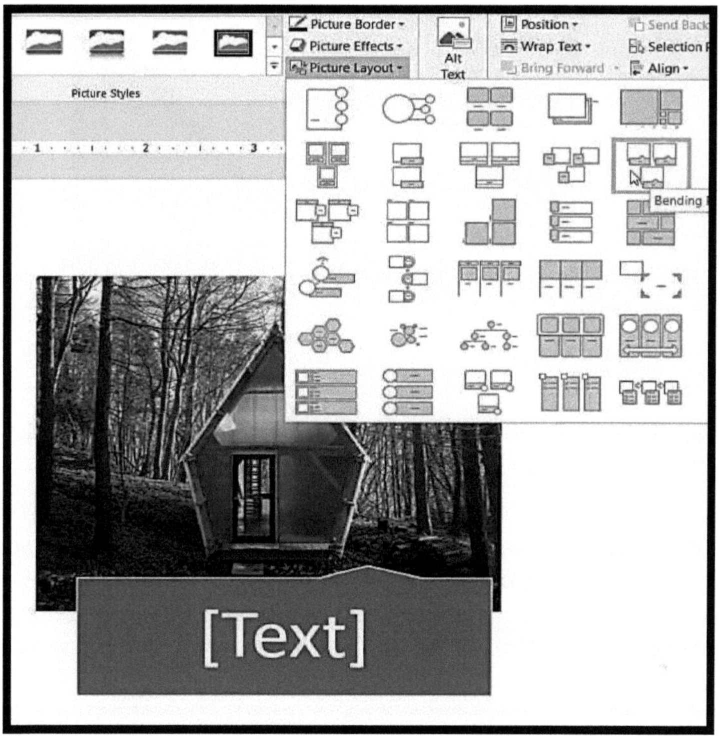

We've previously looked at a number of the other options on this toolbar in prior sections. You can change the position, wrap the text, align the image, and manually edit the height and width of the image, so rather than dragging these corners in, you might put in a precise height and width to resize this image.

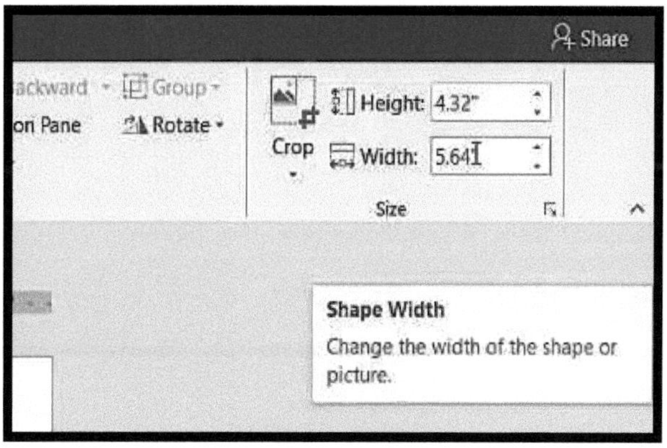

Wrapping Text around an Image

If you resize an image a little smaller, you can see that the text is above and below it, leaving white space on the side.

Now, you might just want to center that image and ignore the text above and below, but you should be aware that there are various text wrapping options available. You'll notice a small square with the words "**Layout options**" when you mouse over it. This allows you to customize how your object interacts with the words it's surrounded by. You can either click there or go up to your contextual ribbon, "**Picture Tools**" and the **Format** sub-tab and select "**Wrap Text**" from there.

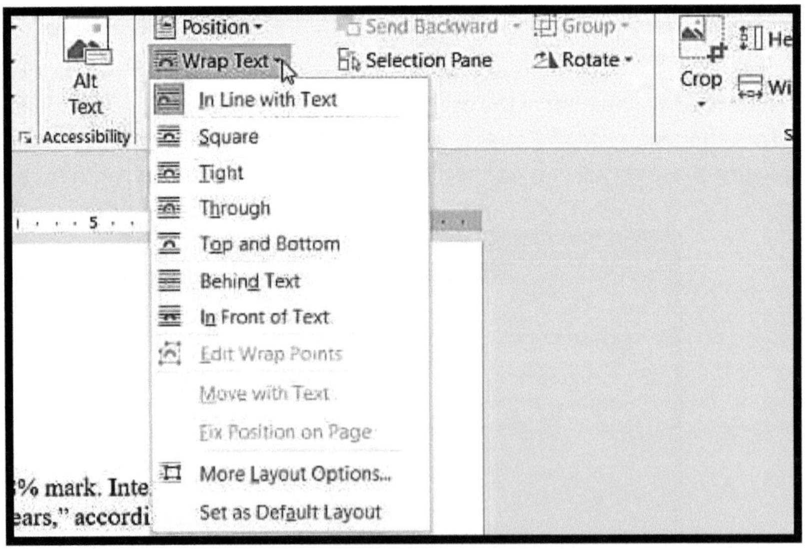

You'll see that the options you see are nearly identical to the Layout Options.

- You may now have "**In Line with Text**" chosen, which implies that the text will be placed above and below the picture. You do, however, have other options here: you have "**Square**," which you can see if you hover over it; with this one, the text flows around the outside of the picture.

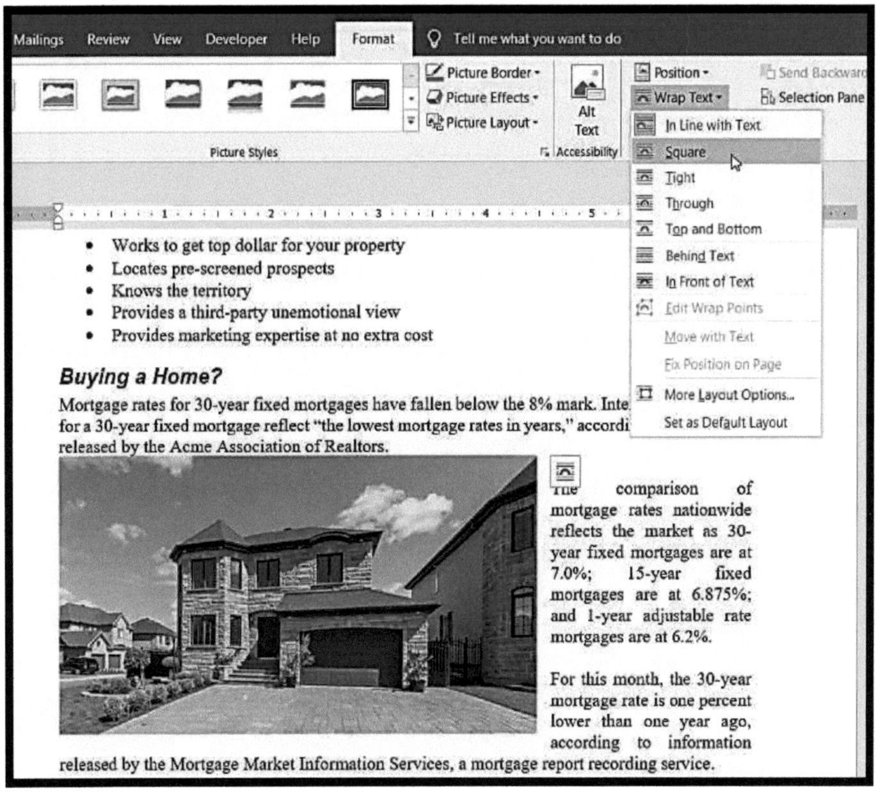

- If you select "**Tight**," the text will be a little closer to the picture edge or a little tighter to the picture.

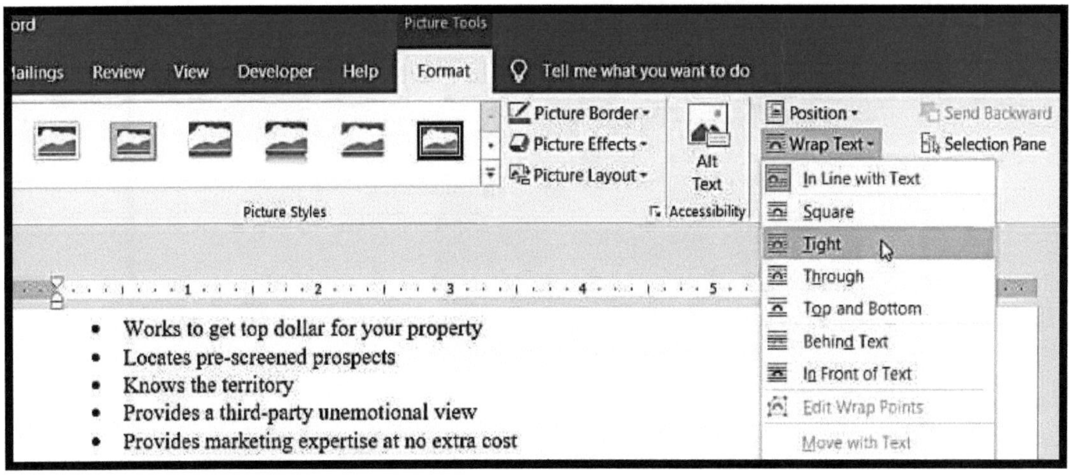

- You also have the option of going "**Through**." If you select this, the text will run through the image if you move the photo slightly to the right. Although this makes it difficult to read, there may be times when you wish to use the "**Through**" text wrapping.

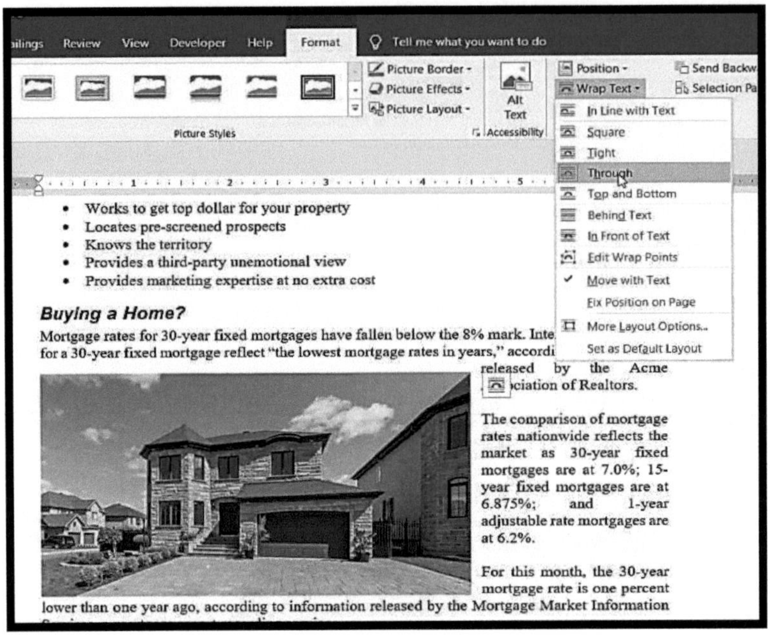

- There's also "**Top and Bottom**," which positions your text at the top or bottom of the image.

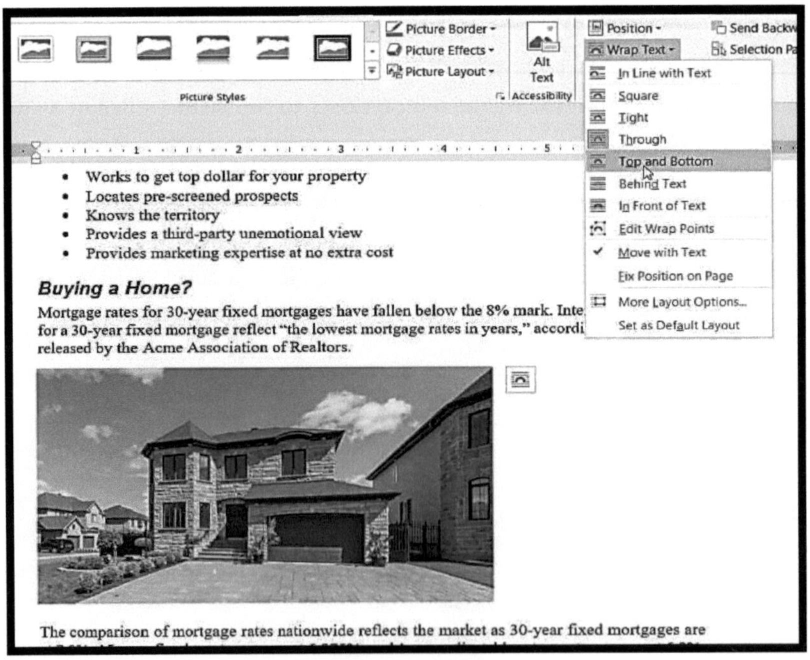

- We have a feature called "**Behind Text**," and if you enable it, the picture will appear behind the text.

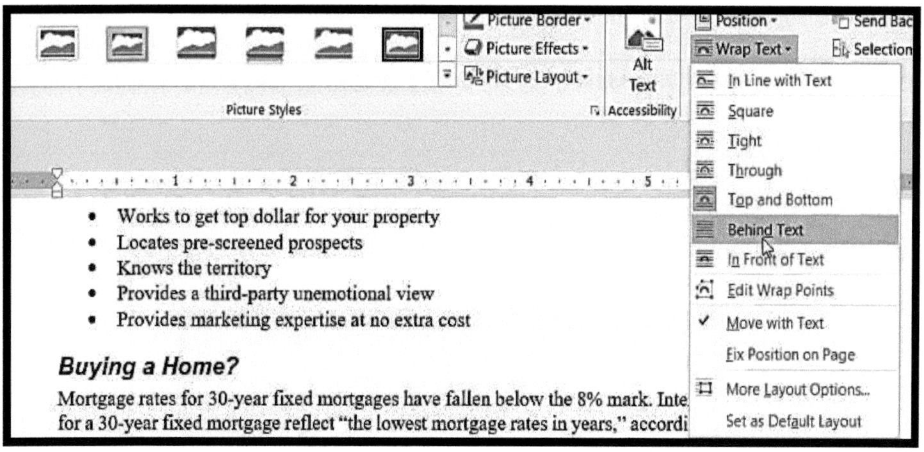

- It can be set to "**In Front of Text**," which means it will hide whatever is behind it.

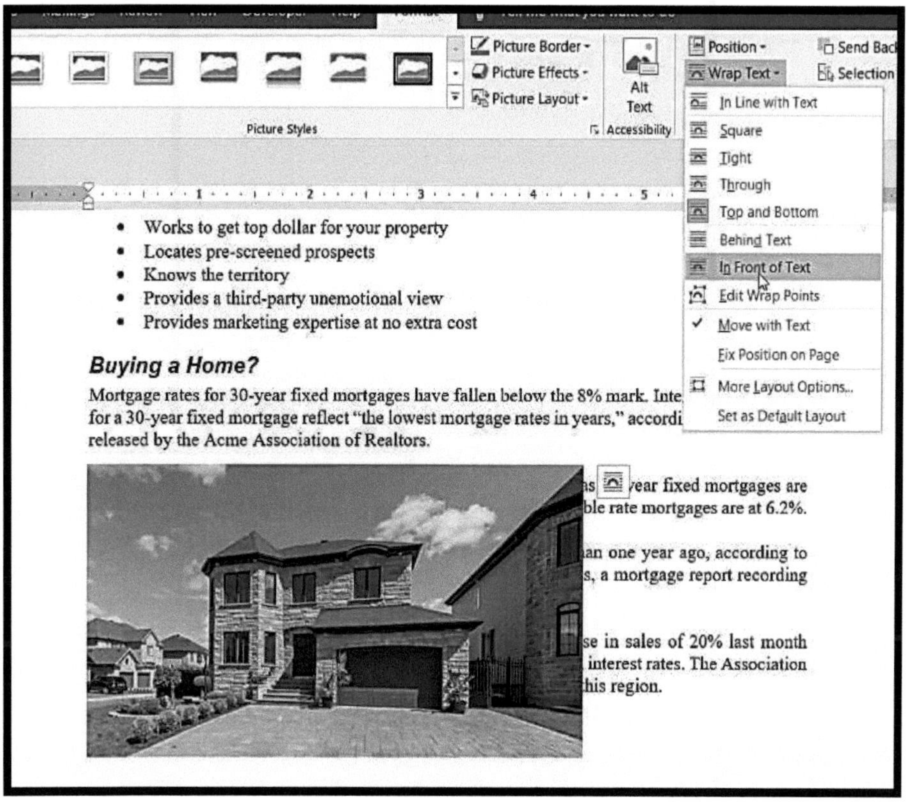

You must ensure that you select the appropriate option based on your documents.

If you're not happy with the way your text looks after you've wrapped it around your image, make some little tweaks. You can remedy this by making the image larger or smaller to fit everything on one line; it's up to you to decide how you want to handle it. It's worth experimenting with them to get your text to flow just the way you want it around, though, or in front of that image.

Now, because there are so many, we'll look at some of the other options you have when it comes to images.

Getting rid of an image's background

Let's return to the first group so you can learn more about the "**Remove Background**" tool.

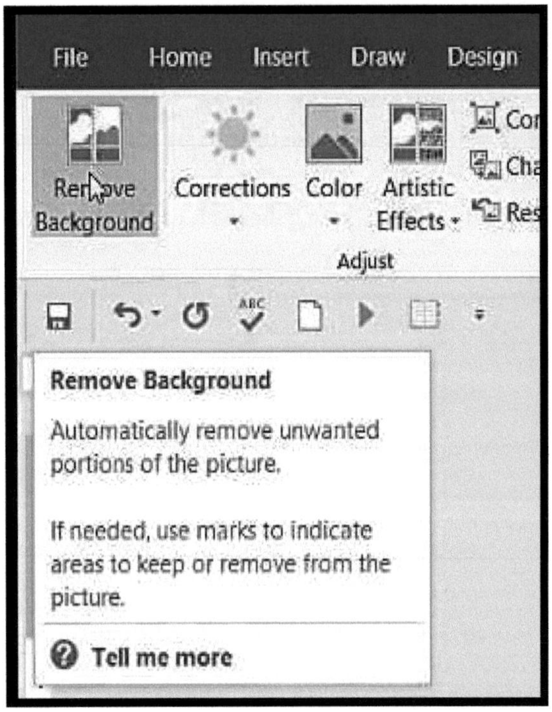

You will notice it reads "**automatically remove unwanted portions of the picture**" if you mouse over it. This would be beneficial if, for example, you wanted to eliminate the background from a picture and only keep the image.

Let's have a look at how it works because it's a little tricky.

- When you select "**Remove Background**," Word will try to figure out what you want to keep and what you want to remove, but it often can't tell what's the background and what you want to preserve, so it takes a guess. In the image below, all the items it's now going to remove are shown in purple, and you may alter this to keep or remove certain regions.

A background removal toolbar will appear at the top and will allow you to mark sections to keep or remove.

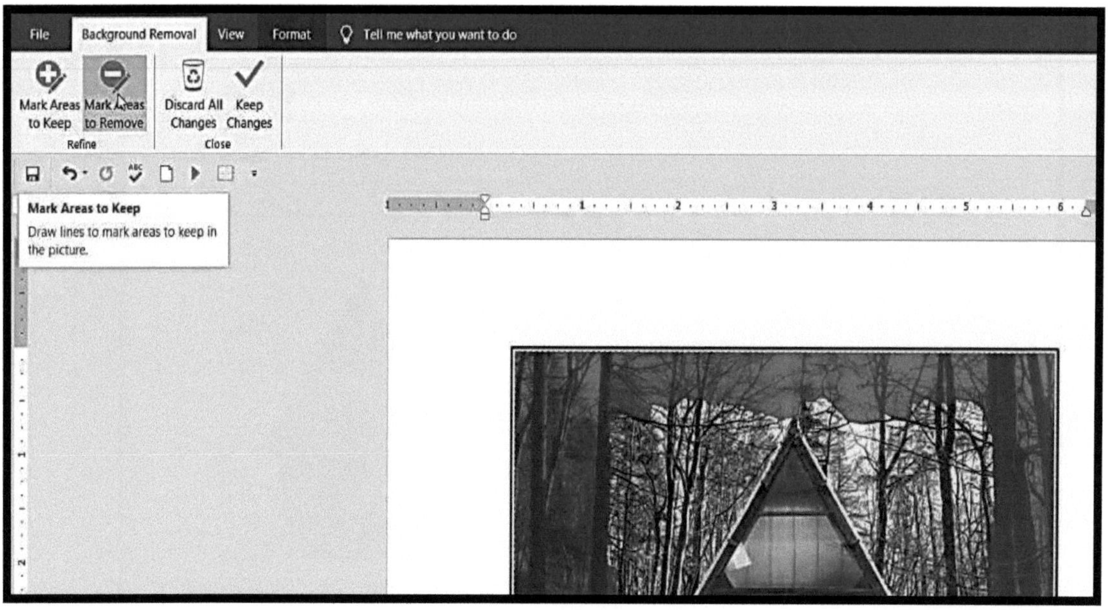

- If you say "**Mark areas to remove or keep**," you simply draw the pencil over the area you want to keep or delete, and as you know, it doesn't always get it exactly right, so you'll have to fiddle around with it a little to get it just right.

- Once you're comfortable with it, click the tick to say "**Keep Changes**," and there you have it, the background has been removed.

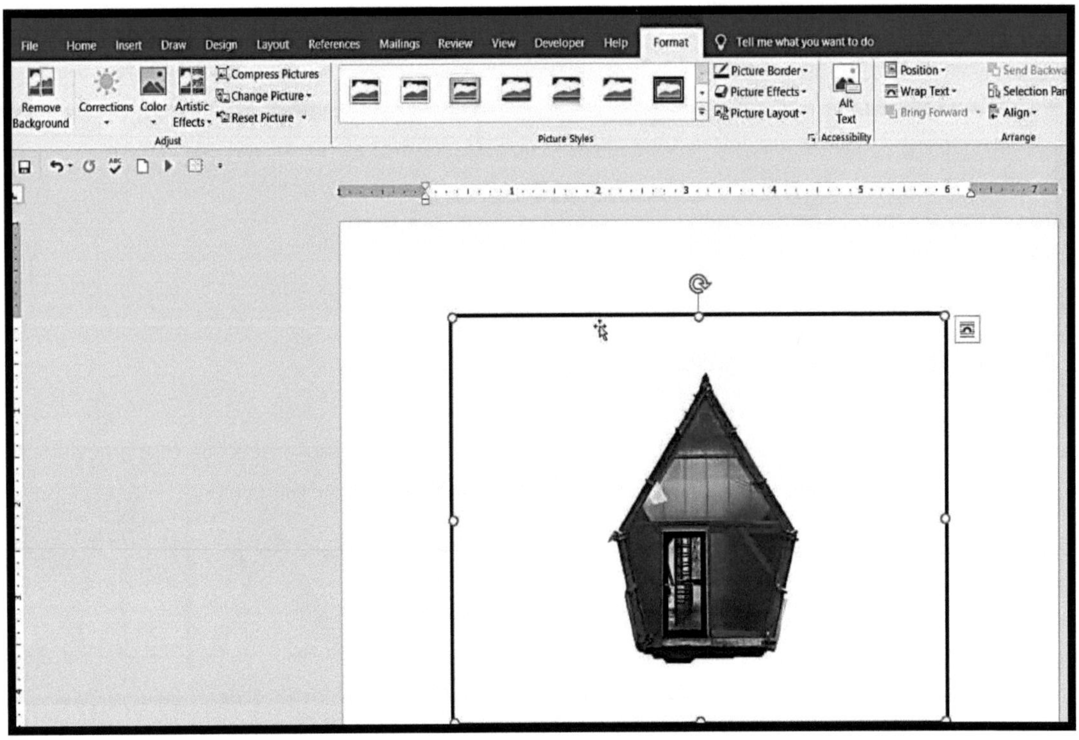

Correcting an Image

If you click a picture and go to the Format ribbon, you'll see some "**Corrections**" options.

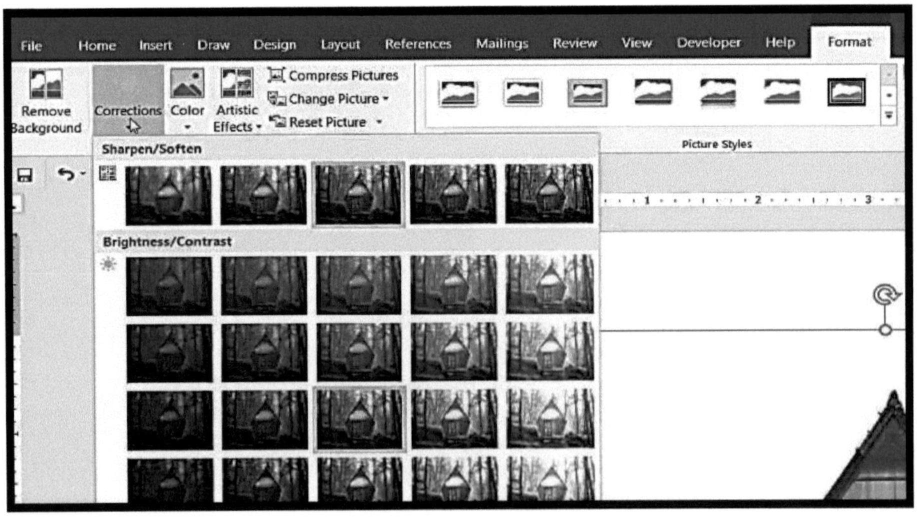

This is commonly linked to your image's color, contrast, and sharpness. You have some **sharpening** options at the top here, as well as some brightness and contrast images, which, again, if you hover over, will modify the brightness of the picture.

Color is the next choice along, so if you want to recolor or modify the color of this picture, you can do so by selecting any of these options, and you can get even more colors or variants by selecting the option at the bottom.

Lastly, we have **Artistic Effects**. You can make your photo look like a drawing, a mosaic, a painting, or some scribble here. All of these are options that you can explore.

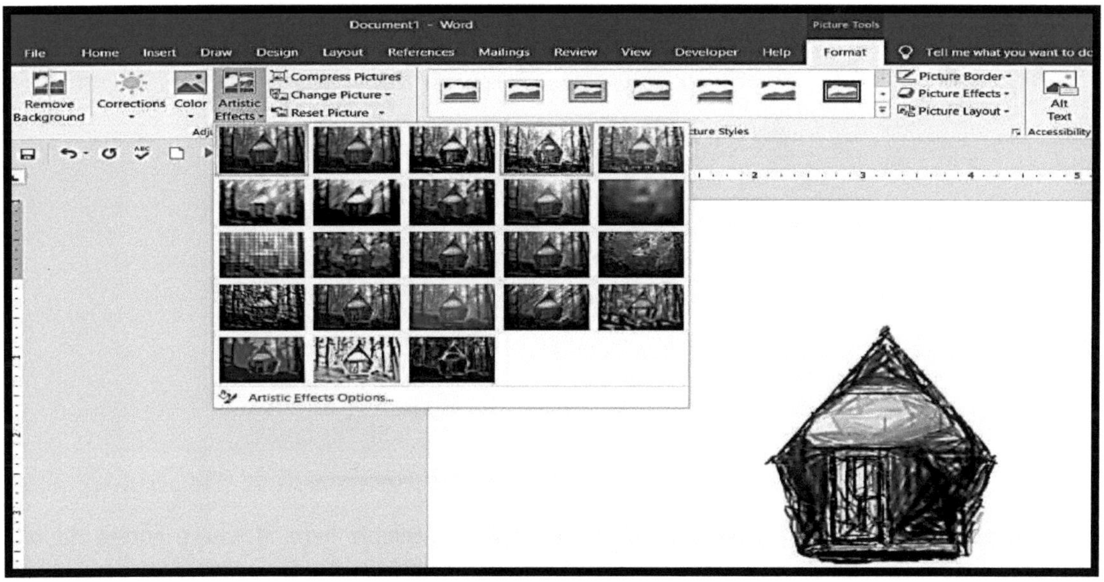

Compressing an image

One thing to keep in mind when using pictures in Word is that if you use a lot of them, your document will grow in size. It's always a good idea to compress your photos in your document to ensure that everything loads as quickly as possible and that you don't have any troubles, and we have the option for that right here.

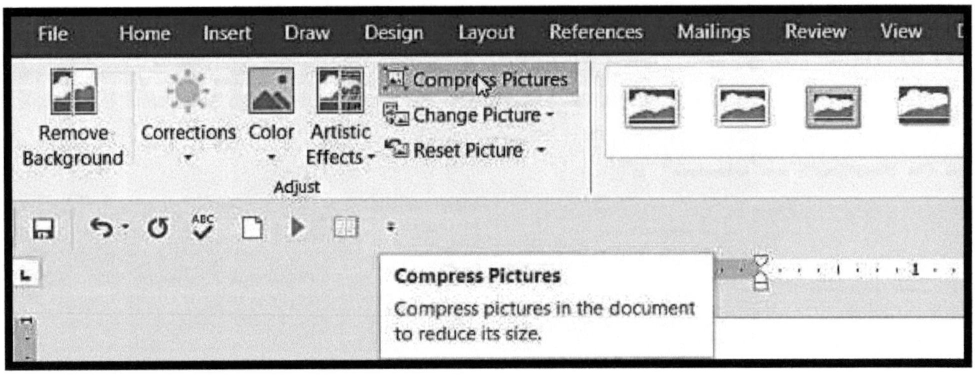

When you click "**Compress Pictures,**" you'll be asked if you want to "**Apply compression only to this picture**" you've chosen. If you only have one picture in the document, this is acceptable; however, if you have a lot of pictures, you should untick this because it will compress everything in the document.

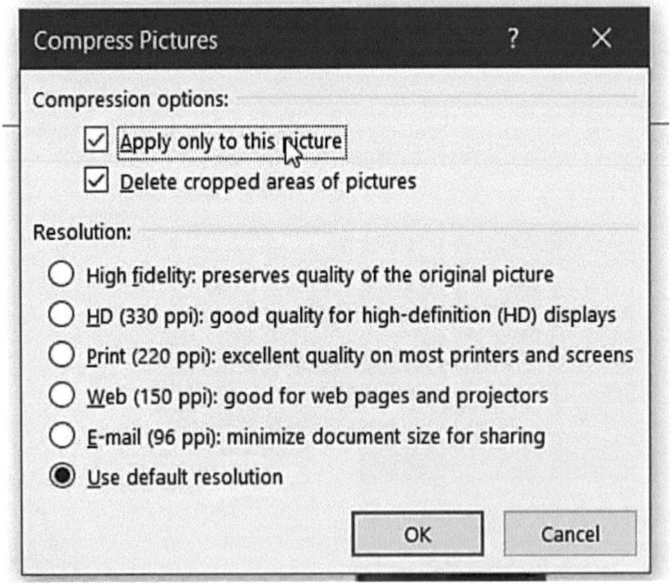

"**Delete cropped areas of a picture**" is another option available here. If you've gotten rid of the background on a picture, you haven't completely removed that from the document, and you could easily undo what you did with the "Remove Background" and restore it, thus the background is still there even though we can't see it. This option will erase any cropped sections of the image, which will assist in reducing the file size.

Changing a Picture

You also have the option to change a photo, so if you want to replace the current picture with something else, you can do so from a file, an online source, or an icon.

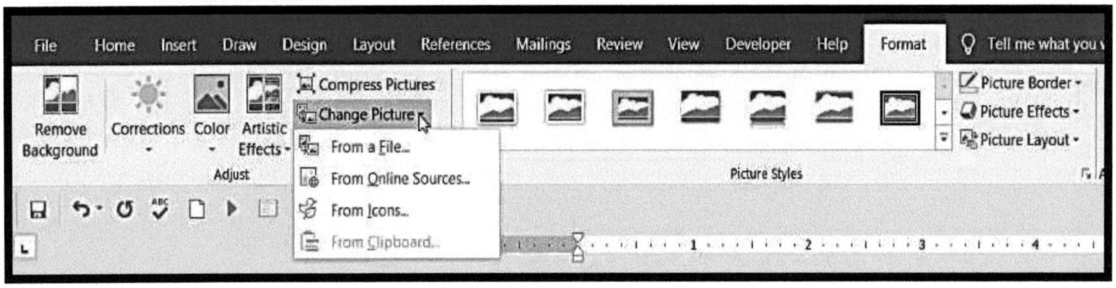

Restoring an Image

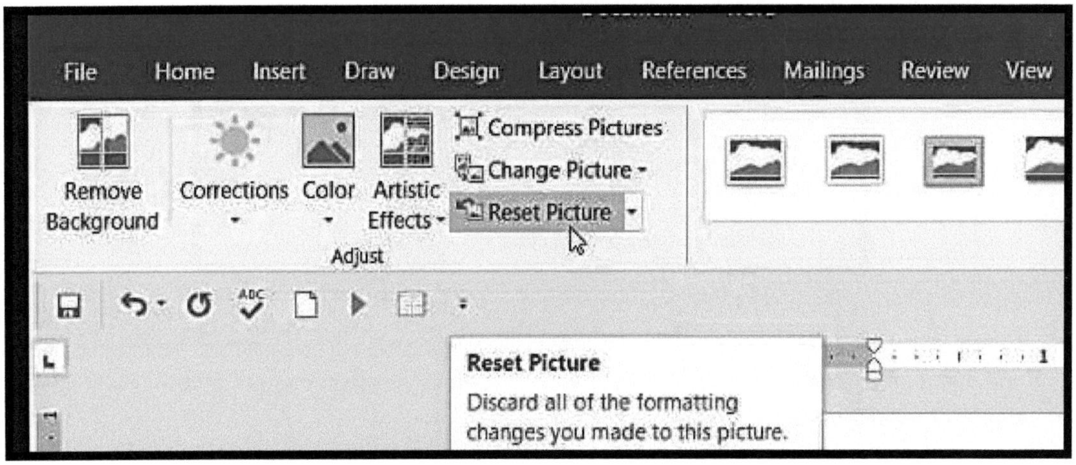

If you've made a lot of modifications to an image and don't want any of them, you can restore it to its previous state by resetting it.

CHAPTER TWENTY-ONE

COMMENTS ON YOUR TEXT

You'll learn the fundamentals of tracking changes and commands in Office 365 in this chapter.

However, with small adjustments to some of the tabs, the same applies to several other versions of Microsoft Word. Turning on track changes, accepting and rejecting track changes, hiding tracked changes, adding and deleting comments, and comparing two documents are all covered in this section. If you printed a copy of your work and asked a friend to proofread it, they would most likely mark the faults in red ink and put any suggestions in the margins. Isn't it more convenient if you could do it electronically?

You can do so with Word by using the "**Track Changes**" and "**Comments**" options.

Enabling Track Changes

We're going to **enable Track Changes** right now so you can see what it means. To do so, navigate to the top-left **Review** tab and select the "**Track Changes**" command.

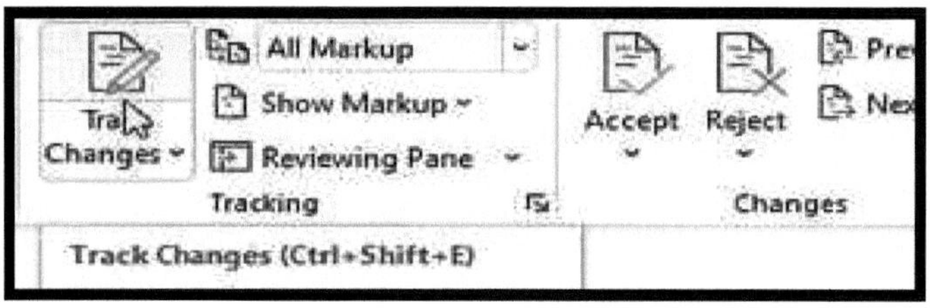

As you can see, it's now grayed out, indicating that any modifications you make will be monitored. With this command enabled, you can now make any modifications to the document and they will display as markup, which is Word's version of the conventional red pen.

When you erase text, for example, it is not removed; instead, it is crossed out, and when you add more text, it is underlined, as shown below.

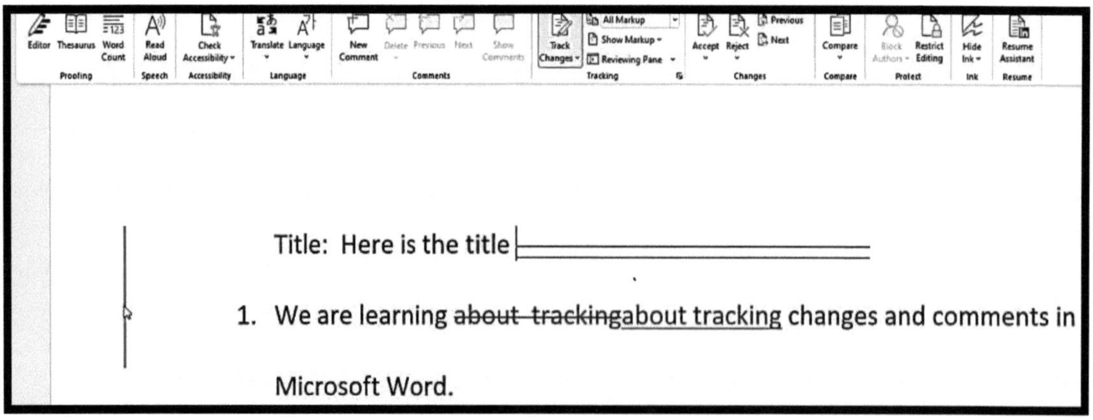

This allows the document's owner to see what modifications you've made before integrating them into the final draft.

Adding a Comment

By selecting the text, you want to remark on, right-clicking, and dropping down to the "**New Comment**" instruction, you can also write comments in the margin.

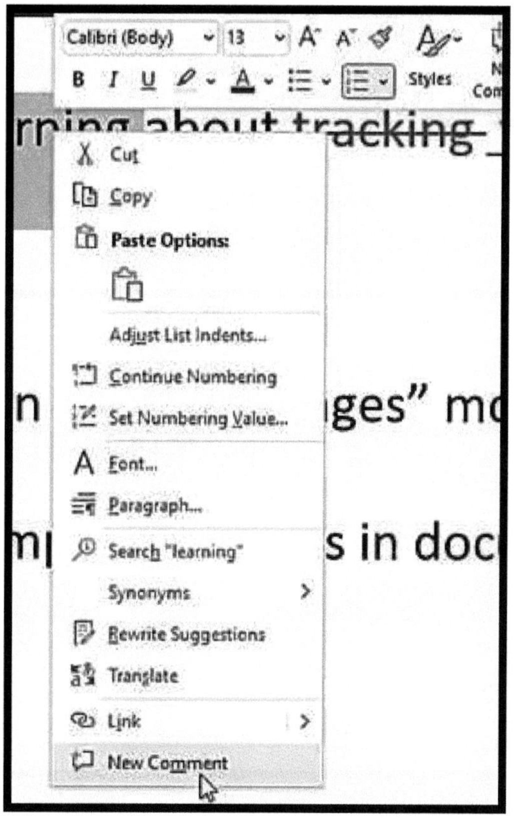

You can also go to the **Review** tab and select "**New Comment.**"

You can now add a comment, which the document's owner will see once you share it with them. When you're finished, simply click anywhere outside the comment to make it part of the others you're making.

Accepting or rejecting a comment

What happens if you're on the other side of the document?

Let's imagine you're sending this document to someone else, that person has the option of accepting or rejecting the changes to make them permanent.

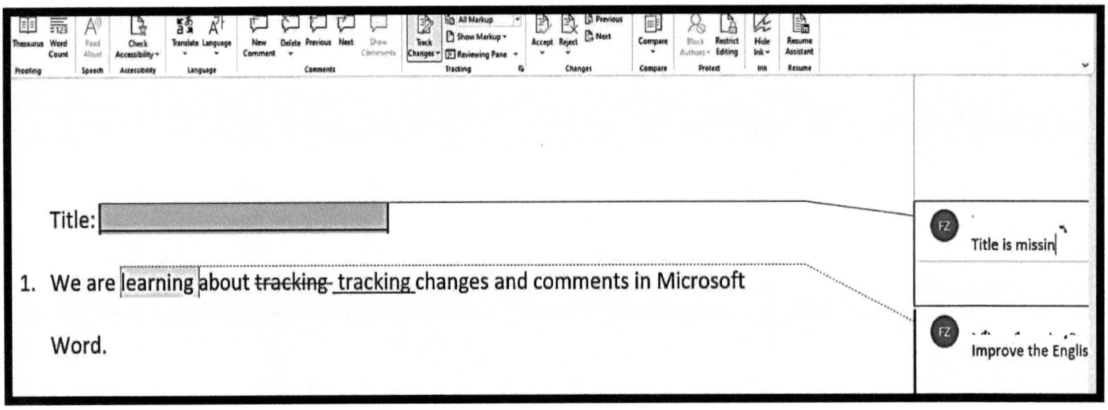

Looking at the changes made from the example above, you can accept or reject the comment by right-clicking on them.

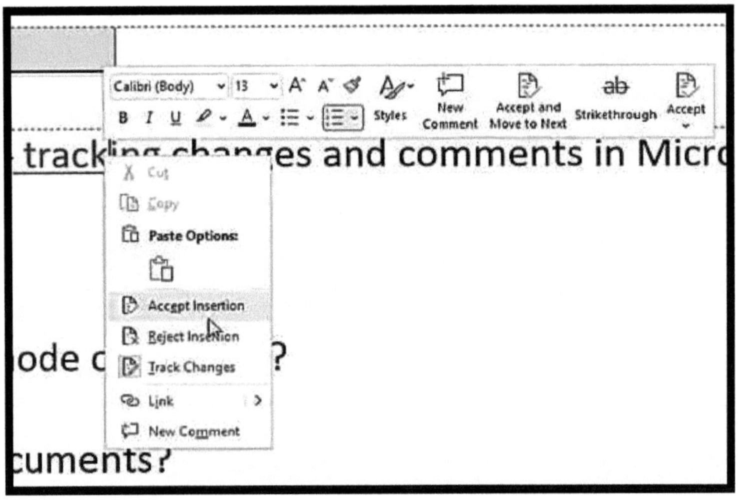

You can either perform this for all of the additions or subtractions at once, or you can do it one at a time. If you want to complete everything at once, select "**Accept all Changes**" from the "**Accept**" drop-down menu.

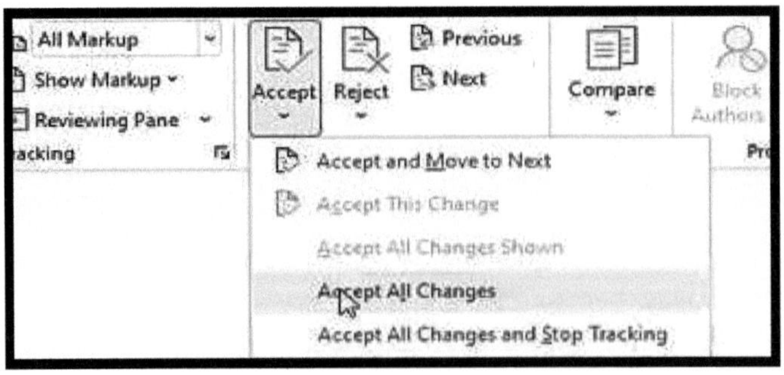

In general, accepting all modifications is not encouraged because the other person may not be as knowledgeable in the topic as you are, therefore you should carefully accept or reject changes. Except for the comments, all additions and subtractions will be done automatically as advised by your collaborator.

Commenting and Replying

You can either respond to the comments or mark the problem as resolved once you've addressed the issue.

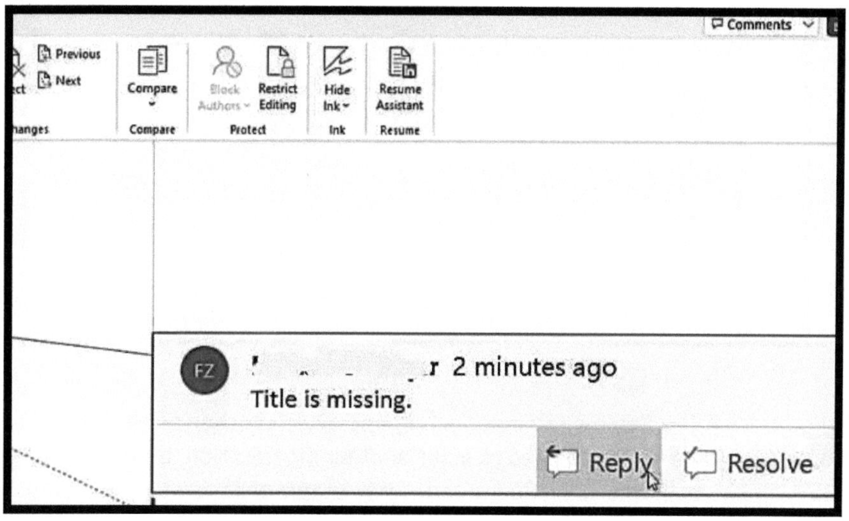

You'll notice that it's now grayed out, indicating that you've fixed the problem. If you need to make any adjustments, you can click on the drop-down and reopen it by choosing "**Reopen**" if you need to add another comment.

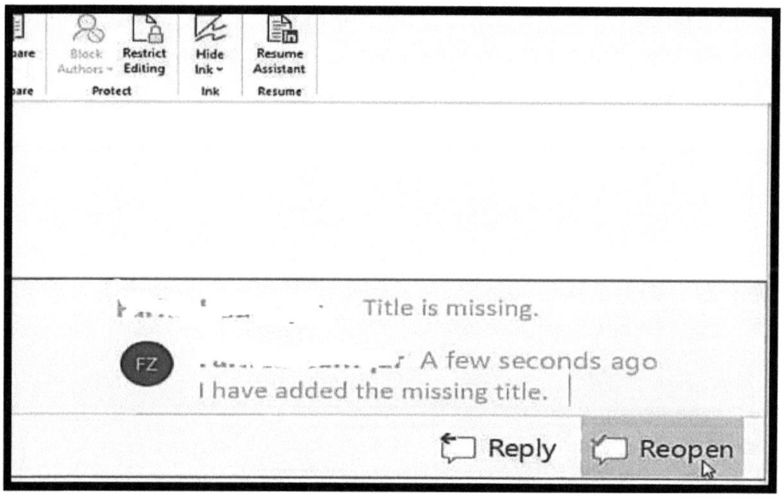

You'll get a hint that there are comments while working on your text. You can erase these comments by clicking on the comment box and selecting the **delete comment** button if you are satisfied that the changes made are acceptable to everyone.

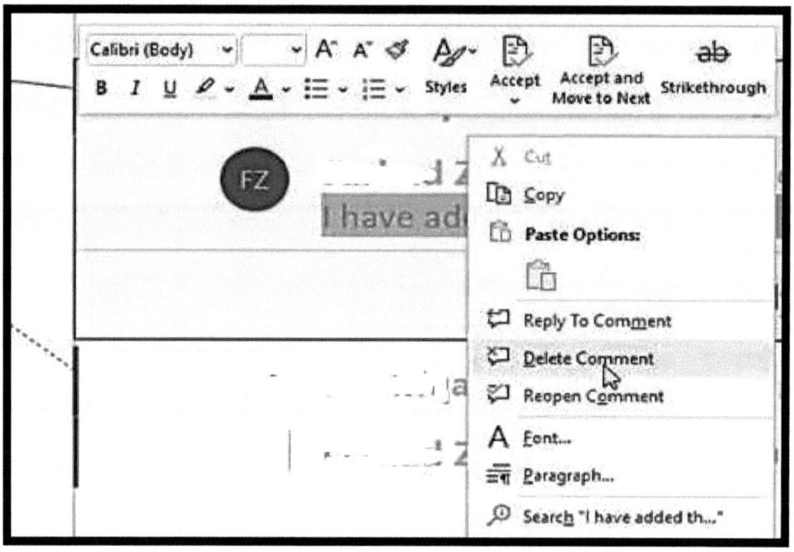

You can also do it from the **Review** tab by selecting **one after another** or **all of the comments** in the document to **delete**. However, if you need to make another modification, there is additional space here where you can simply make the adjustment and notify others that you've done so using the "**Track changes**" modes.

Organizing your document

Now that you've made all of the modifications, you'd like to see the document in its refined state; how can you achieve that? Remember that some markups in your document are visible while you're making modifications. By navigating to the "**Tracking**" group and clicking on the **Markup** options and drop-down, which contains **All Markups**, **Simple Markup**, **no markup**, and **Original**, you can remove these markups. You'll select "**No Markup**," and the document will be displayed in its final form.

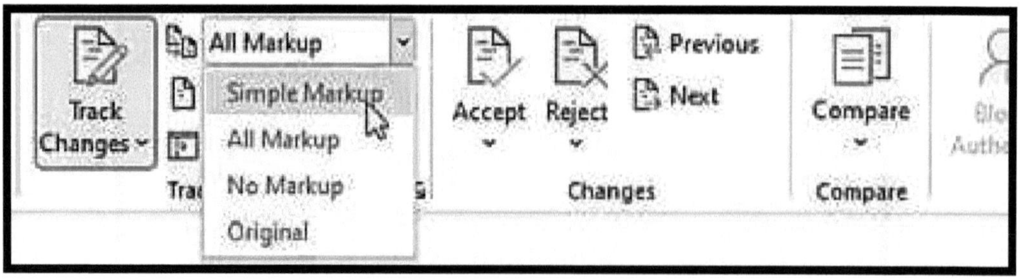

It's crucial to note, though, that the changes are still in effect because you haven't accepted or rejected them yet. As a result, when other people look at these documents, they'll be able to see the markups.

After you've completed all of the modifications, you should accept the document before forwarding it. You can make modifications, edit the document, and add comments in this manner.

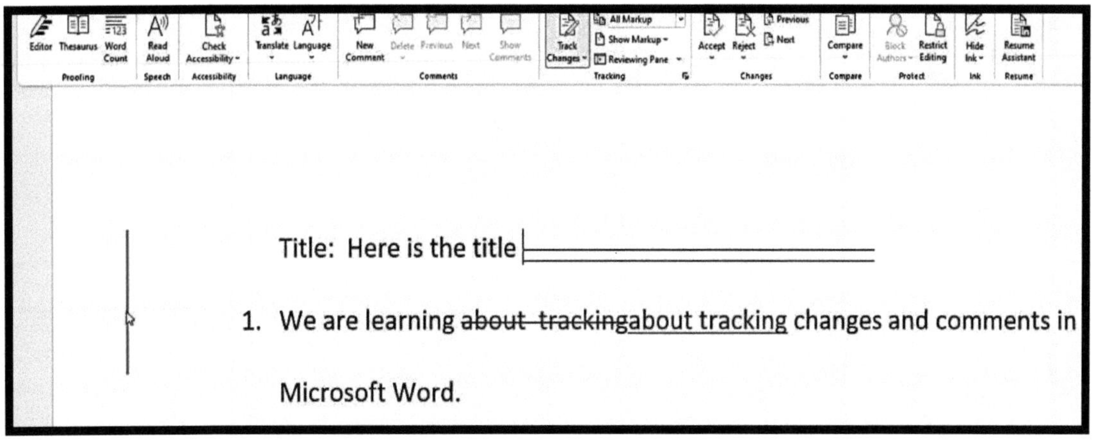

Another thing to keep in mind is that while making changes to your document, you may notice a **small grayed outline**; clicking on it will hide all of your modifications, and when you click, it will change to a red line, indicating that certain changes are hidden here. When you click again, it displays all of the adjustments you've made.

Comparing Two Versions of a Document

The next item we'll look at is how to activate track changes. Let's say you're working on a document with a coworker who made some modifications but neglected to switch on the track changes.

You can use the "**Compare**" option here in such cases. You'll need two versions of the documents to use this "**Compare**" option: one is the original and the other is the changed version. As a result, you must choose between the original document and the edited version, which may have been created by someone else.

When you click "**OK**" after selecting both the original and revised documents, it will display three types of documents: one is **the original document**, the second is **the revised document**, and the third is **the compared document**, which is the version obtained by combining the two and will display the changes made.

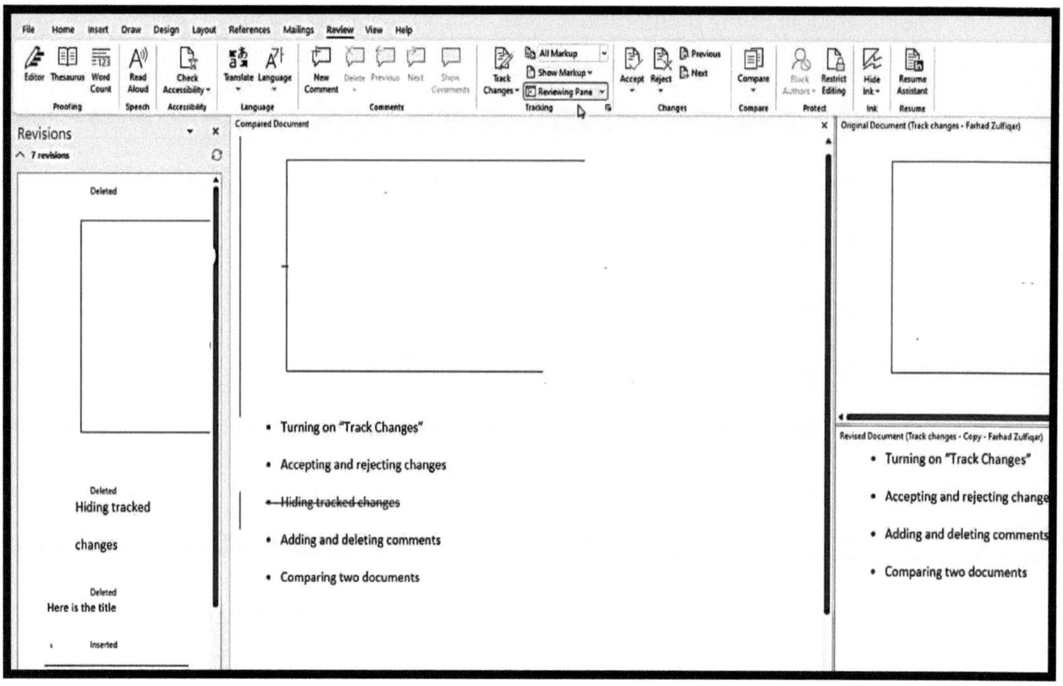

Tracking and Reviewing Changes

When you click on the "**Reviewing Pane**," another box will appear, detailing the changes made in the updated version.

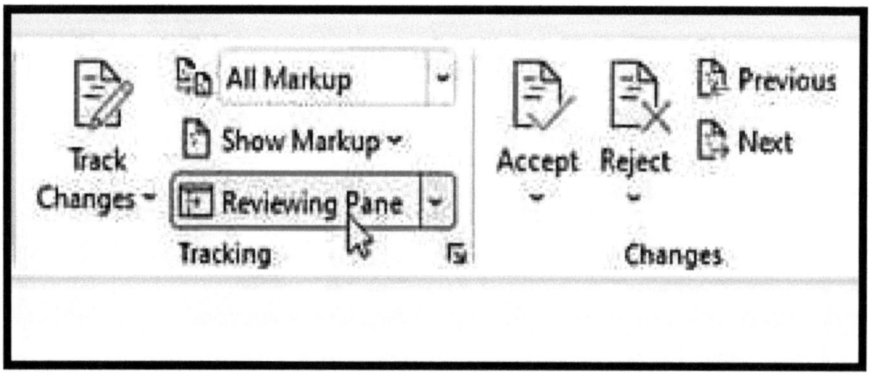

- In the example below, there were five bullet points in the original document that included '**hiding track changes**,' but your collaborator or partner has omitted this one bullet point from the revised version. It indicates that someone deleted the '**hiding track changes**' bullet point in the Compared document.

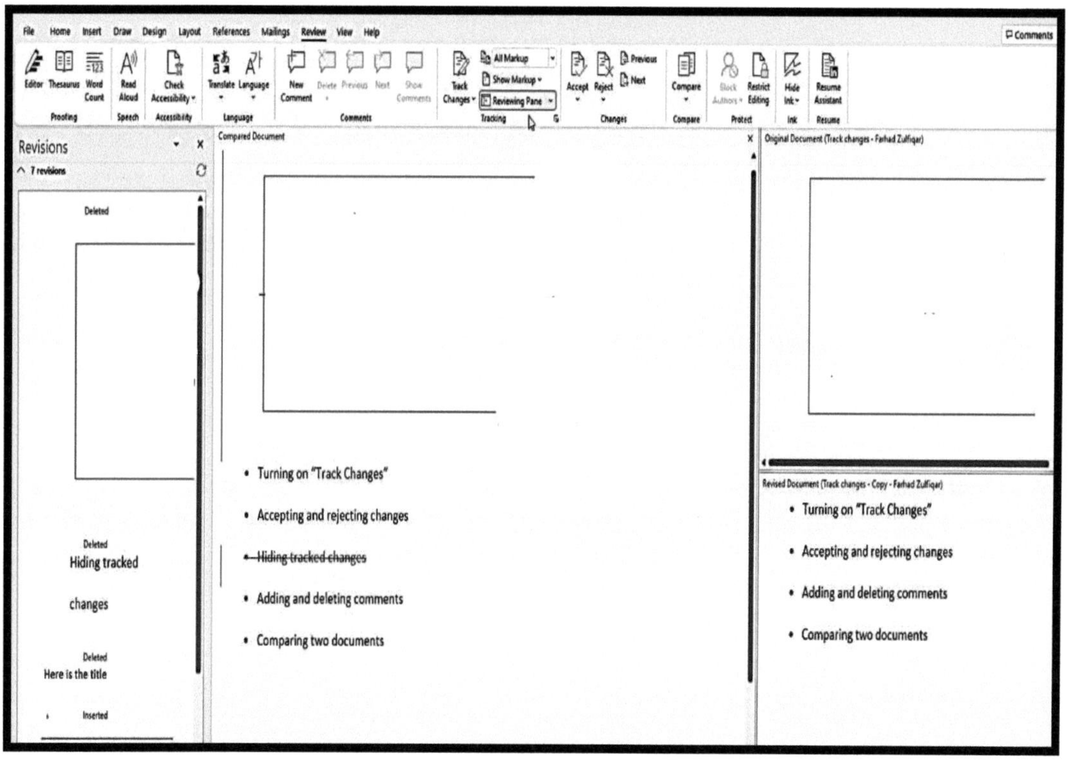

- The original document stated, "**How to Compare Changes in Documents**," whereas the revised version states, "**How to Compare Differences in Documents**," with "**changes**" substituted by "**differences**."

Even if someone forgets to switch on the track changes, you can still study and compare two documents if you have both the original and revised versions.

If you're working on a large project with a large document, such as a 50 or 60-page document, it'll be impossible to look at individual changes without this track change option. That's why, whenever you're working with others, we recommend utilizing Track Change mode to keep track of updates and send ideas back and forth so you can collaborate on the same document with friends, family, or coworkers.

Finally, once you've compared these documents, simply save this newly updated version, and it will appear on your computer after you've saved it.

CHAPTER TWENTY-TWO

MAIL MERGE

We'll look at Mail Merge in this chapter. What is a Mail Merge, and how does it work? Let's say you have a letter that you need to send to a group of individuals, but you want to personalize it with their names and other facts in each letter, and all of these details are stored in another location, such as a spreadsheet or a database. You could either manually generate each letter in an attempt to personalize each one, or you could blend the data from the spreadsheet or database into your word document.

Two things are Involved:

The Time-wasting Process

Take a look at the sample letter below, which is intended to be sent to persons who are attending an adventure camp and for whom we need to write their contact information. To personalize the letter, we'll need to know what class they're in, which team they're on, their team leader, and their campsite, all of which will be based on a database or spreadsheet, and that's what we'll send them.

You can also see that we have a spreadsheet below that covers all of their names, surnames, grades, and classes in one place.

We could do this manually by copying this document and pasting it into another page of the paper, where we can now enter everyone's updated information.

Then we can go through and update these elements one by one by hand; but this will exhaust us because we will be taking so long to complete a basic task. That would be a waste of time if we had to do it for every single person's letter, so how can we do it all at once? Let's have a look at it.

The Easier Way

First and foremost, we'll go to "**Mailings,**" which is where we'll be working. All of our steps will be over here if we click on that. There is a "**Start Mail Merge**" option, as well as a "**Select the Recipients**" option, where we can edit which recipients we want to send to, **insert fields**, **evaluate the results**, and then **finish** them.

There are the stages we will discuss, and they're all right there for us to follow. We'll proceed from left to right, and it'll be really simple.

The Mail Merge process

1. **Starting a mail merge** is the first step. You must inform the Word processor that this document will be merged with another document.

2. Then you must **choose the recipients**, or, in other words, decide where you will acquire the information. Is this information coming from a spreadsheet or a Word document? Data from a text file or a database could be used.

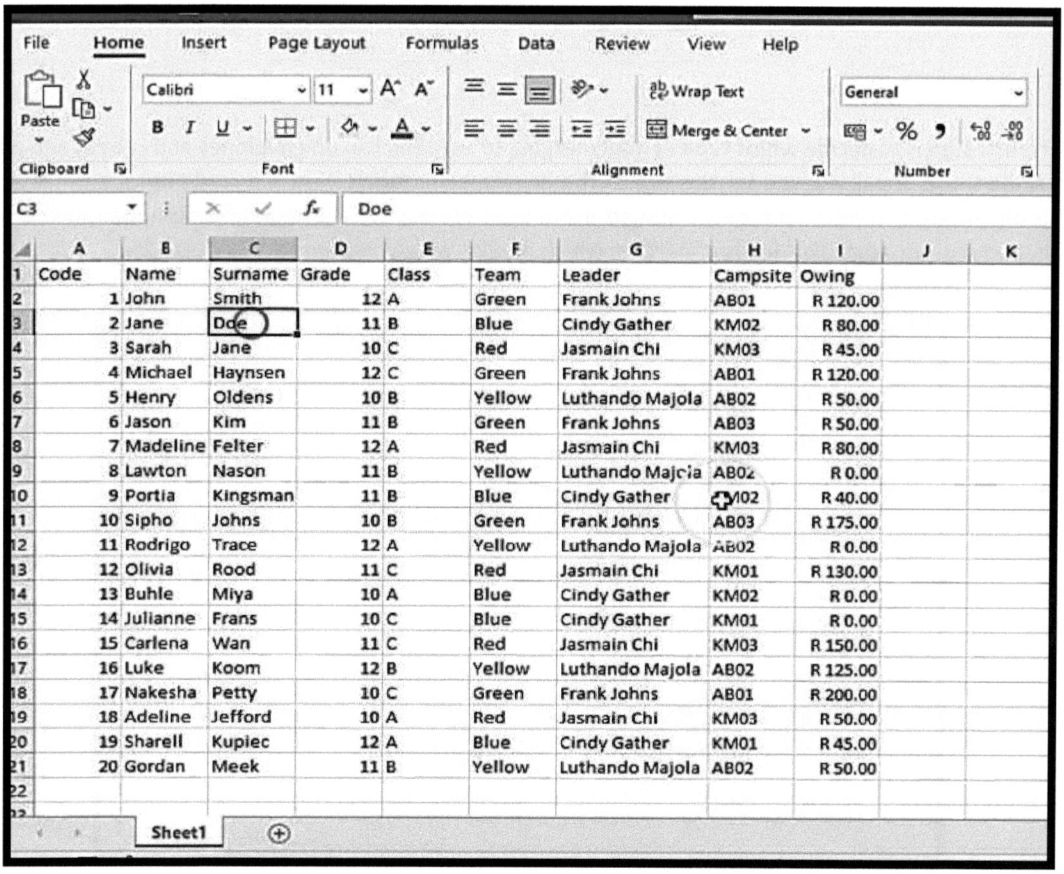

Code	Name	Surname	Grade	Class	Team	Leader	Campsite	Owing
1	John	Smith	12	A	Green	Frank Johns	AB01	R 120.00
2	Jane	Doe	11	B	Blue	Cindy Gather	KM02	R 80.00
3	Sarah	Jane	10	C	Red	Jasmain Chi	KM03	R 45.00
4	Michael	Haynsen	12	C	Green	Frank Johns	AB01	R 120.00
5	Henry	Oldens	10	B	Yellow	Luthando Majola	AB02	R 50.00
6	Jason	Kim	11	B	Green	Frank Johns	AB03	R 50.00
7	Madeline	Felter	12	A	Red	Jasmain Chi	KM03	R 80.00
8	Lawton	Nason	11	B	Yellow	Luthando Majola	AB02	R 0.00
9	Portia	Kingsman	11	B	Blue	Cindy Gather	[illegible]M02	R 40.00
10	Sipho	Johns	10	B	Green	Frank Johns	AB03	R 175.00
11	Rodrigo	Trace	12	A	Yellow	Luthando Majola	AB02	R 0.00
12	Olivia	Rood	11	C	Red	Jasmain Chi	KM01	R 130.00
13	Buhle	Miya	10	A	Blue	Cindy Gather	KM02	R 0.00
14	Julianne	Frans	10	C	Blue	Cindy Gather	KM01	R 0.00
15	Carlena	Wan	11	C	Red	Jasmain Chi	KM03	R 150.00
16	Luke	Koom	12	B	Yellow	Luthando Majola	AB02	R 125.00
17	Nakesha	Petty	10	C	Green	Frank Johns	AB01	R 200.00
18	Adeline	Jefford	10	A	Red	Jasmain Chi	KM03	R 50.00
19	Sharell	Kupiec	12	A	Blue	Cindy Gather	KM01	R 45.00
20	Gordan	Meek	11	B	Yellow	Luthando Majola	AB02	R 50.00

If you don't want to send the information to everyone in your source or the location where all of our data is stored, you can edit the recipients' list or only send it to specific persons on that list.

You can narrow it down a little so that you just utilize specific persons.

3. After that, you must **insert fields**. These are the fields you'll want to include in your document, and they'll come from either your source or your recipients. If you wish to put their names as fields or any other information about them that you've saved in your source document or the recipients', this is where you'll receive your fields and utilize them to insert them into the mail merge.

4. Then you can **preview** your results to make sure everything is working, and it will show you one of the first letters that are compiled from the first document with all the source fields, so you can see if it is in the appropriate format.

5. Once you're satisfied with the results, you can proceed to the **final merge**, which will create a new document containing all of the letters, each of which will be personalized with the information from the source.

With that out of the way, let's get right into it.

Creating The Mail Merge

Get Started

The first step is to **decide which type of mail merging to use**. You can do Envelopes and Labels, and when you do those, it will ask you for the size of the envelopes or labels. You can configure it so that it will automatically determine the paper size to make your life easier. These will be discussed in the next chapter, but for now, we'll focus on letters, which is what we've chosen.

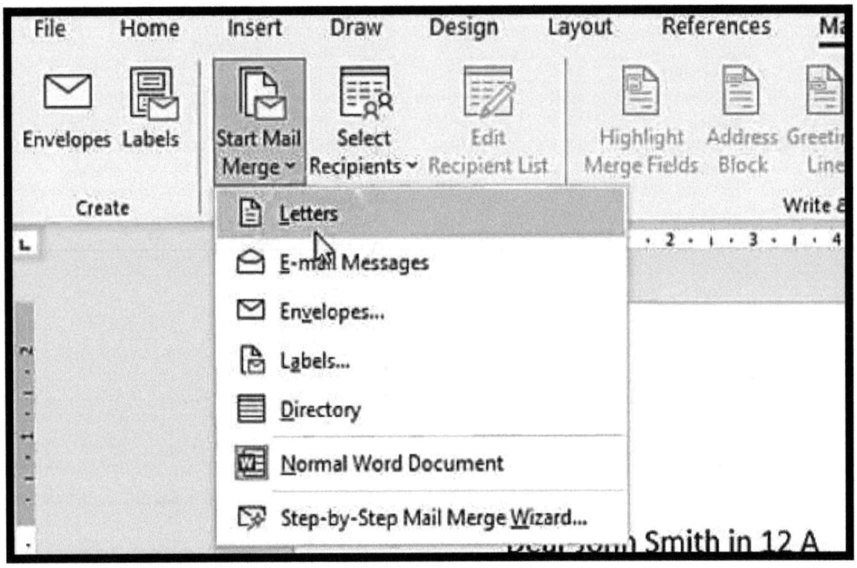

The Recipient List

Step two is to choose our recipients. We'll choose where we want to acquire the information for this letter from here. We can create a new list or make good use of an existing one. As you can see, we already have a spreadsheet with all of the information, which we'll combine with this document. Select "**Recipients**" and "**Use an existing list**" from the drop-down menu.

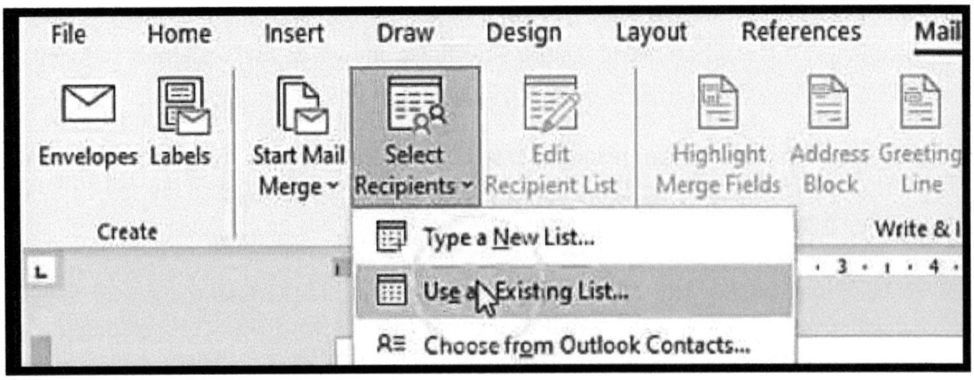

This will open file explorer, and all we have to do now is click on the data we want to open and then click "**Open**." We've now linked this document to that specific spreadsheet.

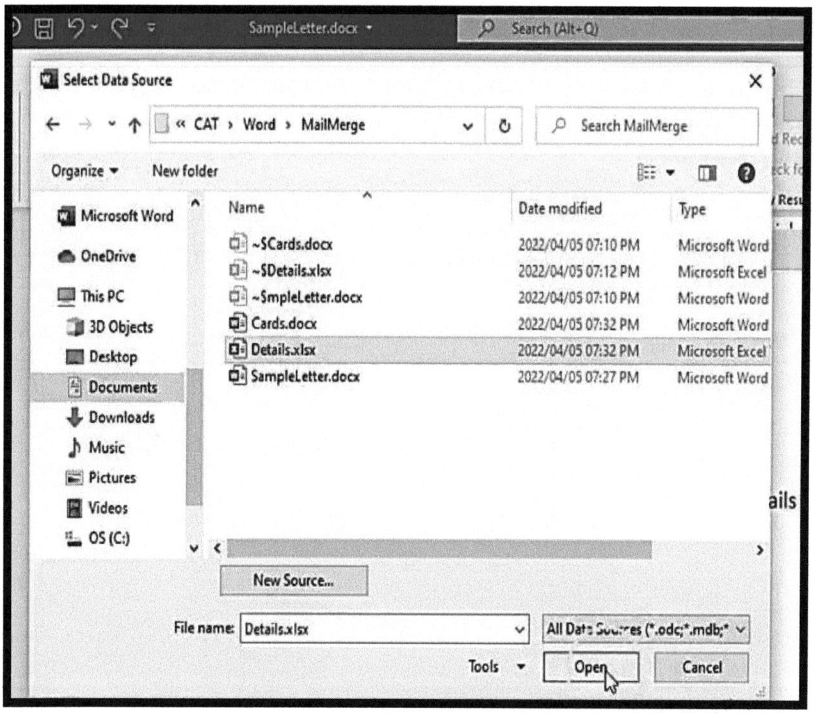

In a spreadsheet, you may have numerous sheets, thus it will ask which one you are using. Please keep in mind that if this were a **database**, it would ask you which **table** or **query** you wanted to use. Your source is dependent on what happens over here. Also, if your first row contains Column Headers, you can click the box at the bottom of this window to indicate that. If you're working with data on a spreadsheet, make sure it's right at the top and that the first heading is in A1 and flows across the page; this will make things much easier. We can now proceed to click "**OK.**"

Making Changes to a Recipient List

If we only wanted to send data in one column, for example, we'd want to use this list but only for those individuals. If that's the case, we can come here and click "**Edit the Recipient List**," where we can see the entire list of names in our source and uncheck the ones we don't want to include.

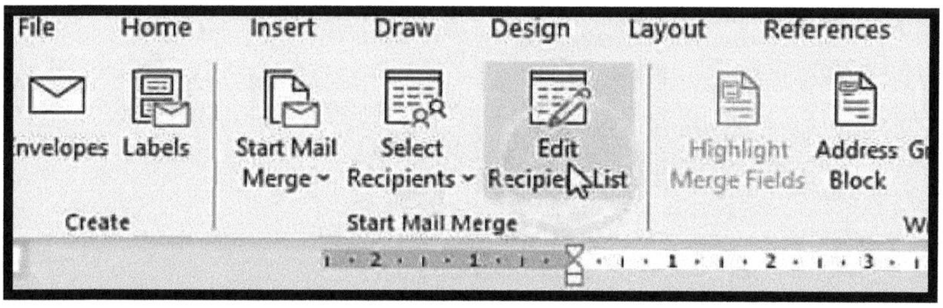

If we wanted to, we could sort it or apply a filter. If you want to change your recipient list, you can use a variety of criteria.

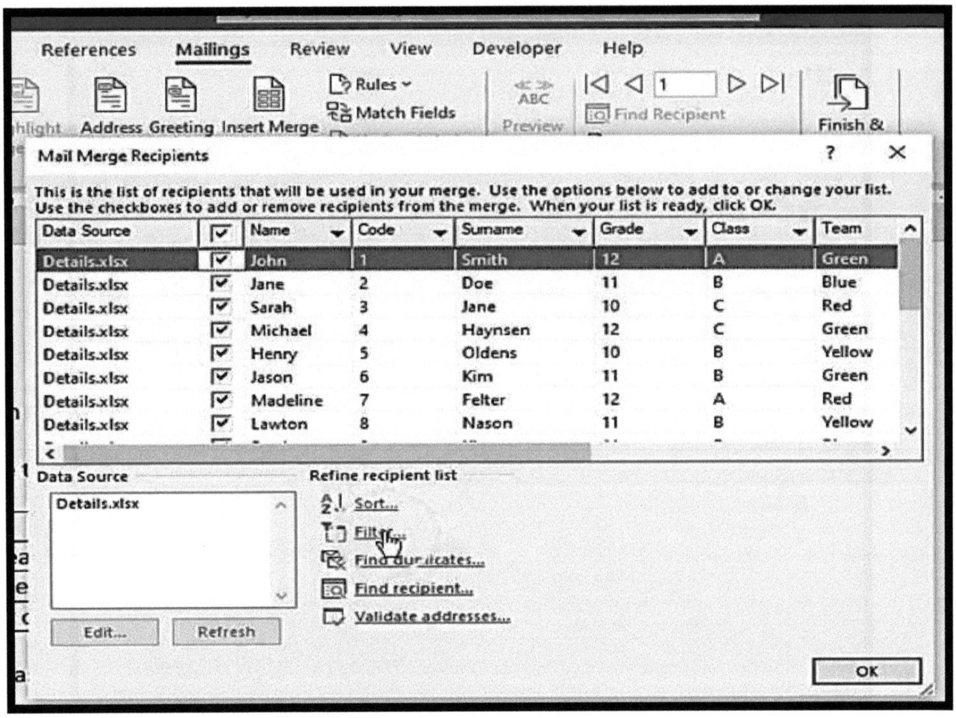

It may be sorting or deleting duplicates, rather than just excluding people. To continue, click "**OK**" after you've finished updating your list.

Inserting the Fields

Now that we have our recipients, we can go on to the third group. First, we need to make a list of all the headers in our recipient list, and now that we've linked this word document to that spreadsheet, when we go to "**Insert Merge Field**," you'll notice that all of those headings are now potential fields that we could enter when we click on the drop-down menu.

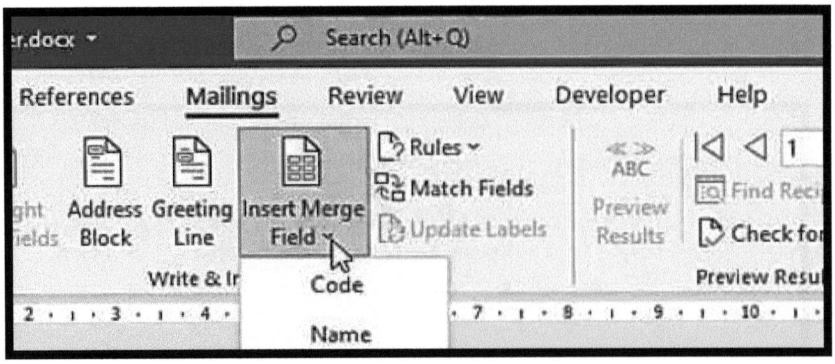

This means that if we want to enter anyone's name, we'll use the name field instead of the individual's name, and for the surname, we'll use the surname field, so whatever's in the surname block for each person will go into that particular location.

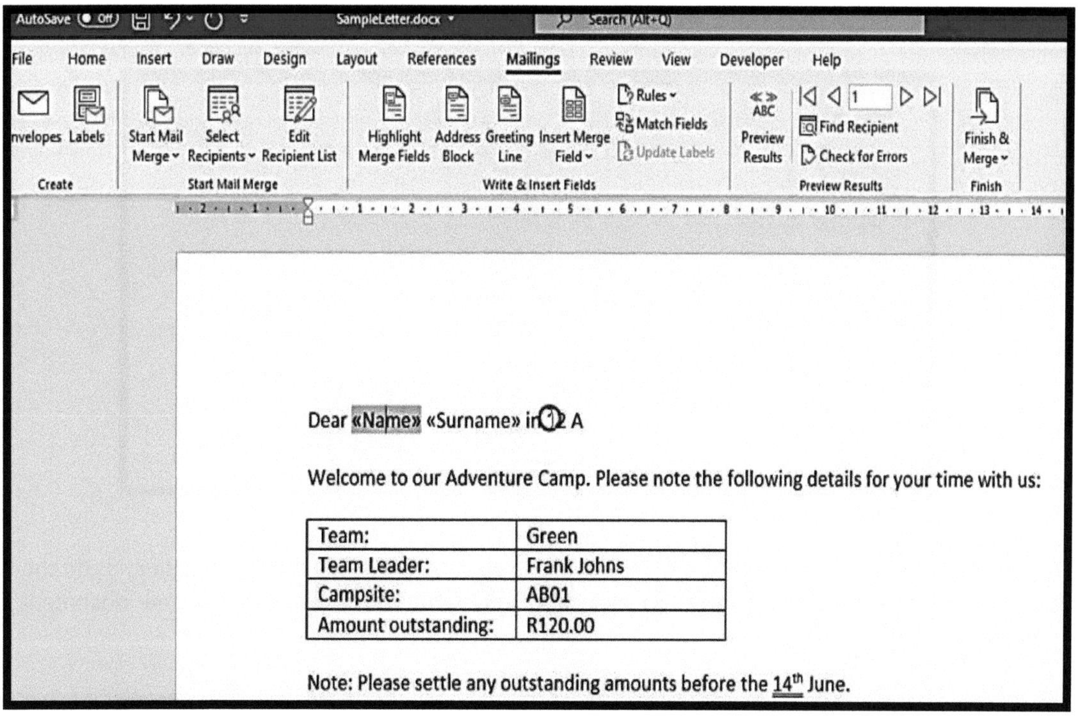

When you utilize the "**Insert Merge Fields**," you'll find that you can't just write in those weird little arrows, but you can see that it's gone gray when you click on it. Now, all we have to do is simply fill in the fields as we require them.

Previewing the Merged Documents

We've built our Mail merge, selected our recipients, and made any necessary edits; we've also entered these fields; we can now evaluate our results before doing the final merging.

If you click on "**Preview Results**," you'll see the page with all those small fields.

Merged at Last

Now that we've reached this point, we can move on to the next step, "**Finish & Merge**." By selecting this option, we get a list of different options for completing the process. We can **send emails** in bulk, **print documents**, or **modify each one individually**.

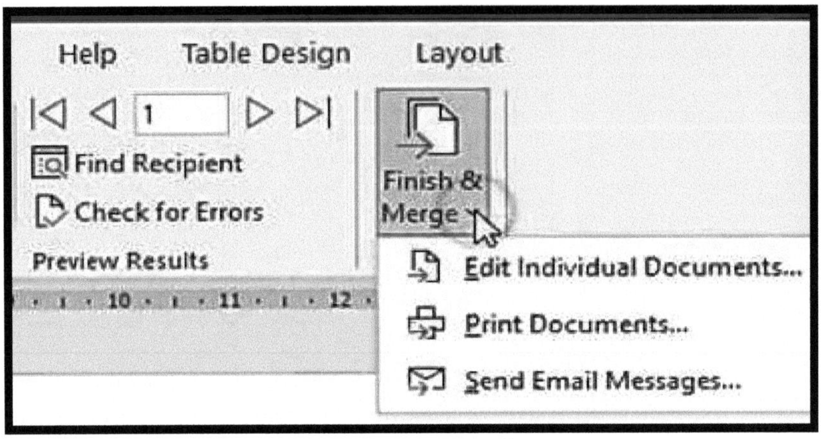

There are still some more options here if we chose to edit them individually. We could either do the first five records or the entire set. Now we can click "**OK**," and you'll notice that a new document has appeared.

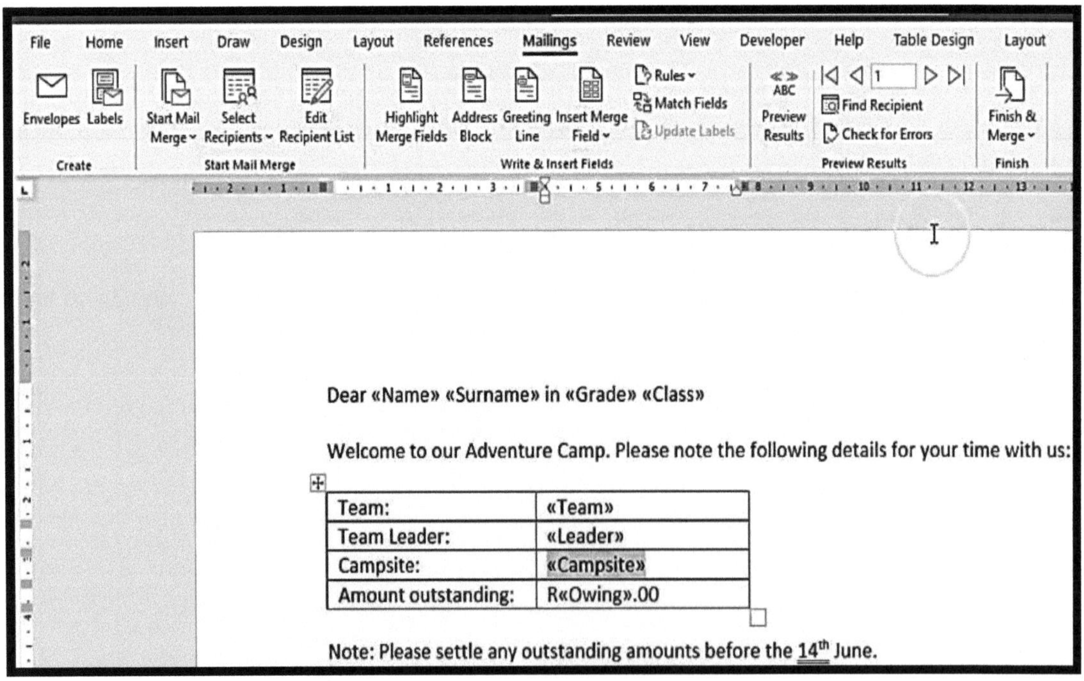

As you can see, all of the fields are filled in, but there are more pages to go down, and on page two, there's another individual with their information; as you continue down this document, you'll notice that it contains everyone's information. Keep your originals and save this as a separate document.

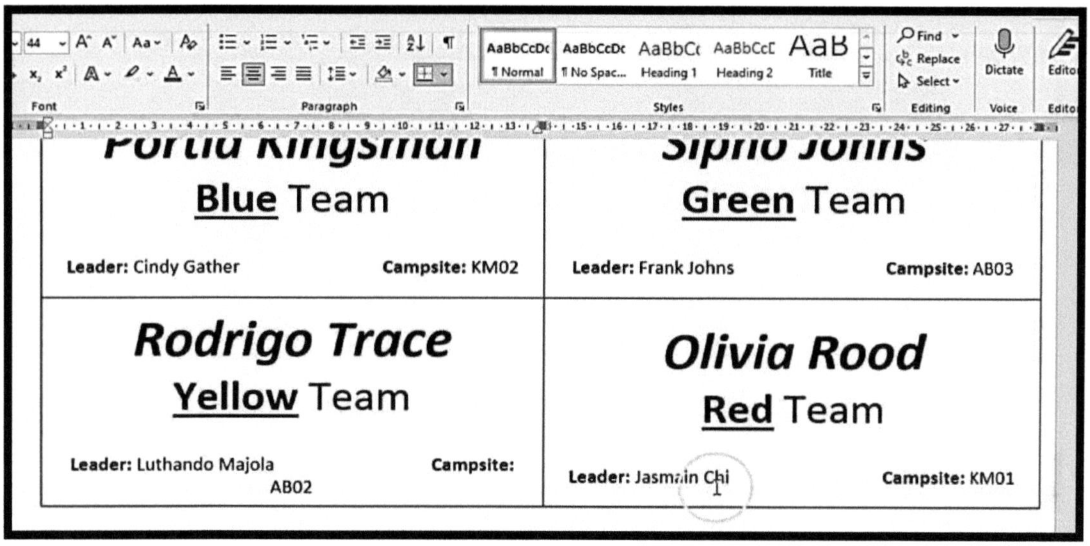

If you have numerous letters on the same page and want to proceed to the next record, you can go to "**Rules**" and play around with all of the rules, but in this case, you just need to click on "**Next Record.**"

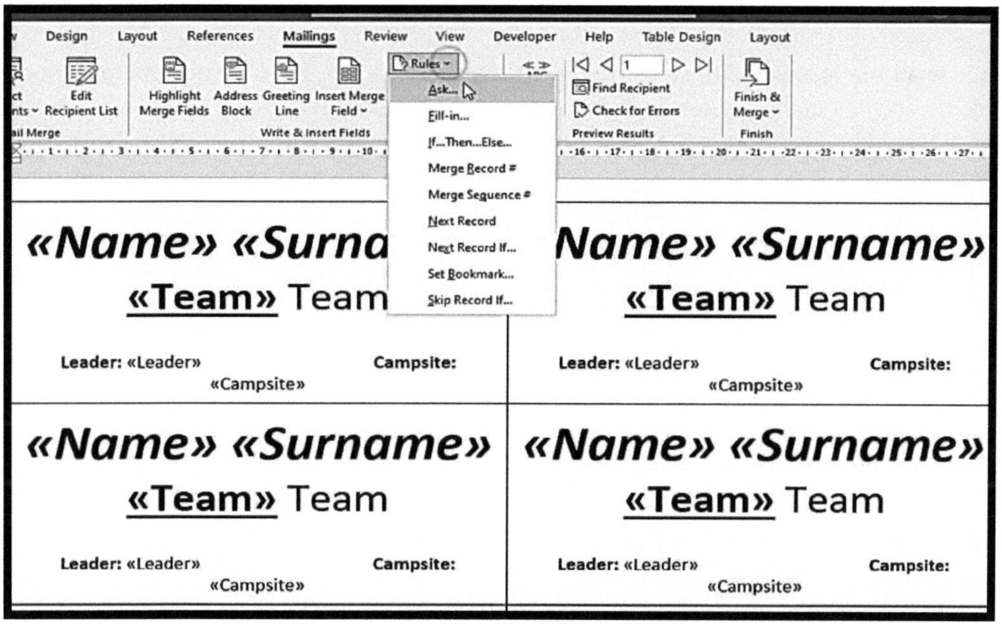

All you have to do now is put the Next Record on the back of each of these. You don't have to do it for the last one because it will be done automatically on the next page, but it must be done for all of them.

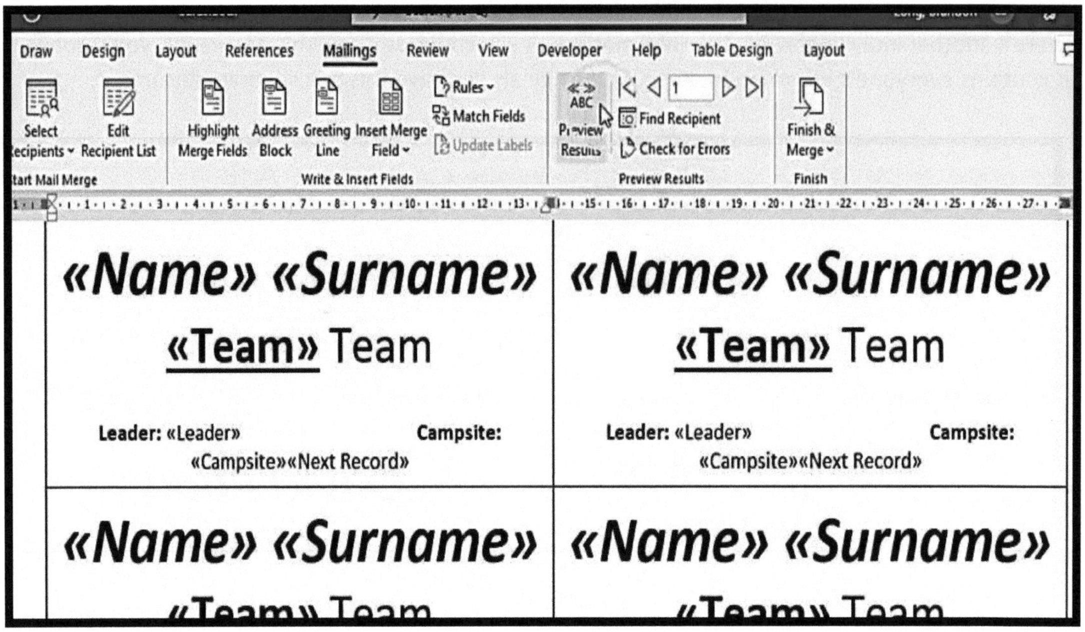

If you go to preview, you'll see a page with everyone's cards on it, and if we run out of records, it'll simply fill in the blanks.

Hopefully, you now have a better understanding of how a mail merge works. If you choose, you can now print and laminate the document.

Keep in mind the following five steps: You go to the **Mailings** tab, **start the mail merge**, **choose your recipients**, and edit them, if necessary, after which you can **search your fields**, **preview** it to make sure it matches everything, and if it doesn't, go back to make the necessary changes, and then **Finish and Merge** and you're done.

CHAPTER TWENTY-THREE

LABELS AND ENVELOPES

Welcome to this chapter, where we'll go through how to work with envelopes and labels. We discussed mail merge in the last chapter, and you became familiar with the Mailings tab. We'll look at the first group in this chapter, which is about creating labels and envelopes.

In this first section of this chapter, you'll learn how to work with envelopes in Word. You'll see how to work with envelopes in Word in the first portion of this chapter. You can print any size envelope, and your Mailings ribbon displays all of your envelope options. We have two options in the first group, the "**Create**" group, if you click on there: **Envelopes and Labels**. In the next section of this chapter, we'll look at labels, but for now, let's begin with envelopes.

Working with Envelopes

After clicking "**Envelopes,**" the Envelopes and Labels dialog box appears; note that this is the same window that you'll use for both envelopes and labels, and the two tabs at the top are for envelopes and labels.

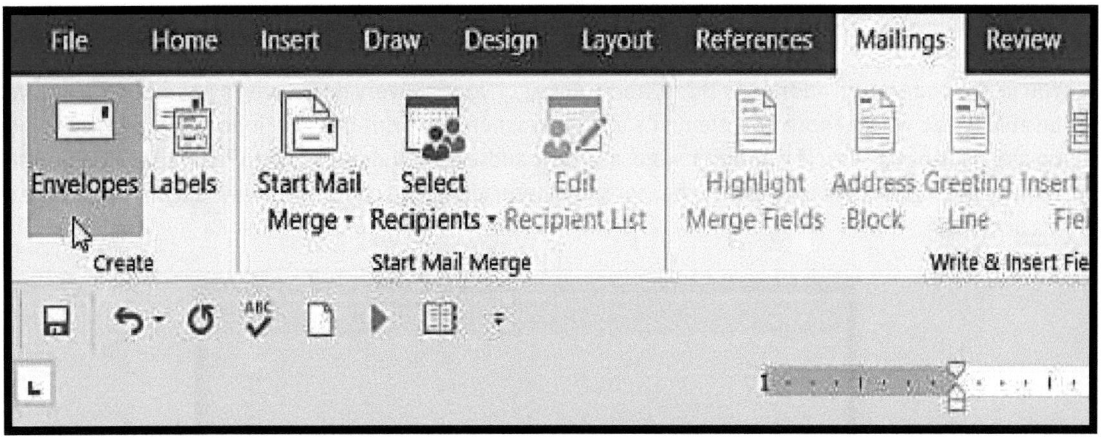

You'll be on the Envelopes tab right now, and you'll notice that the first thing you need to do is type in a delivery address for the envelope's front.

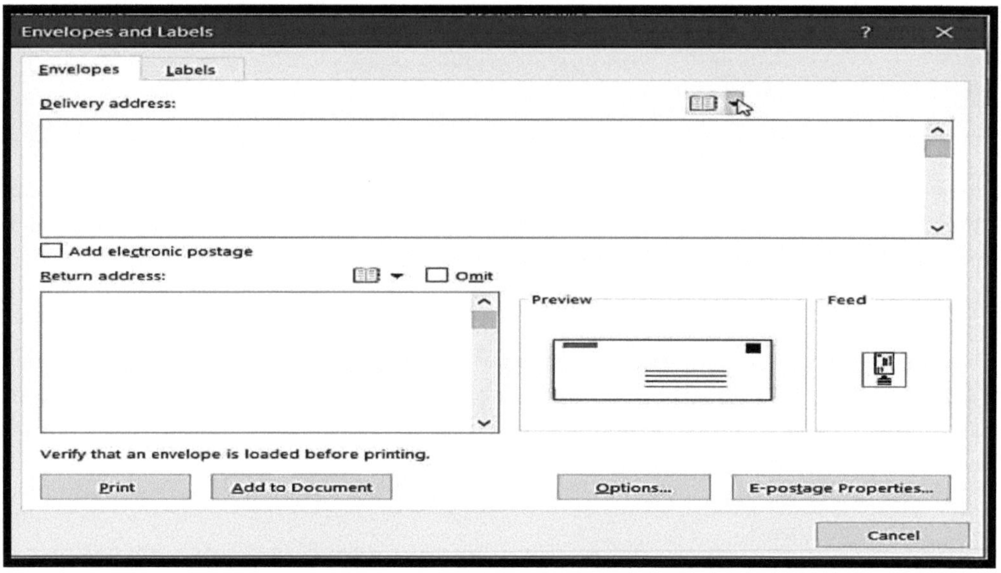

You have a little drop-down just here that will allow you to select something from your Outlook address book, but you can also manually type in a delivery address that you've already copied to your clipboard, and then the second large box we have here is a return address; it's always good to put this in if you can so that if the letter doesn't get delivered for whatever reason, the letter will have a return address.

You've got your return address in there, and you've got your delivery dress, which is wonderful, and you'll be able to see where those two elements will show when you print the envelope in the preview window on the right-hand side. If you don't want a return address, you don't have to have one. If you wish to remove it, you can click the "**Omit**" checkbox just here, and if you check the preview again, you'll see that it's now gone.

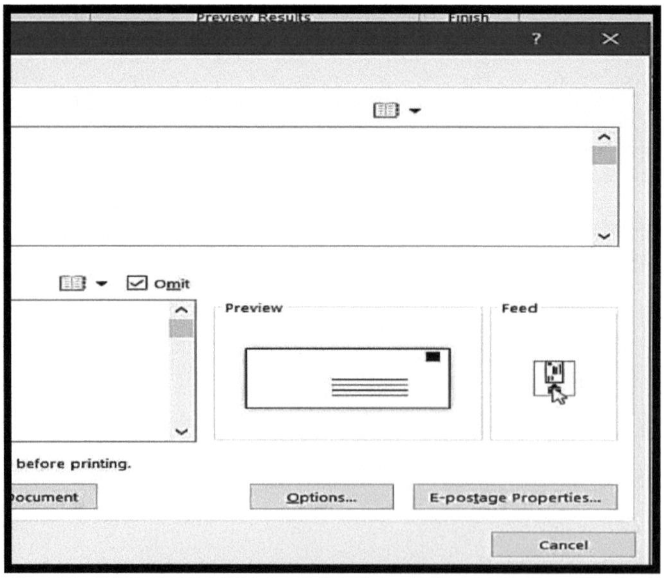

There's also a checkbox for "**Add electronic postage**" here. If you add it, a popup may appear, stating that you'll need to install electronic postage software before you can use this function, which you can get from the Microsoft Office website.

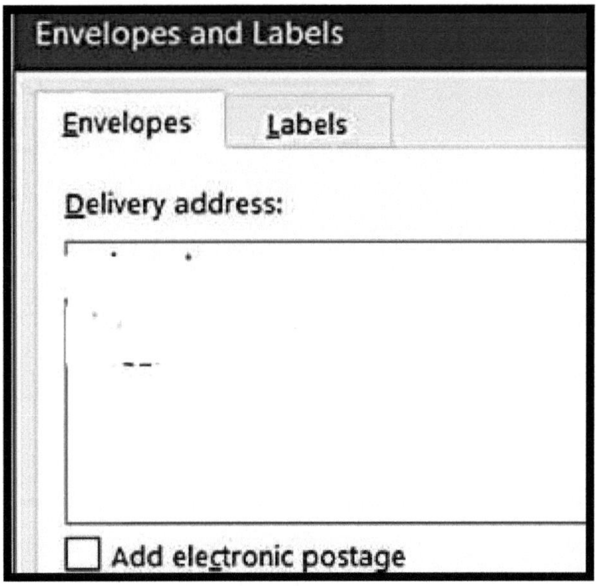

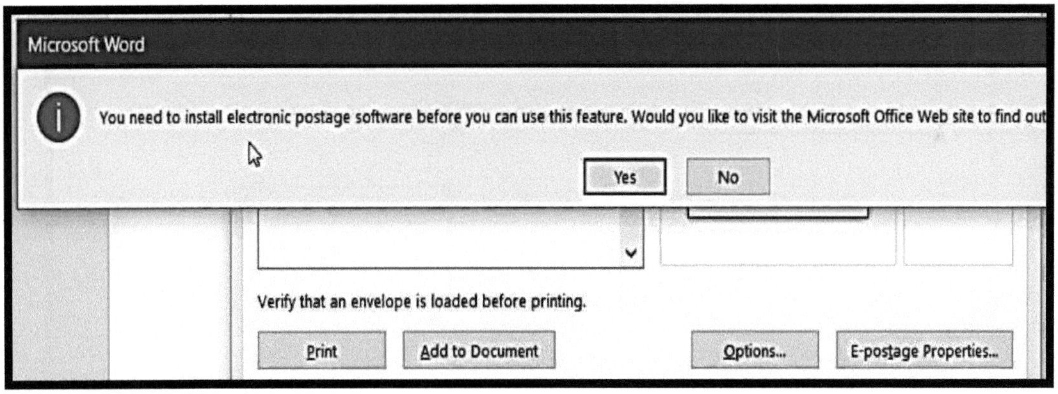

Electronic postage allows you to add a barcode to the top of your envelope, which the post office could scan and use to determine how to route your mail. Just be mindful that the option exists, and if you think it will be useful to you, you might want to look into getting the Word add-in for electronic postage.

We have this tiny button next to the Preview window to the right that reads "**Feed**" and shows you how to feed your envelope into the printer. You can see that you feed it in from the bottom, with the return address on the left-hand side if it is closest to you. That's just a handy little diagram to keep you from putting your envelope in the wrong direction.

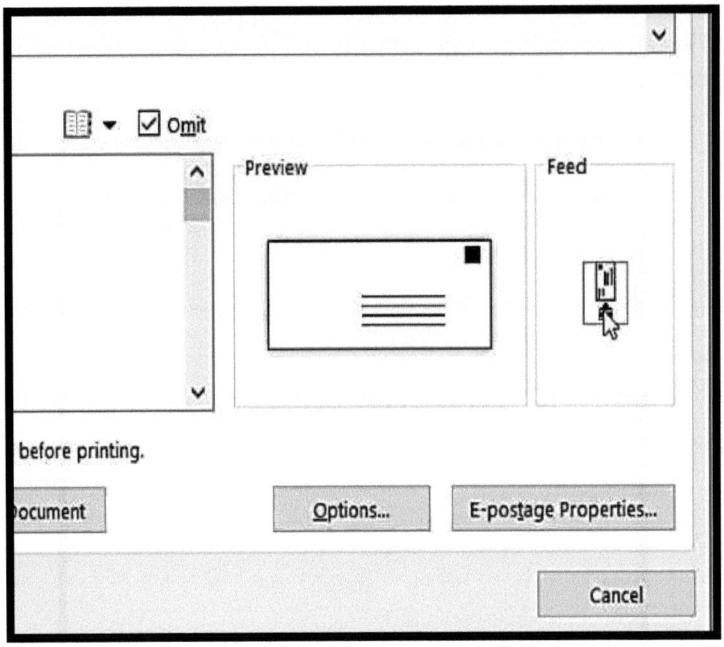

Let's have a look at some of the other options available. You can select the size of your envelopes by clicking the "**Options**" button.

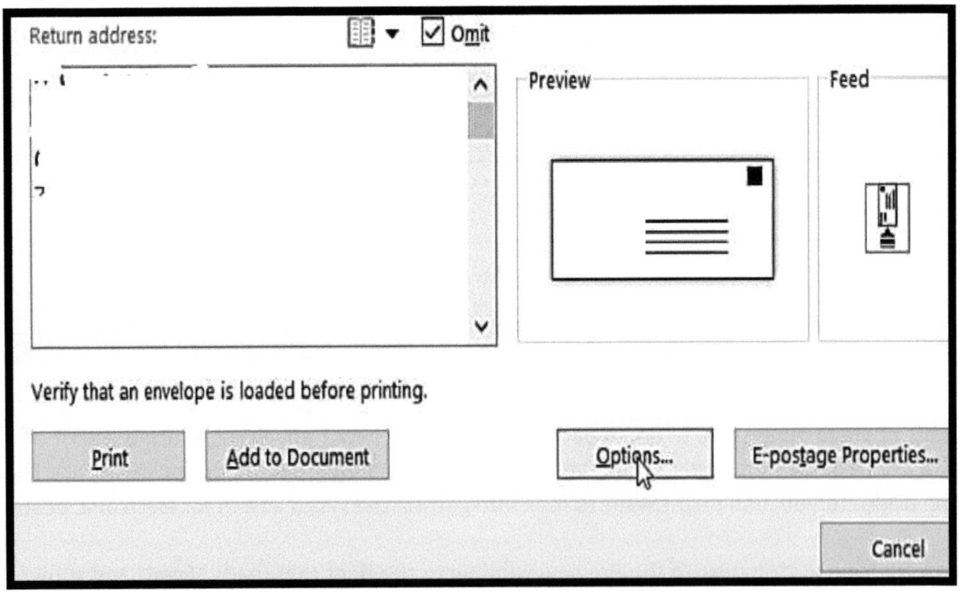

Yours may have a default size of 10, which is a common envelope size, but you can measure the size of your envelope and select the appropriate size from the drop-down menu.

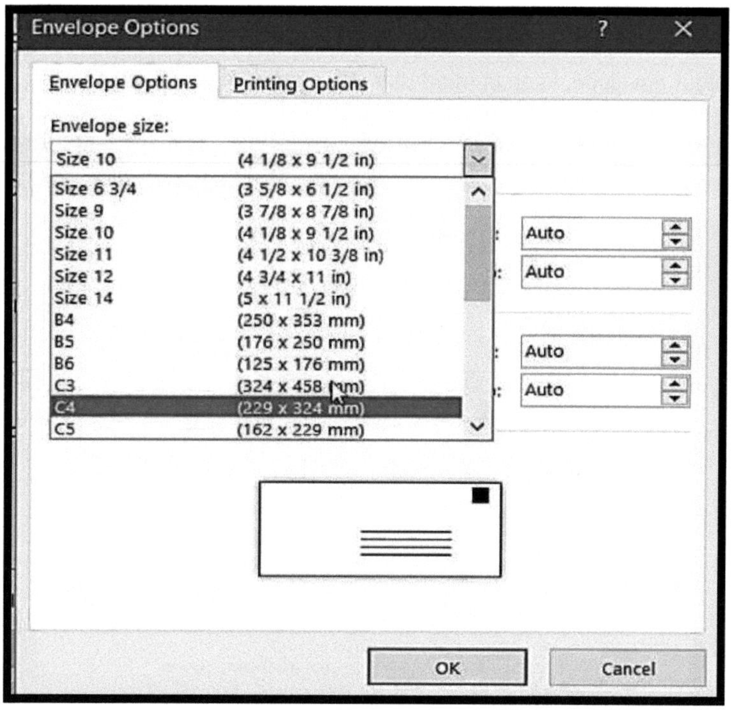

If you like, you could also modify the font for the delivery and return addresses, as well as the spacing, which is how far the delivery address and return addresses appear from the left and top. You'll see a preview of your changes as you make them, so you can see how they'll look.

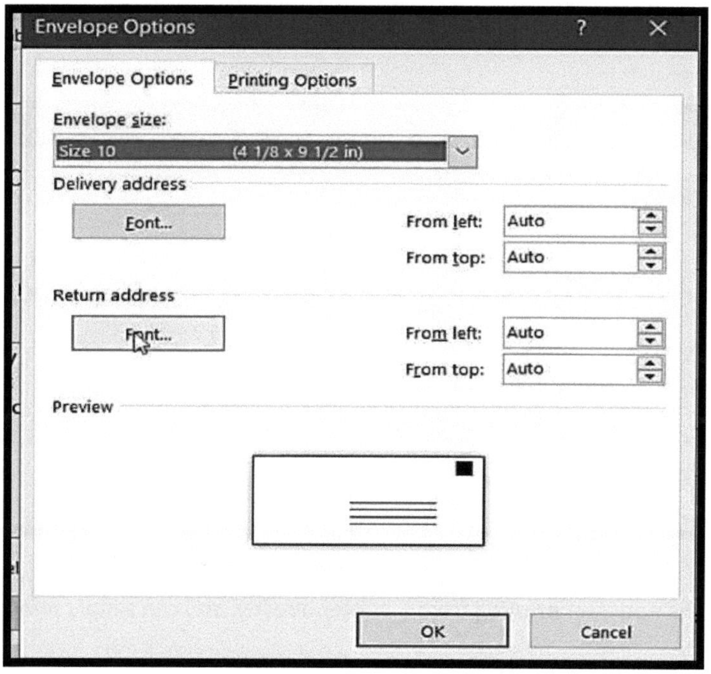

There's also a **Printing options** box, which displays the various feed methods available when feeding envelopes into the printer. You can **rotate clockwise** or **clockwise** and choose "**Face Up**" or "**Face Down**." Before you print your envelope, keep in mind that you have some more customization possibilities.

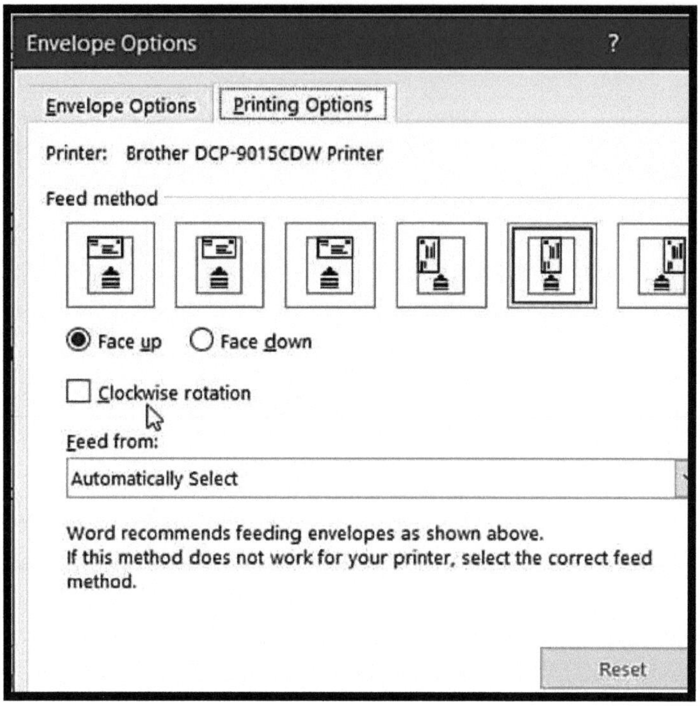

You'll need to add an item for the "**E-postage Properties**" option, so if you've added electronic postage, this will also be active.

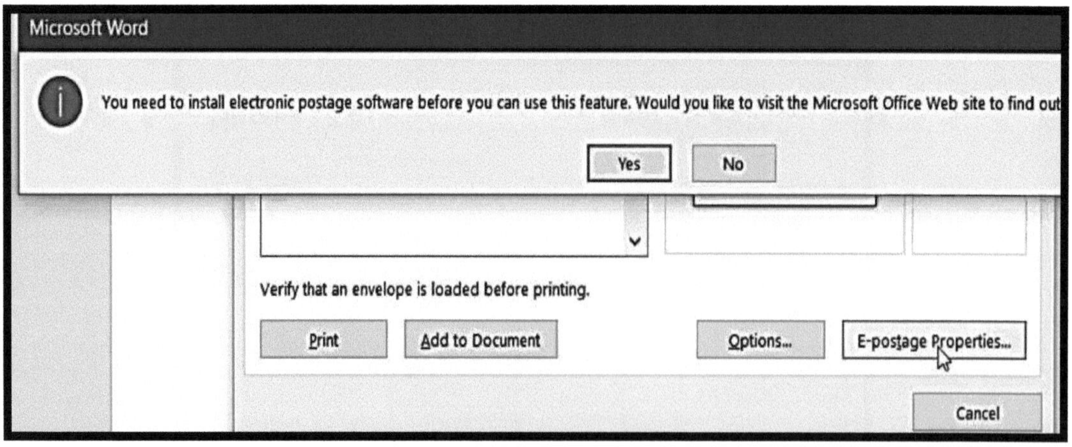

Finally, you have the option of **printing** from a printer directly. You can simply print when you've loaded your envelopes.

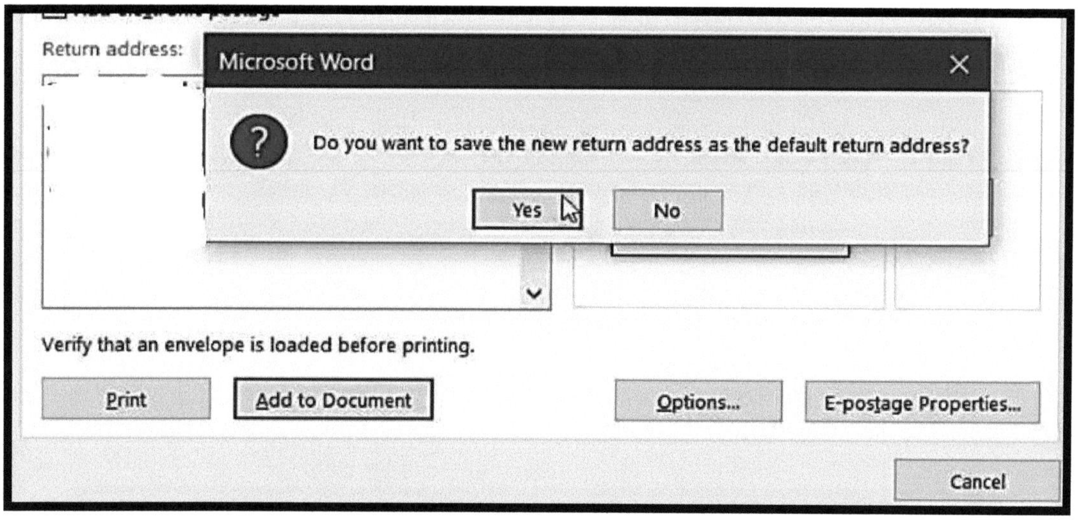

Alternatively, you can choose "**Add to Document**," which will prompt you to save the new return address as the default return address when you click it. This can be handy in some cases, especially if you're printing the same thing over and over. You're going to say yes, and there's your envelope waiting for you.

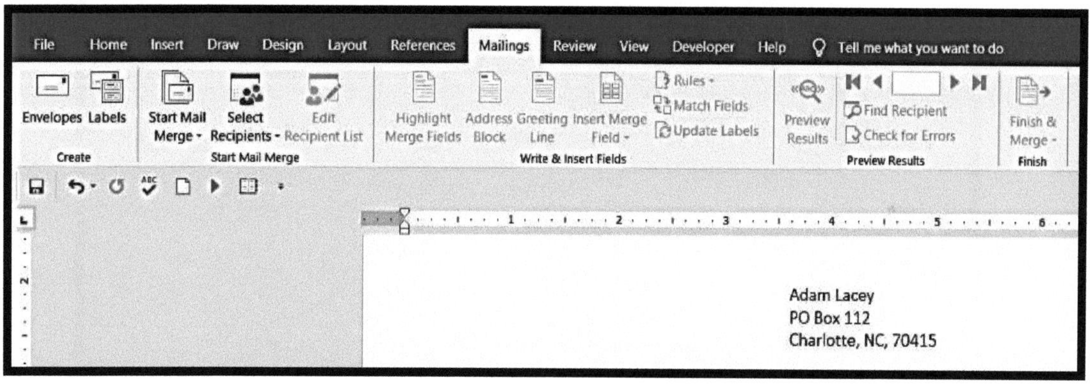

You might want to go in now and make some changes, such as slightly moving that address up, which you can do because it's simply a text field. The good thing about this is that you have your envelope at the top and then space at the bottom to enter your letter or document, so you can print everything at once. You can put your envelopes in one feeder and then put your plain letter paper or A4 paper in another tray and print the document right away. Envelopes are rather straightforward to put up.

Working with Labels

We looked at how to print and customize an envelope in the previous section, and now we'll look at how to print labels. When we talk about labels, we're talking about those adhesive sheets of paper that you can peel off and place onto packages or whatever else you choose, and Word has a great feature for organizing and printing labels. If you go back to the **Mailings** ribbon and look in the first group, you'll see a "**Labels**" option, which you should select.

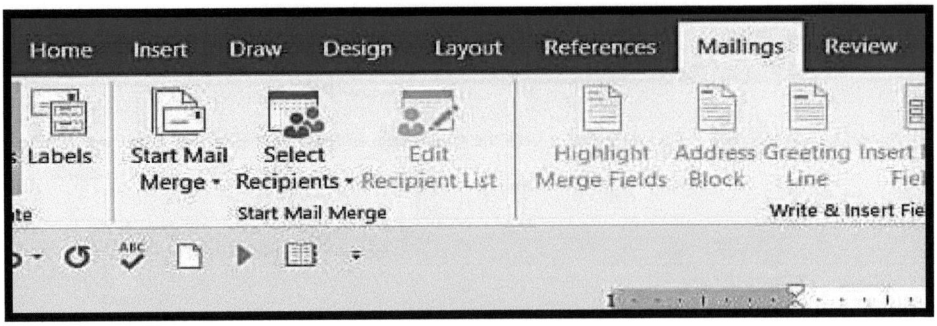

You'll see that this is the identical window that we dealt with in the last section; it's only that this time it's defaulting to the Labels tab.

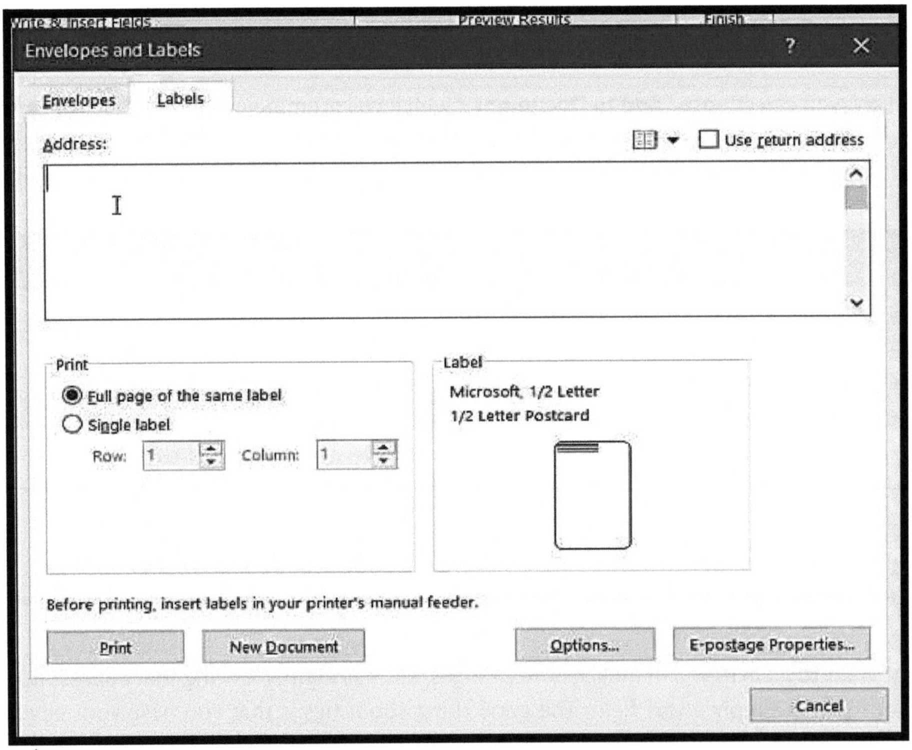

You would type your address in the address field, and then select what you want to print from the drop-down menu. You have the option of printing a whole page of the same label (with this address on every label) or simply a single label. You might be wondering why you'd want a complete page of the same label; well, if you're printing a full-page, having a couple of sheets of pre-printed labels with your company address on them is a fantastic idea. If you choose a single label, you'll need to tell Word where it is on the page by identifying which row and column it belongs to.

Let's have a look at our Label options for a moment. Your printer information is the first part at the top, and this is where you tell Word where your labels are located.

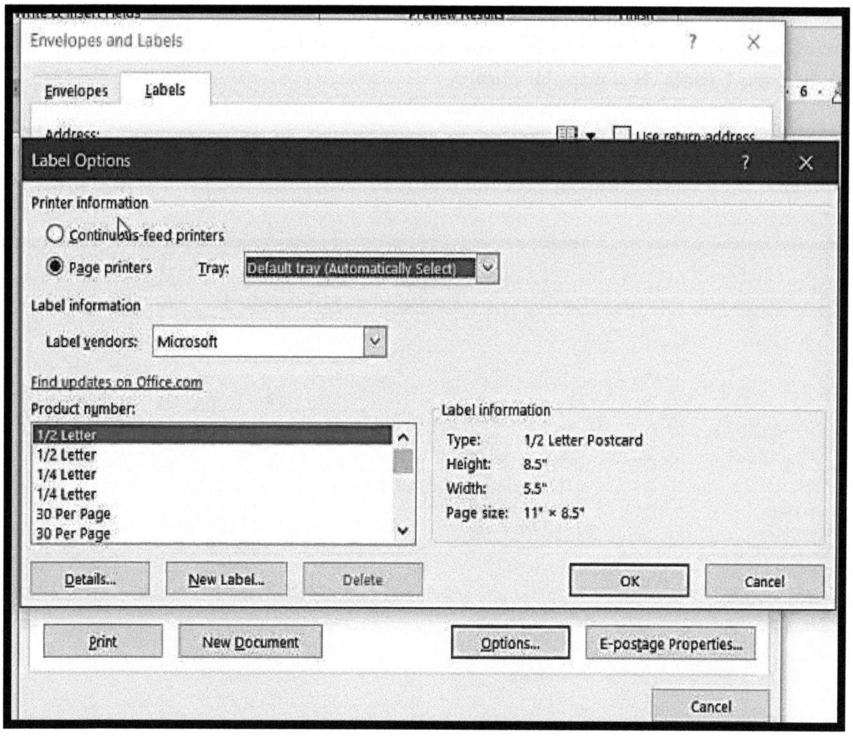

Then you must choose a label from the various options available on this page. The "**Avery US Letter**" is a popular label vendor, and there are a variety of labels to choose from.

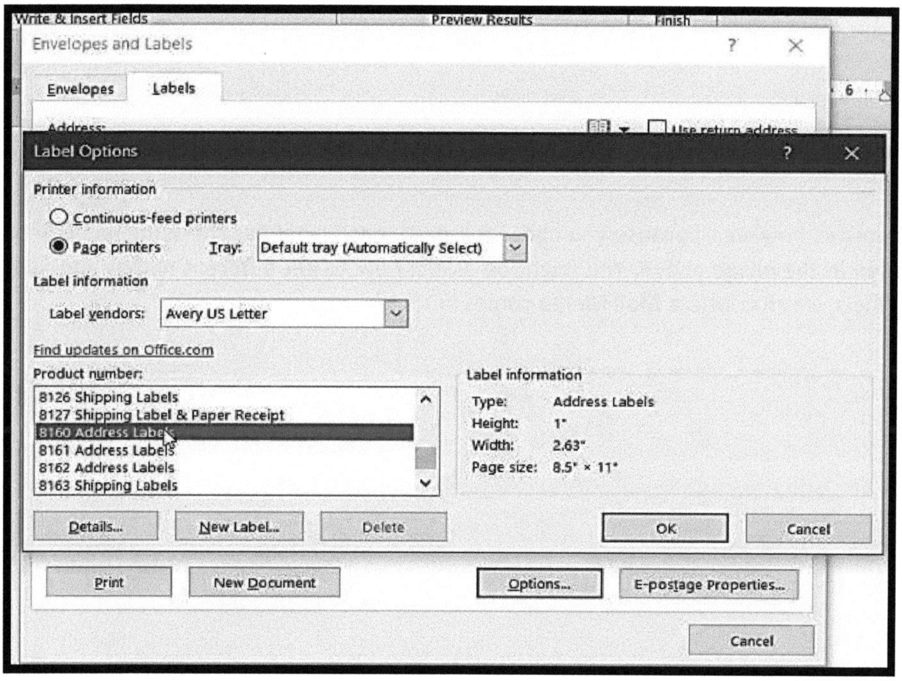

When you click on the **numbers**, you'll see the label information on the side, which will tell you the precise height and width of the label as well as the page size, so you can browse through and locate the one you need. "**8160 Address Labels**" is a popular choice.

You have the option of **printing directly from the printer** or **creating a new document**. Now that you have that address on all of your labels, you can just load them into the printer and print them off.

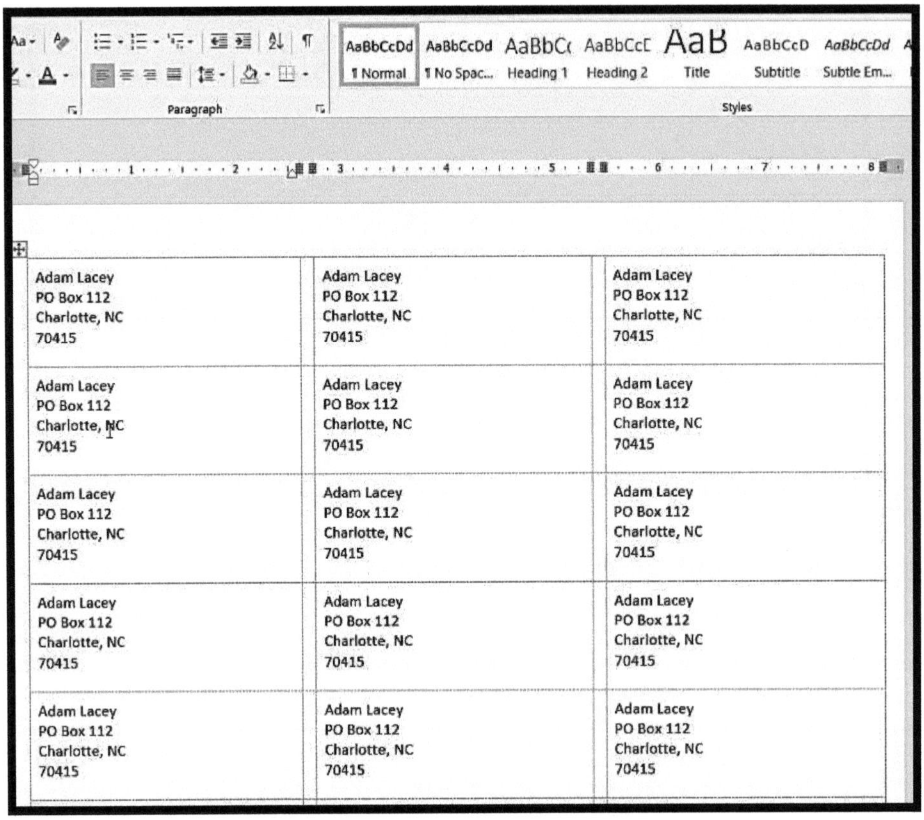

That's fantastic, however, because you chose full-page, each label has the same name and address, as you can see in the image above. You might be asking how to get different names and addresses onto distinct labels, which is where **Mail Merge** comes in.

CHAPTER TWENTY-FOUR

WRITERS' TOOLS

In this chapter, we'll look at various tools that can help you stand out as a writer. As a writer, you'll need to be familiar with these techniques to make your work much easier to read and comprehend for your audience. So, without further ado, let's get started.

Adding a Formula

If your table contains any numbers or numerical data, you can include a formula that adds them all together. First, add a blank row at the bottom, and then go to the Layout ribbon and add a "**Formula**" to your word table under the "**Data**" **group**.

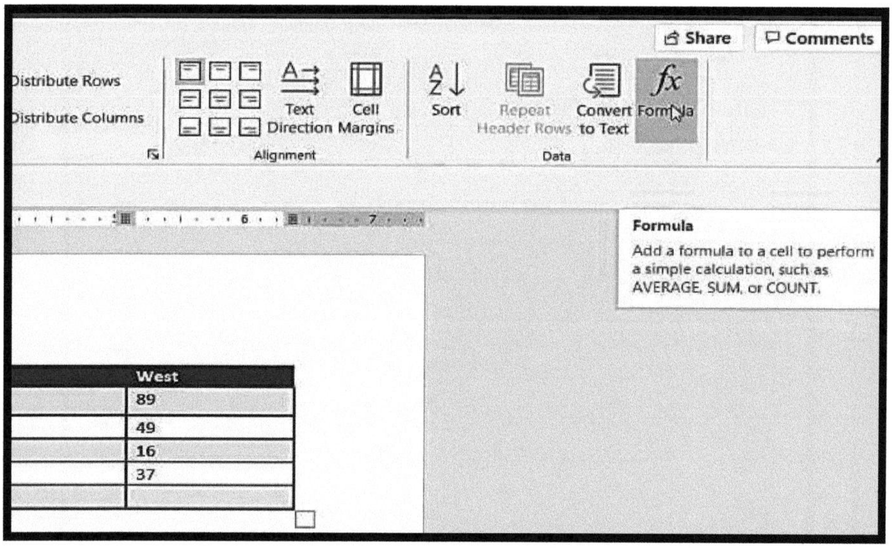

It recognizes that you have numbers above the point where you clicked and automatically enters the formula. So, it equals sum, followed by "Above" in brackets; "**=SUM(ABOVE)**." When you click "OK," the sum is added.

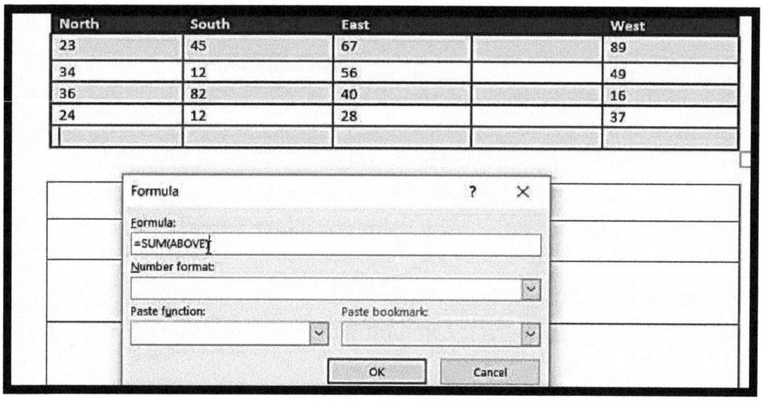

North	South	East	West
23	45	67	89
34	12	56	49
36	82	40	16
24	12	28	37
117	37.75		

In Word, you can utilize roughly 18 distinct functions in formulae, and all of them are listed in the "**Paste Function**" drop-down menu. If you'd rather do an average computation than a total, you can do so by selecting ABOVE, BELOW, LEFT, or RIGHT, depending on where your numbers are present.

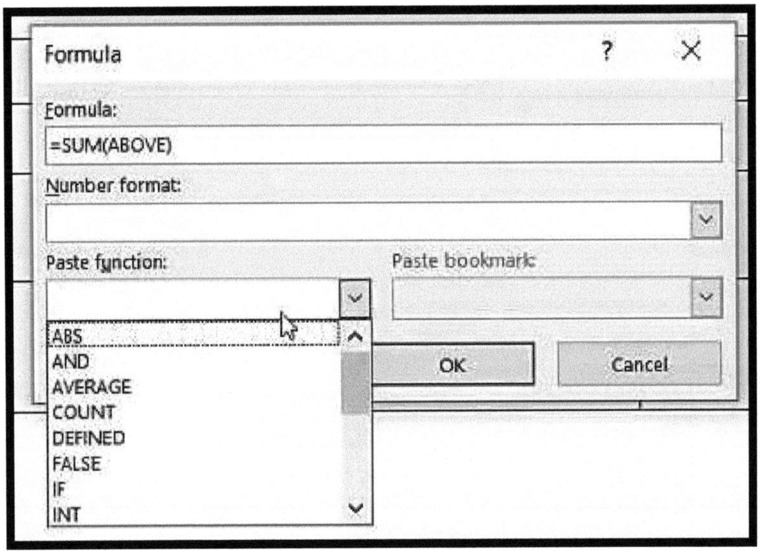

For example, all you have to do is delete the sum function and select **AVERAGE**, then type in "**ABOVE**" and click "**OK**" to get the average.

Collaborating with Others

In this section, you will see how you can **collaborate on a Word document** with other people without needing to send attachments back and forth. In the past, if you've wanted to collaborate on a document with others you needed to send an attachment back and forth and what typically tends to happen is someone might say document one then someone responds with their edits and its document two and maybe someone else had a version of document three and then you have to merge them all and you just end up in this really bad situation of different versions and just a versioning mess.

Luckily with Microsoft Word, you can now work together on a document that's in the **Cloud**. In this section, you will learn step by step how you can do this in Word and these same things apply to the other office apps.

Uploading your File

In the past what you would have done is if you need help on a document, you would have emailed it to someone and then they would have made some edits and they would send it back but if you have multiple people doing that it becomes very complicated. What you want to do is to get other people to help you with this and the best way to do that is to **get this document into the Cloud** so everyone can work on the same version of the file.

The version you have on your desktop is referred to as a **Local copy** because it's located on your computer; a **Cloud copy** is something that everyone can access. That's one of the benefits of putting a file in the cloud, not only is it safer in the cloud, because if something happens to your PC you would lose this file on your desktop, but other people can also access it as long as you permit them to access it and then you can work on that file together all at the same time.

To do that within Microsoft Word there is an option up here in the top right-hand corner that says "**Share**".

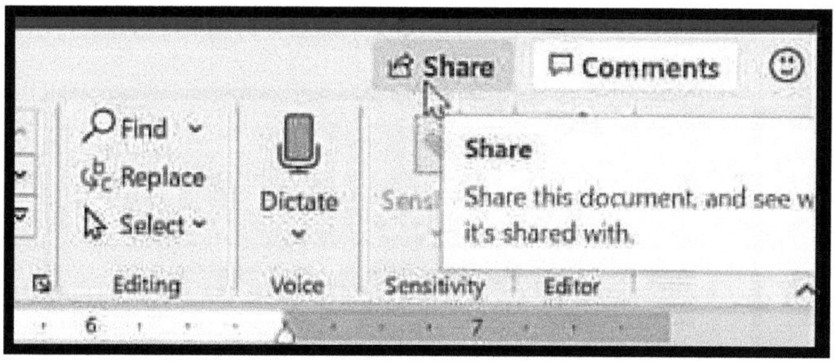

You go ahead and click on that and now what it says is to upload a copy of your document to OneDrive to share it with others so if you want to share this document you need to put it in the Cloud.

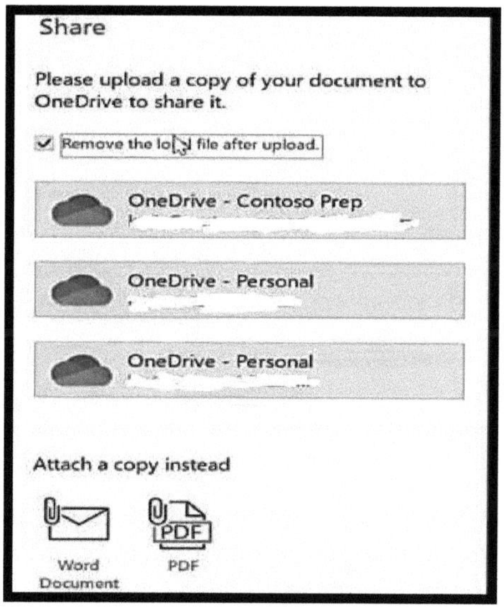

What you could do is once you upload it you could remove the copy or the local file that's on your computer. This probably makes sense because if you have one in the cloud and one on your computer it might get confusing you have multiple copies of the file. You can check that box and get rid of the copy on your computer once you put it in the Cloud.

It also gives you the option to attach the file as either a PDF or as a Word attachment to an email but once again you could end up with version issues where one person makes updates and then that doesn't match your version. Instead, what you should do is get this file in the cloud and to do that you're going to select your account here and what it's going to do now is to upload this file to OneDrive.

Granting Permissions

Now that it's placed the file in OneDrive you can now share this file with others and you have a few different options in terms of the permissions that you can set here. You can allow anyone with the link to edit it or only people in my organization to edit it. You could also allow editing or make it read-only so other people can't edit it. You can also set an expiration date on when they won't have access anymore, you can even set a password to a file or you could prevent downloading of the file; there are lots of different controls you could use.

After applying these settings, you can type in the email address of someone you want to send it to and then Microsoft will send an email to that person with a link to the document. You can also copy the link to the document, paste it where you want to share it, and then they can access the document.

Accessing your Shared Files

Now the file is in the Cloud if you want to know where the file went, on the website **office.com**, what you are going to do is click on **OneDrive**. This is where you can see all of the files that you have saved in the cloud.

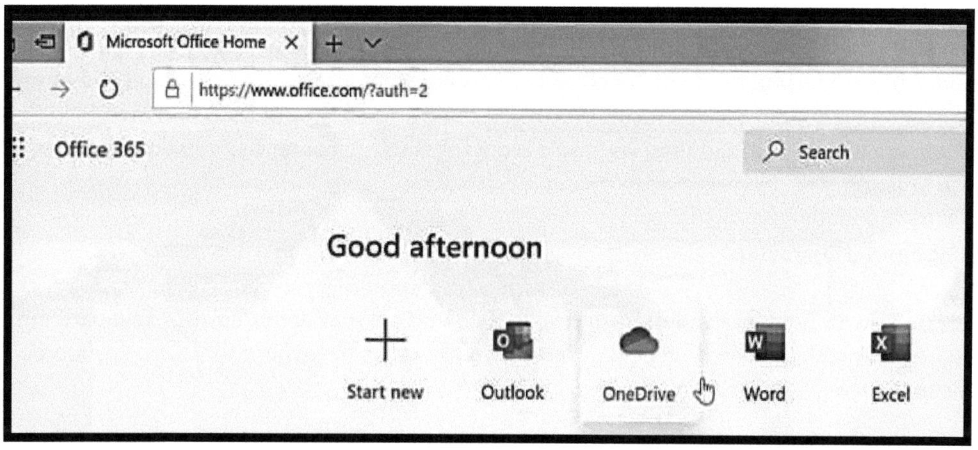

Once it's uploaded your file to OneDrive, not only can you share directly from Word, one of the things you could do is right here next to the name of the file you also can share it from here and the same dialog that you saw in Word pops up here within OneDrive also allowing you to share the file.

Editing a Shared File

Let's say a friend of yours has asked you to help out with a document and shared the file link with you. What you are going to do is paste in the URL of that document and this is going to now open up the Word document and you can start working on the document.

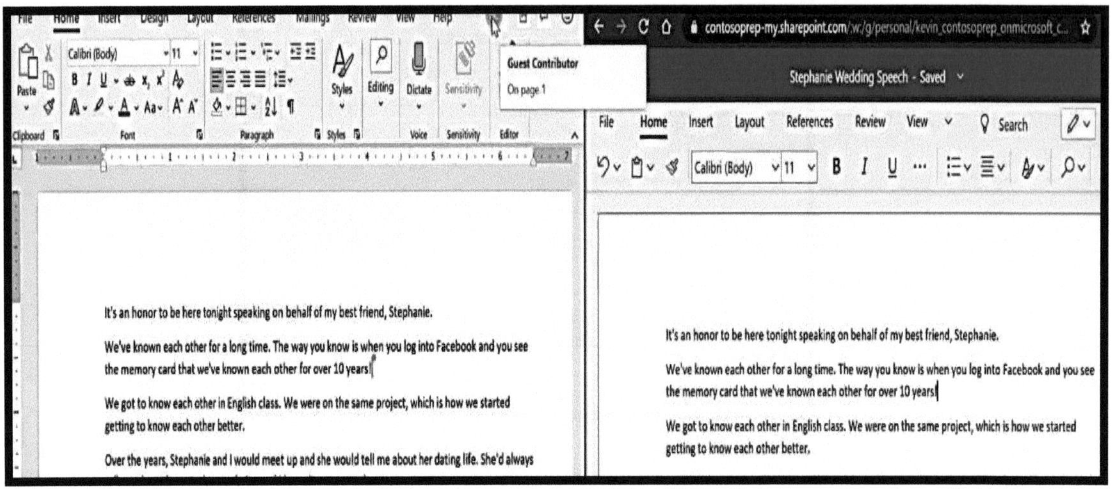

If you share this document with someone, as they start typing on the document from their window, you can see all of the edits that are happening in real-time. Up here in the ribbon as well, you can see who else is within the document.

If they are logged in with their account, you'd also be able to see who exactly is within the document.

This summarizes how you can work together with others on a document in real-time and as mentioned earlier, this isn't just limited to Microsoft Word; you could also use it with Excel, or PowerPoint, you could share your file with others and then you could work together on content so you don't just have to work alone in isolation.

Inserting Special Characters

You'll learn how to type foreign letters and special characters that aren't on your keyboard in this part. We utilize an English keyboard by default unless we are in a place where there are keyboards that are specific to that country's language.

You may occasionally need to enter foreign letters or special characters that are not present on the keyboard you're using at work, school, or even at home, but thanks to **ASCII** (**American Standard Code for Information Interchange**), we can do so. This simply refers to a readable text that is not in binary form, and with the help of ASCII codes, you can overcome the difficulty of writing foreign characters; of course, there are a variety of **ASCII** codes available.

To overcome this difficulty, we will look at options that are already available to us. The **Character map** is what it's called. If you type "**Character map**" into your Windows search area, you'll get a list of results.

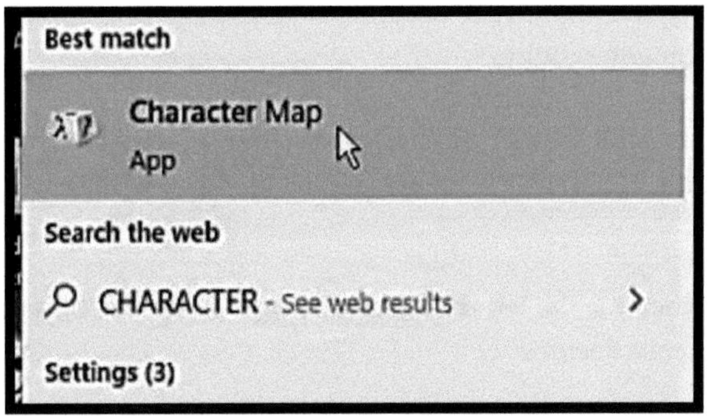

This is a system tool that can be found even in far older versions of Windows, and it provides you with a list of readily available foreign and unusual characters.

Inserting Characters such as Π, Ñ, Ç, and Ũ

Let's say you are trying to type a Spanish character which is the letter N with a special character at the top (**Ñ**). You may need to write a Spanish name with a particular character that isn't available on the English keyboard at work.

To find it, go to the Character map you've already opened and look for it. Now **click on that character**, then on the "**Select**" button here, then on the "**Copy**" button beside it, and this character will be copied to the clipboard once you click on the Copy button. This works for both uppercase and lowercase letters.

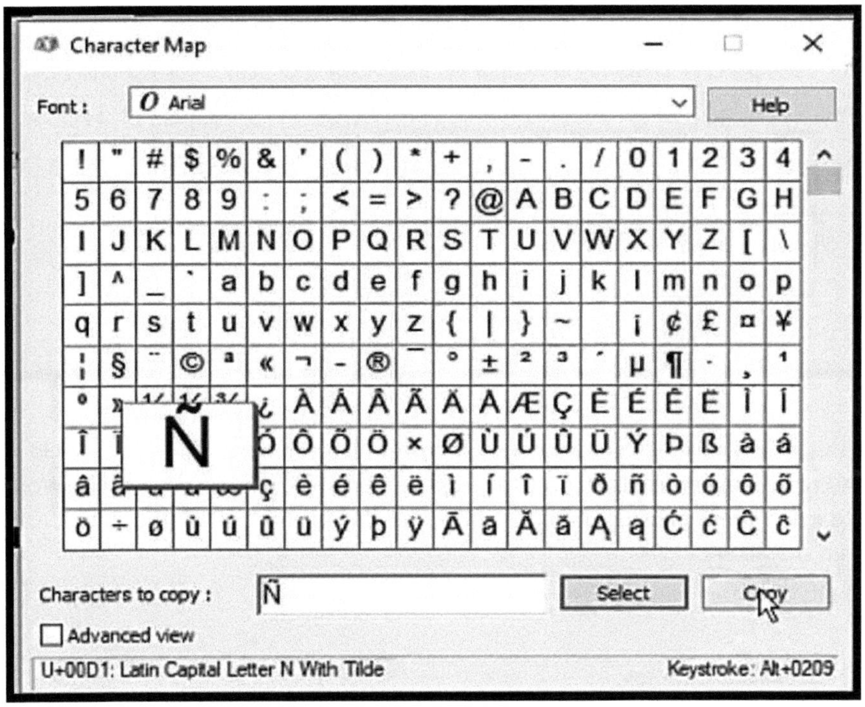

Now all you have to do is paste it into your document. As you can see, you've used this character map to enter the foreign character you require.

This character map offers a lot of lists of characters from various languages if you scroll through it. Consider the letter U with a special character at the top (Ũ). If you wish to use it in your document, follow the same steps.

Another well-known character is the letter C, which has an "S" sound (Ç). After you've found it, select it, copy it, and paste it into your document.

The Russian special letter n (П) is located farther down.

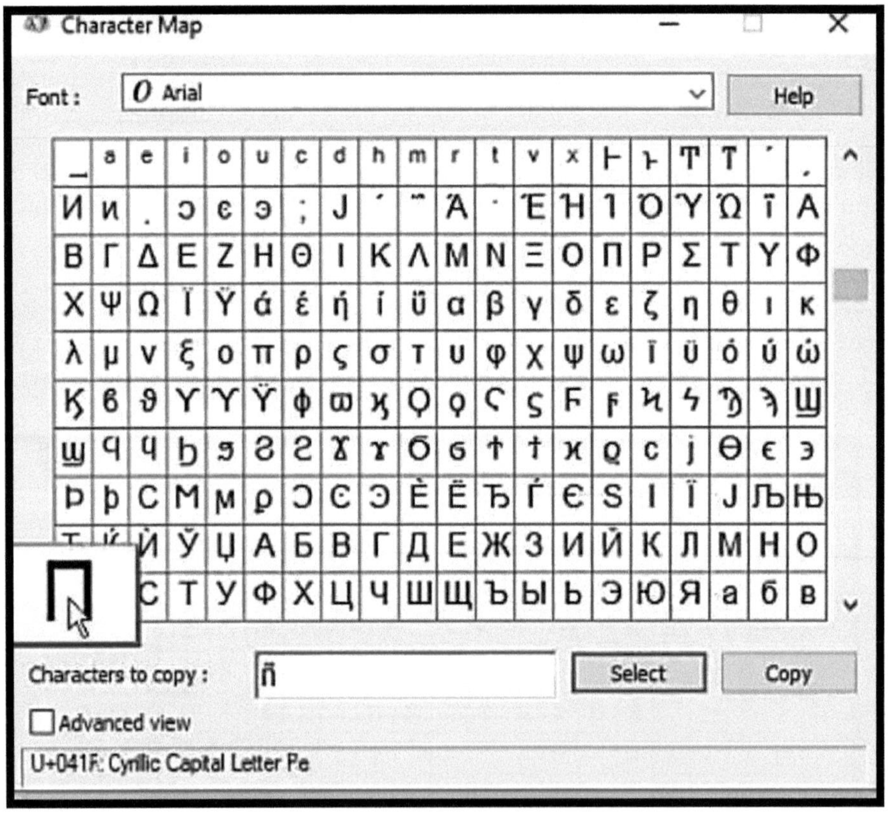

Hopefully, this gives you an idea of what the Character map can do and how it can help you overcome the difficulty of needing to use foreign letters or accented vowels in documents like Word. You've also seen how to insert these unusual characters into your document; explore these options to learn more about foreign characters.

CHAPTER TWENTY-FIVE

PUBLISH YOUR DOCUMENT ON MICROSOFT WORD

In this chapter, we are going to learn how to publish and print a document. Before you print a document, you must look at it and make sure that everything is in order. This chapter will also cover how to export a document and create a PDF

Introduction

After writing, editing, formatting, and proofreading, the last step in document preparation is to put it on the Internet. Publishing is a broad term that includes printing a document on paper and making an electronic document that you can share on the web.

Document On Paper

When you finish your document on Microsoft Word, you can choose whether to make a paper copy of the document, which is also called a printout or a hard copy.

Previewing Before You Print

Before you print, look at what the document will look like. Even if your document is supposed to look the same on the screen as it does on paper, you might find little mistakes like missing page numbers, blank pages that don't belong there, and so on. Therefore, you must get a look at things before you print.

To see your document, first, you have to save your work. Then click on the **Ctrl+P** key (The Ctrl+P keyboard shortcut is the most common way to print). Another way to do this is to click the File tab and choose Print from the File screen's left side. The Print screen comes up, like the one shown in Figure 1.

Figure 1: An image of the Print screen

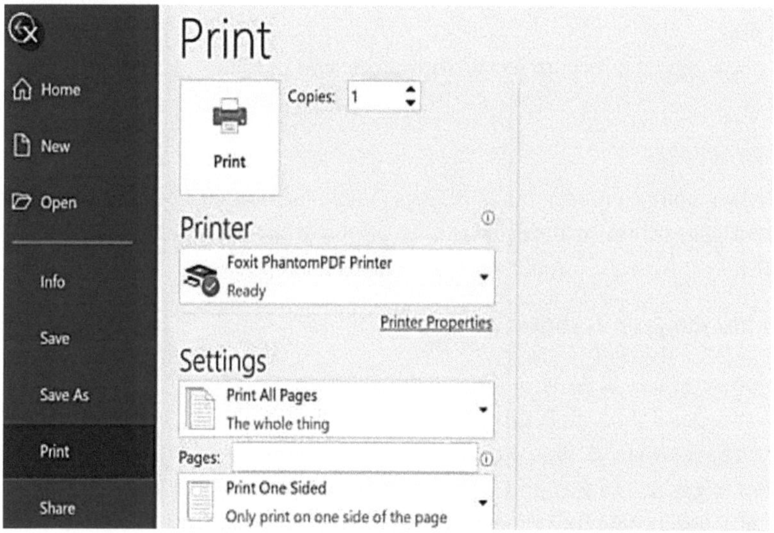

If you want to see a preview of each page, click on the forward and backward buttons at the bottom of the page. To read more text, use the zoom slider at the bottom of this page to make it bigger. The zoom

slider that lets you change the size of the text is shown in chapter 1, figure 5. You might want to look at the layout of footnotes, headers or footers if you're using them in your text. The goal is to find anything that's not appealing before you print. When things need to be fixed, you can click the back button or press the Esc key to get back to your text.

Printing The Whole Document on Microsoft Word

To print a document, ensure the printer is on and ready to print. It is faster to print when the printer is on, so make sure it is turned on.

Save your document and then click on the **Print command** on the File tab or press Ctrl+P. Choose Print from the **File tab**. The Print screen comes up. Look at Figure 1. Click the big Print button. The print screen closes and the document comes out from the printer

Note: Fortunately, you can keep working while the paper is printing. Also, you should not use the Print command again if nothing comes out. It's very likely that the computer is still thinking or sending information to the printer, and that nothing has gone wrong. It's likely that if you don't see an error message, everything will likely print out at some point.

Printing a Certain Page

To print a particular page that you want. Take these steps:

- To print, you need to move the insertion pointer so that it's sitting somewhere on the page that you want to print.
- To make sure you're on the right page, look at the page number on the status bar
- Press the keys Ctrl+P. The Print Range button is next to the Settings button on the right shown in figure 1, it says Print All Pages.
- The menu has a print option called **"Print This Page,"** which you can choose from.
- The Page Range box at the bottom of the Print screen can also help you check the page number.
- Then print.
- That single-page prints with all the formatting you put in.

Printing a Range of Pages

On Microsoft Word, you can print a range of pages including odd number pages, even number pages, or a mix of different pages from your document. To print a group of pages, go to the Print screen or press the keys Ctrl+P.

The Pages text box shows up as shown in figure 2

- To print pages 5 to 9, write 5-9.
- To print pages 3 through 9, type 3-9.
- To print pages 4 and 8, type 4,8.
- To print pages 3, 5 to 9,15 to 17 and 19, type 3, 5-9, 15-17, 19.
- Print All Pages is now turned into Custom Print if you type any value into the box.
- Print when you're ready. Only the pages you tell the printer to print come out of it.

Figure 2: An image of the Print screen with an arrow pointing at the page box

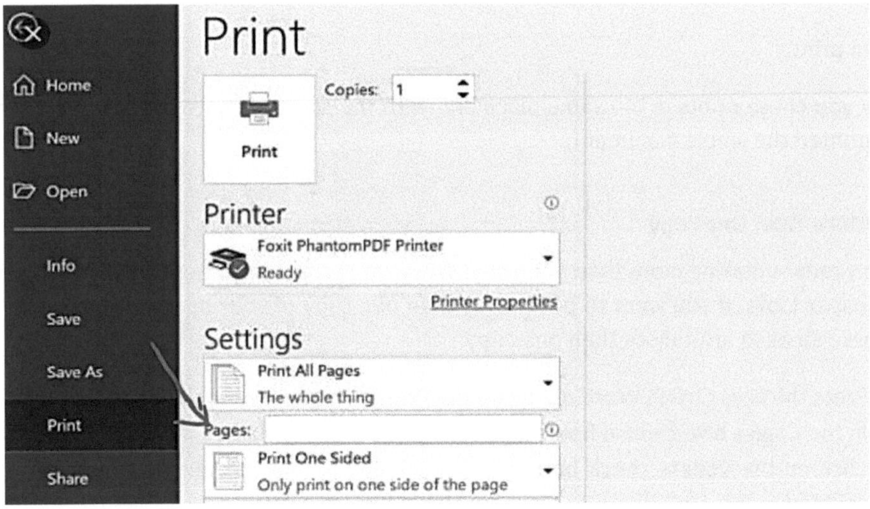

Making Copies of a Page that are Printed on Both Sides

On Microsoft Word, you can print your document on both sides of a sheet of paper. Take these steps:

- Right after you save the document, press Ctrl+P.
- Click the **Duplex Printing button** on the Print screen to make two copies of the same thing.
- Choose "Print on Both Sides, Flip Pages on Long Sides," then click "OK."

Short Sides isn't worth it if you don't plan to bind your document in that way.

Odd and Even Pages are Printed on Both Sides of the Paper

If you have a printer that doesn't print on both sides of the paper, what would happen? If that's the case, you can print all the odd pages in your paper. Turn the paper over and put it back in the printer again. In the next step, print all the even pages. This means that the text will be printed on both sides of the page.

Note: It's a good idea to print only one odd-and-even page first, to make sure that when you put the paperback in, it's in the right orientation.

Printing a Block on Microsoft Word

When you print a document with Word, you have a lot of different options for how it looks. Make sure you have marked a block of text on the screen. Then, ask the Print command to print only that block of text. This is how:

Mark the text you want to print

See Chapter 6 for all the instructions on how to block a text on the word document.

- Press Ctrl+P to bring up the Print screen.
- The Print Range button is next to the Settings button on the right.

- When you go to the menu, choose the Print Selection item.
- To see the Print Selection option, you need to select at least one block in your document.

It's time to print.

The block you chose prints in the same place and with the same formatting (headers and footers) as if you had printed the whole document.

Printing More Than One copy

This simply means making more than one copy. When you print a document with Word, you can choose how the paper looks. If you want to print more than one copy of your document, you can do that, too. Follow these steps to print more than one copy:

- Press the keys Ctrl+P. Word opens up the Print dialogue box as shown in figure 1.
- In the Copies box, choose how many copies you want to print.
- Click on the **Collate check box.** If the check box is selected, this indicates the copies will be collated.
- Click the Print button and click OK.

Note: As a rule, Word prints one copy of a document after the other. This is called collating, and it is how things are put together. However, if you want Word to print five copies of page 2 and then five copies of page 3 and so on, choose the option from the Collated menu button and type in the number of copies you want of each page The collated option is indicated with an arrow in figure 3 below: Then click on print.

Figure 4: The Collation Option

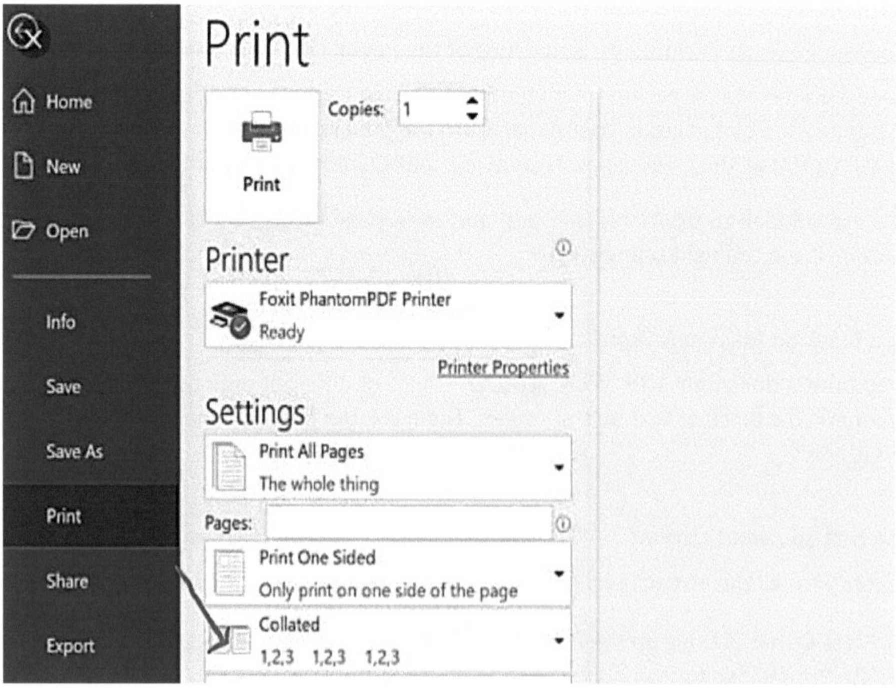

Choosing a Different Printer

Your computer may be able to use several printers, such as one that is connected directly to your computer, as well as some network or wireless printers. To choose a specific printer, such as that fancy color printer, you should follow these steps:

- Press Ctrl+P to bring up the Print screen.
- Make sure you click the "**Select Printer button"** next to the Printer title.
- The button shows the name of the printer that Windows has chosen as the default one.
- Choose a different printer from the list and click on it.

Note: Not all printers have names that are easy to understand. Also, not every printer on the list is a real printer. Several of them are printers that print documents like Microsoft Print to PDF. To add printers, change their names, or set them as the default printer,

Cancelling a Print Job

You can cancel a print job in Word by running to the printer and pressing the **Cancel** button. At times, there is an icon with a red X on it. When you touch that button, the printer will stop. The button stops the document from printing, but it might not stop the printer right away.

The Cancel button doesn't work on some slow printers. You can try to use Windows to stop a print job on a printer with slight issues. Follow these steps:

- D**ouble-click** the little printer icon on the right end of the taskbar to open the printer.
- Select the job that you want to delete from the list.
- Choose either the Document Cancel command or the Document Cancel Printing command to cancel the document or print job.
- Click "**Yes" or "OK**" to stop the job.
- Then shut down the printer's control window now.

Note: It may take a while for the printer to stop. This is because printers have their memory, and even though you cancel it, some pages of the document may be stored and keep printing even after you have cancelled it.

Publishing Your Document Electronically on Microsoft Word

Some documents don't need to be written on paper at all. They are sent to the publisher electronically, and they can be changed right on the screen. When you buy an eBook, you don't need to print it out. And sometimes, documents can be better spread by email than by paper. It's all part of putting your document online and making it public.

Convert a Document to PDF Using Microsoft Word

The best way to convert your document to PDF on Microsoft Word is as follows:

- Open the document you want to convert and then click the "**File" tab.**
- Select the "S**ave As**" option on the backstage screen

- From the Save As screen, select where you would like the PDF to be saved (OneDrive, This PC, a particular folder, or wherever).
- Choose where to save the converted PDF
- Click the drop-down arrow on the right side of the "**Save As Typ**e" box and select "PDF" from the drop-down menu.
- If you want to, you can change the filename at this time. When you're ready, click the "Save" button.
- Change the name of the file (if you wish) and then click the "Save" button

After saving the PDF, you'll be returned to your Word document, and the new PDF will open automatically in your default PDF viewer.

Other Ways to Make a PDF from a DOCX File

If you don't know how to convert a file to or from PDF, use a service like Small pdf, which has free tools for doing so. It's possible to use services like this that have some restrictions or require a fee for full access, but they can be very useful for converting PDFs when you need to do so a few times.

Close Your Document

When you're done you can close your document by simply clicking **Ctrl + W** or clicking the File button and closing the window.

In conclusion, these are necessary things to note about Microsoft word 2025;

- Always remember to save your work on Microsoft Word
- Be very vigilant when using the space bar
- Don't abuse the enter key
- To make your work easy and fast, use most of the keyboard shortcuts.
- Avoid manually numbering your pages and using the bullet in the wrong way
- Do not force a fresh/ new page and always remember the undo icon.
- Make sure your printer is in a good condition before you click on the print button to print any document

BOOK 2

MS-EXCEL

CHAPTER ONE

INTRODUCING EXCEL

Microsoft Excel 2025 is a spreadsheet app and the latest Excel software that allows individuals to arrange, manage, and add data while utilizing formulae. Additionally, the software is included in the Microsoft Office suite, but it is also linked with more office applications.

As the majority of other Microsoft apps are, Excel can be assessed as a cloud-based subscription via Office 365. With Excel, you can perform more functions than you think, as an individual or organization.

The software was specifically made for Mac OS and Windows users. Both users are capable of carrying out functions including creating pivot tables, using graph tools, and forming easy arithmetic. Furthermore, Excel enables us to use the AVERAGE and other functions that will guarantee success in our workplace.

Workers looking to arrange and organize data can use a set of cells formed into columns and rows to do so, and this can be made possible by using Microsoft Excel 2025. Excel also uses histograms, charts, and line graphs to display data.

UNDERSTANDING WHAT EXCEL IS USED FOR

Excel is a popularly used Microsoft Office application that companies and individuals usually use to analyze and save numerical data. Furthermore, it is a spreadsheet program where individuals and companies can record data to create tables. With an MS Excel spreadsheet, it is simple to analyze data. You can summarize data and save it in an orderly manner with the aid of graphs and charts so that you can readily access it whenever you need it. It becomes easy to save data, and you will save a lot of time as a result.

MS Excel also lets users arrange, format, and calculate data while using formulas with a spreadsheet system. There are numerous formulae in MS Excel. By utilizing them, you perform many operations on a huge quantity of data at once, such as computing the sum, average, and so on. As a result, MS Excel is used anytime users need to solve difficult mathematical issues or apply basic mathematical functions to tables with a lot of data.

MS Excel has a plethora of functions that make your job a lot easier and save you time. There are fantastic tools for sorting, filtering, and searching that make your job even easier. You can do your task in much less time if you combine these tools with tables, pivot tables, and other tools.

(NB: Multiple components may be readily found in vast volumes of data to assist in the resolution of a variety of issues and concerns).

Excel allows you to add more complexity to your data. ***(NB: This means you can enhance the data bars, highlight any particular elements for laying emphasis, and quickly make your data more attractive)***. If you have data saved in MS Excel and you want to emphasize something significant, you may do it using the numerous data presentation options provided in MS Excel. You can even make the spreadsheets on which you've placed data more appealing.

MS Excel is useful for budgeting. For example, if a doctor earns $50,000 per month, he will incur some expenses, and if he wants to know exactly how much he is spending each month, he can easily do so using MS Excel. He can enter his monthly income and expenses into excel tables, which will allow him to see how much he is spending and, as a result, control his spending.

Excel is among Microsoft Office suite, and it is compatible with additional applications in the Microsoft Office suite. Like other Microsoft Office products, users can purchase Microsoft Excel 2025 and previous models through the cloud on a subscription basis via Office 365. In addition, this software program uses a cell collection organized into columns and rows to organize and separate data. It is capable of displaying data as charts, line graphs, and histograms.

There are numerous advantages to using MS Excel, which is why it is used by people worldwide for a variety of tasks. Not only does it save time, but it also makes the job easier. It is almost capable of completing any task. For example, you can perform mathematical calculations as well as create graphs and charts to store data. It is simple for a businessperson to compute and save data in it.

LOOKING AT WHAT'S NEW IN EXCEL 2025

Excel 2025 comes in with its new calendar template. Nobody wants to start this year off with a ton of spreadsheets if we are being totally honest. Luckily, this template only has one sheet and it's the only sheet you need.

In the smart calendar sheet, you have a calendar. A cool feature of this template is that when you hover over the interactable aspects of the template, a guide pops out with information on how the tools are used.

This template is flexible because you can adjust out the year to your liking whether you want to skip ahead to 2025 or relieve the good old days of 2019. The template adjusts the dates for you automatically. In the first day label, you can decide what day you want to start your weeks with.

If you want to mark an entry as critical, you need to put an asterisk in front of it so the data shows red. For normal events, you just input values normally and for holidays, you input a forward slash. You get the whole set from January to December and on the top left, you have a full-screen macro button that hides the ribbon for a larger work area and a print selection button that allows you to print the areas you have highlighted.

As always, in the intro sheet, you get the standard items like the content, customization difficulty, explanation of the sheets, and instructions on how to utilize this template.

UNDERSTANDING WORKBOOKS AND WORKSHEETS

The electronic counterpart of a paper ledger is a worksheet. It's a powerful toolset for entering, analyzing, calculating, and manipulating data. A worksheet may be used for simple calculations like addition and subtraction, as well as more complex applications like statistics, audits, and mortgage tables. Worksheets also make it simple to turn your data into useful business reports.

A workbook is a file that contains a collection of worksheets. These worksheets may include a variety of data, but they are generally connected in some way. Each worksheet in a sales workbook, for instance, may include sales data for a single division.

Workbooks may have an infinite number of worksheets, depending on the size of the worksheet and the amount of memory available on your computer. A workbook may also contain chart sheets, Visual Basic modules, dialog box sheets, macro modules, and scenario report sheets in addition to worksheets.

In Excel, a workbook is the same as a file, and a file is the same as a workbook. At the top of the screen, we can see the file or workbook's name.

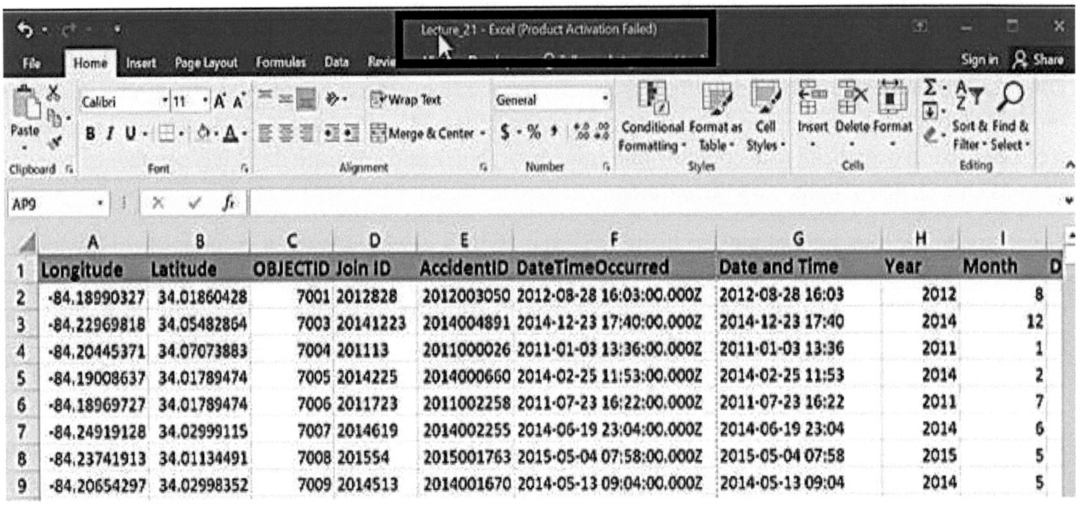

	A	B	C	D	E	F	G	H	I
1	Longitude	Latitude	OBJECTID	Join ID	AccidentID	DateTimeOccurred	Date and Time	Year	Month
2	-84.18990327	34.01860428	7001	2012828	2012003050	2012-08-28 16:03:00.000Z	2012-08-28 16:03	2012	8
3	-84.22969818	34.05482864	7003	20141223	2014004891	2014-12-23 17:40:00.000Z	2014-12-23 17:40	2014	12
4	-84.20445371	34.07073883	7004	201113	2011000026	2011-01-03 13:36:00.000Z	2011-01-03 13:36	2011	1
5	-84.19008637	34.01789474	7005	2014225	2014000660	2014-02-25 11:53:00.000Z	2014-02-25 11:53	2014	2
6	-84.18969727	34.01789474	7006	2011723	2011002258	2011-07-23 16:22:00.000Z	2011-07-23 16:22	2011	7
7	-84.24919128	34.02999115	7007	2014619	2014002255	2014-06-19 23:04:00.000Z	2014-06-19 23:04	2014	6
8	-84.23741913	34.01134491	7008	201554	2015001763	2015-05-04 07:58:00.000Z	2015-05-04 07:58	2015	5
9	-84.20654297	34.02998352	7009	2014513	2014001670	2014-05-13 09:04:00.000Z	2014-05-13 09:04	2014	5

If it's a new file, it'll be called **Book 1**. To open a new workbook, go to **File** > **New** > **Blank Workbook**. At least one worksheet is included in every workbook. You can add more sheets if necessary, by clicking on the Add Sign (+) below the window.

We can remove sheets. We can relocate them. We can copy them. Columns and rows make up every spreadsheet or simply sheet. You can get all the options by right-clicking on the worksheet you want to act on.

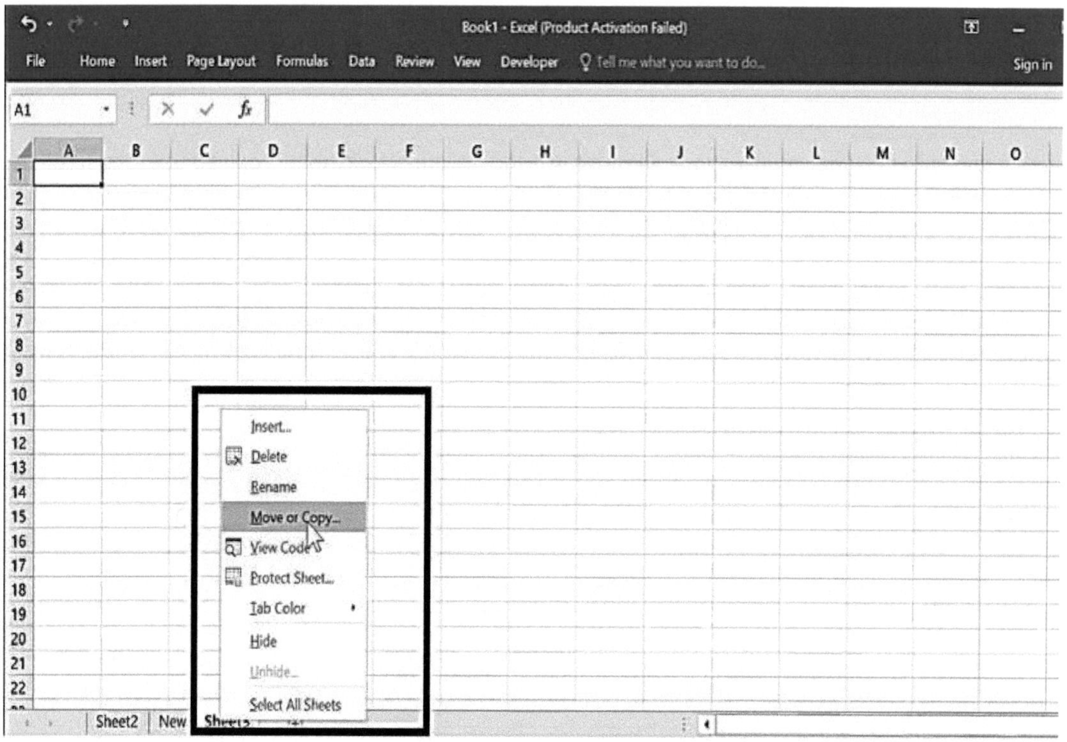

Vertical columns are indicated by letters. The rows are laid down horizontally; numbers are used in identifying them.

There are about 16,384 columns and 1,048,576 rows at the bottom. There is an address in every cell in an Excel spreadsheet. To summarize, a workbook is a file, and a workbook is a file. We have a working sheet with over a million rows and over 16,000 columns on each.

MOVING AROUND A WORKSHEETS

When you open a new workbook in Excel, the worksheet's "**active cell**" is positioned in the top-left corner (**in Cell A1**). You must move about in your worksheet to build it or make changes to it. You can move around your worksheets in Excel in two ways; by using the Keyboard or the Mouse.

Navigating with your keyboard

By hitting the arrow keys and other direction keys on the keyboard, you may navigate around the worksheet. When you move around in the worksheet using the keyboard, the active cell moves with you. The contents of the active cell are then added or edited. Below are the shortcuts you can use;

Up arrow key	Move up the next row (one row)
Left arrow key	Move to the next cell on the left
Right arrow key	Move to the next cell on the right
Down arrow key	Move down the next row (one row)
Control key + End	Moves to the cell that is located at the junction of the column on the right and the bottom-most utilized row.
Control key + arrow key	Moves to the last row or column in the worksheet
Page Up	Moves up one screen page

Page Down	Moves down one screen page
Control Key + Home	Moves to Cell A1

Navigating with your Mouse

Using the wheel on your mouse, you can move around on your worksheet. You can use it to scroll up and down to navigate through the worksheet. You can add more sheets using the mouse by clicking on the add sign below the screen. By clicking on a worksheet tab, it displays the sheet for you to work on/with it.

To navigate quickly to a cell, check the left-hand side of the formula bar and click on the Name box. Type in the cell address i.e. the column letter & the row number. For example, type in **A6**, or **B2**, then press **Enter**.

USING THE RIBBON

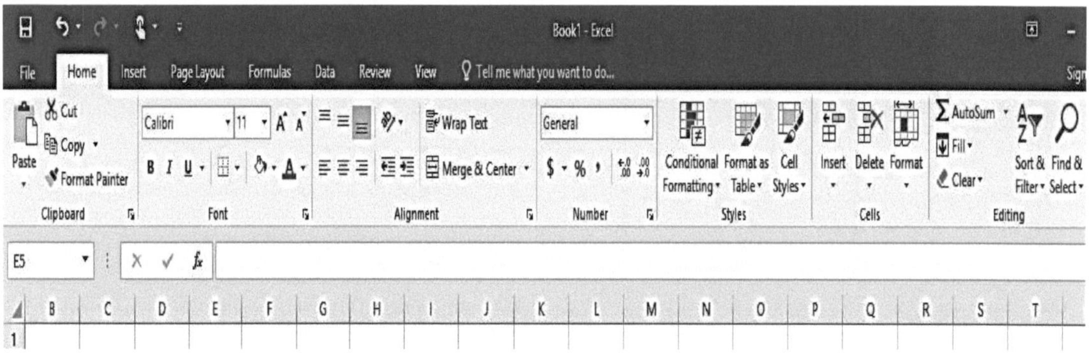

The toolbar has been referred to as the ribbon. It organizes commands into tabs depending on their functionality and shows them as icons in a strip. Meanwhile, before we get started, there are certain things you can't change in Excel. They include:

- Built-in commands cannot be changed or removed. ***(NB: You may, however, conceal a whole group)***.
- The ribbon cannot be resized. The only visual option is to conceal (collapse) it fully.
- Text size, font type, and color selections are available right away. You may also use Excel schemes to change the backdrop of the ribbon across all Office programs.

Customizing the Ribbon

You can customize the ribbon in your Excel. There are different techniques you can use to get to the customization options.

- Select **File > Options > Customize Ribbon** from the File menu.

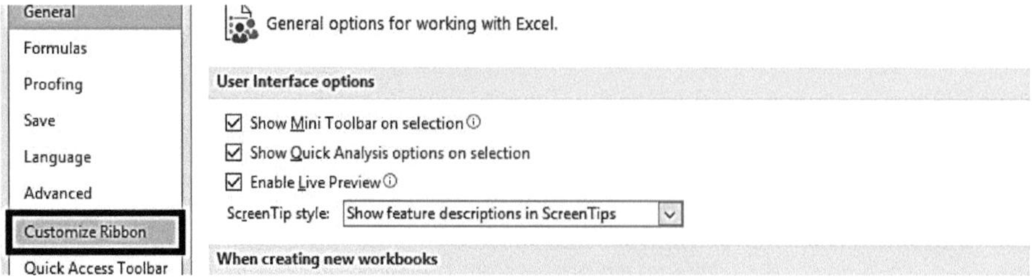

- Right-click on the ribbon and pick **Customize the Ribbon**

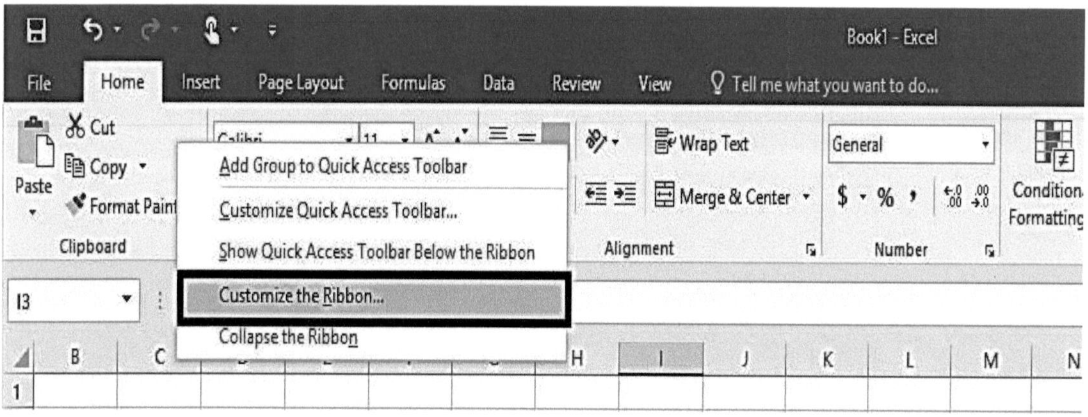

The window gives you lists of instructions to choose from. In contrast, the one on the left provides all popular commands, the right side, lists the ribbon's main tabs, groups, and commands.

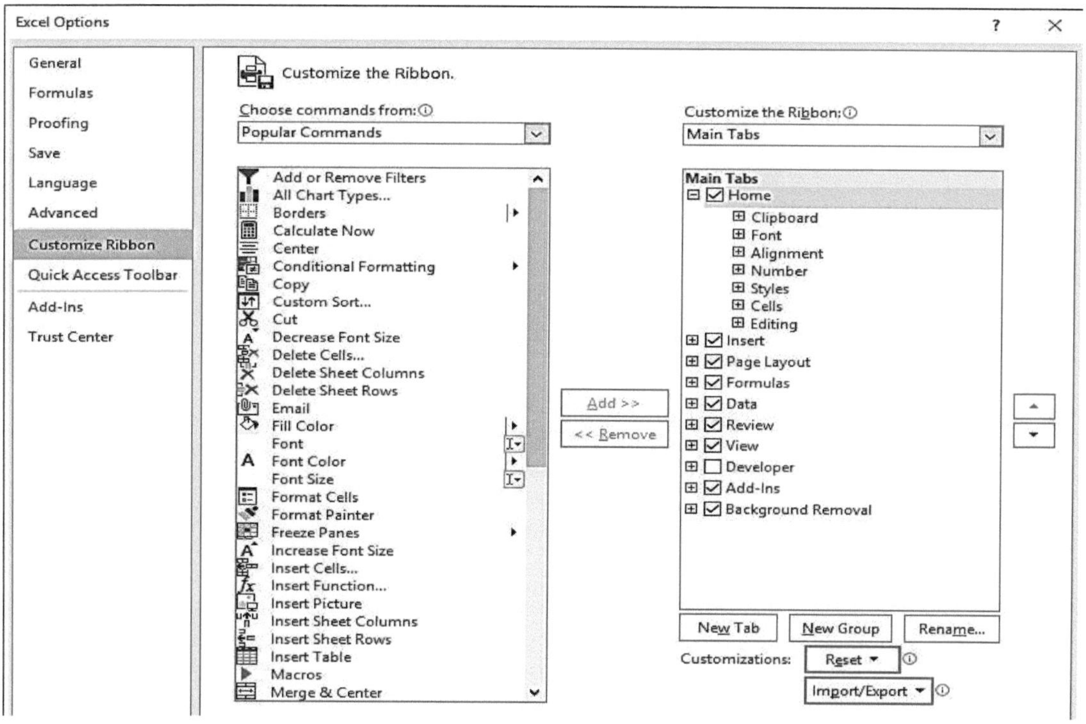

Ribbon Tabs

This is where you make some changes to your worksheets. You can insert a new tab, create a new group, add a command, changes the names and positions of tabs, groups, and instructions.

Inserting a new tab

Go to the right-hand list and choose **New Tab**. The new tab and group are created when you click the button ***(NB: A tab should always have a minimum of one group. Otherwise, you won't add commands)***. Use the Rename button to rename a tab or group.

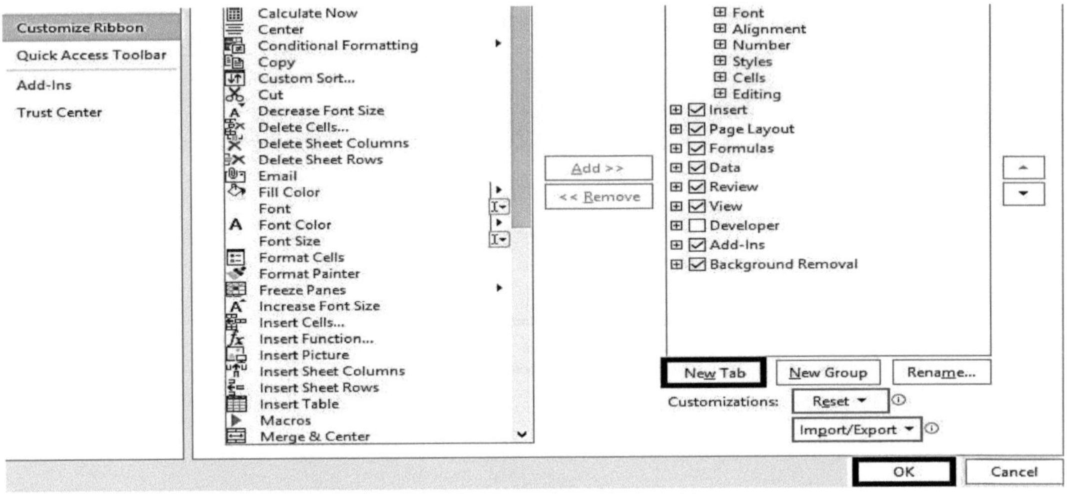

Creating a new group

Choose the tab you want to add a group, then, select **New Group**.

Adding a command

Commands can be added only to custom groups. As a result, begin by picking a custom group from the appropriate list. So, on the left side, choose the command. Choose commands from the menu above to filter the list. To add the command to the specified group, click **Add >>**. When you're finished, click **OK**.

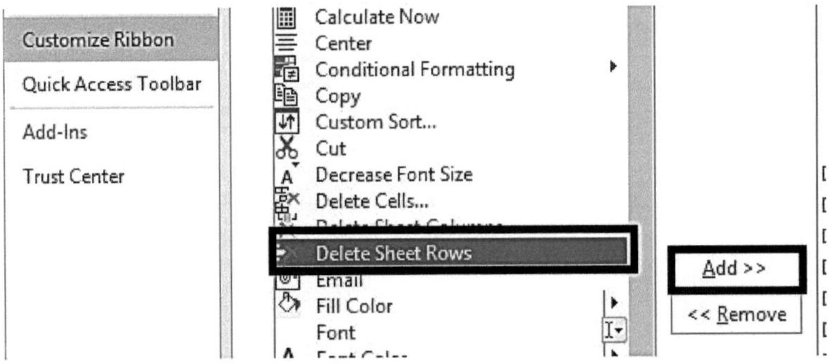

Changing the names of ribbon tabs, groups, and commands

The Rename button, as previously indicated, may be used to rename tabs, groups, and commands. However, when it comes to built-in things, your selections are restricted. Although you may change built-in tabs and groups, none of the built-in commands can be renamed. Except for the File tab, that is. Symbol selection, on the other hand, is not possible for built-in tabs and groups.

Changing the position of tabs, groups, and instructions

You can rearrange the tabs, groups, and instructions on the ribbon in Excel. You may take one of the following approaches:

- On the right side, use the arrow buttons. Select the object you want to move and then click the arrow that indicates the direction you want it to go.
- Make use of drag and drop. Move the mouse by clicking and holding the mouse button on any object.

Contextual Tabs

When you create or update an object in Excel, such as a chart, table, or diagram, contextual tabs appear. The contextual tabs provide you with the choices and tools you need when working with these things, and they are then cleared away when you click out of the object. To restore them, just click on the item, and the tab will appear. It keeps your workspace free of clutter.

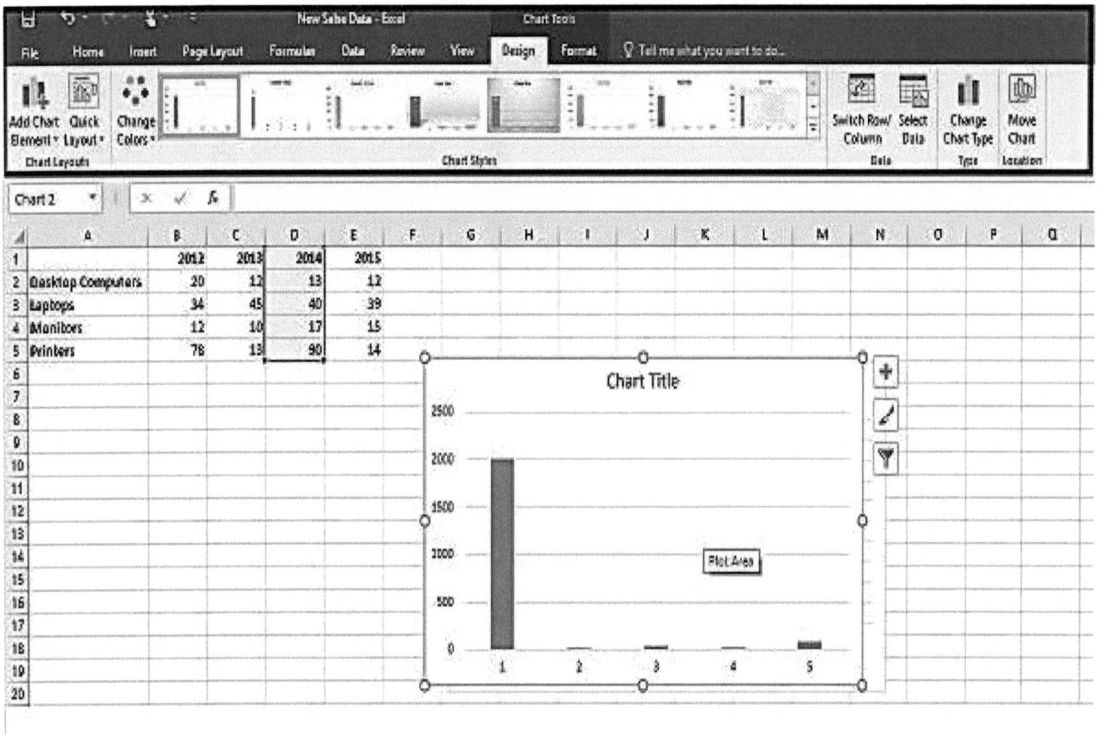

Types of Command on the Ribbon

There are 5 main types of commands on the Excel ribbon. They are; Toggle buttons, Drop-Down buttons, Tick Box, One-Click, and Split Buttons.

Toggle Buttons: With these buttons, you can adjust a cell. When you use this button to a cell, Excel will let you know the button you have applied (you will see the button highlighted with a particular color). Most times, when you have applied different formats to your cells, this button helps you to know the formats you have applied.

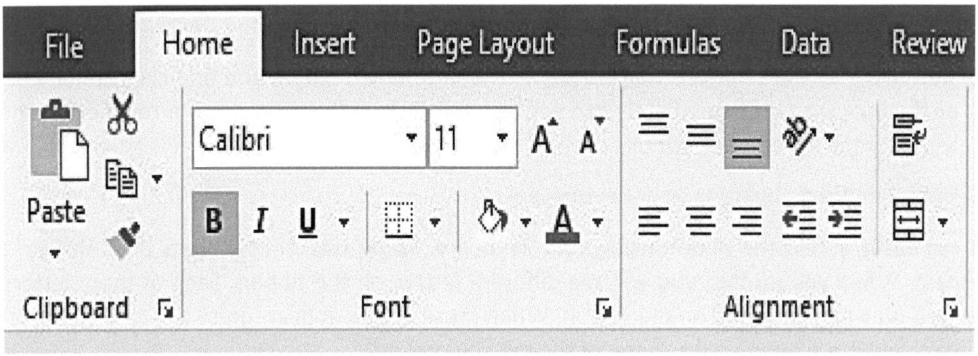

Drop-Down Buttons: These are those little arrows that are next to a particular button. When you click on the arrow, it gives you an extra list of applicable change options that you can use to make changes to a cell. For example, when you click on the arrow next to the Fill Color button, you will see a list of different colors that you can apply to the selected cell.

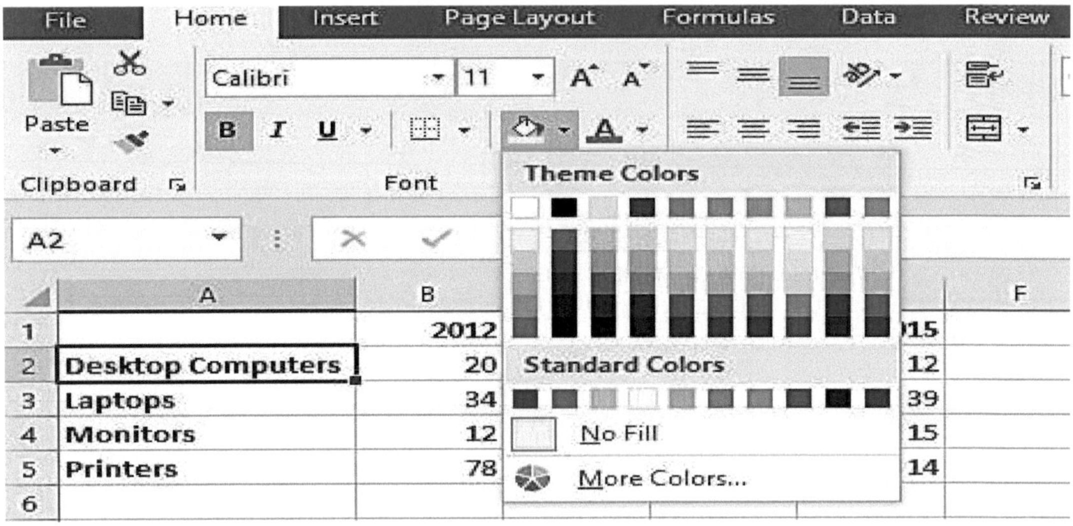

Tick Box: This box is more like the On/Off buttons. On these boxes, you can check and uncheck different box options. For example, when you click on the View tab, you will see some options there with the tick-box beside them. As you can see in the image below;

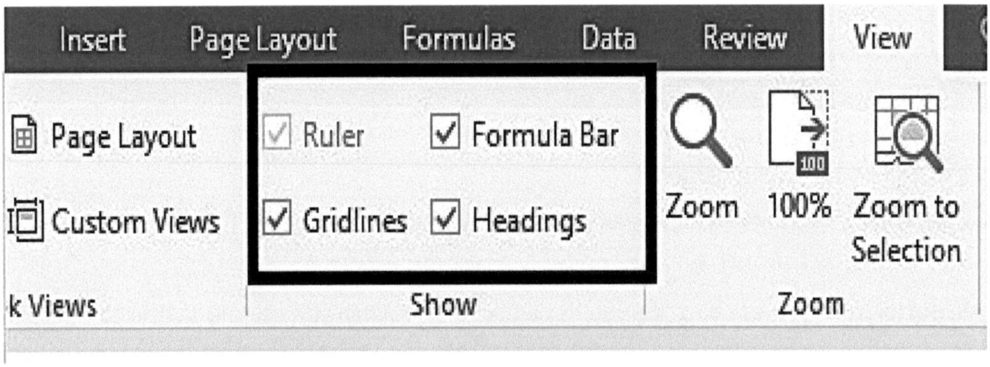

One-Click: These are those commands on the ribbon that when you click on them once, act immediately. Let's look at the **Text alignment** option for an example. Select a cell and click on it, it aligns the cell immediately. Same with the **Bold button**, **Italic button**, or the **Increase** and **Decrease** font size buttons.

Split Buttons: These are buttons that consist of the drop-down option and one-click option i.e. you can click on the button to act immediately and you can also click on the drop-down arrow next to the button to select from the list of other options.

Accessing the Ribbon by using your Keyboard

You can easily access the ribbon using your keyboard. To do this, simply press the "Alt key" on your keyboard. When you do this, you will see different letters on the ribbon. Each of these letters will be displayed on a particular tab on the ribbon. When these letters appear, press any of the letters on your keyboard and it will open up the menu of the tab you pressed.

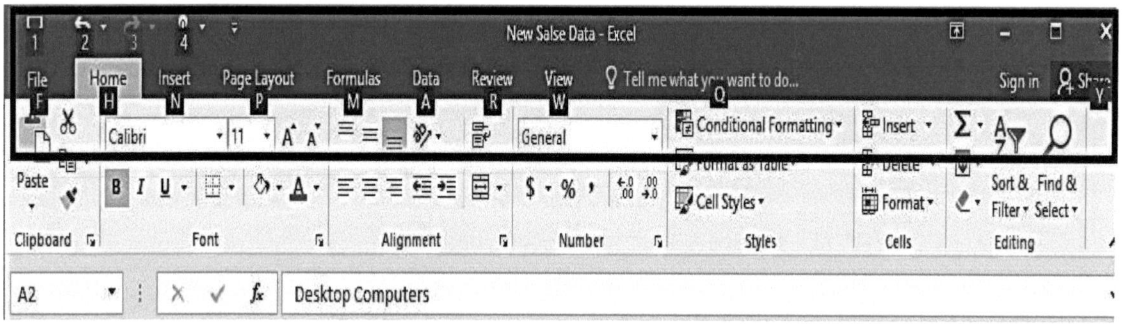

USING SHORTCUT MENUS

Using keyboard shortcuts into your Excel routine is one method to speed things up. Excel specialists seldom use a mouse since it takes more time and is often imprecise. Many key combinations, or instructions, are already incorporated into the application, allowing you to perform actions. Depending on their function, all of the accessible shortcut keys may be grouped into numerous categories.

Shortcuts for formatting data

Combinations of the Control key + Shift key with extra characters may be differentiated as a distinct category of data formatting fast commands.

- **CONTROL KEY + W** — This command will dismiss the current worksheet immediately.
- Save the working document by pressing **CONTROL KEY + S**.
- Create a new working document by pressing **CONTROL KEY + N**.
- **CONTROL KEY + X** – copy the contents of selected cells to the clipboard.
- Open a working document by pressing **CONTROL KEY + O**.
- **CONTROL KEY + V** - this shortcut pastes data from the clipboard into the previously selected cell.
- **CONTROL KEY + P** – brings up a window with printing options.
- **CONTROL KEY + Z** is a command that can be used to undo a previous action.
- **F12** - this key saves the current working document as a new name.
- **CONTROL KEY + T** - This key combination allows you to generate a new worksheet from a single cell and a range of cells around it.
- **CONTROL KEY + 1** – Opens the cell formatting dialog box from the table.

Data entry shortcuts

- **CONTROL KEY + D** – this command copies the contents of the first cell in the marked range and pastes it into all subsequent cells.
- **CONTROL KEY + Y** – the command will, if feasible, repeat the previous action.
- Add the current date using **CONTROL KEY + ;**
- If edit mode is open, ALT + enter inserts a new line within the cell.
- Change the indicated cell using **F2**.
- The Paste Special Docker is opened by pressing **CONTROL KEY + SHIFT + V**.

Data View and Navigation

- CONTROL KEY + G (Go to) – opens the "Go" window on the screen.
- You may navigate to the next worksheet by pressing CONTROL KEY + PgDown.
- CONTROL KEY + END - Go to the end cell of the current sheet immediately.

- The Find dialog box is opened by pressing CONTROL KEY + F.
- Switch between workbooks by using CONTROL KEY + Tab.
- Hide or reveal the ribbon with tools by pressing CONTROL KEY + F1.

Data selection

- SHIFT + Space – Selecting a whole line
- CONTROL KEY + Space – Selecting a whole column
- CONTROL KEY + A – Selecting the whole worksheet.

CUSTOMIZING YOUR QUICK ACCESS TOOLBAR

The Quick Access Toolbar is a resizable toolbar that includes a collection of instructions that are not reliant on the current ribbon tab. You may add buttons that indicate instructions to the Quick Access Toolbar.

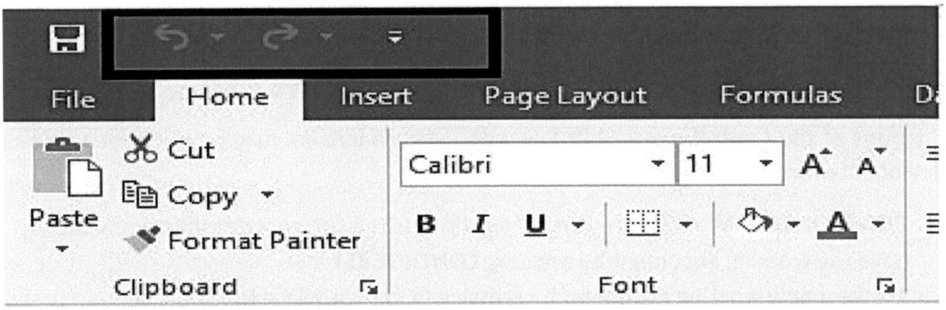

Select the drop-down arrow on the Quick Access Toolbar, it shows a list of commands that you can add to the toolbar. When you click on any of the commands, it will display in the Quick Access Toolbar.

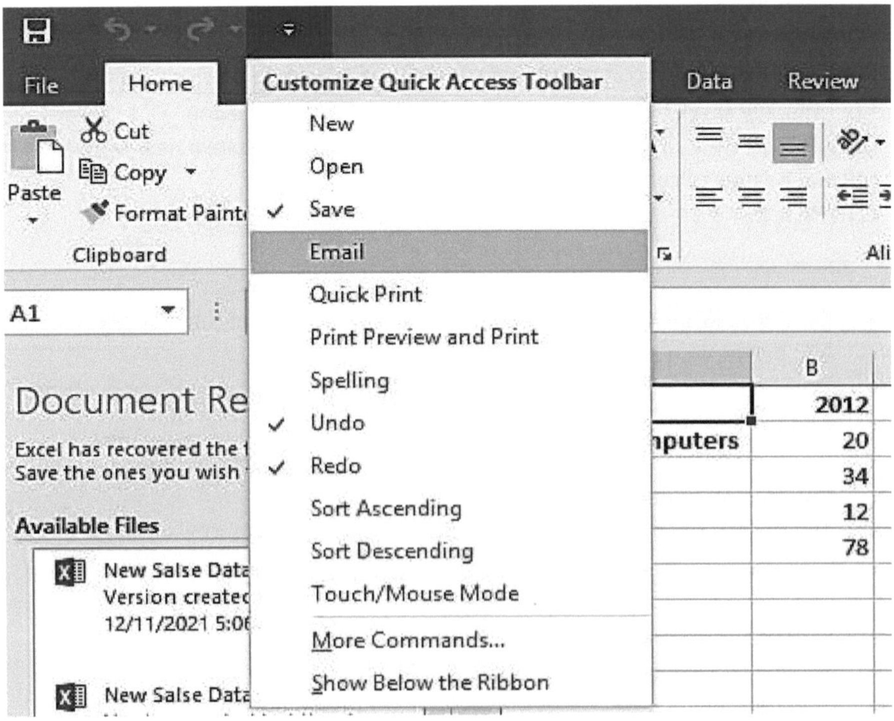

Move the Quick Access Toolbar: You can move it to two places; Below the ribbon or Above the ribbon. To change the location of your quick access toolbar (if it is currently above the ribbon), click on the drop-down arrow, then select **Show Below the Ribbon.**

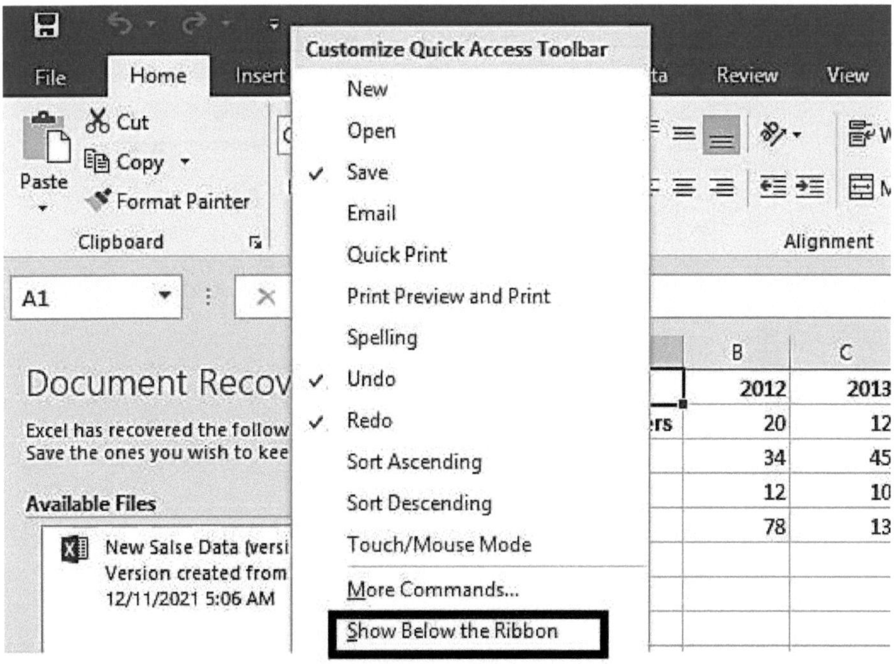

Follow the same steps above then select **Show Above the Ribbon**.

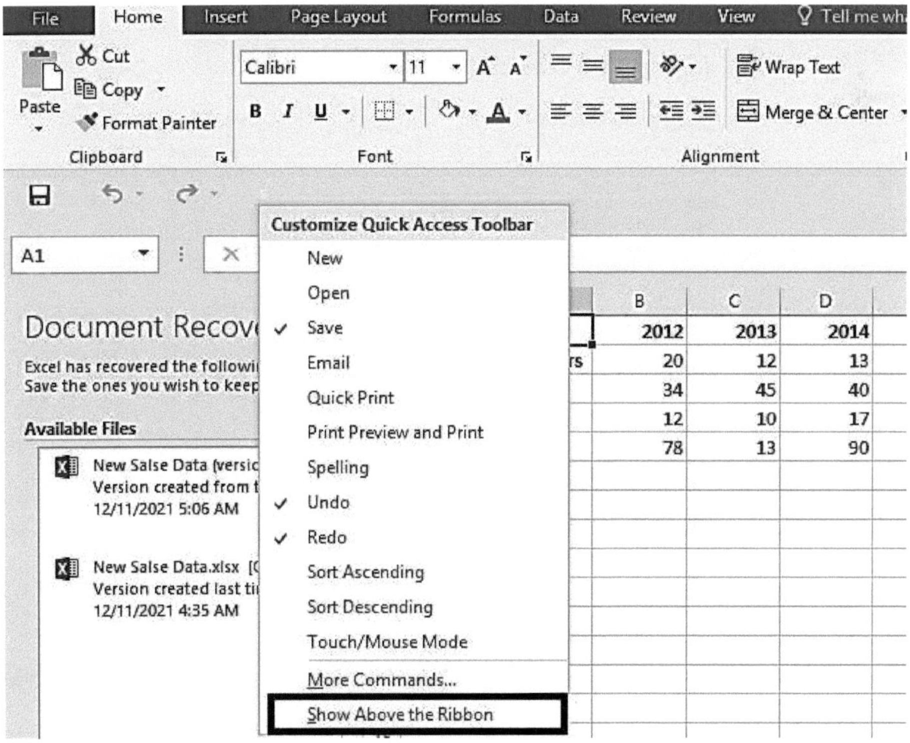

Not all commands are displayed in the lists. So, simply click on the drop-down arrow and select **More Commands** and this will display a window where you will see other commands options.

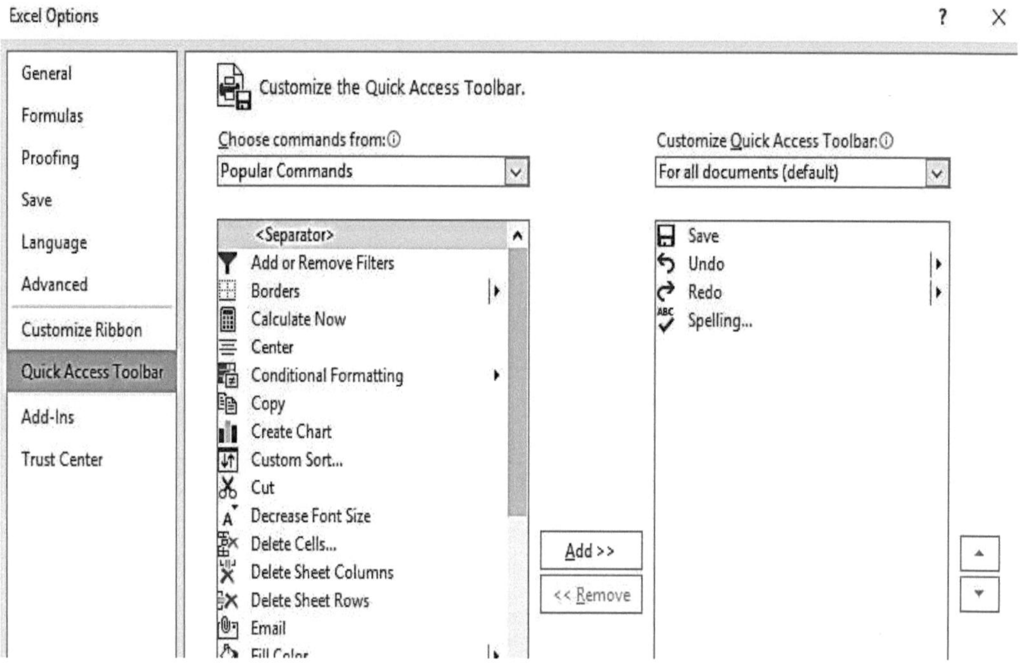

Click the arrow under the "**Choose commands from**" option. This will display different command options.

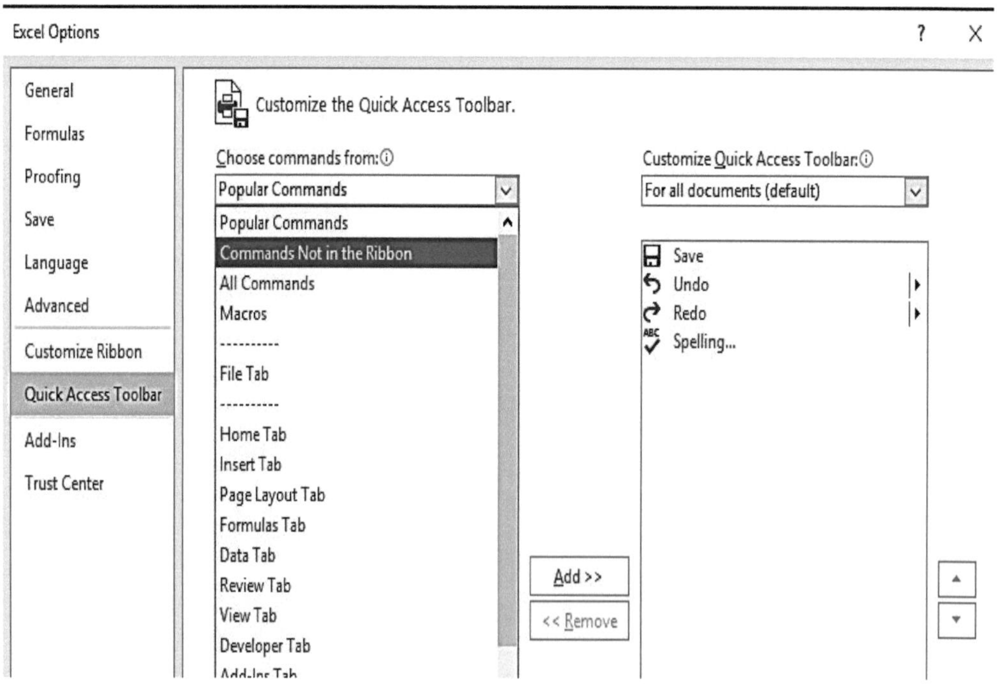

Select an option and choose a command. Then click **Add>>**. You can also remove a command by clicking on the **<< Remove button**.

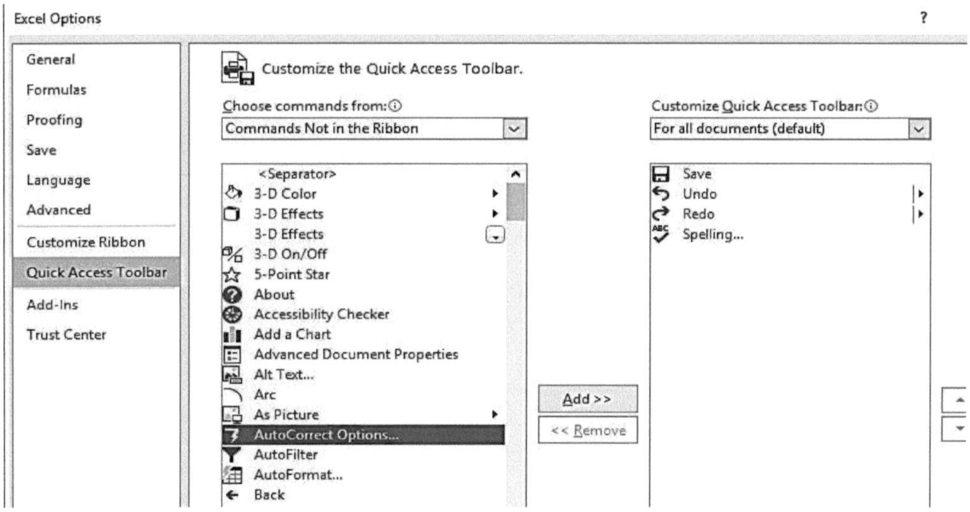

WORKING WITH DIALOG BOXES

The box that appears is known as a Dialog Box. It is a transient window created by an application to collect user input. Dialog boxes are often used in applications to ask the user for further information about menu options.

In Excel, when you click or make use of some commands in the ribbon, it displays a box where you see other options for that command. Excel features a lot of dialog boxes that you'll use a lot, such as Format Cells, Spelling, Past, Find and Replace, etc. There are types of dialog boxes.

Typical Dialog Box: If this box appears on your screen, it doesn't go away by itself until you decide to close it. when it appears, perform your action, then you click **OK**. If you want to dismiss the box, click on the **Cancel** button. This will close the dialog box immediately and no action will take place. The image below is a Typical dialog box.

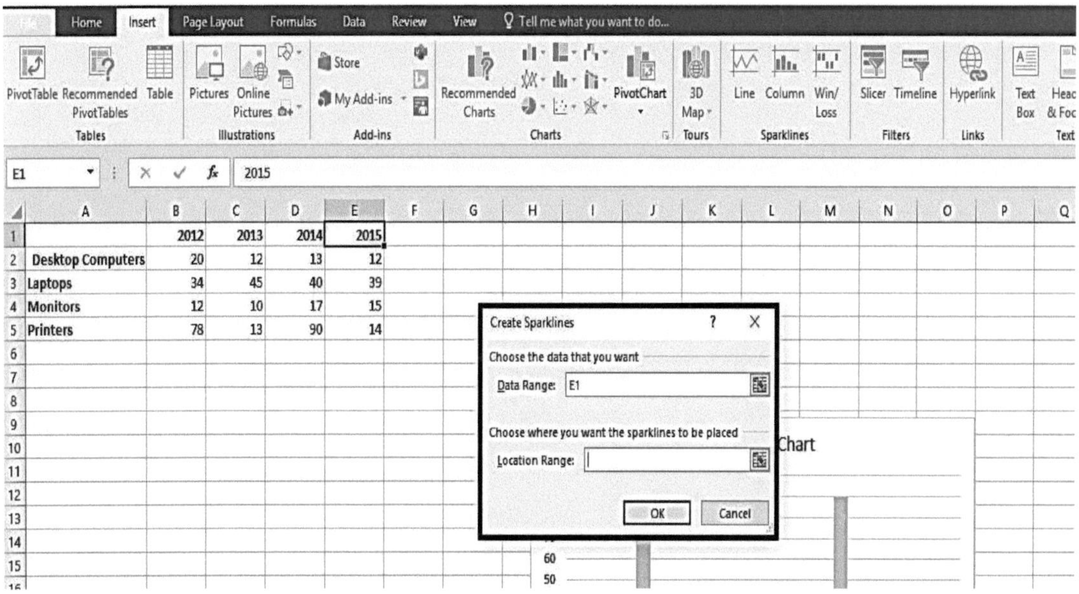

Navigating Dialog Boxes: This box appears when you click on a command that you want to use. You can easily see them from the list of commands on the Insert Tab.

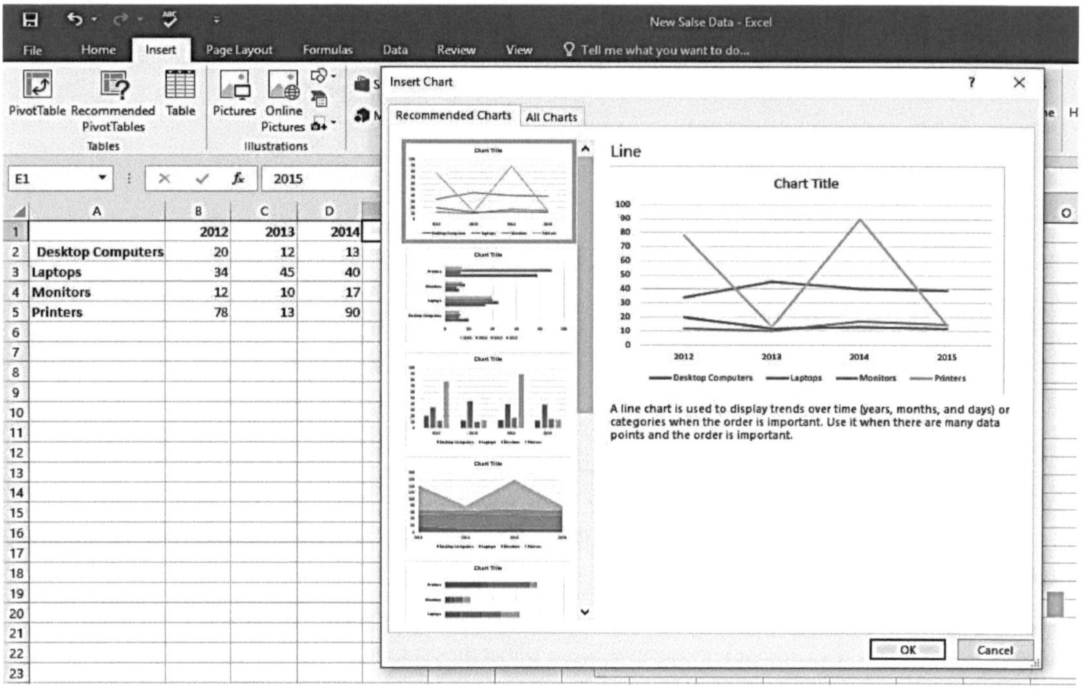

Tabbed Dialog Boxes: Many of the dialog boxes in Excel are Tabbed dialog boxes. It has several commands on it. You make the changes you want, and once you are done, click **OK**.

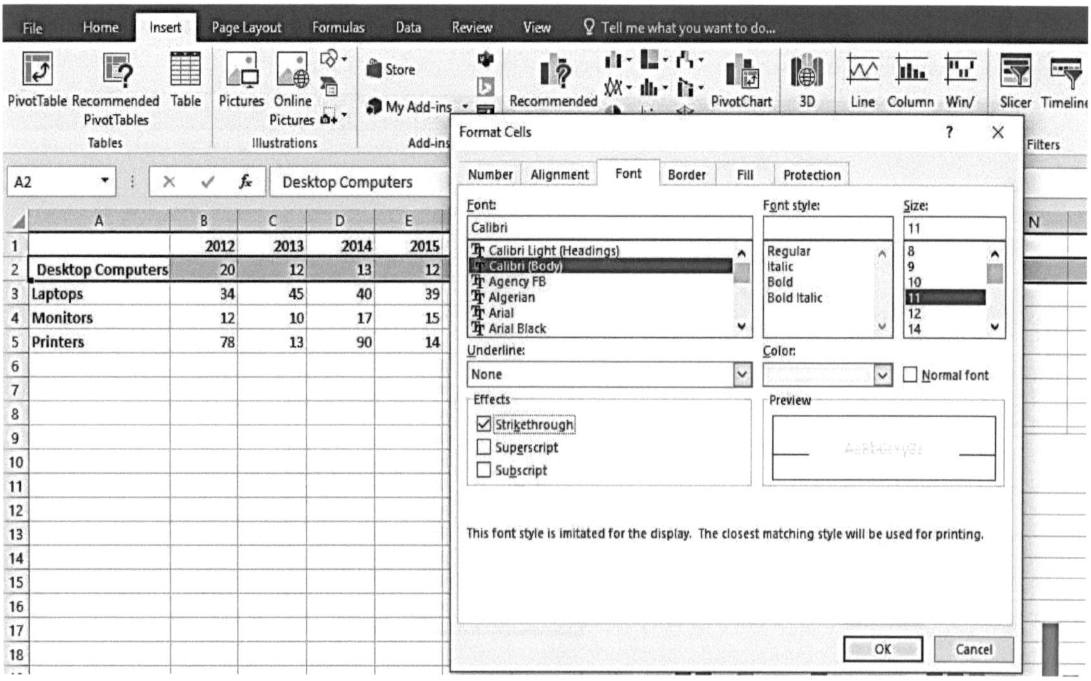

USING TASK PANES

In Excel, task panes are interface panels that display on the right side of the window. It provides users with easy access to common features, data, and instructions. Below is an example.

Open an excel sheet (It should have some details in it).

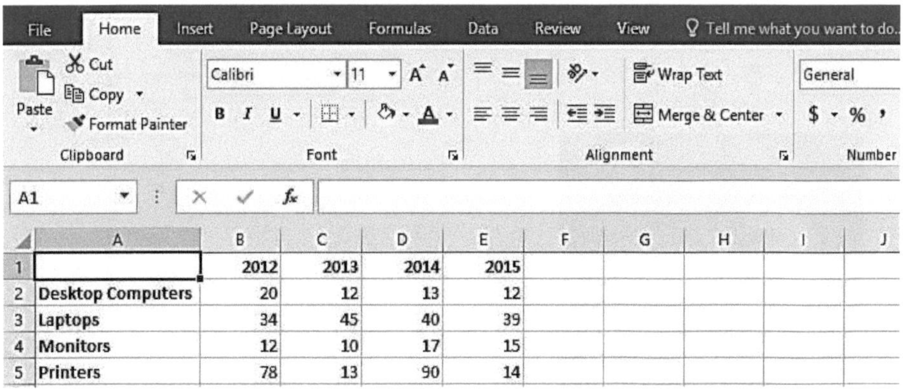

Now, let's insert a picture. Click **Insert** > **Pictures** > Select the picture you want, then click **Insert**. Once the picture has been displayed on the worksheet, right-click on it and select **Format Picture**.

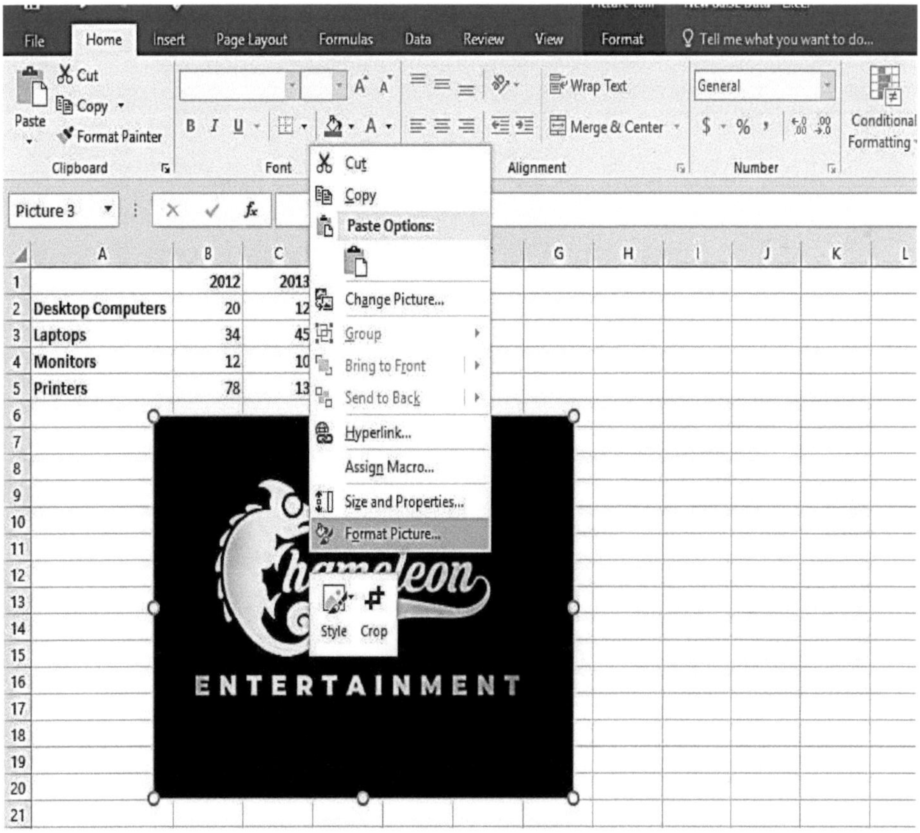

This will open up the Format Picture Task pane. You will see lots of options you can select from in the Task Pane.

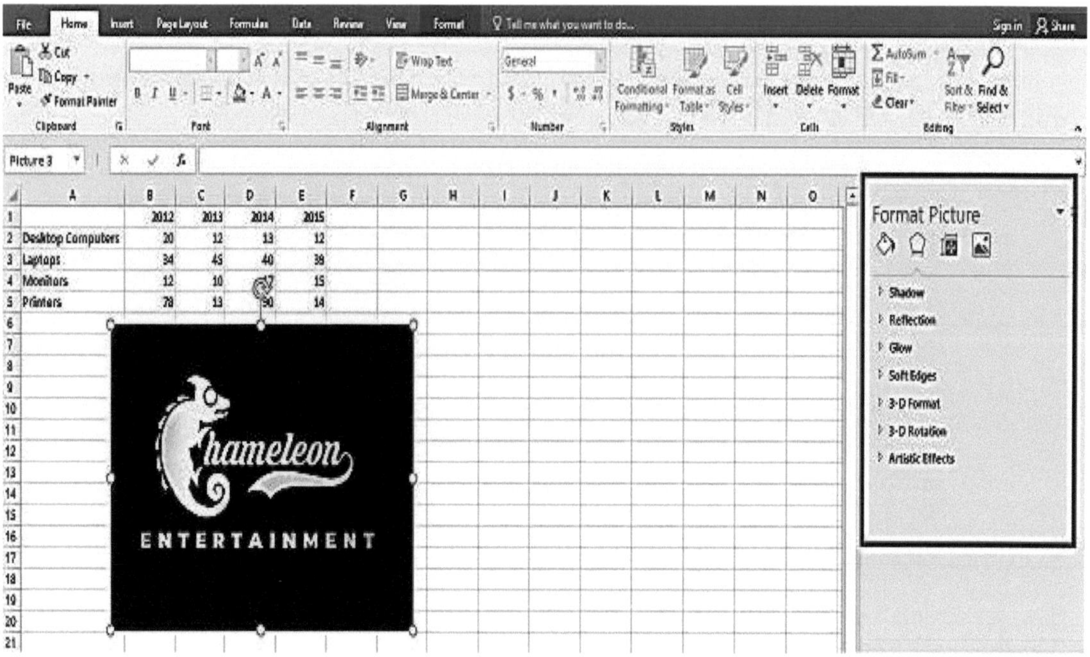

Click on an option. As you make the changes, you will see the picture adjusting to the changes you have made.

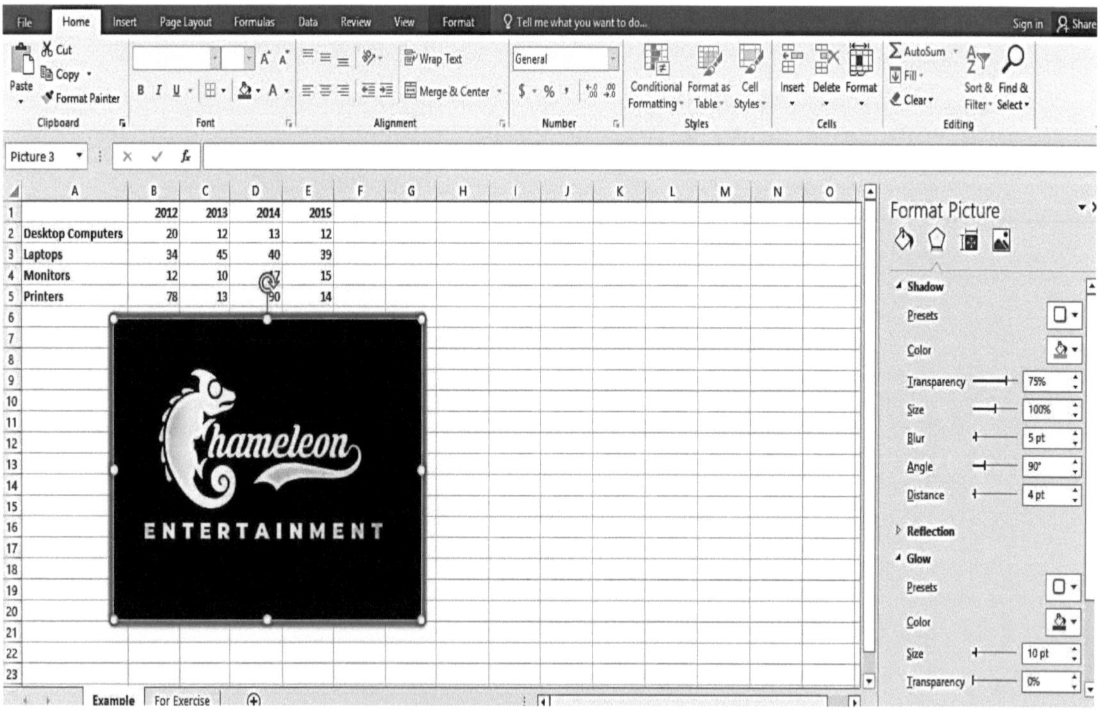

CREATING YOUR FIRST EXCEL WORKBOOK

You'll get a welcome screen when you first launch Excel, where you can select to access an already Excel spreadsheet or you create a new one. You may create a fresh, blank worksheet or a ready-made workbook using a template on Excel's welcome page. For the time being, click the "**Blank workbook**" icon to create a new spreadsheet without any formatting or data.

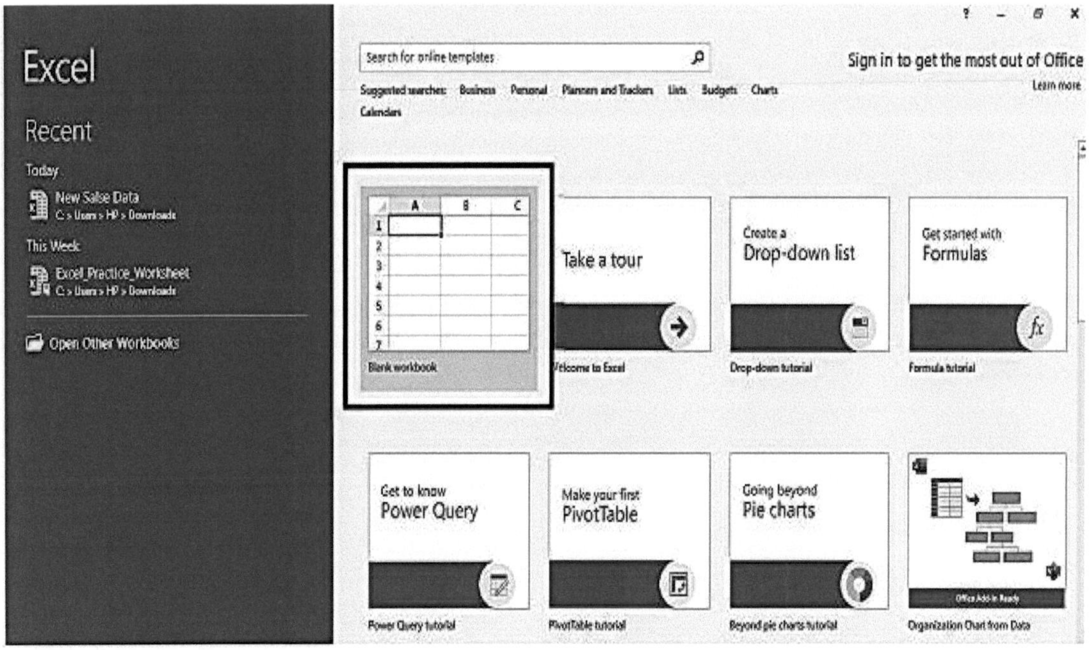

Getting Started on Your Worksheet

A worksheet is a grid of cells into which you may write data and formulae. The grid occupies the majority of the Excel display. It's where you'll do everything from inputting data to developing formulae to evaluating the outcomes. The worksheet consists of rows and columns gridlines and is labeled with numbers and alphabets.

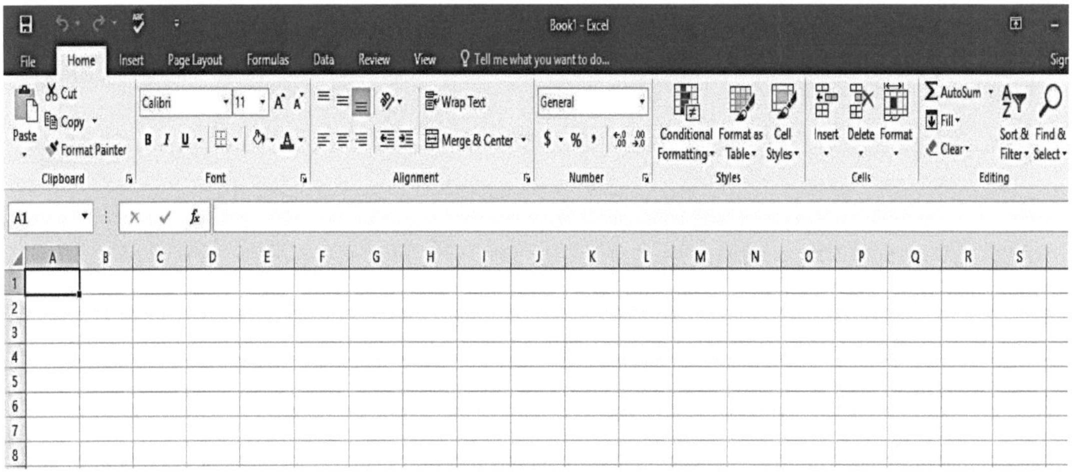

Filling in the Month Names

Now, we are going to fill in some data. First, select **Cell B2** and **Cell C2** at the same time. To do this, click on Cell B2, and why holding the left-click button on your mouse, move the cursor over Cell C2. This will select both cells. Now, on the Home tab, select **Merge & Center** from the ribbon which will merge both cells into one cell.

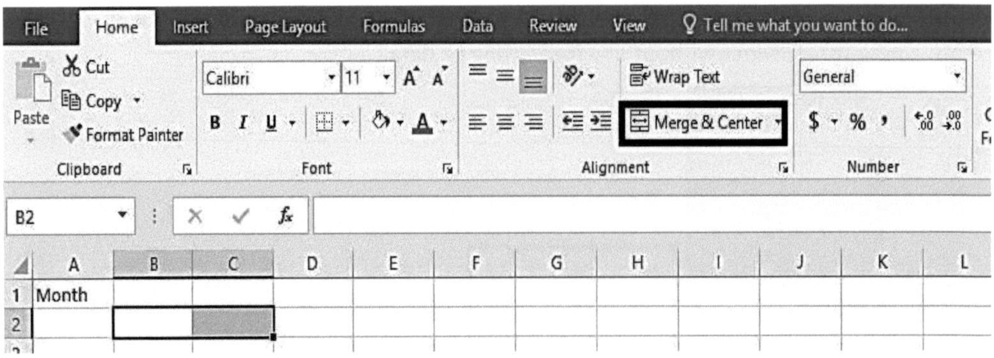

Now, click on the merged cell and type in Sales Report. Now, activate cell B3 and type in the word Month. Activate cell C3 and type in the word Sales. Now, select cell C3. In the alignment group, select the **Center command**.

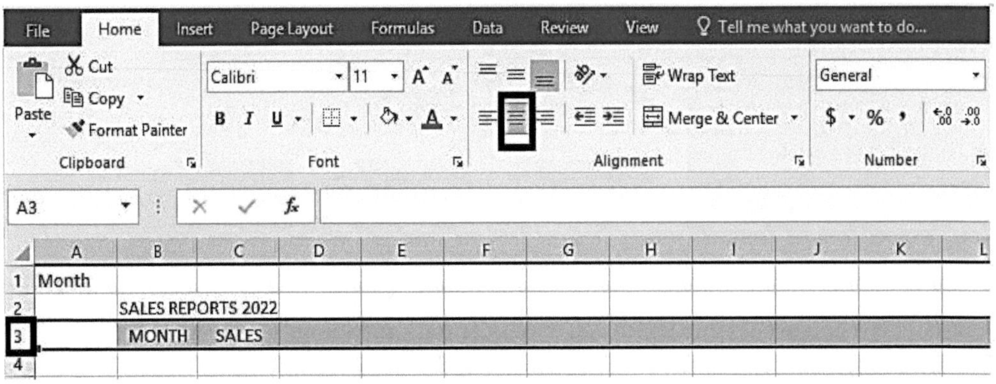

Now, activate Cell B4 and type in **JANUARY**. Then, on the small square box on the bottom right of Cell B4, click and drag down to 11 rows (**B15**). Once you've done this, Excel will automatically complete the rest of the months for you instead of making you type the months because it recognized January.

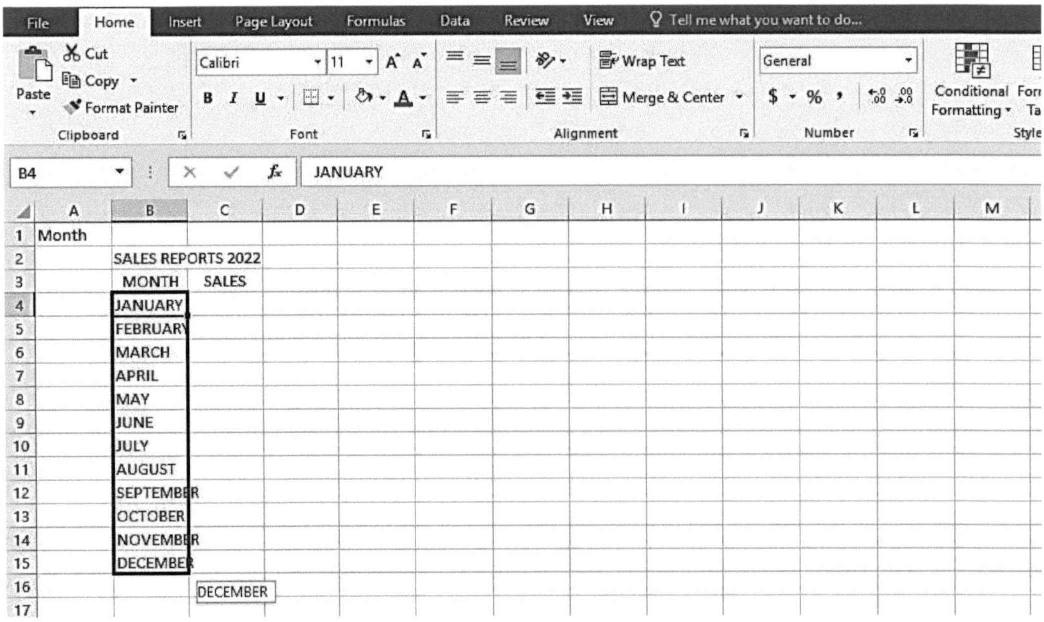

Entering The Sales Data

Now, enter the following numbers (as you can see in the image below) into C4 and C15 in the exact following format without dollar signs or commas. Then, activate cell B16 and type in Total and then center it.

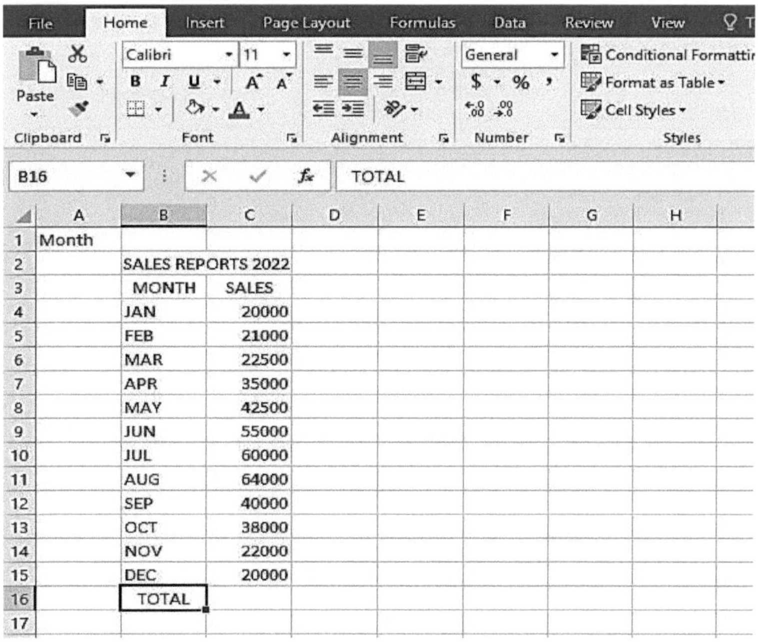

Summing the Values

The SUM function is used for summing values. we will apply it in cell **C16**. Activate cell C16 and enter in the equal sign, sum, and left parenthesis. Then select cell C4 to cell C15 and close the parenthesis. The contents within cell C16 should be as follows: **=sum(C4:C15).**

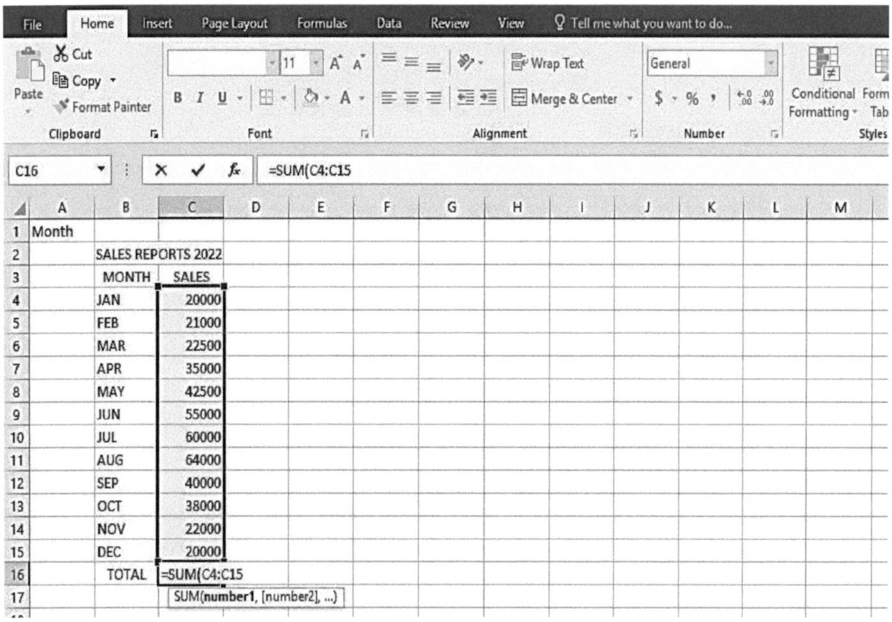

Then, press ENTER, and this sums up the monthly sales.

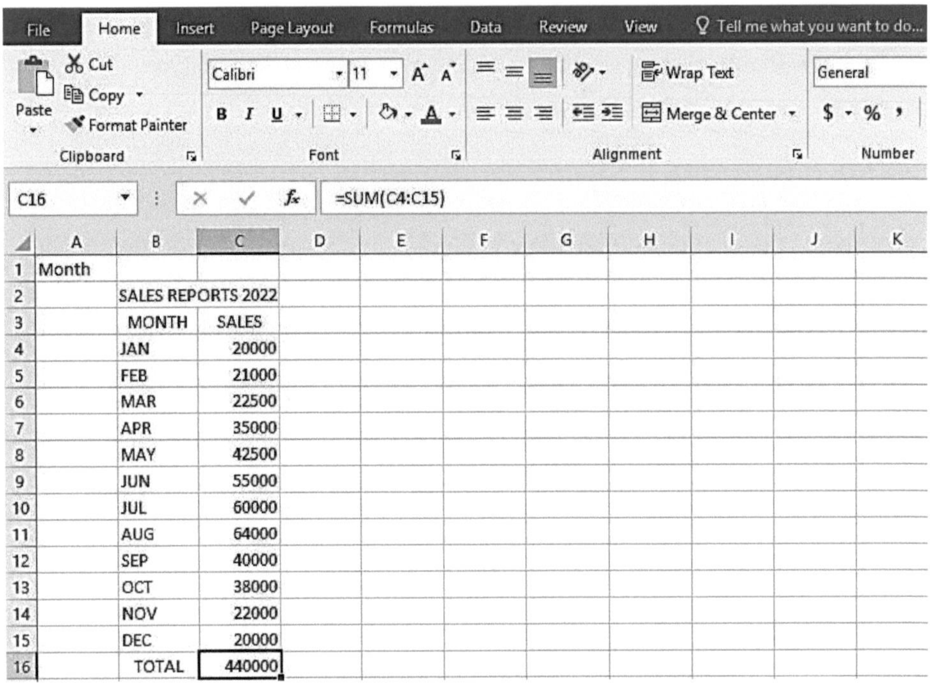

Making Your Worksheet Look a bit Fancier

Select **Cell B2** and **C2**. Click the down arrow on the fill color command on the ribbon and select a color. This will color the title of the table.

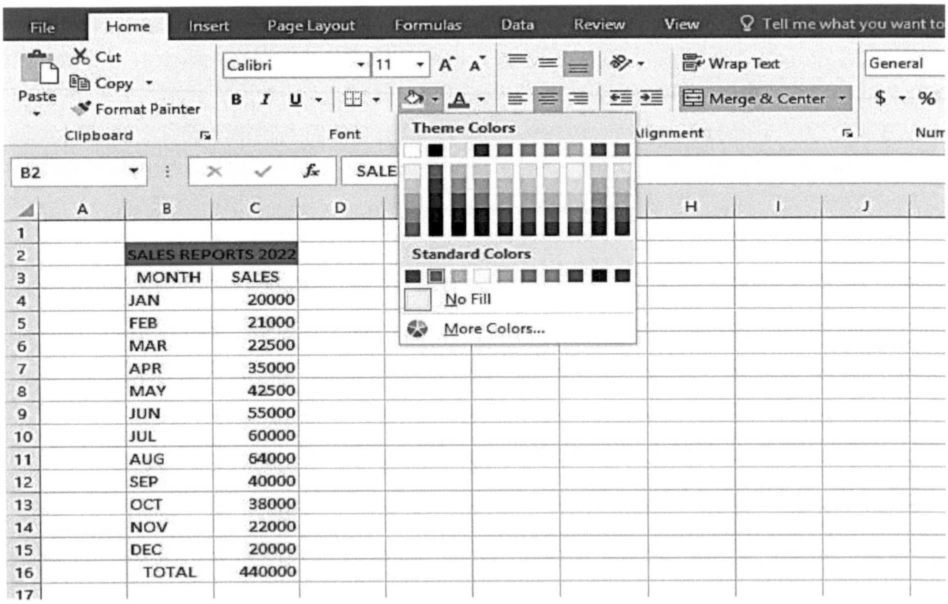

Now, select cells B3, C3, and B16 which are the Month, Sales, and Total. Repeat the same steps above but pick a lighter color. Repeat the same steps for cells **B4** to **B15** which are the Months. Finally, select cell **C16** and fill it with a very light color to differentiate it from the rest of the Sales number.

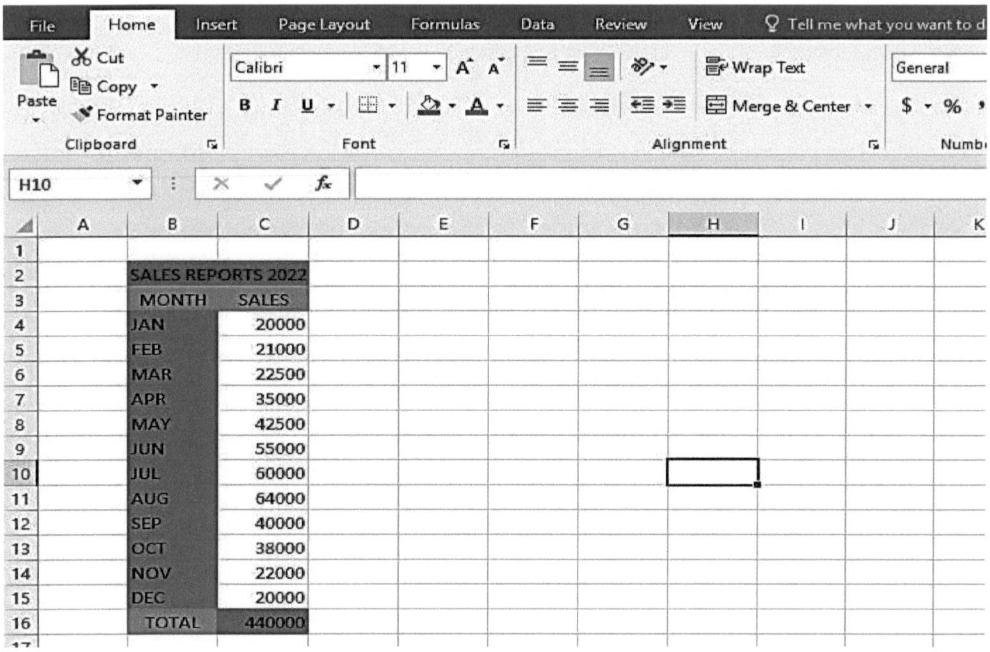

Now, let's add a border

Select cell B2 through cell B16 (the entire **Sales report)**. Select the down arrow of the border command and select **All Borders**. There are other border options there as well.

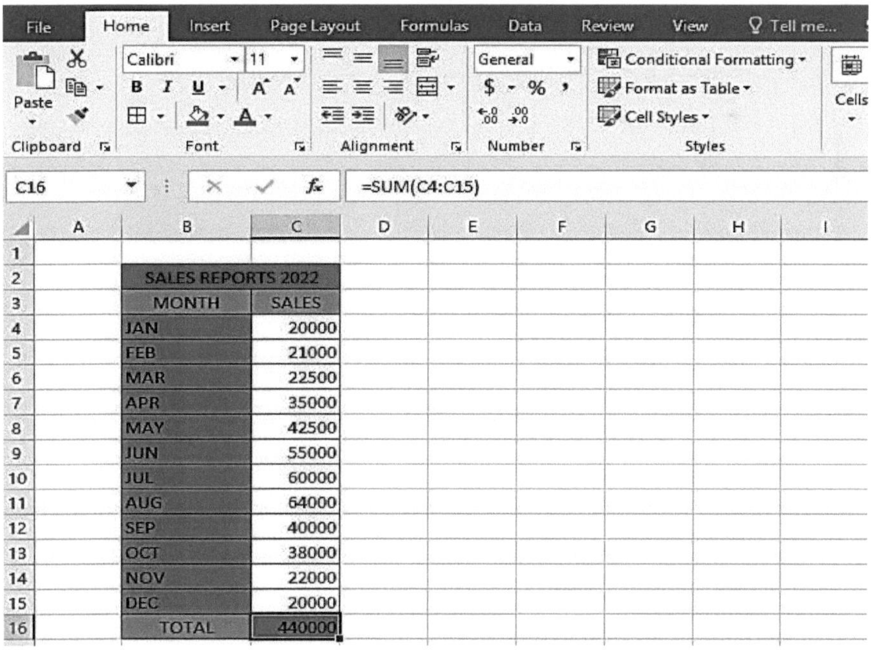

Creating a Chart

Now, choose the cell you want to create a graph for. To do this, click and drag your mouse across the cell. Once you've done this, click on the Insert Tab on the ribbon. You will see the chart command menu on the Insert tab options.

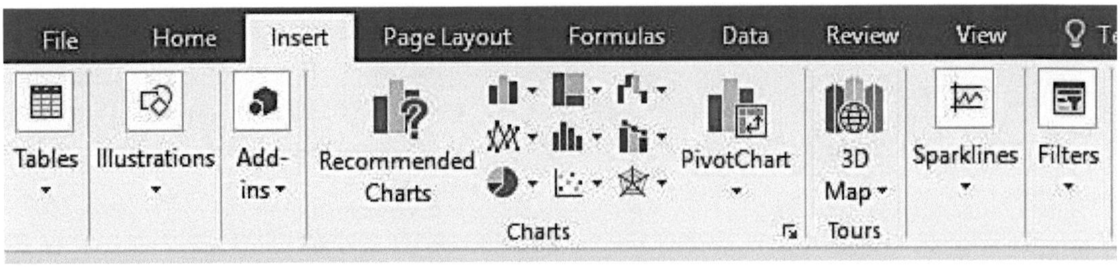

You can create different charts for your work. In this case, we will create a bar chart. Click on **Recommended Charts** > **All Charts**. Then, select **Bar**. Click **OK**.

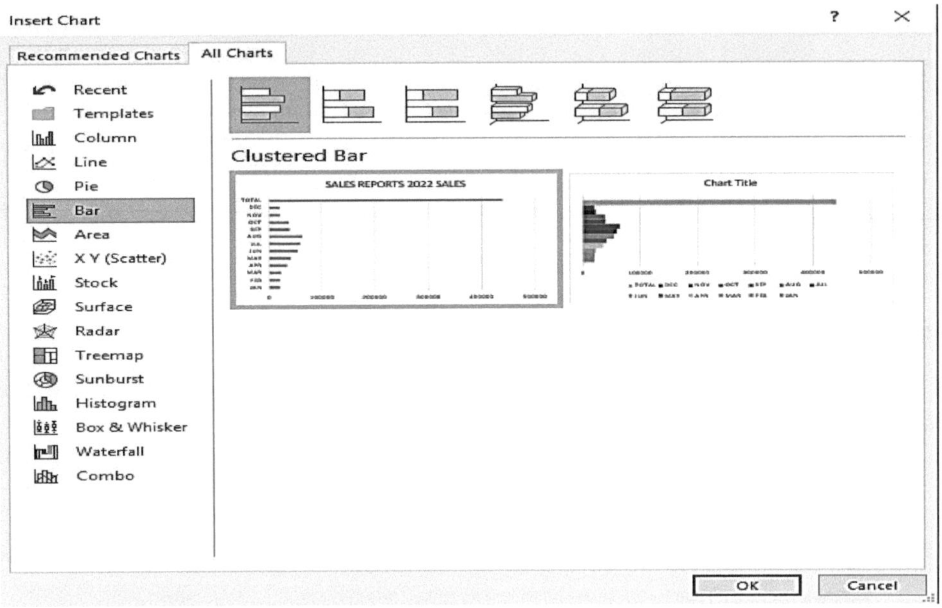

This will immediately create a pie chart for your work.

Formatting the Numbers

Select the cell > click the down arrow on Number group and select any number format option. you want. Click on More Number Formats for more formatting options. On the image below, I formatted the numbers to Currency (Dollar).

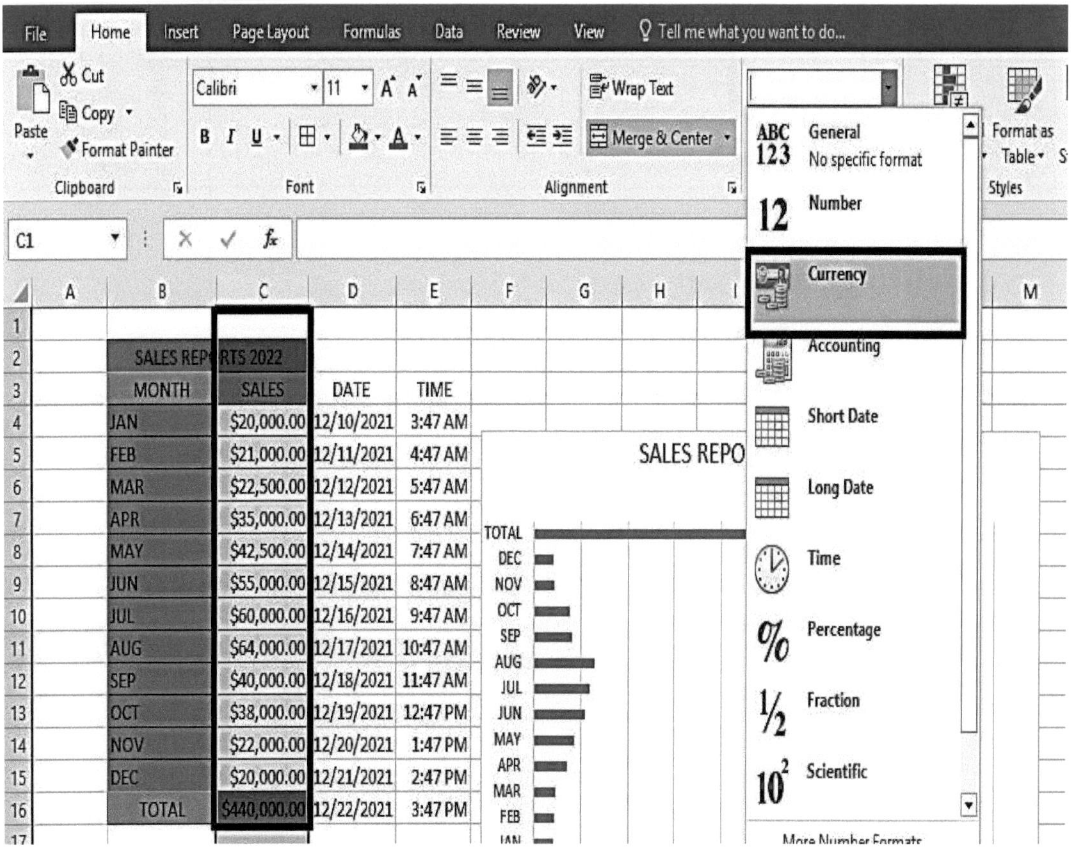

Printing your Worksheet

Excel sheets might have much information, and printing it all at once isn't always practical. By identifying the desired region, navigating to the print options, and selecting the 'print chosen area' option, you may print specific areas of a spreadsheet. Printing selected sheets in a workbook may be done similarly. "**Print Areas**" may also be utilized by people who want to fine-tune their formatting before printing. There are different methods you can use in printing your worksheet.

Printing from a Selection

Highlight the cell(s) you want to print. Click on **File** and choose **Print**.

On the Print Setting menu, hit the down arrow and select **Print Selection**. Then, click on **Print**. This will print out only the selected area in your spreadsheet.

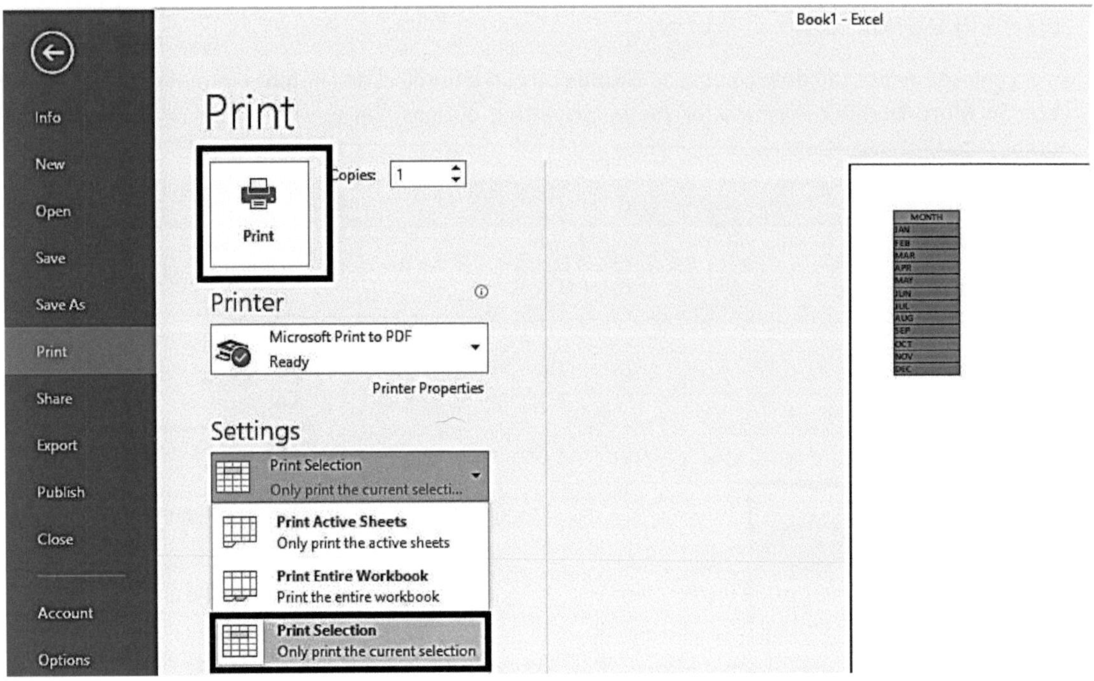

Using a Print Area

Highlight the cell(s). Click on Page Layout. Click on **Print Area**, then select Set Print Area.

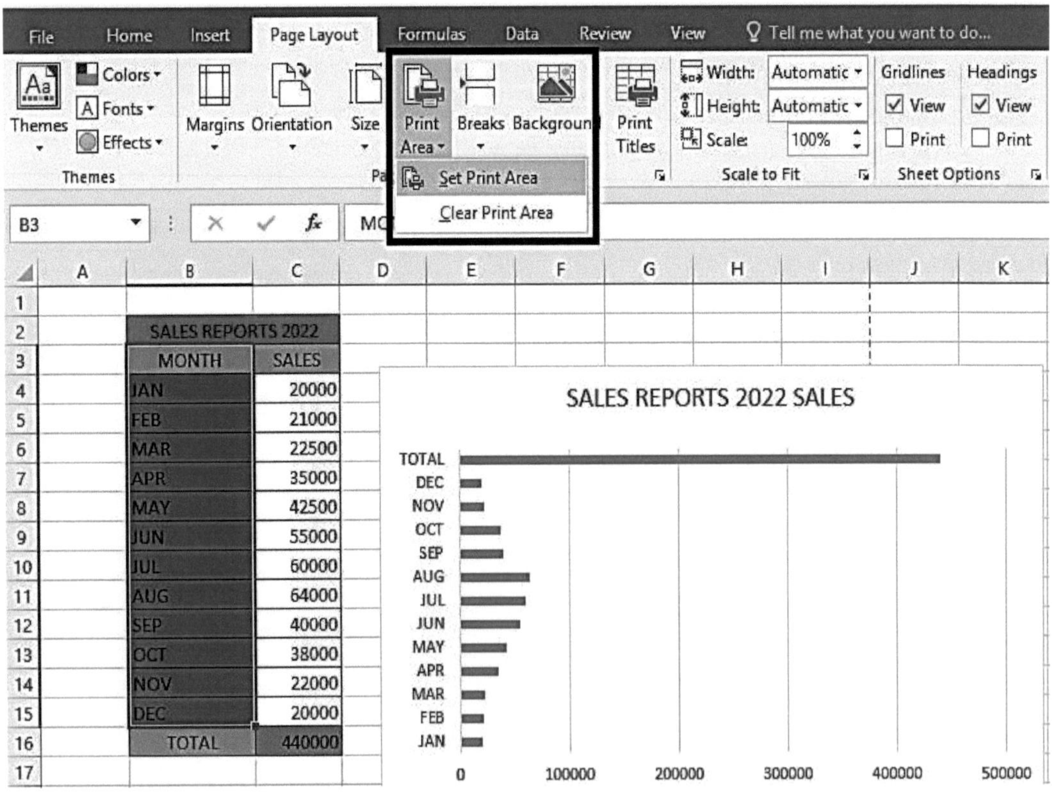

The print area will be identified by the highlighted cells. You may continue working while this area is stored for future printing.

- The "Orientation" button allows you to move between landscape and portrait mode.
- On a printed page, the "Margins" button modifies the margins.
- "Scale to Fit" determines how many pages your printed material will occupy.
- From the same dropdown menu, you may clear, overwrite, or add to the print area.

Click on **File** and select **Print**. On the Print settings menu, select **Print Active Sheet**. Then, click on **Print**.

Saving Workbook

When you save a workbook for the first time, Excel will ask you to name it. Any modifications to the text, numbers, or formulae must be saved using the Save procedure once the file has been given a name.

- Go to **File > Save New Workbook**.
- Choose **File Save As**.
- A dialog window called **Save As** opens.
- Select a location for the file.
- Fill in the File Name: box with a name for your file.
- Select **"Save"** from the drop-down menu.

When working with a spreadsheet, it's a good idea to save it periodically. It's never pleasant to lose data. Using the shortcut key combination **Control key + S,** you may save your worksheet fast.

CHAPTER TWO

ENTERING AND EDITING WORKSHEET DATA

Double-clicking the cell location or using the Formula Bar may alter the data that has been put in it. You may have observed that the data you put into a cell location showed in the Formula Bar as you wrote it. The Formula Bar may be used to input data into cells as well as amend data that has previously been entered.

For example, click on cell B16, and type in the word Tot. Activate cell B16, then, on the formula bar, you will see the abbreviation, **Tot**.

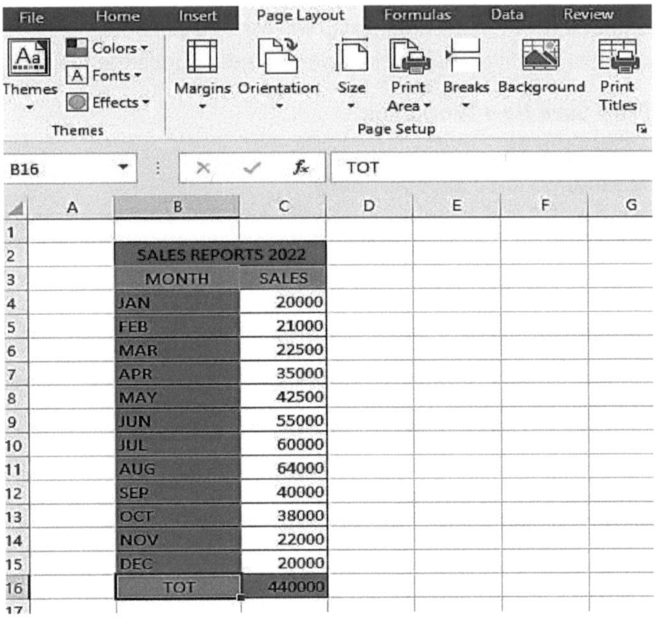

So, click on it and complete the word to Total, then click on the checkmark icon on the left. This will change the word to **Total** in cell **B16**.

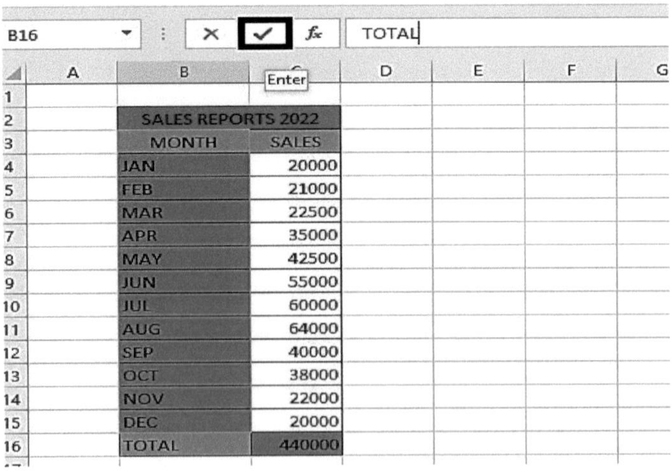

Exploring Data Types

In Excel, there is the Numeric data type, Text data type, Date and Time data type, and Formula data type.

Numeric values

The numerical data which are stored in Excel cells are used in building the majority of formulae in Excel. Numeric data is aligned to the right-hand side. Numeric data may be divided into two categories.

Numeric Data: Numeric data is used to keep track of amounts. 500 bags of cement, for example.

Date and Time: Date and Time values are stored using the Date and Time data type. Technically, Excel saves Date and Time as numbers as well. Date and time data, like numeric data, is aligned to the right-hand side. A date value, a time value, or both may be stored using the data and time data type. Depending on how you structure the Cell, the date and time value might appear in a variety of forms.

Excel's numeric data type is used to record various amounts that may then be utilized in mathematical operations using Excel formulae. Numeric characters (0 to 9) may be used in Excel cells with numeric data types. Aside from numeric characters (0–9), the special characters listed below may be used for many purposes.

+	Positive symbol
-	Negative symbol
()	Negative symbol
%	Percentage symbol
.	Decimal symbol
,	Decimal symbol
E	Exponential symbol

Text Entries

Text is any string of letters, digits, spaces, and characters. This is the most popular data type. Text is the only data type in Excel that is oriented to the left by default. It's mostly used to name table headings, descriptions, and other things. Although text can as well be a number, it should not be utilized in calculations. You may get a **#VALUE!** error if you endeavor to execute any mathematical operation with text data in a Cell. Customer names, customer numbers (using customer numbers in computations would be meaningless), and addresses are examples of the popular usage of the Text data type.

If the text data in an Excel Cell is bigger than the Cell width, the text data will flow over into the empty right-side Cells. If a right-side empty Cell is subsequently filled, Excel hides the spilled text before revealing the data in the new full Cell. See below.

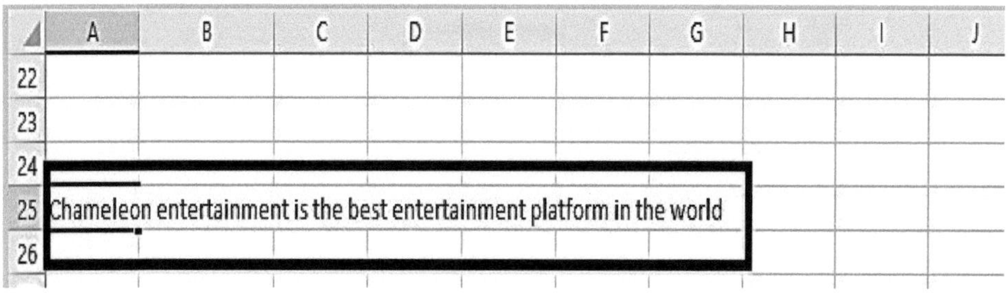

Formulas

A formula is a set of instructions that are entered into a cell to generate a value. It must start with an equal symbol (=). A mathematical equation, cell references, functions, or an operator might all be examples. An expression is another name for a formula. Because MS Excel can build tables with a large quantity of data, you use this function to add formulae to your table and receive faster results. Following are some fundamental Excel functions to get you started, assuming you can now input and function with your favorite formulae.

SUM Formula: First and foremost, the SUM function in Excel is a must-know formula. Values from several columns or rows are often combined. =SUM is a simple selection that adds the column values (A2:A8). Example: Summing up the values of a single row, =SUM(C4-C15).

MEDIAN Formula: The average number of shareholders in a certain shareholding pool is an example of a simple average that the AVERAGE function should bring to mind. Example: The formula =AVERAGE is used to calculate an average (C2:C14). (SUM(C2:C14)/12) is likewise comparable.

COUNT Formula: The COUNT function keeps track of how many cells in a given range contain just numeric values. Example: COUNT is used to count the numerical values in a column (C: C). To count rows, you must change the range of the formula.

IF Formula: The IF function is often employed when you wish to sort your data according to a set of rules. Other formulae and functions may be true, for example, the value is true and false.

Example: =IF(C2D3, 'TRUE,' 'FALSE') – IF(C2D3, 'TRUE,' 'FALSE') – IF(C2D3, ' This method checks the two numbers to see whether C3 is smaller than D3. If the reasoning is right, the cell value should be TRUE; otherwise, it should be FALSE.

COUNTA Formula: COUNTA counts all cells in a given range in the same way as the COUNT function does. Regardless of cell type, all cells are counted. It counts the same things as COUNT: dates, times, strings (including logical values and errors), and empty strings or text.

Entering Text and Values into your Worksheets

Any cell may have either text or number. It might be difficult to input data if you haven't been taught how, so follow the steps below to discover the tips and tricks for efficiently entering data.

Entering Numbers and Text

To input numbers or texts, first of all, activate the cell by clicking on the cell, then input the value and hit **Enter** key. After inputting the value, activating the cell, you will notice that the value is shown in the cell likewise in the Formula bar.

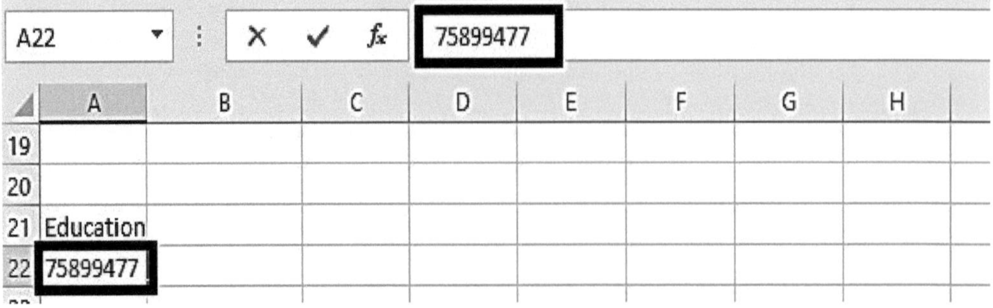

A cell may carry up to 32,000 characters, which is more than enough for a normal chapter in a book. Despite a cell carrying a large number of characters, you'll discover that it's impossible to show them all. The formula bar may not fully display lengthy text. To expand the height of the Formula bar and show additional text, click the bottom of the Formula bar and drag it down.

When inputting values, especially numbers, you may use decimal points, currency symbols, mathematics symbols, etc. excel considers a value to be negative if it is preceded by a minus sign or it is enclosed in parenthesis.

Entering Dates and Times into your Worksheets

According to what you want to perform, there are numerous methods to add dates in Excel. Do you wish to include today's date in a report or invoice, for example? Perhaps you'd want to put a date in Excel that would regularly refresh and always show the current date and time? Or maybe you'd like to have your worksheet auto-fill weekdays or enter random dates? Let's show you how you can do that.

Entering date and time values

Choose the cell. Then, press **Control key + Semi-Colon (;).** This will add the current date to that cell. To enter the time, select the cell, then press **Control + Shift + Semi-Colon (;)** on your keyboard. This will add the current time on that cell.

You can also enter date and time using some functions. To do this, simply Activate the cell you want to add a date on, type in this function on the cell **=TODAY()** and then press **Enter**. For the time, select the cell, type in this function **=NOW(),** and press **Enter**.

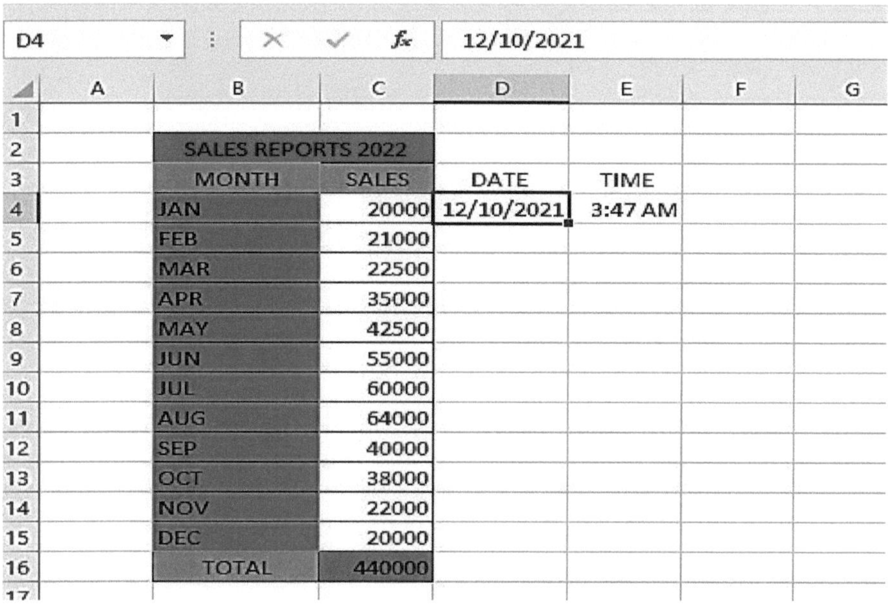

	A	B	C	D	E	F	G
1							
2		SALES REPORTS 2022					
3		MONTH	SALES	DATE	TIME		
4		JAN	20000	12/10/2021	3:47 AM		
5		FEB	21000				
6		MAR	22500				
7		APR	35000				
8		MAY	42500				
9		JUN	55000				
10		JUL	60000				
11		AUG	64000				
12		SEP	40000				
13		OCT	38000				
14		NOV	22000				
15		DEC	20000				
16		TOTAL	440000				
17							

MODIFYING CELL CONTENTS

Deleting the contents of a cell

There are various techniques for deleting data on a worksheet. This is a useful command if you accidentally delete data from your worksheet. The instructions below show you how to erase data from a cell or a range of cells

- Click on any cell you want to delete. Then, on your keyboard, press the Delete key. This will delete the contents of the cell.
- Select cell **B3** to **B16** by highlighting it. To do this, place the mouse cursor over B3, click and drag down to B16. Then, on the Fill handle of B16, click and drag it up to cell B3, then leave the mouse button. All contents of B3 to B16 will be deleted.

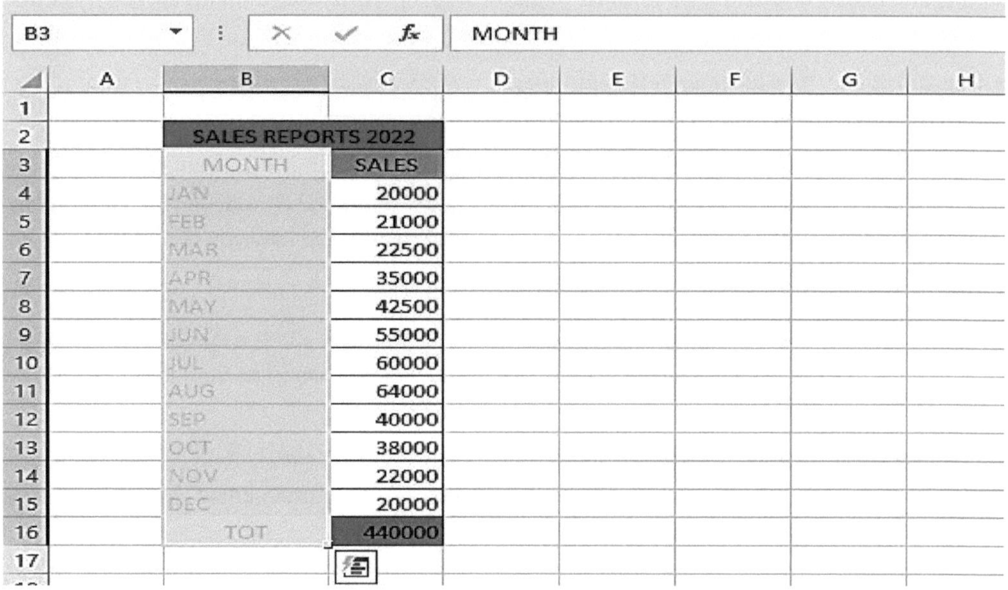

Replacing the contents of a cell

To replace the contents, you can use the Undo button on the Quick Access Toolbar. Click on it and the contents will be back. You can also press the Control key + Z to do this.

Learning some handy data-entry techniques

The core of Excel is data input. If you can't rapidly and properly input data into your spreadsheets, you won't be able to utilize the tools to analyze and report on the data. When it comes to data input, Excel gets a lot of things right, yet certain things aren't straightforward. Even the most seasoned Excel users may sometimes choose more sophisticated procedures than are required. See below.

- If you wish to remain in the same cell but switch to a new line when entering long text, you can't merely hit [Enter]. This just advances you to another cell. Instead, use [**Alt**] + [**Enter**] to make a line break or a new line in the same cell. This technique is often known as a new line or line break.
- Do you use product codes or other ids that start with one or more zeros in your entries? When you input data like this, Excel thinks it's numbered and removes the beginning zeros. Format your worksheet cells as text rather than numbers to persuade Excel to allow you to input values with leading zeros. There are two techniques to consider:

- One: Type an apostrophe (') in front of each item to convert it to a text format. However, if you have a lot of data, this is a time-consuming method since you'll have to add the apostrophe to each record, not just the ones with leading zeros. Your list will not order appropriately if this is not done.
- Format selected cells into the text to prepare for data input. Choose the cells (or a whole column) where numbers will be kept as text. Right-click, then select Format Cells, then in the Category box, select Number Click Text and then OK to apply.

Automatically moving the selection after entering data

As you hit Enter, Excel slides the pointer down to another cell. That isn't always the best course of action. It's aggravating when you're typing data from one column to another. You may now hit the right arrow key instead of hitting **Enter**, however, this will cause you to slow down. Although pressing Tab is simpler than extending for the right arrow key, changing the cursor movement as follows may be the simplest solution

Click on **File**. Click on **Options**. This will open up the Excel option menu. On the left-hand side of the menu, click on Advanced. On the **Editing option**, below the **After Pressing Enter** option, select the direction you want from the down arrow. The direction options are Right, Left, Up, and Down. When you are done, click **OK**.

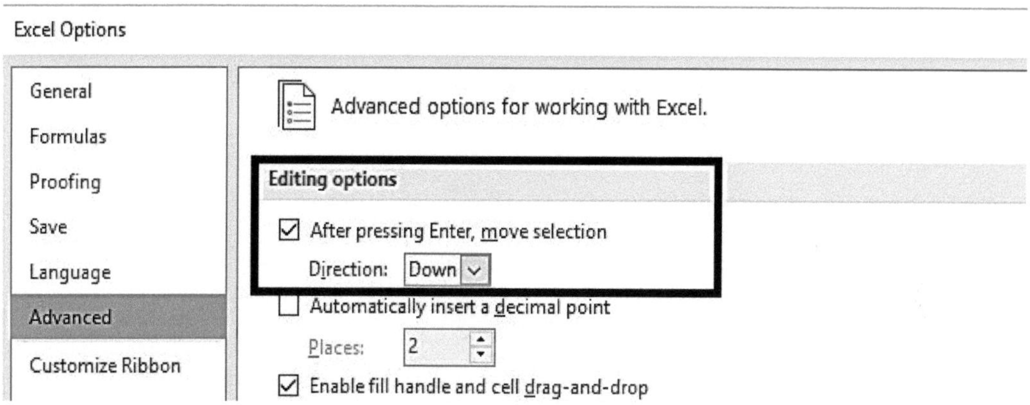

Selecting a range of input cells before entering data

In Excel, working with a whole data set is a regular activity. You could wish to transfer the data, apply a filter or a standard format, or convert it to a table. There are several reasons to pick a data range, but doing so may need to leap through a few selection hoops, particularly if you're dealing with a wide range that covers multiple screens. Fortunately, selecting a whole data range is simple and fast.

Use the **Go To** function to choose a data range as follows:

Any cell in the data range may be selected by clicking it. For example, you might choose any cell from B2 through C15 to choose the data range B2:C15.

[**F5**] is the shortcut key. Press it.

Click the Special button in the bottom-left corner of the **Go To** dialog.
Select Current Region from the pop-up menu that appears.
When you click **OK**, Excel will choose the current data range for you (the current region).

Using CTRL + Enter to place information into multiple cells

In Excel, copying and pasting a single cell is simple. When it comes to copying and pasting a group of cells, columns, or rows, you have several options. Use the shortcut **Control key + Enter key** to activate. Pick a few cells and type any term or value into any of them. Instead of hitting Enter, hold down the Control key and hit Enter. You'll see that the name is copied to all of the chosen cells.

Changing modes

In an Excel spreadsheet, there are four different sorts of cell editing modes. "**Ready**," "**Edit**," "**Enter**," and "**Point**" are the four sorts of Cell modes. An Excel cell's default mode is "Ready." Excel is ready to take data in any of its worksheet Cells while it is in Ready mode.

The default "**Ready**" mode of its Active Cell is changed to "**Edit**" mode by pressing the function key "**F2**" once. If you hit the function key "**F2**" again when in "**Edit**" mode, the Cell mode will change to "**Enter**." The fourth cell mode is called "**Point**." While creating or editing a formula, "**Point**" cell mode enables you to explore a large Excel spreadsheet and choose the required cells.

Entering decimal points automatically

If you need to input many numbers, each of which includes two decimal numbers, as illustrated below, manually typing the decimal point will be a waste of time. Now I'll show you how to quickly add a decimal point to a number in Excel at a precise location.

B	C	D
	34.65	
	77.54	
	32.12	
	224.56	
	0.12	
	9.56	
	55.3	
	23.4	

To do this, click on **File** > **Options**. In the Options menu, select **Advanced**. Under the Editing Option, check the box close to the **Automatically insert a decimal point** option. Then, put in the decimal places in the Places option.

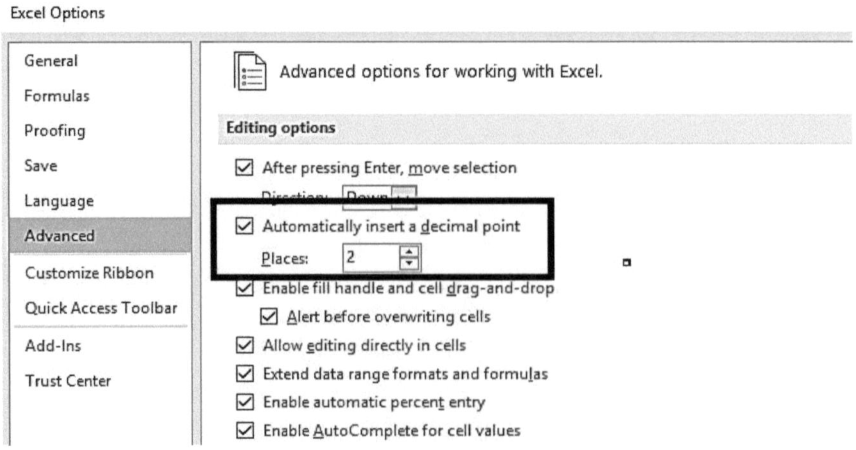

After that, click **Ok**.

Using Autofill to enter a series of values

You may use the AutoFill command to automatically lengthen a predictable sequence (for example, 10,20,30.... day, hours of the day). You may also use AutoFill to propagate formulae - create the formula once, then use AutoFill to spread it to the other cells.

So, select a cell and type the first day of the week; Sunday. Activate the cell you typed in the day. Then, hover the cursor to the little box beside the cell. When the pointer changes to an **Add sign (+),** click and drag down. Excel will fill in the rest of the values i.e., the days of the week.

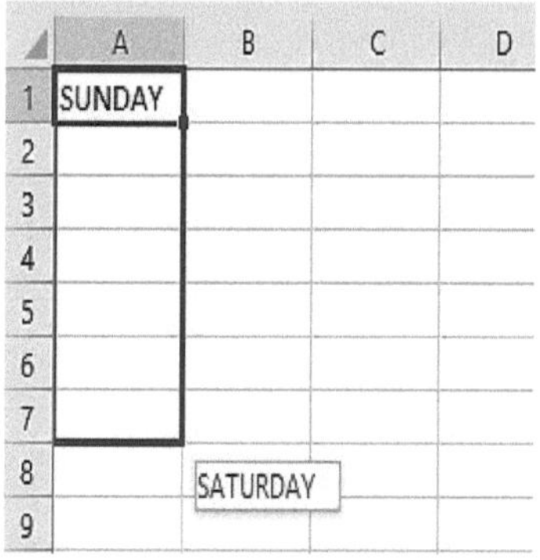

Using Autocomplete to automate data entry

When manually inputting data into a spreadsheet, the AutoFill option comes in handy. This function is useful for a variety of things, but it's especially useful for inputting data in a certain order, such as the digits 2, 4, 6, 8, and so on, or non-numeric data like the days of the week or months of the year. Just like we did in chapter one when we completed the months of the year by dragging down the box beside the month cell. See the example below;

Activate **Cell B4** and type in January. Then, you will see a tiny small box beside the cell which is the Fill Handle.

Click on it and drag down to 11 rows on the bottom right of Cell B4 (**B15**). Because Excel identified January, it will automatically fill the remainder of the months for you rather than making you input them.

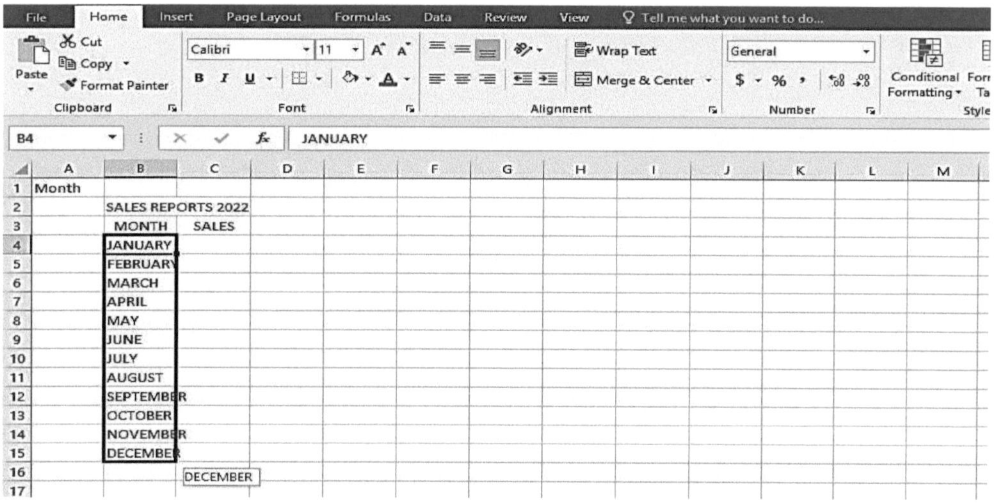

Forcing text to appear on a new line within a cell

By pressing the **Alt key + Enter**, you can insert a line break or start a new line of text or create space between lines and paragraphs. On the cell you want the line break to appear, double-click on it. to break the line, click the desired position inside the cell you have chosen.

Using Autocorrect for shortcut data entry

Excel's AutoCorrect tool may be used to create shortcuts for frequently used words or phrases. Entering lengthy stuff or individual name takes time and often results in mistakes and inaccuracies. To fix this problem, use the AutoCorrect tool to create shortcuts and modify them. Follow these procedures below;

Click on **File** > **Options**. This will open up the **Excel Options** menu.

On the left-hand side of the menu, select **Proofing**. Then, click on **AutoCorrect Options.**

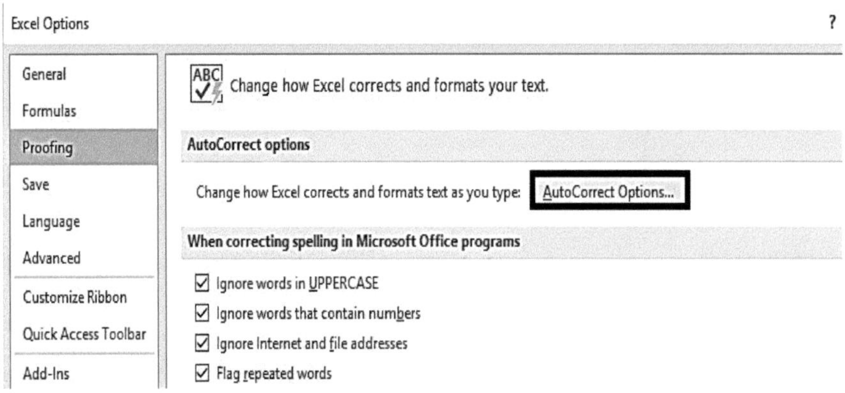

This will open up the AutoCorrect option dialog box. On the Replace text box, type in the shortcuts you want and what you want to replace them with. Then, click **Add**. As you can see in the image below.

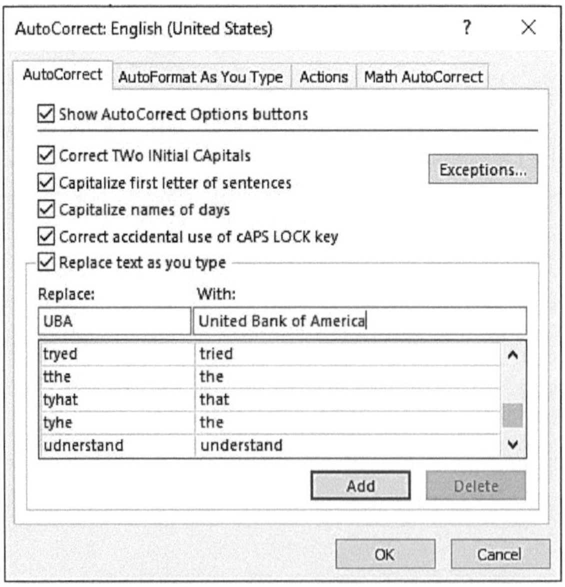

You must give each AutoCorrect item a distinct name after entering or choosing it.

If you're going to name an AutoCorrect entry, don't use a popular term unless you're going to change it.

Don't call an entry Mr., for example. Instead, use an underscore or another symbol to distinguish it, such as *Mr. If you don't make the term unique, Word will automatically insert the AutoCorrect entry anytime you enter it, whether you want it or not. If you've picked a frequent term, reverting each auto-correction will take too much time.

Note that your AutoCorrect list in Excel is shared with other Office products. Any AutoCorrect entries you make in Word will function in Excel and Outlook as well.

Entering numbers with fractions

When we use the slash **("/")** symbol to separate two integers in Excel, the values are converted to decimal format. We may use the Fractions option to retain fractions for such values; first, choose the cell whose value we want to convert fractional and then choose the Format Cells option from the right-click menu list.

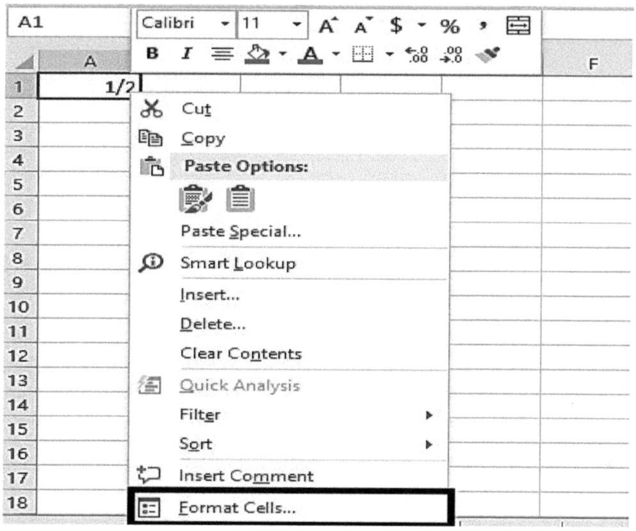

Click on the **Number tab**, pick **Fraction** from the category area, where we have many criteria for fractioning up to one digit, two digits, or three digits like halves, eighths, etc. Then select the **Fraction Type** you want.

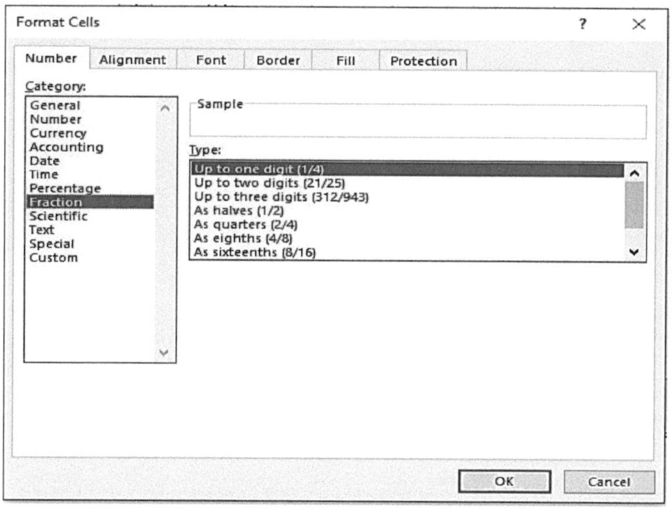

Using a form for data entry

A data entry form can assist by data entry process more efficient and error-free. You use it to put data into your worksheet. By default, there is no data entry form option on the Excel Ribbon. So, you will have to add it to the Quick Access Toolbar so you can access it easily. Follow the steps below to add it;

Right-click on any option on the **Quick Access Toolbar** and select **Customize Quick Access Toolbar.**

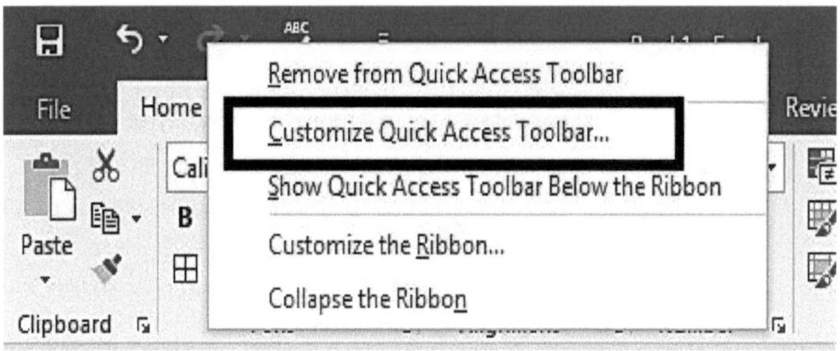

This will open up the Excel Options box. Click on the drop-down arrow on the Choose commands from and select All Commands.

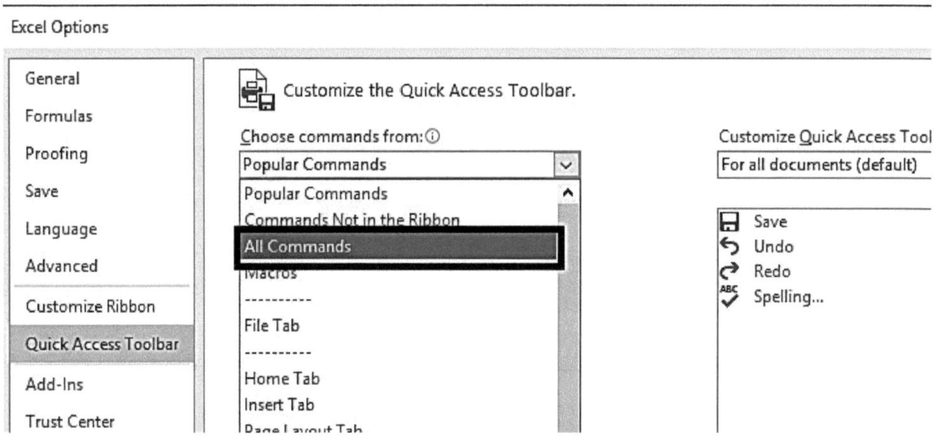

On the All command list, click on Form, then click on the Add button. Then, OK.

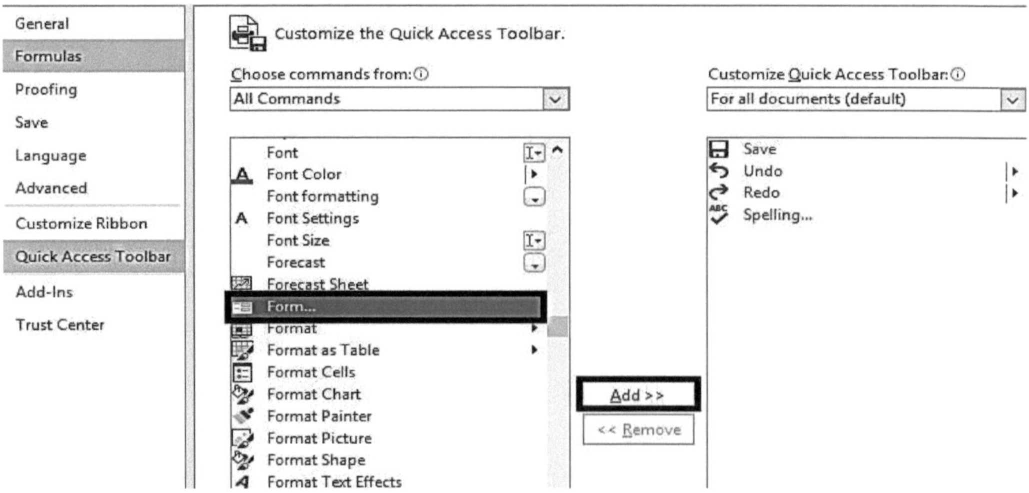

You will see the Form Icon on the Quick Access Toolbar. So, you can easily access it now. Simply click on the cell you want to input data from, then click the Form-Icon.

The Data Entry Form is made up of different options. Below is a brief explanation of what each option does.

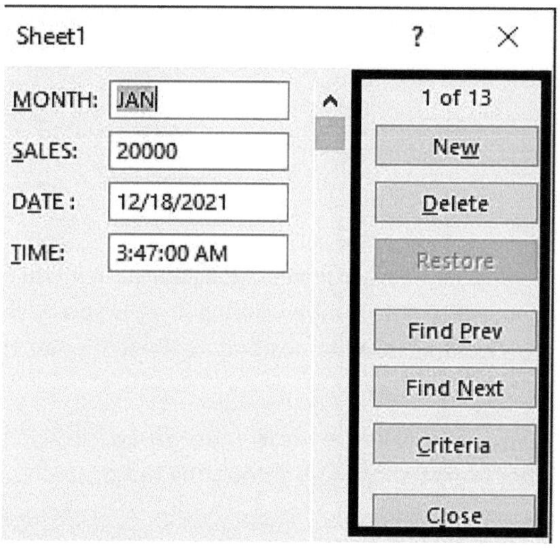

- **New**: This clears any existing data in the form and enables you to start over with a new record.
- **Delete**: You may use this to get rid of an existing record.
- **Restore**: If you're updating an existing entry and haven't clicked New or pressed Enter, you may restore the old data in the form.
- **Find Previous**: This will locate the preceding entry.
- **Find Next:** This will take you to the next page.
- **Criteria**: You may use criteria to discover particular records.
- **Close**: This will bring the form to a close.
- **Scroll Bar**: The scroll bar may be used to navigate through the records.

Let's take a look at all you can accomplish with an Excel Data Entry form.

- Activate the cell for the data.
- Click on the Form-Icon. This will open the data form box. Now type in the data in the fields. Then, press the Enter key. You can also click the New button.

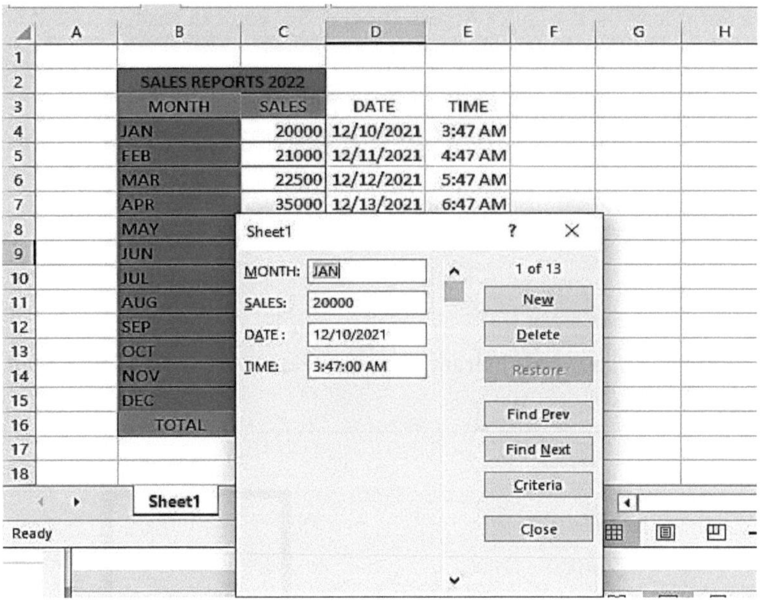

Applying Number Formatting

It's a wise concept to utilize suitable number formats for your data if you're dealing with a spreadsheet. Number formats specify what sort of data you're utilizing in your spreadsheet, such as percentages (%), money ($), times, dates, etc. Number formats not only make it simpler to read your spreadsheet, but they also make it easier to utilize.

To apply a number format, choose the cell > On the Number group, click on the arrow and select a number format. Click on More Number Formats for more formatting options.

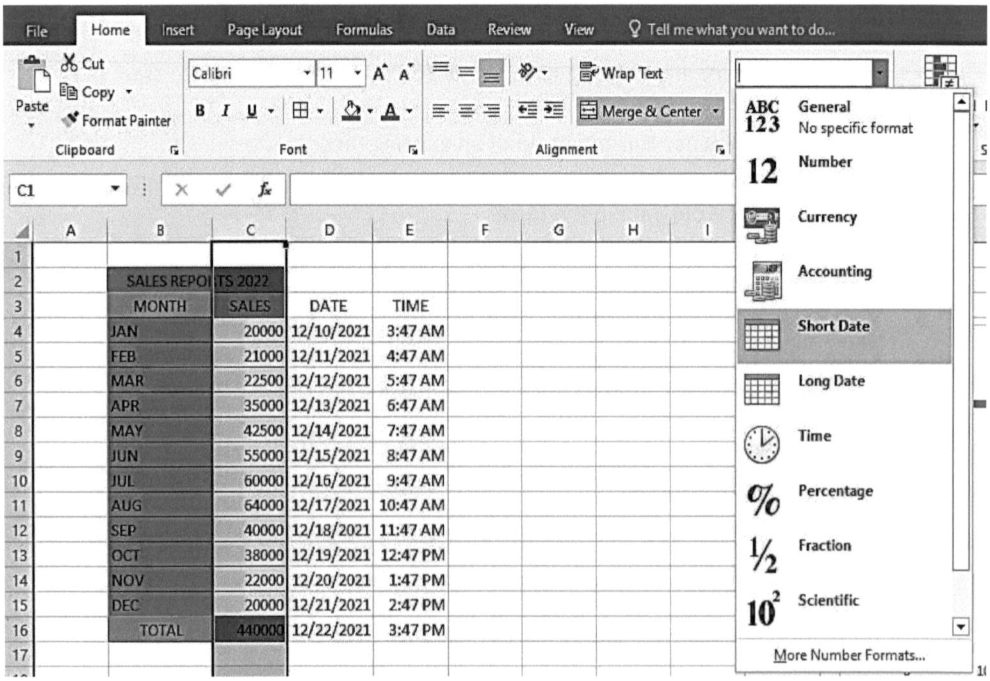

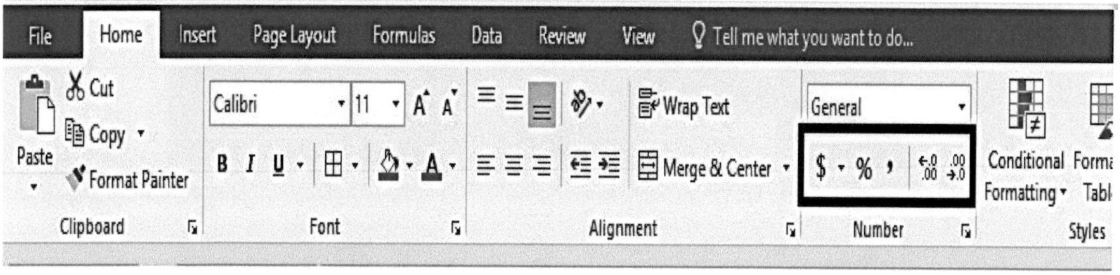

Using automatic number formatting

The AutoFormat option cannot be accessed from the ribbon, so you will have to add it to the Quick Access Toolbar. Right-click on the Quick Access Toolbar and select **Customize Quick Access Toolbar.**

Click on the down arrow on the Choose commands from and select **All Commands**.

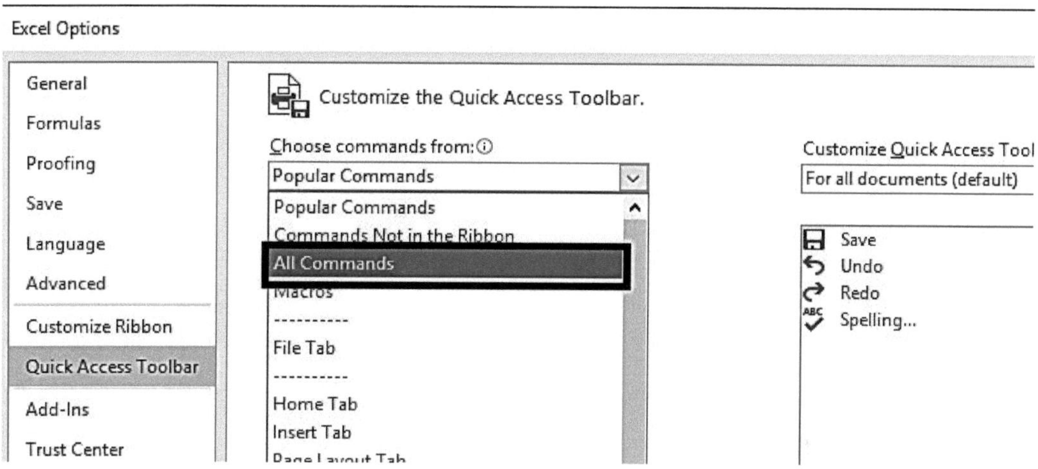

On the All Commands list, click on AutoFormat, then click Add. This will add the AutoFormat Option on the Quick Access Toolbar.

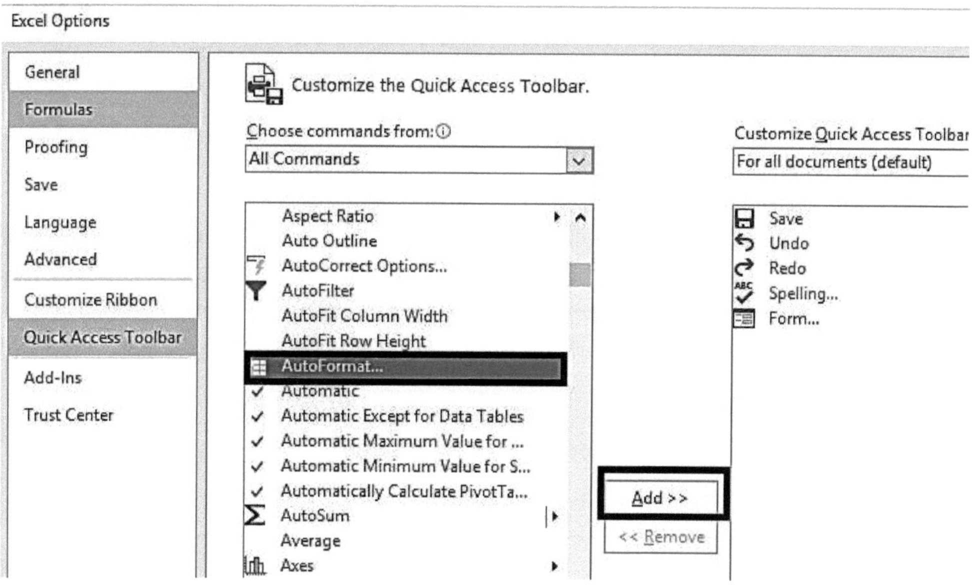

This will add the AutoFormat Option on the Quick Access Toolbar.

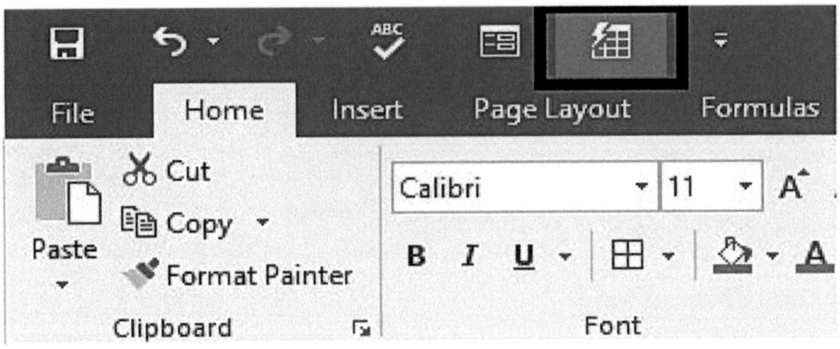

To AutoFormat the numbers, simply highlight the whole data, hit the AutoFormat icon. This will display the AutoFormat menu. Choose any style, then click **Ok**.

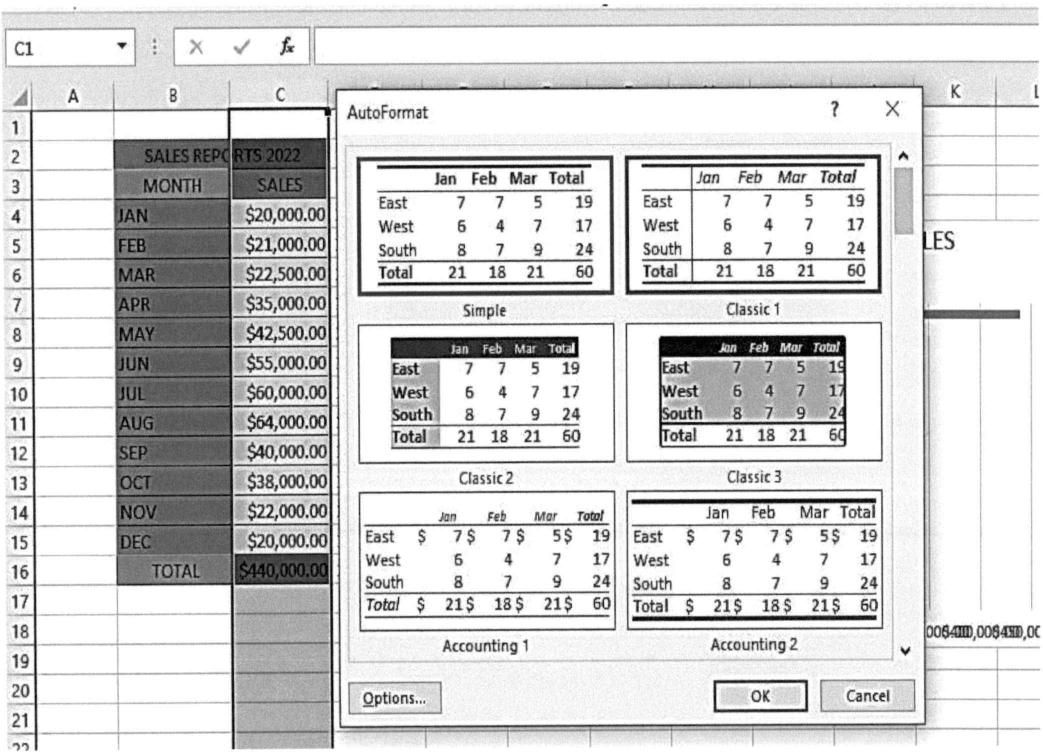

Using shortcut keys to format numbers

There are shortcuts for formatting numbers on Excel. **Control key + Shift key + Number key**. Simply choose the cell(s), then press the shortcut keys. Below is a list of the shortcut keys and their function.

Control + Shift + ` = **General**

Control + Shift + 1 = **Number**

Control + Shift + 2 = **Time**

Control + Shift + 3 = **Date**

Control + Shift + 5 = **Percentage**

Control + Shift + 6 = **Scientific**

Control + Shift + 7 = **Border**

Formatting numbers by using the Format Cells dialog box

Pick the cell for formatting, right-click on it and select **Format Cells**. On the Dialog box, you will see a list of options that you can select from. Choose the one you want and click **OK**.

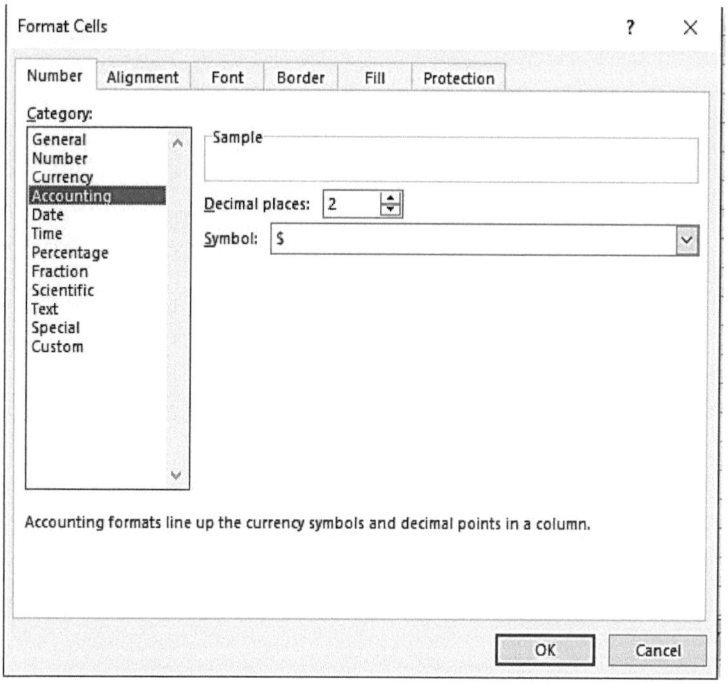

Add your own custom number formats.

You may wish to format a number in a custom manner, such as formatting 421020000 as $421.02 M, and then store this for future use. You can also apply it to multiple cells. To do this, follow the steps below.

Select the cell that has a number value in it, right-click on it, then click on **Format Cells**.

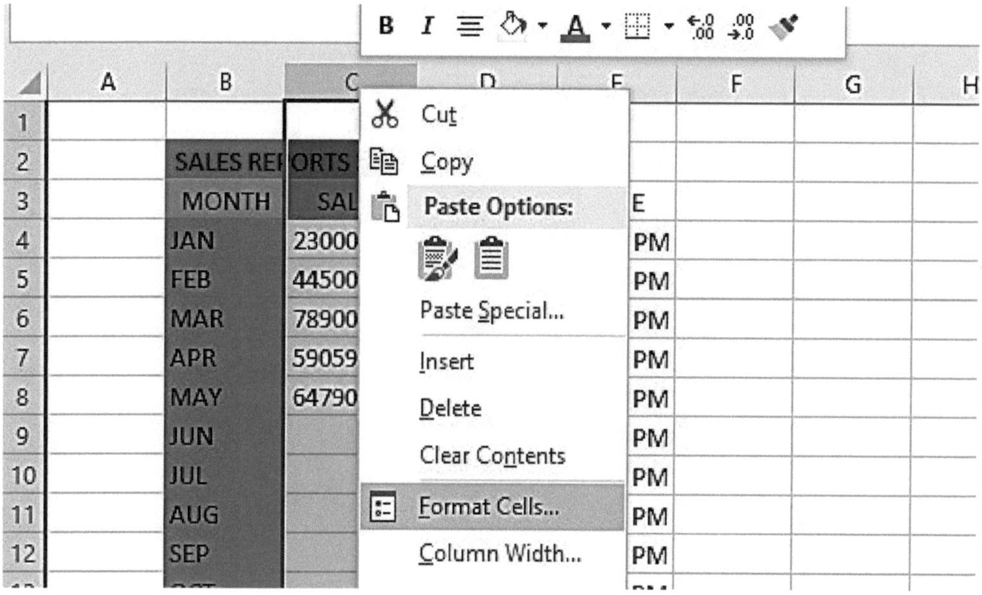

Click on the Number tab, then click on Custom. On the box below the Type option, type in the format code $#.##,," M"; then, click **OK**.

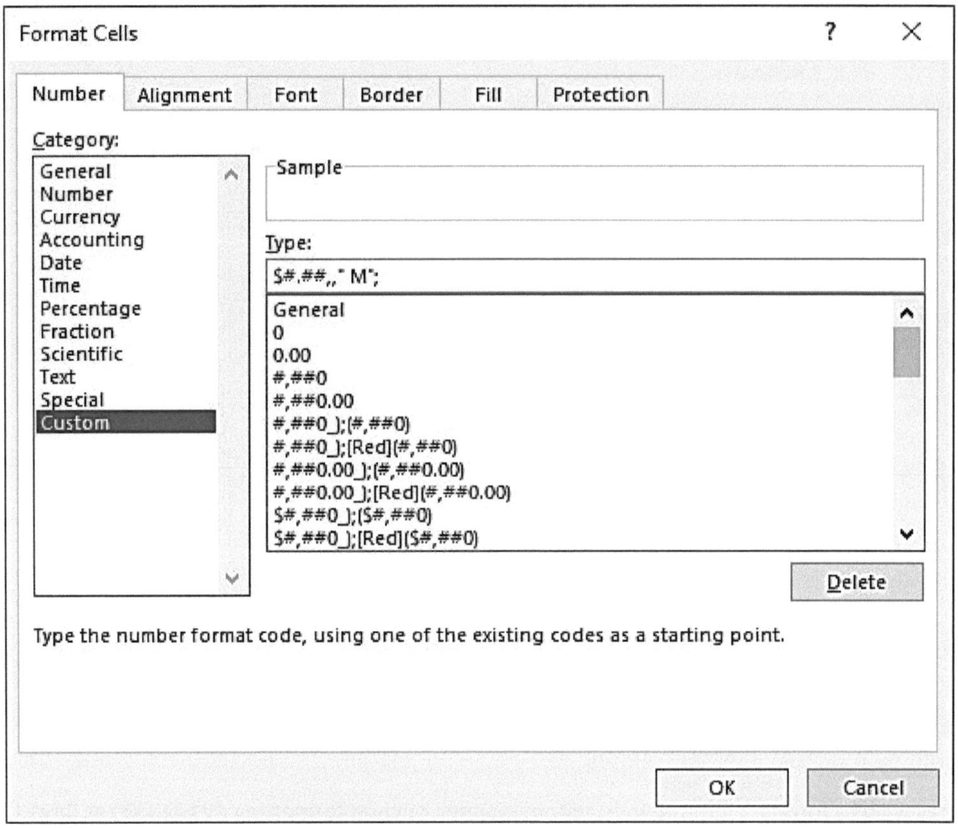

CHAPTER THREE

PERFORMING BASIC WORKSHEET OPERATIONS

Worksheets may be a fantastic tool for organizing your data. Rather than jamming everything into one worksheet, the user may utilize several worksheets inside a workbook.

Learning the Fundamentals of Excel Worksheet

Each file in Excel is referred to as a workbook, and each workbook may include more than one worksheet. Consider an Excel workbook as a notebook, with worksheets serving as pages inside the notebook. You may see a specific sheet, add new sheets, delete sheets, and copy sheets just like in a notepad.

Excel workbook files can contain more than one-sheets. These sheets can be different from each other. It can be chart sheets (containing a chart) or worksheets (which contain rows and columns). Take note that each worksheet has a unique name; by default, a workbook opens with three worksheets named Sheet1, Sheet2, and Sheet3. However, you may add, remove, and rename these spreadsheets as desired.

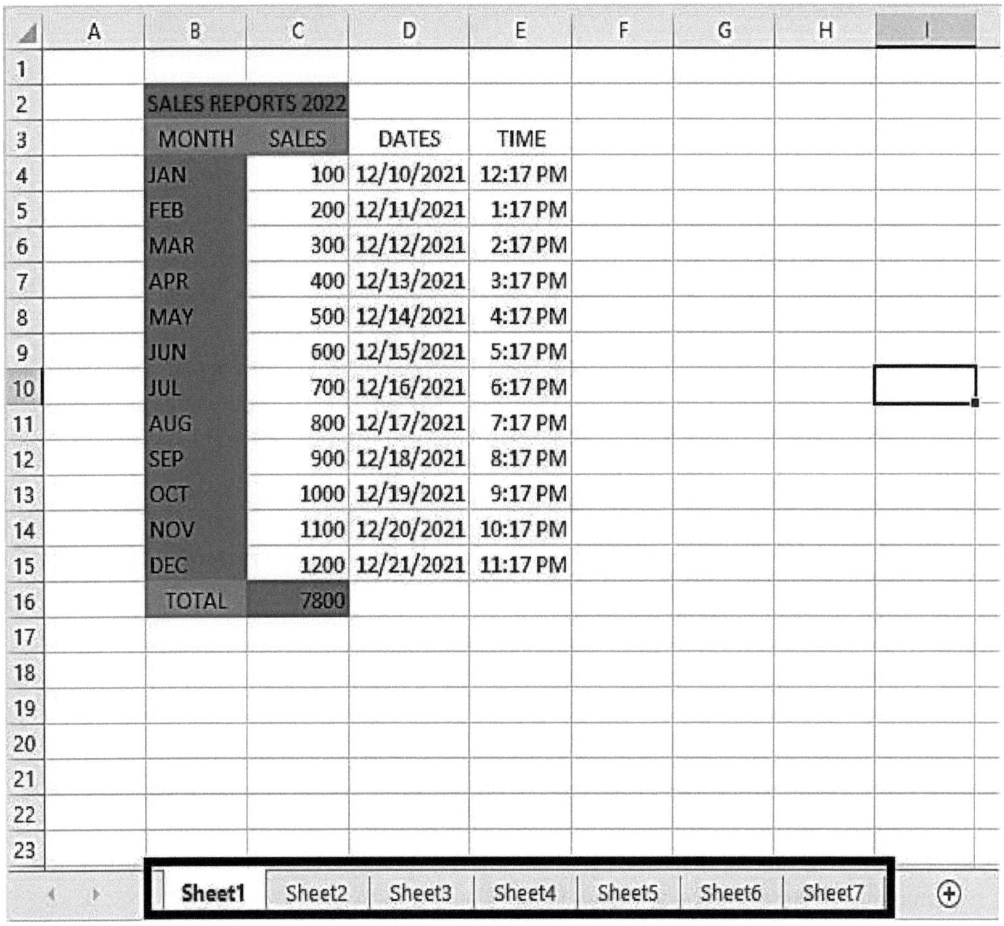

SALES REPORTS 2022			
MONTH	SALES	DATES	TIME
JAN	100	12/10/2021	12:17 PM
FEB	200	12/11/2021	1:17 PM
MAR	300	12/12/2021	2:17 PM
APR	400	12/13/2021	3:17 PM
MAY	500	12/14/2021	4:17 PM
JUN	600	12/15/2021	5:17 PM
JUL	700	12/16/2021	6:17 PM
AUG	800	12/17/2021	7:17 PM
SEP	900	12/18/2021	8:17 PM
OCT	1000	12/19/2021	9:17 PM
NOV	1100	12/20/2021	10:17 PM
DEC	1200	12/21/2021	11:17 PM
TOTAL	7800		

Right-clicking on a worksheet brings up a menu containing some options for making changes to your worksheet. You will see options like **Insert, Move or Copy, Rename, Delete**, and others. You can also change the color of the worksheet.

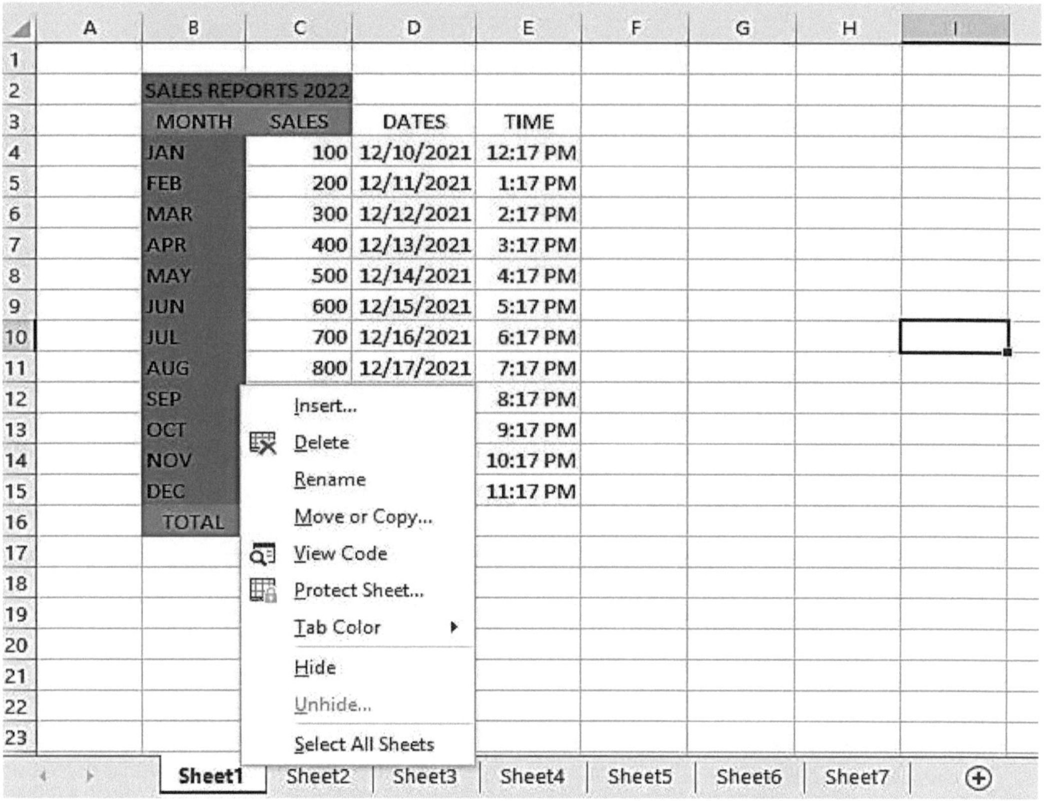

Working with Excel windows

A window is opened for each Excel workbook file that you open. A window is the workbook's container in the operating system. You may open as many Excel spreadsheets as you want concurrently. As with other Windows programs, the Excel window's control buttons are situated in the window's extreme right top corner.

Three window-control buttons are accessible. To minimize the Excel application window, press the left button. The middle button maximizes/restores the Excel software window. To dismiss the Excel application window, click the right button.

The functionalities of the left and right buttons are always identical. They are used to minimize and close the Excel software window. However, the center button's operation is dependent on the present state of the Excel window. If the current state of the Excel window is Maximized, the center button is used to Restore the Excel window to its default state. When the current state of the Excel window is Restored, the center button is utilized to maximize the Excel window. You must comprehend three concepts relating to window state: minimized, maximized, and restored.

- **Minimized state**: A minimized Excel window is not visible on the Operating System screen but is minimized to stay on the Windows taskbar.
- **Maximized state**: The term "maximized Excel window" refers to the state of the Excel window when it has been maximized to fill the whole screen of your Operating System.
- **Restored state**: When an Excel window is restored to its original size, it is said to be restored to its restored condition. When the window state is restored, the size of the Excel window is less than when it is maximized. You may resize the Excel window in this state.

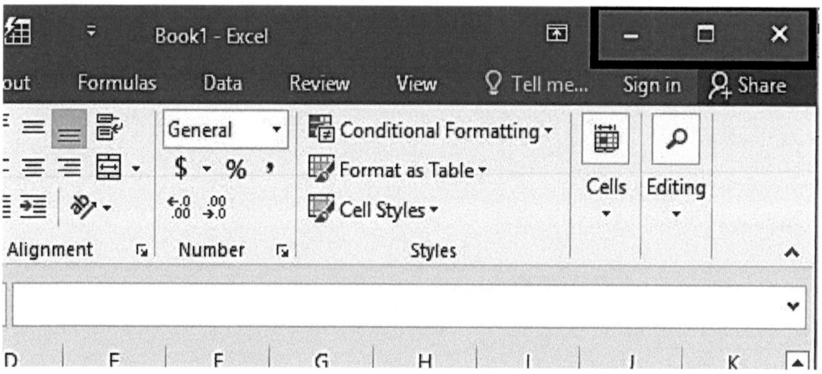

Moving and Resizing Windows

When the Excel window is in the Restored state, drag the Windows pointer to one of the window's sides or corners. As seen here, the Windows cursor transforms into a **double-headed arrow.** Once you see the arrow, click and drag the mouse. The window will begin to resize. Note that when you drag the window, the resizing will affect the height and the width.

To move a window

When the Excel window is in a Restored state, move your mouse pointer to the top of the Excel title bar. Then, click and drag the window to any location on the screen you want.

Switching Windows

You must have at least two Excel workbooks open to switching windows. Click on the View Tab, click on Switch Windows.

Choose the window to switch to.

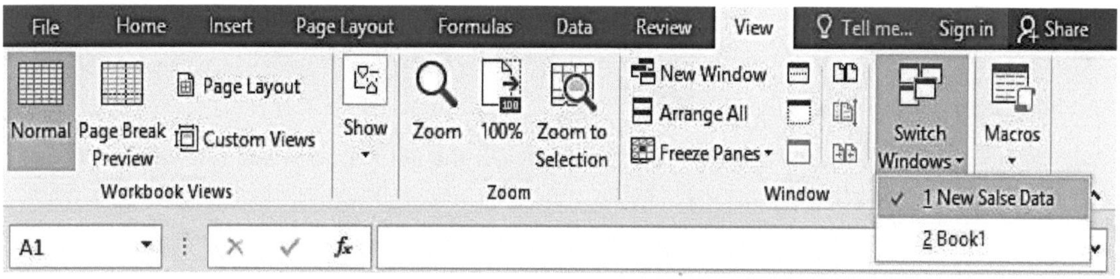

Closing Windows

On the top right-hand side of the Excel window, click on the close button.

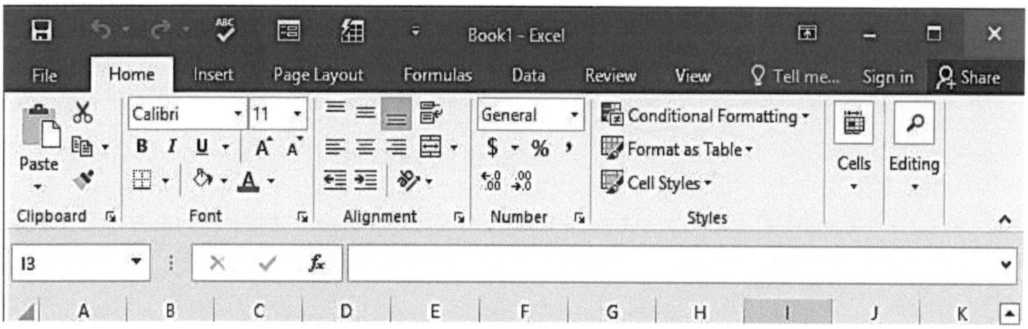

Activating a worksheet

To activate a worksheet, click on the small left arrow at the bottom left of the workbook.

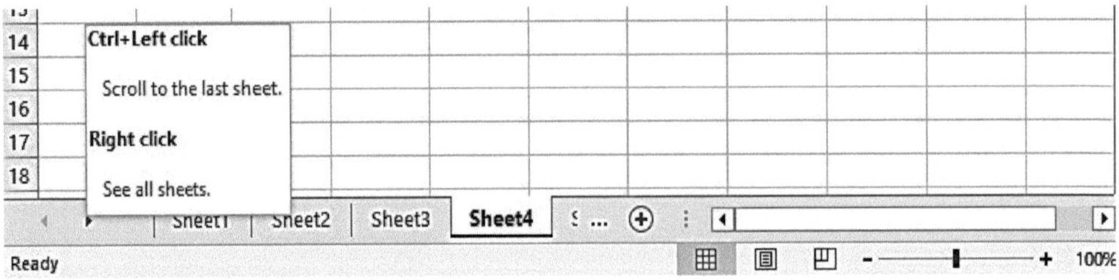

Then, right-click to see all sheets. This will bring up the Activate Sheet menu. Click on the sheet you want to activate and then click Ok.

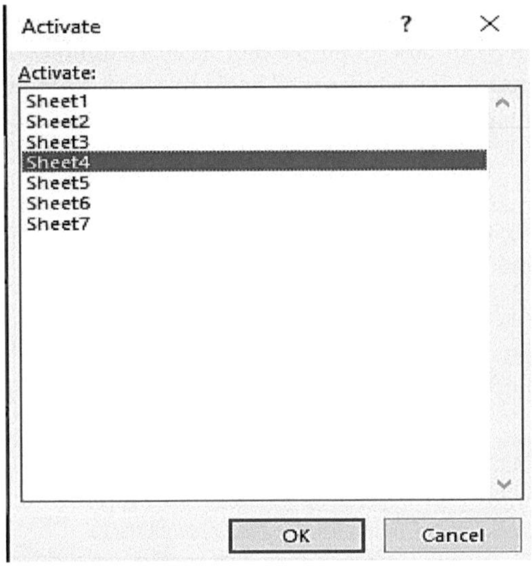

Adding a new worksheet to your workbook

You can do this in two ways. First, click the **(+)** sign below the workbook to add the workbook. You can also right-click on the present worksheet below the menu which should be **Sheet 1**, then click on Insert, click worksheet and then click **OK.**

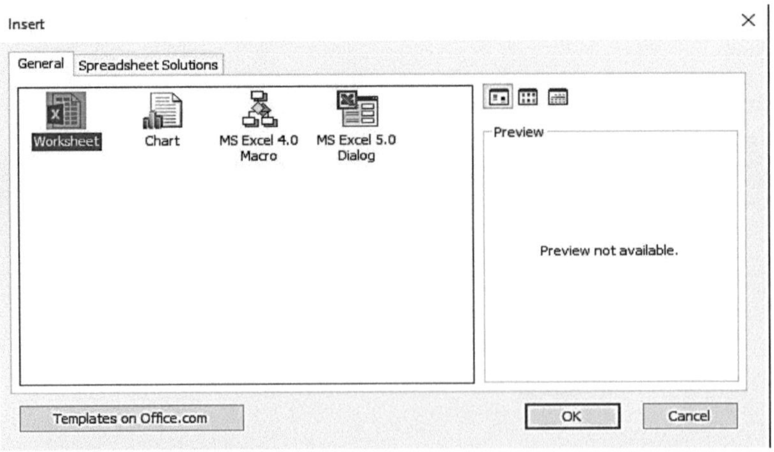

Deleting a worksheet

To do this, simply right-click on the name of the sheet you want to delete, then select **Delete**.

Changing the name of a worksheet

Select the name of the sheet you want to rename by right-clicking on it. when you do so, select Rename. Then, you can now rename the sheet by typing the name on the sheet box.

Changing the tab color

On the sheet, you want to change the tab color, right-click on it. from the list of options there, click on Tab Colour. Then select the color you want to use.

Rearranging your worksheet

You can move or copy a worksheet to another location. Right-click on the worksheet, choose Move or Copy. You can rearrange the worksheet to another position on the same workbook and you can also rearrange the worksheet to another worksheet.

- To reposition a worksheet inside the same workbook, click the worksheet's name.

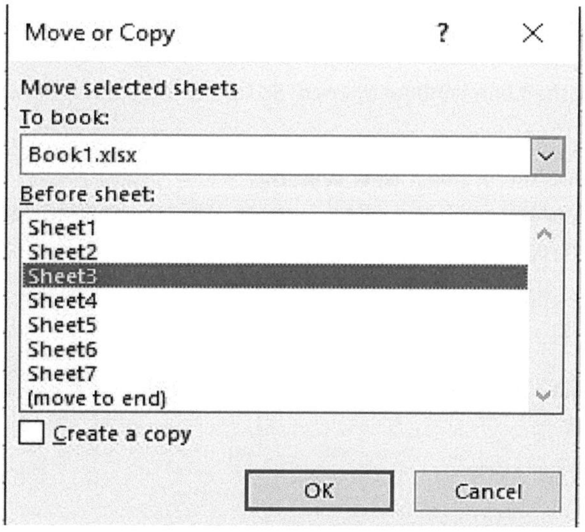

- To reposition a worksheet to a different workbook, choose the new workbook from the "**To Book**" option and click the name of the worksheet you want to move.

Note that your worksheet will vanish from the main workbook when you move it to another workbook.

Hiding and unhiding a worksheet

Select the worksheet, right-click on it and select Hide. To unhide it, select any worksheet from the list and right-click on it, then select Unhide.

CONTROLLING THE WORKSHEET VIEW

Zooming in or out for a better view

By using the Zoom slider in the Microsoft Excel status bar, you may zoom in or out of an Excel worksheet. The status bar is located in the bottom right corner of Microsoft Excel. Slide to the desired % zoom setting. To zoom in and out in small steps, use the - or + buttons. The percentage number associated with the zoom level is also shown in the status bar adjacent to the zoom slider. When you click the percentage, a Zoom dialog box appears, allowing you to choose from preset zoom levels or create your own.

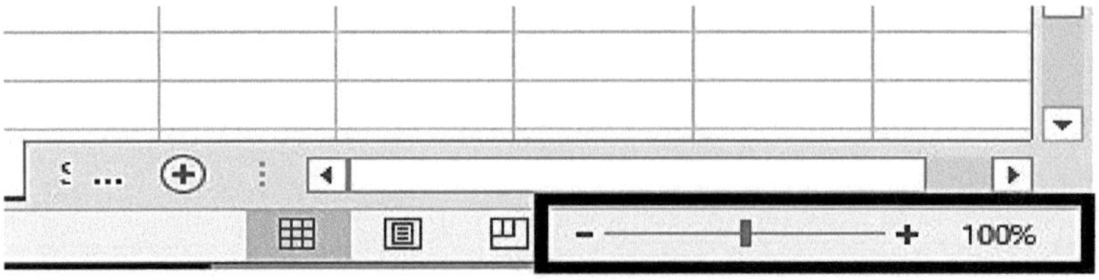

You can also make use of the ribbon. On the ribbon, click on View. On the View tab, you will see the Zoom box with some options in it.

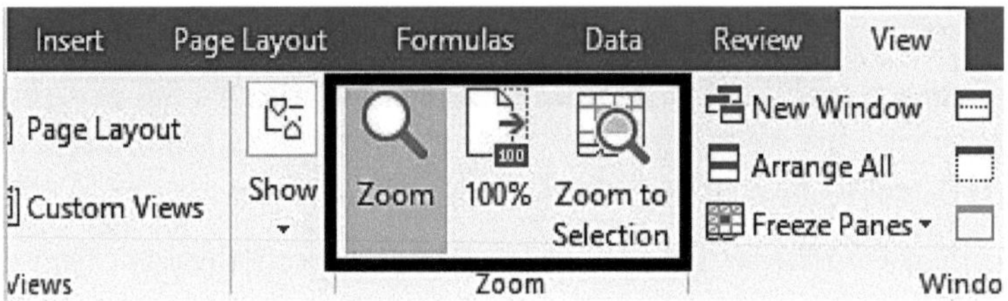

Viewing a worksheet in multiple windows

You need to have more than one window opened. So follow the steps below to do so;

- Click **View** tab.
- From the options there, select **New Window**.
- Click on the **View tab** again and select Arrange All. This will open up a box for you to select how you want to arrange the windows. Once you have selected it, click OK.

The worksheet will now appear on both windows.

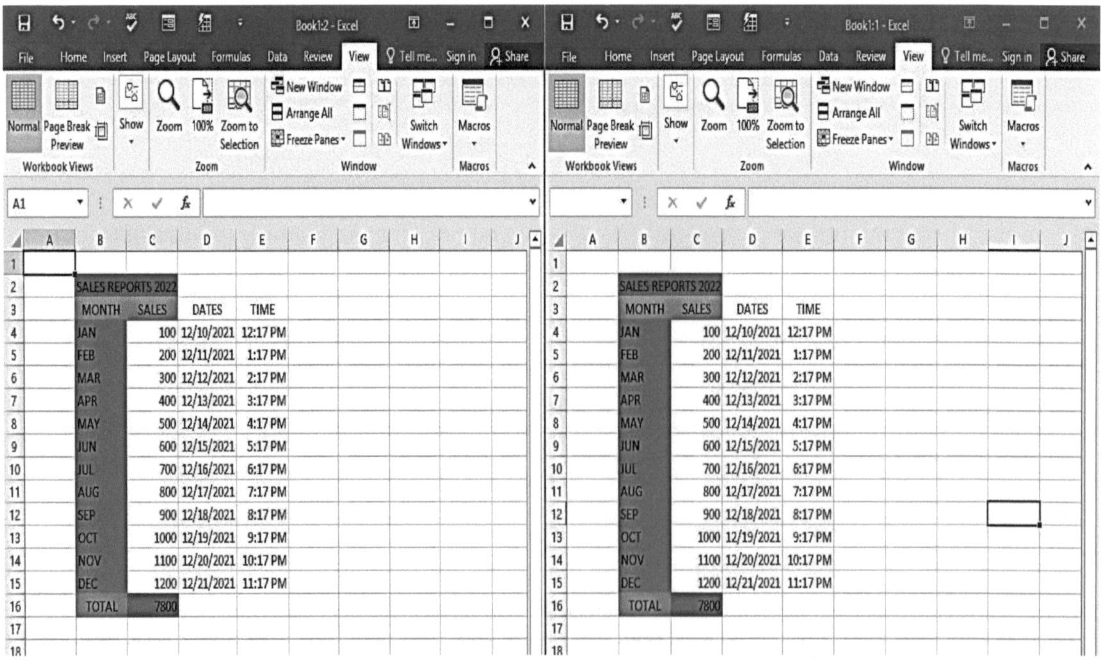

Comparing sheets side by side

To do this, click on the **View tab**, you will see the **View Side by Side option**, click on it.

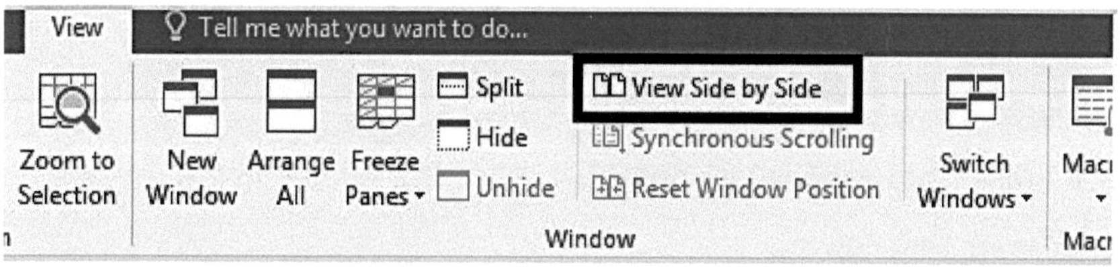

Splitting the worksheet window into panes

To do this, you will make use of the Split Button on the View tab. So first of all, choose the row or column that you will like to insert the split pane. Then click on View on the ribbon, then click Split.

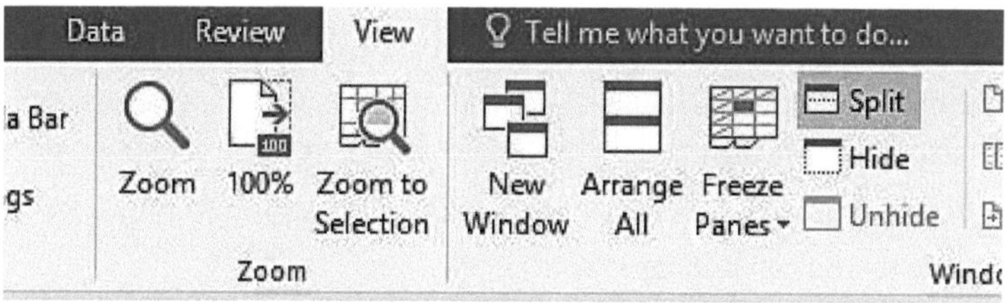

Keeping the titles in view by freezing panes

Freezing panes in Microsoft Excel implies that some rows and/or columns will be constantly visible at the top of a worksheet while scrolling. Below are the steps to do so.

Freeze Top Row

- Select **Freeze Panes** on the View tab.
- Select **Freeze Top Row** from the menu.

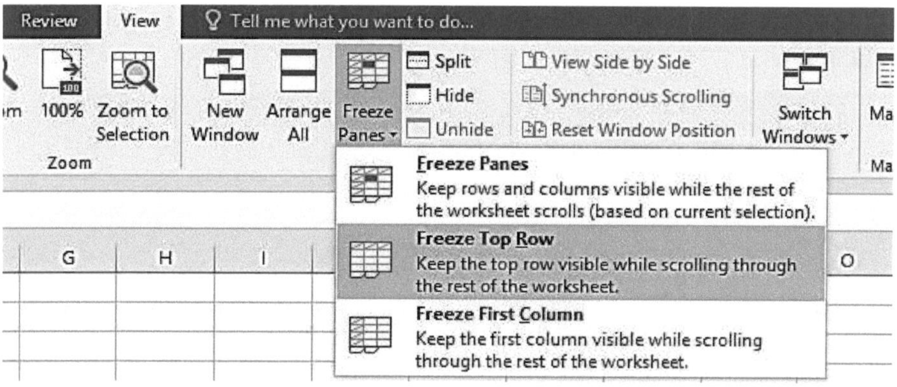

You will see a dark grey horizontal line on the top row.

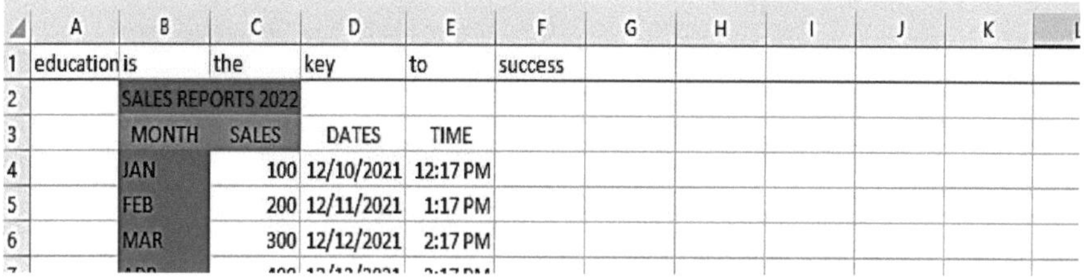

Unfreeze the Panes

- Select Freeze Panes from the Window group on the View tab.
- Select **Unfreeze Panes** from the menu.

Freeze First Column

- Select **Freeze Panes** on the View tab.
- Select **Freeze First Column**
- You will see a dark grey vertical line in the first column.

You can do this to the rows and cells as well using the same steps.

Monitoring cells with a watch Window

In Excel 2013, Microsoft added the Watch Window. It's a tool that displays the value of a cell and its characteristics in another window, including the workbook and worksheet titles, cell or range names, cell address, and formulae. This window may be placed anywhere you like and even docked like a toolbar.

To use the Watch Window tool simply click on Formulas on the ribbon, then click **Watch Window**. This will bring up a box where you can add and remove cells.

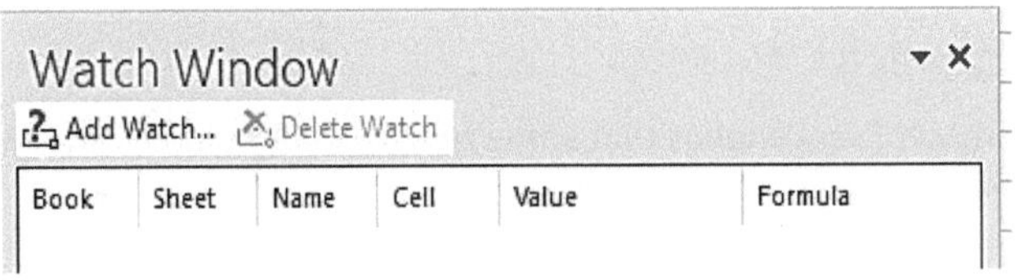

WORKING WITH ROWS AND COLUMNS

Inserting rows

Click on the heading of the row where you want to insert the new row. Then, on the home tab in the ribbon, click Insert.

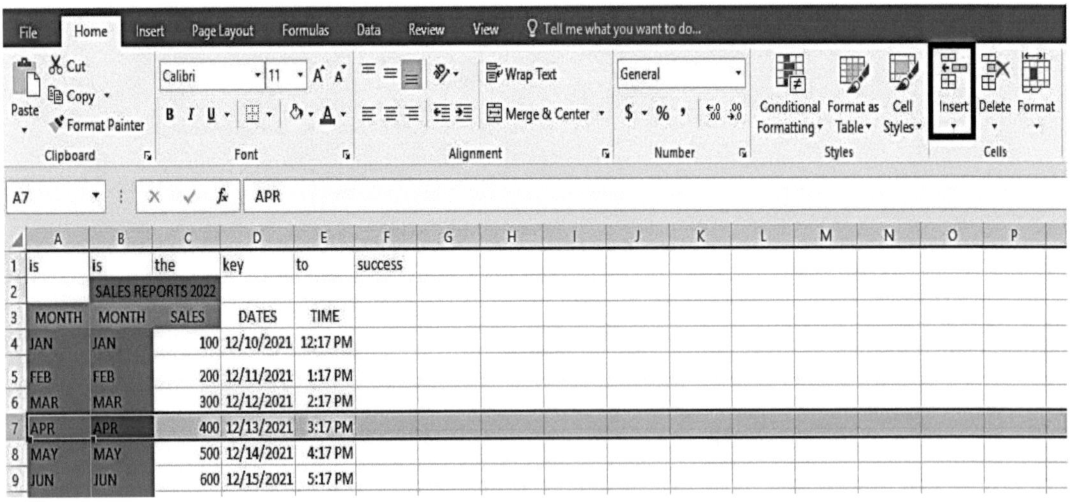

The newly inserted row will appear on top of the row you selected.

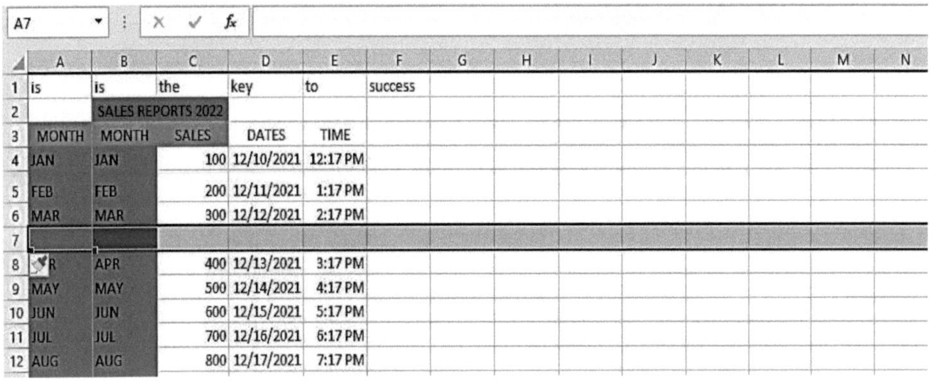

Inserting Column

Click on the heading of the column where you want to insert the new column. Then, on the home tab in the ribbon, click Insert.

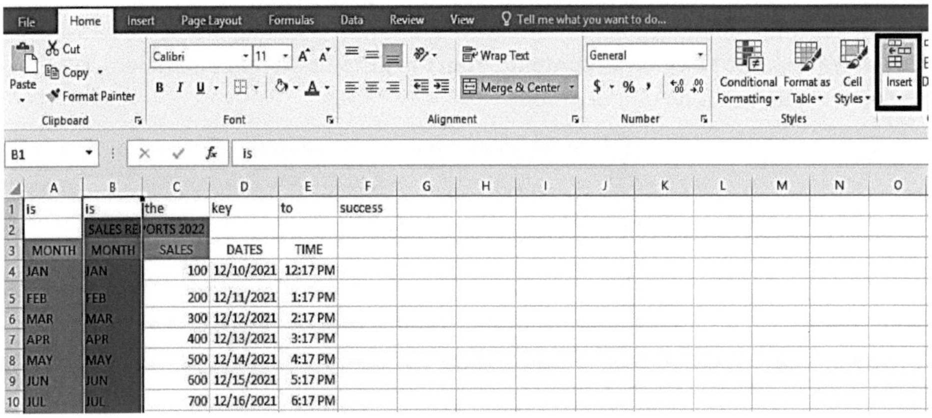

The newly inserted column will display at the left of the column you selected

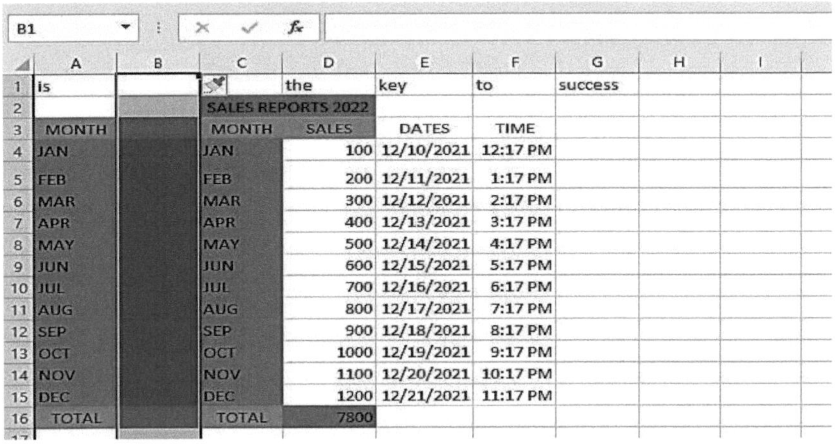

Note: A paintbrush icon appears next to newly added rows, columns, or cells when you insert new rows, columns, or cells. This button enables you to customize the formatting of these cells in Excel. Excel formats added rows by default to match the formatting of the cells in the row above. Hover your cursor over the icon, then click the drop-down arrow to see more choices.

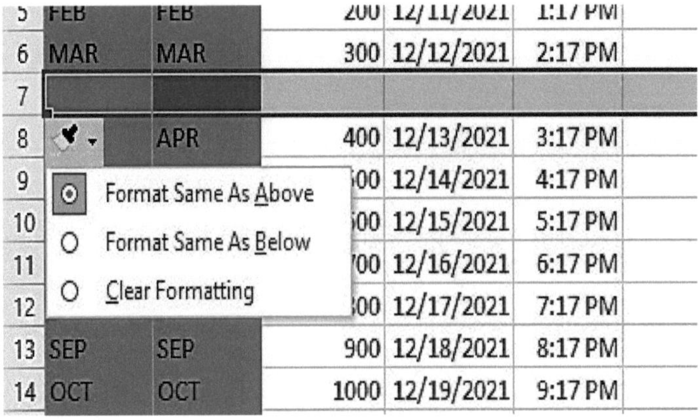

Deleting rows and columns

To delete a row, click on the row you want to delete, right-click and select **Delete**. Also, to delete a column, click on the column you want to delete, right-click and select **Delete**. Note that it is very important to know the difference between removing a row or column and merely erasing its contents. Right-click a heading, then pick **Clear Contents** from the drop-down menu to delete the content from a row or column without causing others to shift.

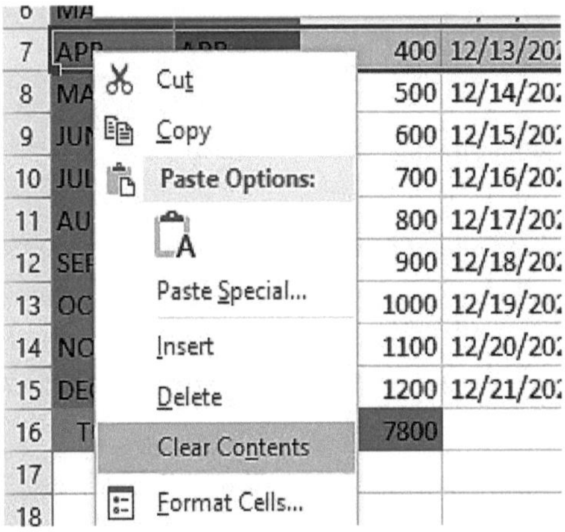

Changing column widths

To do this, simply place your mouse cursor over the column line in the column heading. When you do this, the mouse cursor will change to a double arrow.

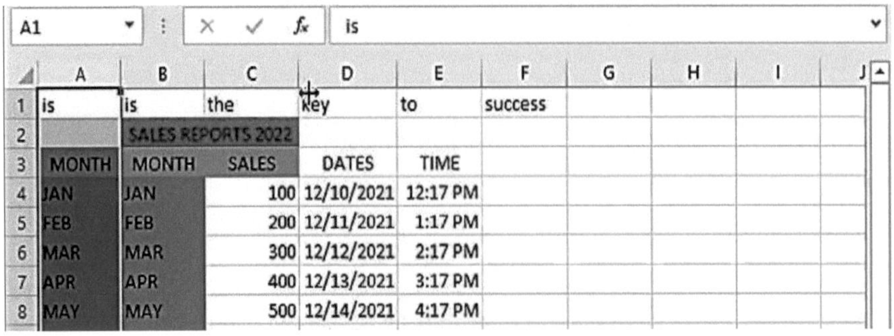

Then, click and drag the mouse left or right depending on how you want it to be. Then, release the mouse.

Changing row heights

Place your mouse cursor over the row line. The mouse cursor will turn to a double arrow.

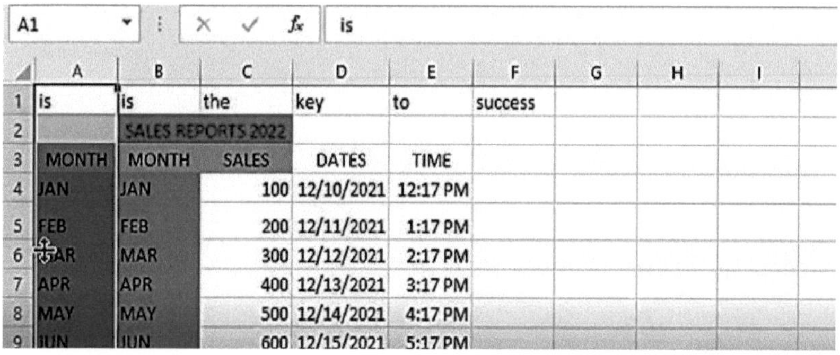

Then, click and drag the mouse up or down. Let go of the mouse.

Hiding rows and columns

Simply choose the column(s) or row(s). Then, right-click on it and select **Hide**.

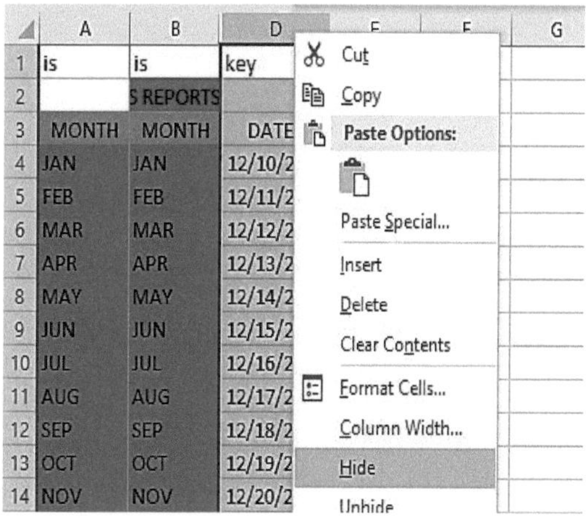

This action will hide the column. You will see a green line on the worksheet after you have hidden the row/column. The green line shows where the hidden column or row is.

To unhide the column or row, move your cursor in-between the two columns where the hidden column was i.e. on the green line, then right-click and click **Unhide**.

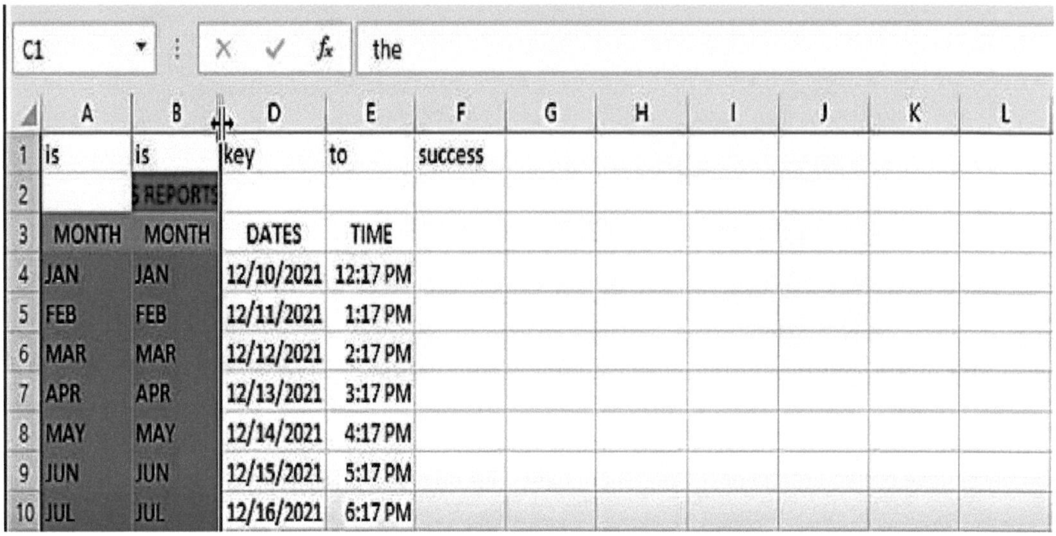

CHAPTER FOUR

WORKING WITH EXCEL RANGES AND TABLES

UNDERSTANDING CELLS AND RANGES

A cell range is a collection of cells that you choose to employ in functions and operations. A cell range is a set of chosen cells. This range is normally symmetrical (square), but it may also be made up of individual cells. A cell range may also be referenced in a formula.

A cell range is specified in a spreadsheet by the reference of the upper-left cell (minimum value) and the reference of the lower right cell (highest value). When different cells are eventually added to this selection, the range is referred to as an irregular cell range. The minimum and maximum values are supplied in Excel. A mathematical range, on the other hand, is a collection of values between a maximum and lowest value.

Select Single Cell Range

The intersection of the row and column in Excel is referred to as a single cell. When you click on any cell on the Excel sheet, you'll see a column name and a row name.

Let's look at the picture below as an example. The intersection of column B and row 3 is the cell chosen below. The cell may be interpreted as B3, which is a mix of row and column names.

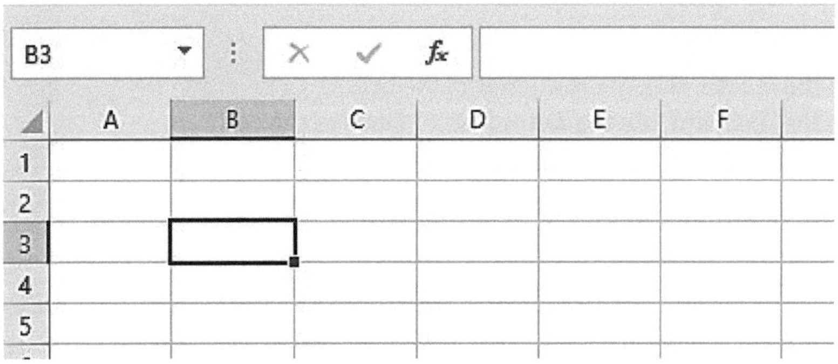

Selecting complete columns

When you select all column cells in Excel, the range is also described. You may use either your mouse or your keyboard to pick all the cells in a column, depending on your preference. The techniques for selecting all column cells are shown below.

Mouse

- Click the column cell name with the mouse.

Keyboard

- Pick the column for the cell.
- Press and hold the 'ctrl' key while pressing the **'space'** key.

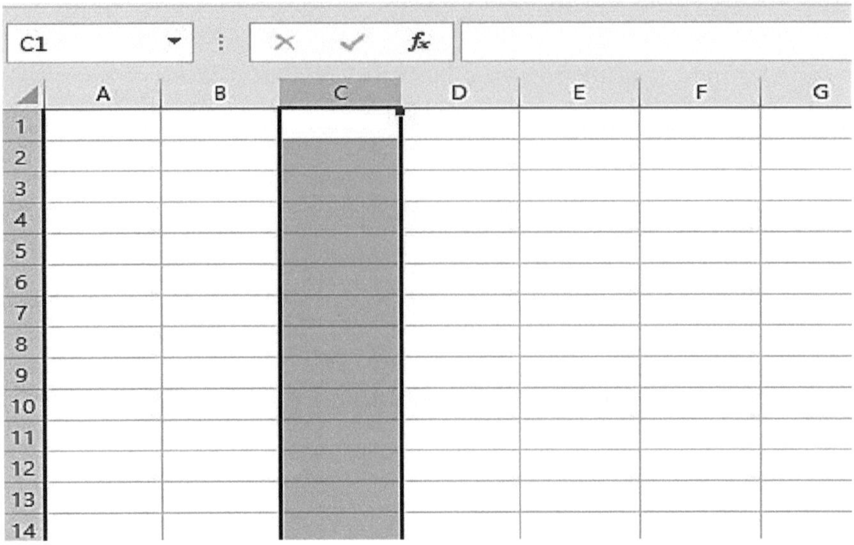

Assume you wish to pick the cells in column C in the picture above. Simply click on the C column name.

With the keyboard, go to column C. Then, while holding the '**control key'** hit the '**space' key** on your keyboard.

Selecting complete rows

- With the mouse, click the row letter of the cell.
- With the keyboard, use the arrow keys to move to the cell.
- Press the '**shift'** key and hit the '**space'** key to select a row.

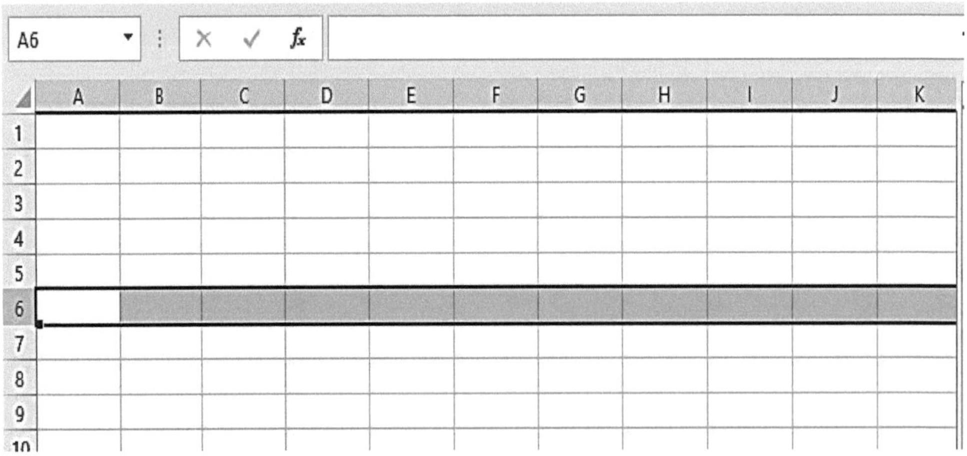

To choose a row, as shown above, do so. You must click the row's name. You may also use your keyboard shortcut to choose a row. Using the arrow keys on your computer, go to any cell in row 6. Now, while holding down the '**shift'** key, press the '**space'** bar on your keyboard.'

Selecting noncontiguous ranges

To choose noncontiguous ranges, follow the procedures outlined below:

- Using the mouse, click on the start cell.

- Press and hold the Control key on your keyboard.
- Select the cells you wish to select by clicking on them.

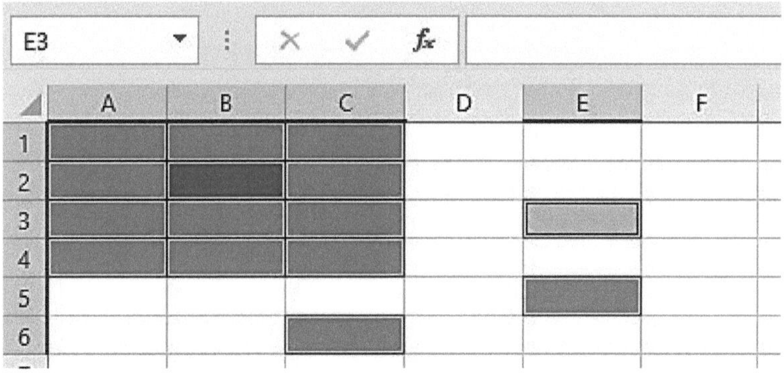

Assume you wish to choose the range (A1:C4, C6, E3, E5) shown in the figure above. Using the arrow keys on the keyboard, click or visit the initial cell. Now press C4 while holding down the 'ctrl' key. With the Control key held down, click C6, E3, and E5 cells again to select them.

Selecting multi-sheet ranges

It's a technique for picking many cells in a certain pattern, such as squares or rectangles. This approach may be used to pick both tiny and big areas of cells. Follow the procedures shown below to choose your pattern.

- Use the mouse or keyboard to go to the first cell.
- Press and hold the '**Shift'** key on your keyboard.
- Select the last cell.

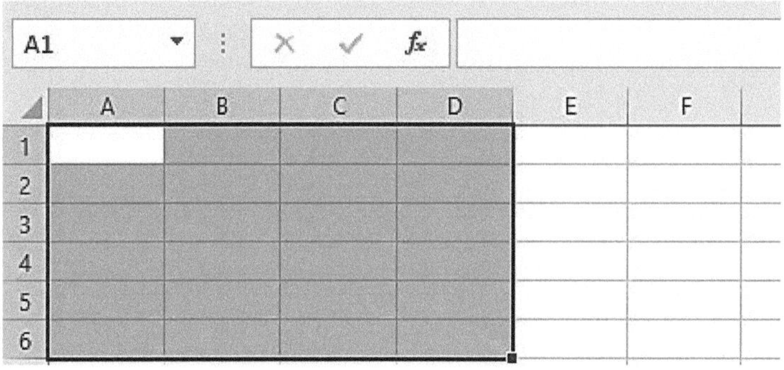

Assume you wish to choose the range (A1: D6), as shown in the figure above. To begin, use your mouse or keyboard to go to cell A1. Now press the '**shift'** key on your keyboard and click cell D6. As seen in the graphic above, this will choose the needed cell range.

Selecting special types of cells

You may copy, move, remove, color, fill, and protect specific cells on a sheet, such as cells holding Constants, Formulas, blank cells, and more. To select particular cells, follow these steps:

- Press Control key + G on your keyboard.
- This opens up the Go to the dialog box. Click on Special.

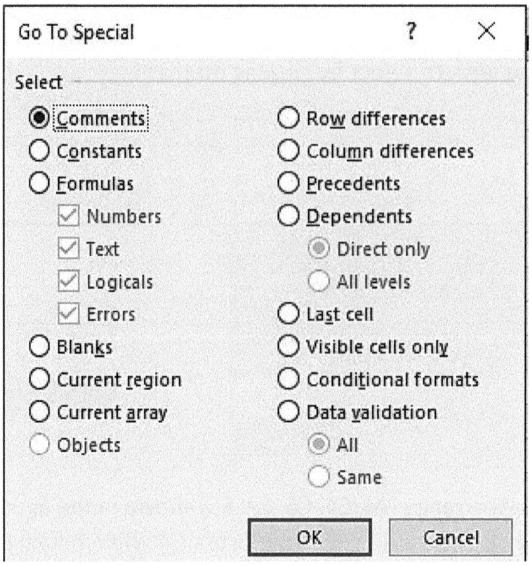

- Click OK after selecting one of the available buttons.

Selecting cells by searching

- Press Control key + G on your keyboard.
- Type in the cell and click Ok.

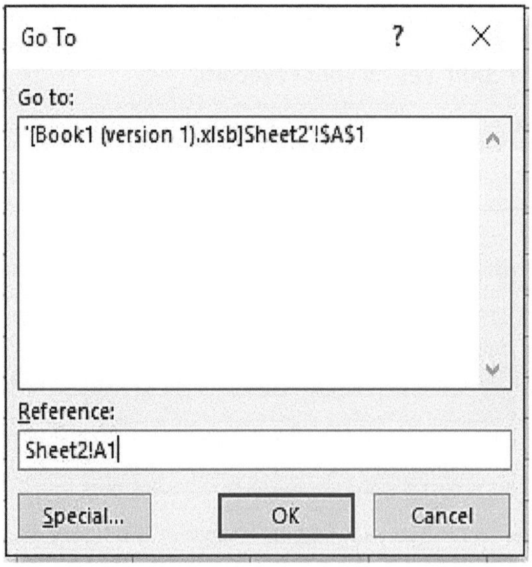

COPYING OR MOVING RANGES

There are different ways to copy and move ranges in Excel. In Excel, copying and pasting a cell is simple. When it comes to copying and pasting a collection of cells, columns, or rows, there are ways for it.

Copying by using Ribbon commands

Click on the cell range heading (row or column). Select Copy from the home tab. To copy as a picture is possible also.

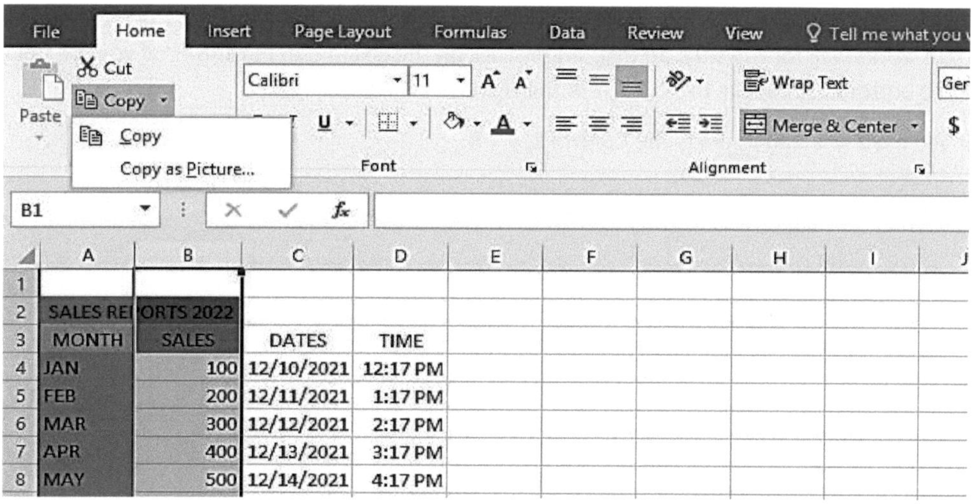

Copying by using shortcut keys

To copy a range of cells, use Control key + C simultaneously after choosing the range of cells. Choose a range of cells in which you wish to paste it again, and then click Control key + V to paste it. This is the most convenient method for copying and pasting several cells at once.

Other shortcut keys are Control key + D and Control key + R. Control key + D is used for copying and pasting data downwards. It can also copy the values of a full column. To do this, choose any range of cells that are in the same column as the cell you want to replicate.

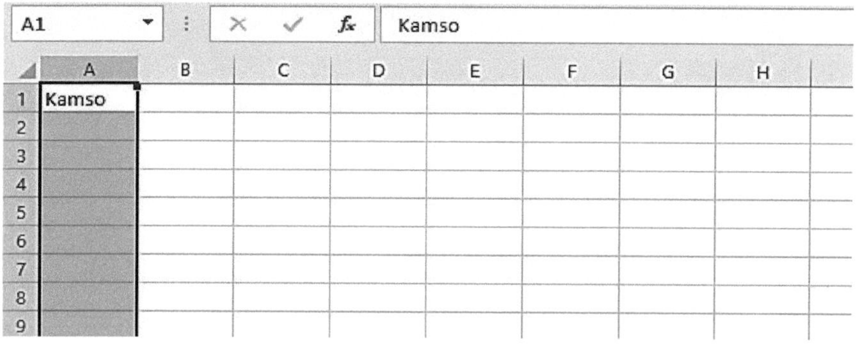

Now, press Control key + D. this action will copy the values downwards.

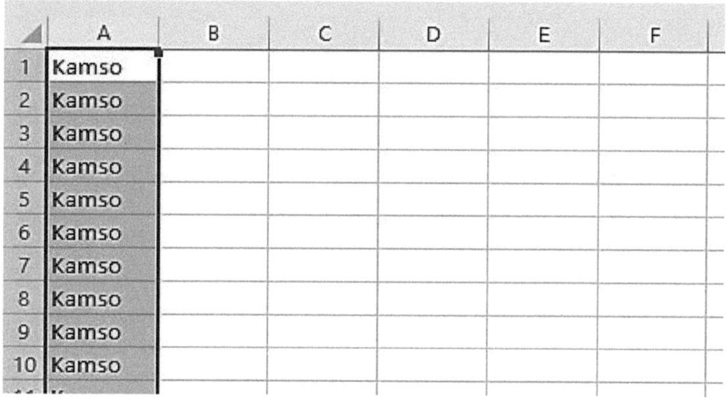

Control key + R also does the same action, but this time, it copies the values to the right-hand side. This shortcut works well for rows. To do this, simply click on the cell range, but this time you are not choosing from the bottom up, choose the cells from the right.

After you have selected it, press the Control key + R.

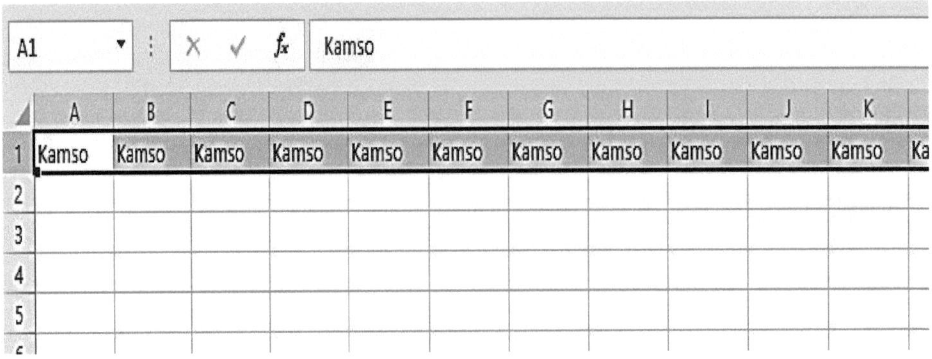

You may also duplicate a single value to several cells using another shortcut. Control key + Enter is the shortcut. To do so, pick a few cells on your worksheet and type any term or value into any of them. After selecting the cells, press Enter to put in a value in any of the cells.

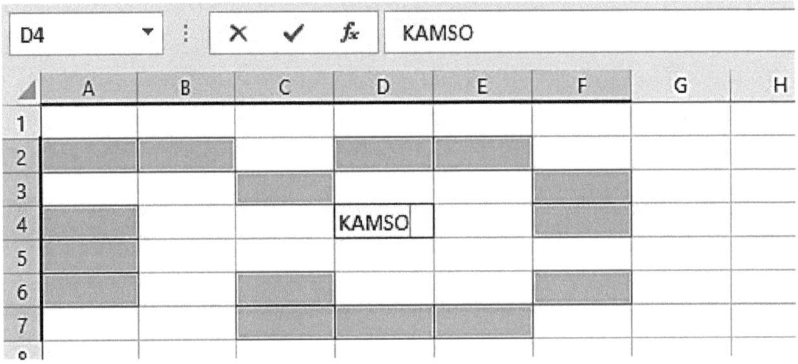

Now, press the **Control key + Enter**.

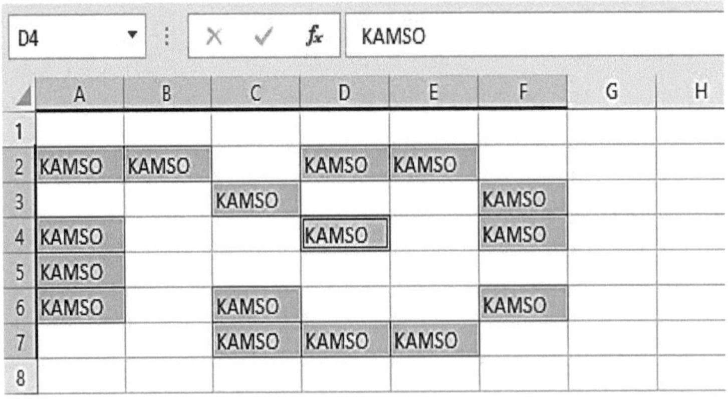

Copying or moving by using drag-and-drop

For copying a set of values for columns, rows, or a range of cells, we will make use of the **Fill Handle** option. To do so, enter some values in the appropriate cells. Select the cells after inputting the data. The value is surrounded by a green box.

In Excel, this green area is called the Fill Handle. Drag the fill handle down to replicate these values for a range of cells by hitting the "+" symbol that appears in the bottom right corner of the fill handle.

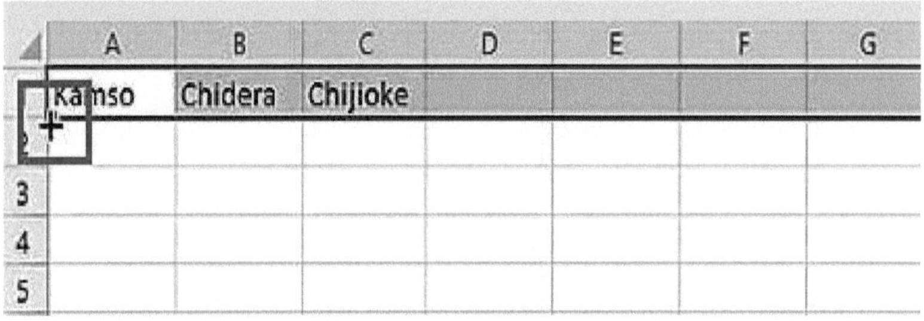

Once you stop dragging the handle, you will get the result, as you can see below.

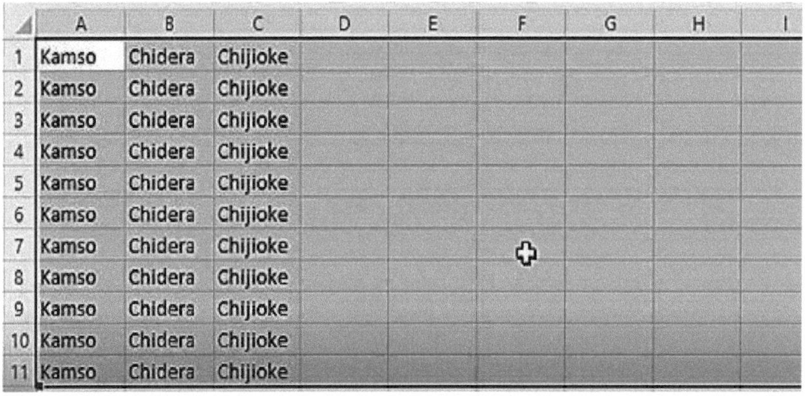

	A	B	C	D	E	F	G	H	I
1	Kamso	Chidera	Chijioke						
2	Kamso	Chidera	Chijioke						
3	Kamso	Chidera	Chijioke						
4	Kamso	Chidera	Chijioke						
5	Kamso	Chidera	Chijioke						
6	Kamso	Chidera	Chijioke						
7	Kamso	Chidera	Chijioke						
8	Kamso	Chidera	Chijioke						
9	Kamso	Chidera	Chijioke						
10	Kamso	Chidera	Chijioke						
11	Kamso	Chidera	Chijioke						

Copying to adjacent cells

So in our example below, there are data in range A1:C7.

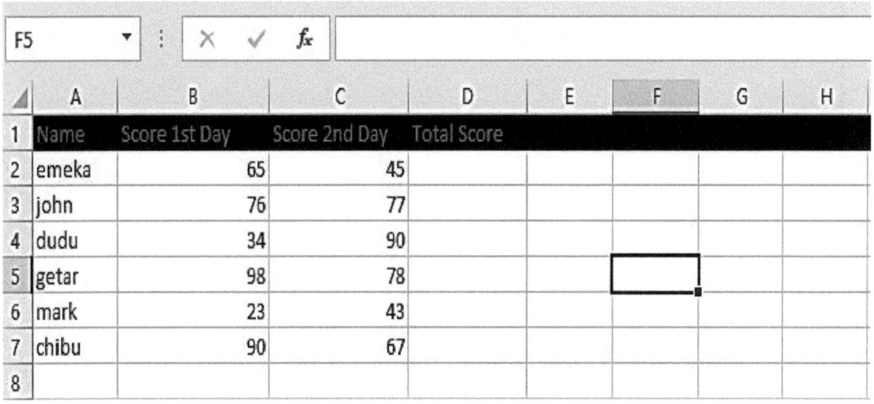

	A	B	C	D	E	F	G	H
1	Name	Score 1st Day	Score 2nd Day	Total Score				
2	emeka	65	45					
3	john	76	77					
4	dudu	34	90					
5	getar	98	78					
6	mark	23	43					
7	chibu	90	67					
8								

Column D is empty and it will consist of the total score of the table. To do this, we are to copy and paste the formula for that. Now, in Cell D2, enter in this formula **"=B2+C2",** then press **Enter**. As you can see in the image below;

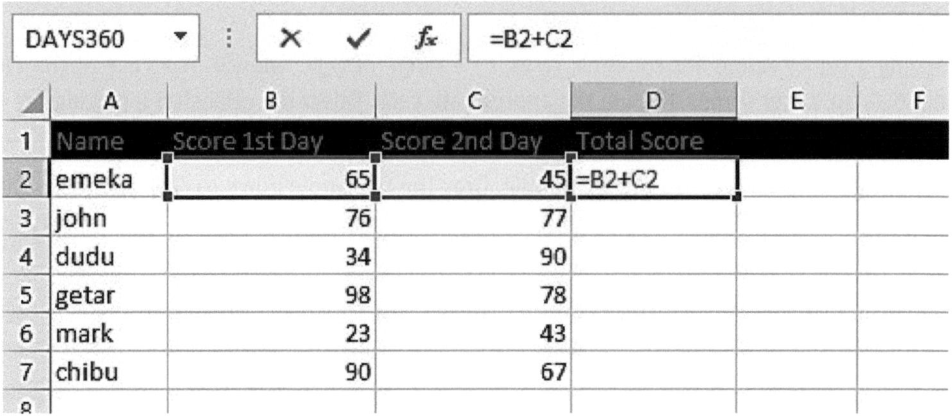

Then, using the fill handle, drag down and release to fill in the empty cells.

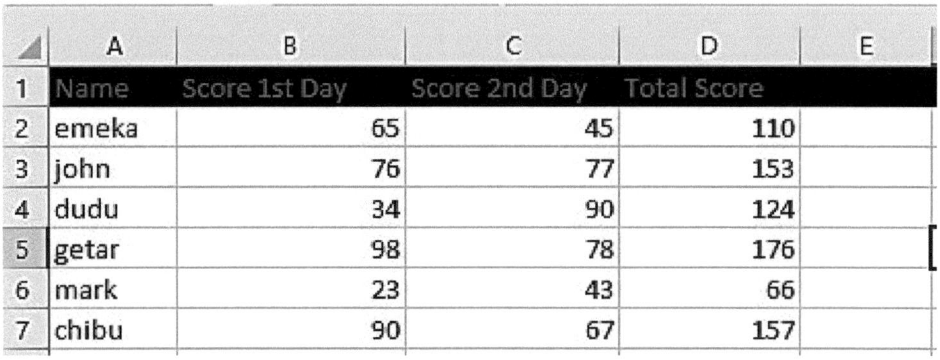

Now, we want to copy a range of cells to the adjacent side. So, select the range of cells. In this example, we want to copy the Total Score column. To do this, we will select the cell on the right-hand side which is Cell E. when you select it, press **Control key + R**.

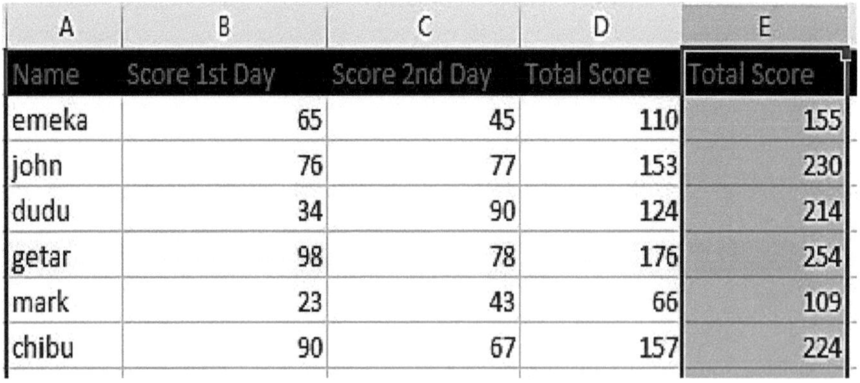

Using the Office Clipboard to paste

You may copy up to 24 things from Office documents or other applications and paste them into another Office document using the Office Clipboard. You can paste text from an email message, data from a workbook or datasheet, and a graphic from a presentation into a document, for example. You may organize the copied objects in the document as you wish using the Office Clipboard.

When you utilize the Clipboard task window, you're not restricted to copying the last thing you copied or cut. Many of the last pictures and text you copied or cut are stored in the Clipboard task window. First, Open the Office Clipboard. On the Home tab, click on the Clipboard dialog box launcher.

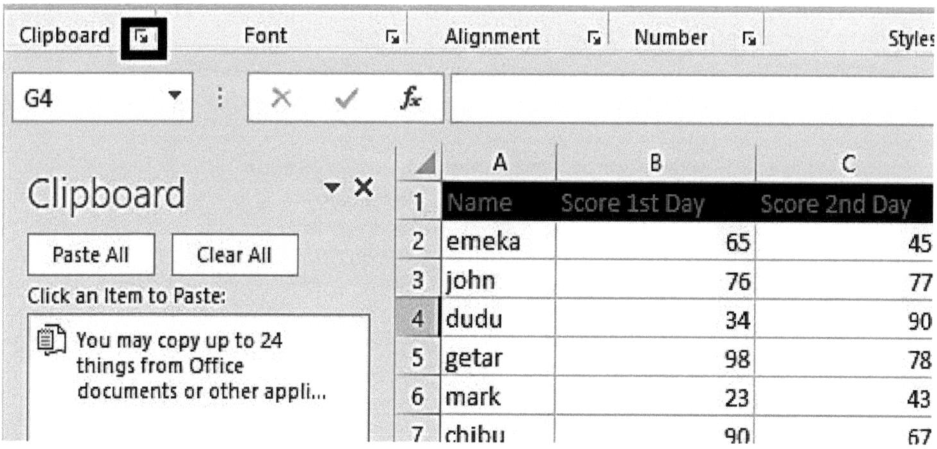

Now to copy and paste, follow the steps below;

- Open the file from which you'd want to copy things.
- Select the first thing you wish to copy and click CTRL+C on your keyboard.
- Continue copying stuff from the same or other folders until you've gathered all you desire. Up to 24 things may be stored on the Office Clipboard. The initial item on the Office Clipboard gets destroyed after you copy the twenty-fifth item.
- Now, open the clipboard pane. On it, you will see all the things you've copied. You can select each item to paste or you can paste all of them. So, to paste one item, click on the item. to paste all items, click on the **Paste All** box on the clipboard pane.

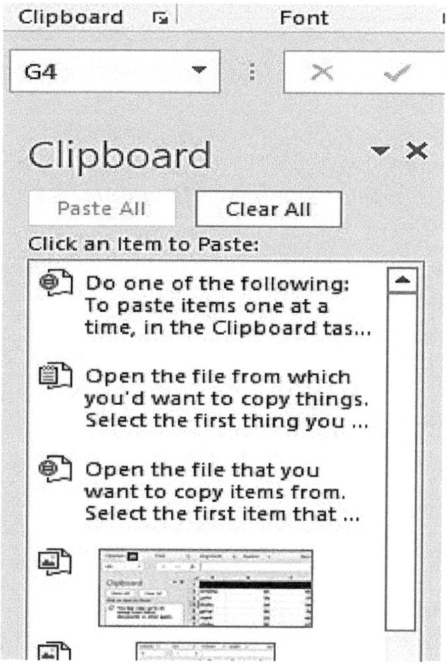

Pasting in special ways

When a conventional copy/paste isn't suitable, Excel's Paste Special provides a variety of choices, such as pasting just certain portions of copied cells or performing a mathematical function on the copied data.

All of the Paste Special options are shown in the image below:

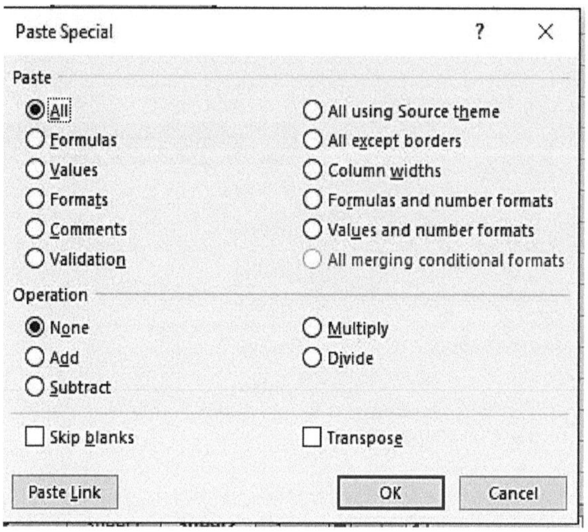

Using the Paste Special Dialog box

This is an easy process. Select the cell or range of cells and right-click on the area you wish to paste the item on the workbook and select **Paste Special**. This will open up the Paste Special Dialog box where you have a list of options on how you can paste the copied item.

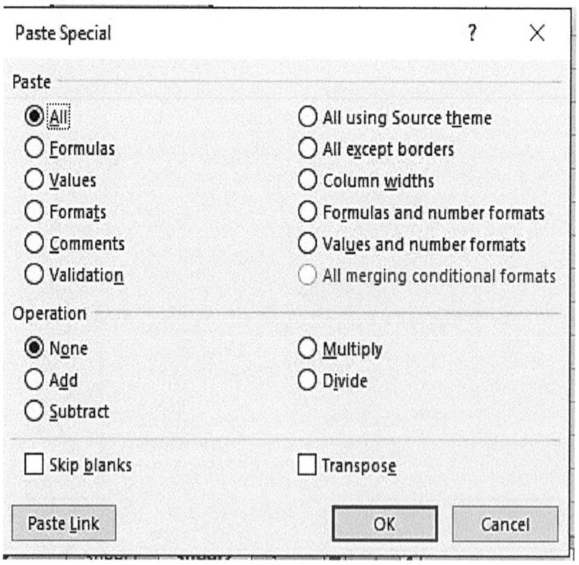

Skipping blanks when pasting

So, we will make use of the Paste Special Skip Blanks Option to do this. As you on the image below, range B1:B10 has some blank spaces in its cells. So, we are going to copy the range and skip those blanks. So, copy Cell B.

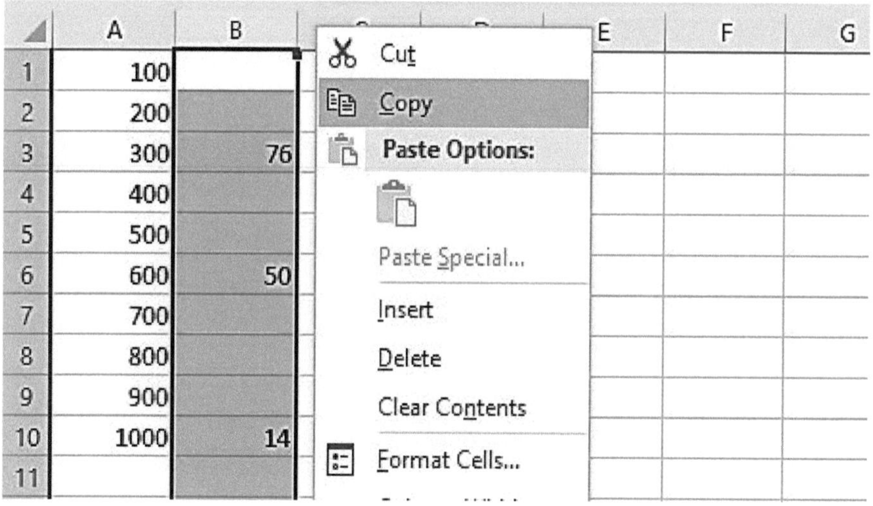

Now, select cell A1. Right-click on it and select Paste Special.

Check the box on Skip Blanks and click **Ok**.

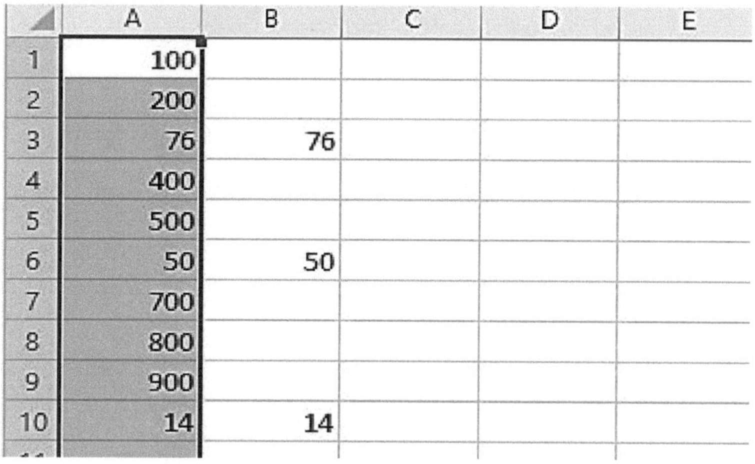

USING NAMES TO WORK WITH RANGES

However, there are certain key requirements for naming named ranges in Excel that you should be aware of before you begin:

- There are no spaces allowed in names. Instead of using a space, you might use an underscore character (such as Annual Total).
- You may name the range with any combination of alphabets and digits, but it must begin using a letter symbol. A number cannot begin a name.
- Except for underscores and periods, no other symbols are permitted.
- Although names are restricted to 255 characters, it is best to make them as brief as possible while still being relevant and clear.

Excel also keeps a few names in reserve for internal use. Although it is possible to construct names that override Excel's internal names, this is something you should avoid.

Creating range names in your workbooks

Choose the cell or range. Click the Formula tab on the ribbon. Navigate to the Define Names section. Click the down arrow and select **Name manager**.

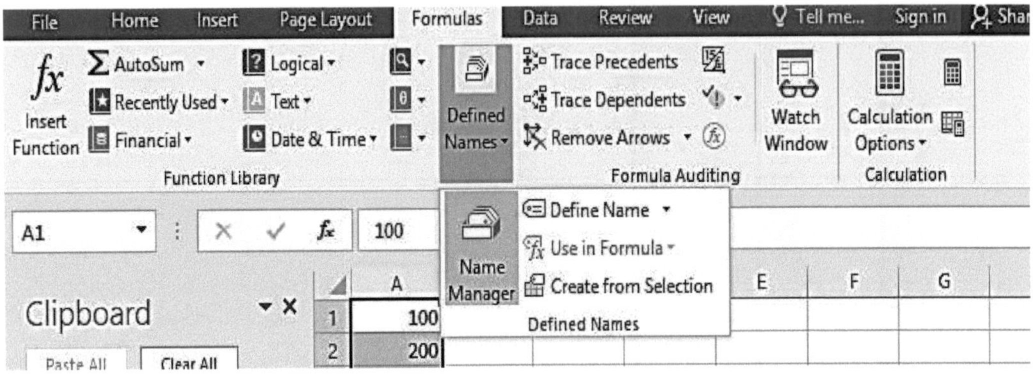

Click on New. This opens up the New Name dialog box.

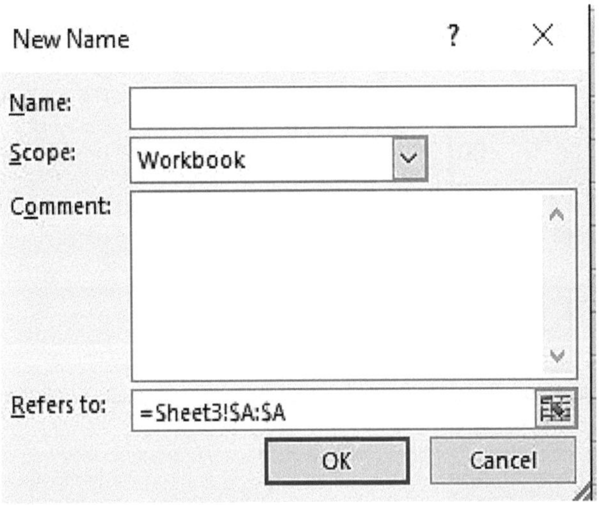

In the Name box, type a name (Excel can display the name if you selected a data range with a heading line). In the area labeled Refers to, the active or chosen cell or range address shows. To add the name to your spreadsheet and exit the dialog box, double-check that the address provided is accurate.

Using the Name box to construct a name is a quicker option:

Choose the cell or range, then input the name. Press Enter. (If you input a name and then click on the worksheet, Excel will not assign the name to the chosen range.) You can't modify the range that a name refers to if it already exists. Instead of selecting the range, attempting to do so choose it.

For instance:

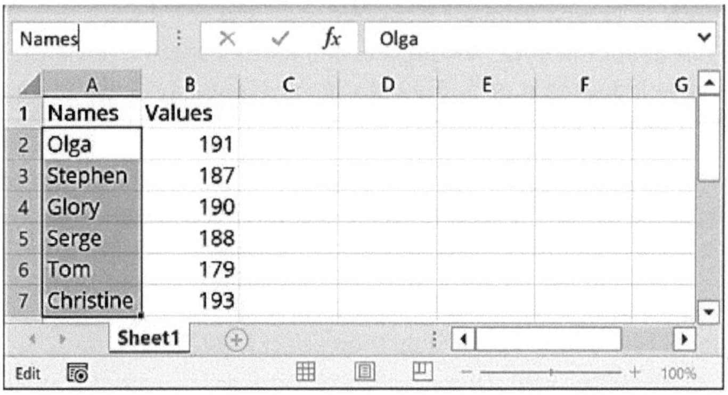

If your formula contains named cells or ranges, you can either input the name instead of the address or select a name from a table, then Excel will automatically insert it. Names and Values are two specified names on the worksheet.

Using the Create Names from Selection dialog box

You may rapidly generate names for numerous cells by using existing row or column labels. Choose the cell range and labels. In the Defined Names group of the Formulas tab, pick **Create from Selection**.

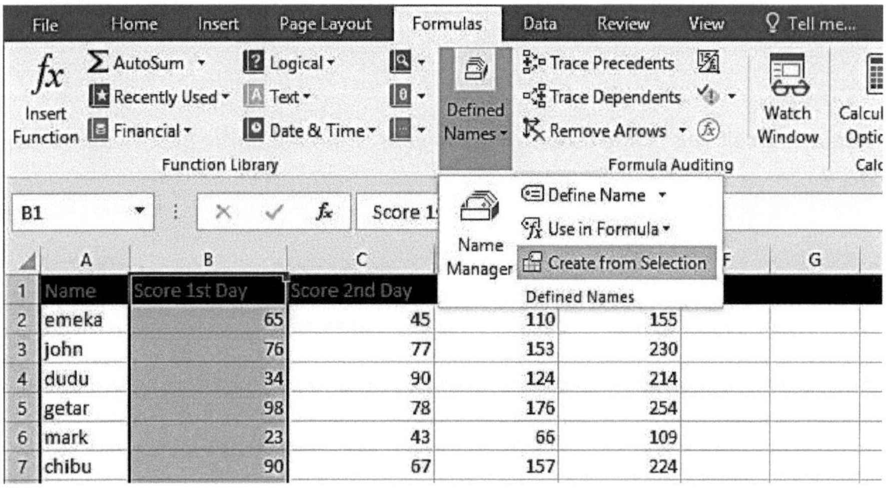

Select the place of the text and click **OK**.

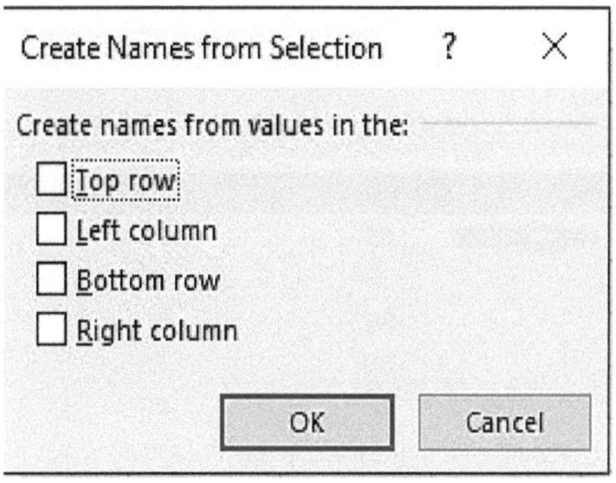

ADDING COMMENTS TO CELLS

Let's say you've gotten an Excel document from someone and you'd want to provide comments, make edits, or ask questions about the data. Writing a comment to a specific cell in the worksheet is a simple way to do this. Because it doesn't modify the data, a comment is typically the easiest approach to add extra information to a cell.

This tool is also useful for explaining formulae to other users or describing a specific value. You may use an image instead of words to describe something in a comment. Follow the steps below to add comments to cells.

- First, choose the cell you would like to add a comment on.
- On the ribbon, click on the Review tab, then select New Comment.

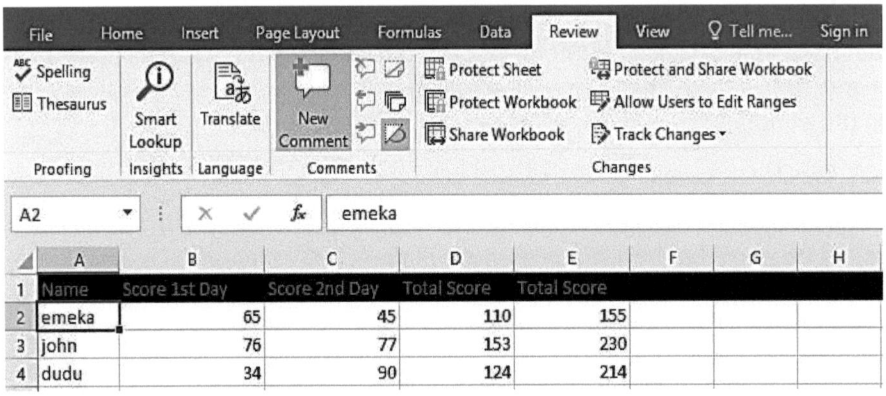

This will display a small box on the screen. Type in your comment in the box. When you are done, click on any cell.

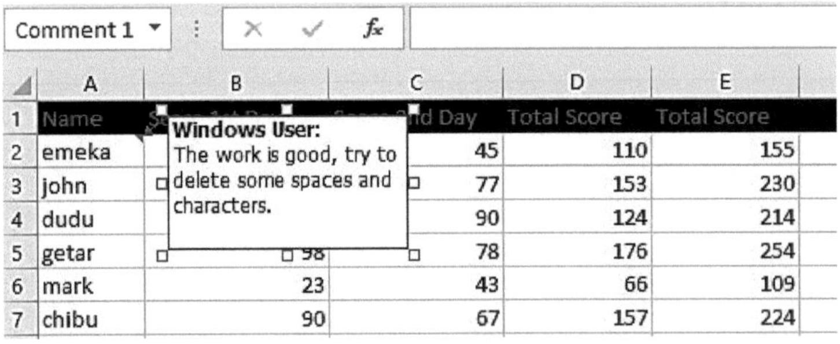

The comment box will disappear. But you will notice a red mark on the column you have just added a comment. That red mark indicated that the cell contains a comment. When you hover your mouse cursor on the cell, it will display the comment.

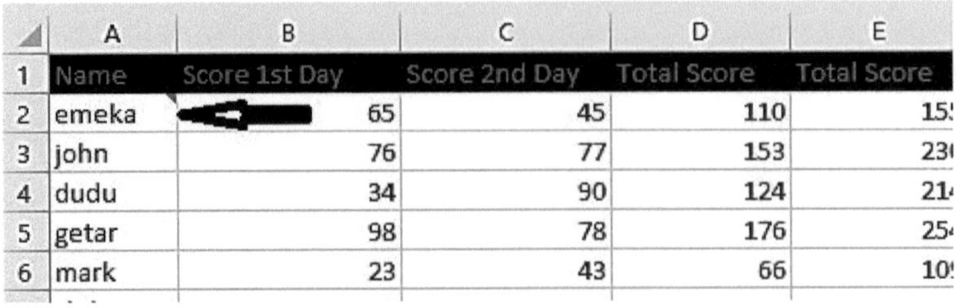

Shift key + F2 will perform the same action here.

Formatting comments

To format the comments, hover your mouse cursor to the cell to display the comment box. Once it has been displayed, right-click on it and select Format Comments. This will bring up the Format Comment dialog box. After formatting, click **Ok**.

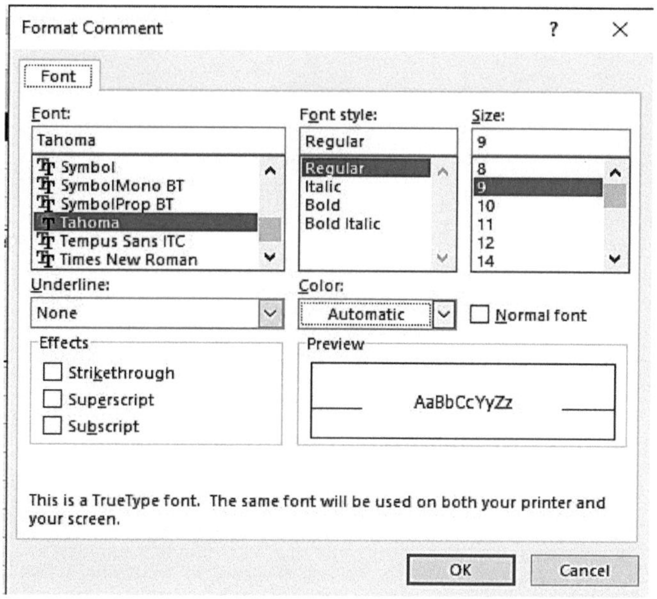

Changing a comment's shape

By default, the shape of the comment box is a rectangle. You can decide to change this shape. You will add a tool to the Quick Access Toolbar. So, click the down arrow on the Quick Access Toolbar and select More Commands.

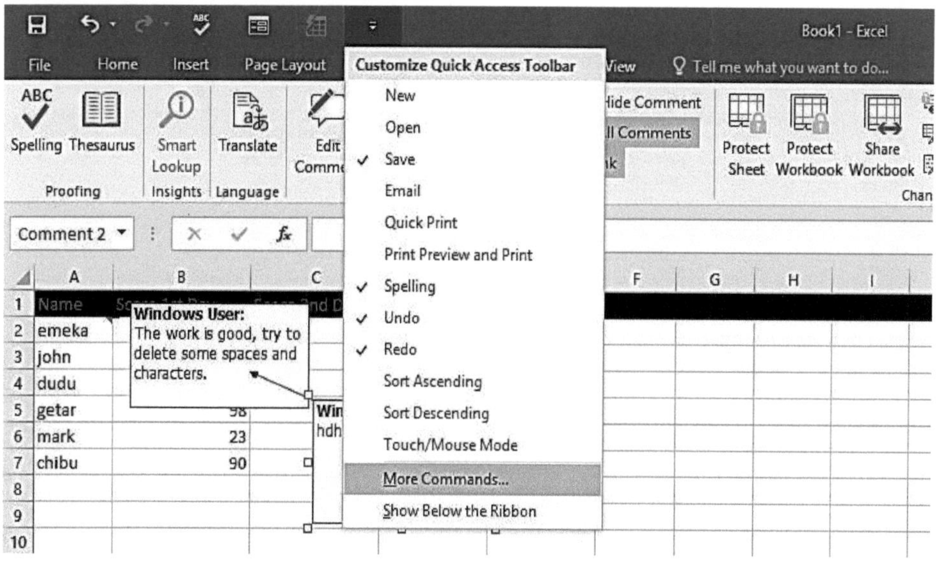

On the **Choose commands from** option, hit the down arrow and select **Drawing Tools | Format Tab.** From the list of commands below the options, select Change Shape. Then, click Add and click Ok.

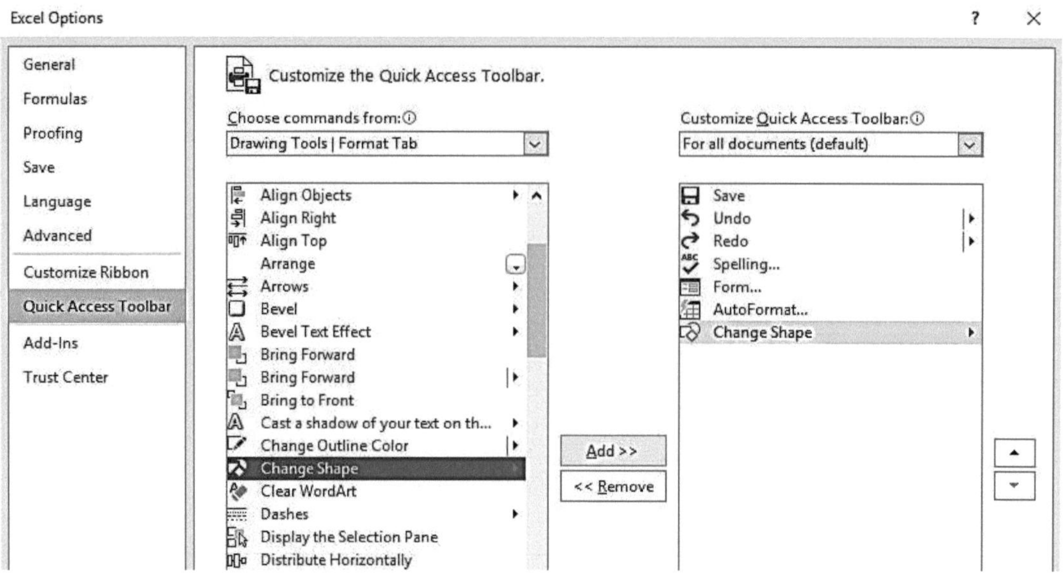

The shape icon will be added to the commands in the Quick Access Toolbar.

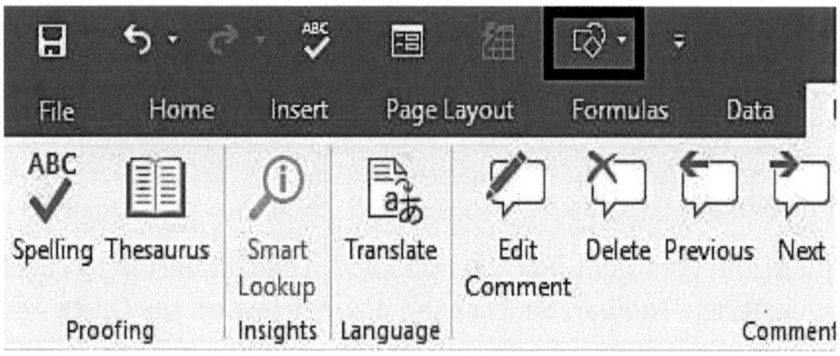

At first, the shape icon won't be accessible. So, you have to first work on the comment box. Click on the comment box and move your cursor around it until it switches to the four-headed arrow. Then, click.

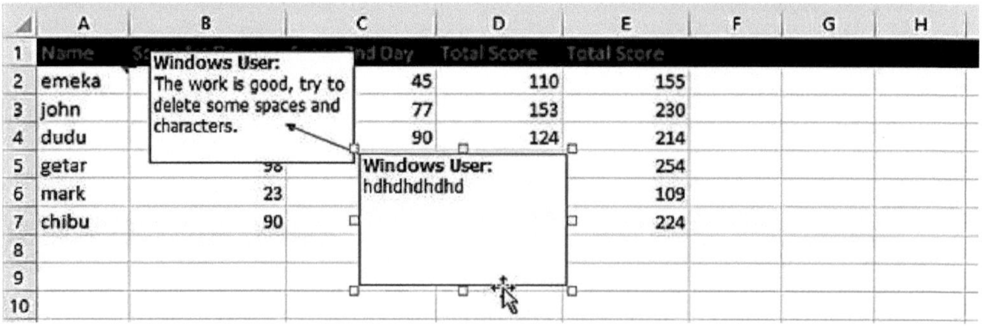

The shape icon will be accessible now. So, click on it. This will display a box that consists of different shapes. Select any shape.

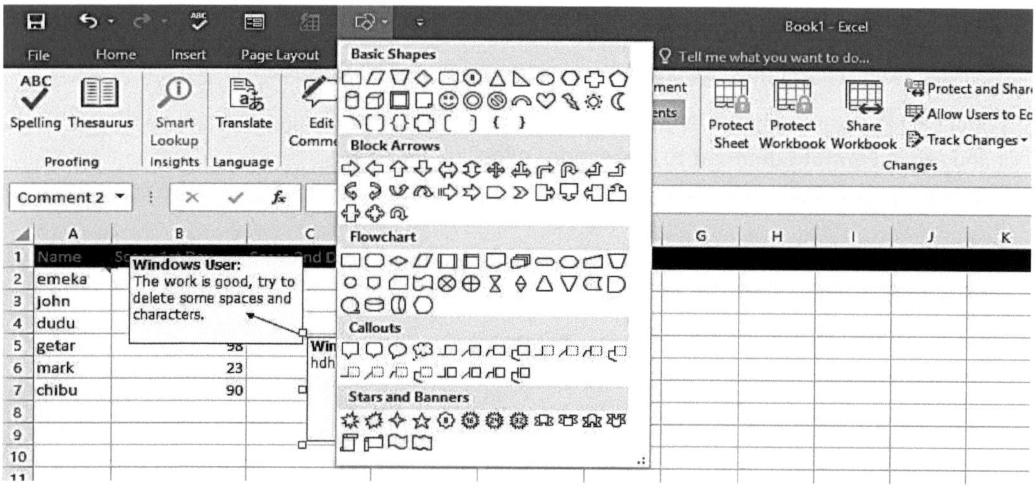

Resizing comments

Most times, when you change the shape of the comment box, the text inside will no longer fit into the comment box. You can solve this issue by resizing the comment box. To do this, click on the comment box, then click on the sizing angles around it (the small transparent boxes) and then drag it to the size you want.

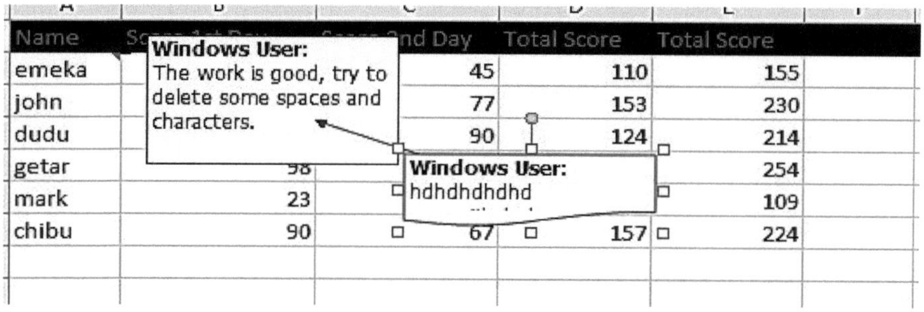

Hiding and showing comments

Simply click on the **Review tab** and select **Show All Comments**.

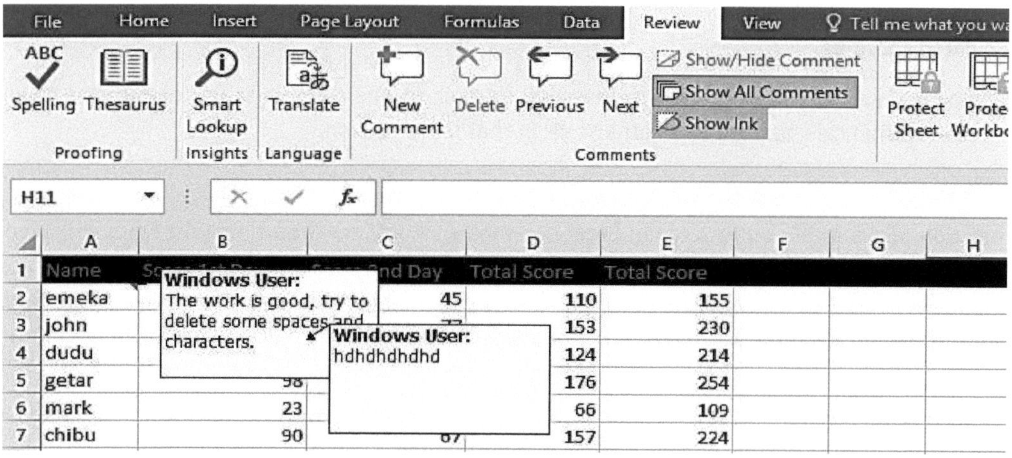

To hide the comments, click on **Show All Comments** the second time.

Editing comments

Right-click on the cell that has the comment on it and selects **Edit Comments**.

This opens up the comment box. Now, you can click on it and start making changes to it or you can right-click and select **Format Comment** to make more edits.

Deleting comments

You may not want a comment to be on a cell again and you want to delete it. Right-click on the cell, then select Delete Comments from the options.

Also, do this with the Review Tab. Select the **Delete icon**.

WORKING WITH TABLES

Microsoft Excel makes it simple to create tables and do computations. Its working area consists of a series of cells that must be filled with information. As a result, the information may be structured and utilized to create graphs, charts, and summary reports.

Working with tables in Excel may seem hard at first look to a novice. It varies significantly from Word's table creation concepts.

Understanding a table's structure

A table in excel consists of many elements which are explained below;

The header row

Every table in Excel consists of a header row. The columns in a table have filtering enabled in the header row. This is to enable you to filter your table as well as sort it easily.

Calculated columns

You may build a calculated column by inserting formula in one cell in a table column and having that formula automatically applied to all other cells in that table column.

The total row

When you add a total row to a table, Excel provides an AutoSum drop-down list from which you may choose among functions like SUM, AVERAGE, and others. When you choose one of these choices, the table will convert it to a **SUBTOTAL** function, which by default ignores rows that have been concealed by a filter. You may adjust the SUBTOTAL function parameters to include hidden rows in your computations.

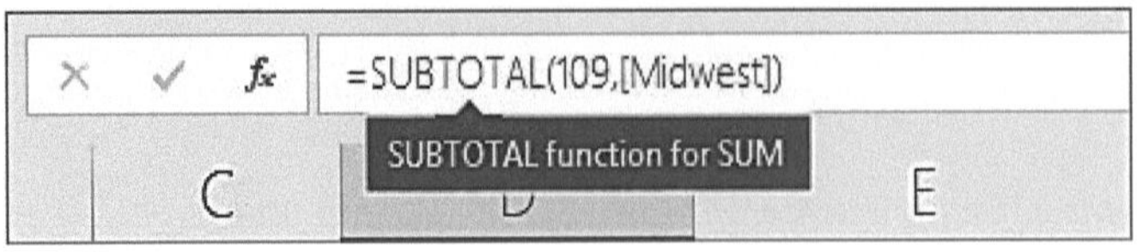

Creating a table

Tables in Excel are useful for providing data collections structure. It contains several useful features, such as data organization, headers, and applied filters. Tables may be accessed via the Insert menu tab or the shortcut key **Control key + T**. All we have to do now is choose the range of cells we want to include in the table. The Design tab, which appears when we pick the table, allows us to adjust table styles. Follow the steps below to create a table.

Choose the cell or range you want to create the table on.

Click on the Home tab, then click on **Format as Table**. This will display a menu that consists of some table styles. Choose any style you want.

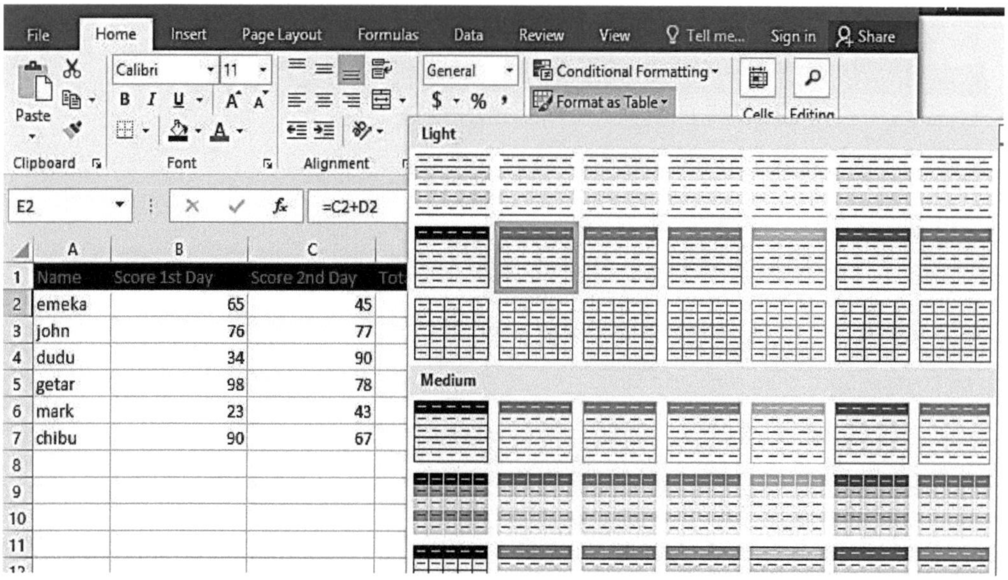

When you click on the style you want, a dialog box will open. Check the box on the My table as header (Checking this box makes the first row of the range the header row) and click Ok.

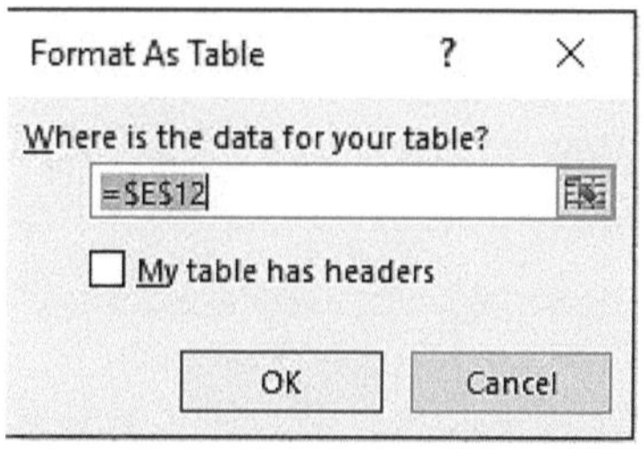

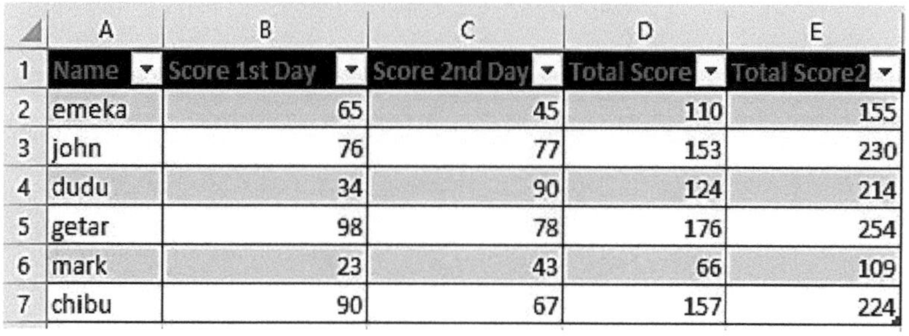

	A	B	C	D	E
1	Name	Score 1st Day	Score 2nd Day	Total Score	Total Score2
2	emeka	65	45	110	155
3	john	76	77	153	230
4	dudu	34	90	124	214
5	getar	98	78	176	254
6	mark	23	43	66	109
7	chibu	90	67	157	224

Sorting a table

One of the most frequent data management tools is sorting. You may sort your table in Excel by one or more columns, ascending or descending order, or by performing a custom sort. First, click on the cell. Then, click on Sort & Filter.

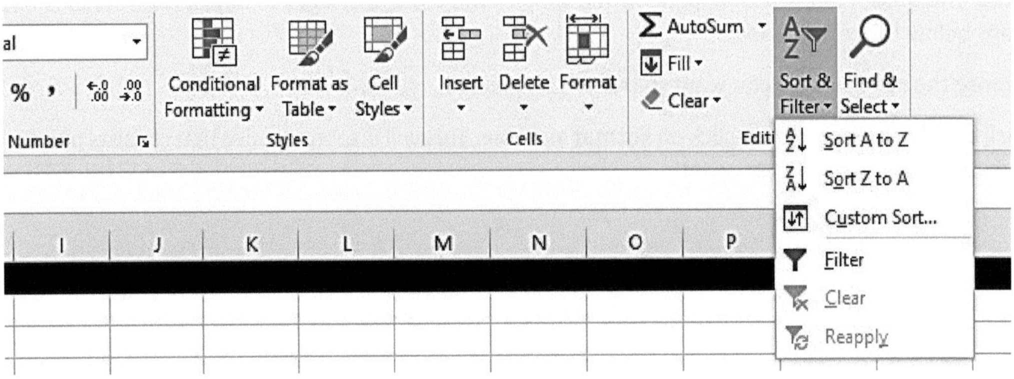

You will have different sorting options.

- **Sort A to Z:** This is to sort in ascending order
- **Sort Z to A:** This is to sort in descending order
- **Custom Sort:** This is for applying various sort criteria in multiple columns.

Click on **Custom Sort**, then click **Add Level.**

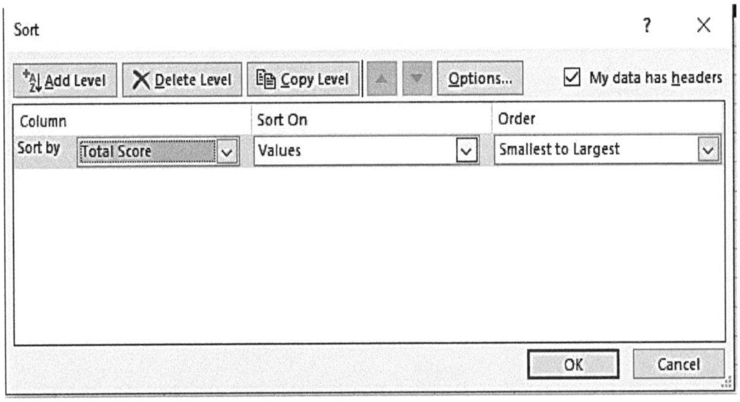

Then, enter in how you want it to be sorted. Once you are done, click Ok.

Filtering a table

Click on cell on your worksheet. On the ribbon, click on the Data tab, then select Filter.

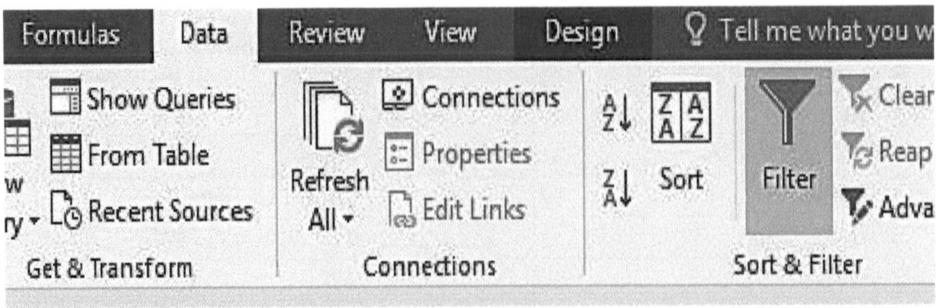

Now, click on the arrow on the column header.

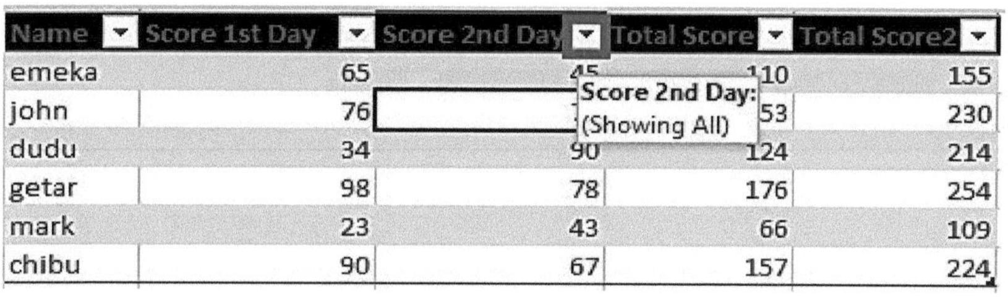

Then, click on Number Filters. Select any filter option.

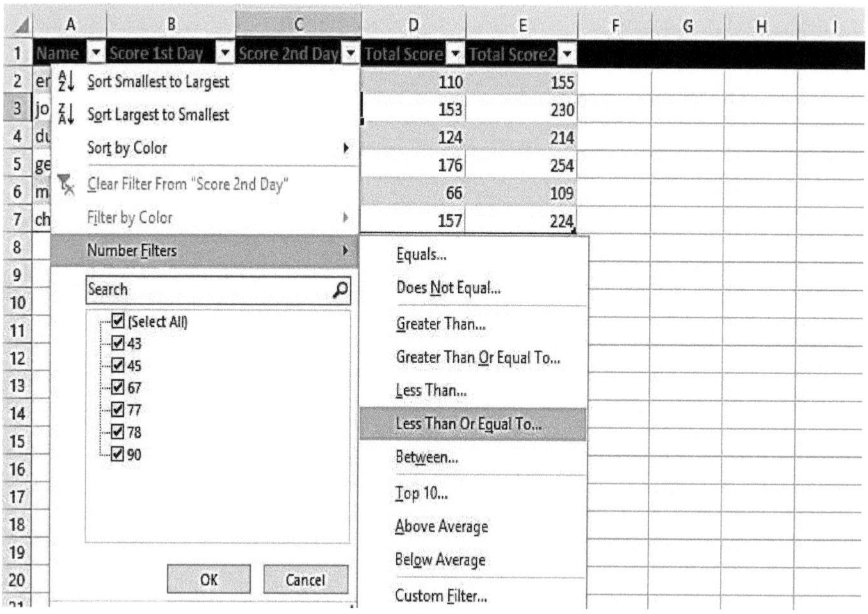

Filtering a table with slicers

Adding Slicer filters to your tables may dramatically boost the usability of tables by allowing you to filter table data more quickly and simply. Choose the table for filtering. Then, click on Insert Slicer. You can find the Insert Slicer on the Design tab on the ribbon.

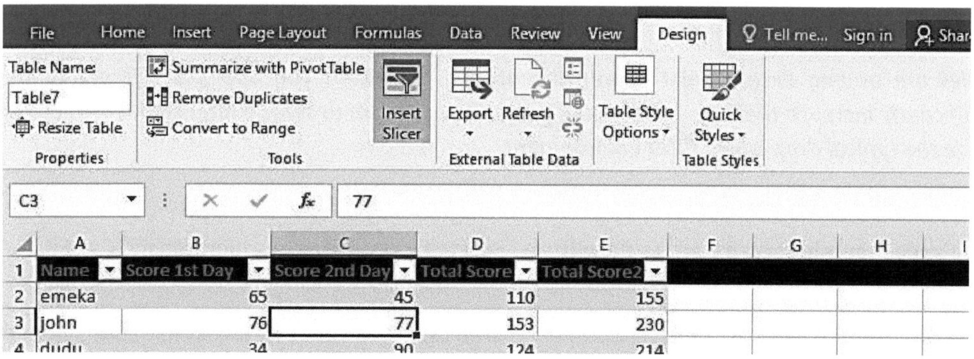

When you click on Insert Slicer, it opens up a dialog box. On it, you are to check the boxes of the field in the table in which you want to filter with a slicer.

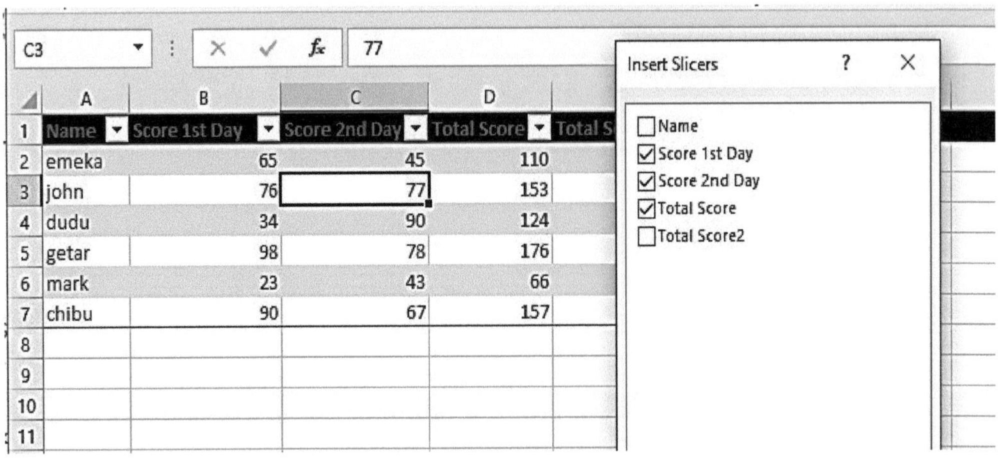

Then, click Ok.

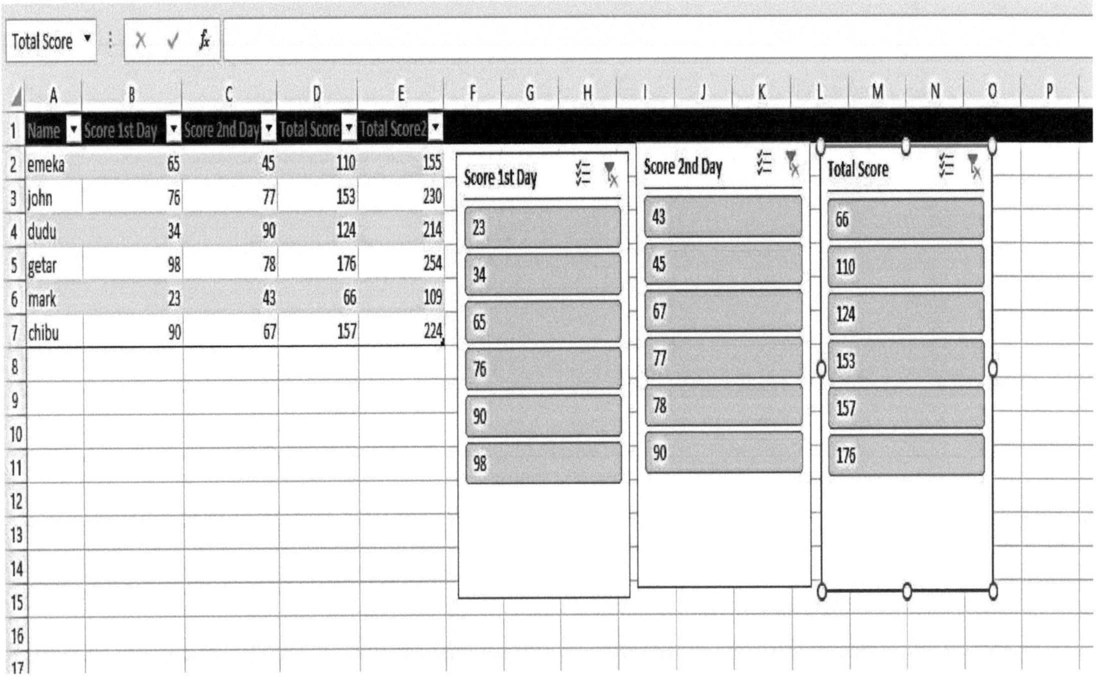

Tables are among Excel's most powerful features, but when you add Slicer filters to them, you significantly increase their use by allowing you and your team to swiftly filter tables without having to utilize the typical drop-down filter environment.

Changing the table's appearance

What's the first thing you'd want to do with an Excel table once you've generated it? Make it appear just how you want it to!

To do this, simply select the table, click on the **Home** tab, then click on **Format as Table**. This will display a menu that consists of some table styles. Choose a style.

CHAPTER FIVE

FORMATTING WORKSHEETS

GETTING TO KNOW THE FORMATTING TOOLS

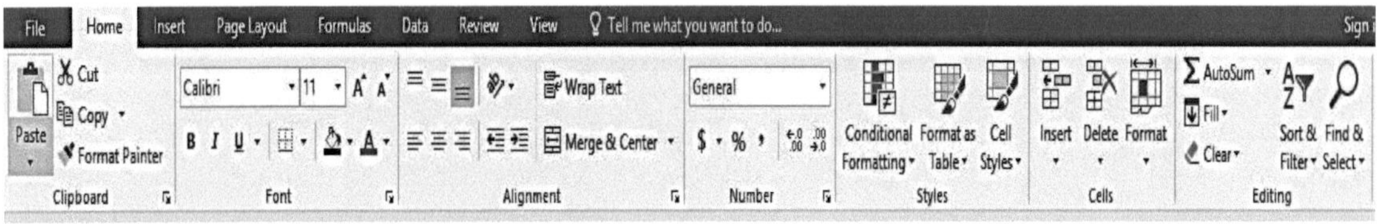

Using the formatting tools on the Home tab

When you're thinking about how to design a spreadsheet, it's helpful to know what tools you have at your disposal. What tools alter the appearance of a worksheet?

Italic, Underline, and Bold: These three tools are fundamental options. You see them almost in any text editing tool. Just highlight the text or sentences and click on any of these tools.

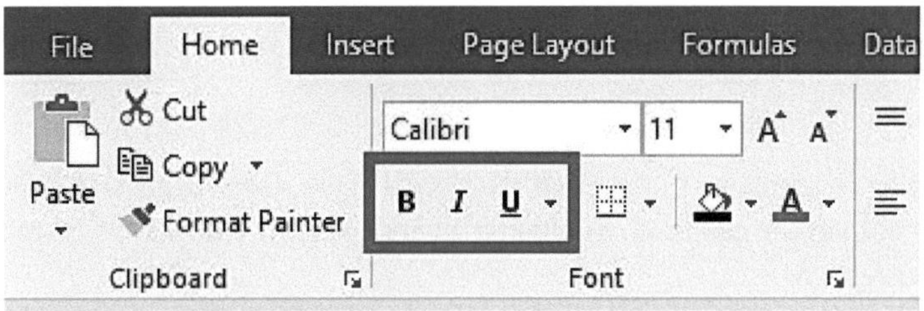

Borders: Borders are useful for segmenting your data and separating it from other portions in your worksheet. The border tool in Excel can create a variety of borders, but it might be difficult to use at first. Begin by highlighting the cells to which you wish to apply a border. Select one of the built-in styles from the Borders dropdown menu.

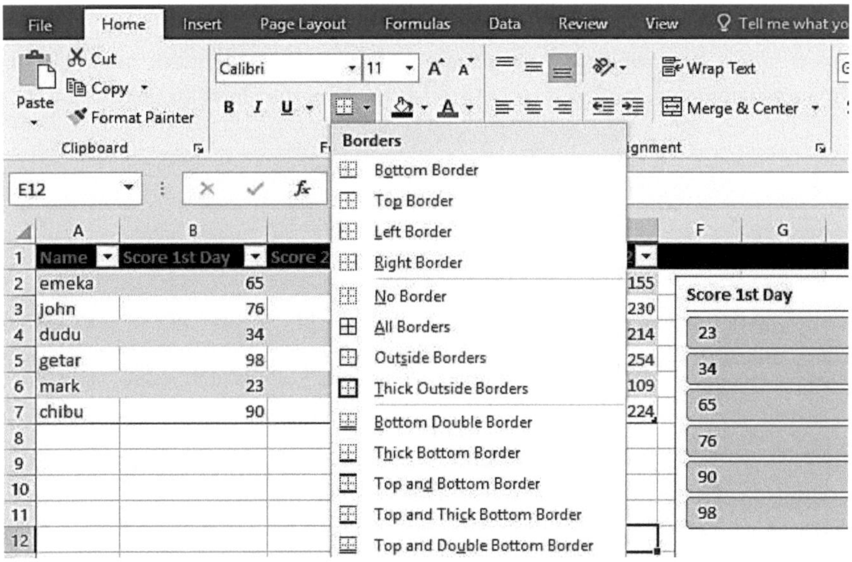

Shading: Shading, commonly known as **fill**, is a color applied to the cell's backdrop. To apply shading to a cell, click and highlight the cells you wish to shade.

Then, on the Font tab of the Home ribbon, click the arrow next to the paint bucket option. To apply a color to a cell, choose one of the several color thumbnails. I'll also utilize the **More Colors** option to launch a full-featured color choosing window regularly. The easiest way to make writing legible is to use light tones.

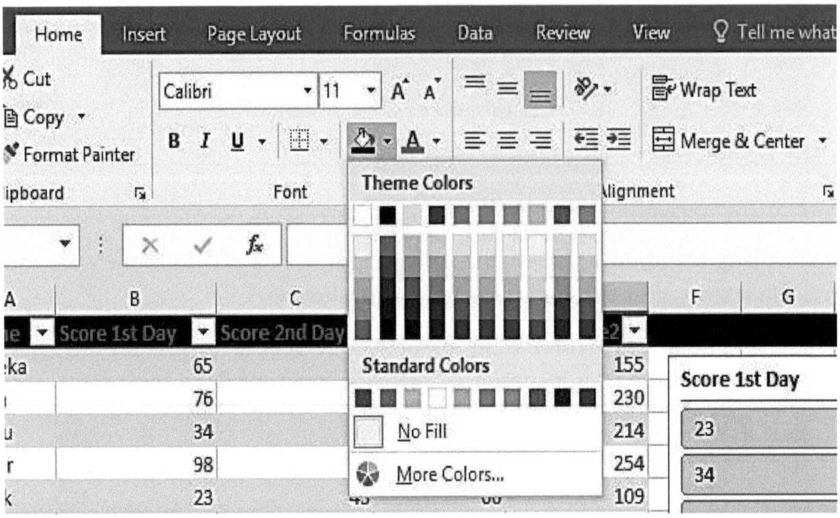

You may use shading to emphasize important data once again. One suggestion is to utilize a constant fill depending on the data of the cell, such as blue for every "input" field where you manually enter data, as I indicated previously.

Don't go overboard with the shading. When you apply too many of these to your cells, it detracts from the information provided in the spreadsheet.

Alignment: How the contents of a cell are aligned to the edges is referred to as alignment. Text may be aligned to the left, center, or right. In a cell, content is left-aligned by default. When working with huge data sets, you may wish to experiment with alignment to improve readability.

Text on the left border of a cell is a regular change I make, whereas numeric quantities should be right-aligned. Also, when column headings are centered at the top, they look fantastic.

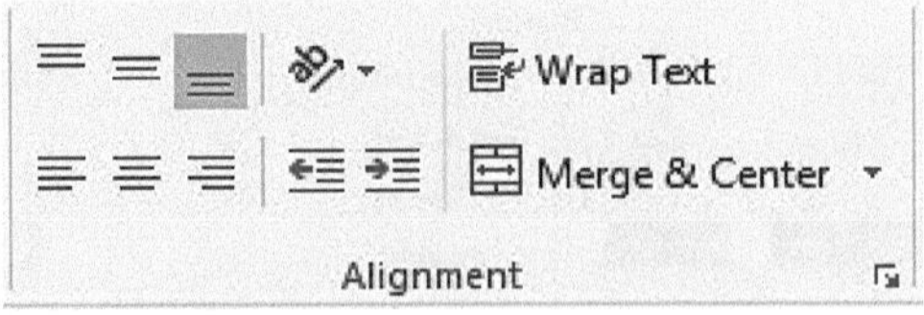

The three alignment buttons on the Alignment tab of Excel's Home ribbon may be used to change alignment. You may also modify whether the content aligns towards the top, middle, or bottom of the cell by aligning it vertically.

Using the Format Cells dialog box

When formatting a cell value from General to Number, Accounting, Currency, etc., Format Cells is mostly used to do this. Format Cells allow us to modify the formatting of a cell number without affecting the actual number. We may modify the quantity, alignment, font style, Border style, Fill options, and Protection in Format cells.

Also, use **Control key + 1** to open it. You can also use the **Shift key + F10** as well as right-clicking on the cell you want to format and select Format Cells. On the Format Cell box, you will see six tabs which are **Number, Font, Border, Fills, Protection, and Alignment.**

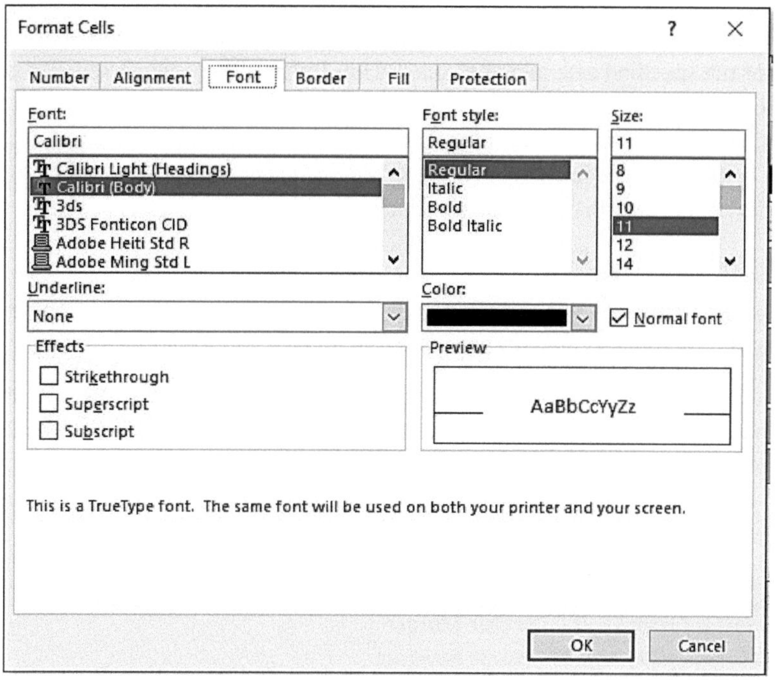

Simply select the cell or range. Press the shortcut keys listed above, open the box, and then click on any of the tabs depending on your format style. Once you are done, click Ok.

Formatting your worksheets

Worksheets are often dismissed as dull, utilitarian tools. They are, without a doubt, quite helpful for organizing data and doing computations. That doesn't mean we can't add some flair to our spreadsheets with some Excel styling.

When we format a worksheet correctly, it gives a second layer of significance to it. Formatting isn't a haphazard process; it's a technique of applying certain styles to indicate the kind of data in a cell.

Using fonts to format your worksheet

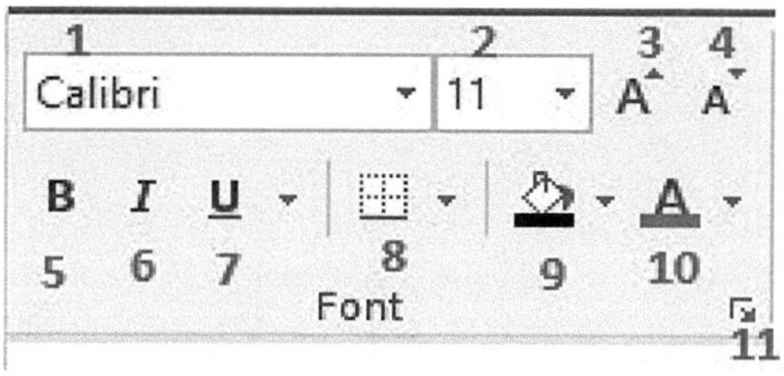

1. **Font** – Changes the font of the currently selected cell (s). Fonts are several methods of displaying the same letters.
2. **Font Size** – Controls the text size (the font). Font size is proportional to the size of the number. In this box, you may write a custom size. You may use the numbers 1 through 409 in Excel, including half sizes.

3. **Change Font Size** – Increases the font size.
4. **Reduce Font** – Reduces the font size.
5. **Bold** – Highlights the chosen cell(s). Control key +-B and Control key +-2 are shortcut keys.
6. **Italic** – Italicizes the chosen cell(s). Control key +-I and Control key +-3 are shortcut keys.
7. **Underline** — Highlights the chosen cell(s). Control key +-U and Control key +-4 are shortcut keys.
8. **Borders** – For the specified cell, add and remove borders (s). The down arrow will bring up a large list of border options. Click **More Borders**... for more border options.
9. **Fill Color** – Changes cell background color. The cells have "**No Fill**" by default.
10. **Font Color** — Modifies the font color
11. **Format Cells** — This button brings up the dialog box.

Wrapping or Shrinking text to fit the cell

When text does not fit inside a cell and overflows to nearby cells, or even when text extends to the next cell, this is one of the problems encountered by Excel users.

Furthermore, when there is material in the neighboring column, Text is sometimes cut-off. Because of the column's narrow width, all of the contents of the cells are not visible in the image below, however, you may always change the width. You'll also come across hyperlinks that don't fit in the cell.

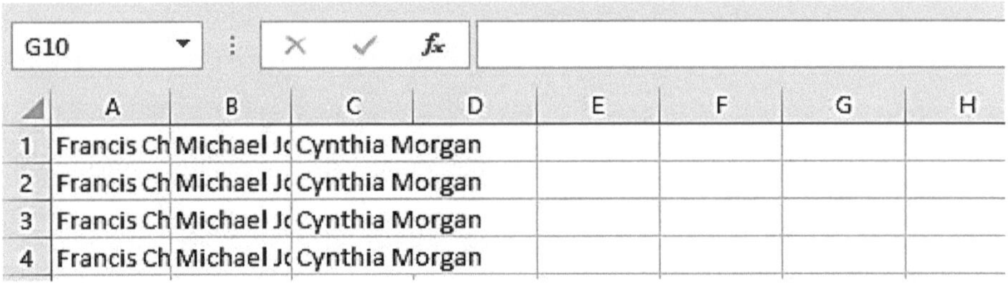

Purpose of Wrap Text

Wrap Text ensures that all of the texts in a cell remain inside the cell's bounds, ensuring that all of the cell's contents are visible to readers. So, to do this, simply

Click on the Cell(s) or Range. On the Home tab, on the Alignment group, click on the Wrap Text option.

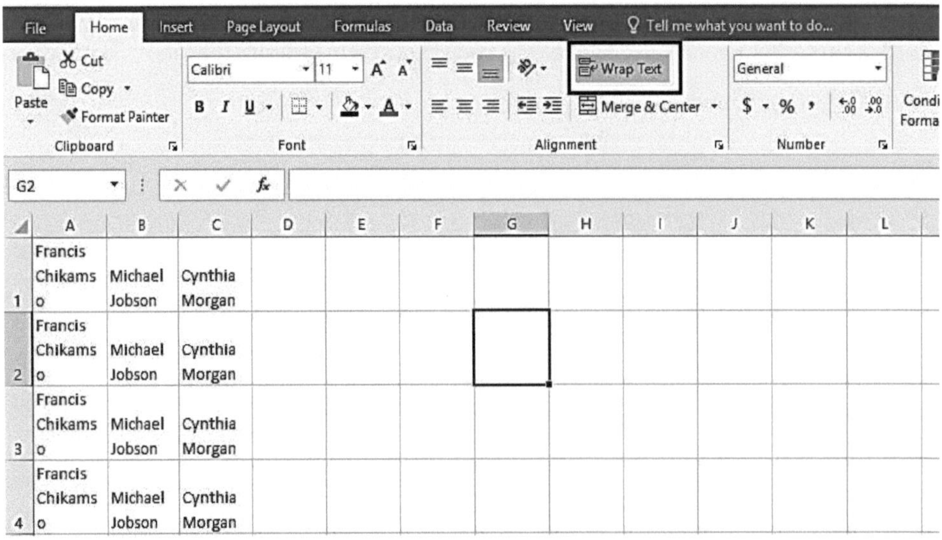

The breadth of the cell remains constant, but the cell is made taller than previously to ensure that all contents are contained inside the body of cells. As a consequence of this operation, the text in Excel will move down rather than across.

Shrinking text

Consider the case when you don't want to raise the cell size horizontally or vertically to fit the text inside the cell boundaries. You may utilize Excel's **'Shrink the text'** option in such cases. To use this, the first thing you are to do is to unwrap the text you have wrapped. Select the cell and click on **Wrap Text.**

Right-click on the cell or range.

Click on **Format Cell**. On the Alignment, tab selects the box **Shrink to fit.**

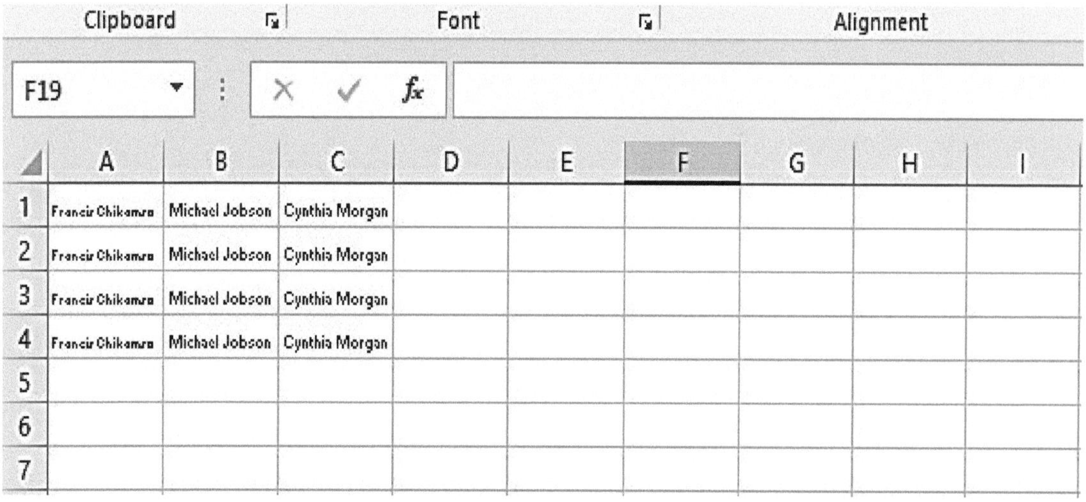

Do you notice how the type size of the contents in the cell changed? This effect will not affect the cell boundaries; width and height of the cell.

Merging worksheet cells to create additional text space

Once you have opened your worksheet, choose the cells for merging.

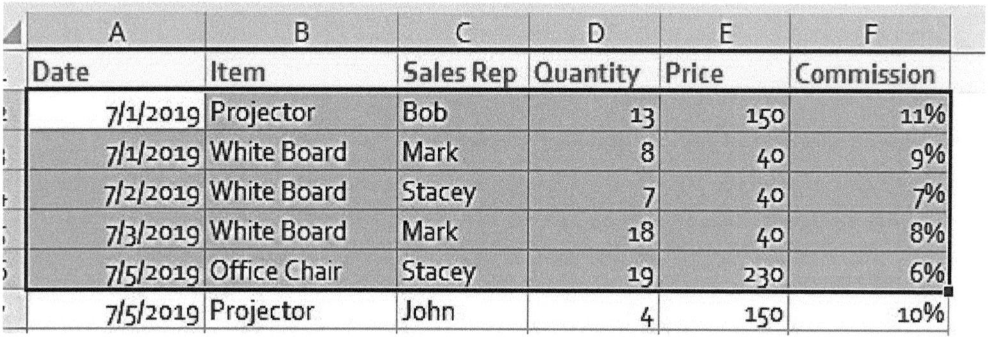

Date	Item	Sales Rep	Quantity	Price	Commission
7/1/2019	Projector	Bob	13	150	11%
7/1/2019	White Board	Mark	8	40	9%
7/2/2019	White Board	Stacey	7	40	7%
7/3/2019	White Board	Mark	18	40	8%
7/5/2019	Office Chair	Stacey	19	230	6%
7/5/2019	Projector	John	4	150	10%

Select **Merge & Center** on the Home tab. This will merge the cells and will center the content. Click the down arrow on the **Merge & Center** option to see other merging options.

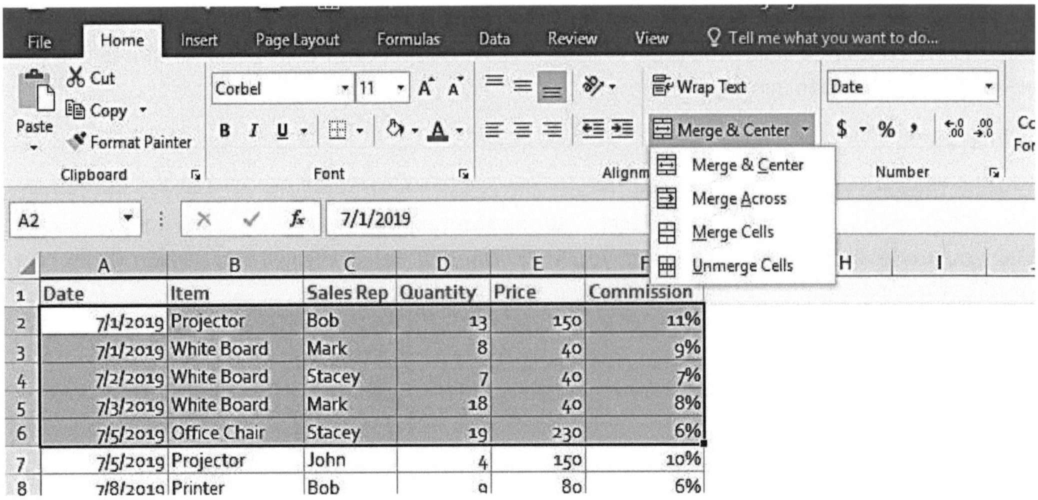

Displaying text at an angle

Using an angle for your text, particularly for column and row headings, maybe aesthetically pleasant. You may swiftly rotate the text in either a clockwise or counterclockwise direction. Simply follow the steps below;

- Click on the cell or range of cells.
- Click on the down arrow near the **ab** (with an arrow below) symbol in the alignment group.
- Choose either **Angle Counterclockwise** or **Angle Clockwise** from the top two choices.

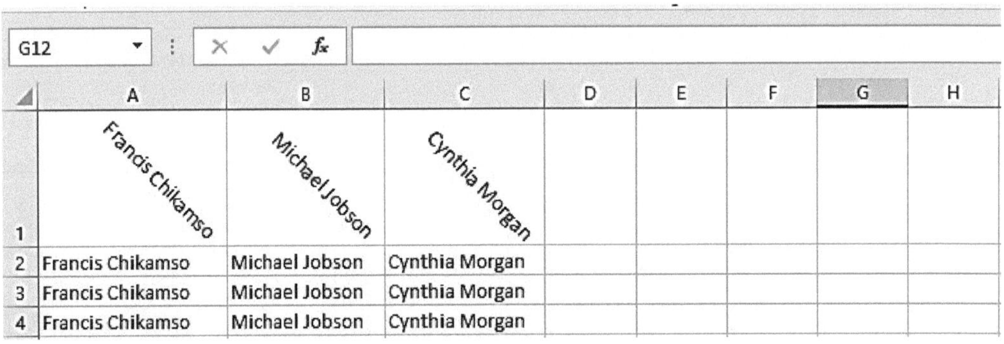

Using colors and shading

First, select the cells, then press **Control key + 1** to launch the Format Cells dialog.

Click the **Font tab.** After choosing font color, click ok. In the image below, I changed the color of the Total Score 2 column to blue.

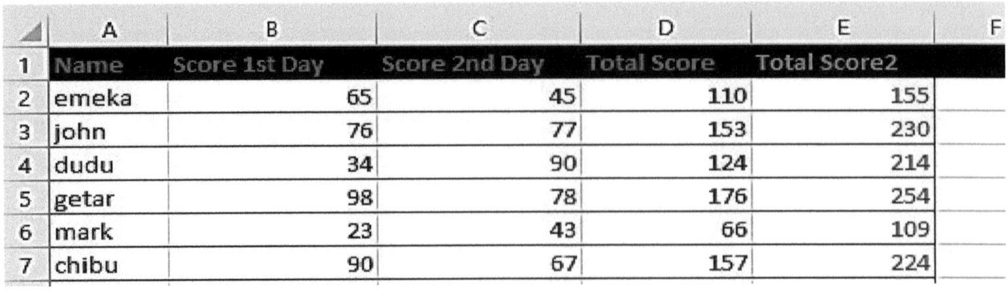

To fill the cells with color, select the desired cells. Launch the format cells dialog box. Click on the **Fill** tab and select a color, then click **Ok**. This formatting draws attention to the column headings and the data in the total score column.

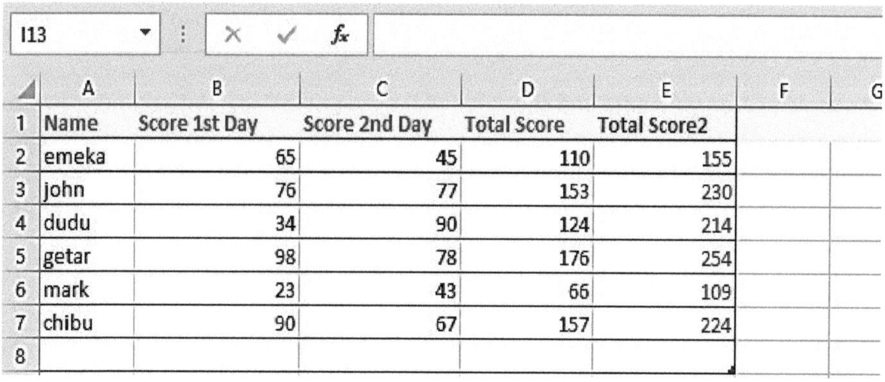

Adding border lines

Highlight the cells. Click on the drop-down arrow.

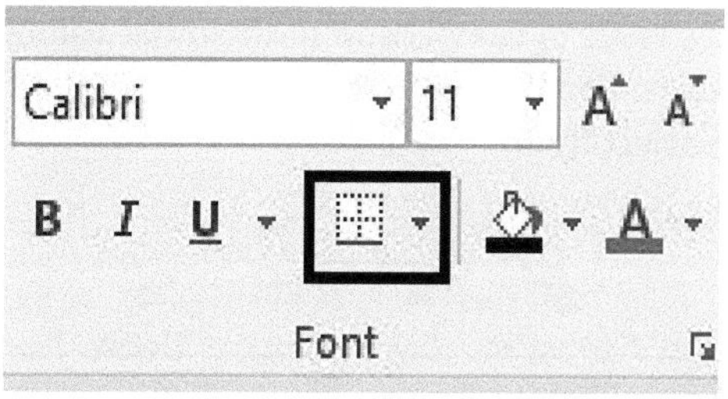

So, select the kind of border you want to add and where you want to add it.

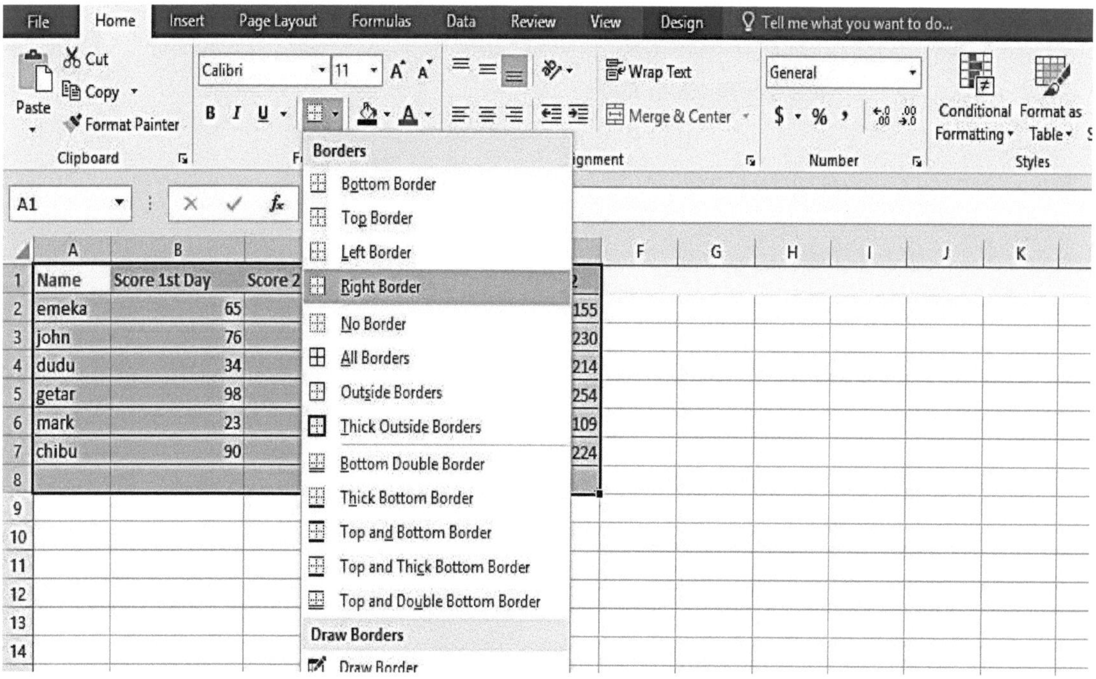

Using conditional formatting

In Excel, conditional formatting is used to emphasize data based on certain conditions. It would be tough to see different patterns merely by looking at your Excel worksheet. In Excel, conditional formatting allows you to visualize data and make spreadsheets more understandable. It enables you to add formatting to cell values such as colors, icons, and data bars based on the cell values.

It makes it simple to scan your data and check for vital signs graphically. With numerical data, conditional formatting works best. Simply choose a column of data and make sure you're on the Home tab of Excel's ribbon to get started. From the Conditional Formatting dropdown menu, you may choose from a variety of styles. Each of the formats your cells in a different way, but they all adjust to the cells you've highlighted.

Using graphical conditional formats

Data bars, color scales, and icon sets are the three conditional formatting choices for displaying graphics. These sorts of conditional formatting might help you see a range of numbers.

Using data bars

Horizontal bars are shown directly in the cell using the data bars conditional format. The length of the bar is determined by the cell's value in comparison to the other values in the range. To do this, simply select the column or row. Click on **Conditional formatting** and select Data bars which will display a list of different data bars for you to select from.

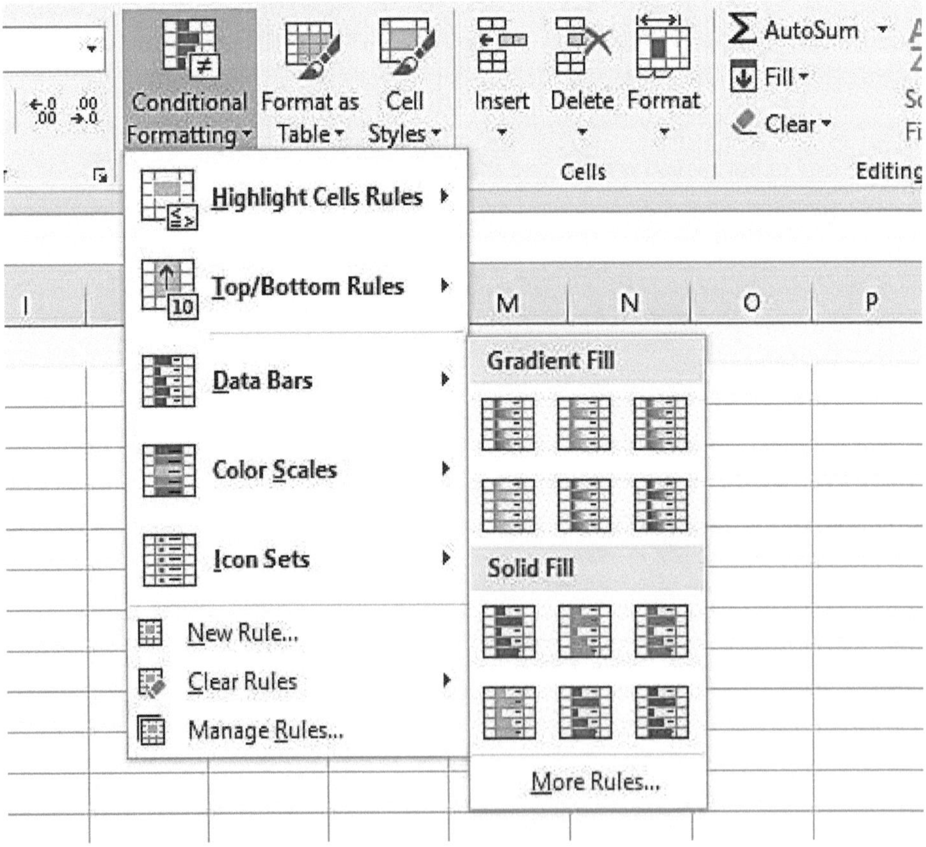

Pick one and the effect will be applied on your worksheet.

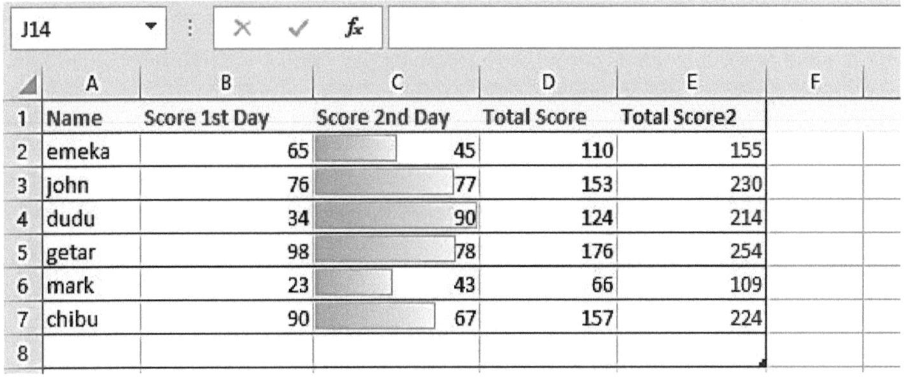

	A	B	C	D	E
1	Name	Score 1st Day	Score 2nd Day	Total Score	Total Score2
2	emeka	65	45	110	155
3	john	76	77	153	230
4	dudu	34	90	124	214
5	getar	98	78	176	254
6	mark	23	43	66	109
7	chibu	90	67	157	224

Surprisingly, the colors used for data bars are not theme colors if you pick one of the 12 data bar types. The data bar colors do not change when you alter the document theme. However, if you use the New Formatting Rule dialog box to add the data bars, the colors you pick are theme colors.

Using color scales

Select the column or row. Click on **Conditional formatting** and select Color Scales which will display a list of different color scales for you to select from. Select one and it will be applied to your worksheet.

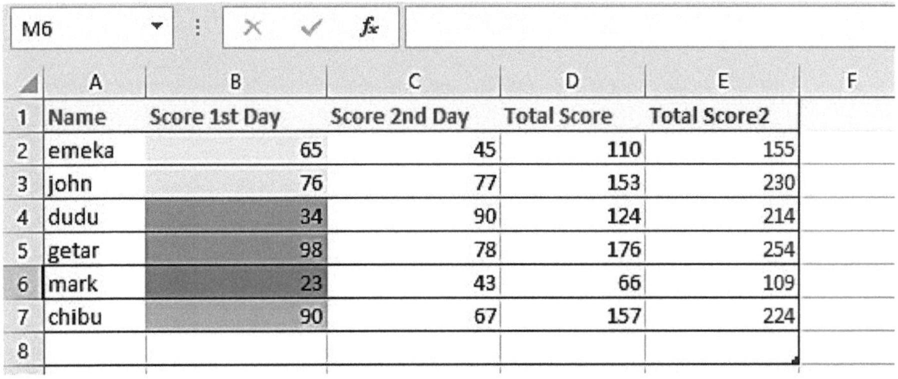

	A	B	C	D	E
1	Name	Score 1st Day	Score 2nd Day	Total Score	Total Score2
2	emeka	65	45	110	155
3	john	76	77	153	230
4	dudu	34	90	124	214
5	getar	98	78	176	254
6	mark	23	43	66	109
7	chibu	90	67	157	224

Using icon sets

Select the column or row. Click on **Conditional formatting** and select **Icon Sets** which will display a list of different icon sets for you to select from. Select one and it will be applied to your worksheet.

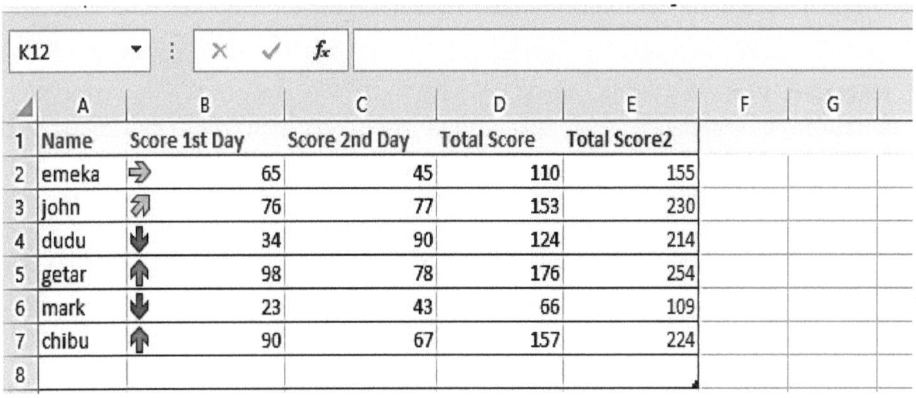

	A	B	C	D	E
1	Name	Score 1st Day	Score 2nd Day	Total Score	Total Score2
2	emeka	65	45	110	155
3	john	76	77	153	230
4	dudu	34	90	124	214
5	getar	98	78	176	254
6	mark	23	43	66	109
7	chibu	90	67	157	224

Creating formula-based rules

Formulas are a more powerful technique to apply conditional formatting since they enable you to apply rules based on any reasoning. Excel comes with a plethora of "presets" that makes it simple to establish new rules without having to use formulae.

Although Excel comes with several "presets" for conditional formatting, they are restricted. You may, however, construct rules using your unique formulae. You may take over the situation that activates a rule and apply precisely the reasoning you need by creating your formula. Formulas provide you with the greatest amount of power and versatility.

To create formula-based rules, simply choose the cells. Then, on the Home tab, click on Styles > Conditional Formatting > New Rule. This will open up the new formatting rule dialog box. Click on the **Use a formula to determine which cells to format** option, and then type in the formula.

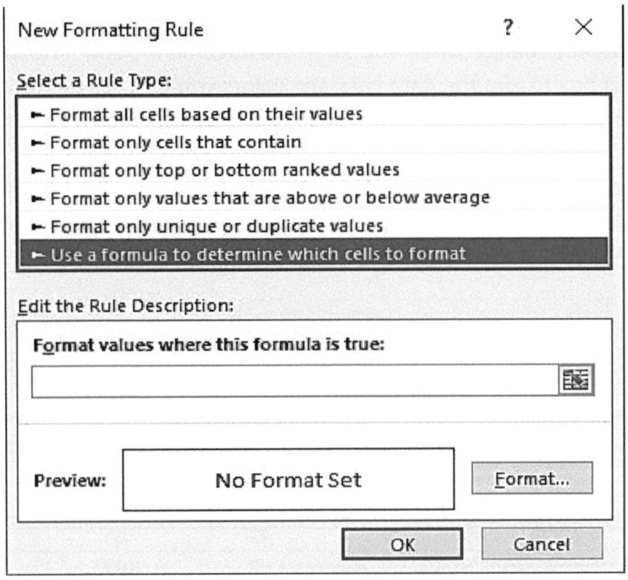

The formula has to be logical. It must return either TRUR or FALSE. When it is true, the conditional formatting will apply but when it is false, it will not apply. Note that the formula must start with an **equal to (=)** sign.

Understanding relative and absolute references

There are several types of cell references in Excel formulas, some of which we have listed below.

- **Absolute cell references**: (Those preceded by a $ sign, such as $A$1) are always consistent, regardless of where they are copied.
- **Relative cell references**: (Those without the $ sign, such as A1) alter depending on the relative location of the rows and column when replicated over several cells.
- **Mixed cells references**: This is the absolute column and relative row ($A1), or relative column and absolute row (A$1). These are mostly used in conditional formatting rules to indicate that a column letter or row number should be fixed when a rule is applied to all other cells in the chosen range. In conditional formatting rules, cell references are tied to the specified range's top-left most cell. Simply create a formula for the top-left cell when establishing a new rule, and Excel would "**replicate**" your formula to all of the other cells in the given range.

- **Relative references**: Every cell references are a relative reference by default. When you copy them on many cells, they will adjust depending on the relative location of the columns and rows. Now, let's create a formula and copy it using relative references.

From the image below, we will create a formula that will multiply the student's score on the first day by the student's score on the second day. So, we will create the formula in cell E2. So first, select cell E2.

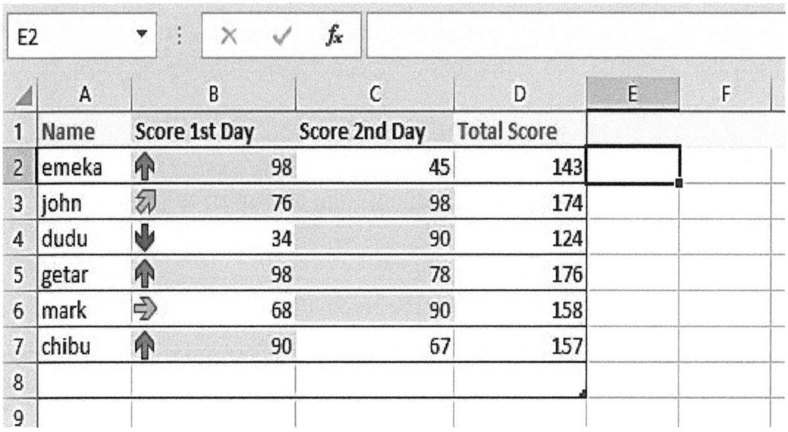

	A	B	C	D	E	F
1	Name	Score 1st Day	Score 2nd Day	Total Score		
2	emeka	98	45	143		
3	john	76	98	174		
4	dudu	34	90	124		
5	getar	98	78	176		
6	mark	68	90	158		
7	chibu	90	67	157		
8						
9						

On the cell, we will put in the formula which will calculate it for us. The formula here is **=B2*C2**.

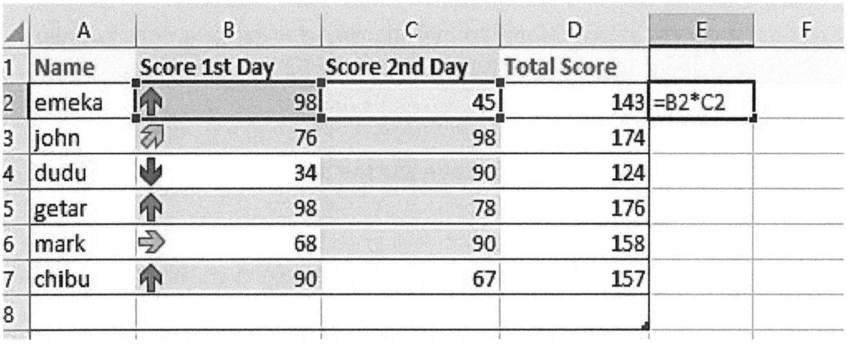

	A	B	C	D	E	F
1	Name	Score 1st Day	Score 2nd Day	Total Score		
2	emeka	98	45	143	=B2*C2	
3	john	76	98	174		
4	dudu	34	90	124		
5	getar	98	78	176		
6	mark	68	90	158		
7	chibu	90	67	157		
8						

After typing in the formula, press Enter. Now, drag the fill handle to the last cell on the column. This will copy the formula to the cells with relative references and fill in the rest of the column with the calculated answer.

When you double-click on the cells, you will see their formula. The relative cell references will differ from each other (that depends on their rows).

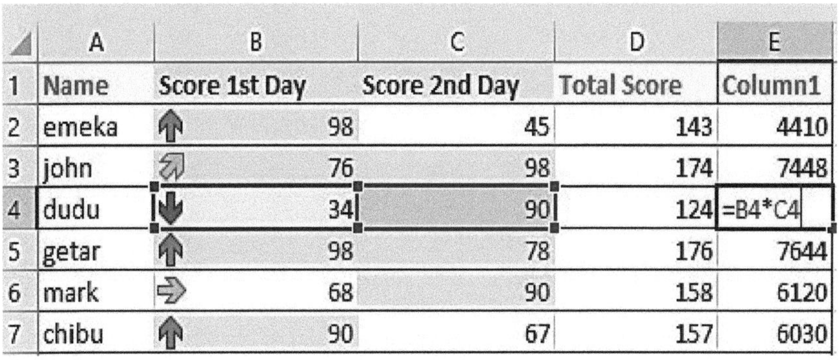

	A	B	C	D	E
1	Name	Score 1st Day	Score 2nd Day	Total Score	Column1
2	emeka	98	45	143	4410
3	john	76	98	174	7448
4	dudu	34	90	124	=B4*C4
5	getar	98	78	176	7644
6	mark	68	90	158	6120
7	chibu	90	67	157	6030

Absolute references

Maybe you may not want your cell reference to change when it is copied to other cells. In relative references, it changes, while absolute references don't change when it's copied. You use this to make your row or column remain constant.

Conditional formatting formula examples

Assume you choose the range A1:B10 and want to apply formatting to any cells in the range that are greater than the value in cell C1. Fill in the following conditional formatting formula:

=A1>$C$1

The reference to cell C1 in this example is absolute; it will not be altered for the cells in the chosen range. To put it another way, the conditional formatting formula for cell A2 is as follows:

=A2>$C$1

The absolute cell reference is not changed, but the relative cell reference is.

After choosing the Use a Formula to Determine Which Cells to Format rule type, each of these examples employs a formula written directly into the New Formatting Rule dialog box. You get to choose the sort of conditional formatting you want to use.

Identifying weekend days

Although Excel has a variety of conditional formatting rules for dates, it does not allow you to identify dates that fall on a weekend. To get weekend dates, use the following formula:

=OR(WEEKDAY(A1)=7,WEEKDAY(A1)=1)

This formula supposes that you've chosen a range and the active cell here is cell A1

Highlighting a row based on a value

In my example below, I want to highlight the records that contain everything about Bob.

	A	B	C	D	E	F
1	Date	Item	Sales Rep	Quantity	Price	Commission
2	7/1/2019	Projector	Bob	13	150	11%
3	7/1/2019	White Board	Mark	8	40	9%
4	7/2/2019	White Board	Stacey	7	40	7%
5	7/3/2019	White Board	Mark	18	40	8%
6	7/5/2019	Office Chair	Stacey	19	230	6%
7	7/5/2019	Projector	John	4	150	10%
8	7/8/2019	Printer	Bob	9	80	6%
9	7/10/2019	Printer	Laura	16	80	2%
10	7/10/2019	Office Chair	Mark	15	230	9%
11	7/10/2019	Diary	Bob	15	16	1%
12	7/10/2019	Office Chair	John	7	230	2%
13	7/13/2019	Diary	Laura	23	16	11%
14	7/17/2019	White Board	Bob	20	40	5%
15	7/17/2019	Office Chair	Mark	9	230	3%
16	7/20/2019	White Board	Stacey	23	40	6%
17	7/20/2019	White Board	Stacey	4	40	5%

So, I highlight the entire worksheet, select Conditional Formatting (Home tab) and select New rules. Choose the **"Use a formula to determine which cells to format"** option. On the box, below it, type in this formula **=$C2=" Bob"**

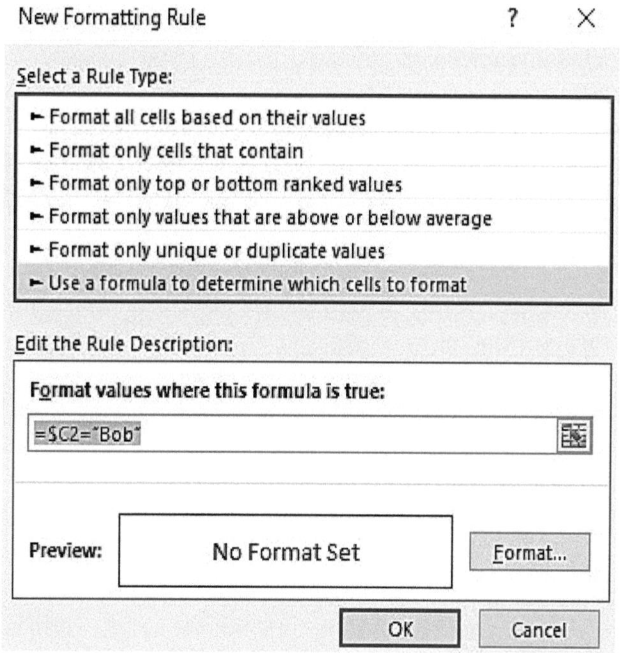

Then, click on the Format option. Click on the Fill tab and choose a color that you want it to be highlighted with. Then, click Ok.

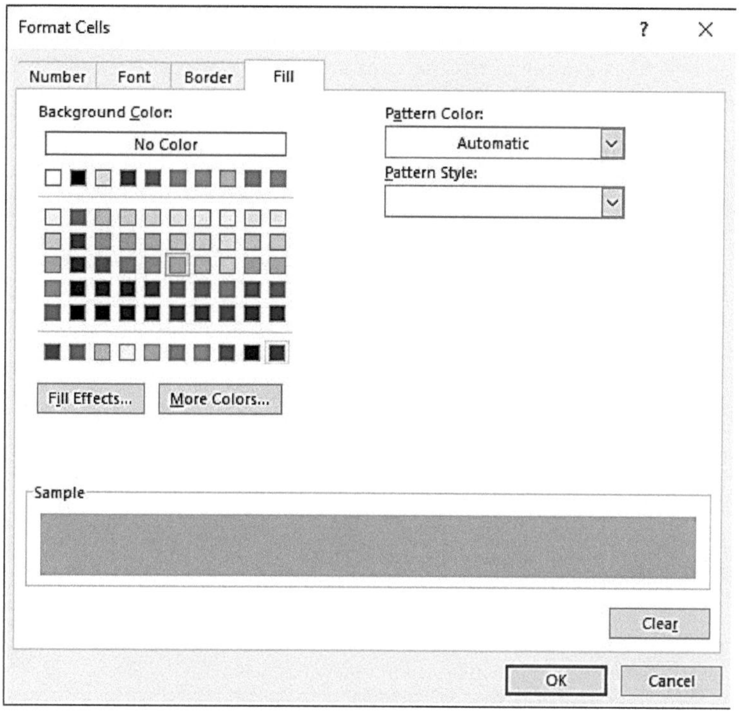

This will be the result.

Date	Item	Sales Rep	Quantity	Price	Commission
7/1/2018	Projector	Bob	13	150	11%
7/1/2018	White Board	Mark	8	40	9%
7/2/2018	White Board	Stacey	7	40	7%
7/3/2018	White Board	Mark	18	40	8%
7/5/2018	Office Chair	Stacey	19	230	6%
7/5/2018	Projector	John	4	150	10%
7/8/2018	Printer	Bob	9	80	6%
7/10/2018	Printer	Laura	16	80	2%
7/10/2018	Office Chair	Mark	15	230	9%
7/10/2018	Diary	Bob	15	16	1%
7/10/2018	Office Chair	John	7	230	2%
7/13/2018	Diary	Laura	23	16	11%
7/17/2018	White Board	Bob	20	40	5%
7/17/2018	Office Chair	Mark	9	230	3%
7/20/2018	White Board	Stacey	23	40	6%
7/20/2018	White Board	Stacey	4	40	5%

Explanation

Conditional Formatting examines each cell for the condition we've set, which is =$C2=" Bob" in this case. As a result, it will check whether cell C2 contains the name Bob or not while inspecting each cell in row A2. If it does, that cell is highlighted; if it does not, it is not.

It's important to note that the dollar symbol ($) comes before the column alphabet ($C1). By doing so, we've ensured that the column will always be C. When cell A2 is tested for the formula, it will also check cell C2, and when cell A3 is examined for the condition, it will also check cell C3. By using conditional formatting, we can highlight the whole row.

Displaying alternate-row shading

Pick a range. Click on Conditional Formatting and select New Rule.

Date	Item	Sales Rep	Quantity	Price	Commission
7/1/2019	Projector	Bob	13	150	11%
7/1/2019	White Board	Mark	8	40	9%
7/2/2019	White Board	Stacey	7	40	7%
7/3/2019	White Board	Mark	18	40	8%
7/5/2019	Office Chair	Stacey	19	230	6%
7/5/2019	Projector	John	4	150	10%
7/8/2019	Printer	Bob	9	80	6%
7/10/2019	Printer	Laura	16	80	2%
7/10/2019	Office Chair	Mark	15	230	9%
7/10/2019	Diary	Bob	15	16	1%
7/10/2019	Office Chair	John	7	230	2%
7/13/2019	Diary	Laura	23	16	11%

Click on "Use a formula to determine which cells to format"

Type =**MOD(ROW(),2).**

Click **Format** to choose a style. Click Ok.

	A	B	C	D	E	F
1	Date	Item	Sales Rep	Quantity	Price	Commission
2	7/1/2019	Projector	Bob	13	150	11%
3	7/1/2019	White Board	Mark	8	40	9%
4	7/2/2019	White Board	Stacey	7	40	7%
5	7/3/2019	White Board	Mark	18	40	8%
6	7/5/2019	Office Chair	Stacey	19	230	6%
7	7/5/2019	Projector	John	4	150	10%
8	7/8/2019	Printer	Bob	9	80	6%
9	7/10/2019	Printer	Laura	16	80	2%
10	7/10/2019	Office Chair	Mark	15	230	9%
11	7/10/2019	Diary	Bob	15	16	1%
12	7/10/2019	Office Chair	John	7	230	2%
13	7/13/2019	Diary	Laura	23	16	11%
14	7/17/2019	White Board	Bob	20	40	5%
15	7/17/2019	Office Chair	Mark	9	230	3%

How does it work?

The residual of a division is returned by the MOD function. The ROW() method returns the number of rows in a table. For instance, MOD(7,2) = 1 for the seventh row since 7 divided by 2 equals 3 with a leftover of 1. MOD(8,2) = 0 for the eighth row since 8 divided by 2 equals 4 with a residual of 0. As a consequence, all odd rows will be darkened if they return 1 (TRUE).

Creating checkerboard shading

First, choose a range of cells and follow the steps above. On the formula box, type in this formula **=MOD(ROW(),2) =MOD(COLUMN(),2).** Click on Format and select a color. Click Ok.

	A	B	C	D	E	F
1	Date	Item	Sales Rep	Quantity	Price	Commission
2	7/1/2019	Projector	Bob	13	150	11%
3	7/1/2019	White Board	Mark	8	40	9%
4	7/2/2019	White Board	Stacey	7	40	7%
5	7/3/2019	White Board	Mark	18	40	8%
6	7/5/2019	Office Chair	Stacey	19	230	6%
7	7/5/2019	Projector	John	4	150	10%
8	7/8/2019	Printer	Bob	9	80	6%
9	7/10/2019	Printer	Laura	16	80	2%
10	7/10/2019	Office Chair	Mark	15	230	9%
11	7/10/2019	Diary	Bob	15	16	1%
12	7/10/2019	Office Chair	John	7	230	2%
13	7/13/2019	Diary	Laura	23	16	11%
14	7/17/2019	White Board	Bob	20	40	5%

Shading groups of rows

To shade rows in excel, formulas to be used will be based on CEILING, ROWS, and **ISEVEN**. The formula for it is **=ISEVEN(CEILING(ROW() –offset, n)/n).**

- **n** stands for the number of rows in a group
- offset stands for the number used to normalize the first row.
- So to shade group pf rows, simply follow the steps above and on the formula box, type in this formula **=ISODD(CEILING(ROW) -5,3)/3).**

Working with Conditional formats

We'll go through the basics of building a new conditional formatting rule in the example below. Although there are several forms of conditional formatting, the core processes are the same for all of them. Let's have a look at what we've got.

In the image below is a table that consists of the scores of a group of students. Now, we will create a conditional formatting rule which will highlight the test scores that are over 70. We will make it highlight it with the green color. To do this, follow the steps below;

First, choose the cells you want to format. On the Home tab, click on **Conditional formatting**. From the list of options there, click on the **Highlight Cells Rules**, then select **Greater Than.**

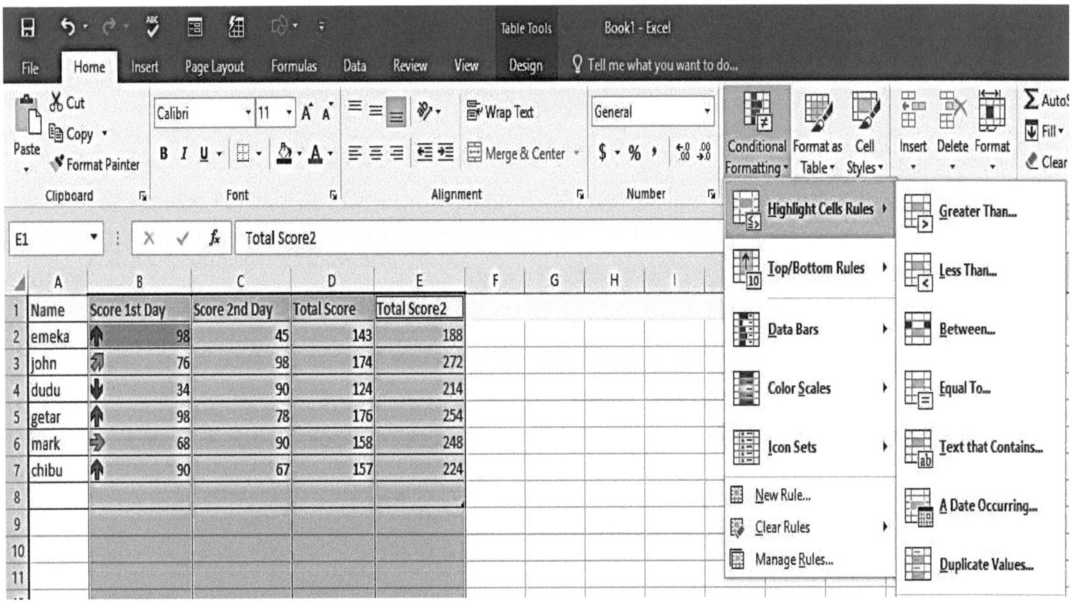

Type in 70 and select the green color. Click Ok.

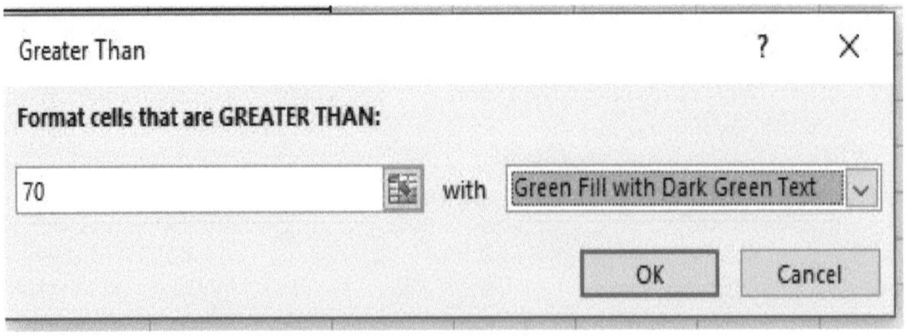

After you have clicked Ok, the rule will be created and you will see the result on your worksheet.

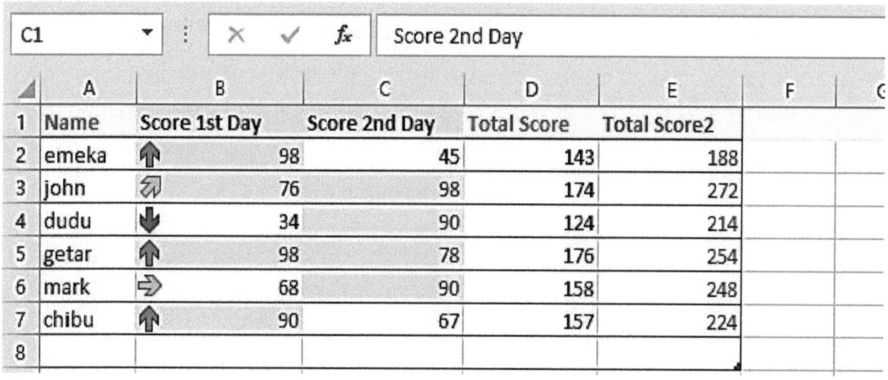

To access the conditional formatting rule, which you have created, click on **Manage Rules** on the conditional formatting menu. if the selection has cells that have conditional formatting applied to it, the rule will display in the Rules Manager window.

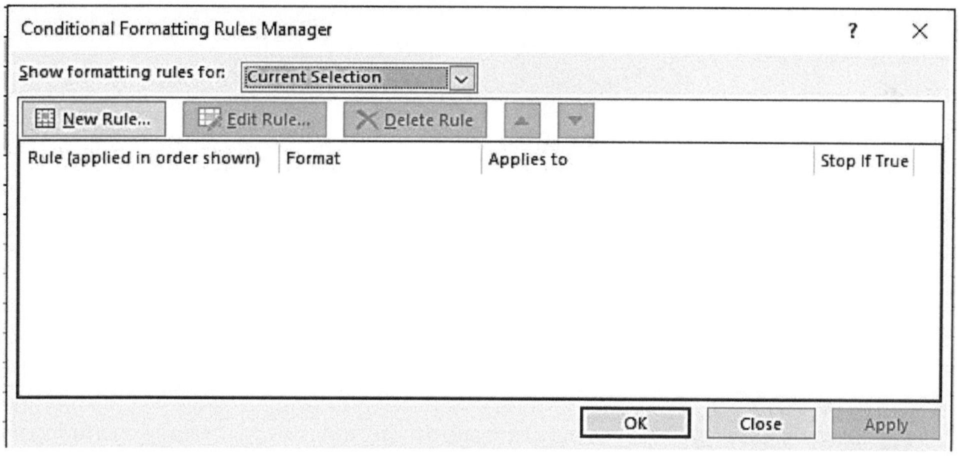

But if the current selection has no cells which have conditional formatting applied to it, then, the rules will not display. So, to make it display, click on the drop-down arrow on the "**Show formatting rules for"** option and select "**This Worksheet**". It will display all the rules and conditional formatting you have applied to that worksheet.

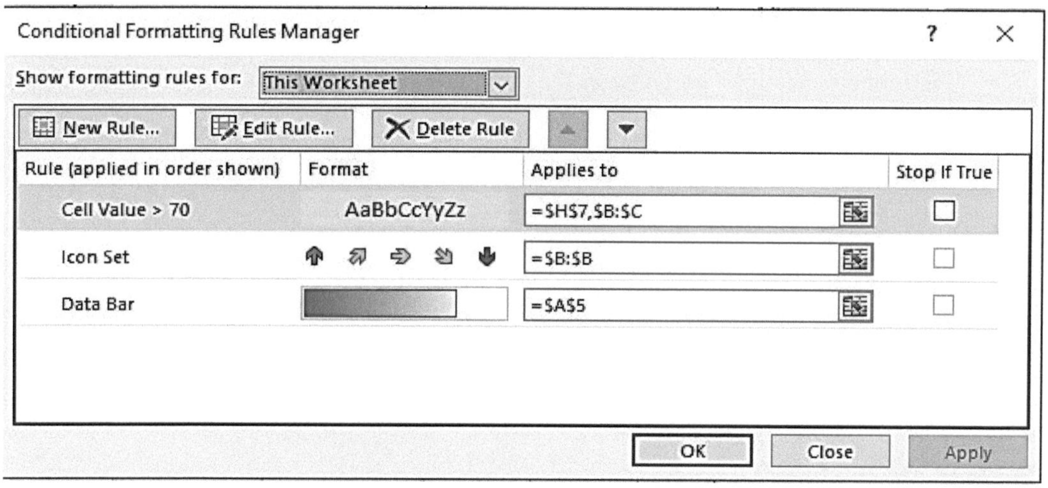

You may use the following steps to develop an excel formula for conditional formatting:

- Select cells or columns to format. If you wish to apply formatting to an entire row, however, you may pick numerous columns or the whole table. More data may be added by transforming cells into tables using the Insert tab and then choosing some empty rows below one's data.
- Select New Rule from the Conditional Formatting menu.

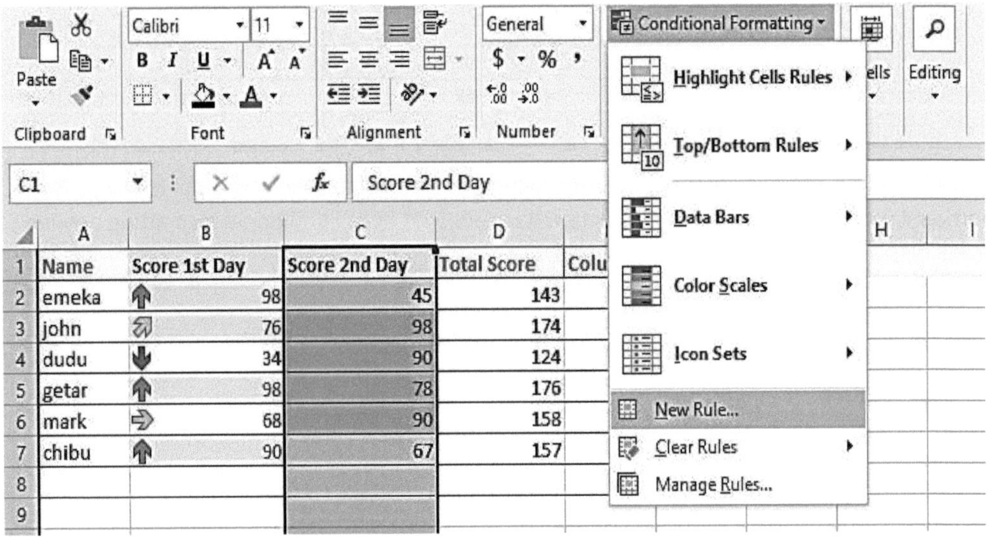

- Select Use a formula tab to decide which cells to format under New Formatting Rule.

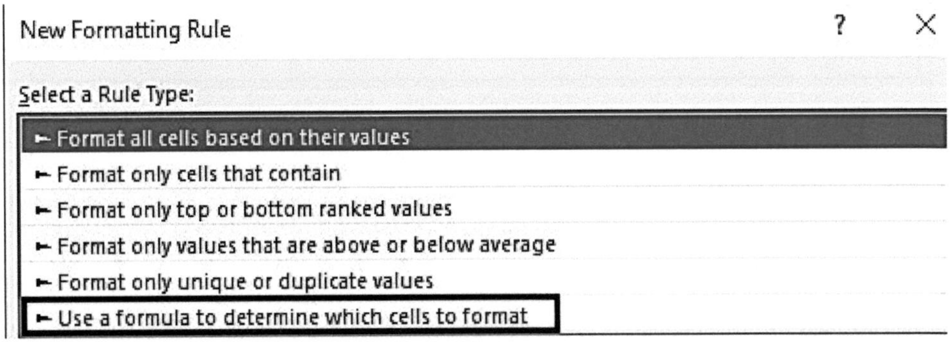

- The formula may then be typed in the appropriate area. Pick the Format option to select a custom format.

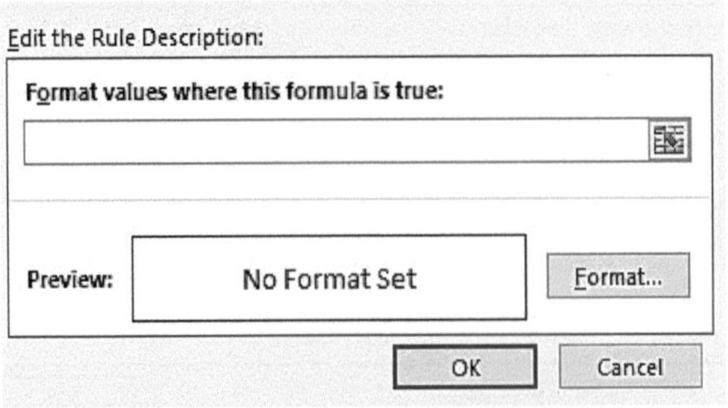

- The Font, Border, and Fill tabs may be switched to experiment with choices such as font, style, and fill color. Under the More Colors option, one may also choose a chosen color from the RGB or HSL charts and then click Ok.

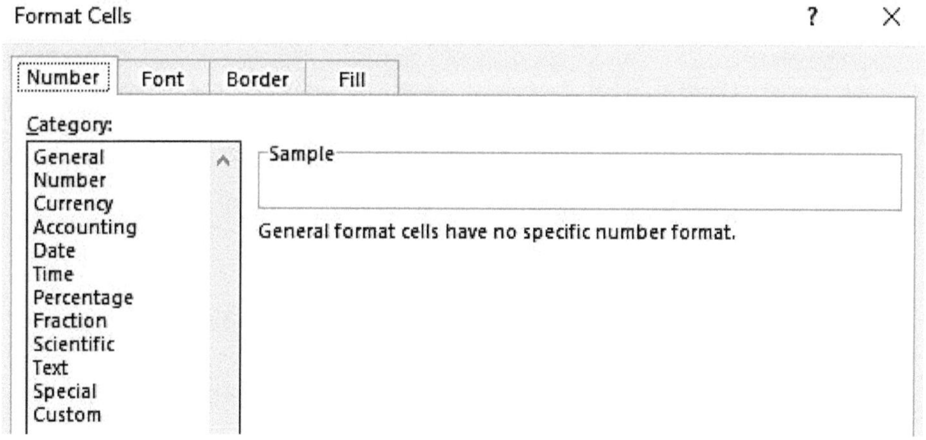

- Once the preview section has shown the required format, select the Ok button to save the rule. If the results aren't what you're looking for, click Format and start modifying.

When changing a conditional formatting formula, always hit the F2 key, then use the arrows to go to the desired location inside the formula. F2 is pushed a second time while adding a cell, and then the cell is clicked.

USING NAMED STYLES FOR EASIER FORMATTING

Applying styles

First, choose the cell or range of cells. Below the Conditional Formatting option, click on Cell Styles. On the menu that opens, select a style.

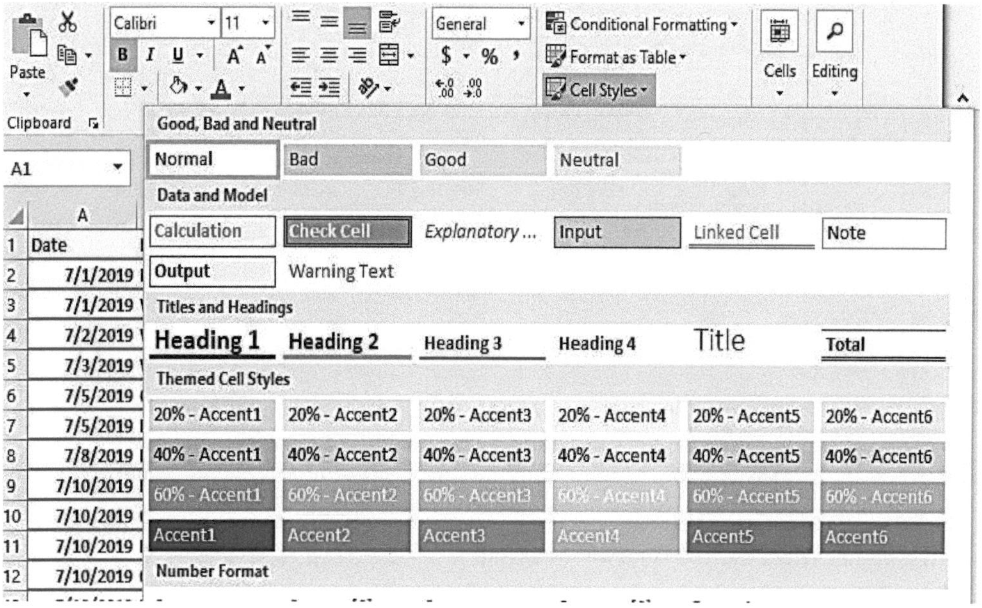

Modifying an existing style

Right-click on it and select **Modify**.

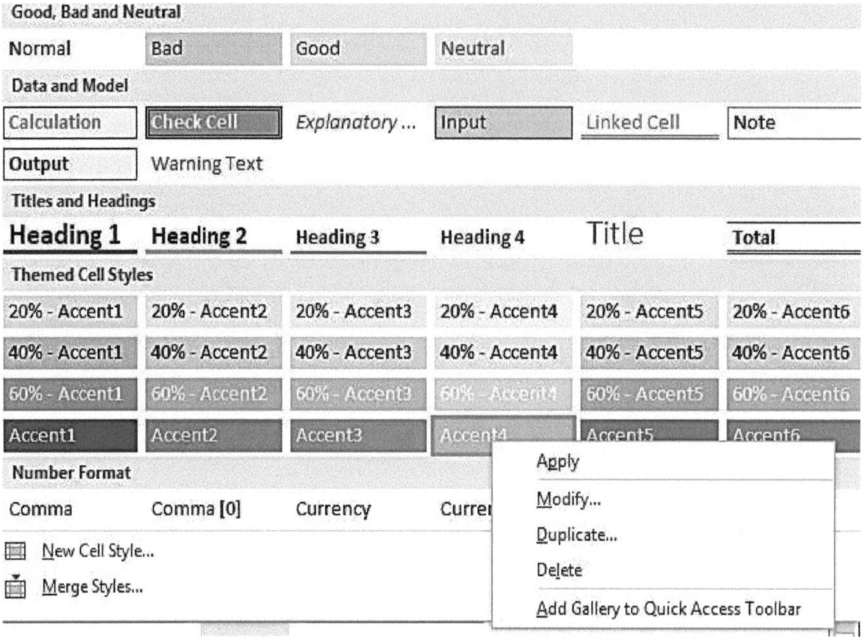

A box appears containing different modifying options. Click **Format** to see the rest of the options.

Creating new styles

Choose the cell, click on **Cell Styles**. Pick **New Cell Style**.

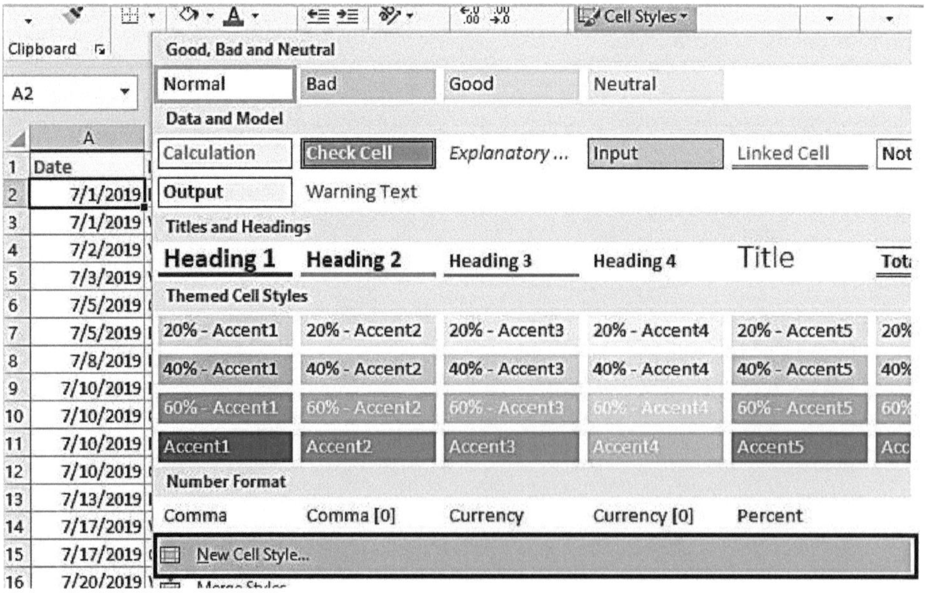

Enter style name. Select Ok. The style will be saved.

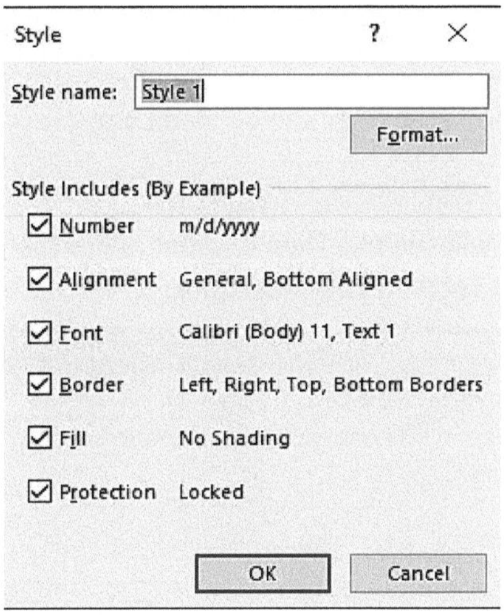

Merging styles from other workbooks

Most times, when people create styles, they like to use them in other workbooks. You can easily do that. First, launch the workbooks that have cell styles. Also, launch the other workbook that you will like to merge the style with.

Click on Cell Styles and click on Merge Styles. This opens up the Merge Style dialog box. Choose the workbook and click **Ok**.

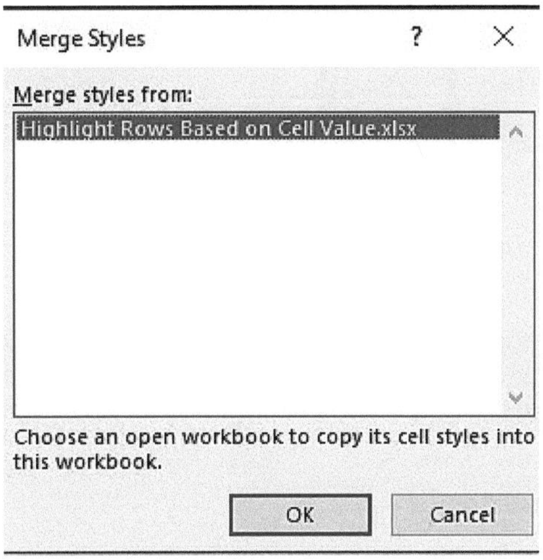

Controlling styles with templates

In your Excel Start folder, save the workbook as a template. After that, click File, click New to select a template for the new workbook. Template files may also hold additional named styles, which is a great method to keep your workbooks looking consistent.

UNDERSTANDING DOCUMENT THEMES

Themes are built-in and customizable functionalities in Excel that allow you to customize the look of your worksheet. In Excel, there are numerous themes, each of which has 12 colors and two fonts (heading and body). The themes make it simple to coordinate colors, fonts, and visual formatting so that they may be utilized and modified quickly.

A standard color theme may be chosen, a custom color theme can be created, theme fonts can be changed, a specific theme can be converted to a new theme, a custom theme can be saved for reuse, and numerous adjustments to the preset themes can be made using Excel themes.

Applying a theme

Click the **Page Layout** tab. You will see the Theme option. Click the down arrow on the Theme option. You will see different themes to apply to your workbook.

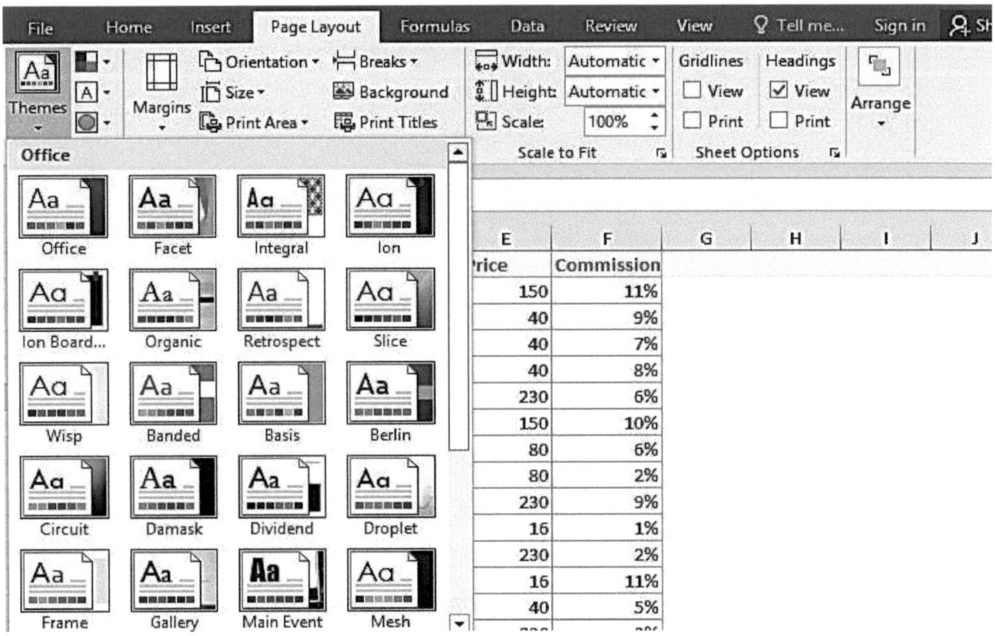

Customizing a theme

To customize a theme, click the down arrow on any of the color, fonts, or effects options and select **Customize**.

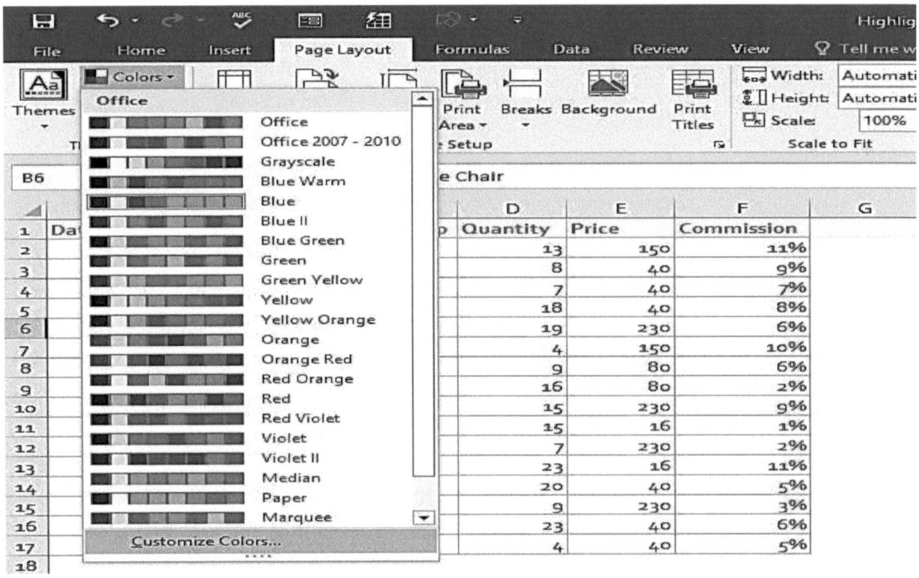

From the Create New window, make your theme using the options there. You can add text background, Accents, Hyperlink, etc.

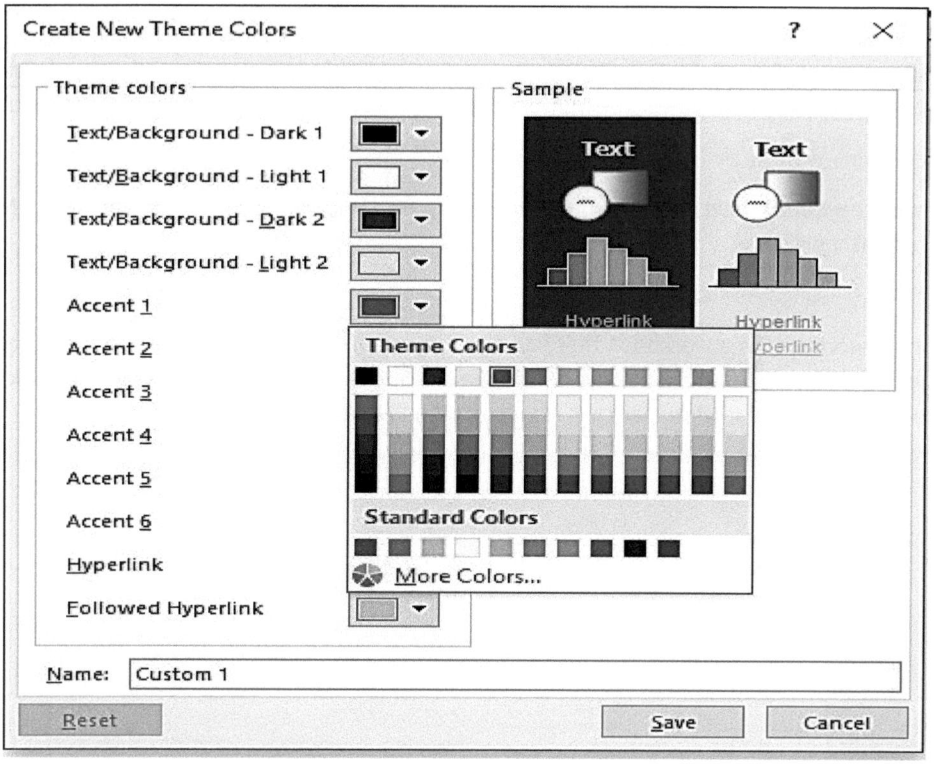

Once, you are done, click **Ok**.

CHAPTER SIX

UNDERSTANDING EXCEL FILES AND TEMPLATES

Users may build a new workbook from a blank document in Excel 2025. They also have the option of creating a new worksheet from an existing one. A new workbook comes with three worksheets by default. We may, however, adjust the number of worksheets in a workbook to meet our needs.

Creating a new workbook

To create a new worksheet, simply launch Excel on your computer, click on File and select New. Then, click on Blank workbook.

Opening an existing workbook

An existing workbook is a workbook that has been saved already and stored in the computer or cloud. You can open it from your drive or online. To open the file, click on File and select Open. Click **Browse**. A pop-up menu will appear where you will search for the file. Select the file and click **Open**.

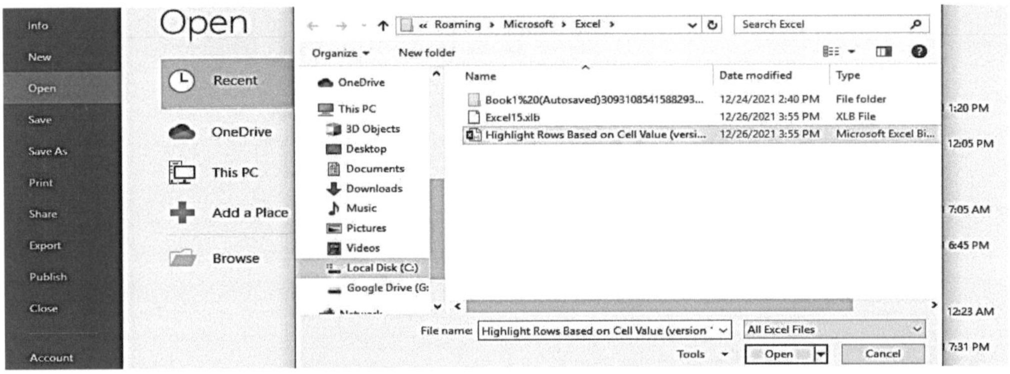

Choosing your file display preferences

Click on File, then select Browser View Options. Click on the **Show tab**, then on the down arrow and select **Sheets**.

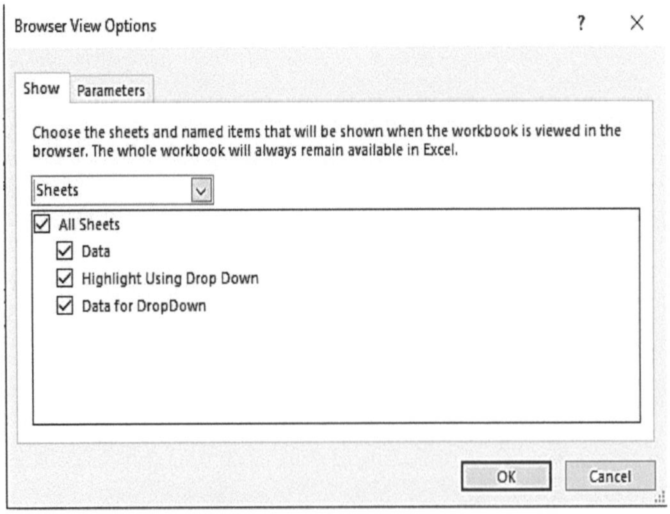

Pick the worksheets from the list, press **Ok**.

Saving a workbook

Click on File > Save. Select a location where you want to save the workbook. Type in the name for the workbook, then click Save.

Using AutoRecover

When Excel stops response or it shuts down, AutoRecover automatically recovers the last saved work and will open the file. It will give you two options to choose from which are: to maintain the changes in the documents or dismiss them if we have already saved the data. This helps a lot because it prevents us from losing important data.

It only retrieves files that have been saved. The workbook must have been saved in the system or computer at least once.

Enabling AutoRecover in Excel

Select File, select Option. On the options menu, click on Save. AutoRecovery options are on the right.

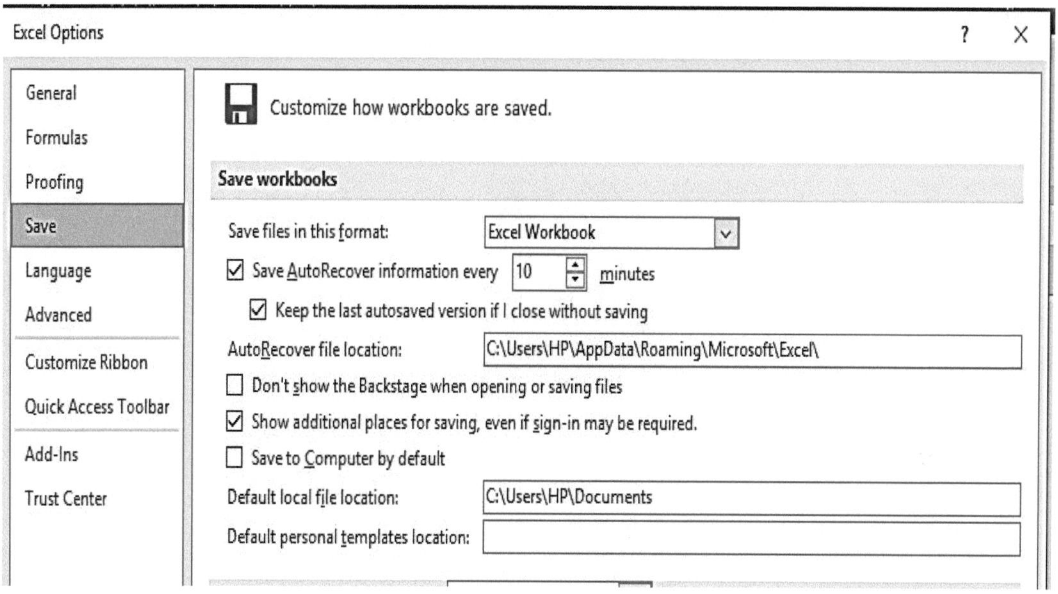

By default, it is already enabled. If yours is not, enable It and click Ok.

Recovering versions of the current workbook

So first of all, launch the workbook which you want to recover. Select **File** > **Info** > **Manage Versions.** On it, you will find the automatically saved versions with the time beside it.

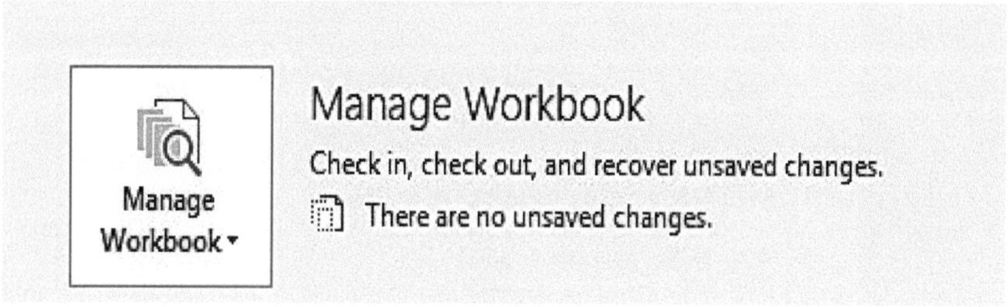

Recovering unsaved work

Click **File** > **Open** > Recent Workbooks. Scroll down, then click Recover Unsaved Workbooks.

Click on the workbook for recovering.

Configuring Auto recover

By default, your files are backed up by Excel every ten minutes. You can decide to change these settings. simply click on File > Options > Save > Save Workbooks.

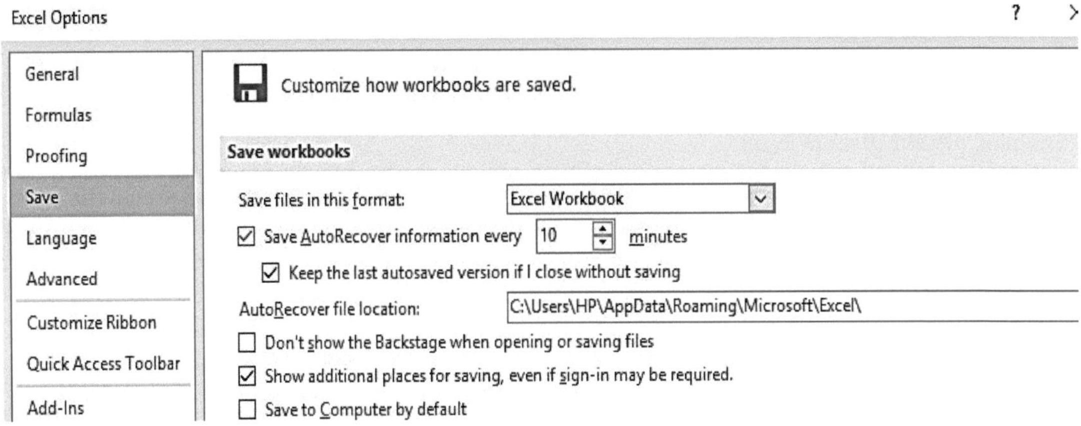

Password-Protecting a Workbook

Excel, however, provides three distinct levels of data protection. This prevents data from being accessed or edited without a password. Follow the steps below to protect your workbook.

Click on File > Info. Click on Protect Workbook. Then select Encrypt with a password.

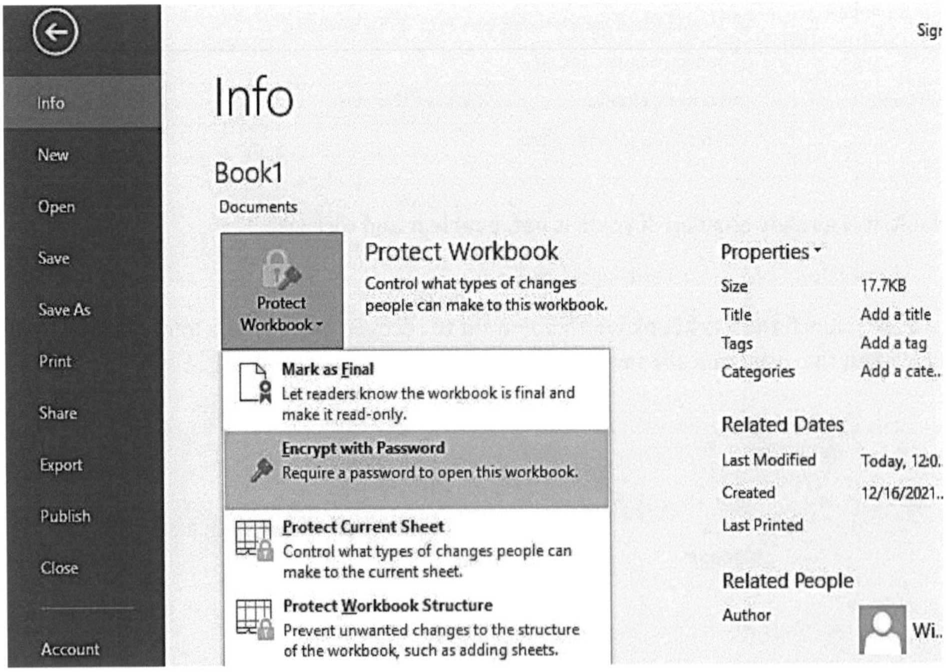

This will open up a password box. Enter in the password and click Ok. You will be asked to confirm the password again. Click Ok. Once you are done, you will see the Protect Workbook option as it has been highlighted stating that a password is needed for the workbook.

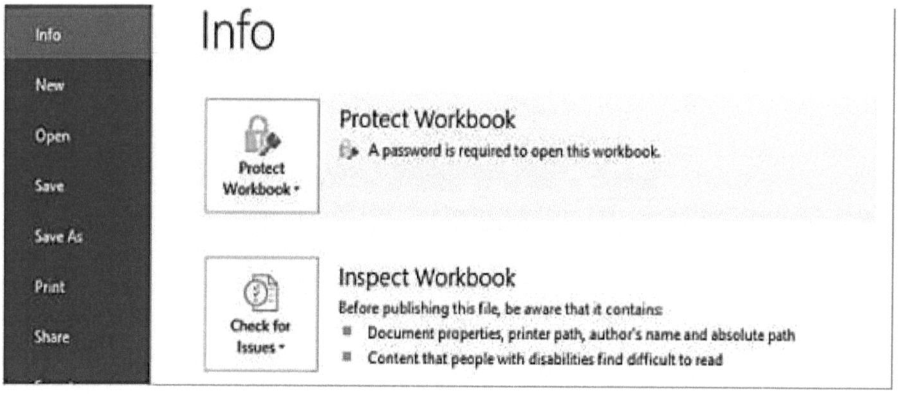

Organizing your files

Microsoft Excel is a spreadsheet-based tool that employs functions and formulae to help you manage figures and data. There are ways to discover a solution to assist you to organize your worksheets.

- **Templates**: You shouldn't spend time re-creating Worksheets from start every time you need to evaluate data. Instead, Microsoft Excel comes with a large number of user-created templates. Simply select one of the numerous templates available in Excel by going to File > New. Of course, selecting a template that exactly matches your data analysis method may be tough.

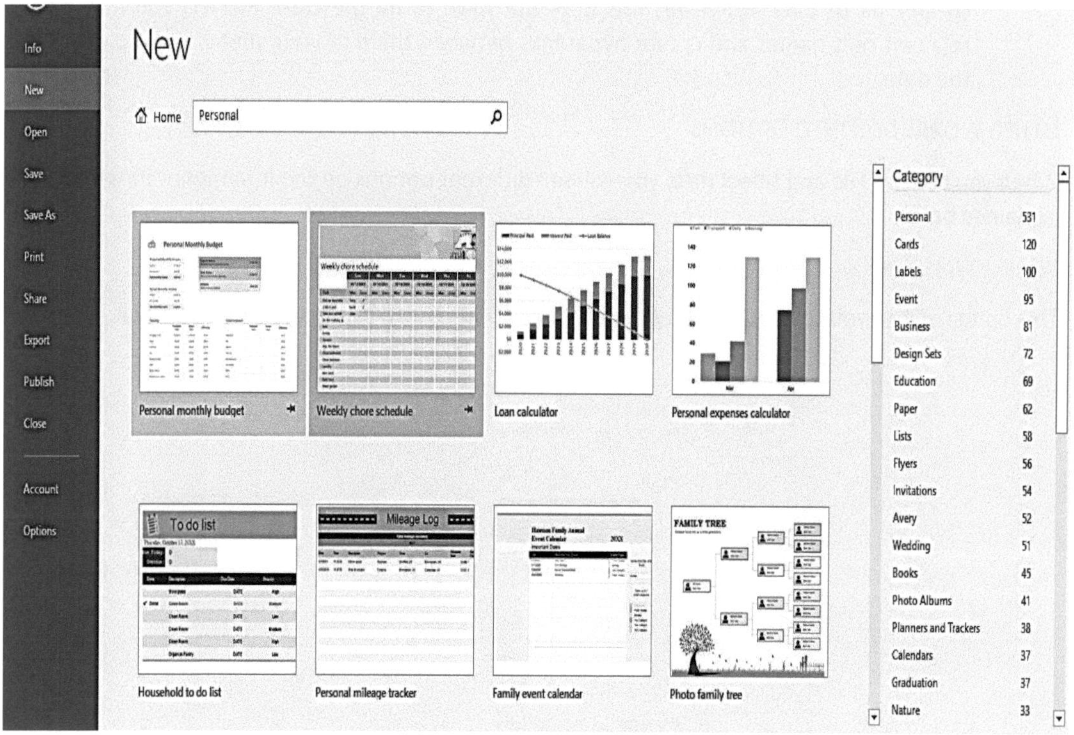

- **Multiple Sheets**: When dealing with anything complicated or with much data, use multiple sheets for your works. Divide your data into different worksheets and name each one appropriately. This will make it simple to locate the information you want.

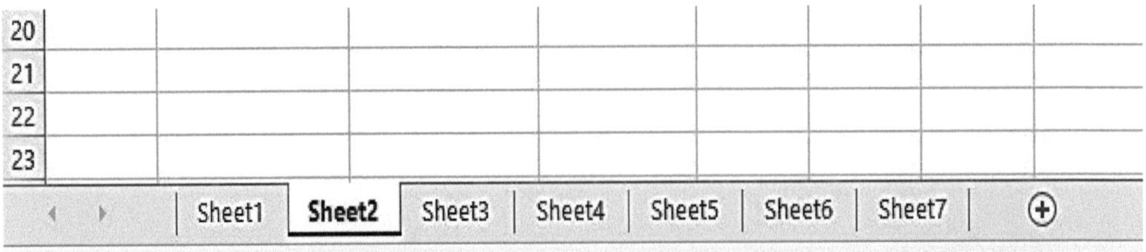

- **Highlighting the essential data**: You always should emphasize the most critical information while using any management application. When working with Excel, this is particularly true if you have a full team working on a single spreadsheet. One of the simplest ways to achieve this is to construct a dashboard sheet that summarizes your major data elements. Furthermore, you should always maintain your vital information in a legible font and use conditional formatting to provide heights and colors to important cells.
- **Sorting Data**: Knowing how to categorize/sort your data is an important element of data analysis. It makes no difference whether you wish to order the names alphabetically or list the goods from cheapest to most expensive. Sorting your data can aid you in better understanding and visualization of your data. You may arrange your data alphabetically, numerically, by built column list, formatting, an icon set, or by time and date in Excel.
- **Hyperlink cells**: A times, you will often spend time digging through all of the worksheets for a specific bit of data. However, this does not have to be the case. Rather, you may give your relevant cells names and create hyperlinks between them in your sheets to help you navigate the data.

OTHER WORKBOOK INFO OPTIONS

When you click on File and select Info, you will see different options on the Info menu. those options are explained below.

Protect Worksheet options

This option allows you to control the type of changes that people can make to a workbook.

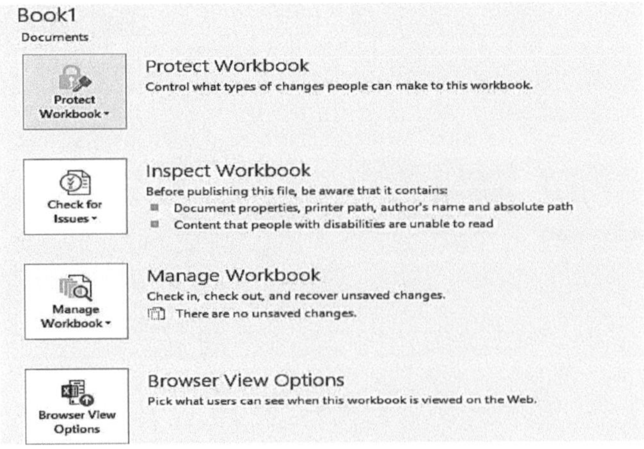

Check for issues options

This is to inspect your workbook, check accessibility, and check compatibility.

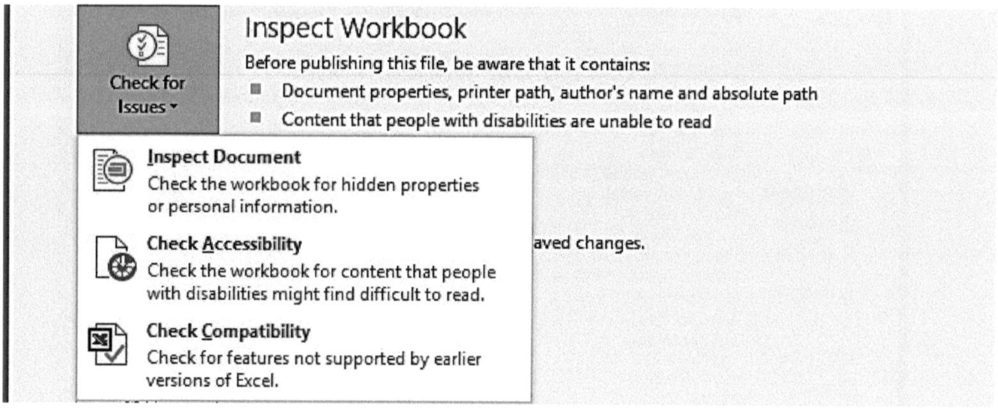

Manage workbook option

This is to check in, check out, and recover unsaved changes on your workbook.

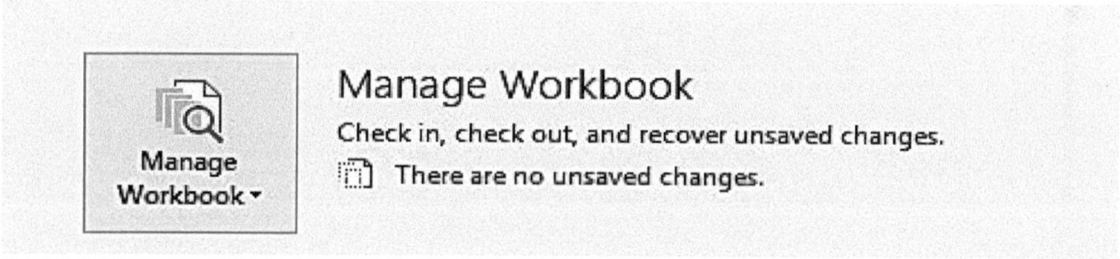

Browser view options

This determines what users can see on your workbook when viewed on the internet.

Compatibility mode section

This helps you to create documents that can be viewed by everyone. Since Excel has been there for years, it has a variety of versions. As a result, documents created in later versions might not be compatible with older versions. It is dependent on the content of a document, which is a difficult factor to consider. For instance, a feature developed in Excel 2021 will almost certainly not work with Excel 2013. The majority of Excel users, however, are unaware of this.

Microsoft implemented Compatibility Mode to Excel to address the problem. Your workbook in compatibility mode will be readable in former versions of Excel. Compatibility Mode guarantees that workbooks created in recent editions may be viewed even if you're using an earlier version. Some documents may show improperly or not open at all if you don't use Compatibility Mode. When software upgrades affect the core of a program over time, this is typical.

To save in compatibility mode, simply open the workbook, click File > Save As >. Then, click on the down arrow on the **Save As Type** box. Select the version you want.

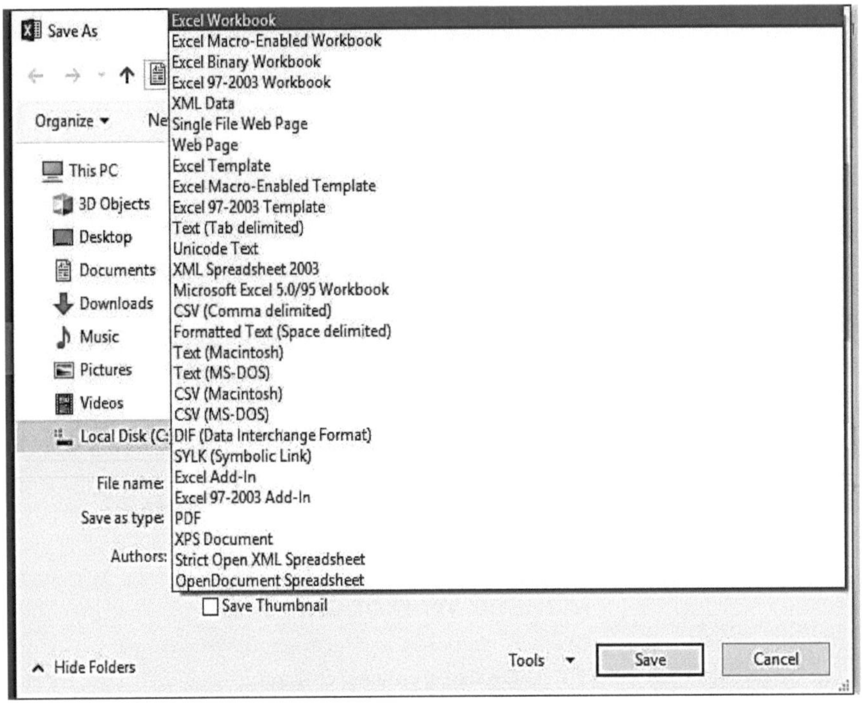

Closing Workbooks

When a worksheet is open, and you want to close it but not the Excel program, simply click on **File** > **Close**. Use **Control key + W** also. To close every workbook that is open when you have more than one workbook open, simply press down the Shift key as you select Close from the File menu.

Safeguarding your work

Nothing is more frustrating than working for hours on a complex Excel spreadsheet only to have it destroyed due to a power outage, a hard drive crash, or even a human mistake. Protecting oneself against these tragedies, however, is not difficult.

We examined the AutoRecover option earlier in this chapter, which causes Excel to store a backup copy of your workbook at regular intervals. It's a fantastic concept, but it's far from the sole safeguard you should use. If a file is critical, more precautions must be taken to protect its security. Your work may be safeguarded using the following backup options:

- Save copies of the documents on the same disk as a backup. As you are saving your work, pick the **Always Create a Backup** option, this is essentially what occurs. Although this option provides some security if you make a mess of the spreadsheet, it is useless if your whole hard drive collapses.
- Keep a duplicate on a separate disk for backup. This assumes that your computer has more than one hard disk. Because the chances of both hard drives failing are slim, this technique provides greater protection than the previous strategy.

- Keeping a backup copy on the internet (cloud) is a good idea. Assumes that your computer is linked to a server where you may upload and download files. This procedure is relatively risk-free. However, if the network server is in the same building as you, you are in danger if the whole building burns down or is destroyed in some other way.

Working with templates

Microsoft Excel templates are a valuable feature of the Excel experience and a time-saving tool. Once you've produced a template, all it takes is a few simple modifications to make it fit your current needs, and it can then be utilized in a variety of circumstances and reused over and over again. Excel templates may also assist you in producing consistent and appealing papers that will wow your coworkers or superiors while also making you appear your best.

Excel schedules, budget plans, receipts, inventory, and dashboards are all examples of commonly used document types that benefit from templates. What could be cooler than selecting a fully prepared worksheet with the design and feel you desire and that you can quickly customize to fit your needs?

Exploring Excel templates

To explore the Excel templates, click on File > New. You will see a search box. On it, you can search for templates on the web.

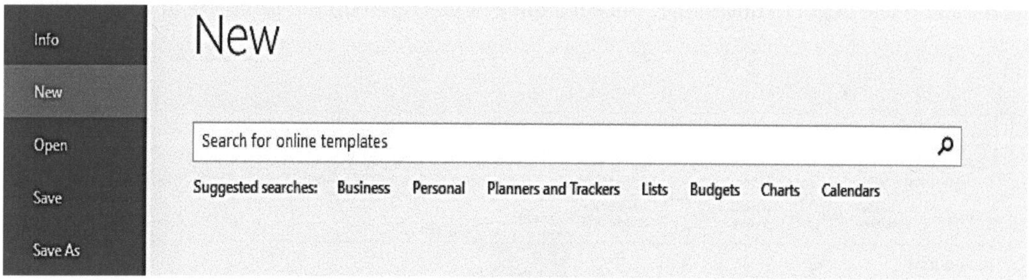

Viewing templates

Choose **File** > **New** to bring up the available templates screen in Backstage View, to see the Excel templates. You can get templates maybe from your drive or Microsoft Office Online. In the right panel, when you pick a template thumbnail, you will get a preview.

Creating a workbook from a template

Click on **File** > **New**. On the search box, type in the template you want to use or you click on any of the options below the search box which are the suggested searches. You must be connected to the internet.

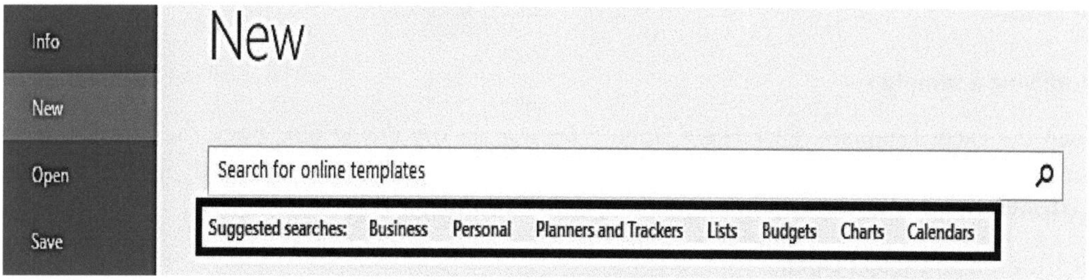

This will open a box, where you will see the information about the template. Click Create.

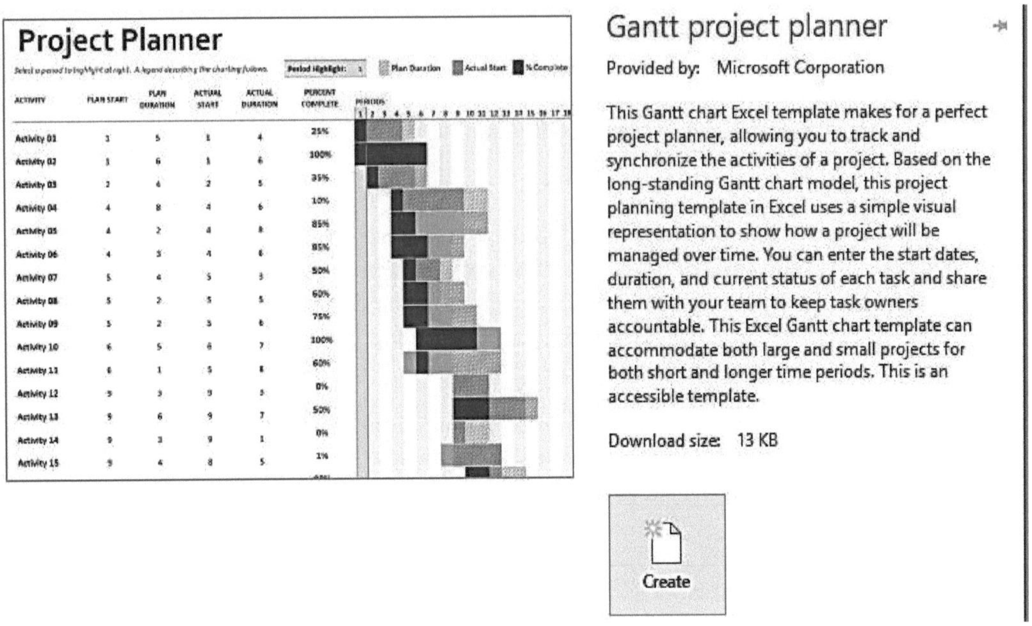

The template will begin to download. Once it's done, it will open up the template in your worksheet.

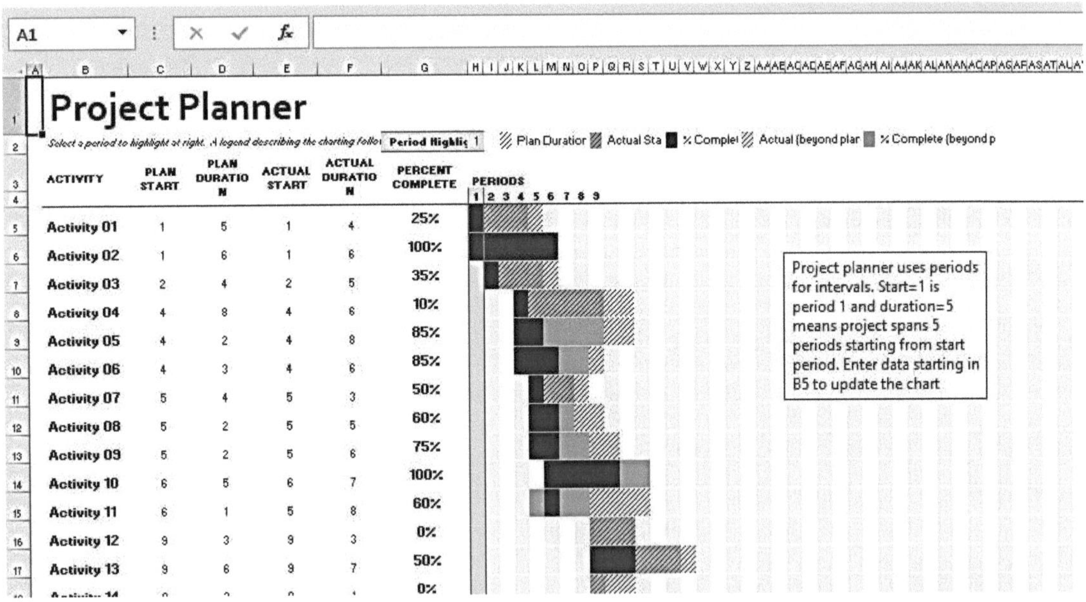

Modifying a template

Open the Excel Template. Click File > Open > Browse. In the dialog box, open the location where the template is saved. When saved in the default save location, put this path in the Address box **C:\Users\%Username%\Documents\Custom Office Template**. Press **Enter**.

Click the file and select **Open**.

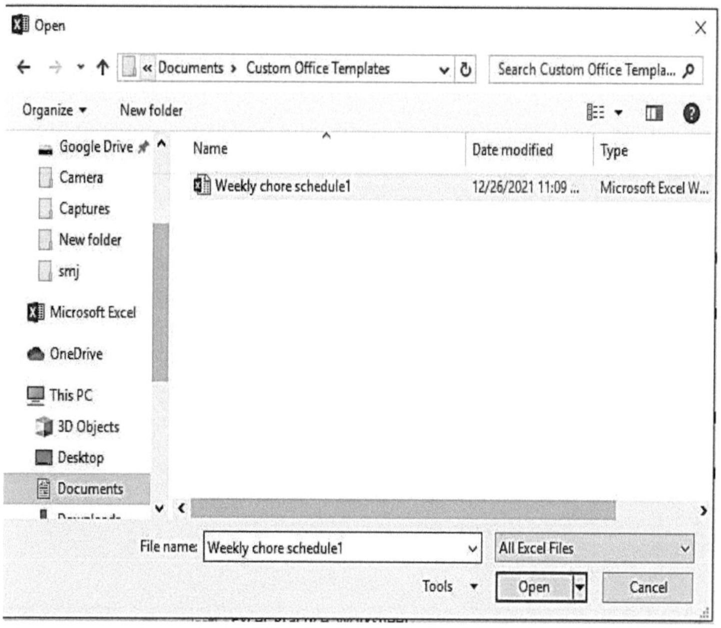

Creating a worksheet template

Create a blank workbook. Apply some data into it (if you want) then click on File > Save > Browse. Select a location where you want to save the file. On the Save as type box, click on the drop-down arrow and select Excel Template.

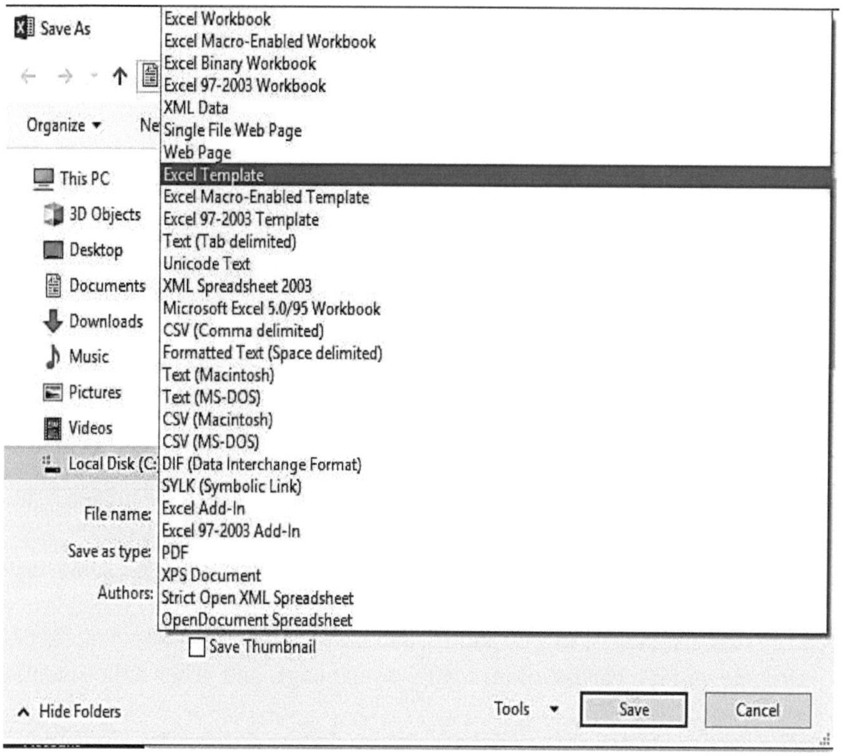

Click on Save to save the worksheet. By default, this file will be saved in the Template folder.

Editing your template

Click on File > Open > Browse. Click on Documents and select Custom Office Templates folder.

Select your Template, then click Open. apply the edits you want to your templates.

Resetting the default workbook

Click on File > Options. On the General tab, below When creating a new workbook option, choose the options best suitable for you. When you are done, click Ok.

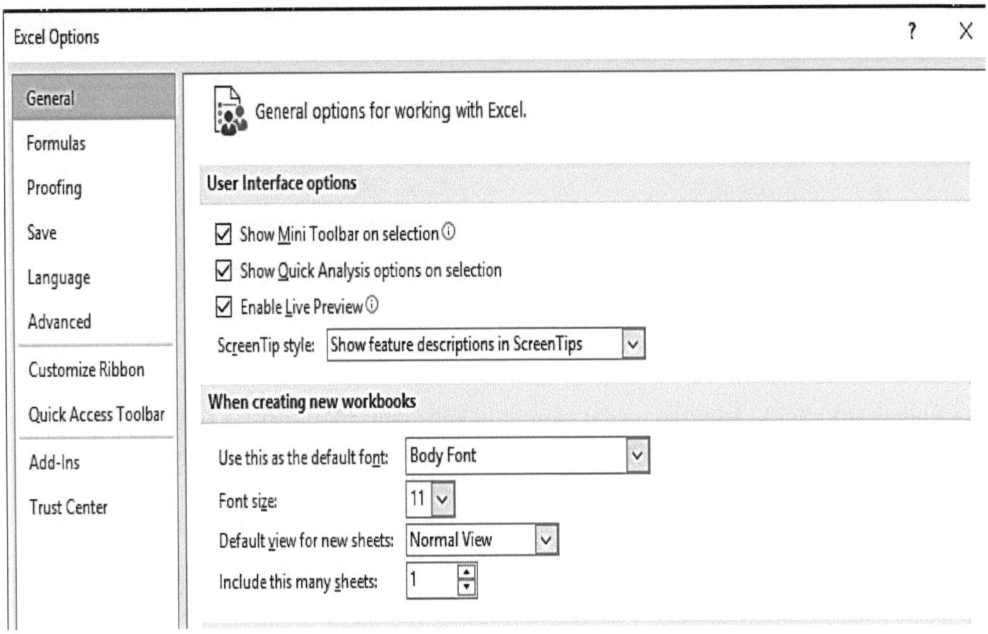

Using custom workbook templates

It's simple to create your own Excel templates. You begin by creating a workbook using the normal method, and the most difficult aspect is getting it to appear exactly how you want it to. Because whatever formatting, styles, text, and pictures you use in the workbook will appear in all subsequent workbooks based on this template, it's worth spending some time and effort in both the design and contents. You may save the following settings in an Excel template:

- The number of sheets and the kind of sheets
- Formats and styles for cells
- Each sheet's page layout and print regions
- To make particular sheets, rows, columns, or cells invisible, use hidden regions.
- Protected zones to keep some cells from changing.
- Text that should appear in all workbooks generated using the same template, such as column labels or page headers
- Formulas, hyperlinks, charts, photos, and other graphics are all examples of graphics.
- Drop-down lists, validation messages or warnings, and other data validation features are available in Excel.
- Options for calculations and window views, such as freezing the header row
- Custom forms using macros and ActiveX controls

Creating custom templates

After creating the workbook, click on File > Save as > select a location. In the dialog box, enter the name of the template. On the Save as type box, choose Excel Template **(*.xltx).** Select Excel Macro-Enabled Template **(*.xltm)** if the workbook consists of Macro. As you select these templates, the file extension in the **file name** field will change to the corresponding extension.

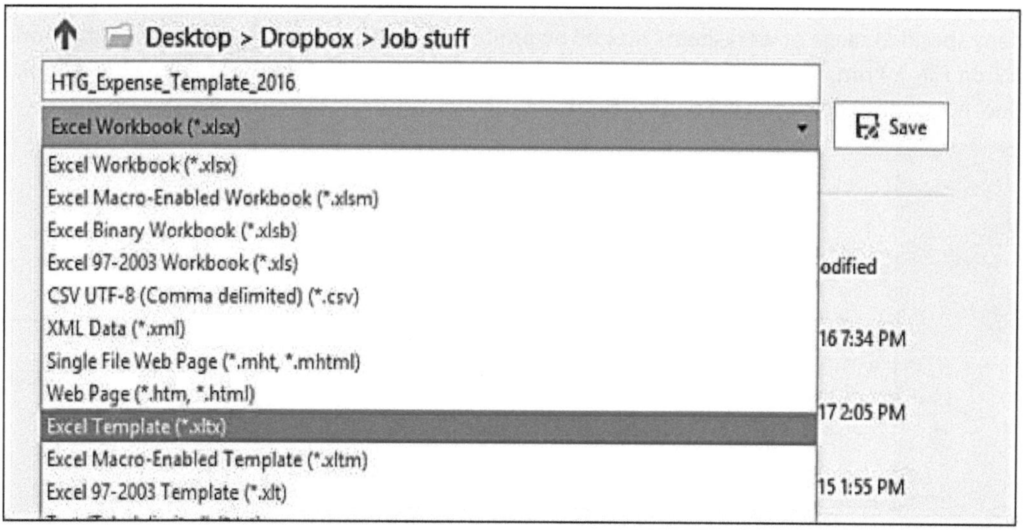

Saving your custom templates

When saving your work as an Excel Template, the location will be changed by Excel. It will be saved to the default templates folder:

(C:\Users\<User Name>\AppData\Roaming\Microsoft\Templates). So, if you want another location for the file, try to change the location after choosing Excel Template **(*.xltx)** as a document type. Even though you choose another location, the copy of the file will be stored in the default template folder. So, click on **Save** to save your custom templates.

CHAPTER SEVEN

PRINTING YOUR WORK

Doing Basic Printing

Print in Excel is for printing the data in the worksheet, but only to the width that the printer option allows for chosen and available pages. The present worksheet, active sheet, whole workbook, any chosen table, or any specified range of worksheets may all be printed. To print your work, simply open the worksheet, click on **File** > **Print**. On the **Print setting** menu, click the arrow and pick **Print Entire Workbook**. Then, select how many copies you want. Click **Print**.

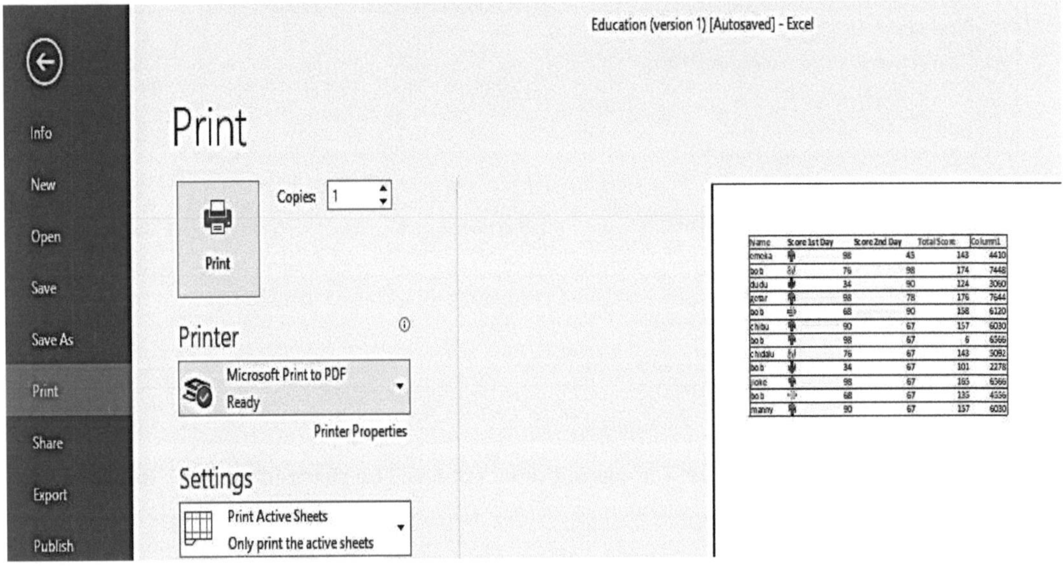

Changing your Page View

This is an excellent way of knowing how your worksheet appears when printed. On the page view, the functionalities of the normal view will be there but this time you will have a few more tools like header, footer, layouts, and more. This will help you complete your page perfectly.

Normal view

The standard view of your worksheet. How your worksheet looks when you've not made any view edit on it.

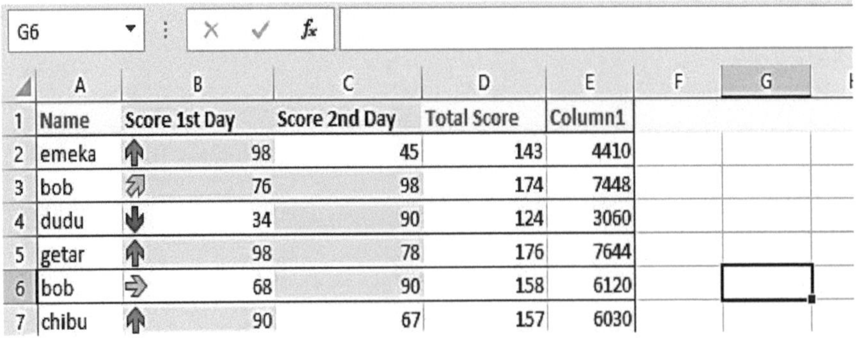

	A	B	C	D	E	F	G
1	Name	Score 1st Day	Score 2nd Day	Total Score	Column1		
2	emeka	98	45	143	4410		
3	bob	76	98	174	7448		
4	dudu	34	90	124	3060		
5	getar	98	78	176	7644		
6	bob	68	90	158	6120		
7	chibu	90	67	157	6030		

Page layout view

Select the worksheet that you want to change the view. Click the View tab on the ribbon, then select **Page Layout View**. Your workbook will be displayed in the page layout view as you can see in the image below.

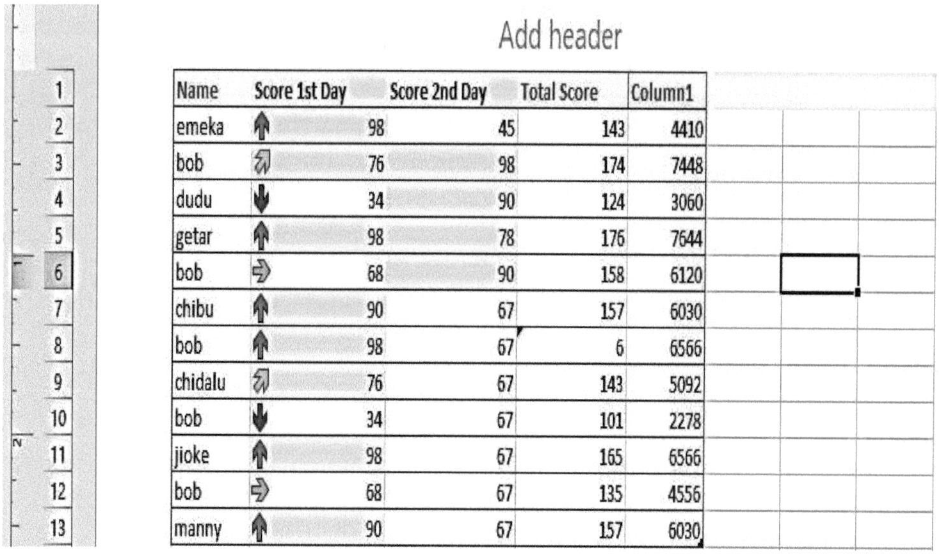

	Name	Score 1st Day	Score 2nd Day	Total Score	Column1
1	Name	Score 1st Day	Score 2nd Day	Total Score	Column1
2	emeka	98	45	143	4410
3	bob	76	98	174	7448
4	dudu	34	90	124	3060
5	getar	98	78	176	7644
6	bob	68	90	158	6120
7	chibu	90	67	157	6030
8	bob	98	67	6	6566
9	chidalu	76	67	143	5092
10	bob	34	67	101	2278
11	jioke	98	67	165	6566
12	bob	68	67	135	4556
13	manny	90	67	157	6030

Page break preview

Select the worksheet. Click View, then **Page Break View**.

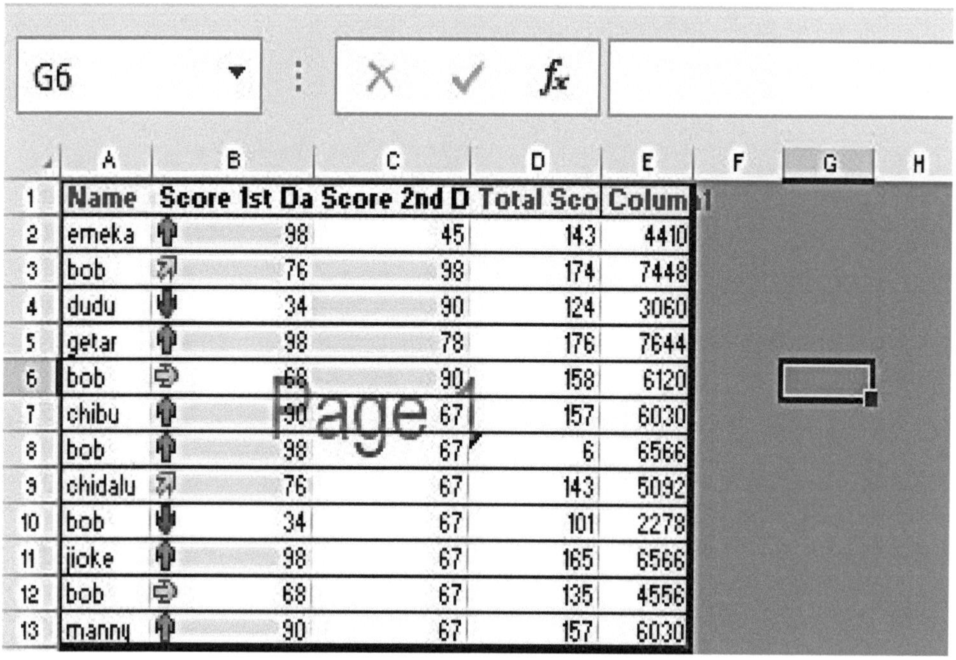

	A	B	C	D	E
1	Name	Score 1st Da	Score 2nd D	Total Sco	Colum
2	emeka	98	45	143	4410
3	bob	76	98	174	7448
4	dudu	34	90	124	3060
5	getar	98	78	176	7644
6	bob	68	90	158	6120
7	chibu	90	67	157	6030
8	bob	98	67	6	6566
9	chidalu	76	67	143	5092
10	bob	34	67	101	2278
11	jioke	98	67	165	6566
12	bob	68	67	135	4556
13	manny	90	67	157	6030

ADJUSTING COMMON PAGE SETUP SETTINGS

Choosing your printer

To print, your work, you need a printer. So, on the print page, click on the drop-down arrow below the Printer section. You will see a list of printers, select one from there. If you want to add a printer, click Add Printer.

Specifying what you want to print

When printing a workbook, you can decide what you want to print from that particular workbook. When you go to the print setting menu and click on the drop-down arrow, you will see a list of options in which you can select how you want to print your workbook.

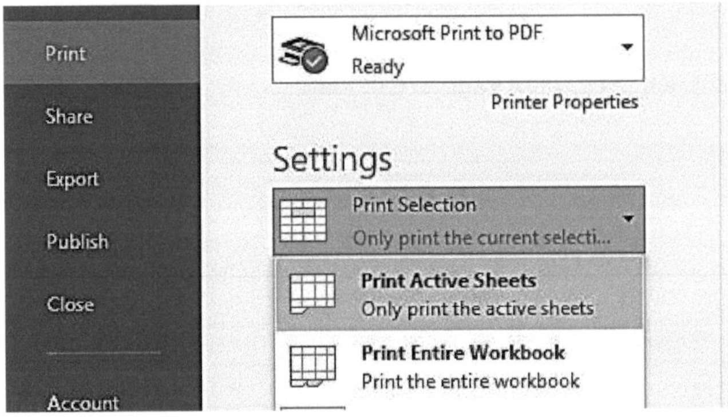

Below are the options you will see and what it means;

- **Print Entire Workbook**: To print the whole workbook. If you have many sheets, use this option to print them all.
- **Print Selection**: For printing just the current selection in your workbook. So, to print some portion of your worksheet, simply highlight the areas, then select this option.

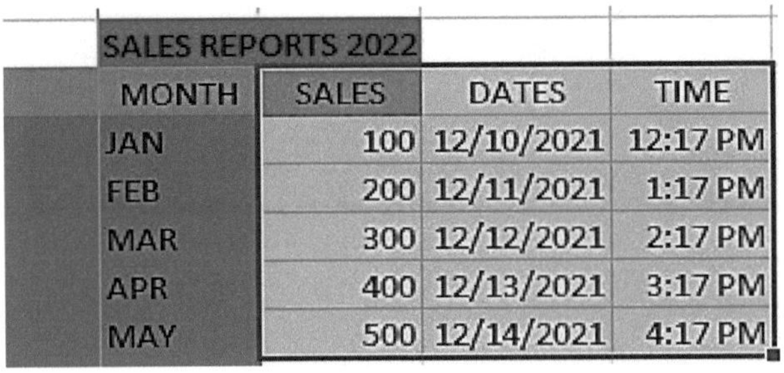

SALES REPORTS 2022			
MONTH	SALES	DATES	TIME
JAN	100	12/10/2021	12:17 PM
FEB	200	12/11/2021	1:17 PM
MAR	300	12/12/2021	2:17 PM
APR	400	12/13/2021	3:17 PM
MAY	500	12/14/2021	4:17 PM

- **Print Active Sheets**: To print just the active sheet i.e. the sheet that's being displayed on the screen. In the image below, the active sheet is Sheet 1. So that is what you will print.

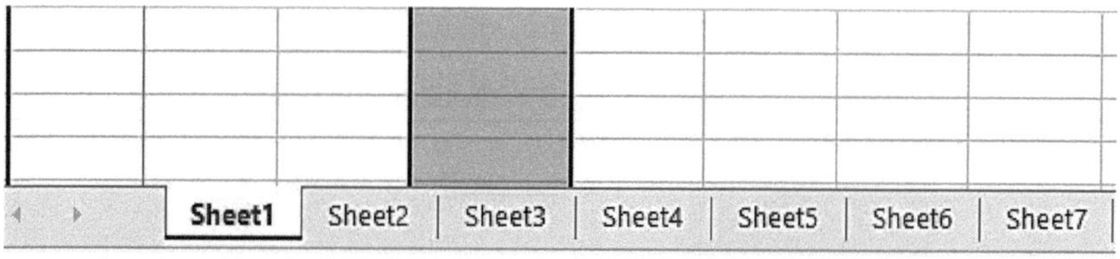

- **Print Selected Table**: This option is to print the selected table on your worksheet.

Changing page orientation

There are two types of page orientation in Excel which are Landscape orientation and Portrait orientation. The Portrait orientation is mostly used when you have more rows and lesser columns on your worksheet while Landscape orientation is used when you have more columns and lesser rows on the worksheet. You will find this page orientation on the Print setting menu.

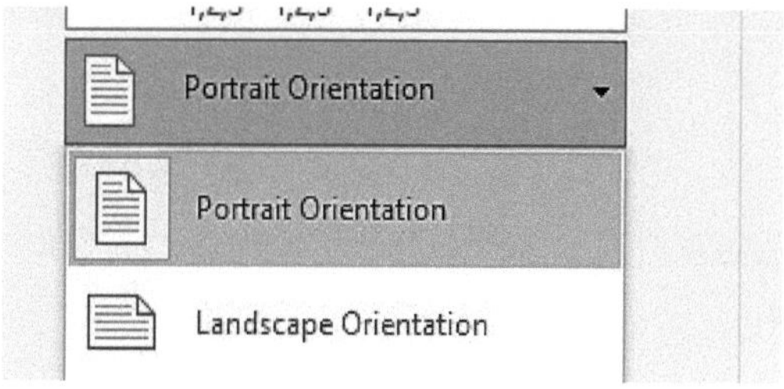

Specifying paper size

Click on Page Setup below the Print Setting option.

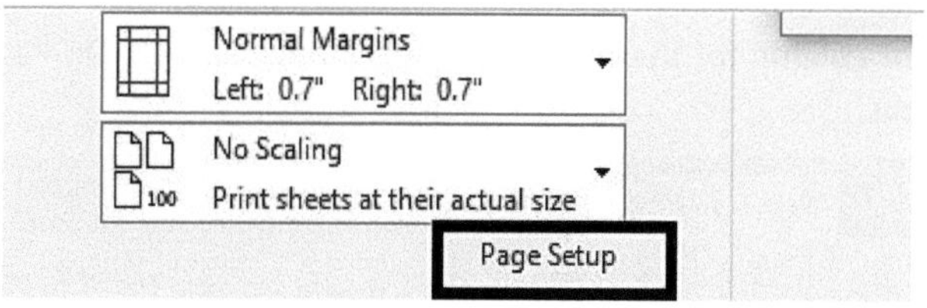

In the dialog box that opens, click on the drop-down arrow on the Paper Size option. Pick the paper size for your workbook.

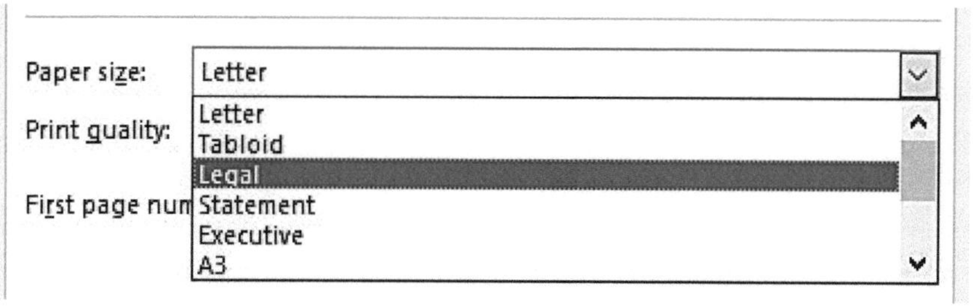

Printing multiple copies of your reports

If you want to print more copies, click on the arrow next to the Copies option. Type in the number of copies you want and then click **Print**.

Adjusting the page margins

to change the page margins of your worksheet, click on the arrow on the Normal Margin option. You will see different options for margins. Select the one you want to use.

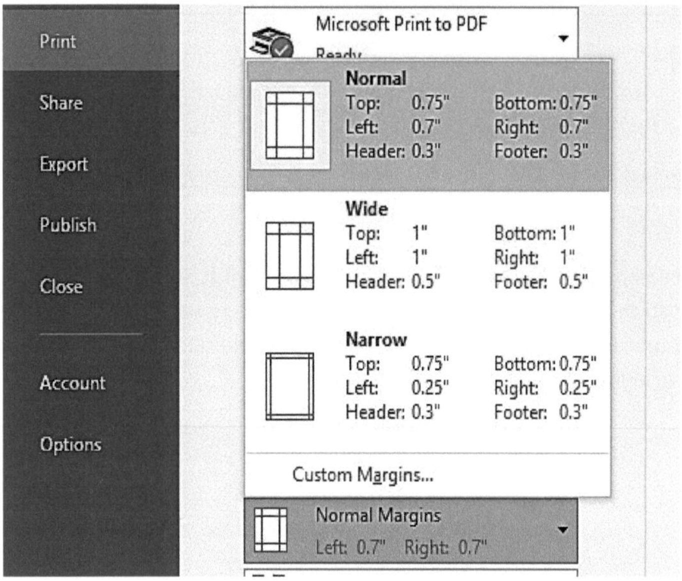

Understanding page breaks

As the name sounds, Page Breaks is used to break pages into different forms. In Excel, it is used to reduce the length of a page to minimize data misalignment when printing the work.

Inserting a page break

First, choose the row or column for the page break. Click Page Layout. On-Page Setup, click Breaks and select Insert Page breaks.

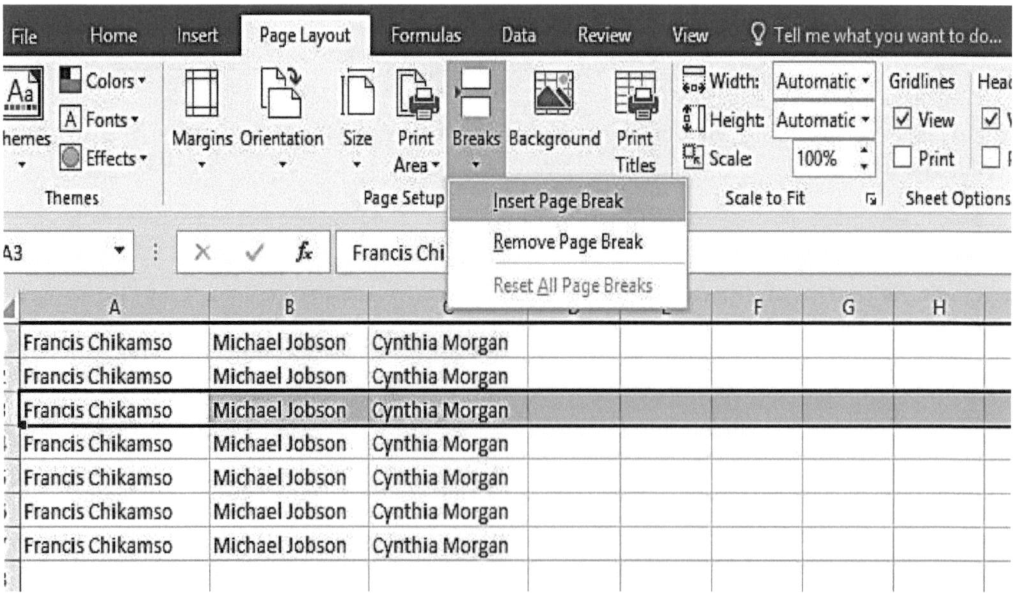

This action will display a thick line on the worksheet to let you know where the new page starts from.

1	Francis Chikamso	Michael Jobson	Cynthia Morgan	
2	Francis Chikamso	Michael Jobson	Cynthia Morgan	
3	Francis Chikamso	Michael Jobson	Cynthia Morgan	
4	Francis Chikamso	Michael Jobson	Cynthia Morgan	

Removing manual page breaks

To remove the page breaks, click on a cell that is below the cell you inserted a page break, click on Breaks, then select **Remove Breaks**.

Printing row and column titles

Click the Page Layout tab, select **Print Titles**.

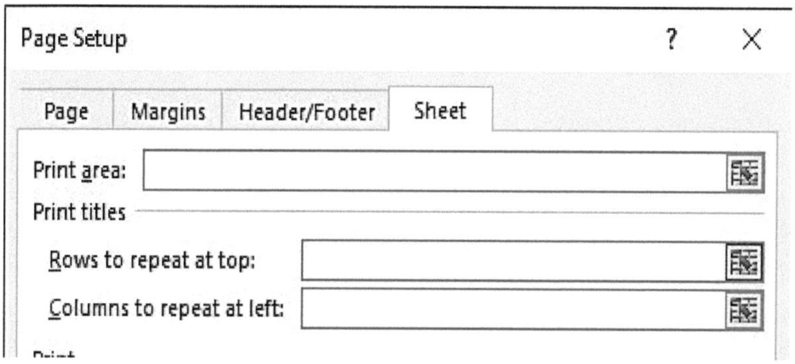

On the Rows to repeat at the top and Columns to repeat at the left box, choose the row or column you wish to print. After that, click Ok.

Scaling printed output

Scaling is used to determine how you want to print the sheets. Click on the Scaling arrow to see the list of options just as you can see in the image below. You can decide to print the actual size of the sheets, fit the sheets on one page, fit all columns on one page, or fit all rows on one page.

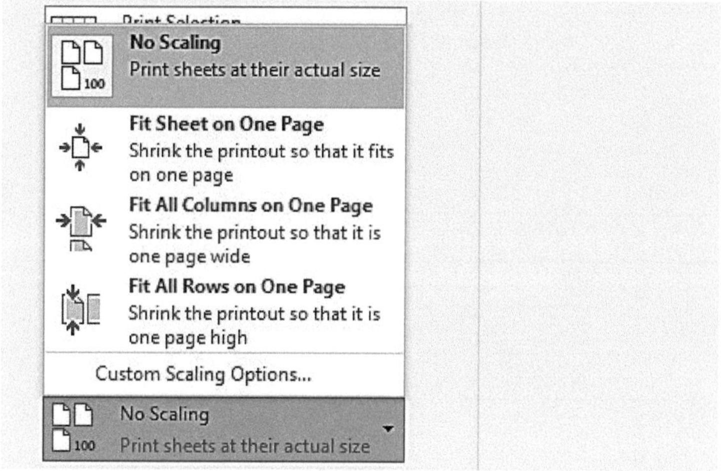

Printing cell gridlines

By default, the letters and numbers that function as row and column headings do not print. Neither do the gridlines on the worksheet. If you need to print them also while printing, then you can turn them on. So, on the ribbon, click Page Layout, then select Print Titles. On the Page Setup box, check the box next to Gridlines. If you also want to print the row and column headings, check the box close to the option. Once you are done, click Ok.

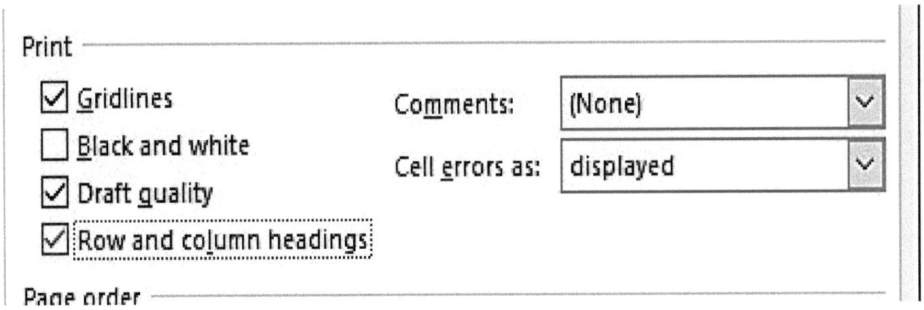

Using a background image

By default, when printing a worksheet, it only prints the cells which contain the data. If there is any other content on the worksheet, it won't be printed. However, you can add a background to your excel printouts. You can add a background by clicking on Page Layout, then, select Background. The issue with this step is that excel will not be able to print any backgrounds that have been added this way. So you can make use of shapes, images to work this out. So, let's insert a shape and use it as our background image.

Click **Insert** and on the illustrations group, select the **Shape icon** and click on the arrow. Select any shape. Here, I choose a shape from the rectangles shape group.

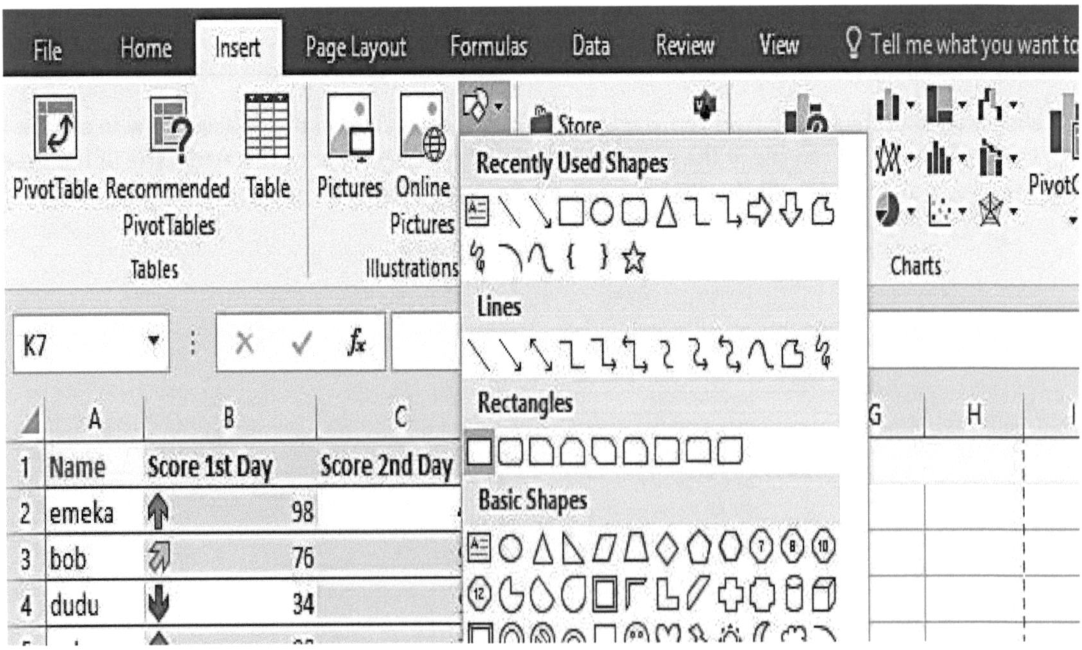

Then, draw how you want the shape to be on the worksheet. The Drawing Tool tab will now be active. Click Shape fill. Select White. Then, click Shape Outline. Then, select White as the outline color.

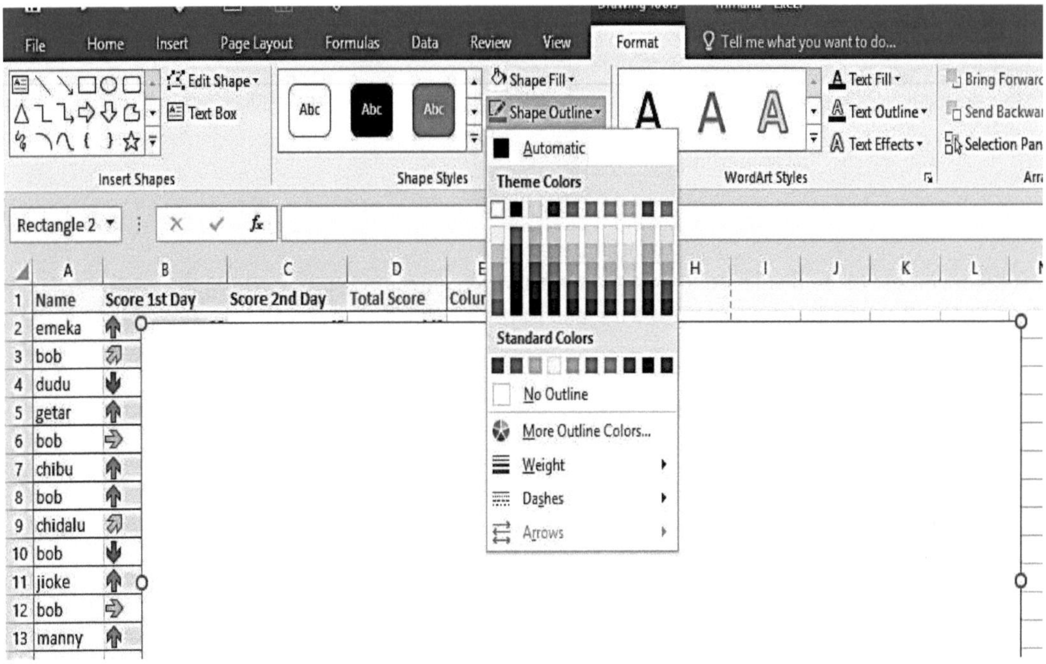

Now, on the shape, right-click. Chose Format Shape. This opens up the Format Picture pane

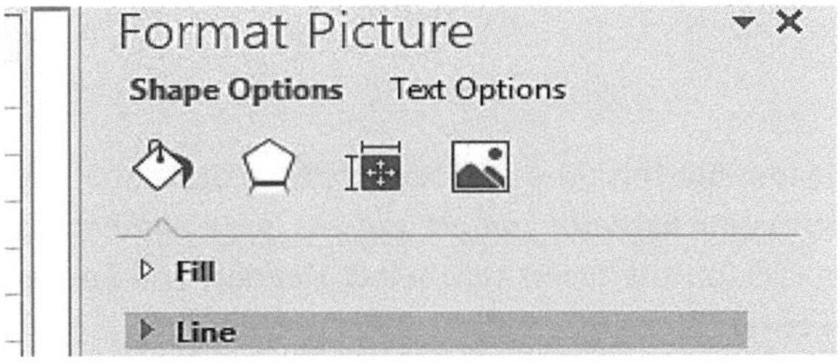

On the pane, click on Fill. Then, choose the Picture or texture fill option. Click on File. This opens a menu where you will select the picture you want to use. Click on the picture and click **Insert**. Check the box next to the Tile picture as a texture option. On the transparency slide, set it to 75%.

Now, click on the size and properties icon (the third icon on the Shape Options). Click on Properties, check the box next to Move and size with cells, and also the box next to Print Object. Then, click **Close**.

Now, click File > Print. You will see the preview of the page with the background image on it.

Adding a header or a footer to your reports

A document's header is a text or picture that appears at the top of every page. A header may make your content stand out. Including a header in your papers can save you time when it comes to creating templates since it will display on every page.

A well-crafted header may pique your audience's interest right away. If you want to be heard, you must first engage your reader. And having a terrific header in your reports and/or articles is one of the most effective methods to hook them.

Inserting a header

Open the Excel file. On the ribbon, click on the Insert Tab. Then, select Header and Footer from the Text Group.

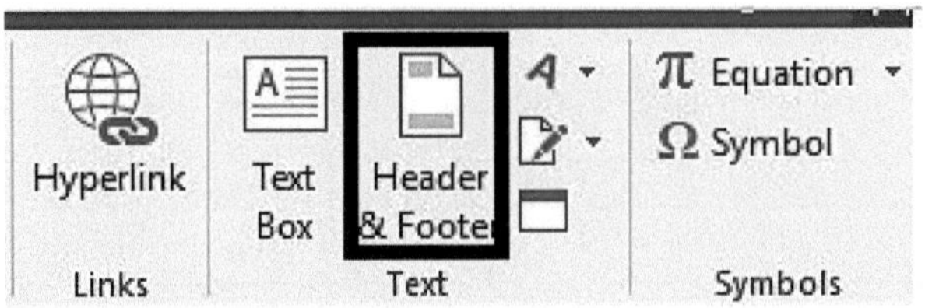

Your worksheet will be displayed in a **Page Layout View**. Now, on the box above the worksheet, enter in the header. By default, you write your header in the center box, but you can also write on the other boxes. Just click on them and write.

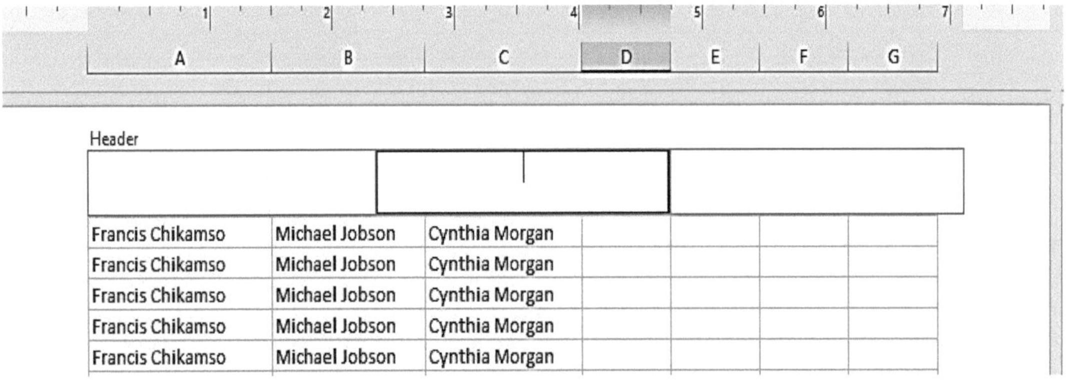

Inserting a footer

A footer is more like a header but this time it is located at the bottom of the page. It is as important as a header. You can add information like your contact address, page number, homepage link, and so on your footer. To insert footer, click on the Insert tab, select **Header, and Footer,** then go to the Navigation group and select **Go to Footer**.

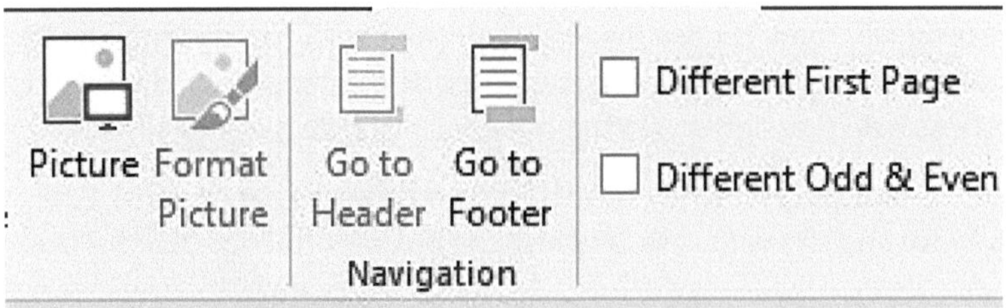

This will lead you to the footer below the page.

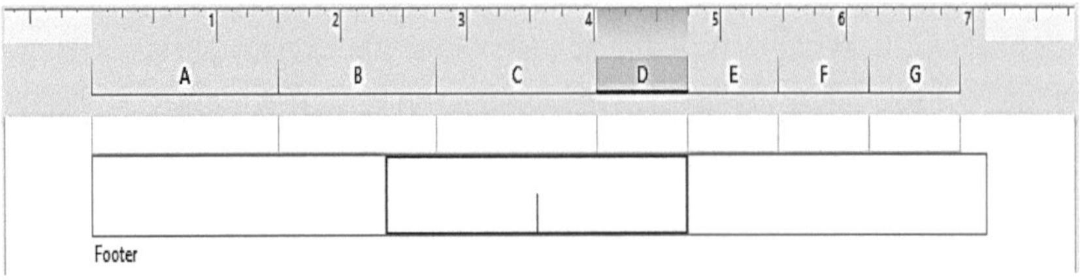

Understanding header and footer element codes

There are elements that you can apply to the header and footer of your page. Some of the elements are File path, Picture, Page Number, Current Date, File Name, Sheet Name, Current Time, and so on. To apply any of them, click on the icon of the elements you want to use.

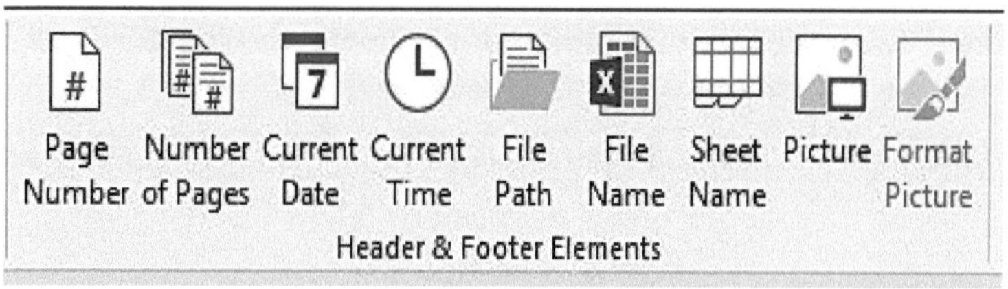

Let's say you want to insert the current date on the header, just click on **Current Date**, you will see this **&[Date]** written on the header.

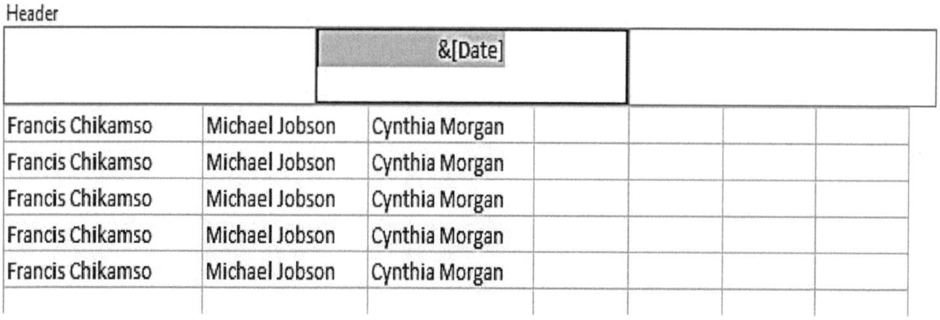

Click on any area on the worksheet to see the current date written there.

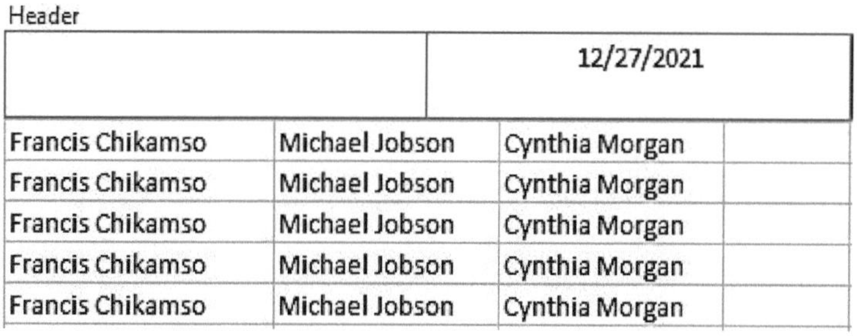

Exploring other header and footer options

There are also other options you can make use of. On the Design tab, hit the down arrow on the Header and Footer icons to see the list of headers or footer options.

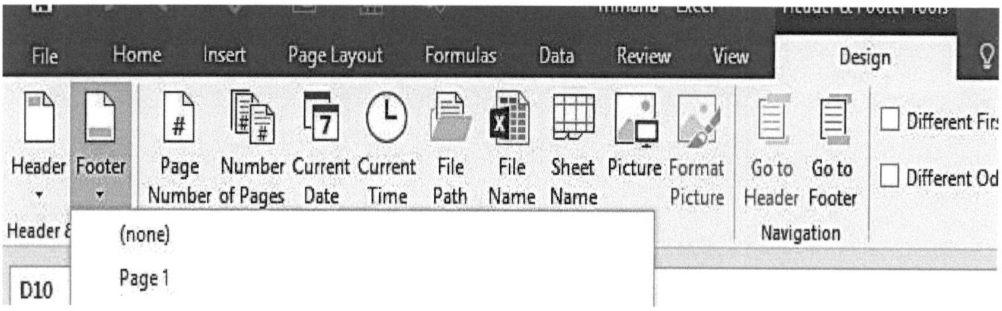

EXPLORING OTHER PRINT-RELATED TOPICS

Copying page setup settings across sheets

Say you've got a workbook containing many worksheets, but you wish to make an identical page setup based on a sheet. If you simply have one worksheet, you may quickly set the page setting for it. Also, if you wish to create the same page configuration for additional worksheets in your workbook, doing so one at a time is not a smart idea. Follow the steps below to do so;

Click on the worksheet tab in which you want to copy the page setup. Right-click and click on Select **All Sheets**.

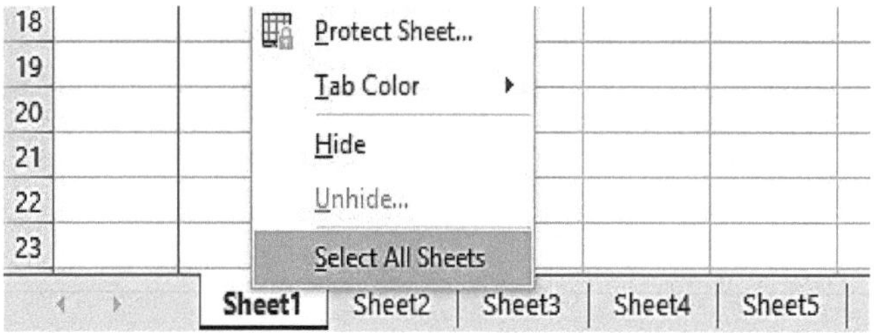

Click the Page Layout tab. On the Page Setup group, click on the small arrow located at the bottom right of the group menu.

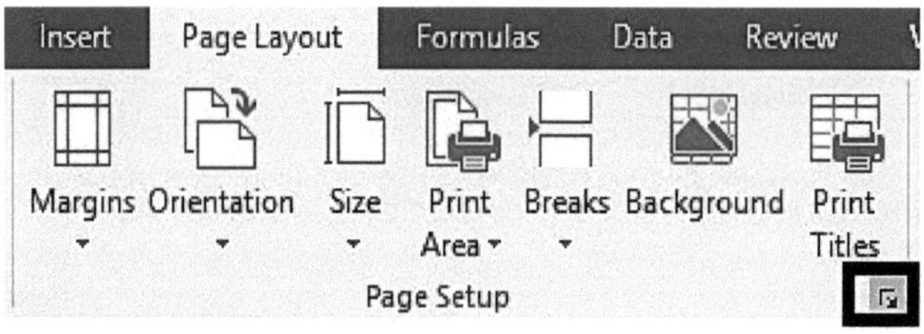

This will open up the Page Setup Dialog box. Then, click Ok. This will copy the page setup of the active worksheet to the selected worksheet.

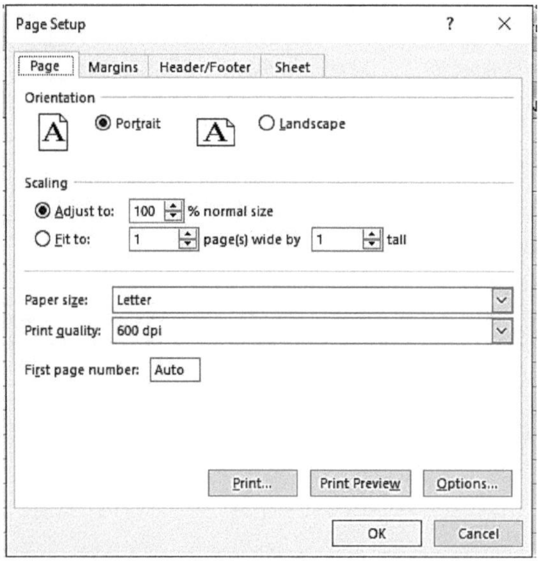

To ungroup the worksheet, right-click on a tab, then, click Ungroup Sheets.

Preventing certain cells from being printed

First, click on the cells. Right-click and select Format Cells. You will see the Format Cell dialog box, then click on the Number tab and select Custom. In the type option box, type in this "";"";"";"". Then, click Ok. This will hide the contents of that cells.

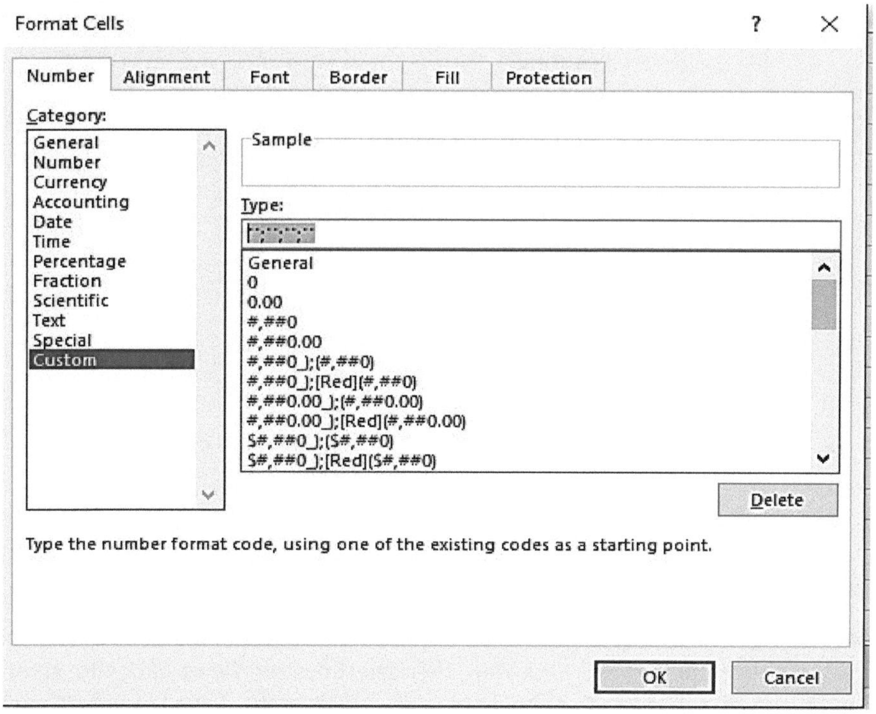

Preventing objects from being printed

You may have objects like boxes, pictures, shapes, etc. on your workbook but don't want to print them alongside other data.

To prevent them, first, click on Page Layout and select **Print Titles**.

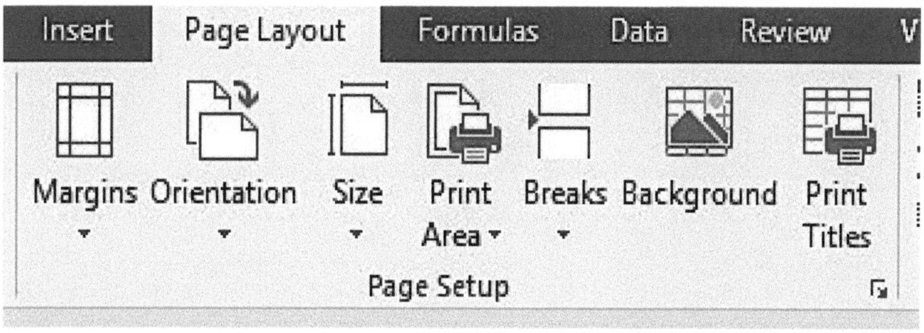

Click on the Sheet tab and check the box on the Draft quality option. Click on **Print Preview** to see if the object is on the preview. It will not be available. Then, click Ok.

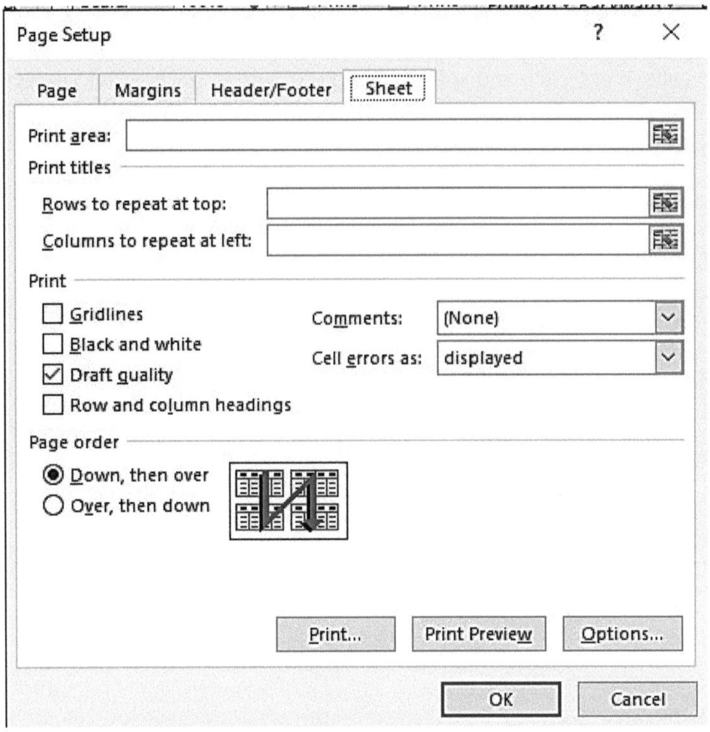

Creating custom views of your worksheet

The custom view helps in making your presentations smooth. To create custom views, simply choose the area on your worksheet for the view. Click View, then select Custom Views. Click Add, enter the name for the custom view. Then, select Ok. To navigate to a view, click on the down arrow on Custom View, then pick a view.

Creating PDF files.

Click on **File** > **Print**. On the Printer option, click the down arrow and select Microsoft Print to PDF. Then, click Print. A menu will appear where you will choose the location for the file and the file name.

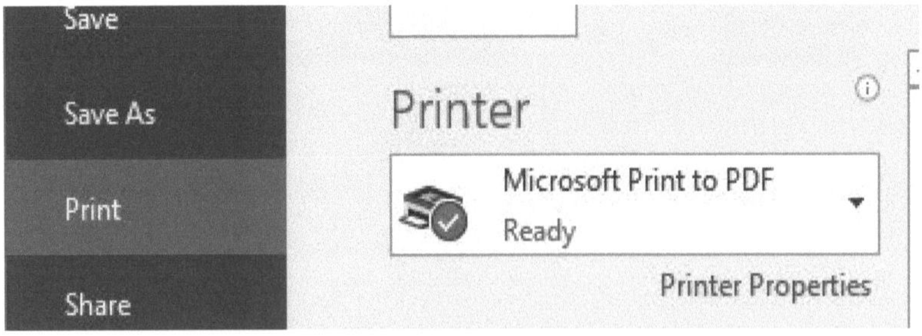

Excel print's Limitations

- When your printout is illegible, Excel does not notify you.
- Adjusting the margins from the Preview Pane is difficult.
- It will be difficult to discern cell borders when a user produces a worksheet or a table without gridlines.

CHAPTER EIGHT

CUSTOMIZING THE EXCEL USER INTERFACE

About the Quick Access Toolbar

The Quick Access Toolbar is located at the top left-hand side of the Excel window (above the ribbon) by default. You easily access some tools from there. It saves you time.

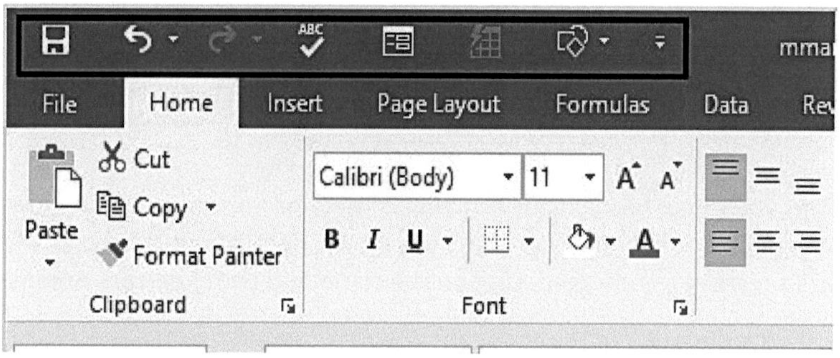

Customizing the Quick Access Toolbar

You can make some changes to the default settings of the Quick Access Toolbar. You can add more tools, delete, and even change the position.

Adding new commands to the Quick Access Toolbar

On the Quick Access Toolbar, click the down arrow to see the list of commands. Click on any to add it to the toolbar.

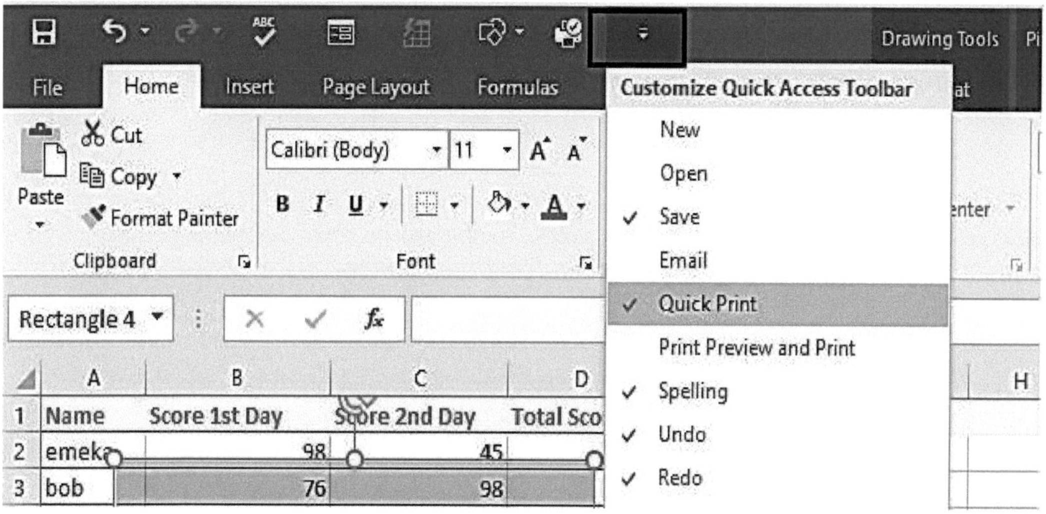

If the command you want to add to the toolbar is not on the list, click on More Commands. This will open up the Excel Options window. On the Quick Access Toolbar menu, you will be the Popular commands listed for you. You can also find other commands. Simply click on the arrow below the Choose Commands from option and select any of the options such as Commands Not in the Ribbon, All Commands, Macros, and others.

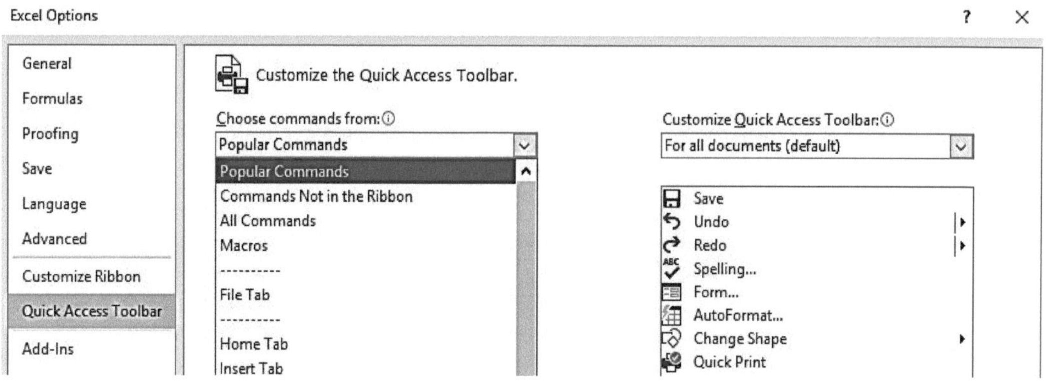

Any of the options you select has a list of commands that you can choose from. Choose the command, then click on **Add>>.** The command will be displayed on the right side of the menu. After you have clicked **Add>>,** click **Ok**. To remove a command, click on the command and then click **Remove**.

You can also rearrange the order of the commands. Simply click on a command, then click the up or down arrow at the right-hand side to move the command above (to the left) or below (to the right).

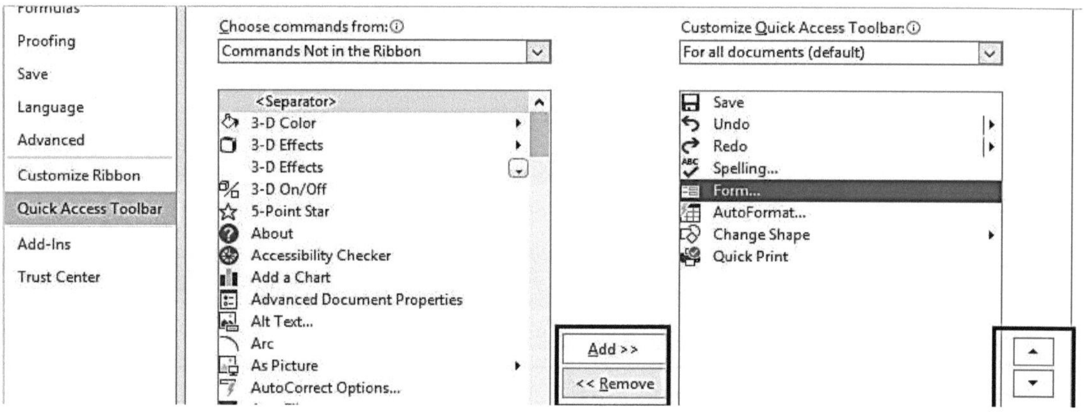

You can also change the position of the Quick Access Toolbar. Simply click on the drop-down arrow and click on Show Below the Ribbon. The Quick Access Toolbar will go down below the ribbon.

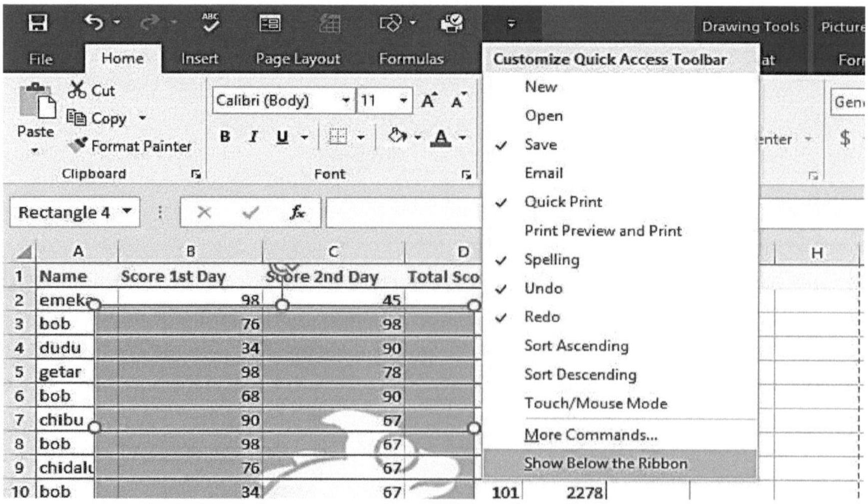

Other Quick Access Toolbar actions

You can reset the customizations you made on the Quick Access Toolbar. When you reset it, all the commands you added will be gone. To do this, right-click on the Quick Access Toolbar and select customize the ribbon. Below the menu, at the right-hand side, click on the drop-down arrow on the Reset option and select **Reset All Customizations**.

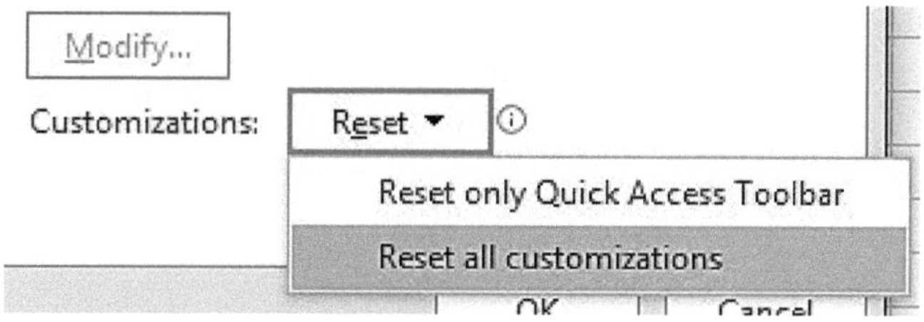

A dialog box will appear for you to confirm your action. Click Yes. Then, click Ok.

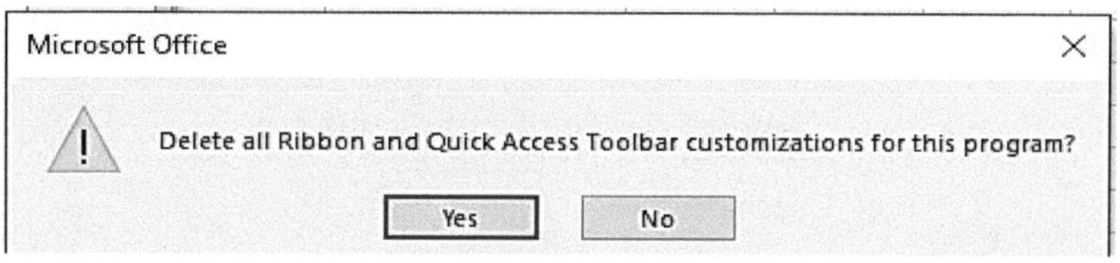

You can also Import or Export customization files. Simply click on the **Import/Export** option and select an option. When you click Import, it opens up a menu where you will choose the customization file you want to import.

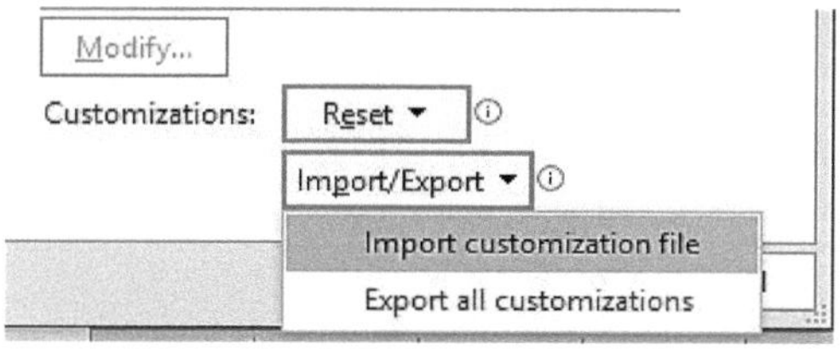

Customizing the Ribbon

This is where you get access to all the commands you are to use when working on your workbook. You can change the settings of the ribbon. You can hide the tabs or show them, add your tab or group to the ribbon, rename groups and rearrange them, and lots more.

Why you may want to customize the ribbon

You may like to have your tab or group on the ribbon. You will like to include those commands that you often use in the ribbon by adding them to a group and building a tab for them so that you can easily access them. You may also want to not display some tabs that you rarely use. Whatever the reason may be, I will guide you on how to do all these.

What can be customized

Yes, the ribbon is customizable but it is not everything on the ribbon that you can customize. Things you can customize in the ribbon are: Showing and Hiding tabs, renaming tabs, adding and removing groups from the ribbon, exporting and importing your customized ribbon, rearranging tabs, custom commands, and groups, and creating a new tab and group.

What cannot be customized

The color of the ribbon cannot be customized, however, customizing the color scheme of the whole Office is possible. Ribbon size, text size, and default icons are not customizable but it is possible to hide the ribbon or customize it to display just the names of the tabs. You cannot change the names, icons on the ribbon. You cannot remove the built-in commands.

How to customize the ribbon

There are many ways you can customize the ribbon. You can do so using the Customize the Ribbon window. You can create a new tab on the ribbon, create a new group, add commands to the ribbon, and also add commands to the new group you have created.

Creating a new tab

To create a new tab, simply right-click on the Quick Access Toolbar and select customize the ribbon. This opens up the Excel Options menu. on the Customize Ribbon option, on the bottom right-hand side of the window, click on **New Tab**.

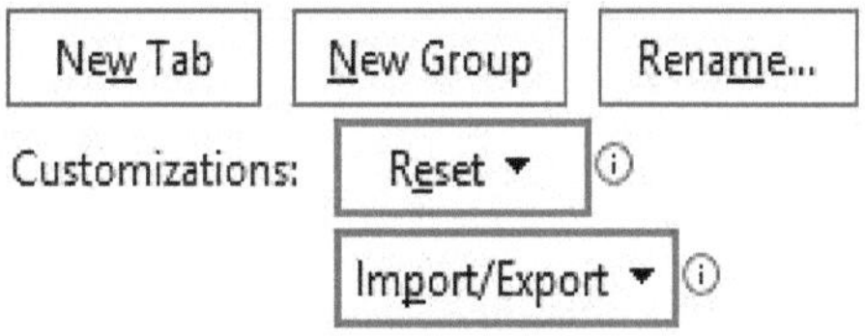

A New Custom tab and a Custom group will be added. The Custom group was added because you can only add commands to custom groups.

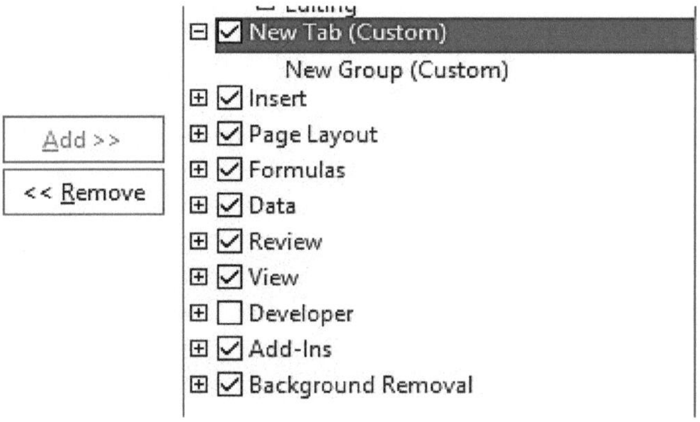

Click on the New tab you just created, then select **Rename** to give the tab a name of your choice.

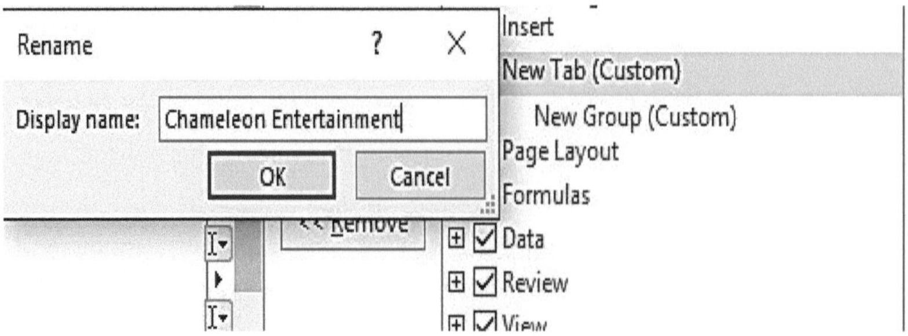

Then, click Ok. You will see the newly created tab on the ribbon.

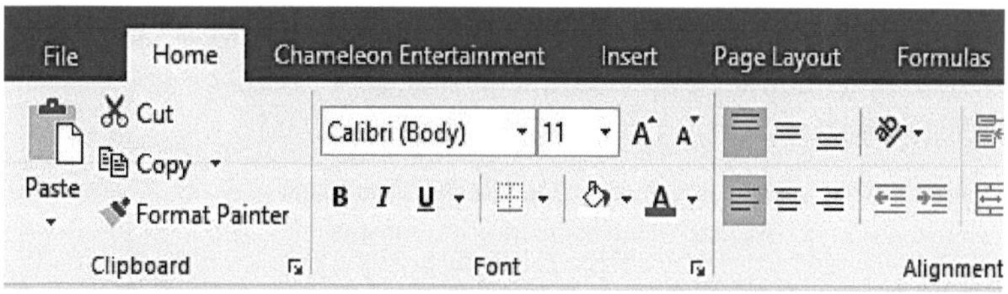

Creating a new group

As you can see in the image below, the new tab we created has no group in it. Let's create a group on it.

Right-click on the Quick Access Toolbar and select Customize the Ribbon. On the right side of the window, click on the tab for the group (Here, I clicked on Chameleon Entertainment). Click New Group. Click on the Custom Group and select Rename to give it a name and symbol to represent the group when the Excel window is in Restored mode (narrow). Click Ok.

After creating the group, you will first add some commands to the group.

Adding commands to a new group

From the **Choose command**, click on any of the commands you want to add. Click the Add>> button after selecting each command.

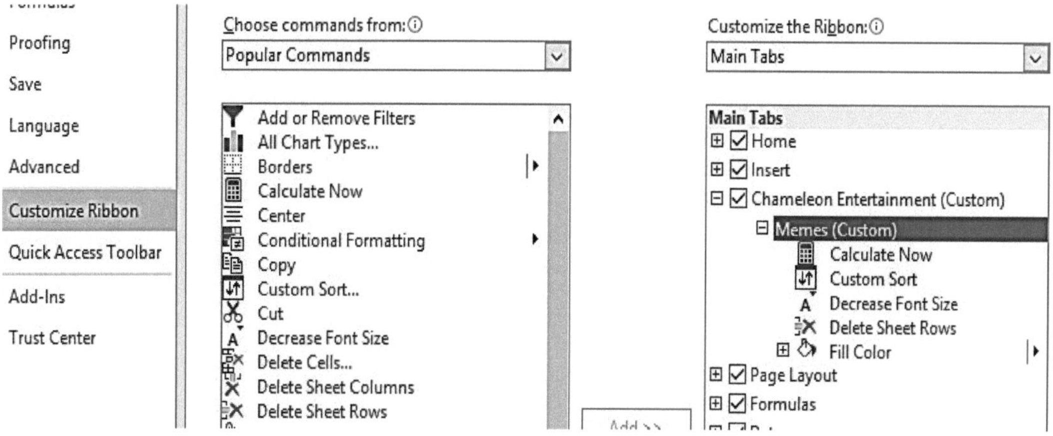

Click Ok when you are done. Now, you will see the new group and the commands in it on the tab.

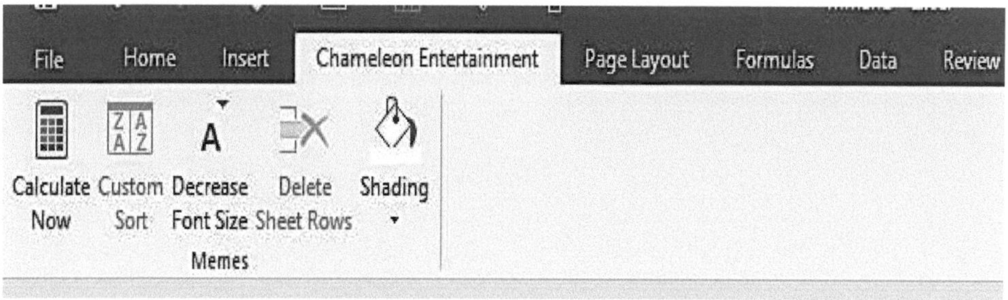

Resetting the ribbon.

After making some customizations, you can change back the ribbon to its default settings. Simply navigate to the Customize Ribbon Window. At the bottom left side, click the down arrow on the **Reset** option and select **Reset all Customizations.**

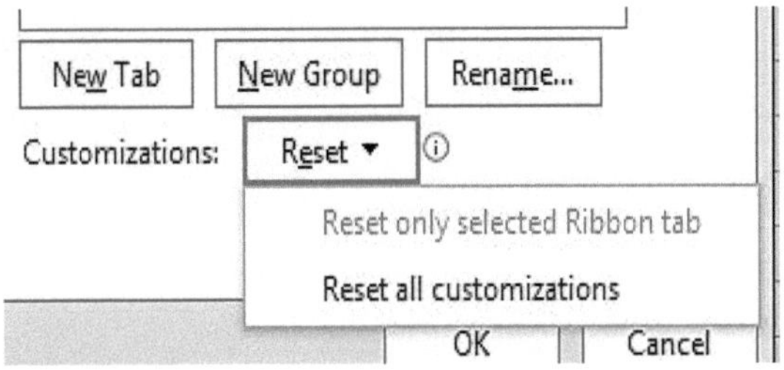

CHAPTER NINE

GETTING STARTED WITH EXCEL CHARTS

What's Chart

The chart is commonly referred to as a graph in Microsoft Excel. It's a visual depiction of data from a spreadsheet that could help you comprehend the information better than simply looking at the numbers.

A chart is a tool for graphically displaying data in different chart layouts like Line, Pie, Surface, Bar, Scatter, Column, etc.

How Excel Handle charts

Charts provide a pictorial depiction of any data collection. A chart is a visual description of data that uses symbols like columns in a column chart to describe the data. You may pick from the list of chart kinds in Excel, or utilize the Excel Recommended Charts option to see charts that have been created for your data.

In Excel, the data which you are to use in creating a chart is the one in your worksheet. However, you can use data from other worksheets or another workbook. The Charts in Excel are not static i.e. you can make changes to them anytime and anyhow you want.

Embedded charts

The charts you make in an Excel Worksheet are embedded charts. It stays on the worksheets. On a worksheet, is see an invisible drawing layer, which is where embedded charts are displayed.

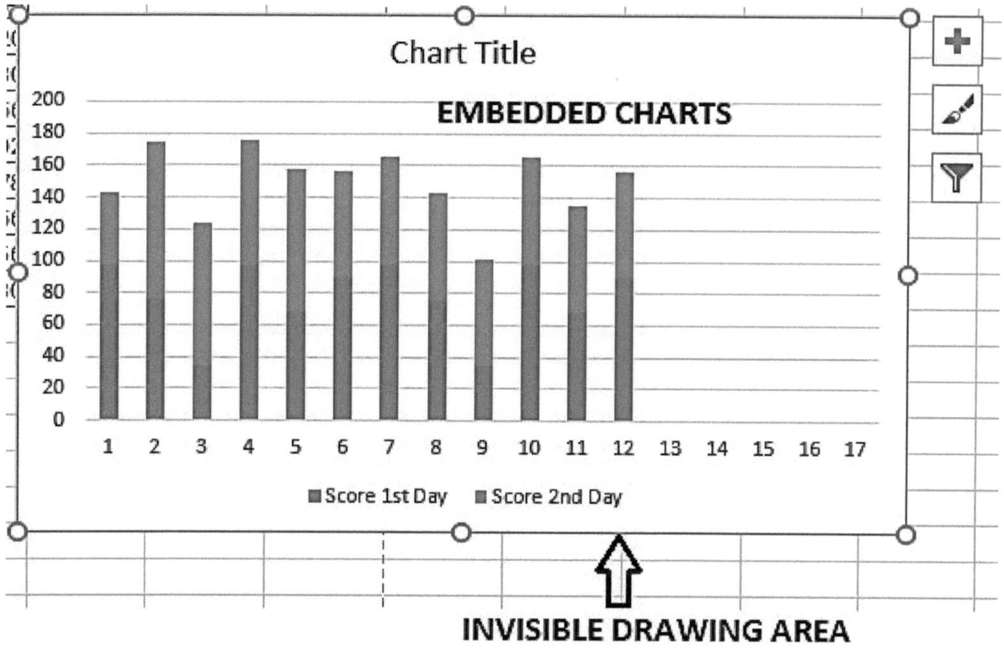

Just like other objects, you can change the position of an embedded chart, resize an embedded chart, change the properties, make changes to the borders, and so on. To make changes to the chart, simply click on the chart. The chart will be activated, then on the ribbon, the Design and Format tab comes up. With those tabs, you can make changes to the embedded chart.

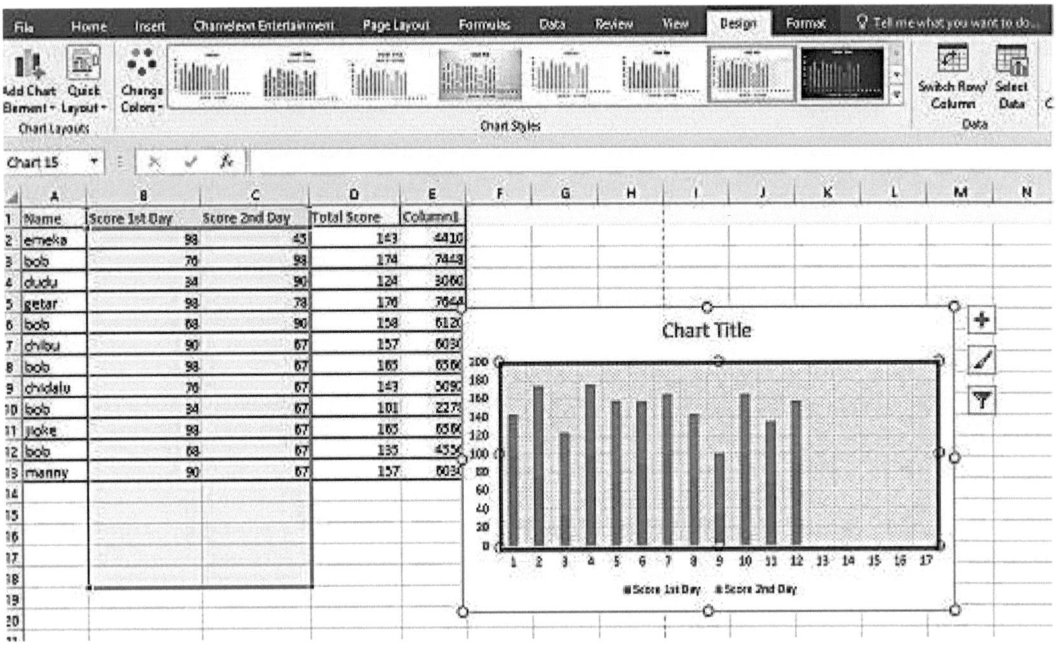

When you create a chart in Excel, by default, it is created as an embedded chart.

Chart sheets

Chart Sheet is the sheet that holds a chart in your worksheet. It is just like a worksheet just that worksheets contain both the data and your chart while chart sheets contain just a chart. An embedded sheet is always on the chart sheet. Just like how worksheets are called Sheet 1, Sheet 2, so are chart sheets called Chart 1, Chart 2, etc.

You can change the position of an embedded chart to a chart sheet. A chart sheet can be an embedded chart. Simply click on the embedded chart, click on the Design tab and select Move Chart. A dialog box will appear. Check the New Sheet option. Put in a name for the chart sheet if you want. Then, click Ok.

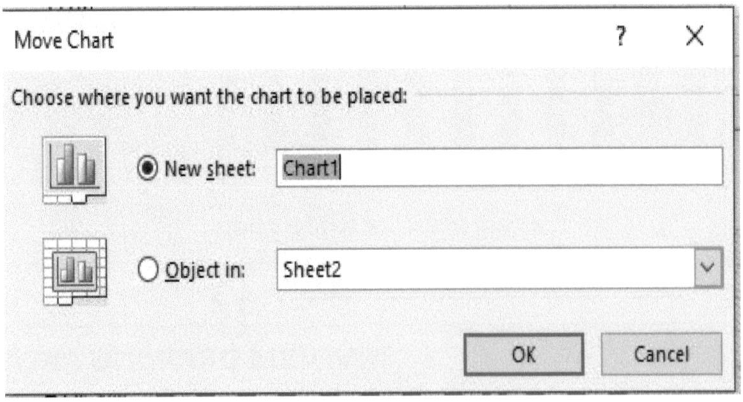

Parts of a chart

A chart is made up of different parts. When you create a chart in Excel, you will see the different parts of it located on the right side of the chart.

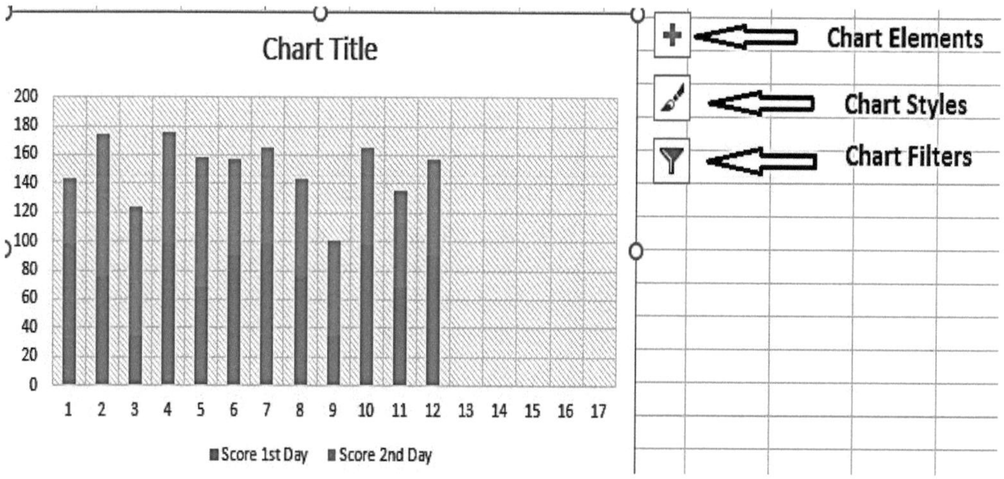

Chart Elements: These are parts of a chart that give more description to your chart. They make your data have detailed meaning. When you click on the chart element icon, you will see a list of different elements such as Axes, Axis title, Chart title, Gridlines, Legend, etc.

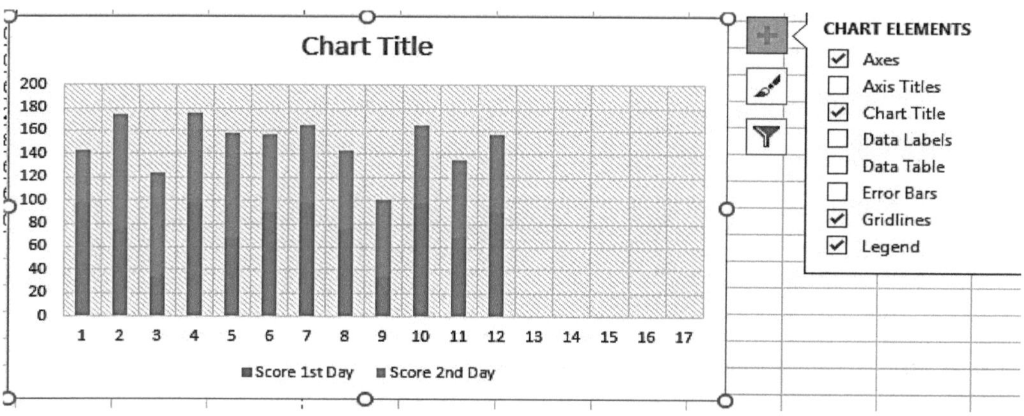

When you check the box beside each of the elements, the preview of the element will be displayed on the chart. As you can see in the image below, I checked all the boxes and the preview of it is displayed on the chart.

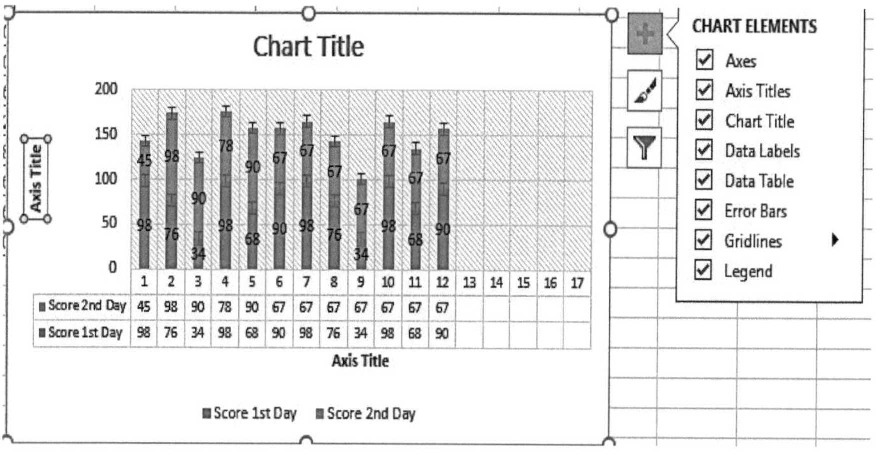

Axes: A chart has two which are the **Vertical axis (Y-axis)** and **Horizontal axis (X-axis)**.

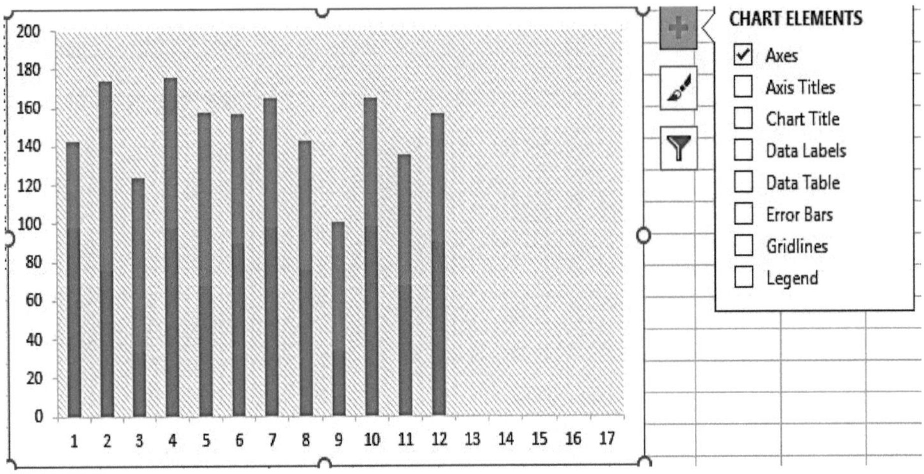

Axis Titles: This gives more understanding feature of the data in the chart. You can add them to any vertical, horizontal, or depth axes in the chart. You can't add them to charts that do not have axes like doughnut charts or Pie charts.

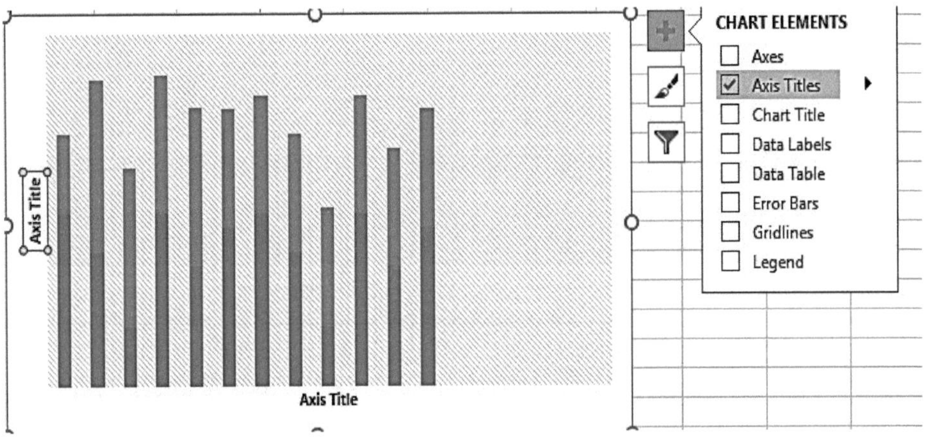

Chart title: This is the title of the chart you have added. It appears by default on top of the chart sheet. Click on the chart title to edit it.

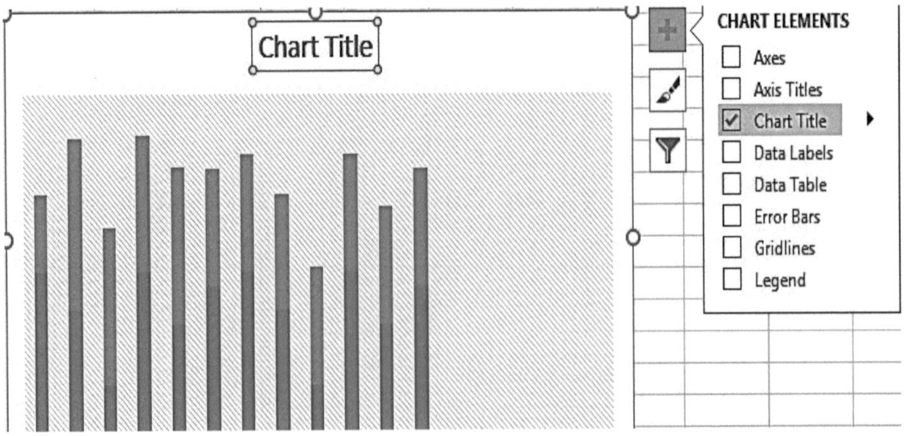

Data labels: This shows the details of a data series or data points. It makes your chart to be easily understood by the readers or viewers.

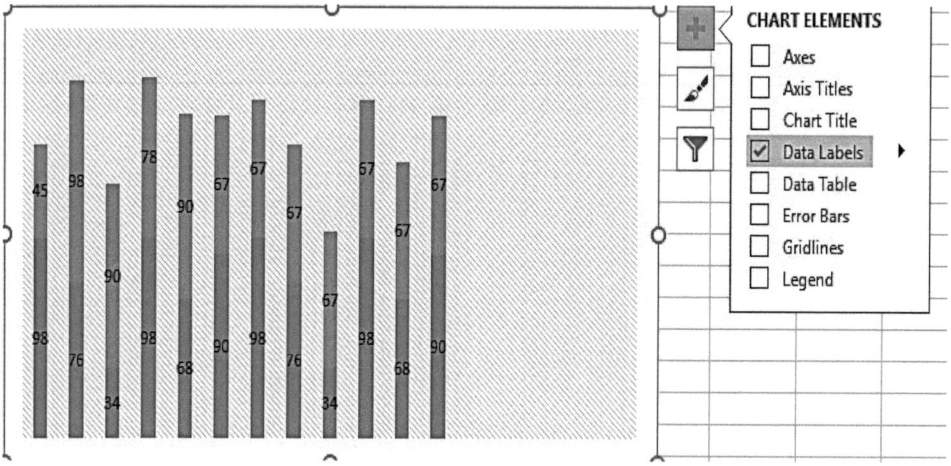

Data table: This displays the table of the data for the chart.

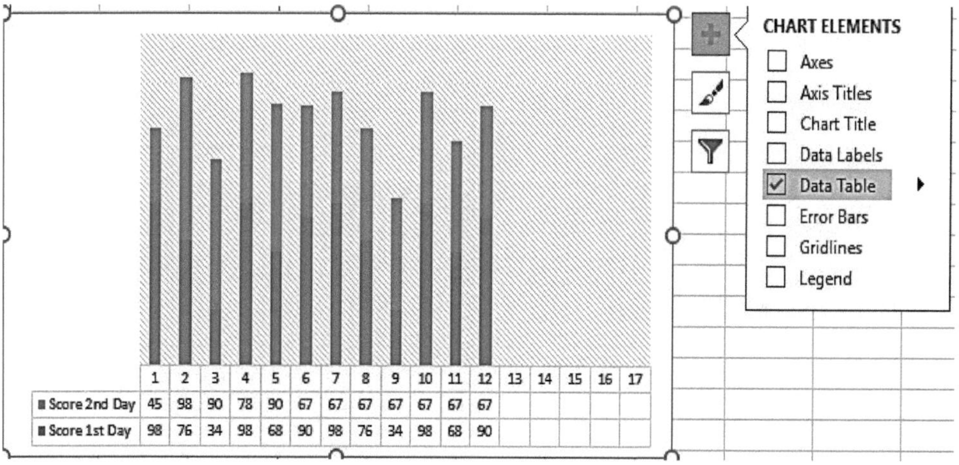

Error Bars: Displays the possible error amounts comparatively to every data marker in a data series.

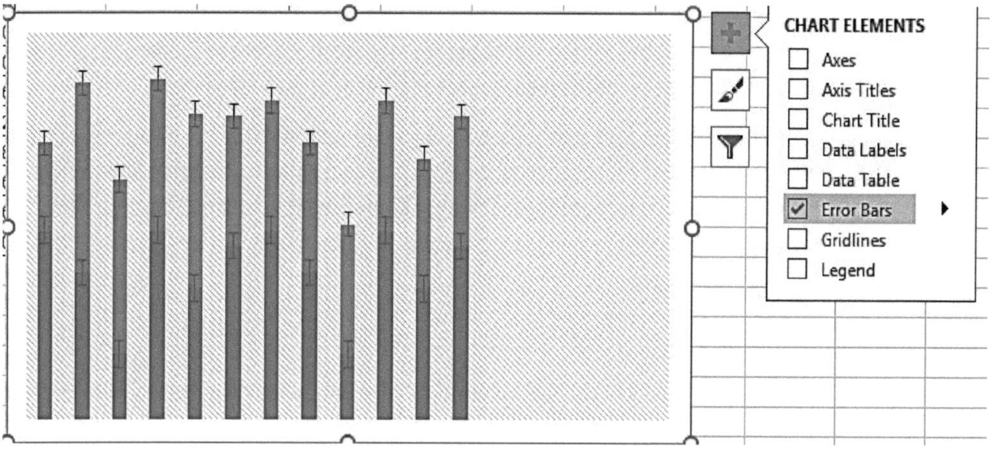

Gridlines: This makes the data easy to comprehend.

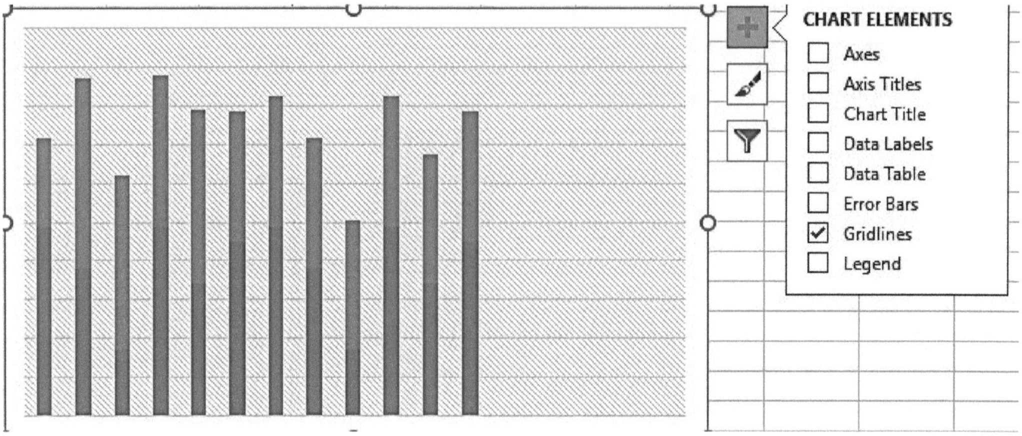

Legend: Displays the rows or columns on which the chart is created on. It also gives it a special representation with the use of different colors to describe them.

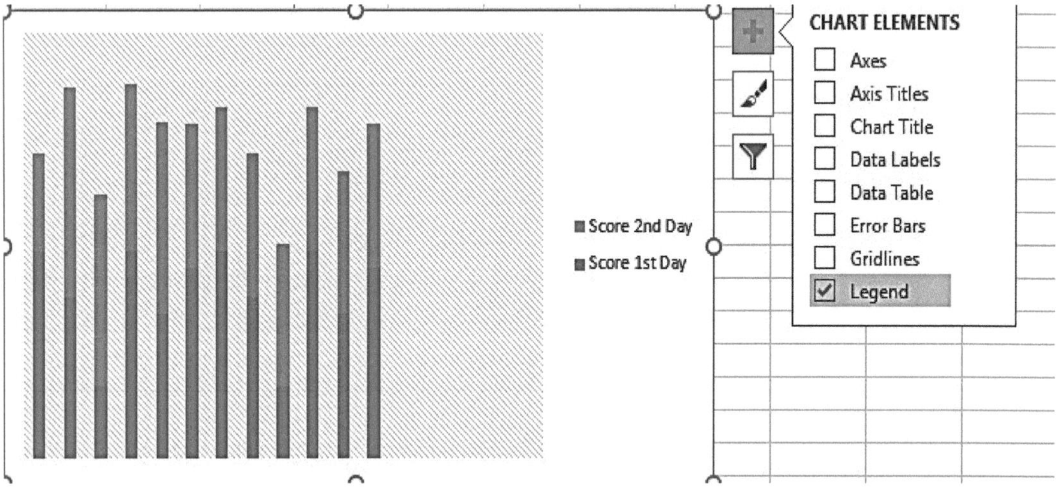

Chart limitations

The amount of series available in a chart in Excel 2010 has stayed the same as in previous versions, 255, but the number of points in a series has expanded substantially, from 32,000 to 1,048,576, allowing whole columns in Excel 2007 and 2010 to be utilized as the source data for a chart series.

Because the prior restriction of 256,000 points is less than the new maximum of points per series, the total number of points permitted in a chart has risen. I didn't have time to test this limit since drawing a series with a million points takes over a minute. (Hmm, I'm not sure whether the help files have been updated with the new restrictions yet.

BASIC STEPS FOR CREATING A CHART

Creating the chart

To create a chart, simply select the cell (s), row, or column that you want to make a chart for. Then, click Insert on the ribbon. You will see the Chart group in the Insert tab.

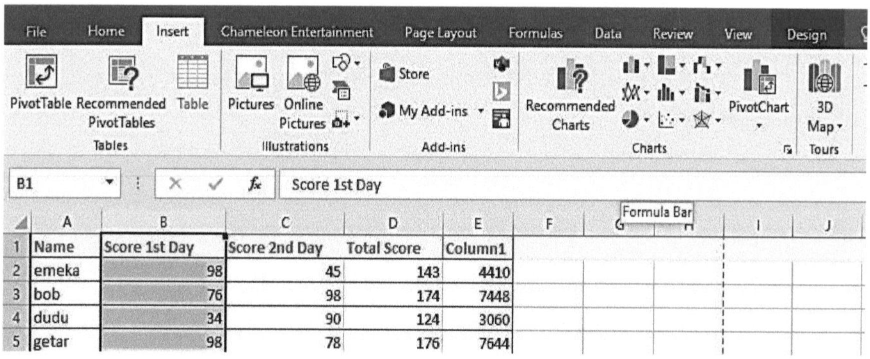

So, select the type of chart you want to use from the list of charts there. Click on the drop-down arrow for each chart icon to see the different chart styles. You can also click on Recommended Charts to select another chart style. So, click on a chart and it will be displayed on your worksheet. The chart represents the data of the selected area in the worksheet.

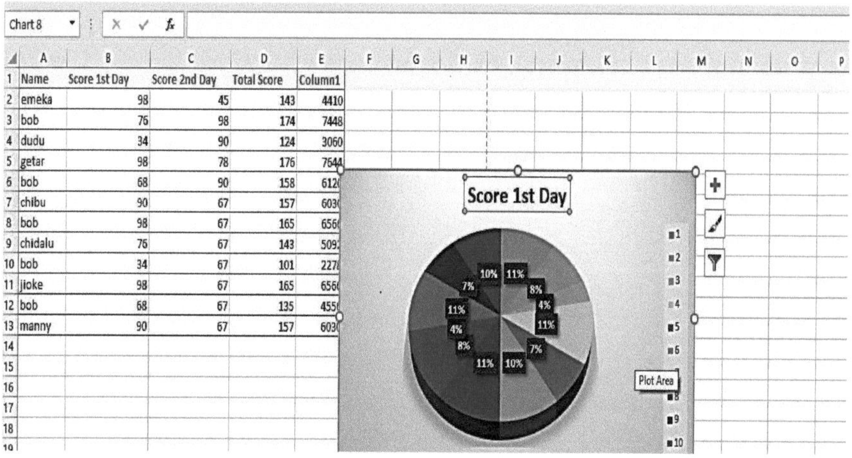

On the ribbon, you can change the chart style for the chart on your worksheet.

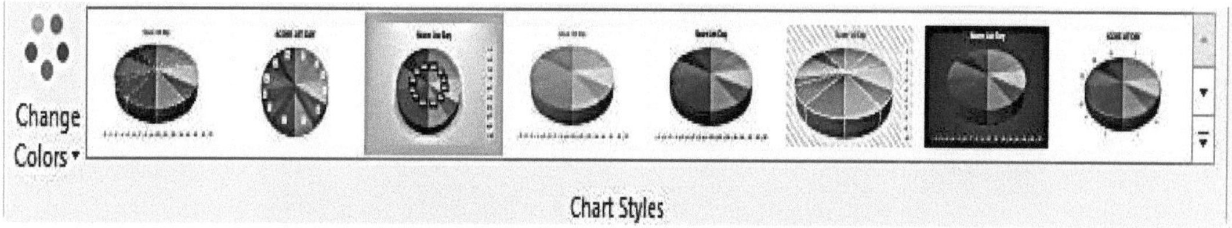

Switching the row and column orientation

After adding your chart, you can switch the row and column. Simply click on the Switch row and column icon on the ribbon.

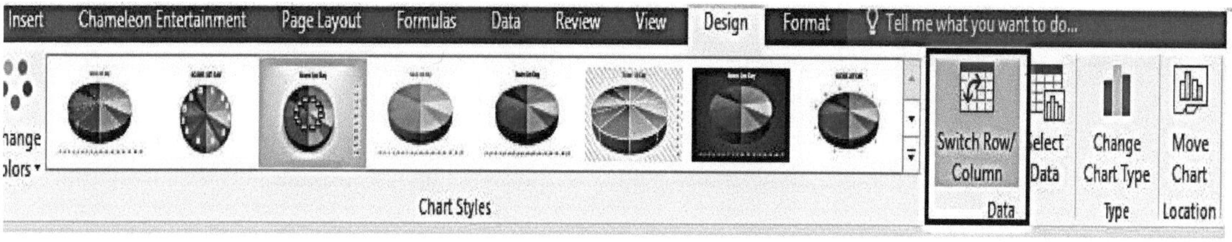

Changing the chart type

Click on Change Chart Type in the ribbon. This opens up the Change Chart Type menu. on it, you will see the list of different chart types.

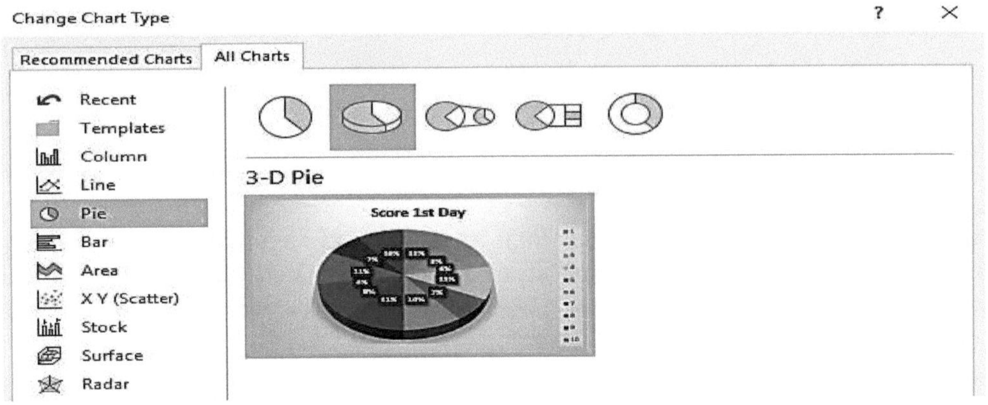

Click on the one you want to change the chart type to and select Ok. The chart will change on your worksheet.

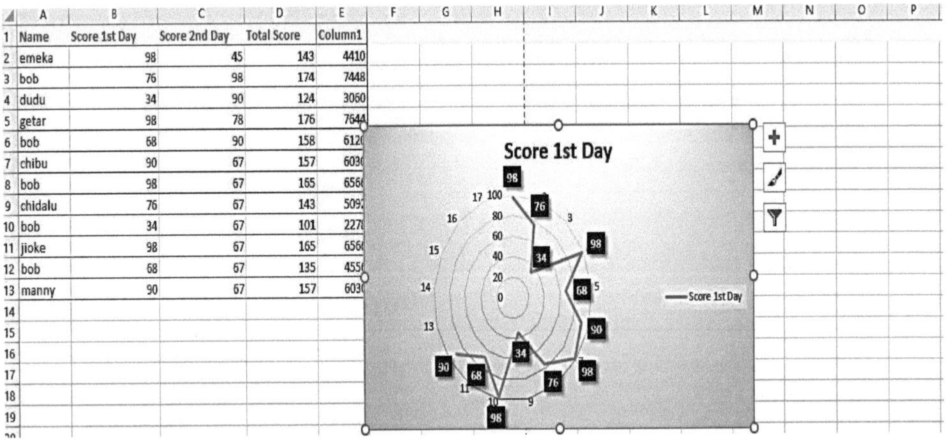

Applying chart layout

At the left side of the ribbon is the Chart Layout group. Click on Quick Layout, then select a layout.

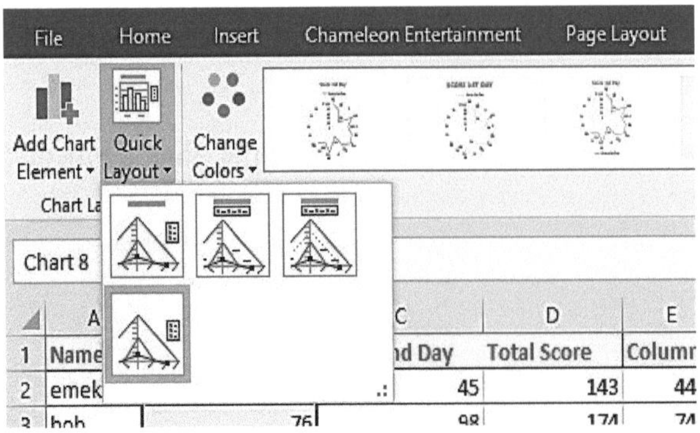

Adding and deleting chart elements

Chart elements provide extra information to your charts, making them more relevant and attractive to the eye. Click on **Add Chart Elements** and select an element.

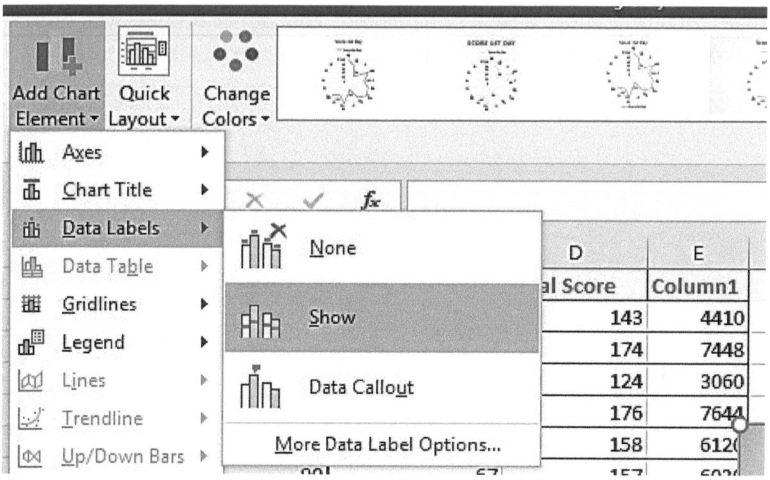

To delete, right-click and select **Delete**.

Formatting chart elements

To format the elements, right-click and select Format (the name of the element you added comes after Format).

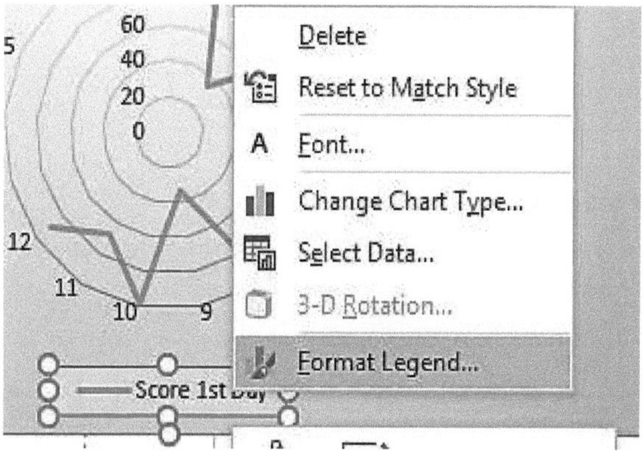

On the Format Pane menu, click on an option.

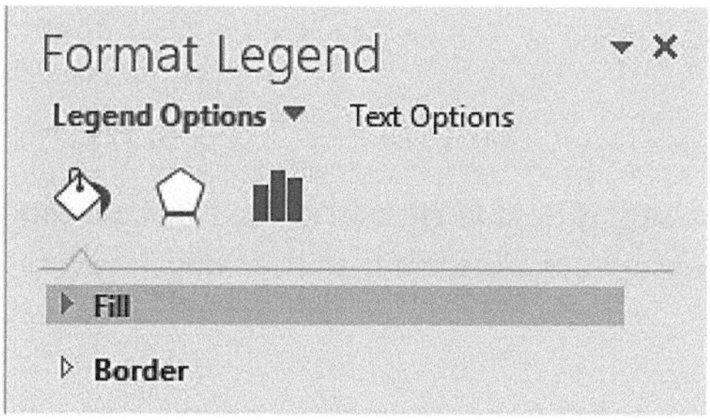

MODIFYING AND CUSTOMIZING CHARTS

Moving and resizing a chart

To move a chart, click the chart, then drag it to anywhere to want to place it. To resize it, put the cursor on the chart edge, then drag it up, down, or sideways. You can also right-click on the chart, then select **Move Chart**. This opens a dialog box.

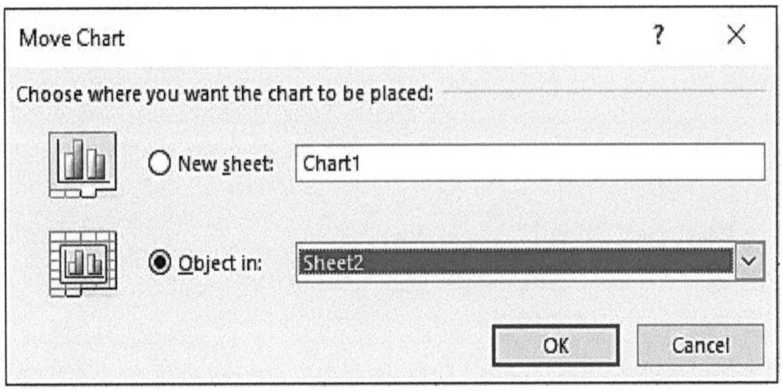

Converting an embedded chart

We can connect a chart in Excel and the data in a Word document together using Embed in Excel. When linking them, any changes you apply in the chart will also be applied on the Word document automatically only if they are saved in a similar folder or location. Follow the steps to do so.

On the Insert tab, then select **Object** from the Text group.

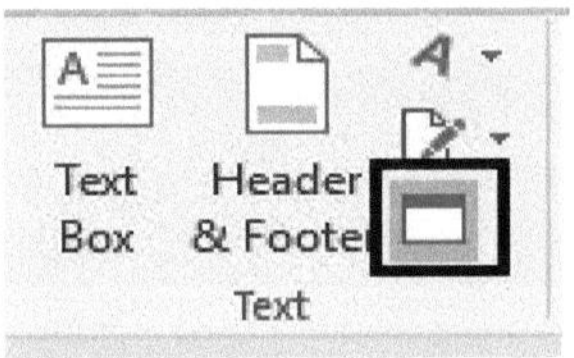

Click on **Create from file** tab and select **Browse**.

Find the file (chart) and choose Insert. Check the box on **Link to File**. Then click **Ok**.

To edit the embedded chart, simply double-click on it. the Excel worksheet which contains the chart will open. make your edits and ensure you save them after editing in Excel.

Copying a chart

Simply right-click on the chart, select Copy. Then, navigate to where you want to paste it and right-click, then select **Paste**.

Deleting a chart

Click on the chart, then Press the **Delete key** on your keyboard.

Copying a chart formatting

First copy the chart. You can click on the chart and press **Control key + C** or you right-click and select **copy**.

Then, click on the chart you want to format. On the clipboard group, click the down arrow on the Paste icon and select Paste Special.

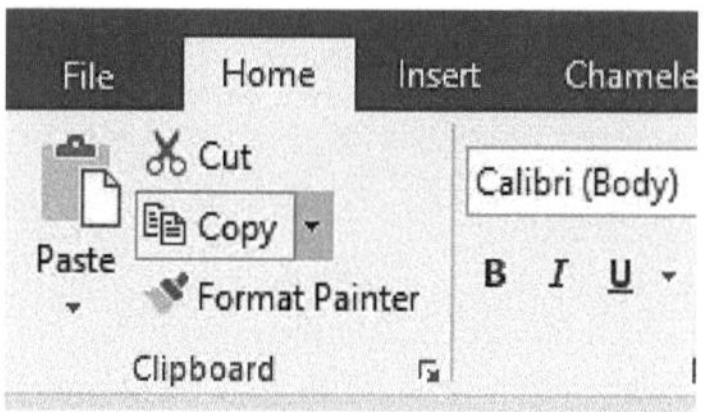

Check the **Format box**. Click Ok.

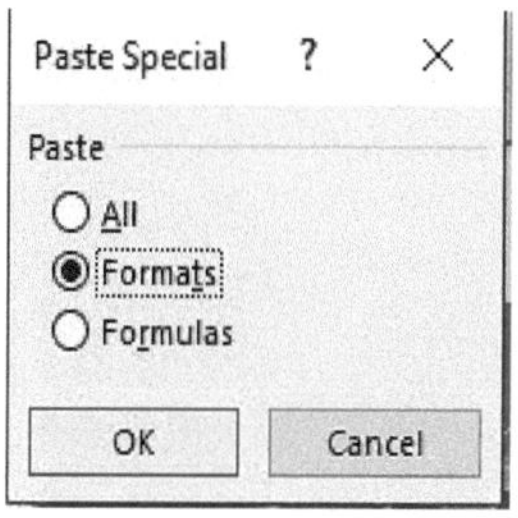

Renaming a chart

You might well have observed that when Excel generates a chart, it assigns it a specific number. It is found in the Name Box on top of the grid in the left corner. You have the option of changing the name to something more informative to you.

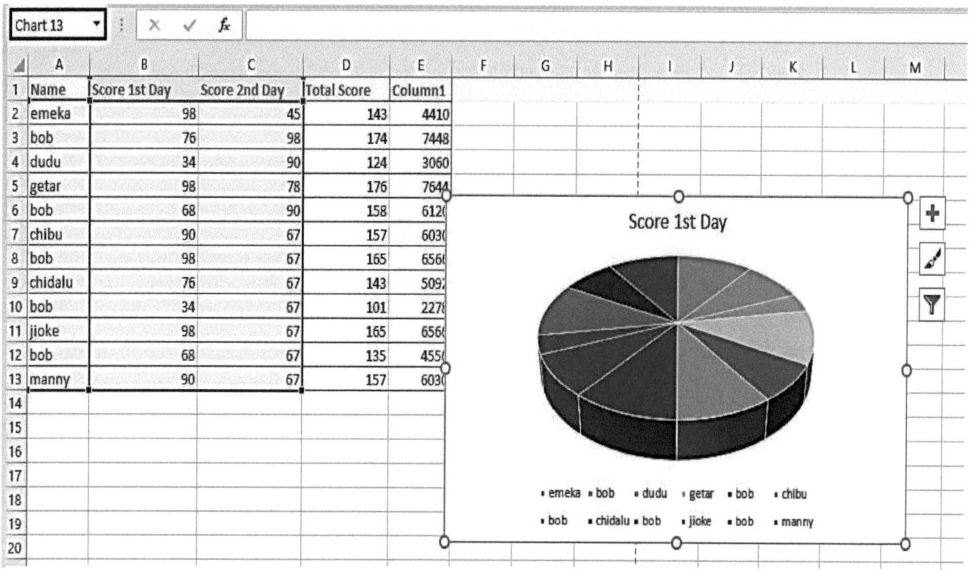

There are ways to rename a chart. First, click the **Name box**, then, type in the name for the chart.

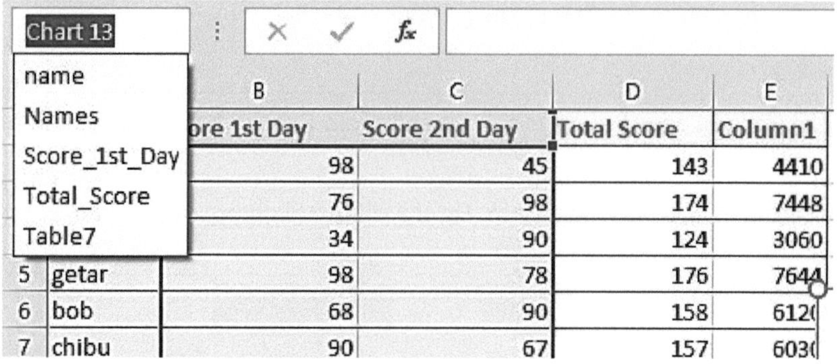

Renaming a data series in a chart

Simply right-click on the chart and click on Select Data.

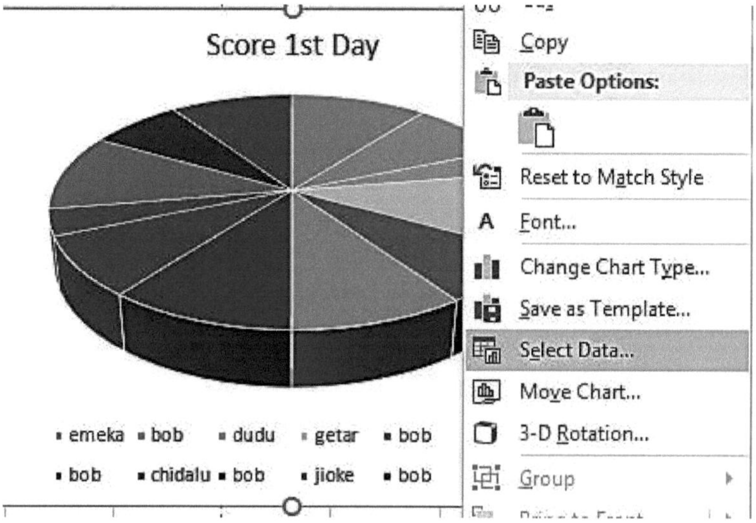

This opens up the Select Data window. So, highlight the data series you want to rename. Click Edit.

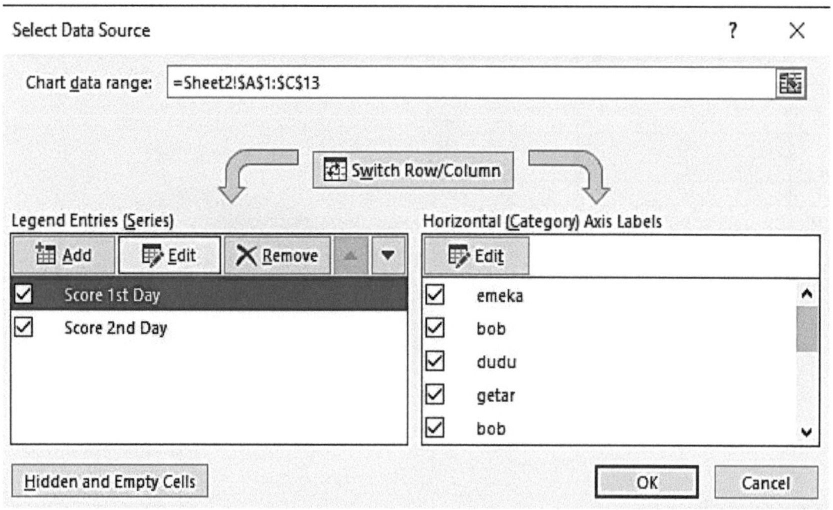

The Edit Series dialog box will appear. On it, remove the names on the type-in box and put in the name you want for the chart. Click Ok.

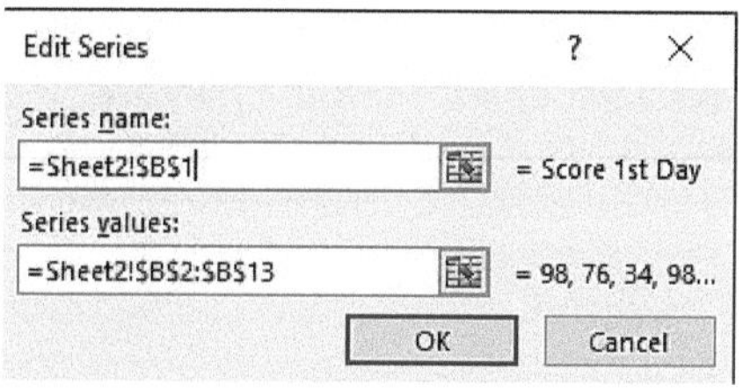

Click **Ok** again.

Printing charts

Choose the chart. Click on File and select Print. On the **Printer** option, select a printer. Then, on the **Print Setting option**, make sure it is set to Print Selected Chart, though it will be there by default. Click **Print**.

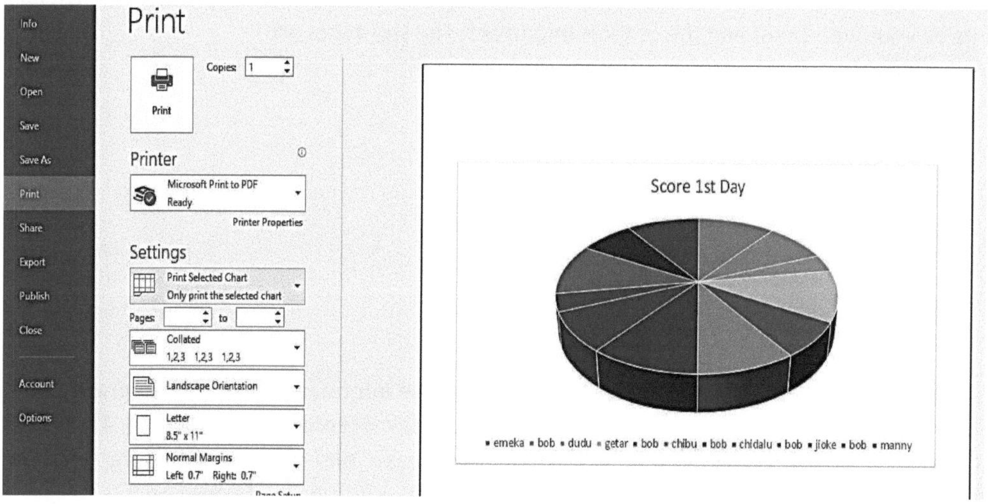

UNDERSTANDING CHART TYPES

Choosing a chart type

In Excel, there are different kinds of charts. You have a list of charts that you can use when working on your worksheets. There are also sub-types of these charts. Below are the charts;

Column charts

In a Column Chart, the horizontal (category) axis is used to show the categories, while the vertical (value) axis is used to display the values. Arrange the data in columns or rows on the spreadsheet to make a column chart. The sub-types of the column chart are;

- 3-D Column
- 3-D 100% Stacked Column
- Clustered Column
- 100% Stacked Column
- 3-D Clustered Column
- 3-D Stacked Column
- Stacked Column

Bar charts

Individual item comparisons are shown using bar charts. The data are grouped along the horizontal axis and the categories are arranged along the vertical axis in a Bar Chart. Organize the data in columns or rows on the Spreadsheet to Make a Bar Chart. The sub-types are

- Clustered Bar
- Stacked Bar
- 100% Stacked Bar
- 3-D Stacked Bar
- 3-D Clustered Bar
- 3-D 100% Stacked Bar.

Line charts

Line charts are used to show trends over time. It can be over years, months, and days. It is also used to display categories when the order is not important. You can utilize this chart when you have lots of data points on your worksheet and the order is important. The sub-types are

- Line
- Stacked Line
- 100% Stacked Line
- Line with Markers
- Stacked Line with Markers
- 100% Stacked Line with Markers
- 3-D Line

Pie charts

This is a circular chart. It is a sort of graph that depicts the information in a circular graph. The slices of pie illustrate the data's apparent size and are a sort of graphical representation of data. A list of categories and numerical variables is required for a pie chart. The phrase "**pie**" refers to the entire, whereas "**slices**" refers to the individual components of the pie.

It is divided into different sectors in which each of them represents a part of a whole. The number of elements in one data series is equivalent to the total of the elements in a pie chart. In a pie chart, the data points are shown as a proportion of the whole pie. Arrange the data in one column or row on the spreadsheet to Make a Pie Chart. The sub-types are

- Pie
- 3-D Pie
- Pie of Pie
- Bar of Pie

XY (scatter) charts

XY (Scatter) charts are often used to display and compare quantitative quantities, such as data from science, statistics, and engineering. There are two Value Axes in a Scatter chart.

- Axis of Value Horizontal (x)
- Axis of Vertical Value (y)

It merges x and y values into a singular data point and shows them in clusters of irregular intervals. Organize the data in rows and columns on the spreadsheet to make a Scatter chart. The X values should be in one row or column. The y values should be in the adjacent rows or columns.

When in doubt, use a scatter chart when:

- You wish to adjust the horizontal axis scale.
- You should use a logarithmic scale for that axis.
- The horizontal axis values are not equally distributed.
- On the horizontal axis, there are a lot of data points.
- To expose additional information about data that comprises pairs or grouped sets of values, you wish to modify the independent axis scales of a scatter chart.
- Instead of showing disparities between data points, you want to illustrate commonalities across big amounts of data
- You wish to compare a large number of data points over a long period.
- The more data you put in a scatter chart, the more accurate your comparisons will be.

Area charts

Area charts are useful for plotting changes over time and highlighting the entire value throughout a trend. An area chart displays the connection of parts to a whole by displaying the total of the plotted data. Arrange the data in columns or rows on the spreadsheet to make an Area Chart. The sub-types are

- Area
- Stacked Area
- 100% Stacked Area
- 3-D Area
- 3-D Stacked Area
- 3-D 100% Stacked Area

Radar charts

The average values of many data series are compared using radar charts. Organize the data in columns or rows on the worksheet to create this chart.

Surface charts

A surface chart comes is helpful when you want to combine two sets of data. Colors and patterns, much as on a geographical map, identify regions with similar values. To make a Surface chart, make sure that the categories and data series are both numeric values and organize the data on the spreadsheet in columns or rows.

Bubble charts

A Bubble chart is like a Scatter chart, only that it has a third column that specifies the size of the bubbles that indicate the data points therein the data series.

Stock charts

Stock charts, as the name indicates, may depict price movements in stocks. A Stock chart is used to display changes in other data, such as rainfall intensity or yearly weather.

Organize the data in columns or rows in a specified arrangement on the spreadsheet to make a stock chart. To make a basic high-low-close Stock chart, for example, organize your data with High, Low, and Close as Column headers in that sequence.

NEW CHART TYPES FOR EXCEL

Histogram charts

A histogram is a graphical illustration of numeric values distribution. A histogram is a kind of column chart that displays the frequency of data in a certain range in a more straightforward manner. It uses the number of data points that fall inside a specific range of values to visualize numerical values. It uses vertical columns to display the frequent increase and fall in data. In Excel, a histogram chart is divided into five sections which are **Title, X-axis, Y-axis, The bars,** and **Legend.**

The **title** of the histogram describes the information contained inside it. The **X-axis** is a clustered interval that depicts the range of values in which the measurements are taken. The **Y-axis** is a scale that displays the number of times the values happened inside the intervals defined by the X-axis.

The **bars**: There are two dimensions to this parameter: height and width. The number of times the values happened inside the interval is shown by the height of the bar. The interval, distance, or area covered is indicated by the width of the bars.

Pareto charts

A Pareto chart, also known as a Pareto diagram, is a graph that uses the Pareto principle as its foundation. It's a kind of sorted histogram in Microsoft Excel that includes both vertical bars and a horizontal line. The relative frequency of values is represented by the bars, which are drawn in decreasing order, while the line shows the cumulated percentage.

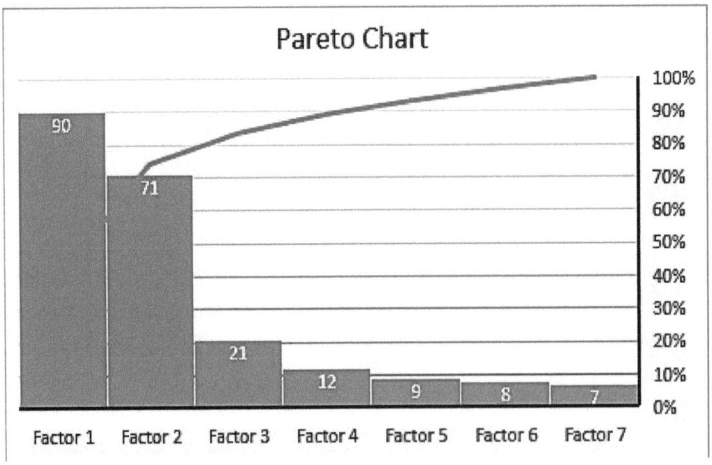

Pareto graph emphasizes the main pieces in a data collection and displays their relative relevance to the whole.

Waterfall charts

A waterfall chart is a kind of graph that is often used to visualize financial information, compare profits, and analyze sales or product value over time. It's also used to visualize inventories and analyze profit and loss. They originally gained popularity in the late twentieth century, when McKinsey & Company used them in a client presentation.

Box & whisker charts

A box and whisker chart, also known as a box plot, is a statistical analysis tool in Excel that shows you how values are dispersed in a collection of data. For example, you may use a box and whisker chart to illustrate statistical information on test results across topics to determine which subjects need greater focus from pupils.

Treemap charts

A treemap chart is a style of data representation that excels at portraying hierarchical data. Each element is displayed as a rectangular shape on a treemap, with smaller rectangles representing sub-groups. The color and size of rectangles are usually associated with the tree structure, making the groups and sizes simpler to perceive. Treemap charts are excellent for emphasizing each object's contribution to the overall hierarchy.

Sunburst charts

A sunburst chart, like treemap charts, is a style of visualization that works well for visualizing hierarchical data. A sunburst chart is a circular diagram in which each circle symbolizes a group hierarchy level. The high-level groups are plotted in the inner circle, while the sub-categories are plotted in the outer rings. The segments' sizes are proportionate to the values they represent. A sunburst chart is divided into three sections: Plot Area, Chart Title, and Legend.

Plot Area: The plot area is where the graphic expression occurs. Like pie and donut charts, a sunburst chart illustrates parts of the full data set. A top-level group is represented by each hue. **Chart Title:** Make an effort to be detailed and succinct. The **legend** is an indication that aids in the differentiation of data sets. Each color denotes a category at the top level.

Funnel charts

Funnel charts are comparable to their name in that they are used to display data behavior at each step specified, and when the numbers decrease, the chart takes on the form of a funnel, thus the term funnel chart.

Map charts

These are used to depict a certain Key Performance Indicator (KPI) and demonstrate its representation over different geographical locations for a given category, such as a business, sector, or commodity.

CHAPTER TEN

USING ADVANCED CHARTING TECHNIQUES

SELECTING CHART ELEMENTS

When you are done creating a chart, you can add some chart elements to your chart. You can select the chart elements in different ways.

Selecting with the mouse

Simply click on the chart element. Also, click directly on a bar in your chart to choose the data series. Double-click on an element for it to display.

Selecting with the Keyboard

You can use the up and down keys to select chart elements in your worksheet. Simply, click on the chart, press down the Control key, then use the up and down keys to navigate through the main chart elements.

Some chart elements have sub-elements. To access those sub-elements, use the right and life arrow keys. Click the chart, use the up arrow key to pick the element, use the right/left arrow keys to pick the sub-elements.

Selecting with the chart element control

You can select chart elements from the chart element control on the Format tab. When you click on a chart, the list of the main Chart Elements will be displayed on the menu.

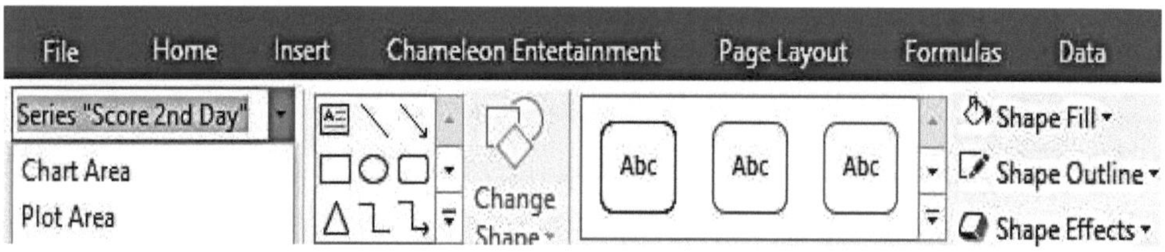

With this menu, you can go to the chart title, plot area, and so on. When you click on an element, the name will show there.

EXPLORING THE USER INTERFACE CHOICES FOR MODIFYING CHART ELEMENTS

Using the format task pane

Pick the chart element that you want to modify, then right-click on it and select Format (the name of the chart element comes after format). This will open up the Format pane. On it, you will see a list of options for the selected chart element.

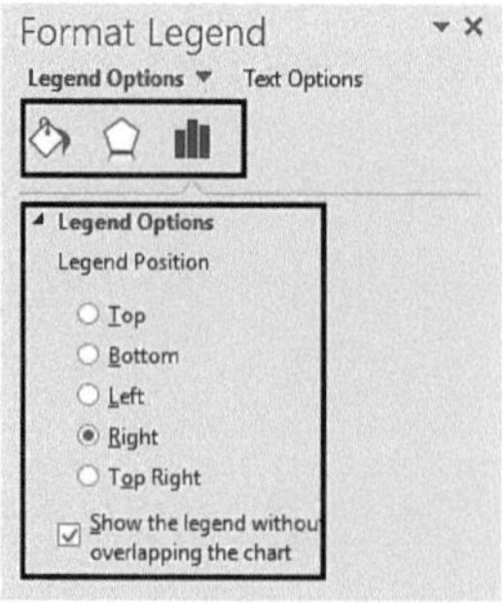

Using the chart customization buttons

Click the chart you have created on your worksheet, then select a customization button on the right side.

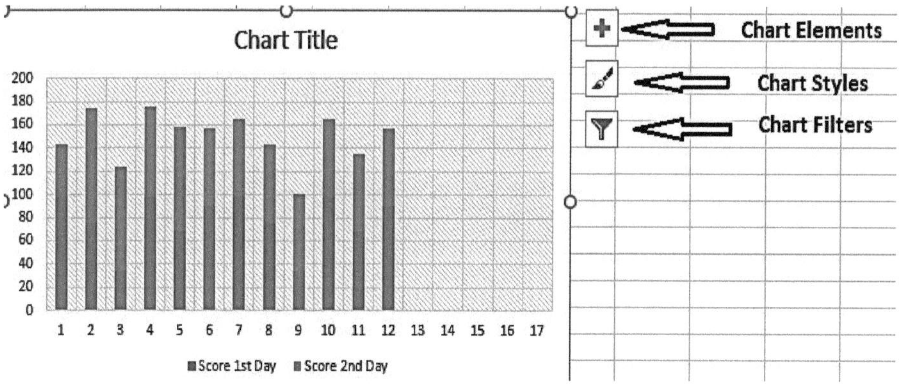

Using the ribbon

First, pick the chart element, navigate to the Format tab which is located below the Chart Tools. You will see options for modifying your chart elements.

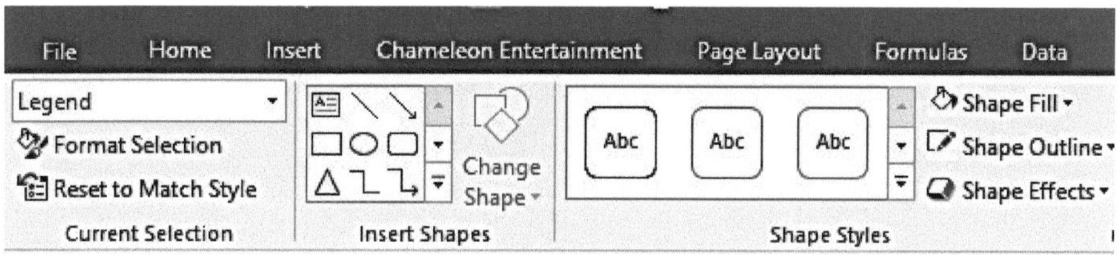

Using the Mini toolbar

Right-click on the chart element to see the mini toolbar. On it, you will see some chart modifying options like fill color, style, and outline. Click on any of the options to modify your chart.

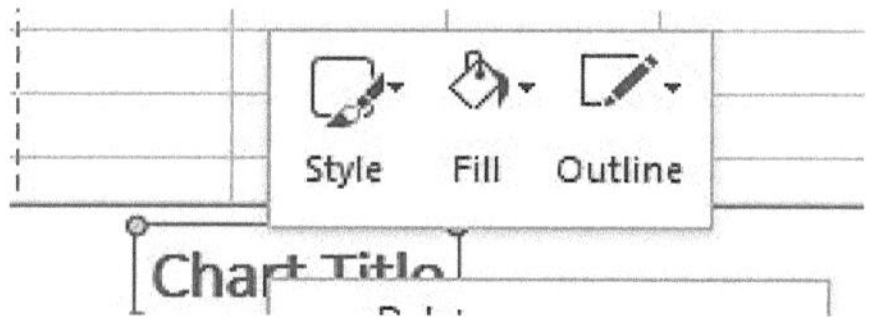

Modifying the chart area

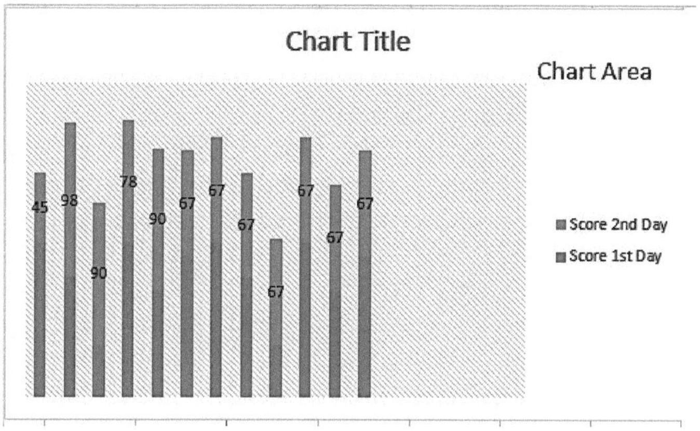

The chart area is the part of a chart that encompasses all elements of a particular chart. You may change the appearance and position of your chart. A few options are shown here, but there are many more! To access **Chart Tools** in any situation, you must first pick the chart. When you click on the chart, you can use the Design tab and the Format tab to modify it.

To add labels such as title, axes, etc. Select the **Add Chart Element option** and select a label. The **Chart Tool tab** contains many modifying options. You can change colors, chart styles, move charts, add borders, add shapes, etc.

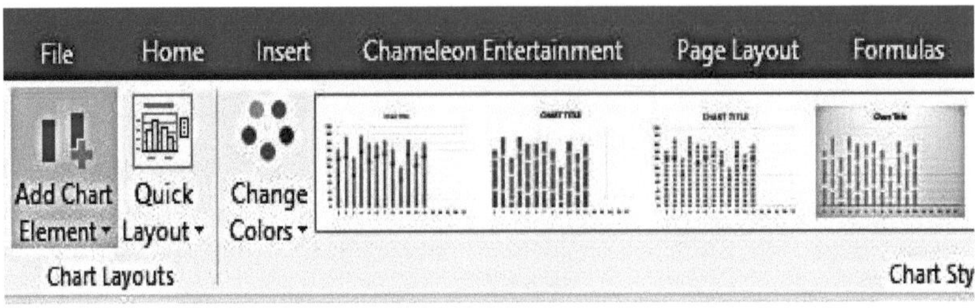

Modifying the Plot area

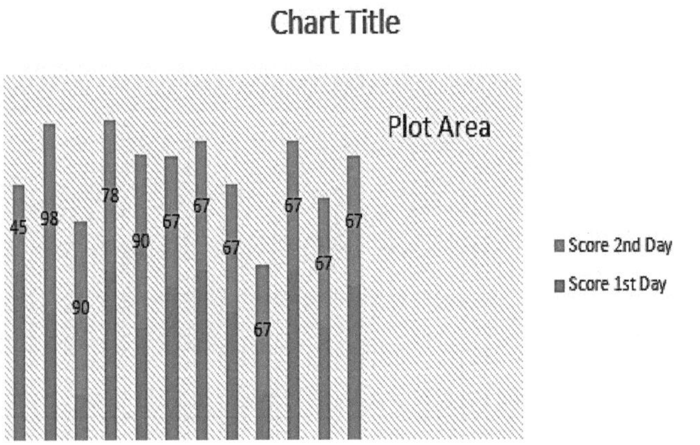

The plot area of a chart is found inside of a chart. It consists of the actual chart. Just like the chart area, you can change the border and fill the plot area. When you add elements to a chart, it changes the plot area size.

When you right-click on the plot area, you will see some options for modifying the plot area.

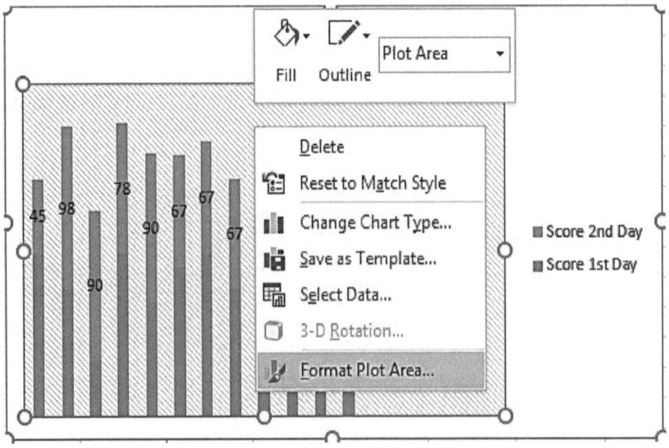

Working with Titles in a chart

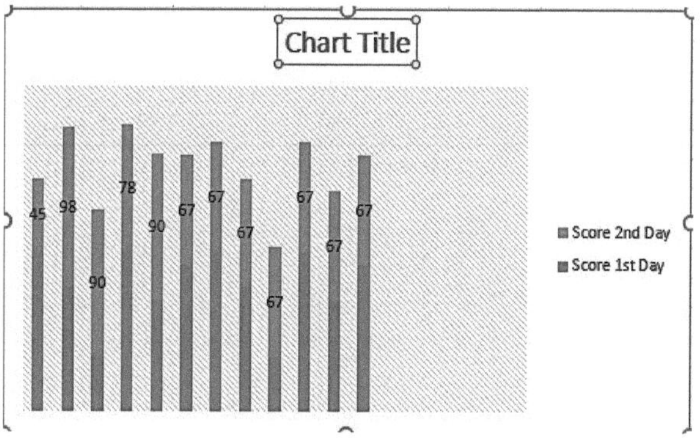

The chart title is located at the top. When creating a chart with just a column, the column heading becomes the title of the chart. When you create a chart with a group of columns or rows, the title is **Chart Title**. Click on it to edit the name.

You can add or delete the chart title if you want to. Simply click on the chart, select the chart element icon, then uncheck the Chart title box. You can also change the position of the chart title.

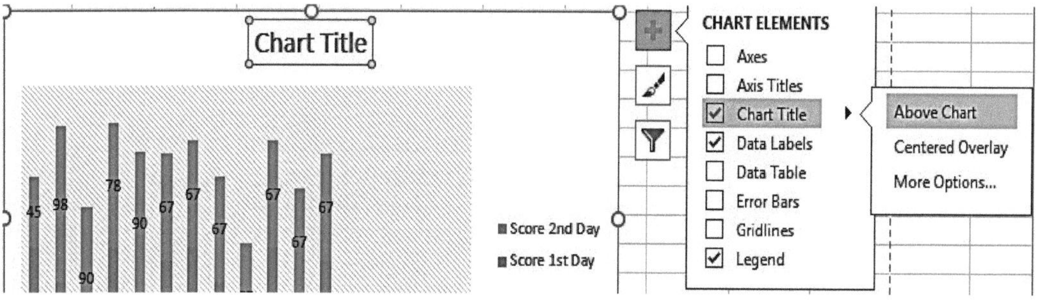

Working with Legends

As said in the previous chapter, the legend is displayed by default at the right side of the chart when you create a chart. If you want to hide it, click on the **Chart Elements** icon, then uncheck the box beside the Legends option.

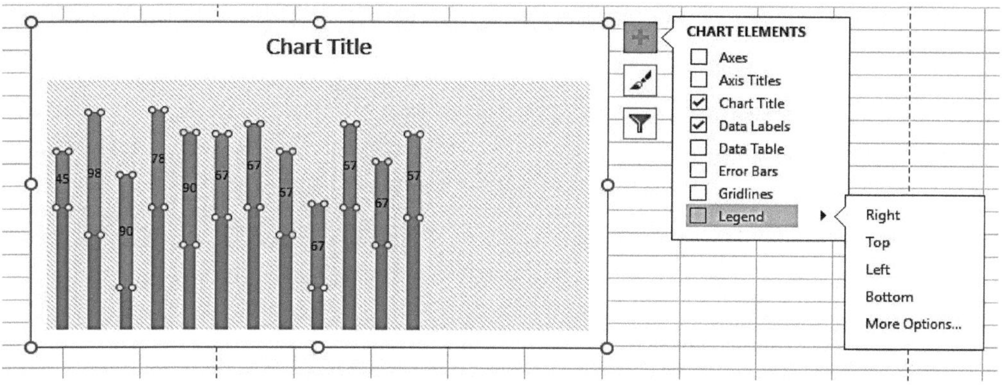

To change the position of the legend, simply click on the chart element icon, then, click on the arrow next to **Legend**, then select where you want to move it to.

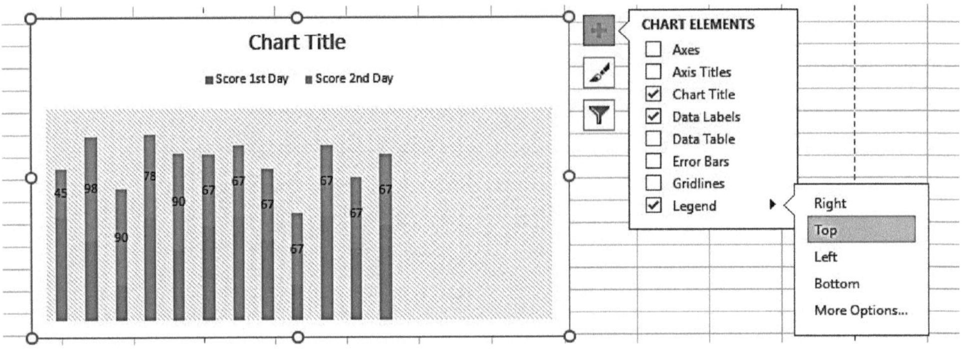

Working with Gridlines

Click the Chart Element icon on the chart and check the box beside Gridlines. You can select the type of gridlines you want to use on your chart. Click on the arrow next to Gridlines and select an option from the box.

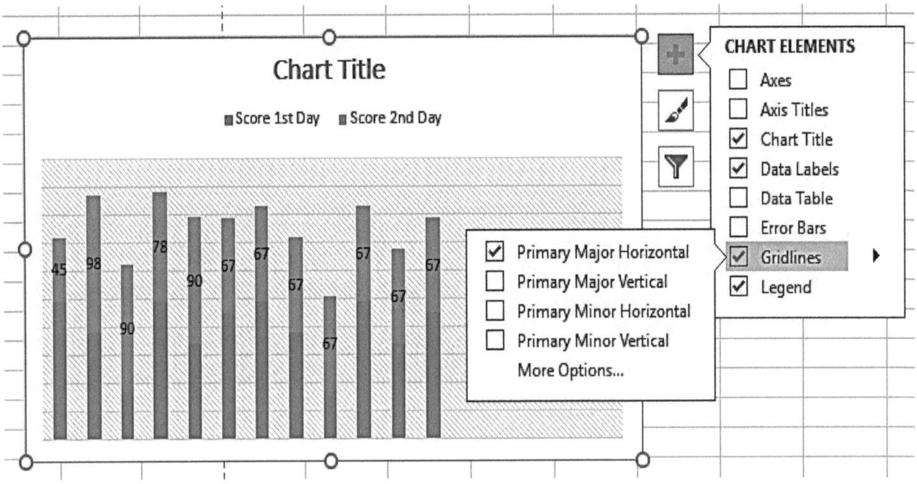

Modifying the Axes

You may have added an axis on your chart but the scale of the axis is small or too big to display the units in the chart. You can modify it to your choice. Simply right-click on the axis you want to modify and select Format Axis.

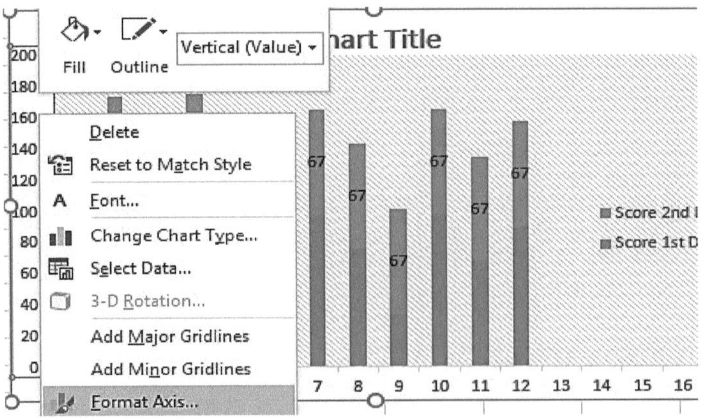

This will open up the Format Axis pane. On the pane, select any icon to modify the axis. Each of the icons has its modifying options.

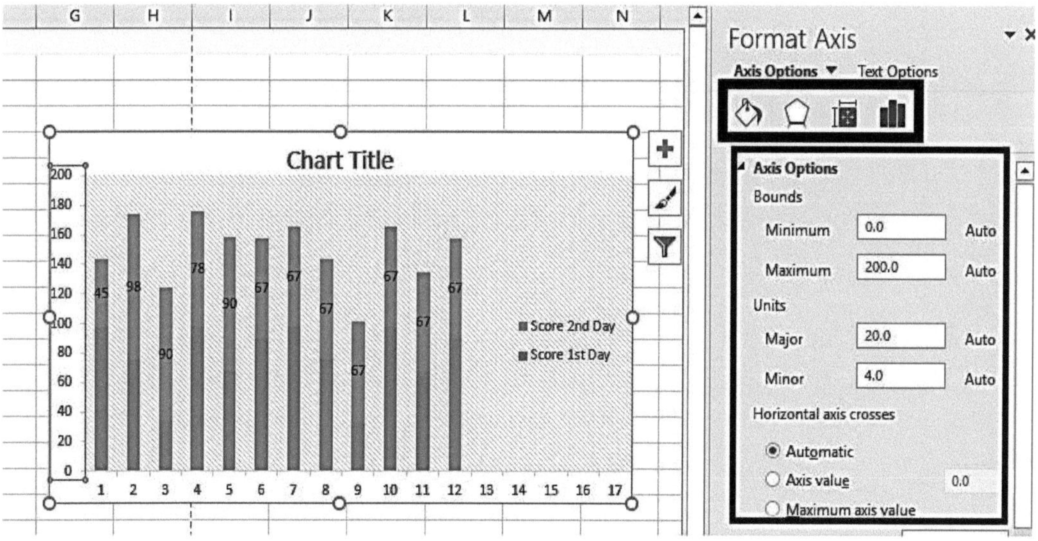

WORKING WITH DATA SERIES

Deleting or hiding a data series

You might want to delete some data on your worksheet. Simply click on the Chart Filter icon, then uncheck the box beside the data series you want not to display.

Adding a new data series to a chart

Right-click on the chart, then click on **Select Data**. The Select Data Source box will open. On the left side is where the data series are listed, such as Add, Edit, and Remove buttons. So you can add, remove and edit data series with these buttons.

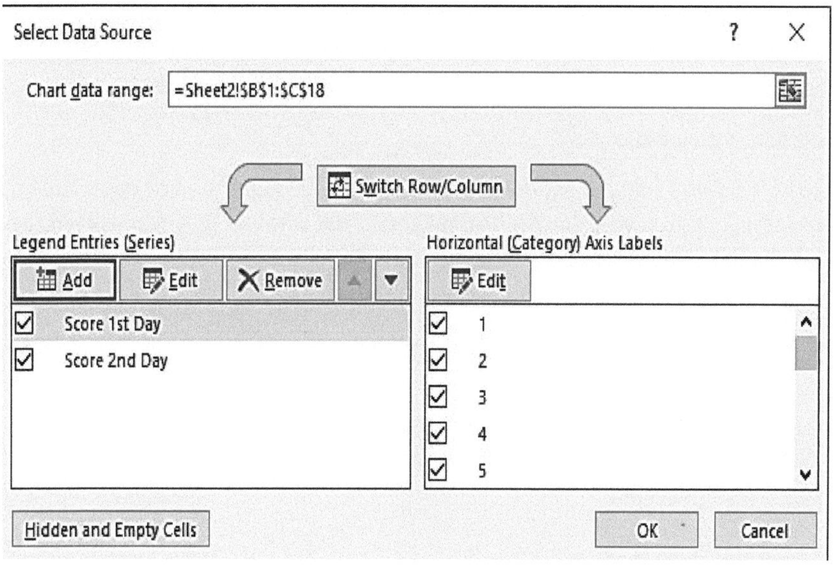

Click on the **Add button**. The Edit Series box opens, then enter the name of the series and its value. Then, click Ok.

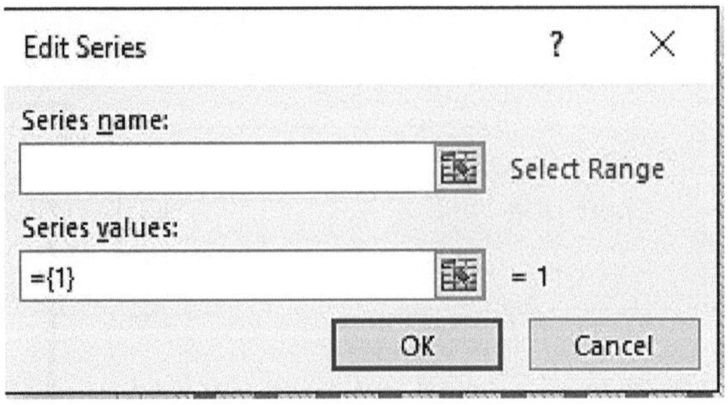

Changing data used by a series

To change the data in a series, go to the **Select Data Source menu** using the steps above. In the data series box, select the series that you wish you change, then click on the Edit button. Then, apply your changes.

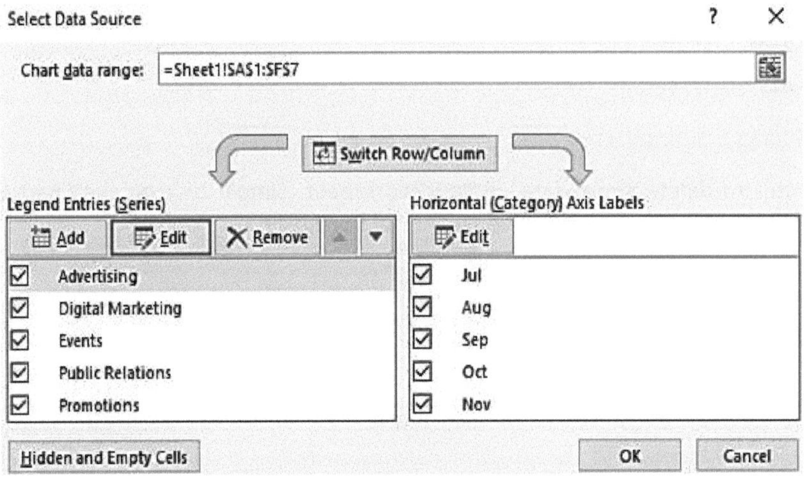

Note that these changes are likely to break connections to the source data in your spreadsheet.

Using the Edit series dialog box

If you want to edit a data series, select the Chart Filter icon. This opens a box containing data series. Click on the Edit series icon at the right side of the data series you want to edit. You will see the icon when you hover your cursor on the data series.

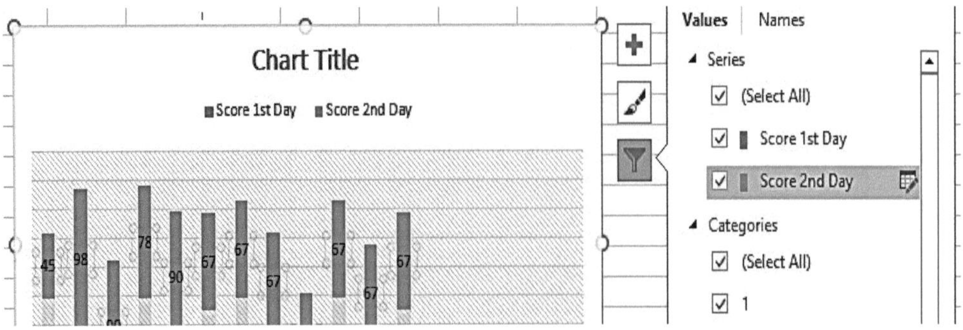

When you click on it, it opens the **Edit series** dialog box. You can now edit the series name and value from the box.

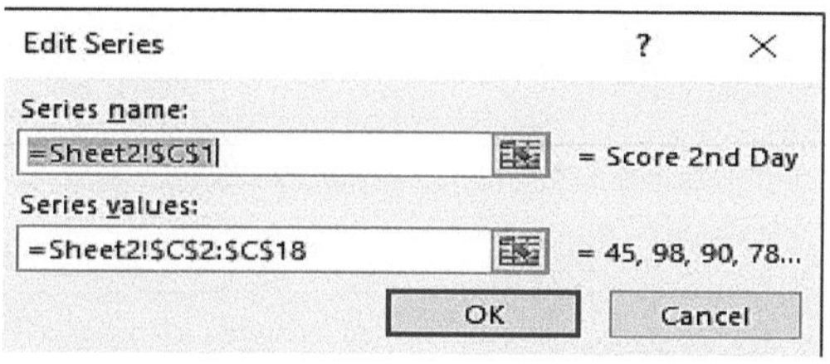

Editing the Series formula

The data in your chart is connected with the series formula. This formula is only available for a chart. When you click on a series in your chart, you will see the series formula on the formula bar. The formula is generated by default. It is written by Excel after creating a chart or adding a series.

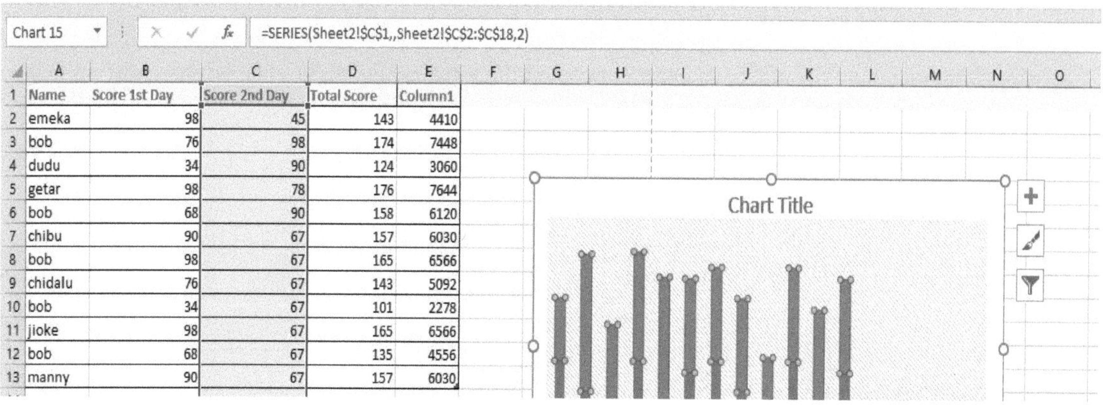

	A	B	C	D	E
1	Name	Score 1st Day	Score 2nd Day	Total Score	Column1
2	emeka	98	45	143	4410
3	bob	76	98	174	7448
4	dudu	34	90	124	3060
5	getar	98	78	176	7644
6	bob	68	90	158	6120
7	chibu	90	67	157	6030
8	bob	98	67	165	6566
9	chidalu	76	67	143	5092
10	bob	34	67	101	2278
11	jioke	98	67	165	6566
12	bob	68	67	135	4556
13	manny	90	67	157	6030

The formula for the data series I just selected is written in the formula bar as **=SERIES(Sheet2!$C$1, Sheet2!$C$2:$C$18,2).**

So, you can add a new formula. You can change the "**C**" in the series formula and add another alphabet like "G". The chart will apply the series name and Y values in Column G.

Displaying data labels in a chart

You may add data labels to your Excel graph to highlight information about the data series, making it simpler to interpret. You may add labels to one data series, all data series, or individual data points, based on where you'd like to concentrate your users' attention.

Simply select the data series. If you want to add a label to just a data point, select the data point after choosing the series.

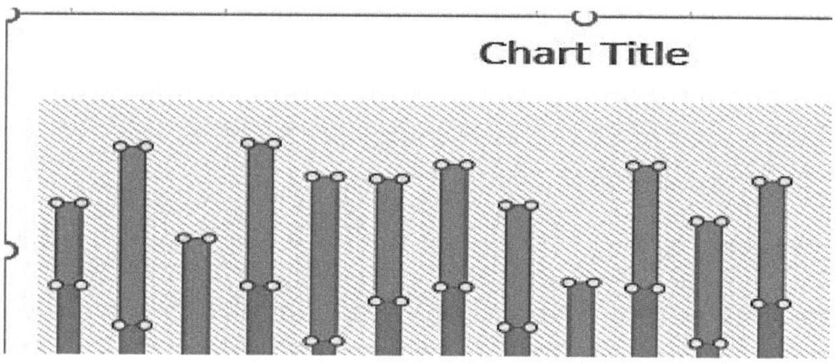

Then, click on the Chart Elements icon, then, check the box beside Data Labels. You can click on the arrow to decide the location where you want to place the data series.

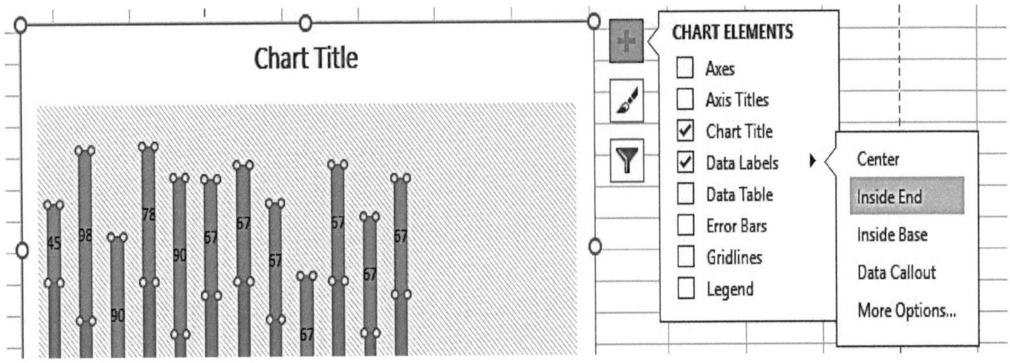

Handling missing data

The data you're graphing may be missing a data point or more. Excel has some solutions for dealing with missing data, including the ability to chart concealed data in a range. Right-click on the chart and choose **Select Data**. Click on Hidden and Empty Cells.

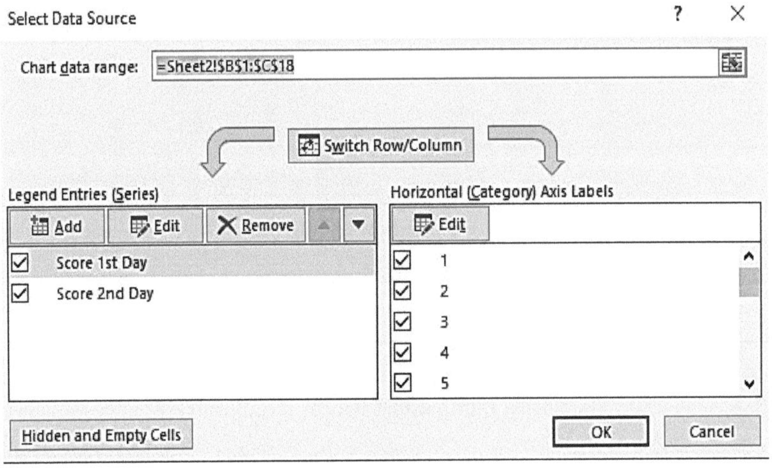

The Hidden and Empty cells box will open. Then, decide how to handle the missing data.

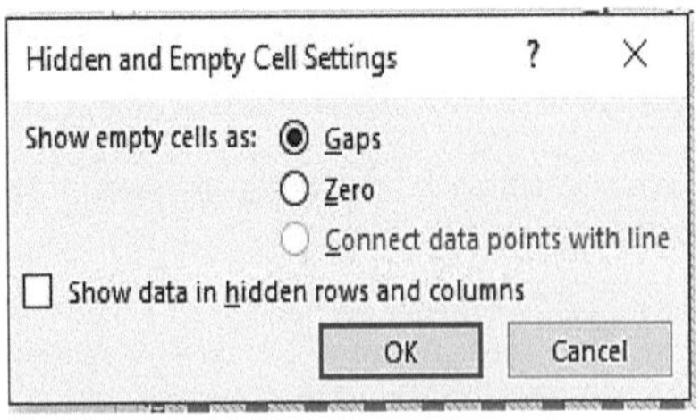

Gaps: This is the default option of Excel. The missing data is not included in Excel. A gap is given by the data series for every missing data point.

Zero: The missing data is seen as zero.

Connect Data missing data as zero: The missing data is calculated with the data on either side of the missing point(s). This only works for some of the X Y scatter subtypes and line charts.

Adding error bars

To add error bars on your chart, simply click on the chart, then select the Chart Element and check the Error bars box. You can click on the arrow to select the kind of error bar you want to add.

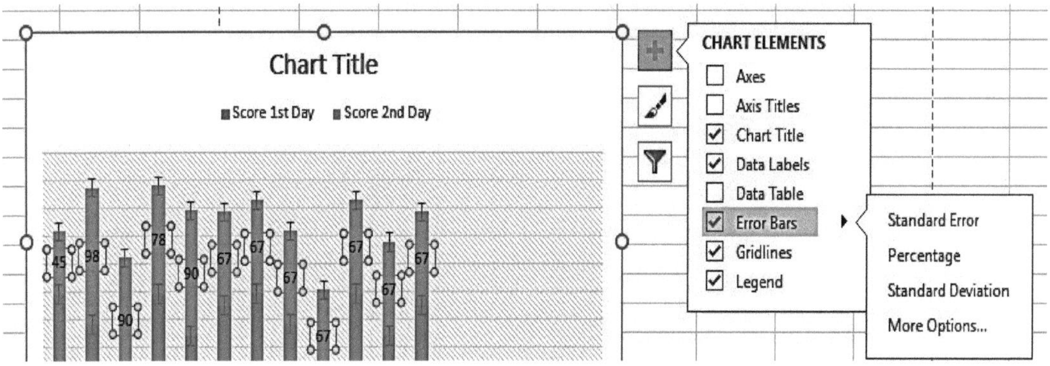

Creating combination charts

A chart that consists of two or more chart types in a chart. Follow the steps below to do so;

Simply select the cells/range. Click the Insert tab and select the Combo symbol.

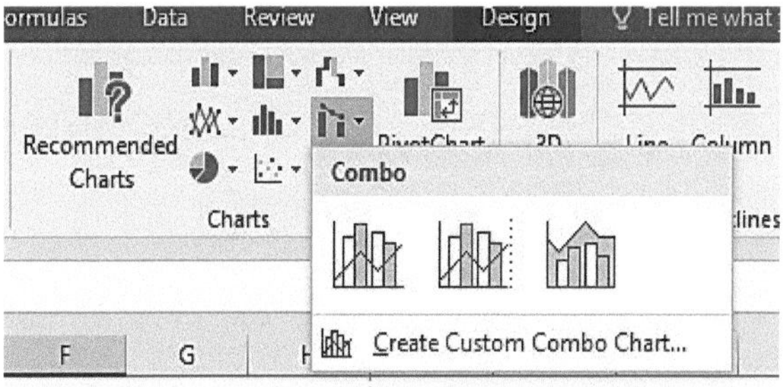

Select Create Custom Combo Chart. This opens the Insert Chart menu. Below the menu, select the chart type for the series name you have selected. You can select two different types of chart types. You can check the box on the secondary axis options. Then click Ok.

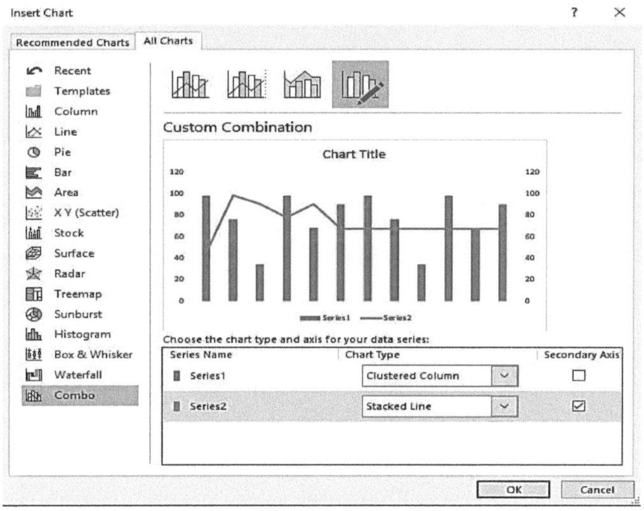

You will get the chart on your worksheet.

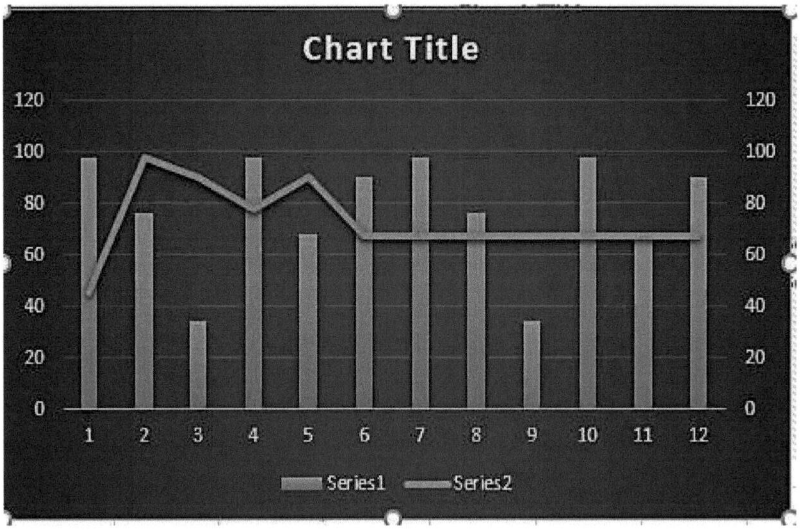

Displaying a data table

Select the chart you would like to display its data table. Click on the **Chart Element** icon, then check the **Data Table** box. Click on the arrow beside it to choose if you want to display it with legend keys or not.

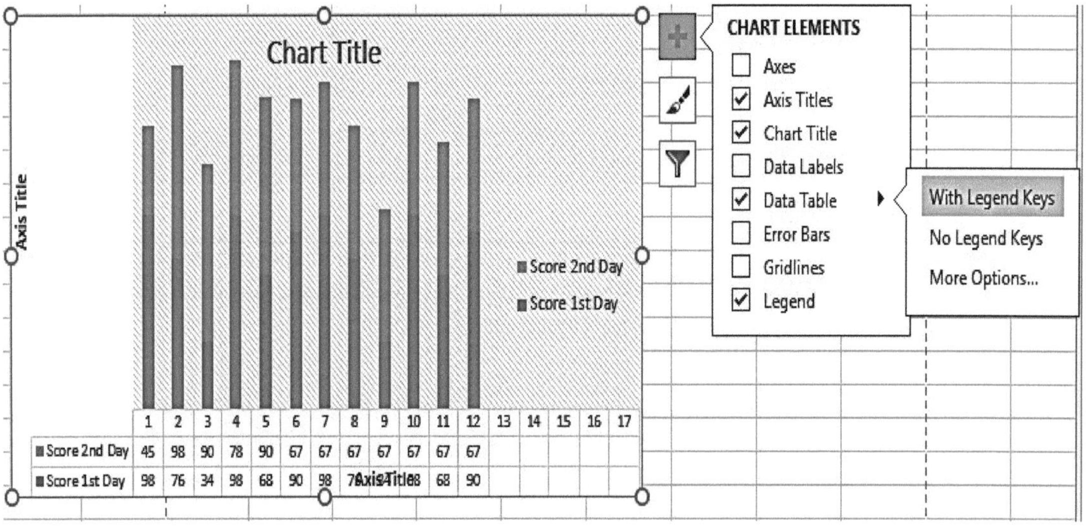

Creating Chart Templates.

After creating a chart, right-click on it and select Save as Template. The Save Chart Template dialog will open. Type in the name of the template then selects Save. Chart Templates are saved in the Charts folder. They are included in the Templates folder in the Insert Chart and Change Chart Type dialog.

CHAPTER ELEVEN

VISUALIZING WITH CUSTOM NUMBER FORMATS AND SHAPES

Visualizing with Number Formatting

Number, currencies, proportion, accounting, dates, and times are all designed formats in Microsoft Excel. However, there are times when you need something quite special. You may develop your number format if none of the built-in Excel formats satisfy your requirements.

Number Formatting is a tremendously effective technique. The goal of this lesson is to walk you through the components of Microsoft Excel number formatting so you can understand custom number formatting.

Doing basic number formatting

So, highlight the cells containing the numbers for formatting. On the Number group, the default format for your number is General format. So, click the down arrow close to General and select the format you want.

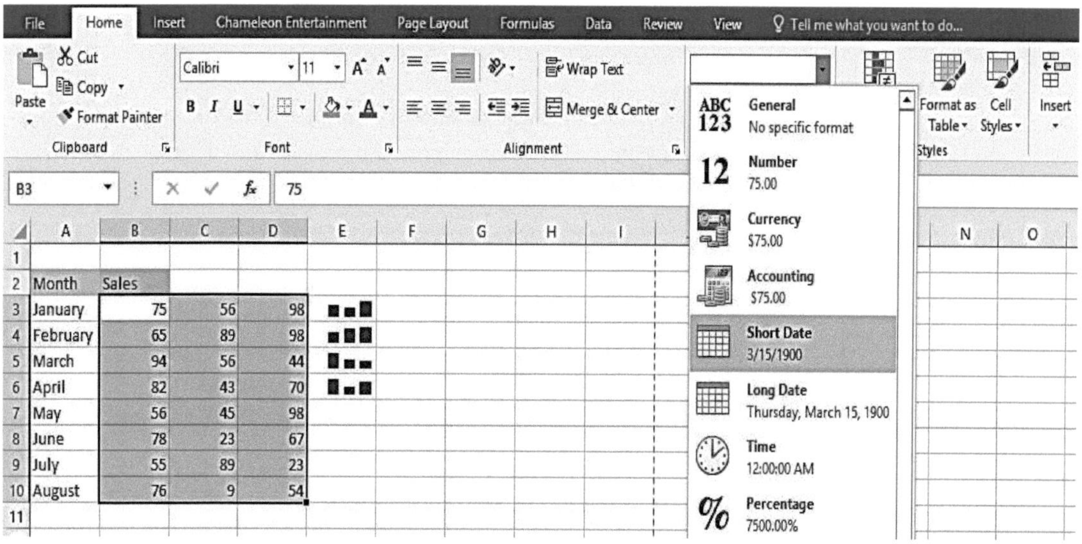

I formatted the numbers using the **Accounting** format option.

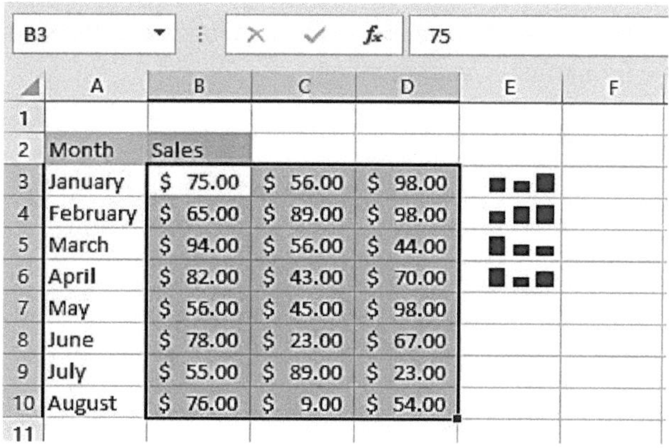

Using the formal cell dialog box to format numbers

You may not find the formatting option on the menu. So, to find them, simply highlight the cells containing the numbers, then on the Numbers group, click on the down arrow which is the dialog box opener.

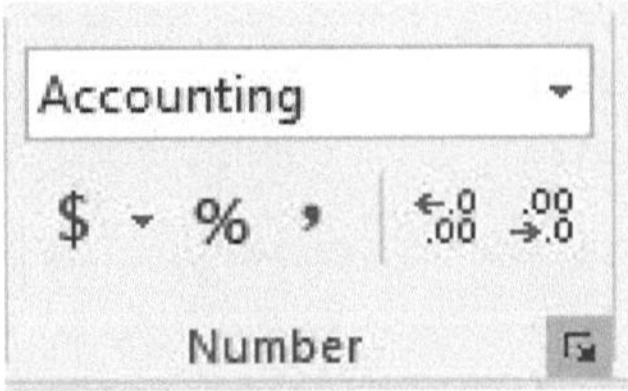

This will open up the Format Cell dialog box. On the Number Tab, you will see lots of number formatting options. Click on the one you want to use, then click **Ok**.

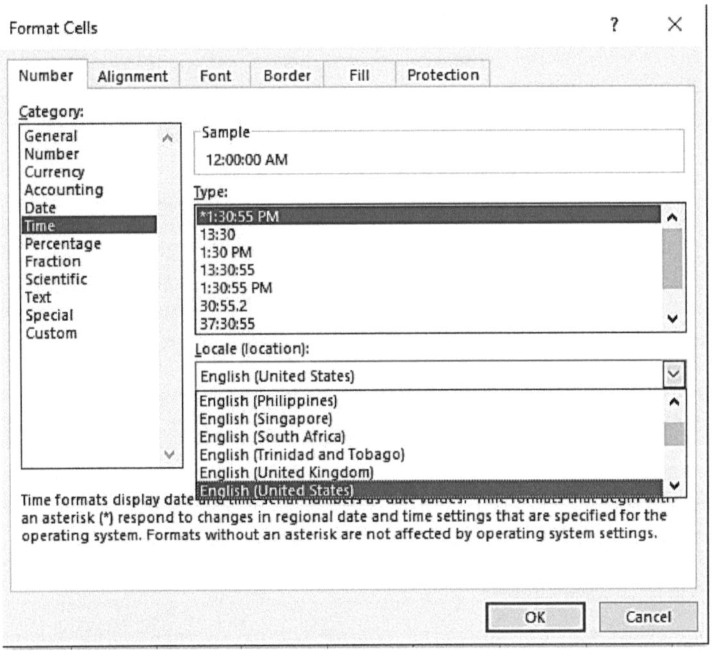

Using shortcuts keys to format numbers

With shortcuts, you can easily format the numbers in your worksheet. Use **Control + Shift + Number key**. With it, you apply some formatting to your numbers.

Simply select the cell(s), then press the keys below;

Control + Shift + ` = **General**	Control + Shift + 5 = **Percentage**
Control + Shift + 1 = **Number**	Control + Shift + 6 = **Scientific**
Control + Shift + 2 = **Time**	Control + Shift + 7 = **Border**
Control + Shift + 3 = **Date**	

So, press and hold down the Control key and Shift key, then press any of the numbers on your keyboard.

GETTING FANCY WITH CUSTOM NUMBER FORMATTING

Formatting numbers in thousands and millions

In Excel, you may wish to format a number in a custom manner, such as formatting 421020000 as $421.02 M, and then store this for future use. You can also apply it to multiple cells. To do this, follow the steps below.

Select the cell that has a number value in it, right-click on it, then click on **Format Cells**.

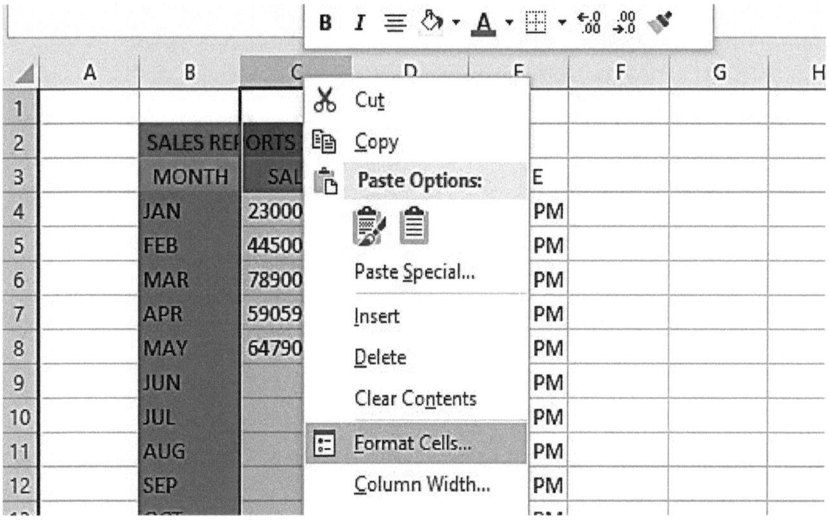

Click on the **Number tab** on the Format cell box, then click on Custom. On the box below the Type option, type in the format code $#.##,," M";

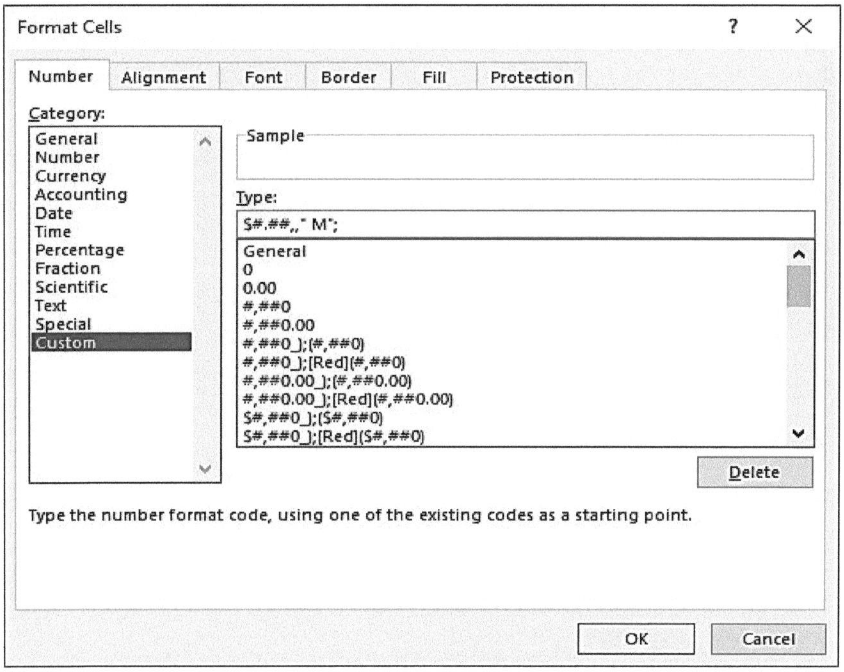

Then, click OK

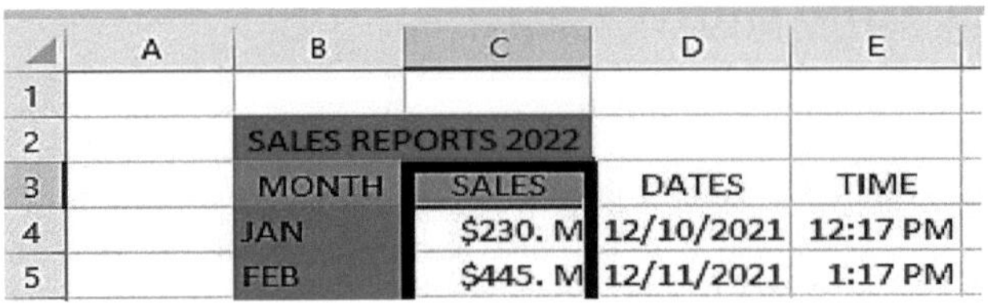

Hiding and suppressing zeros

The number format can hide the zeros in the values of the cells. The hidden values are shown in the formula section. When any number in the cells changes to a nonnegative value, the value will be displayed in the cell with a format similar to that of a general format. Follow the steps to do hide zeros;

Click on the cells that have zero value in them. On the Home tab, select Format then select Format Cells.

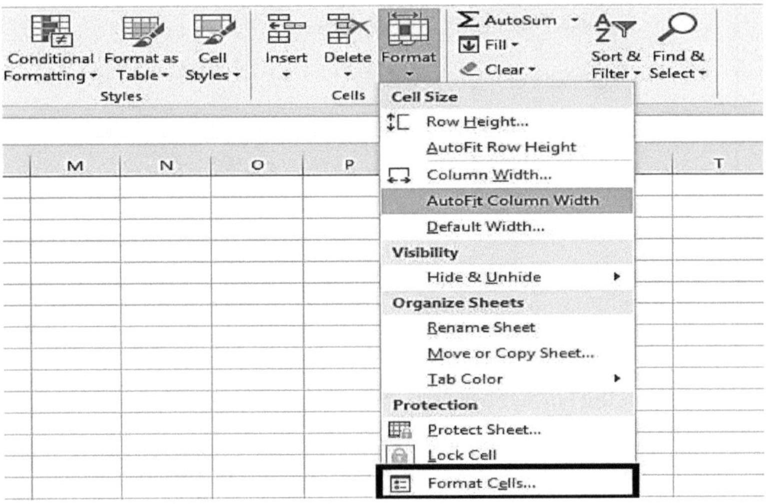

On the Format Cell box, on the Number Tab, click on Custom. On the Type box, put in this format code **0;-0; @**. Click Ok. The Zeros will be hidden.

To display them again, select the cells, navigate to the Format cell box, then on the Number tab, click General, and click Ok.

Applying custom format colors

So to apply custom format colors, simply highlight the cells, then open up the Format dialog box with the steps we explained before.

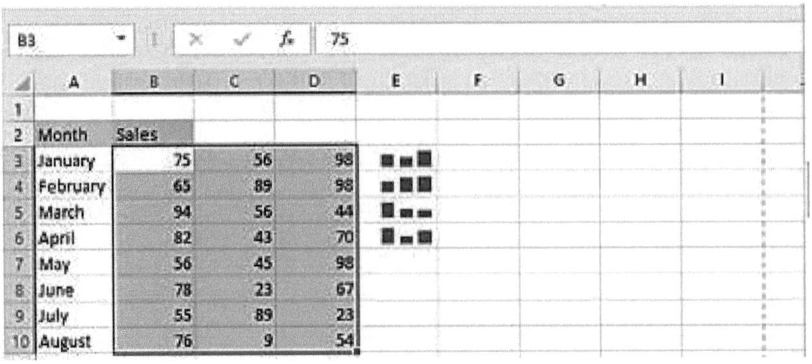

Then, on the Formal cell box, click **Custom**. On the Type box, click any format option there. In my case, I go with **"#,##0".** So, I want to create a custom currency format. So, we add a dollar sign. You can see the preview of what you are doing in the **Sample** box.

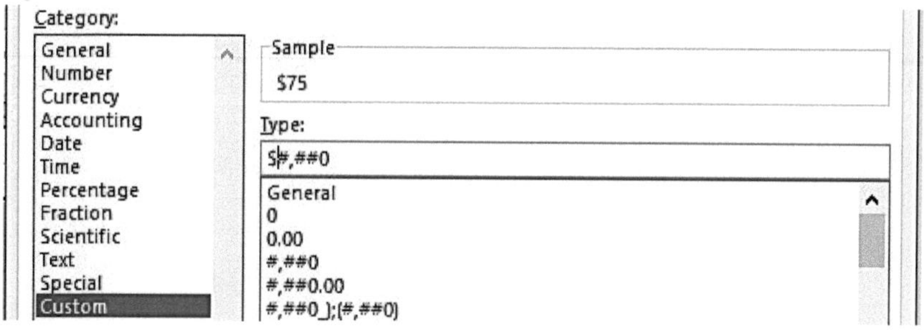

So, to apply color to the custom format, type in the color name at the front of the format and it should be written inside a square bracket.

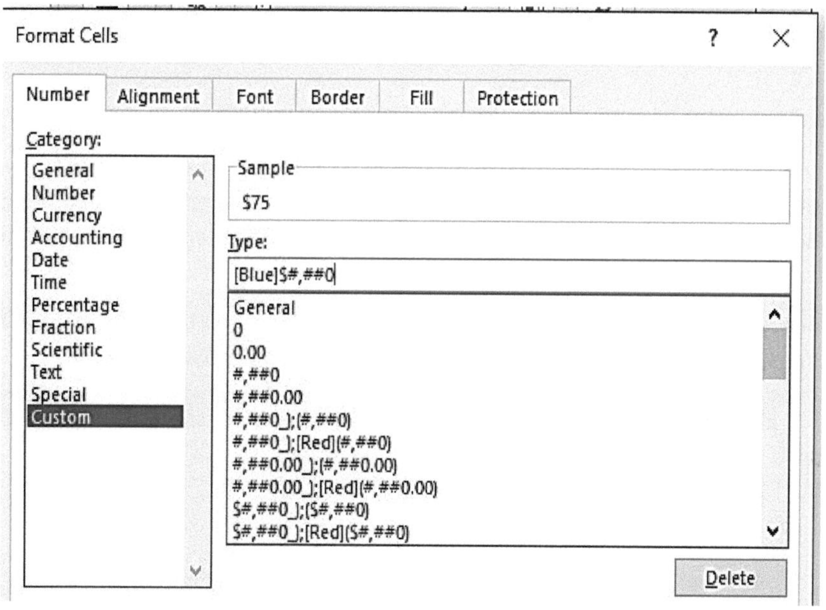

You cannot see the color preview in the Sample box. Click **Ok**. It displays on your worksheet.

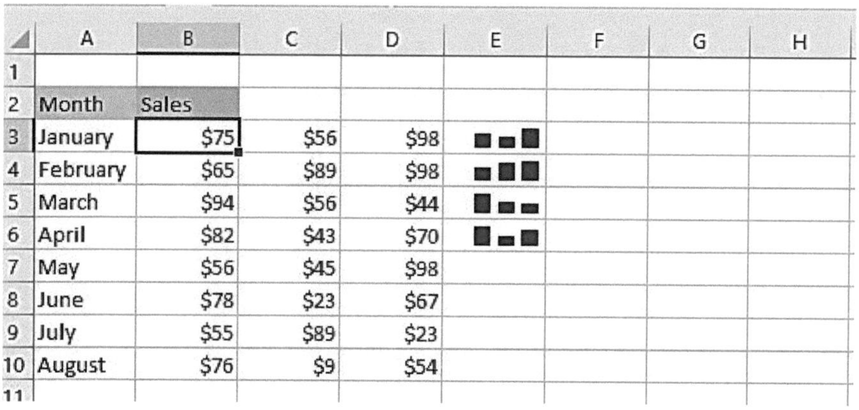

Formatting dates and time

Right-click on the cell, then pick **Format Cells**. On the Number tab, pick Date or Time. Select an option. Then, click **Ok**.

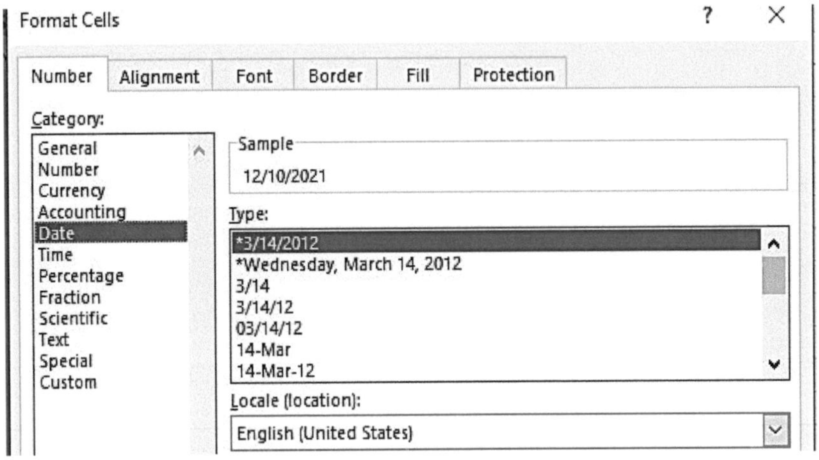

Using a symbol to enhance reporting

Including symbols in your worksheet may help you visualize your data. Here, I will go with the up and down arrows in our number formatting to indicate the increase and reductions in our data. The symbols we'll employ are Unicode symbols, which are a platform-agnostic standard that may be found on any language PC.

First, we will access the symbols by inserting them. So, click on a cell that has no data in it. Click Insert and select the Symbol icon on the Symbols group. This opens up the Symbol menu.

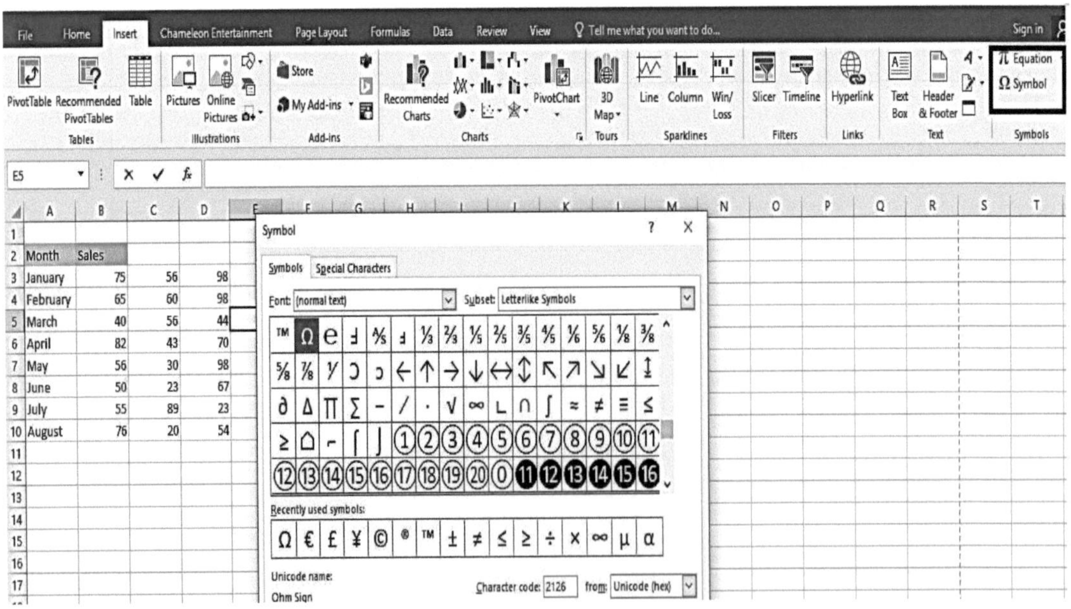

Select the font you want from the **Font box**. On the **Subset box**, choose **Geometric Shapes**. From the list of shapes, choose the **up arrow** shape and click **Insert**. Choose the **down arrow** shape and click **Insert**. Then, close the menu.

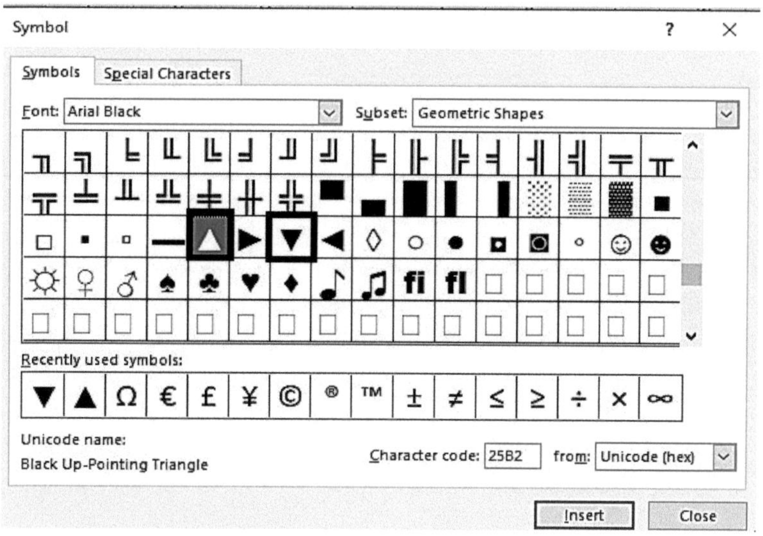

Now let's use the symbols in our formatting

Choose the data to include the symbols. Then, right-click and choose Format Cells. Select Custom on the Insert tab. In the type box, type in this format code **[Green] ▲ 0.0%;[Red] ▼ 0.0%.** Click **Ok**.

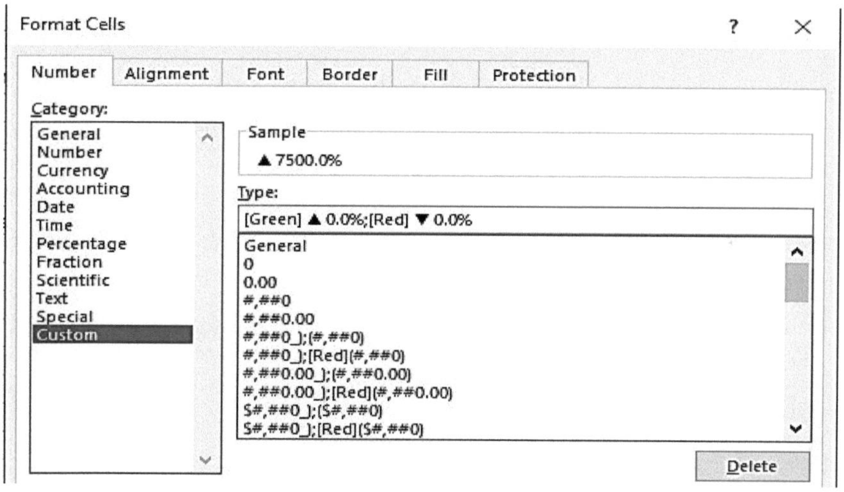

The positive numbers will change to green and the up arrow will be there. The negative numbers will change to red and the down arrow will be there.

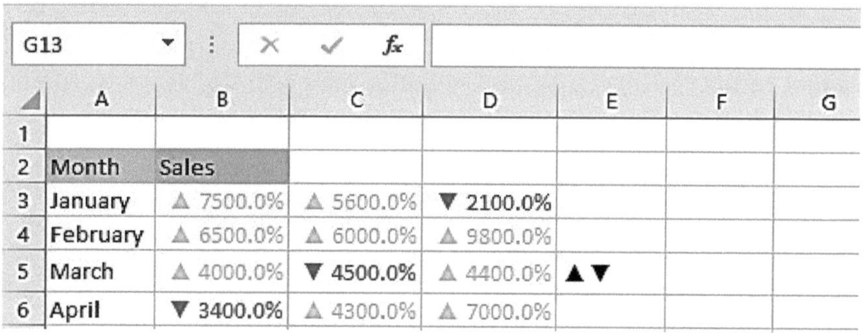

USING SHAPES AND ICONS AS VISUAL ELEMENTS

Inserting a shape

Using shapes in your worksheet helps to make your work neat and understandable. To insert a shape, click on the Insert Tab. Click on the **Shape icon** on the illustration group. Select a shape.

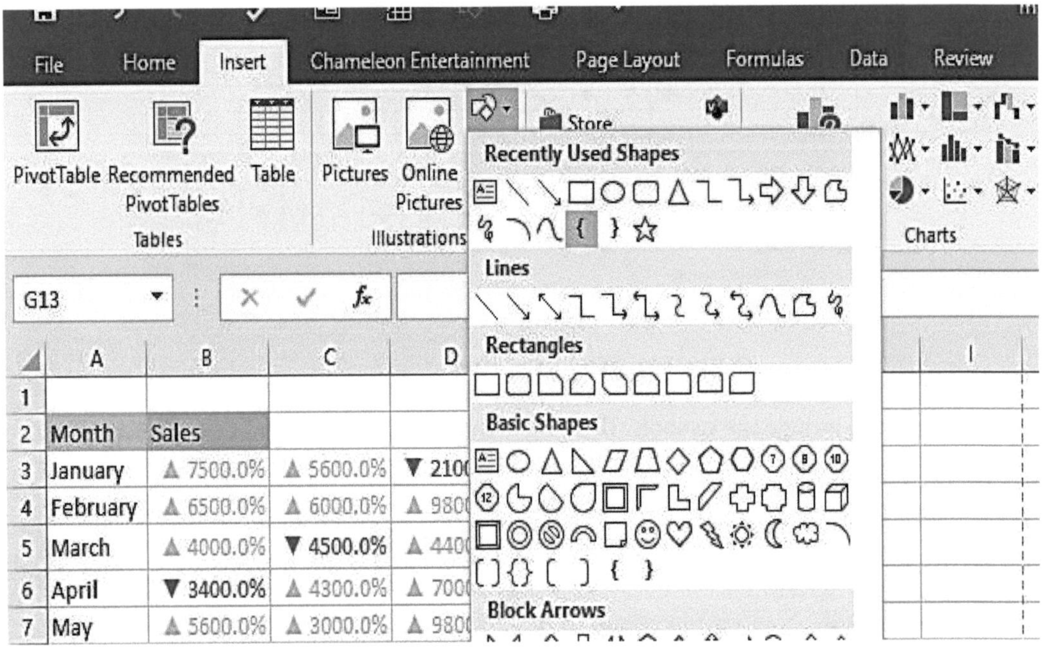

Inserting SVG icon graphics

SVG means Scalable Vector Graphics. To insert them, click Insert Tab. Click on Icons. This will display different icon categories. Click on any category and select an icon. Then, click **Insert**.

You can insert more than one icon at the same time by clicking on them. You can also search for icons. Click **Insert**, the icon displays on your worksheet.

Use the Graphic Format tab to modify the icons.

Enhancing Excel reports with shapes

With Shapes, you display a perfect presentation of your work. You can use shapes to show any common chart in Excel. Excel shapes are used to enhance the appearance of your dashboards and reports, as well as for realistic and efficient applications. We'll look at ways to make your worksheets more appealing with shapes.

Creating Custom Button Links: Creating custom button links with shapes is a wonderful way to move across worksheets, connect to reports on shared servers or just link to a webpage.

To create a Hyperlink out of a shape, simply insert the shape. After adding the shape to your worksheet, add text to it by right-clicking on the shape, then, select **Edit Text**.

Once you have added a text, press **Control key + K**.

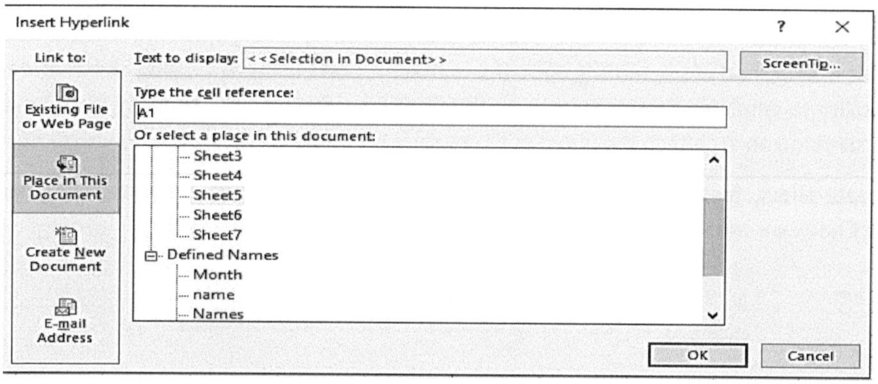

Here, I use the Place in this document option. We will choose the sheet for linking. So, choose the "**existing file or web page**" option. Then, get the location of the document or you put the address on a website.

Building Dashboards: Excel shapes most times, are used to enhance and update the appearance of data in a worksheet. Flow charts and aesthetically pleasing spreadsheets are often the first things that spring to mind when people think about shapes in Excel. Smaller forms and other shape attributes may be used to create some nice-looking dashboards.

Layering shapes to save space

The image below has different shapes on my worksheet like text box images, charts slicers, etc.

To align your shapes, use the **Align command** to align them to the top, bottom, right, or left to save some space on the worksheet. Simply select the shapes by holding down the Control key and clicking on the shapes. Then, on the Format tab, click **Align**, select an option.

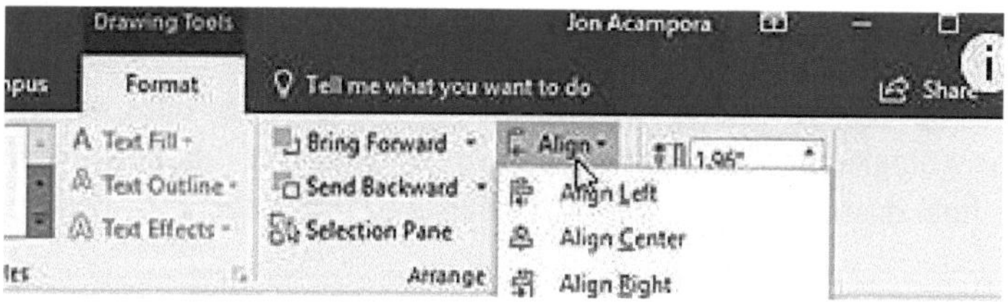

Constructing your infographic widgets with shapes

Infographic is a method of presenting data or summary reports using appealing charts and pictures. They are the ability to employ external objects or visuals to visualize images. They are not like dashboards in that they demand an architectural mindset to create your infographics.

So, to create them, first select the cells, then on the Insert tab, click on Bar chart, then choose 2D Clustered Chart. we will have a chart like the one below.

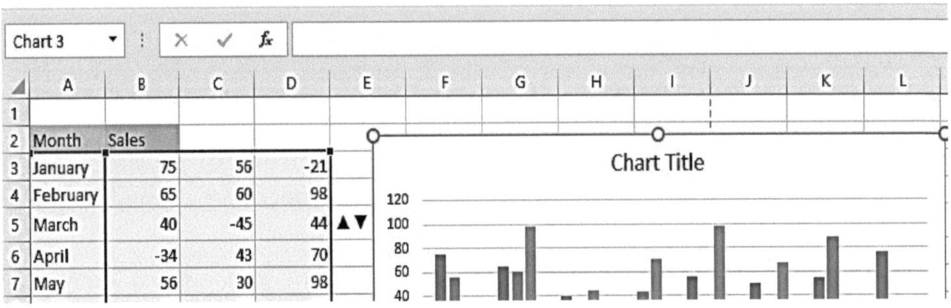

Right-click and select Format Data Series. On the series option, set the Gap width to 0%. Make the chart background to be light white using the Shape Fill icon. Adjust the chart size for the image to fit in.

Creating dynamic labels

To progress beyond your basic worksheet abilities, creating dynamic charts can enhance it. The idea is to pick dynamic range as the source data. Changes and additions to the source data will be instantly reflected in the chart.

We will first create a table. So, select the data range, then, on the Insert tab, navigate to the Tables' group and select **Table**. Select the table. If your data has headers, check the box next to **My Table Has Header.** Then, click Ok.

The chart built on the table is going to be dynamic. Now, let's create a chart. Highlight the table. Then on the Insert tab, on the Chart group, click on **Column Chart** and select **2-D column chart**. When you update the table. It will reflect on the chart automatically.

Creating linked pictures

Picture link makes an image of grouped cells in form of a picture. You can move it around the worksheet as well as resize them. The picture will update as the source cells changes.

First, pick the cells and copy them by pressing **Control + C**. Pick the cell for creating the link picture. Click **Paste** (Ribbon) > select **Paste Special,** and pick **Picture Link**. The picture link will display on your worksheet.

Some do ask, the importance of creating a linked picture and when is it needed?

At first glance, picture links may appear to be a pointless function. They are, nonetheless, rather strong. Here are a few examples of how picture links may be used:

In dashboards and reports, we usually need to mix charts, tables of data, conditional formatting, and other elements on a single page. Build the different dashboard components in different spreadsheets first. Then, in the final dashboard, provide picture links to these sections. Resize and position them as desired.

In Dynamic Charts: Because picture links are just photos with a formula applied to them, you can quickly create dynamic charts and dashboards with them.

Picture link helps as proof to your data just in case someone makes some changes to the data.

USING SMARTART AND WORDART

SmartArt basics

SmartArt helps you to add diagrams, captioned pictures, and visual lists in your work presentation. It comes in different forms, colors, arrangements, etc. To insert a SmartArt, click on the Insert tab, then on the Illustration group, click on SmartArt.

This will open up the SmartArt dialog box. On it, you will see different SmartArt options.

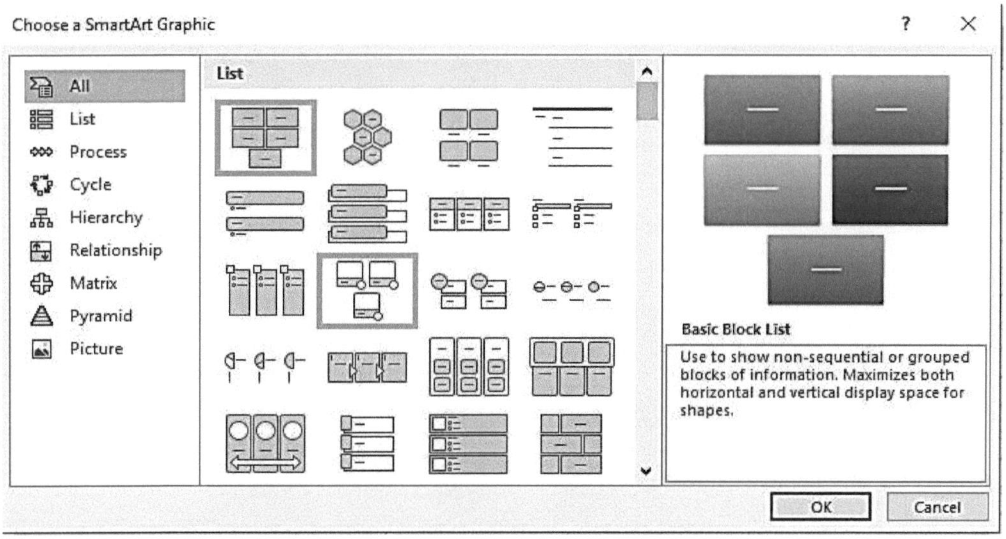

Pick any SmartArt graphics, then click Ok. Check for the image in your worksheet. On the [Text} placeholders, you can type in titles, captions, pictures.

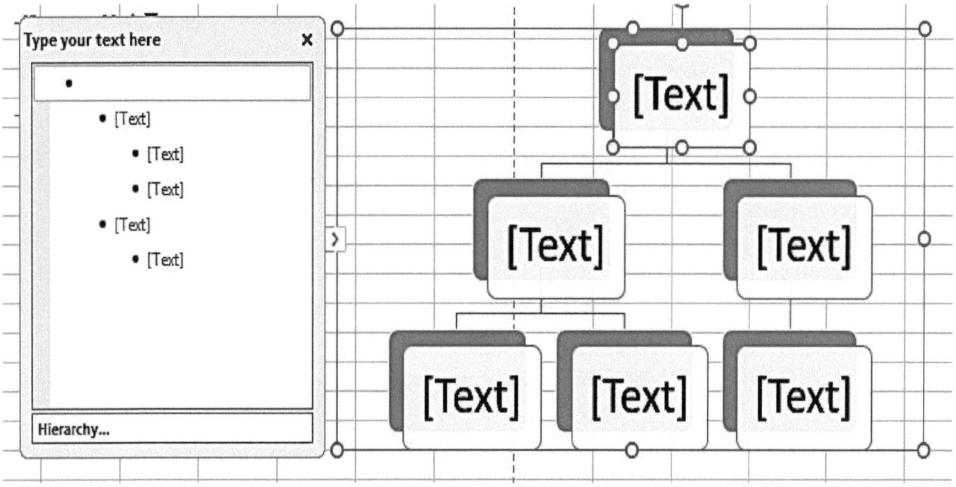

Modify the SmartArt Graphics such as changing the layout, styles, color, etc. using the SmartArt tab.

WordArt basics

Select WordArt from Insert Tab. Select a style.

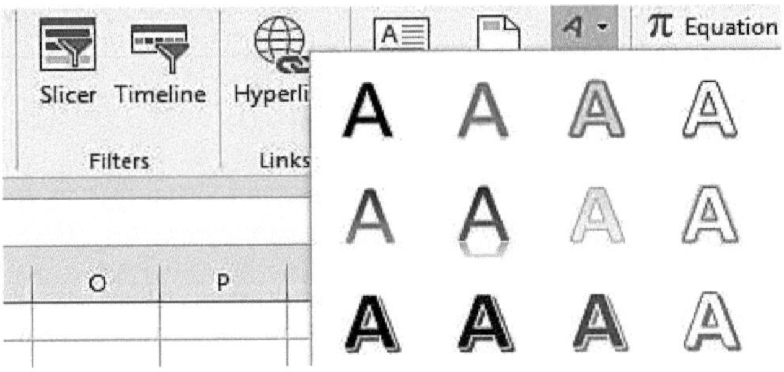

Click and type in your text.

Modify WordArt with the Format menu. You move it by dragging the edges.

WORKING WITH OTHER GRAPHICS TYPES

About graphic files

When most people think about worksheets, the last thing that springs to mind is art. Graphic images, when utilized with care, will not only draw attention to otherwise dull tables and lists of statistics but also improve the readability of their material. Hand-drawn visual shapes that you may add to the spreadsheet to bring attention to unusual data points.

There are types of graphics in Excel for enhancing your worksheet. They are Clip art images provided by Microsoft, hand-drawn graphics (shapes, lines, text, WordArt), and graphics files that consist of digital images or digital photographs.

Inserting screenshots

Click Insert, then click on Screenshot. This will display a menu where you are to take a screenshot of any open window on your desktop or take a screen clipping of just a part of your screen

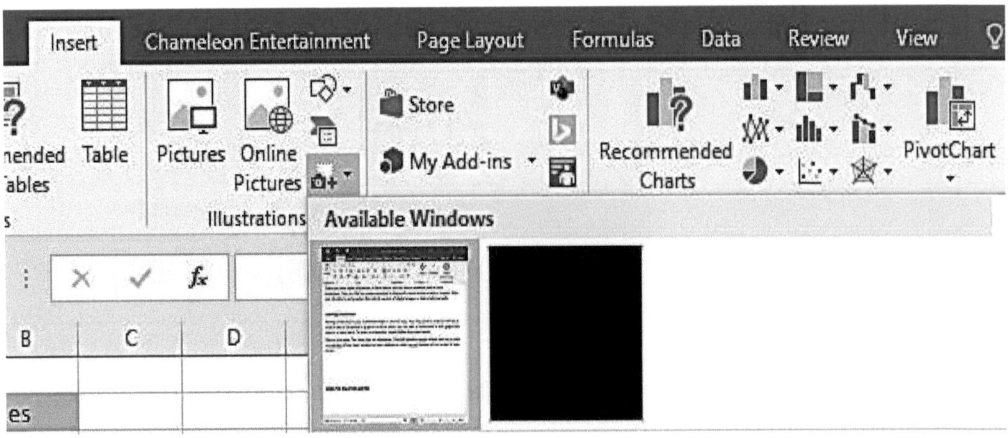

Choose any of the options and the screenshot will display. You can modify the screenshot using the Format tab.

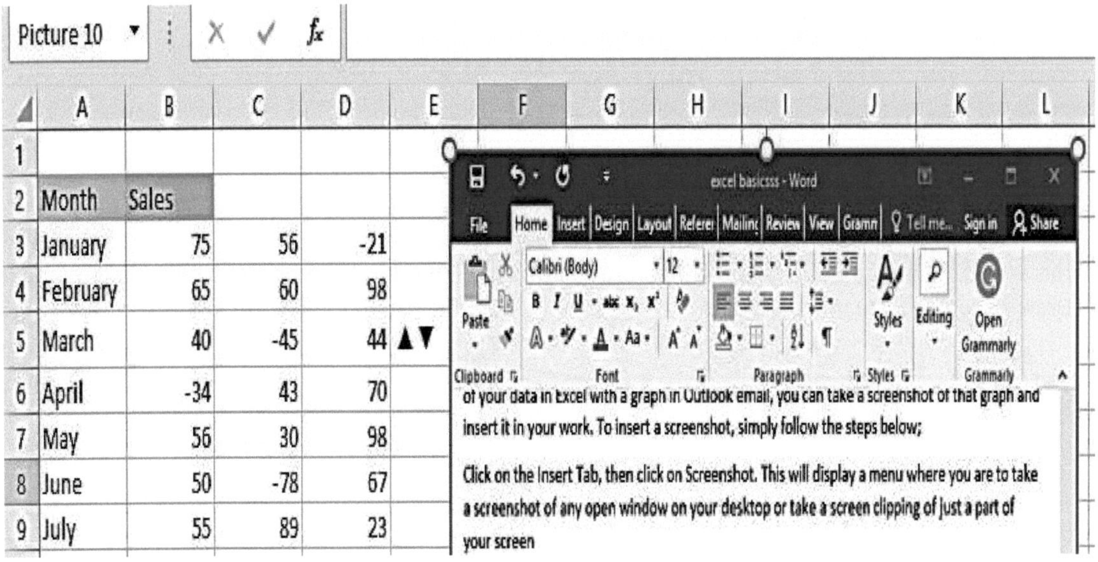

USING THE EQUATION EDITOR

The Equation editor puts in equations or format equations. To insert equations with the Equation editor, simply;

Click Insert, under the Symbols group select **Equation** and a drop-down menu will appear. This will display some of the preinstalled equations in Excel. Click on any of the equations.

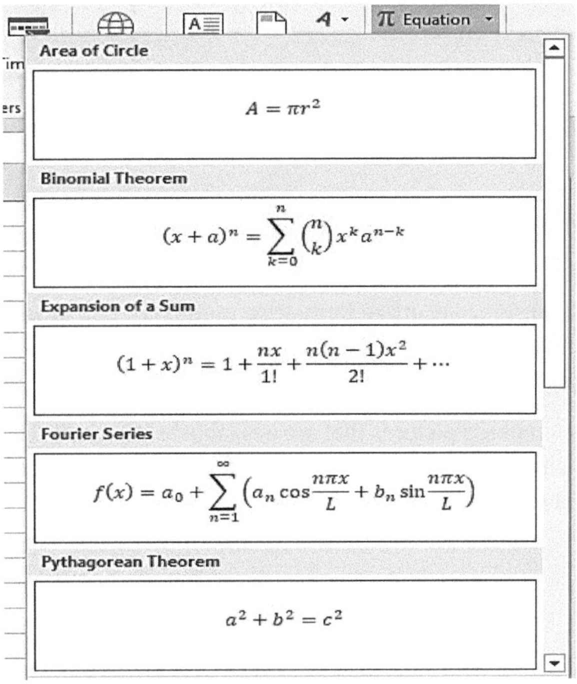

Here, I selected the Expansion of a Sum Equation. Then click Ok.

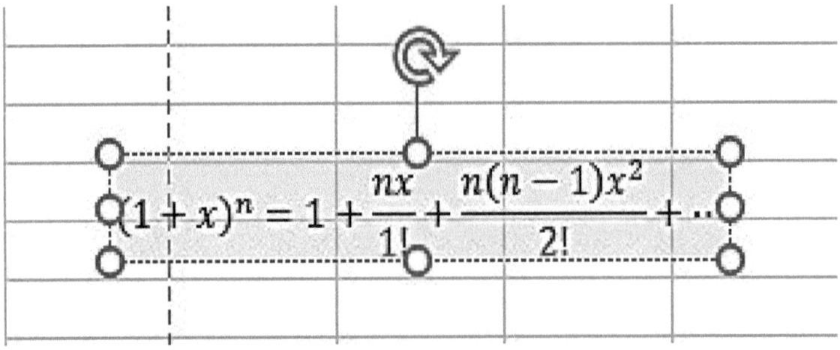

You have just inserted an Equation. You can also make your equation. To do this, click on Equation (not the arrow). A text box will appear with a caption of **Type Equation here**. Modify the equation with the Drawing too and Equation tool tab.

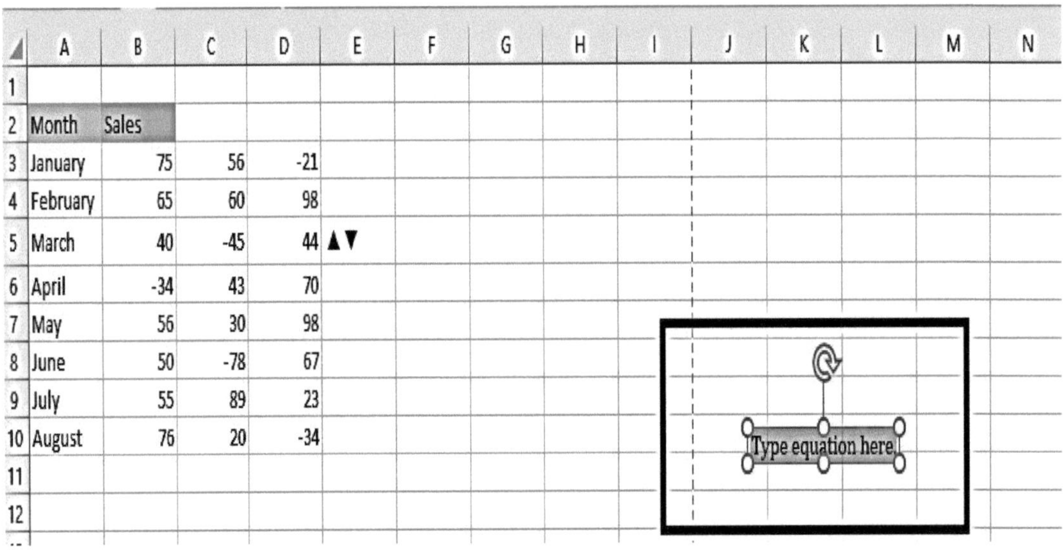

You can right-click on the box, then select Format Shape for more modifying options.

CONCLUSION

Number formatting, like all of Microsoft Excel's tools, has a multitude of options. This chapter just skims the surface of what's possible. Number formatting may be something you're already familiar with.

This chapter looks at how users may utilize formatting methods to construct layers of visualizations and transform their data into useful representations. It explains how to develop custom number formats and gives several examples that users may use as-is or modify to meet their requirements.

Shapes are configurable graphical pictures that are available in Microsoft Office, including Excel. Scalable Vector Graphics (SVG) icons are included in Excel 2019's new icon collection. Without sacrificing picture quality, SVG graphics may be resized and formatted.

BOOK 3:

MICROSOFT OUTLOOK

INTRODUCTION

Microsoft Outlook is a personal resource management system software created by Microsoft and is also available as part of the Microsoft suite. Though it is primarily an email client, it also has other functions which include calendaring, managing tasks, managing contacts, note-taking, web browsing, and journal jogging, amongst others.

While Outlook can be used by individuals as a stand-alone application, organizations make use of it as a multi-user software (through the use of Microsoft Exchange Server and also SharePoint) for the use of shared functions such as mailboxes, folders, data gathering, and the scheduling of various appointments.

Microsoft has designed applications for use on most mobile platforms, which include Apple iOS and Android. Furthermore, all data in Outlook Mobile can be synced from Windows Phone devices. With the use of Microsoft Visual Studio, developers now know how to create their unique custom software that can be used with Outlook and other Office packages.

Just like that of the desktop version, Outlook mobile provides an aggregation of attachments and files that are saved on cloud storage platforms; a "focused inbox" that highlights messages from contacts that are used frequently; the addition of calendar events, files, and the inclusion of locations to messages without the need to switch apps. The application also provides support for various emailing platforms and services like Outlook.com, Microsoft Exchange, and Google Workspace, formerly known as the G Suite, amongst others.

Outlook Mobile was created to bring together functionality that was normally supposed to be found in different applications on mobile devices that are close to the personal information managers on PCs. It is created around four major hubs, which include mail, calendar, files, and people. The people hub regularly enlists contacts and aggregates communication that is used often with them, and the files hub also has a way of aggregating messages and also allows you to integrate with other storage devices online like Dropbox, Google Drive, and OneDrive.

Outlook provides support for various email services and platforms, which include Exchange and iCloud. Google Workspace, Outlook.com, and Yahoo Mail. The application also supports the use of different email accounts at once. Just like the desktop version of Outlook, Outlook Mobile enables users to see appointment details and also reply to exchange meeting invites and schedule meetings. It also adds the three-day view option and lovely calendar options from Sunrise.

As of March 2020, Microsoft announced the launch of a series of new products to appeal to business customers of its Team platform, in addition to the products that were introduced the month before. The chat and collaborative modules now have more efficient and integrated waypoints specifically created to make group work for organizations simple and to also give encouragement to such an organization to use the Microsoft platform to become the preferred company chat platform.

With the very first version of Outlook being Outlook 97 version 8.0, released on the 16th of January, 1997, Microsoft has now released a better and more packed version. Outlook 2019, version 16 was released on the 24th of September, 2018, and is also included in the 2019 Office Suite and also the online version, i.e., Office 365.

Some of the features of Outlook 2019 include a focused inbox, the inclusion of multiple time zones, the option of listening to emails rather than just reading them, a much easier and faster way of sorting email, an automatic download of attachments to the cloud, and also the True Dark Mode that helps to reduce the brightness of the screen and also increases the life span of the battery.

CHAPTER 1

OUTLOOK FEATURES YOU NEED TO KNOW

Explaining Why So Many People Use Outlook

Most people prefer to make use of Outlook rather than Gmail accounts. Outlined below are some of the reasons:

Its compatibility with email, contacts, and calendars. The main mode of communication, especially in business, today remains email. Most users opt for the option that Outlook has just one place for effective communication at work for things to get done effectively. It doesn't matter if it is to have meetings organized, the need to gather more information about a contact, having to put a call through a conference call, or jump into an online meeting, Outlook seems to be the perfect hub. Users say Outlook helps them get through their daily tasks without having to think too much. Each Morning Outlook has been designed in such a way as to wake them each morning by reminding them of their activities that have been scheduled for the day.

Offline access to email messages

Outlook provides users the option to access their emails when they are not connected to the internet. This means users can catch up with work-related mail even when they are on a plane and have the airplane mode on their phones turned on. Most of them also work on Outlook throughout their journey, since they can read and respond to email just like they would if there were an internet connection. Once the internet connection is established, the messages will be sent automatically.

In total contrast, Gmail only gives access to email messages from the previous months once there is no internet connection, which can only be accessed in an offline mode on Chrome and Safari browsers. Outlook does not have such restrictions.

A customized mode of organizing email

Not everyone likes to work alphabetically sorting through their folders. Some like to make use of the search box to find their email, while others enjoy the use of specific bins to search for their mail. This makes the use of Outlook more rampant. In comparison to Outlook, users have fewer options for organizing their emails on other platforms. They are limited to the use of just labels to organize their email.

Grouping email items for better organization

With the use of Outlook, users can attach groups to different items like calendars, appointments, etc. An item can be used to attach to different groups to have them classified in more than just one way. At the slightest glance, users can see how their time is spent across various areas by just looking at the color-coding of the events on the calendar. Those who make use of groupings and color-coding to keep their work organized won't be able to imagine how they would cope without them. Google has rendered color-coding useless due to its inability to differentiate items in their different boxes through the use of colors.

Rules to reduce the clutter of email

With the use of Outlook and the consistent push to upgrade the app, Microsoft has found a way to better manage the clutter in the box even as research is ongoing to make progress in that regard every day. Gmail is equipped to filter, but the actions that should be taken after the filtering of the emails are quite limited. Setting up very important rules to flag important emails based on the sender of the message is much easier with the use of Outlook.

Access to rich contact information

The experience shared when being able to locate a person's office with the use of their contact card is simply amazing. The contact card in Outlook provides a lot of rich information about a person. The picture in the contact card also helps users link people's faces with their names. "Presence" information displays the status of the user if they are either busy, online, or offline. Users can also start a chat, book an appointment, have a quick chat over the phone, or start an online conference. Unlike Google mail and other platforms, the contact card in Outlook has a person's job title, department, and also their location information.

Scheduling meetings

With the use of Outlook, users can schedule and reschedule resources such as meeting rooms, projectors, etc., just by including them as a resource in the meeting. The resource will then perform like any other attendee and also includes the ability to respond automatically to the meeting invite and the ability to see busy schedules and those that are free. Reports state that attempting to perform this function on other platforms, especially Google Mail, can be very frustrating as it is a cumbersome multi-step process that involves primary and secondary calendars.

The ability to ignore conversations

Not all conversations are worth responding to; some are meant to simply be ignored. Most such conversations are possible outcomes of a person's oversight or who has the last say attitude. Outlook has a very unique way of ignoring such conversations. With a simple click, all present and new emails in that conversation will be moved to the deleted items folder. This will also help to remove clutter from the inbox and also help focus on more important tasks.

Users also like the cleanup option. This way, only the latest email thread with all comments will be kept. The rest of the emails will then be moved to the deleted items folder. All unique conversations will still be kept in one place, but with Gmail, users will only have the option to mute conversations. This ability is not discovered easily, and it will not work if the recipient gets re-added to the line of the email.

Sharing and delegating calendars

Users who are administrative assistants have revealed how tedious it is to have to work with Google Calendar. They believe that with the use of Google calendar, there is no other option apart from sharing all the details on the calendar or just the free/busy schedule. It cannot share the free/busy with the subject of the events because it does not have the level of control as to the details that should and shouldn't be shared.

When they switched over to Outlook from Gmail, they had much better control over the management of calendars. For example, they could decide if they need to share a person's free or busy status as well as the subject of the meeting, attendees, the location, etc. Once the calendar has been assigned, the administrative assistants can also manage the assigned calendars with their own calendars in a very simple and unique way, which will also reduce the confusion of adjusting the wrong calendar.

The very important thing Outlook offers is the ability for it to be used in a way that brings about more productivity in the job, be it as an administrative assistant, helping the team lead, or even for personal use. The availability of various choices and flexibility is what is missed the most with Outlook.

The Easiest Way to Do Anything in Outlook

Outlook manages email, calendar, contacts, and other tasks all in just one hub. Hence, its steady growth since 2017 doesn't come as a surprise. Here are some ways to carry out major tasks in Outlook easily. Allow Outlook to communicate with you.

When we say Outlook talking, it's not really as you may presume, it's simply having to engage Microsoft's Cortana productivity assistant. Once Cortana has been set up in Outlook, there is an option to listen to new emails as opposed to reading them. By speaking only, messages can be listened to, flagged, deleted, and also replied to. Cortana doesn't only help to read emails; he or she can also help to keep things organized. The AI will feed you in on the changes to events and also pick out schedules that clash with others on the calendar.

Using a focus box

Cluttered inboxes can be an uphill task to scroll through. Scan through messages easily with the use of the focused inbox. There is an option to make two identical inboxes, one focused and the other unfocused. You can then toggle between the two with the use of a simple button slide at the top of the inbox.

Outlook can help get mail sorted, moving important messages to "focused" and transferring junk mail to "others". There is always an option to reclassify messages by moving them from one section of the inbox to another. The "Always move to focus" option can also be selected or otherwise if there is a need for communication from that particular sender to appear at a particular place. The interesting thing about the focused inbox is that the more it is used, the smarter it gets, i.e., the ability to depict junk from real messages. It is sensitive to how emails are sorted with the aim of better classifying incoming messages and helping to focus only on what matters.

Share Calendars

Oftentimes, a lot of time is wasted in meetings. While that is being worked on with various organizations, Outlook removes the inefficiency of trying to fix another meeting again.

With the use of shared calendars in Outlook, get to the calendar settings and choose the people you would love to share it with, such as friends, family, or colleagues at work. You can use the color-code option for the different people that you share with and also control their permissions to let them either share, view, edit, or manage the calendar. Details that will be shared can also be chosen between all, only title and locations, or only when I am busy.

Make use of the scheduling assistant

When sharing calendars, there is an option to make use of the schedule assistant instead of having to always send emails to come up with a time to get together. Make an event on the calendar, then add the people that should be invited. The schedule assistant time picker can then be used to drag and drop to a time on the calendar that turns green. This means everyone is available for the meeting.

Get people's attention by mentioning them

When working alongside someone or in a team, saying "Hey Jane", i.e., mentioning the person's specific name, will draw the person's attention. The same can be done with the use of Outlook's @ mentions. When you type "@ Jane" in an email, the person will see that they have been called into a conversion. Furthermore, when people @ mention you, the inbox will show the important sentences that surround the @ mention in the message excerpt. With this, you will know what you need to do at a glance

Customizing the swipe options

In a dating app, for instance, it is expected that you swipe either left or right to pick a potential match. You can be decisive in an Outlook inbox as well. Open Outlook settings and choose from the selections below; delete, mark as read, mark as unread, flag, or archive.

Using Email: Basic Delivery Techniques

If you are new to Microsoft Outlook, follow the instructions listed for the "initial setup of an email account."

Initial Setup of an Email Account

- When Microsoft Outlook 2019 is opened for the first time, the select profile window will be displayed. Confirm the profile when done by selecting the OK button.
- The welcome to Outlook 2019 will then be displayed. Click on the Next button.
- Insert the email address.
- Click on the advanced options button.
- Select I'd like to set up my account manually.
- Click on the connect button. Then choose the account type window that will open.
- If you prefer to check your email with the use of IMAP (recommended), choose the IMAP account type. If there is a need to check your email using POP3, choose the POP account type.

Reading email

To read email messages with the use of Outlook, simply follow the steps below.

- Click the Inbox. This will display a list of messages.
- Click on the messages you want to read. The message text will then be displayed in the reading pane on the right side or lower part of the screen. As each of the messages is clicked, the contents of the messages will then show up in the reading pane.

Note: There is also an option to make use of the arrow keys to move from one email message to the next.

Click on the icon that appears to be like a gear (on the far-right side of the ribbon) to make adjustments to the mail settings.

The reading pane can also be opened on the right or in the lower part of the screen, and it can also be completely closed. If the reading pane is closed, there will be a need to double-click any message to view it in another window.

Answering email

- In the message list, select the message that should be replied to.
- Right click on the message > Choose the reply or reply all option.
- Type the response and then click on the "send" button.

Creating new email messages

- Navigate to the top of the page and choose the new email option.
- On the To line, insert the name or email address of the person whom the message is meant for.
- In the Add, a subject line, insert a brief description of what the message should be about.
- To have a file attached to the message, click on the attach button.
- Type the message and then click on the "send" button.

Schedule sending mail; send it later.

A new feature known as "send later" is now being introduced to Outlook. This feature will allow users to be able to defer sending an email and then choose a particular date and time when the email should be sent. To do this, simply follow the steps below.

- Compose the email message.
- Click on send and send later.
- Choose the preferred date and time when the message should be sent.
- Finally, click on the send button.

Note: The message will always remain in the drafts or sent items folder (based on the device) until it is time for it to be sent. There is also an option to either edit or delete the message at any time until it is meant to be sent. After making adjustments to the message, select the send and send later option. Note that it is not compulsory to have access to the internet when the message is scheduled to be sent.

Forwarding an email message

"Forwarding a message" simply means sending an already existing message without making any adjustments to it.

To forward a message, simply follow the steps below.

- Locate the message list and select the message that should be forwarded.
- Navigate to the top right-hand corner of the message pane > choose the select button or select and then select forward.
- When the message that should be forwarded is already in the message box, select the send button.

It is worthwhile to note that only one message can be forwarded at a time. When a message is forwarded, the original message will remain in the mailbox, and a copy will be sent to the new recipient.

Sending a File

Most of the daily work is likely done in programs other than Outlook. Documents might be made in Microsoft Word or spreadsheets created in Excel. When there is a need to send a file by email, Outlook becomes involved, though there are times it works in the background. A file can be shared by sending a link to a file that has been saved on OneDrive. If the file cannot be found on OneDrive, there will be a prompt to have a copy saved on it before the file can be shared.

To send a file as a link, follow these steps:

- Open the document that should be sent in Microsoft Word.
- Once the document has been opened on the screen, click on the share option in the upper right corner of the Word window.
- Once the share dialog box is displayed, select the One Drive option.

The share dialog box will be displayed if the file has not been saved on OneDrive. Note that the file has to be saved on OneDrive before the link can be shared (if there is no need to save the file on OneDrive, follow the next steps below to share the file).

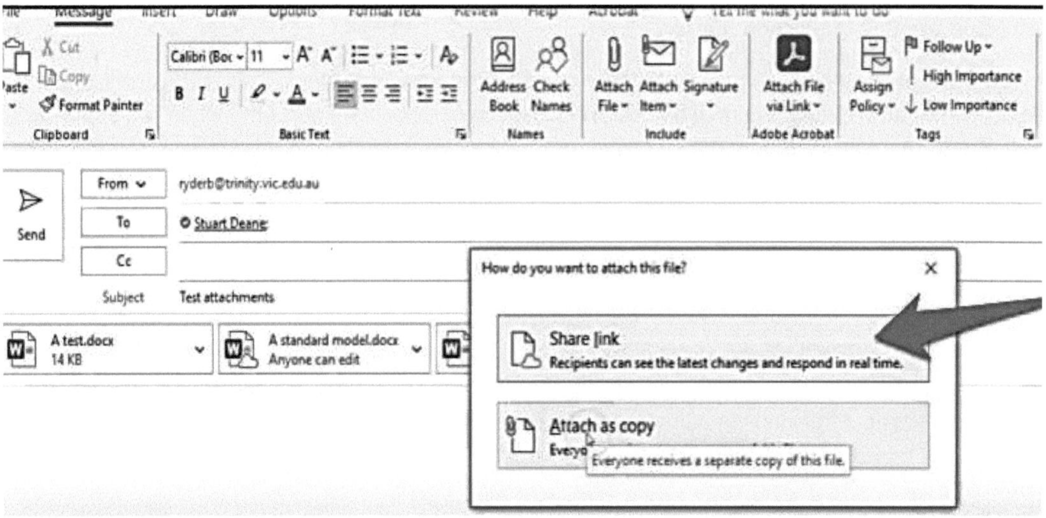

- Locate the share task pane and then insert the email address of the person that the file is intended to be sent to. To insert more than one address, separate the addresses with semicolons.
- If there is no need for the recipient to have control over editing the file, click on the "can edit" drop-down menu and select the "can view" option. These settings will determine if the recipient will be made to make adjustments to the files.

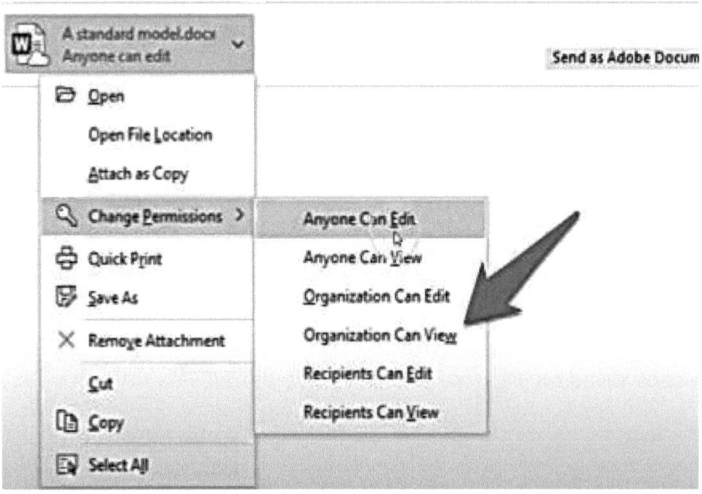

- Fill in the message box with a message for the recipient. The message will then be displayed as the email message body.
- Finally, click on the share option.

To send a file as an attachment, follow the steps below.

- Open the document that should be sent in Microsoft Word.
- Once the document opens on the screen, click on the "Share" button in the upper right corner of the Microsoft Word window. This button offers a very quick way to share. There is also an option to use the file share option.

- If the share dialog box is displayed, click on Word Document, or rather, if the share task pane displays, click on the send as attachment option and then click on Send a copy. Notwithstanding the method used, the new message form will be opened with the document displayed in the attached line. If you prefer to type a message in the body of the screen, you can go ahead, but it is not compulsory.
- Insert the email address of the person the message is intended for. To insert more than one address, separate them with a semicolon.
- You can tweak the subject line if need be.
- Finally, click on the "send" button.

Maintaining Your Schedule

Get much more done within the next 24 hours if the Microsoft Outlook 2019 calendar is kept current. Appointments can be entered and managed from right within Outlook.

Opening an appointment in Outlook

If at any point in time you have had to use the old method of planning with the use of paper, the Outlook calendar will be very familiar to you.

- Select the calendar button in the navigation bar.
- Click on the Day button on the Home tab and a grid in the middle of the screen lines that shows each segment of the day will be displayed. The length of these segments can also be tweaked from as little as 5 minutes to as much as 60 minutes.

To enter an appointment at a certain time, follow the steps below:

- If need be, click on the calendar located in the navigation bar to change to the calendar module if needed.
- Enter a name for the appointment.
- Click on the next line at the very time the appointment is supposed to begin.
- Finally, click on the enter button.

Time management has a lot to do with more than just inserting appointments. In any case, if you are busy, there might be a need to manage the time by slicing and dicing the list of appointments to see when you are free to add additional appointments.

If there is a need to have a broader collection of calendar views, select the view tab located on the ribbon and then make a choice from the views listed under the change view button.

Adding a contact

Typically, Microsoft Outlook refers to the information stored about people and organizations as "contact information." It also stores this information in a folder known as Contacts. To gain access to the contact folder, select the contacts icon (which has an image of two people) located in the lower-left corner of the Outlook window. The contact folder then shows each of the contacts that have been entered into Outlook.

To store a person's contact information, follow the following steps:

- Locate the contacts window, select the "New contact" button located in the upper left corner. A new contact window that is untitled will then be displayed.

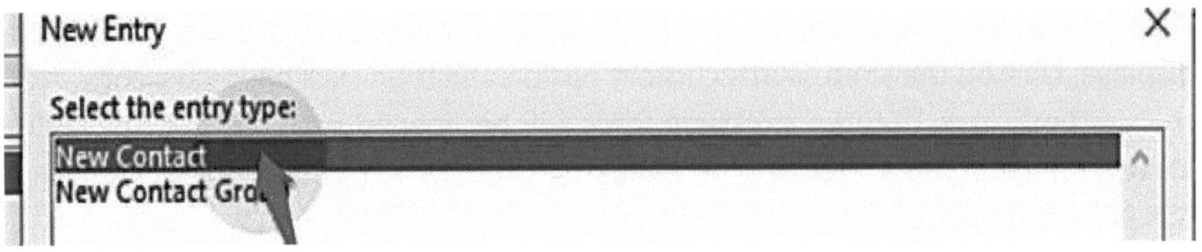

- Fill in the information that needs to be saved for the person or organization. You can fill it in as much or as little as you want.
- Finally, click on the save and close button located on the ribbon. The contact will then be added to your contact list.

Here are some tidbits that should be considered when entering contact information:

- Anything that is typed in the Full Name Field will immediately replace the word "Untitled" in the dialog box title bar.
- In the File As field, the mode in which the entry is alphabetized is defined. The default is by the last name, and it will reverse whatever name that is entered in the full name field. For example, if the name entered is Mark Joe in the full name field, it will appear as Joe, Mark. For people's names, this can be the best way to go about it, but for a company's name like ACME Corporation, it might probably not be the best for it to be alphabetized as Corporation, ACME. To have that fixed, open the File As drop-down list and select ACME Corporation.
- Multiple email addresses can be saved for one person. Notice that email is not just a field label, it is also a drop-down list. Open the list and select Email 2, Email 3, and so on for the additional addresses.
- All phone number fields also have drop-down lists that are in association with their labels. Up to four phone numbers can be saved for one person, and a label for each of the numbers can be carried. For example, the label "mobile" can be assigned to one of the phone numbers.
- There can be up to three addresses for one person: business, home, and other. You can toggle between these addresses with the drop-down list in the addresses section.
- When an address is entered but wasn't entered in the proper mailing format (address, city, state, and zip code), a dialog box will pop up so that that information can be filed in. This is to protect you, as it will ensure that every address entered is usable.
- Make use of the Notes pane to store any other additional information about the person that doesn't correspond with any of the fields stored.
- The default fields that are displayed are known as "general fields." There are also more fields available. Navigate to the contact tab, move to the show group, and select details to see the other fields.
- To customize the appearance of the contacts, select from sets of fields rather than the defaults.

Entering a Task

It is not enough to know what to do; there is a need to always know what to do next till the task is completed. When you are working on a lot of demands, there is a need for a tool that helps in showing what should be done next at a simple glance so that the work can keep moving without an abrupt stop.

Microsoft Outlook has a lot of task management tools that can be of use in the organization of lengthy to-do lists in order to obtain maximum performance. The tools include the tasks module. The to-do lists and the to-do bar below are a quick way to get started.

To enter a new task, simply follow the steps below:

- Locate the navigation bar and select Tasks to change to the Tasks module if there is a need for it.
- On the Home tab of the Ribbon in the Current View group,

Select Simple List to ensure the task list is in the Simple List view. New tasks can be created from just about any view, but the wording of the step that follows is a little different based on the view that is being used, so the choice of views is just to avoid any form of confusion when getting started with Outlook.

- Select "Click here to include a new task box" and then type in the name of the task. In some of the views, the wording used is "Type a New Task," and in other views, the box for adding a new task with this method is not present at all. This is why it is best to start in the Simple List view to completely avoid any form of confusion.
- When all has been completed, tap the enter key and the new task will be moved down to the task list with the other tasks.

With the use of Outlook, you can manage just about anything perfectly. From a simple shopping list to a complex business project.

Taking Notes

Almost everyone makes use of sticky notes to set reminders for themselves, from bits of information to website passwords or some memorable quotes. Outlook also has notes, which are the electronic equivalent of sticky notes. Anything can be stored on a note, but they are best used for the storage of little pieces of information like confirmation codes, pins, or membership ID numbers. A note can be left open as long as Outlook is also open. This way, you will always be reminded each time it is seen or have the note closed so that it won't be in your way.

To make a note, simply follow the steps below:

- Show the notes portion of Outlook. To get this done,

Click on the More icon in the bottom left corner of the window, and in the menu that is displayed, click Notes.

- Select Home, then click on "New Note" or use the keyboard shortcut and press Ctrl + N. A new blank note will then be displayed. Type in whatever you want on this new blank note. Note that the first few words of the first line will be displayed in the form of an icon title. Try as much as possible to be descriptive there. But if you are trying to be secretive about the information, like a password, there might be a need for some misleading words in the first line. You can decide to keep passwords in a note that is titled "Family Birthdays." With this, if anyone wants to snoop or steal passwords, the person wouldn't care about "Family Birthdays" and would simply pass that by.
- Close the note by simply clicking on the X button located in the upper right corner, and the note will be saved automatically.

Below are some useful Outlook note tidbits.

- To open a note again, simply place the mouse arrow on the note and double-click on it. The note will then remain open until it is closed again or Outlook is exited.

- To move a note, drag it around with the use of its title bar (the colored bar at the very top of the note), placing it wherever you like on-screen, even beyond the boundaries of the Outlook window. The note will remain there until Outlook is closed.
- To make adjustments to a note, open the note by double-clicking and then make the necessary changes to it.
- To change the size of the note, click on the note and drag its bottom right corner.
- To have a note deleted, click on the note and follow these steps:
- Select Home, then Delete.
- Press the delete key on the keyboard or right-click on the note.
- Select the Delete option from the menu that is displayed.

To restore a note that has been deleted, open the Mail portion of Outlook, show the contents of the Deleted Items folder, find the note, and then move it to the Notes icon that can be found in the bottom left corner of the window.

There is also an option to;

- Right-click on the note.
- Select the Move to Folder option, which then indicates the very part that it should be moved to.

Categorizing Outlook notes

Given that a note is an incredibly versatile item, creating categories for it might be very useful. This way, one will be able to differentiate one type of note from the other. By assigning a color to a note, the type of information the note should contain can be specified. For example, you can decide to make financial notes green in color and make family notes blue.

Follow the steps below to apply a color to a category note.

- Right-click on the note and select the option to categorize. A menu of color choices will then be displayed.
- Click on the preferred color.

If this category has been used once, that is all at this point. The notes icon and background will change to the color chosen. However, the first time a certain color is used, the Rename Category dialog box will be opened. This way, a name can be assigned. Outlook does not assign any special meaning to color; it is best done based on personal preference. If there is a prompt, change the name in the Name box and then click on the OK button.

Keep in mind that if the default email account is of the IMAP type, the Categories option won't be displayed. Should this option not be displayed, do not panic. Just check the email account.

If there is a need to rename a category, simply follow the steps below:

- Click Home, then Categories, and then All Categories. The color categories in the dialog box will then be open.
- Click on the preferred category and click on the rename button. The name then becomes editable.
- Finally, type in the new name and press the Enter key.

You can also add and delete categories and alter the color associated with a category. For instance, this might be useful if you wanted to have the color used to represent a particular category changed, but you don't want to lose the category information already attached to existing notes. Note that it is possible to have up to 25 categories.

A note can have different categories assigned to it. Each category is an on or off switch for each note. A note icon will show only the color of the category that was last assigned to it, but it will still keep other categories as well.

To remove a category from a note;

- Click on the note and select "home."
- Categorize, and then click on the specific category to switch it off for that particular note.

Categories are also shared by the calendar and tasks features in Outlook. As categories are being created, it is necessary to keep in mind how you might prefer to categorize appointments and events also.

CHAPTER 2

GETTING MORE DONE WITH LESS EFFORT

In this chapter, we will delve more deeply into how Outlook works, its components, and how much can be achieved without spending a lot of time trying to figure out how things work. Carefully go through the points outlined below to gain more knowledge.

Outlook and Other Programs

Lots of users prefer to use Outlook as their email client and personal information manager. Over the years, Outlook, as part of the Microsoft Office Suite, has proven to be a standard solution (in conjunction with the Microsoft Exchange Server) for both public and private organizations.

Even though without a doubt, Outlook is the option to beat, many freelancers and small companies cannot afford solutions intended for larger businesses, but there are lots of email management programs aside from Outlook that can be used for free. Some of which are

Postbox

This program was created by one-time Mozilla employees and hence was based on Thunderbird. About ten years later, the software has developed into an autonomous and extremely effective mail client. The design of its interface is like that of other solutions, and it is also easy to use. Users who want a unique look can adjust themes or design their templates.

For a more efficient operation, Postbox makes use of different shortcuts. Users can access the Quick Bar via hotkey so that messages can be moved or categorized quickly. In addition, when creating an email, a signature can be entered with the use of the Quick Bar without having to use the mouse.

There are lots of benefits when writing emails in the postbox. The software has various templates and text blocks that can be used to write cover letters and replies in a placeholder that can be added and created where the name of the receiver is always added automatically.

Thunderbird

For both private users and those in companies, Thunderbird is one of the most preferred options as an alternative to Outlook. The open-source solution is also available for free. The free version of the program is rather streamlined and offers only the most basic functions. One major advantage is the addition of various add-ons. This means that there is room for expansion for the email program. However, the add-ons and extensions are made to suit their respective versions. If there is a need to update your version of Thunderbird, the add-ons must be updated as well.

Spike

Spike was released in the year 2013 and it combines certain functions of classic email programs with those that are used in modern messenger apps. Immediately a private mailbox is linked to the application, elements like subjects or signatures are no longer necessary, while the basic mail client functions, which include the central inbox or contact management, will be integrated into the modern messenger environment. It is also not coincidental that the creator of Spike described it as a conversational email application. Note that both audio and video calls can also be made via the software.

The spike depends on modern standards in terms of security. With just a single click, communications that also include attached files can be encrypted. This way, you can be sure your messages are fully protected against unwanted access. Spike is free for private users; however, monthly fees are charged when business email accounts are added.

Mailbird

Mailbird is an Outlook alternative that is only free in the test version. This email solution allows the unification of messages and contacts from different accounts into just one box. With different free themes, the interface can be designed as it best suits you.

Mailbird offers different interfaces to various applications and also enhances the mailbox with helpful features for better interaction and teamwork. For instance, Twitter, Whatsapp, Calendar, and Dropbox can be integrated into the mail to change it to a multi-functional program.

Outlook's Main Screen

Using the Folder pane

The folder pane shows all the folders in Outlook. If this pane is minimized, other folders will not be displayed and you will not be able to gain access to them. There are two different ways this folder can be viewed.

The first is by;

- Clicking on the left side of the screen. This will help in expanding the folder pane, making other folders visible.
- The second option is to click on view, then the folder pane, click on normal so as to see things in a clearer way.

The folder pane in Outlook is said to be the main navigation tool between mailboxes, folders, and different modules like mail, calendar, and contacts. Note that the folder pane has a couple of different options and tips that can help make the pane fit more into your kind of style and help you work effectively and efficiently.

You can choose to enable or even disable the folder pane by Pressing the ALT + F1 buttons.

To change from one module to the other, make use of the icons enlisted in the lower area of the Folder pane. If you also prefer to see the names of the modules alone instead, disable the Compact Navigation through the Folder Pane Options dialog box.

Note that with the use of the Folder Pane, you can configure the way modules are shown and in the very order they are shown; they can also show module icons or names for easy navigation; add shortcuts to modules; and lots more.

To get the best out of this folder, you should spend some time clicking on various options, including the ones explained above.

The Information Viewer: Outlook's hotspot

The information viewer is the very place where most of the action in Outlook takes place. If we can assume the folder pane as being the channel sector on a TV, the information viewer will be more like the TV screen.

The information viewer is the very place where emails are read, contacts are searched for or added, and contact names are also displayed. If you also wish to do a whole lot of other fancy things like sorting contacts, tasks, and so on, the information viewer is the perfect place to get all of that and more done.

Based on the fact that lots of information, more than what can be seen at a glance, can be stored in Outlook, the information viewer helps to show a preview of the information available. This way, you are up to date with all you need to know. The calendar, for example, can store dates as far back as the 16's and as far ahead as you can imagine. The smallest calendar review that can be displayed is a day, and the largest calendar view is a month.

The information viewer also helps to arrange what it displays into smaller units known as "views." There is an option to create your own views and also save them, but you can also decide to use the view that comes with Outlook.

Move through the different previews of the information displayed by Outlook by clicking on various parts of the information viewer. Some people love to say they are browsing the information viewer when moving around it; it seems more like just scanning through the pages of a notebook.

You can also browse through the calendar data in the information viewer to keep you abreast of things to come.

To do that, simply follow the steps below:

- Click on the calendar in the navigation bar or make use of the keyboard shortcut and press the Ctrl + 2 buttons.
- Next, click on the workweek button located on the Home tab of the ribbon. The workweek view of the calendar will then be displayed. Note that a workweek means 5 days if the regular calendar week shows 7 days.

To further spice things up, you can decide to change the appearance of the information viewer in several ways. For example, there might be a need to see the schedule for just a day or just the items that have been fixed for a particular category. Views can ensure that you get a glance at the very preview of information needed.

While checking out the calendar, you can also decide to check the To-Do bar located on the right side of the screen. The To-Do bar shows the appointments you have and also reminds you of things you need to do. To turn on this feature,

- Click on the View tab located on the ribbon > click on the To-Do Bar > click the calendar button.

In Outlook, every module (mail, calendar, people, tasks, etc.) has its own version of the Ribbon, arranged specifically to meet the needs of the module. Most of all the buttons are visibly labeled with the very actions they are used for, such as replies, business cards, new appointments, and lots more.

A little button known as "properties" can be found in the lower-right corner of some groups.

- Click on it if you want more information displayed than what is shown on the ribbon. Properties are also known as "dialog boxes" as they open some sort of dialog box launcher related to the group when it is clicked.

The Ribbon Ties

Viewing Screen Tips

Each button located on the ribbon shows a little popup known as a "ScreenTip" when the mouse is placed over it. The ScreenTip informs you about the name of the button and also tells you things that will happen if you click on the button.

Some buttons have a small arrow that points downwards or to the right side of the button.

Click on the arrow to have a menu or list open. A very popular example, pertinent to almost all the modules in Outlook on the Home tab, is a button known as "Move." When you click on the Move button, it opens a menu and shows all the various places an Outlook item can be sent to.

Using the New Items Button

Each module in Outlook has a "New Items" button that enables the creation of an item in any module. For instance, if you are checking the name and address of a customer whose name was also mentioned in a very interesting article in one of the daily newspapers, you will want to remember to refer to it at any point in time. You can do this by creating a new item from the new contact option, selecting the particular date, and then saving it with the name of the customer or something unique that can always make you remember the incident in the daily diary.

Finding Things in a Flash with Instant Search

With the use of instant flash, you can find out through anything you want to do in Outlook, simply

Insert the first few words or alphabets and Outlook will bring up suggestions for you.

Taking Peeks

One very unique feature in Outlook is a tiny pop-up window known as a "peek," which is displayed when you place your mouse over the modules such as People, Calendar, or Tasks in the navigation bar. This little but unique feature offers great help when giving a reply to an email about an event that needs to be scheduled. Feel free to take a quick peek at your calendar while you continue to work on that particular email. If a broader view with more information is needed, there is an option to make the peek window larger by clicking the button at the top right corner of the peek screen or by double-clicking on the calendar, people, or tasks in the navigation bar.

Getting Help in Outlook

The help feature in the Office applications, which includes Outlook, goes beyond just rendering help as it tries to get things done for you. Does that sound awkward? It isn't, it is just amazing and very helpful.

When working in the help feature, a lightbulb icon and a textbox are located at the top of the screen with the inscription, "Tell me what you want to do." Once that box is clicked, type in what you might need help with and it will display a list of things that starts with a list of things that can be done. For example, if you enter the word "delete," the help feature will bring up a link to the Delete and Delete All commands and also the folder with deleted items. When one of the commands is selected, it will delete the very Outlook item that has been chosen. When you click on "Deleted Items," it will take you straight to the deleted items folder.

Note that it is important that you request only the things that can be done in Outlook. If you type something like "what can I eat?", the choices Outlook will offer you might disappoint you. But if you are trying to do something that involves email, appointments, or tasks, Outlook should provide you with the very important links you need to get things done quickly.

Scroll beneath the list of links to view other choices that might be displayed by Outlook. Once you point at the "Get help" option, it will open a submenu of various help topics that are in some way related to what you have asked. You would have to click on that yourself, as there is no option to help you do that.

The final option in the help menu is called "Smart Lookup," which helps to open the Smart Lookup task pane and makes use of Microsoft Bing, which is the search engine for Microsoft, to search for the phrase that has been entered. When using smart lookup, there may be a need to activate the intelligent service. For example, if you type "marry a millionaire" and click on the smart lookup link, a list of marriageable millionaires will be displayed, and if not, at least At the very least, you will get the definition of the word or phrase that has been typed and probably a link to Wikipedia also.

CHAPTER 3

ON THE FAST TRACK: DRAG TILL YOU DROP

You don't always have to type. In fact, having to type will soon fade away with various inventions in technology. In this dispensation, we seem too stuck with the keyboard. By making use of the mouse, trackball, or touchpad, you can touch and drag an item rather than having to type all the time.

Most people understand that a tool such as Outlook can improve their productivity dramatically, but most of them ignore the productivity of some of its tools, which include tasks and calendars and the ability to sync all your information with ease. Some of these tools can be very powerful weapons in the battle to make the most out of your time.

Dragging

If there is a need to work quickly and easily in Outlook, there is a simple trick called "drag and drop" that offers the opportunity to get things done accurately. This feature is very easy to use, but most people do not take advantage of it. Dragging is simply the process of moving an item from one place to another easily and quickly. It can be done easily with the use of the mouse or the touchpad of a laptop.

Most touchpads have two buttons located just below the touchpad itself that acts similar to the two buttons on the mouse pad. These two buttons might look like the real buttons on the laptop case, or just outside the touchpad, or on the touchpad itself. If that is the case, you might not notice they are there until you tap the lower-left corner or lower right part of the touchpad.

- Click on the left button once or twice to get just the same effect as though you were clicking the mouse the same number of times.

Dragging with a touchpad needs more skill than dragging with the use of a mouse.

You can click twice and then slide your finger to do just the same thing as dragging a mouse, or you can choose to hold down the left touchpad button with just one finger while the other finger slides or drags.

The mysteries of using a touchpad can take quite a while to get used to. It is best to add a mouse to the laptop to ensure you are not slow at the task you are trying to complete. It's best you learn how to make use of the touchpad though, as there might be a time when the mouse won't be available and the touchpad at that time will be a lifesaver.

Dispatching Tasks in a Flash

In the business world today, everyone seems to be doing less talking and more instant messaging and emailing. When your boss needs you to get certain things done, he most often communicates via email. The downside, though, to this is that all the messages can clutter the email box so quickly that you might lose track of what you are supposed to do.

Most productivity experts often propose that emails be converted into To-Do list items immediately after they're received. This way, you are sure you won't be losing track of important details. Make tasks from email messages by moving the messages to the Task label button located in the navigation bar. The other information, such as date and category, can then be added later. Note that a single move and drop is all that is needed.

Making Time Stand Still with Calendar Wizardry

The most common method of cheating plans with other people is the use of email. It is very cheap, fast, efficient, and reliable. It doesn't matter if you are including people at lunch, throwing a party, putting up a show, or arranging an exhibition, you must already have an idea of how easy doing all of these can be with the use of email.

When a plain email announcement about a certain event is received and you want to insert the details of the event into your calendar, you can do that in Outlook by following the steps below;

- Click on the mail icon in the navigation bar or make use of the keyboard shortcut by pressing the Ctrl + 1 button to change to the mail module. Once this has been done, a list of emails that have been received will be displayed.
- Choose the message for which you want to make an appointment.
- Move the message that you have chosen to the calendar in the navigation bar. An appointment form will then be opened with a text from the message that has been moved into the section bearing a note of the new appointment form.
- Perform any changes that should be added to both the start time and end time. This feature was made editable as the default start time might not be what you want.
- If there is a need to add more information about the event, enter the information in the appropriate box that can be found on the new appointment form.
- Once all of the above has been completed, click on the "save and close" button. This way, all the information regarding the event is now safely saved in your calendar for future reference.

The most amazing thing about creating an appointment from an email message is that all the details embedded in the message might just end up in your calendar. If there is a need for driving directions, agenda details, and any other information that was added in the message, simply double-click on the appointment in the calendar to get the details. If you use a smartphone with Outlook, such as an Android, Windows, or iPhone, all of the information on the Outlook calendar will be synced to your mobile device. With this, you are sure to have your appointment details with you wherever you go. If you are working with any organization that makes use of Microsoft Exchange for email, organizing meetings can be very easy as you can take advantage of certain features that are much more powerful.

Keeping Friends Close and Enemies Closer

You can move an email message to the People button in the navigation bar to create a contact record that also has the email address embedded in it. If this method is followed, the risk of misspelling the email address will be eliminated, and the work will also be saved.

To add a new contact from an email message, simply follow the steps below:

- Locate the Mail module > choose the various messages to which a new contact record should be added.
- Move the selected message(s) to the People button in the navigation bar. Once this is done, the new contact form will be displayed and will include the name and email address of the person who sent the message. Also, note that the original email sent will be displayed in the Notes area.
- If there is a need to add more information, enter the specific information into the relevant box on the new contact form. Existing information can be changed. You can also add more information such as the company the person is working for, the postal mail address, other phone numbers if available, and also personal details like if there is a need to send a complimentary gift, get the person a pet, and so on. If the body of the email message has certain information

that can be useful as contact information, click on that information and move it to the necessary box of the New Contact form.

- Finally, click on the "save and close" button. Now you are certain that the email address and any other detailed information for the new contact have now been saved for future reference.

As an alternative, you can also capture an email address from an incoming message. This option seems to be very quick and much easier. To do this, simply right-click on the name of the person that sent the mail in the incoming message field (in the reading pane; note that this option will work if you right-click on the message from the message list). It may not seem like much, but right-clicking on it does. After this, a shortcut menu will be displayed. You can then choose the Add to Outlook Contacts option to open the New Contact Form and then follow the last two steps of the above-listed method.

Creating Instant Email Messages

Whenever an item is moved to the inbox, Outlook will immediately change it to an outgoing email message. If the item moved to the inbox has an email address, e.g., a contact. Outlook will automatically compose the message with the person's email address already filled in. If the item moved to the inbox has a subject, e.g., a task, Outlook will also create the message with the subject automatically filled in.

Creating an email from a name in your contact list

Having to add messages can be one of the most productive drag-and-drop operations in Outlook. Dealing with email addresses can be very difficult and tedious, and if by mistake, a letter from the email address appears missing, the intended message won't go through. The best approach to this is to keep the email address as a contact in the contact list of the people to whom you send email messages and also create new messages with the use of the email address.

To create an email address from the contact list, simply follow the steps below:

- Select people in the navigation bar to change to the People module. The contact list will then be displayed. You can choose to work with any view, but the address cards view is the easiest to work with. Click on the first letter of the person's name to view the person's card.
- Move a name from the contact list to the mail button in the navigation bar. This will then bring up the message form with the address of the contact filled in.
- Enter a subject for the message and ensure to keep it simple and straightforward; a few words will be enough.
- Select the text box and enter your message. There is also an option to change the text to bold, italics, or any other effect by choosing the necessary button located on the toolbar.
- Click on the "send" button. The display will then go back to the contacts list and the message will be sent.

Outlook Workspace Expansion

Views offer various ways an item can be seen in a given folder. Every folder in Outlook, like the inbox and calendar, offers users the opportunity to change the view and rearrange the way certain settings like fonts and others are organized.

One of the very common changes that can be made to a view is changing the size of the font in the reading pane and the message list or when composing a new message.

To change the font or font size in the messaging list, follow the steps below:

- Click on the view option >select View Settings.
- Click on other settings in the advanced view settings box.
- Choose the column font or the row font. Note that to make changes to the size of the font in the message preview, the subject in the default inbox view, or the sender's name, make use of the row font option.
- Choose the font, the actual font style, and the font size that you prefer> click on the OK button three times to ensure the settings are saved and hence also applied. Take note: if all you want to change is the font or font size for the message preview (the line of the message text that can be seen beneath the subject and sender), click on the font beneath the message preview.

Changing the font or font size located in the reading pane

Outlook's reading pane will not allow you to make changes to the default font or font size. Nonetheless, there is an option to zoom out or zoom in with ease. There is also an option to command Outlook to show all of the email messages in plain text, and also take more control over the size of the font. On the other hand, Microsoft 365 subscribers (online version) can choose a certain zoom percentage that cuts across all of the messages read.

- Locate the lower right corner of the reading pane and choose the percentage (most times 100%) to show the zoom option when reading the dialog box.
- Select one of the default percentages or insert the percentage of your choice. Choose the "remember my preference" option in the checkbox. This way, the zoom level will be kept the same across all messages received.

To make changes to the zoom percentage, make use of the following steps:

If you want to zoom in or zoom out, find the magnification slider, which is located at the lower right corner of the reading pane.

Drag the slider to either left or right to make changes to the size of the text in the reading pane. Bear in mind that if any changes are made to the zoom level, they will only persist when the message is being viewed. If another message is chosen, the zoom percent will go back to 100. If you return to the initial message, the zoom percent won't be automatically saved.

Apply changes to the font size for messages when creating new messages, replying to messages, and forwarding messages.

To apply changes to the size of the font for messages when creating, replying, and forwarding, make use of the steps below.

- Click on File>Options > Mail, and then the Stationary and Fonts option.
- Choose the font button for the New Mail Messages option or the Replying or Forwarding message option to make changes to the default font, font color, and font size when creating or replying to messages.
- Finally, click on the OK button twice to ensure the changes are effective.

CHAPTER 4

THE ESSENTIAL SECRETS OF EMAIL

Email is very important for all professional communication in the office. If the organization you work with makes use of Outlook, learning how to make judicious use of it can be very vital to your professional development since the skills learned will help you arrange meetings with those you work with and also send email messages easily. Furthermore, you will be able to show your colleagues, boss, or team lead that you have learned well and improved over time when you make use of the features in Outlook with so much ease. You can even offer to teach them what they want you to know about the use of these tools.

Front Ends and Back Ends

There are basically two things needed to send and receive an email; a program that helps with the creation, saving, and management of messages and a program that actually sends the messages to people and also receives replies from them (exchanging messages). A few people in tech call these two parts the "front end" and "back end," as the case may be. Outlook, on the other hand, is a front end for email messages. It helps with the creation, formatting, storing, and management of messages, but it doesn't really do much about getting the message to the intended destination. Ensuring that messages get to their intended destination is the work of the back-end service (like the Microsoft Exchange Server in your office), the Internet Service Provider you use, and also by an online email service like outlook.com or gmail.com.

Email messages cannot be sent or received anywhere in the world without the use of an internet connection. The phone company you use most of the time provides internet services that can be used for this purpose. To ensure your email messages are sent quickly and you also get replies as quickly as possible, make sure you choose the best Internet Service Provider out there. Remember though, that the easiest choices aren't always the very best choices. There are several companies ready to offer internet services. Shop around to get the best value for money.

Creating Messages

In a lot of ways, electronic mail (email) is much better than the normal paper mail commonly known as snail mail. Email is delivered at a much faster pace (almost instantly) than paper mail. The speedy delivery can be of great help when closing a last-minute deal, sending across vital information towards the close of working hours, or for last-minute birthday greetings. Email is also very cheap to use; it is actually free most of the time.

The quick and dirty way

Creating a new email message can be very easy. Simply follow the steps below.

- Open Outlook The mail module will then be displayed, opening up the inbox.
- Click on the New Email button

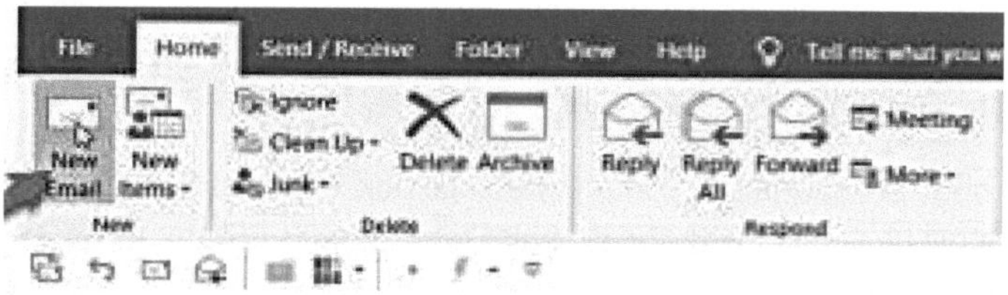

- Insert an email address in the To box.
- Insert a subject in the subject box.
- Insert a message in the message box.
- And finally, click on the send button.

Most of the time, Outlook starts in the Inbox only. This will only change if the settings have been changed. If there is a need to start up in an entirely different folder than your inbox,

- Click on the File tab and choose Options.
- Click on the advanced option and in the Outlook Start and Exit part, change the Start Outlook. Click on the "browse" button to check all the folders in the Outlook data file. To begin in another module, select a folder that is similar to that module, e.g., Calendar or Notes.

The slow and complete way

You might like the more comprehensive way of creating an email message. If you love to create fancy emails, especially if you want to take advantage of every key feature Outlook has to offer, follow the steps below:

- Locate the Mail module > select the New email button or press Ctrl + N. The new message form will then open up.
- Click on the "To" text box and insert the email address of the person you are about to send a message to. If you are sending to more than one person, separate their email addresses with the use of either the comma or the semicolon. There is also an option of clicking on the To button itself, locating the names of the people you intend to send messages to in your address book, and double-clicking on their names to ensure they are added to the To text box.
- Click on the Cc text box and insert the email address of the person whom you would like to have a copy of the email message you are about to send. You can also click on the "cc" button to include people from your address book.
- Click on the Check Names option located on the New Message form's ribbon or press the Ctrl + K button. If you don't know the exact email address of those you want to send messages to, there is a feature known as "Check Names" that allows you to insert a portion of the address and then check it in the address book. Ensure to check what the feature enters to be double sure it is correct. There are times it will just automatically enter a similar name in the address book, which can lead to an embarrassing result if it is not corrected before the message is sent. An example would be a romantic message intended for your spouse being sent to your boss instead.
- Insert the subject of the message into the subject box. The subject should be very brief and simple, as it makes someone enthusiastic about reading your message rather than a very lengthy subject line. If for any reason, you forget to insert the subject, Outlook will open a window that asks if you really intend to send the message without a subject. Click on the "Don't Send" button to return to the message and include a subject. If there is really no need for a subject, click on the Send button to send the message.
- Type the body of your message in the message box. If Microsoft Word is the word processor you use, you must be familiar with the graphics table, modes of formatting, and all the tricks in Word to ensure your email appears more attractive. These very same tricks can be found in Outlook

by making use of the tools at the top of the message form on the Format Text tab. There are times when you might not need to insert anything into the message textbox, like when you are forwarding a message or sending an attachment. If this appears to be the case, just skip this step and move on to the next.

Be extremely careful when formatting your email. This is because not all email systems can deal with graphics or text that has been formatted like boldface or italics. This way, the text you send to your client will not look like gibberish. Furthermore, most people read their email on phones or phablets, which can make the text you send look odd. If you are not certain about how your recipient will receive your mail, stay off of adding graphics or formatting text. You should only include them if you are very sure the recipient is also using Outlook.

- Click on the Review tab > select the Spelling and Grammar button at the upper part of the message screen. You can also choose to press the F7 button. Outlook will then run a spell-check to make sure your message is free of spelling and grammatical errors.
- Finally, click on the send button or press Ctrl + Enter or Alt + S. Outlook will then move your email to the outbox. If you are connected to the internet, Outlook will send any message in the Outbox immediately. If the message was composed when you were not connected to the internet, click the F9 button to send the message again when you are connected to the internet. When a message has been sent, it will move to the sent folder automatically.

Another option for telling Outlook to send messages from the Outbox is to tap the little button looking like two envelopes overlapping that can be found on the Quick Access Toolbar at the top left corner of the Outlook window, which can be seen from any module in Outlook. If the mouse pointer is moved over this button, a screen tip will be displayed notifying that it is the send/receive all folders button. Whenever messages are sent by clicking on the send button in a message or by tapping the F9 button, it is also a way of telling Outlook to receive all messages coming in.

Setting priorities

Certain messages can be more important than others. Sending a report to your boss is not the same as sending a friendly message to a teammate or colleague. Setting the importance level of a particular message to "high" informs the recipient that the message needs urgent attention.

There are basically three levels of importance you can choose from:

- Low
- Normal
- High

Setting sensitivity

There are times when there is a need for your message to be seen by only one person. Alternatively, you might just want to ensure your message is not altered by anyone after sending it. The sensitivity settings in Outlook help to put a restriction on what any other person might be able to do to your message after you have sent it. They also help you decide who that person can be.

In applying the sensitivity settings to a message;

- Click to open the properties drop-down menu for messages.
- Click on the list box arrow closest to the word "sensitivity" and any of the displayed levels as briefly described.

Most of the messages sent via Outlook have just the normal sensitivity, and this is what Outlook makes use of if the settings are not changed. Settings such as Private, Personal, and Confidential only inform those receiving the message that there might be a need to treat the message in a different way from the way they treat other normal messages. There are even certain organizations that employ the use of some strict measures in dealing with confidential messages.

Applying the sensitivity settings of a message to either private or confidential does not make it any different from other messages; all it does is inform the person receiving the message that the message has some information that might be delicate. If you use Outlook at work, make sure to double-check with your system administrators before assuming that the information you're about to send via email is secure.

Setting other message options

Upon clicking the Properties dialog box, you might see that there are quite a number of odd-sounding options. A few of these options are Request a Read Receipt for This Message (which informs you when the receiver reads the content of the message) and the Expires After option (this marks a message as expired if the receiver does not open the message before a stipulated time). These options are quite useful. The only glitch is that your email and that of the email receiver must support the setting, else it might not work. If both of you are on the same network, making use of the Microsoft Exchange Server, all should work just fine. If you are both not using Outlook or an Exchange Network, it can just be a gamble.

Adding an internet link to an email message

All Microsoft Office programs will automatically identify the address of certain things on the web. For example, if you insert the name of a certain webpage like www.facebook.com, Outlook will change the color of the text from black to blue and also underline the address, thus making it look like a hyperlink that is clicked on to move to another web page entirely.

This feature makes sending information about an exciting blog or website very simple; simply copy or type the address into your message. Note that Outlook might not identify the text as a web address if it does not start with "www." In that case, simply add HTTP:// at the front of the text. Based on what the receiver uses in reading the email, he or she ought to be able to

simply click on the text and display a web browser forthwith, opening the page mentioned in the message.

Reading and Replying to Email Messages

Outlook has a number of methods to notify you when an email message has been received. The status bar located towards the left side of the Outlook screen will inform you of the total number of email messages you have in your inbox and the number of messages that are still unread. The word "inbox" will change automatically in the folder pane to a boldface form when there is an unread email. The titles of unread messages will also be bold.

To click open and read an email message, follow the steps below:

- Locate the Mail Module: click on the title of the message you want to read twice. This will open up the message in its window.
- Select the close tab option (X) or press the escape key to close the message pane when you are through.

Viewing previews of message text

At the time you begin to receive plenty of emails, some of them will be important, while others might not be very important, if not totally useless. Upon receiving the mail in your inbox, it can be very useful to know which is important and which is not. This way, you can focus on the very important one. You cannot depend on those sending the mail to tell you how important or otherwise it is. Outlook offers help by giving a glimpse at the first few lines of the message.

When launched, the message preview is always on and set to 1 line. This option can be changed to 2 or 3 lines, and you can also choose to turn it off totally. The Message Preview option becomes totally redundant if the reading pane is turned on, but if it is not, the preview can be a very useful feature.

To gain control over the previews of messages that have not been read, follow the steps below:

- Find the Mail Module>click on the View tab option in the ribbon.
- Click on the Message Preview option.

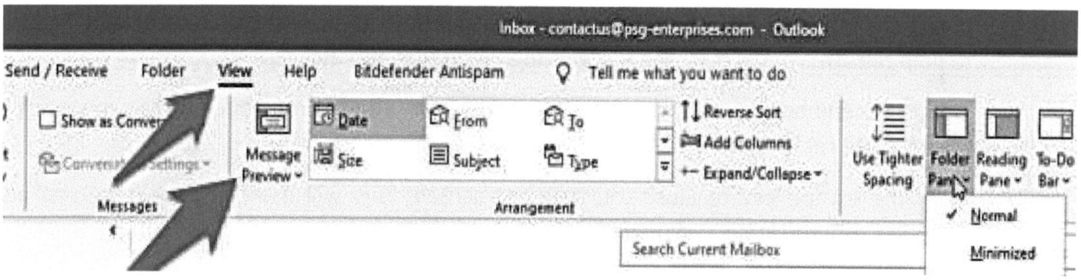

- Click on the setting you prefer (1 line, 2 lines, 3 lines, or off).
- If, for any reason, there is a prompt to alter the settings found in all mailboxes or just this folder, select the option that works best for you.

All the modules in Outlook have a lot of viewing options that can be used to make information much easier to use. The Message Preview option is a very good way to scan through email if you don't want to make use of the reading pane.

Follow the steps below to set up the Reading Pane;

- Locate the Mail Module >click on the View tab located on the Ribbon.
- Select the Reading Pane option.
- Select from the options on the right, bottom, or off.

Either of the options you decide to choose, you cannot be wrong. You can also switch from one choice to another if the first choice chosen is not comfortable for you. When the reading pane is turned on, you can skim through it by using either the upward or downward button.

Sending a reply

One of the amazing benefits of using email is that replying to a message is very easy. There is no need to know the person's address when replying; just click on the reply icon and Outlook will sort the remaining.

To reply to a message, follow the steps below:

- Locate the Mail Module>select the title of the message that you want to reply to. If it is enabled, the selected message will be displayed in the reading pane.
- Select one of the options below

To reply to those located in the From Field, select the **Reply** button.

To reply to those in the CC field and the From field, select the Reply All button. A reply screen will then open in the reading pane area where the initial message was before. Email sent to a bunch of people can also be received at once. Ideally, it is expected that one person at the least will be named in the To field and more than one in the Cc field, which is for those you are sending just a copy of the email to. There is no need to keep replying to those in the CC field. You can also decide to reply to some of them if you wish.

- Enter your reply into the message box. Do not be dismayed if you find out that the message box already has some messages in it; they are part of the message that you are replying to. The cursor blinking will be in the upper part of the screen. With this, anything you type will appear before the other message. This means when you reply, the recipient can view the initial message to help refresh his or her memory about the discussion at hand.
- Select the send button, and your message will be sent with the message form disappearing and the message replying to reappearing.
- Press the escape key to close the message screen. This will make the message replied to disappear and the inbox will reappear.

Resending messages

Asking people to do just what you want is another amazing feature of email. Nonetheless, most people overlook things because they receive too much email. When you discover you are making the same requests again, it is high time you took advantage of the resend feature in Outlook. With this, there will be no need to type the original request again; just find the original message and send it again with a lovely reminder about when the initial message was sent.

To send a message again, follow the steps below:

- Locate the Mail Module and click on the sent items folder in the folder pane.
- Search for the message with the initial request and click it twice. This will leave the original message unopened. Doing this is essential because resending a message is not done from the reading pane.
- Select the Actions Button located on the Message Tab and then click on the Resend This Message option. This option will create a new copy of the former message automatically.
- Insert a reminder or make certain changes to the message if need be.
- Finally, click on the "Send" button.

Don't get caught by phishing

Some cunning people are always looking to trick and deceive you, and most of the time, it is always on the internet. Recently, a very common scam known as "phishing" has caused people to lose their time and money, causing them grief after responding to an email from an impostor who makes a claim to be a representative of a bank or certain financial institution.

If you receive an email claiming to be from a bank or any other business and you are asked to click on a link for verification of your personal data, don't do it.

Oftentimes, the link will take you to a website that seems genuine, but the data you are asked to enter can be used for theft or fraud. Reach out to the organization directly to be sure the email isn't fake. If you are unsure of what to do, it is best to delete such an email.

If, on the other hand, you actually want to check in with the sender of the email, visit your browser and check out the organization's website if it is one you are quite familiar with. If the message looks very old, please do stay away from it.

That's not my department: Forwarding Email

When you don't have the answers to an email you received, you may need to forward it to someone else who can.

Follow the steps below to forward a message;

- Locate the Mail Module>select the title of the message you intend to send. The selected message will then be displayed in the reading pane.
- Click on the Forward button. The forward screen will then be opened, replacing the reading pane. The subject of the initial message will also be the subject of the new message, with the exception of the letters FW, which means forward and is entered at the beginning.
- Select the To text box and insert the email address of the person you intend to forward the message to. If the person is already in your address book, insert the person's name into the search box and Outlook will detect the email address.
- Select the Cc textbox and insert the email address of those you want to send a copy of the message to. A lot of people forward very trivial things to their colleagues via mail. A host of recipients are always added as cc addresses.

Blind Copying for Privacy

Whenever a message is sent to a very large group, all those who get the message can view the email addresses in the To and Cc fields. This means email addresses that some might want to keep private have just been given out. Everyone might have received too many bizarre, unsolicited messages before, and most of them will get peeved at the broadcast of their messages without their permission.

Blind copies offer the best of both worlds. If all the email addresses were inserted into the Bcc field, no one's privacy would be compromised. By making use of Bcc addresses, addresses that need to be kept secret can be kept secret.

Deleting Messages

It is possible to disregard an email message without having to think twice. There might be no need to read it at all. By taking a glance at the inbox list, you already have an idea of the sender and the content of the message. Hence, there is no need to waste so much time reading some unnecessary jokes. Simply take it off.

If by accident, a message you still need is deleted, to undo the action very quickly, simply press the Ctrl + z button instantly. If some other actions have been committed in error, click on the Deleted Items folder in the Folder pane; the messages that have been deleted in the past will be found there. To bring back a message that has been deleted, simply move it from the Deleted Items folder to the icon of whichever folder that it should be in.

To have a message deleted, follow the steps below:

- Locate the Mail Module > select the title of the message that should be deleted. It is not a must to read the message. You can delete it from the list immediately.
- Select the Delete button located on the Home tab on the Ribbon or simply click on the Delete key that is on the keyboard or press the Ctrl + D button.

The Delete button can be recognized easily; it is marked with a very big "X" sign, which simply means "Make this message disappear."

When messages are deleted, Outlook will not take off the deleted items; it simply drags them into the Deleted Items folder. (In some other mailing accounts and mail systems, it is known as trash rather than deleted items, though they mean the same thing.) If there are messages that have been unread in the Deleted Items folder, the name of the folder will be followed by the items that have not been read. It is possible to do away with the deleted messages forever by right-clicking on the Deleted Items folder located in the Folder pane and then selecting an empty folder. After the Deleted Items folder has been emptied, the messages in the folder will be gone forever.

Saving Interrupted Messages

If it happens that you get disturbed while composing an email, don't give up, as all is not lost. You can still get back to the mail. Whenever you begin to compose a message and then switch away to do other things without first sending the message, Outlook will save it automatically in the drafts folder.

To make sure it does this,

- Press the Ctrl+S keys before it saves automatically. When you are ready to continue working on the message being composed.
- Click on the message and complete it by clicking on the send button.

You can also click on the "discard" button if you decide not to continue with the message.

Saving a Message as a File

There are times you receive an email that is so wonderful or disheartening and you want to have the email saved. You can decide to either print out the message and show it to someone else, save the message to a disk, or simply send the message to a desktop publishing program.

To save a message as a file, go through the following steps:

- Locate the Mail Module with the message box open, choose the File tab on the Ribbon and then choose the Save As Option or select F12. This will open up the dialog box.
- Make use of the navigation pane located on the left side of the save as box to select the drive and folder in which the file should be saved. Outlook will at first choose the document folder by default, but there is an option to save the message on any drive and folder as it suits you.
- Select the File Name text box and insert the specific name you intend to give the file. Insert just any name you want. If you type a name that is not compatible with Outlook, it will bring up a window that states that the filename is not valid.

- Select the triangle at the end of the save as type box and select text only as the file type. There are various file options to choose from, but the text-only file format is the one that is read the most by other applications.

The various file type options are:

Text Only: This file format is very simple and easy to use in that it helps to remove all of the message formatting. As the name implies, only text messages are saved.

This is a format that is used to save messages that will be frequently used in Outlook. It helps to save not just message formatting but also attachments.

Outlook Message Format: This format ensures all message formatting and attachments are kept but can be read by Outlook only.

This is just the same as the former file format, but it makes use of the international characters that can be read by any version of Outlook that makes use of various languages. This is Outlook's default setting.

This helps to save a message in a file format option that can be shown in a web browser like Edge or Firefox. Or any application that can show HTM or HTML files. File attachments are basically not saved, although the message formatting is always kept. Furthermore, in addition to saving a copy of the message with the HTM file extension, another different folder is created which has the supporting files that the HTM file needs.

This is also known as the HTML file format; the only exception is that an additional folder is not created due to the fact that all of the content is saved in just one file. Applications that can show HTM and HTML files should also be able to show MHT files.

- Click on the save button and the message will be saved to the folder specified.

CHAPTER 5

EMAIL TOOLS YOU CAN'T DO WITHOUT

Outlook has the capacity to perform lots of tricks with the email messages that are sent out and the ones that are received. Messages can be flagged with reminders, customized with signatures, or have special formatting added to the messages sent as replies.

Nagging by Flagging

Over the years, flags have become the most used feature in Outlook. If you are the type that receives hundreds of messages daily and needs help with remembering those you need to reply to so that they won't get lost in the mix, it is best to flag that message as soon as it is read. This way, you are sure you will get back to the sender. A flag can also be planted in a message that you send to others in order to remind them of a particular task they have to carry out for you and the person on the other end using Microsoft Outlook.

One-click flagging

If flagging a message serves as a reminder of what needs to be done in relation to a specific message, you should be aware of the quickest way to do so.

When you move your mouse over any message in the inbox, towards the very end on the right side, a gray outline flag will be displayed, looking more like a shadow flag. When that shadow is selected, the color will change from gray to red, which means it has been flagged. When you then check through your list of messages, you can easily recognize the one that needs more attention. This way, the tracklist of flagged messages can be maintained even if they are below the bottom of the screen.

Once you have sorted the messages you flagged, click on the flag again. This will replace the flag with a checkmark indicating that the message has been taken care of.

Setting flags for different days

If you select a message once to include a flag, a copy of the message will be displayed in the task list, and various tasks scheduled for that day will also be displayed. There are times when you might not be in the mood to deal with certain messages. You might feel it is better to wait till the next day or the next week. All you have to do is

- Right-click on the flag and a list of possible dates for a flag will appear, including the next day, this week, next week, No date and customs will be displayed. Once a due date has been picked, it can still be changed if the need arises by moving the item from one due date to another with the use of the TO-DO bar. You can also choose to double-click on the item to have it reopened and then choose a different due date.

If the due date comes and goes and the flag remains unchanged, the message heading in the Inbox and To-Do bar will turn red.

Changing the default flag date

For those who are constantly busy or simply enjoy procrastinating, the default due date of flags can be changed by following the steps below;

- Click on the "Follow Up" button located in the Tags group on the Ribbon. This will then have the flag shortcut displayed.

- Select the Set Quick Click and a dialog box will be opened. The list in the box will provide different options for a due date.
- Lastly, select the date that best suits you. The date will then become the default flag due date.

If you have issues committing yourself to a certain date, choose the "No Date" option and wait until someone raises an alarm.

Adding a flag with a customized reminder

Without a doubt, flags can actually do a whole lot more than just stand for a week or more. Flags in Outlook can pop up as a reminder, reminding you of what needs to be done. Flags can also be used to pester someone else when a reminder is attached to a message sent to someone else. Adding a reminder to a flag cannot be completed with a single click.

Follow the steps below to do that;

- Locate the Mail Module and right-click on the message that needs to be flagged. The flag shortcut menu will then be displayed.
- When this is done, the custom dialog box will then open up and if you click on the OK button at this point, the message will be flagged and ready for a reminder at 4 pm. This reminder can be adjusted, especially if the time is too close.
- Select the list box arrow at the right side of the Flag to Text box and click on one of the menu items. A handy flag means "Follow Up" and reminds you to affirm a certain arrangement or appointment, as the case may be.
- Insert the dates in the Start Date box, Due Date box, or all of the boxes. The date and time inserted will determine when a reminder will pop up to help jog the memory of the appointment. Ensure the dates are typed in a way Outlook will remember.
- Click on the OK button. Once the reminder date that has been entered into the custom dialog box is up, the reminder dialog box will then help to offer a nudge.

Changing the date on a reminder

Don't get nagged with a reminder; you can always pull it off and do it much later.

To change the date on a reminder that was sent to you by someone else, follow the following steps:

- Locate the Mail Module and select the message that has the reminder that needs to be changed. The message will be displayed as highlighted when it has been selected. To open the custom dialog box, simply right-click on the message.
- Click on the Home tab and select the Follow Up on the Ribbon option, then include a reminder. As an alternative, press the Ctrl+Shift+G buttons.
- Click on the Reminder checkbox and, if it has not yet been selected, choose a new date for the reminder flag to show up. If the checkbox has already been selected, leave it as if you click on it again, it will then leave it deselected.
- Finally, click on the OK button.

There is also an option to click on the snooze button in the reminder dialog box to turn off the reminder flag for some time when it pops up, the same way it is done with the alarm clock.

Saving Copies of Your Messages

There is nothing handier than having to know what has been sent and when exactly it was sent. All outgoing email messages in Outlook can be saved. This way, you can always go back and check the message that has been sent. Immediately after the Outlook program is installed on a computer, it starts saving sent messages. This feature can, however, be turned off. Before attempting to do so, check your message folder to be sure it contains sent messages.

To save copies of the messages, simply follow the steps below.

- Choose the File tab > select the Options button. This will then open up the Outlook dialog box.
- Select the Mail button and locate the navigation window on the left side. The mail settings will then be displayed.
- Move down to the Save Messages section and click on the Save copies of the messages in the Sent Items folder check box if it hasn't already been selected.
- Then finally click on the OK button.

Setting Your Reply and Forward Options

The look of a forwarded message can be controlled as well as the replies. If you use Microsoft Outlook, your text can be made to look really incredible in your messages by the addition of graphics, special effects, or some wild-looking fonts. If mail is being sent to people who use other programs and not Microsoft Outlook or to those using web-based email services like Gmail, some of the effects might not be well translated.

To set your options, follow the steps below:

- Click on the File tab, locate the ribbon, and click on the Options icon. The options dialog box for Outlook will then open up.
- Select the Mail button located in the navigation window on the left side and the mail settings window will then be opened.
- Move downwards to the Replies and Forwards section and select the list box arrow at the right side of the Reply to a Message box. A menu of options will then be displayed, including the original text as the default option. The diagram located on the left side of the menu will show how the message will be displayed when each option is chosen.
- Select the preferred style to use for replying to messages. A little diagram on the left side of the menu will change instantly to display what your choice will look like. You can try another if you don't like the choice you made earlier.
- Click on the list box arrow that can be seen on the right side of the When Forwarding a Message box. This menu option looks almost the same as the one above, with just one fewer choice.
- Select the style preferred to be used for forwarding messages.
- Click on the OK button. The Outlook Options dialog box will then be opened.

By making use of Outlook, there is so much room to explore and do all sorts of fancy and even useful tricks with email. If the advanced options menu appears confusing, you can easily ignore them and just click on the Reply button and insert your reply.

You can also choose to delete the original message when a forward is created or replied to, but it's best to include at least a part of the original message. This way, it will make the response easier to understand. There is also an option to choose and delete some parts of the main text that are not relevant to your reply.

Adding Comments to a Reply or a Forward

When a message is forwarded or replied to, it is okay to type the reply above the original message if the original is short and simple. But if the original message is quite complicated and embedded with so many questions, it is best to type the reply in the original message, replying to each question as it appears.

The only issue that might arise with typing replies within the original message is that it is not always very clear what the format of the new reply is and the original message. To sort that out, Outlook provides a means by which you can preface any in-line replies that are being made with your name or whatever is needed. This also makes the reply text bold and clear, making it stand out more.

To include the in-line replies with your name, go through the steps.

- Click on the File tab located on the Ribbon and select the Options icon. The Outlook Options box will then open.
- Select the mail button located in the navigation window on the left side to open the mail settings window.
- Move downwards to the Replies and Forwards part and choose the "Preface comments with the checkbox.
- Within the preface comments, With the text box, enter the text you deem fit to be used as the prefix to all of the text that will be typed whenever you reply to messages.
- Finally, click on the OK button.

Sending Attachments

If you have created a document in another application and you don't want to send it just yet, there is no need to type the same message all over again in Outlook; simply send the document as an attachment to an email message. Word processing documents, spreadsheets, and other presentations such as PowerPoint can all be attached. You can also choose to include pictures and music; any type of file can be sent as an attachment.

The Send From menu is the simplest way to send a file from a Microsoft Office program such as Microsoft Word.

- Open up the file in the program it has been created in.
- Choose the File tab option located on the ribbon.
- Select the "share" button.
- Choose the Word document from the share dialog box and write the message in Outlook.

If you don't want that option, you can have a message sent directly from Outlook by following the steps below.

- Locate the Mail Module and select the New Email button that can be found on the Ribbon. You can also press the Ctrl + N button if you prefer to use a shortcut.
- Choose the Attach File button found on the New Message form's Ribbon.

A list will then drop down to display the names of the files that have recently been worked upon.

You may be lucky enough to find the name of the file there, then just click on it. If the name of the file is not there, select the Browse This PC option located at the lower part of the screen. This will then open up the Insert File Dialog Box.

Click on the name of the preferred file and tap the send button. The name of the file will then be displayed in the Attached box in the message form's message header. When the email message is sent, a copy of the selected file will be sent to the recipient.

- Enter a message if you have one to send. There might be no message and you just want to send an attachment. Take note that the content of the attachment will not be displayed on the screen of the recipient until the attachment is opened.
- Click on the "To" button in the message form. The dialog box for choosing names will then be displayed.
- Choose a name from the contact list and click on the button in the Select Names dialog box. This will then open up the name of the person that has been selected. This process can be repeated if you want to include more than one recipient.
- Select the OK button. The name of the person can now be found in the To box of the message.
- Click on the Subject text box and insert a subject for the message. The option of a subject is not compulsory, but if you want the message to look important, a subject can be helpful.
- Finally, click on the "send" button.

Emailing Screenshots

It is often said that pictures are worth more than a thousand words. Most of those words become four-letter words when the computer begins to act up. This makes it very difficult to describe the type of problem in an accurate manner. When problems like this arise, Outlook can be of help.

A screenshot is simply a picture of the screen of a computer that is captured to show what is being done on the computer at that point in time. This eBook contains lots of screenshots to make certain steps and procedures more comprehensible. The exact same thing can be done with the screenshot feature in Outlook. A screenshot can be sent to help a person solve a problem with his or her computer. A screenshot can be sent off just about anything, including pictures and documents. The possibilities of this feature are endless.

To include a screenshot in an email message, simply follow the steps below.

- When an email message or a reply is being composed, select the Insert tab on the ribbon. If you notice the screenshot button is grayed out, ensure the cursor is inside the body of the email message.

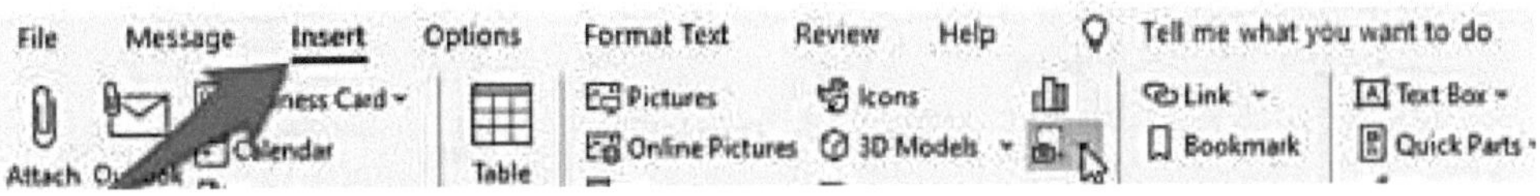

- Tap the screenshot button. This will then display a gallery of thumbnail images.

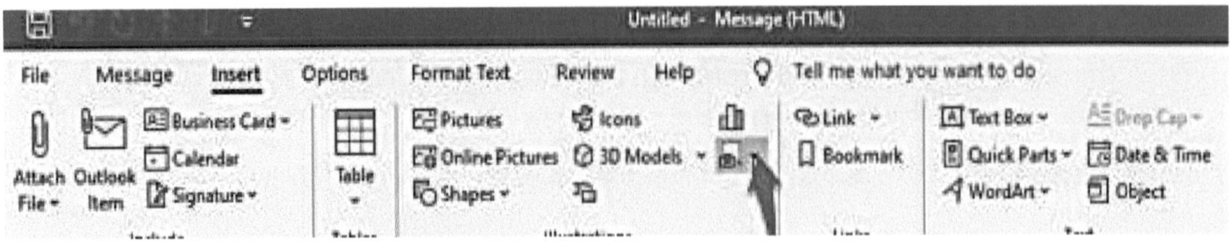

- Choose any one of the screens from the gallery. The screenshot that has been chosen will then be displayed in the body of the email message.
- Conclude with your email message and send it to the recipient.

Creating Signatures for Your Messages

Most people love to include signatures at the very end of the messages they send. A signature, most of the time, is just a few lines of text that show you to all those that read your message and also states certain things you want them to know. Lots of people include their names, the names of their businesses, and also the web addresses, their motto, and a little personal information. You can set Outlook to automatically include a signature in all outgoing messages, but a signature file must be created first. To create a signature

- Choose the File tab found on the Ribbon > Click on the Options button. This will then open the Outlook Options dialog box.
- Select the Mail button in the navigation window towards the left side. This will open up the Mail settings window.
- In the Compose Message section, click on the signature button. The signature and stationery dialog box will then be open.
- Click on the New Button icon.

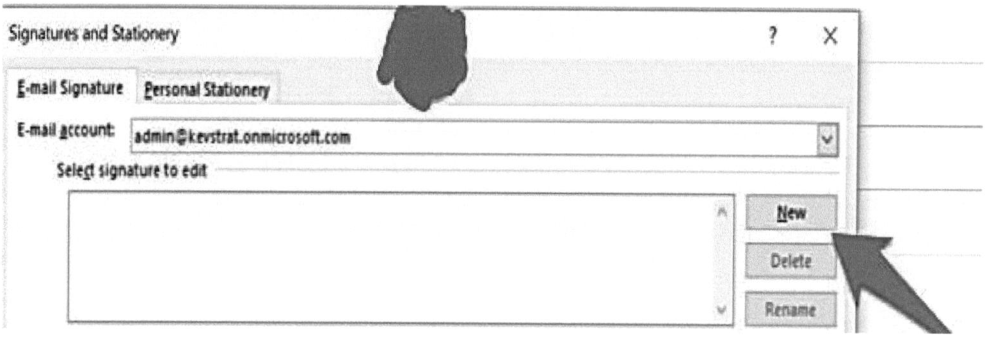

The dialog box for a new signature will open.

- Enter a name for the new signature. The name that is typed will then be displayed in the New Signature box.
- Click on the OK button to complete this process. The dialog box for the new signature will then close.
- Enter the text of the type of signature that you prefer in the Edit Signature box and include any formatting that suits you. To apply changes to font, color, size, or other text characteristics, make use of the button in the text box. You can choose to make the signature in Word and then copy and paste it in the Edit Signature box.

CHAPTER 6

DEALING WITH MESSAGES

Most people spend a lot of time on email messages. Some experts also estimate that an average business employee spends about 2 hours per day on emails. Very soon, more time will be spent on email than on actual work.

Outlook has some tools for dealing with the influx of flotsam and jetsam that finds its way into your message inbox. Different folders can be created for filing mail, and you can also make use of Outlook's View feature to help slice and dice incoming messages into smaller groups. Old messages can also be archived to save your inbox from becoming cluttered.

The Rules Wizard is even considered a better feature than the View feature. The Wizard responds to messages coming in just the way you want. All messages can also be moved from specific senders to the folder you choose.

When talking about spam, the use of junk email filters in Outlook is a better way to deal with aggressive or offensive messages. Ensure the filters are turned on already. You can also crank up the settings so that there will be less junk mail cluttering the message inbox.

Organizing Folders

By now, you are most likely used to arranging items into folders. Windows helps to arrange all the other documents into folders. Outlook will also do the same. Simply create a folder and move your stuff into it.

Creating a new email folder

The simplest and easiest way to cope with messages coming in is simply to file them. Before a message can be filed, at least one folder wherein the file will be stored will be created. The folder is there for life unless you decide it is no longer needed and you delete it. You can create as many folders as you need.

To create a folder, simply follow the steps below:

- Locate the Mail Module and choose the inbox option in the folder pane or press the Ctrl + **Shift + I** buttons. If you were not doing some other thing before, the inbox should be selected by default when Outlook is opened.
- Choose the Folder tab and select the New Folder button located on the Ribbon. This will open the Create New Folder dialog box.
- In the Name text box, enter a name for the new folder. You can choose to use any name that suits you. You can also make as many folders as possible. There shouldn't be too many folders either, to avoid confusion.
- Click on the OK button. The new folder will then be displayed in the Folder pane.

Moving messages to another folder

Filing messages can be as easy as moving them from the folder they are into another folder where you prefer them to be. All you need to do is open the box when they arrive and move each message to the folder where you want it. For another method of moving the messages to a different folder, follow the steps below:

- Locate the mail module > click on the title of the message that should be moved. The message will then be highlighted.
- Choose the Home tab option and click the Move button on the ribbon. The move drop-down menu will then be open.
- Choose the name of the preference folder to which the message should be moved.

Organizing Your Email with Search Folders

The search folders feature in Outlook was created to help arrange messages in the inbox and in other folders. The Search Folder offers one space where a specific type of message can always be found. A search doesn't move messages; it just creates a kind of an imaginary location for messages. This way, you only have to look at one type of message at a time.

When Outlook is initially opened, search folders are not displayed in the folder pane. If there is a need to use search folders, there will be a need to include one of the default search folders or make a search folder of your choice.

Setting up a search folder

To set up a search folder, follow the following steps:

- Locate the Mail Module and select the inbox to choose the folder in the Folder pane.
- Choose the Folder tab and then select the New Search Folder button located on the Ribbon. As an alternative, you can also press the Ctrl + Shift + P buttons. This will open the New Search Folder dialog box.
  - Choose the type of search folder that should be added from the list in the New Search Folder dialog box. Lots of folders are available and will be displayed. You can choose to use an existing folder or create a folder of your choice.
  - If a button showing "Choose" is displayed in the lower part of the New Search Folder dialog box when you click on a search folder, click on the button and enter the information requested.
  - Click on the OK button and the New Search Folder dialog box will be closed, displaying the new search folder in the navigation pane.

Some very useful predefined folders include:

Email marked for Follow up: This folder displays messages that have been flagged only. When the flag is removed from a particular message, it will no longer be seen in this folder but can still be found in the inbox or any folder it primarily is.

Large Mail: This folder arranges messages by the amount of storage space they need. Normally, you aren't concerned with the amount of storage required by messages you receive but don't be surprised if the system administrator where you work insists on not storing too many messages. If you have lots of messages that include attachments, your inbox might get filled up quickly.

The large mail folder can be used to discover the messages taking up so much space and delete the largest ones. The messages that will be displayed in the large mail folder will be characterized by size, beginning with "large," then "huge," and "enormous."

Unread Mail: This folder displays only unread messages. When a message in this folder is read, it leaves the folder, but it will still be displayed in your inbox.

Using a search folder

There is no need to do anything special to make use of a search folder. Simply

- Click the name of the preferred search folder that you want to search for in the Folder pane, and a list of those kinds of messages will be displayed.

When you are set to get back to your inbox,

- Click on the inbox button located in the folder pane to view the collection of all the messages again.

Deleting a search folder

Once the search folder has served its purpose, there is no need to keep it again.

You can delete the search folder by:

- Locate the Mail Module in the Folder pane > right-click on the search folder that should be deleted.
- Click on the Delete Folder button.
- Choose Yes. The search folder will then disappear.

Using the Reading Pane

If all you want to do is skim through your bunch of messages swiftly, the reading pane can be of great help. The reading pane is shown by default in Outlook. If somehow it gets closed,

- Click on the View tab option.
- Select the reading pane located on the ribbon.
- Click on the right icon to then open it.

When the reading pane is fixed to the right, the inbox screen will then show a list of messages on the left of the center of the folder pane is shown and the content of the chosen message is also on the right side. When you want to move from one message to the other, press the upward or downward arrow key. Messages can also be viewed in the inbox when the message title is selected. If you wish to have the message displayed on the lower part of the screen instead,

- Select the "View" tab option.
- Choose the reading pane button on the ribbon
- Click on the bottom option. Note that you won't be able to see many messages using this method. It is best to set the reading pane to show on the right side.

The reading pane shows a lot more message content than when the message preview option is used. If you receive messages that have formatting or images, the graphical view will be better appreciated when the message is viewed in the reading pane.

Playing by the Rules

Rules are just another of the amazing features in Outlook. This feature helps you make Outlook act on a specific type of email message automatically. For example, if you get lots of messages that can easily waste your time when you are trying to sort them, you can simply set up rules in Outlook to sort incoming mail into various folders. There are so many rules that can be created with the use of the Rules Wizard.

Creating a rule

There comes a time when you will discover the need to make a rule. By making a rule, you can set those you want to read their messages and avoid reading some mail that might make you feel bad or set the day on a bad note for you.

The Rules Wizard is termed a "wizard" because of the way the feature leads you to create each rule step-by-step. The process can be very simple.

To create a rule to move a message coming in from a particular person to a specific folder, follow the steps below:

- Locate the Mail Module and choose the inbox icon located in the Folder pane.
- Move to the home tab located on the Ribbon and select the Rules button > Click on the Manage Rules and Alert option. Do not click on the option to "Create a Rule" on the Rules button's options menu, as the options will be limited based on the message that is currently selected. Choosing Manage Rules and Alerts will open the Rules and Alerts dialog box, though you will still be just a click away from the Rules Wizard.
- Select the Create Rule button. This will then open the Wizard dialog box, which has a list of the types of rules you can make.

Select the type of rule you want to make.

The Rules Wizard provides various common types of rules you might want to make, like:

- Drag messages from a person to a folder.
- Drag messages with a certain type of word into the subject folder.
- Drag messages that were sent to a public group to a specific folder.

The collection of suggested rules is split into useful groups like "Stay Organized" and "Stay Up to Date." When you have made your choice, click on the Next button, and then you will see a message in the rule description box.

- Locate the select conditions box, ensure that the From People or Public Group selection has a checkmark sign at the front of it, and click on the first piece of text underlined in the rule description box which says People or Public Group. This will then open the Rule Address dialog box.
- Click twice on the names of each of the people whose messages should be moved to another folder. The email address of each of them will then be displayed in the From text box at the lower part of the Rule Address dialog box.
- Once all the people whose messages should be moved have been chosen, click the OK button. This will then close up the Rule Address dialog box and the names of the people that have been chosen will be displayed instead of "people" or "public group" in the rule description box.

- Choose the next piece of text underlined in the rule description box that says specified. This will open a dialog box providing you with a choice of folders that the messages can be moved to.
- Click twice on the name of the folder in which you want to move the messages. This will close the dialog box and the name of the folder that has been selected will appear in the sentence in the rule description box. More rules can be added by choosing them from the Select conditions box. If the next button is selected a few more times, actions like clearing the message flag and exceptions like "except if sent to me alone" can be added to your rule.
- Click on the Finish button. This will open up the Rules and Alerts dialog box again, offering a list of all the rules. Each rule has a check box next to it, and by selecting or deselecting the checkbox, you will either turn on or off the rules. If a checkmark is shown next to the rule, it means it is turned on and, if otherwise, it means it is turned off.
- Select the OK button to have the Rules and Alerts dialog box closed.

Rules can do a lot more than just sort messages that are coming in. Rules that can reply automatically to certain messages, flag messages with a specific word in the subject, and also have messages from certain people deleted can be created.

Putting a rule into action

Immediately after messages enter the inbox, rules swing into action. When a rule to move messages from a particular person to a specific folder is created, the message that arrives after that rule must have been created will move automatically, but the messages that were already in the inbox before the creation of the said rule will remain in the inbox.

If you want the rule to affect the messages already in the inbox, follow the actions below.

- Locate the Rules and Alerts dialog box and choose the specific rule you want. If the Rules and Alerts dialog box is not already open, click on the Rules button located on the Home tab of the Ribbon and then click on Manage Rules and Alerts.
- Select the Run Rules Now icon located at the top of the dialog box. The dialog box for the Run Rules will now be opened.
- Select the name of the rule you want to run and place a checkmark near it.
- Choose the "Run Now" button. The Rules and Alerts dialog box is currently covering the screen; hence, the result of the rule cannot be seen yet. Once the close button and OK button have been selected, it will be obvious that the rule has already been carried out.

Filtering Junk Email

Are there times when you feel overwhelmed by the influx of junk email in your inbox? You are not alone. Lots of junk email is now being sent over the internet instead of genuine ones. Outlook has a feature that checks all your incoming messages and moves the ones that look like junk email (spam) automatically to a special folder. Anything that gets moved into the junk folder can be immediately deleted over and over again after you must have checked to be sure that Outlook has not made a mistake in moving a real email to the junk folder.

There is just no such thing as a perfect machine, and there is no program running on a machine that can be perfect. This means Outlook can make certain mistakes when catching junk mail, but it can still be relied upon. Some people, however, prefer to make use of software that works with Outlook to sieve out junk mail.

Fine-tuning the filter's sensitivity

Based on your settings in Outlook, automatic filtering might be on already and probably set to low, or it might as well be turned off. The low setting is somewhat conservative, looking for the most visible spam but not stopping most emails from entering is at least kind of plausible as legitimate mail. The high setting can be very aggressive and sometimes takes off messages you want to see. To make some changes to Outlook's junk email settings, follow the steps below:

- Find the Mail Module > click on the Home tab > select the Junk icon in the Delete group located on the Ribbon

Choose the Junk Email Options. This will open up the Junk Email dialog box with the Options tab placed at the top.

Choose the option you prefer from the following options:

No automatic filtering With this setting, every sleazy message will get straight into your inbox.

Low: Only the junkiest of junk will be moved, though some nasty mail will make it to your inbox.

High: This option is so aggressive that it will sometimes move some legitimate mail. If you choose this setting, check the Junk Email Folder frequently to be sure that important messages do not get marked as junk by mistake.

Safelists only: This setting will remove all messages from your inbox except the ones from certain people or organizations that have been designated in the Safe Senders lists.

The checkboxes at the lower part of the Options tab also provide a range of other choices:

Instead of moving it to the junk email folder, permanently delete the suspected junk email: This option might be a little too aggressive. There is just no perfect junk email filter yet. It might be better to drag junk messages to the junk email folder and manually empty the folder after it has been properly checked. If, however, you work with an organization that limits the amount of email you should store, this option might be just perfect for you.

Disable links and other functionality in phishing messages: The term "phishing" is used for an email message that attempts to impersonate a business or a financial institution in a bid to steal your personal information or have your laptop infected with a virus. This is most of the time the first step when it comes to operations of identity theft. As such, Outlook will try to get false emails and also make the web links ineffective. On no occasion should you release your personal information to any financial institution over the phone, it is best you do so over the phone or directly log on to its website.

Warn Me About Suspicious Domain Names in Email Addresses: There are some locations that have bad reputations both on and off the internet. If you get an email from a location that is suspicious, Outlook can give you a warning so you don't get into trouble.

- Click on the OK button and the Junk email dialog box will close.

Filtering your email with sender and recipient lists

The junk email feature in Outlook helps you make decisions if you would prefer to set up a safe and blocked list of your own. There is an option to make a list of people whose messages should be moved to this folder all the time. Take a look at the tabs of the Junk Email Options dialog box for a description of the specific types of senders that can be entered.

Whenever you receive a message from an email address or domain that is specified in this option, Outlook ensures it does not treat the message as a junk email, irrespective of the content of the message.

If you get a message from a mailing list online, the message might come from many different people, but it is always addressed to the list. In this case, the name of the list should always be on your safe recipient's list.

Blocked Senders: This is the direct opposite of the other choices above. Messages from the addresses or domains in this list are expected to always be treated as junk emails.

International: There are a lot of spam messages that come from abroad. There are times when you receive a long and endless list of spam from senders whose email addresses end up in strange letters like herte@spam.ru. The odd letters at the very end of the address are known as top-level domains, and they also show the country of origin of the sender. For example,.ru is Russia's top-level domain, which is a common source of spam these days.

Furthermore, if you receive lots of spam messages in foreign languages, you can also set Outlook to ban messages like that. Locate the international tab, select the Blocked Encoding List icon, and choose the various languages.

If you also receive legitimate emails regularly from senders whose messages make use of a particular top-level domain, and you don't want to block that domain even if you receive a lot of spam from it, do not block the languages that are used by these legitimate senders.

To add a person to your Blocked Senders list, follow these steps:

- When you receive a message from someone you don't wish to hear from anymore, choose the message.
- Select the Junk button located on the Home tab of the Ribbon.
- Select the Block Sender option.

This method also works for adding people, domains, or groups to the Safe Recipients and Safe Senders lists. To be more precise, there is an option to go directly to the appropriate tab located in the Junk Email Options dialog box and insert the addresses or domains you prefer to sieve.

You could also save time by using the following junk email options:

- At the bottom of the Safe Senders tab is the option labeled "Also Trust Email From My Contacts." By checking this box, all messages from anyone in your address book will be treated as safe.
- Recipients: In Outlook, if you check the box labeled "Automatically Add People I Email to the Safe Senders list," messages will be accepted from the people to whom you have already sent messages.
- Import and Export: Adding a long list of people to your Safe Senders or Blocked Senders list can be done in Notepad and then imported into Outlook. A company with a large client list may offer this option to all its employees.

Filtering domains

When it comes to junk mail choices in Outlook, you have one rather powerful option you should be aware of. The option has to do with sieving domains. By following these steps, you can enter the entire company in your Safe Senders list if you do business with someone at that company.

- Go to the Mail module > select the message.
- From the Home tab on the Ribbon, select the Junk button.
- Set the option to Never Block the Sender's domain

Archiving for Posterity

It doesn't take long for your message inbox to overwhelm you, faster than you can deal with it. Several people simply delete their messages after reading them. It is best to keep some messages for reference's sake.

Storing too many messages slows Outlook down when you have a lot of them stored. Not only is a large collection of messages challenging to manage, in a large organization you may not be allowed to store many emails to prevent them from clogging up the server.

Outlook has an archive feature that you can use to store messages and other items that you no longer need to view but that may be useful in the future. By reducing the number of emails you store in the email system, you can easily get along with your system administrators if you use Outlook on an Exchange network at work. You may see that your Outlook items are disappearing if you use the AutoArchive feature to automatically archive items.

Below you will find instructions on finding items that Outlook has archived for you to keep. People tend to archive email messages most often, but nearly all Outlook items can be archived, including calendars and tasks.

Setting up Auto Archive

The AutoArchive setting of Outlook prevents Outlook from automatically archiving your items unless you change it. However, some businesses may allow their users to use it.

You should check your company's email retention policy before you make any changes to the AutoArchive settings. Some companies use an auto-delete service to delete old messages.

Follow these steps if you want to enable AutoArchive, see how Outlook archives items, or change how Outlook works:

- Click the File tab and choose Options. The Outlook Options dialog box will then be opened.
- In the navigation pane on the left, click the "Advanced" button.
- A menu of options appears for working with Outlook pages.
- Go to the AutoArchive section and click the AutoArchive Settings button.

It's best to avoid making radical changes to the AutoArchive dialog box, at least not until you look at what's already set up.

A couple of important tidbits that are normally provided by the AutoArchive dialog box are:

- Is the Auto Archive feature enabled or not?
- What is the frequency at which Outlook archives items?
- How old do items have to be in order to be archived by Outlook?
- Identify the location and name of the archive.

AutoArchive turns on automatically every 14 days without changing any of the other AutoArchive settings, sending items older than 6 months to the archive file shown in the AutoArchive dialog box.

At the top of the Auto Archive dialog box, you can select or deselect the "Run AutoArchive Every" check box to turn on or off the AutoArchive process. Replace the 14 in the text box with any number between 1 and 60 to change how often AutoArchive runs.

Setting AutoArchive for individual folders

Having the AutoArchive settings set up for each folder separately might be a better option since you will be able to select exactly which files will be archived and which won't.

- From the Mail module, select Inbox from the list of folders. The inbox will then be highlighted.
- In the Properties group of the Ribbon, click the AutoArchive Settings button under the Folder tab.
- A dialog box with Inbox Properties appears, showing the AutoArchive tab.
- Using these settings, archive the selected folder.
- Make sure the drop-down menu is set to Months.
- Messages that are very recent from your inbox can be auto-archived if you choose "Weeks."
- Clean Out Items Older Than Type 6 if it doesn't already appear in the Clean Out Items Older Than text box.
- You should now see a notification that items older than six months will be deleted.

- It is possible to auto archive messages from the inbox that are anywhere between one day old and 999 months old by entering any number between 1 and 999.
- Choose to Move Old Items to the Default Archive Folder.
- It is likely that this setting is already selected, but make sure the option to permanently delete old items is not selected.
- Choosing this option would remove all old inbox messages instead of archiving them.
- To proceed, click the OK button.
- Even though you're only setting the AutoArchive settings for a single folder, you must enable Outlook's AutoArchive option.

To use AutoArchive with each folder, repeat these steps. You should at least check what the current AutoArchive settings are for any folder you do not want to auto-archive.

By enabling Outlook's AutoArchive feature, you probably enabled AutoArchive for some other folders that you might not want to archive; there is no way to auto archive a folder without also turning on Outlook's AutoArchive setting.

Whenever you create a new folder, it's automatically set not to auto-archive, even if you previously applied your auto-archiving settings for all folders. If you want your new folder to autoarchive, go through the previous steps for that folder. Also, when you turn on Auto Archive for Outlook, the Deleted Items folder is set to auto archive using the default settings. If you don't clean out your Deleted Items folder, all the emails you thought you'd never see again will instead be archived for posterity. You should consider setting the Deleted Items folder to not auto-archive.

Starting the archive process manually

Here are the steps you can follow to archive messages at any time:

- Navigate to the File tab.
- In the left-hand navigation pane, click the Info button.
- Go to Tools and click on it.
- The Archive dialog box appears when you click "Clean Up Old Items."
- Then click OK once you've selected your settings.

Manually starting an archive will give you a better sense of control over the process and enable you to:

- Establish a deadline for archiving items (say, the first of the year).
- Archive the appropriate folders and place the items in the archive.
- Create different archive files for different Outlook folders.

One disadvantage of all these controls is that it's possible to accidentally archive something and not be able to find it again.

If you want to keep track of your archived items, don't change the name or location of the files to which your archived items are sent. Outlook doesn't provide much assistance here.

Finding and viewing archived items

AutoArchive can sometimes seem like magic. Objects from the past are mysteriously filed away without any action from you.

It seems simple enough, doesn't it? You may not notice it, but when you suddenly need one of those items that appeared in your archive, you will. Then you have to figure out where it went and how it got there.

Using these steps, you can open a data file containing archive items:

- Navigate to the File tab, click on Open & Export, then select Open Outlook Data File. A dialog box for opening Outlook data files appears.
- Choose the file you wish to open. The filename of the selected file appears in the File Name text box. For example, to open the default archive file, select "archive."
- Click on the OK button. In the Navigation pane, below your normal set of folders, you can see the name of the data file you opened.

The following steps will help you determine the name of the archived data file to open:

- Navigate to the File tab.
- Click on "Info".
- Select the Tools button.
- Click on the "Clean Up Old Items" button.
- Look in the text box for the Archive File.
- The archive file is located here. Please do not change the information; otherwise, Outlook might start sending your archived items to another location.

In the Archive File text box, you will usually find a lot of complex information with colons and slashes and other things that everyone can't understand.

Closing the archive data file

Archive data files can remain open in the Folder pane for as long as you like, although most people close them once they locate what they're looking for.

You can speed up Outlook by closing unnecessary data files.

Follow these steps to close an archived data file:

- Right-click on the archive data file's name in the Folder pane. An options menu appears.
- The Archive can be closed by selecting "Close." If the archive data file is called something else than an archive, the name of the archive data file will replace the term "archive."

In the folder pane, your archive folder disappears.

Arranging Your Messages

There is no one who gets a little email anymore. When you get one message, you get a lot of them, which quickly fills up your inbox. The next thing you know, you're scrolling through a never-ending stream of messages, trying to find the one needle in the haystack that you needed a week ago.

Thanks to Outlook, you have a bunch of options for organizing your inbox so you can figure out what is important, what can wait, and what can be ignored. Whenever Outlook displays the reading pane on the right side of the screen, four labels appear at the top of the list of messages.

Two of the leftmost labels are titled "All" and "Unread." When you click on any of your messages, they are all displayed. Messages you haven't read yet will appear if you click Unread; once you read a message, it will disappear from this view, although you can see it again if you click All.

On the right, you can see the system Outlook uses to organize the order in which your messages are displayed. Your messages are displayed in chronological order if the "By Date" label is displayed. Most of the time, you'll want to see your messages that way.

Below are other ways in which messages can be arranged:

Date of Conversations:

Your inbox will look like this when you first set up Outlook. Messages are displayed in the order of receipt, just as you would expect.

From:

The arrangement, as you might guess, organizes your message collection according to the person who sent the message.

While setting up a search folder can sometimes be faster than using the From arrangement, a search folder can still be the best way to track messages from specific people.

In order to:

Not all the messages you receive are addressed to you. In some cases, you receive messages addressed to a list of people, so you don't see your name in the To field. The To field of each message is separated according to whether your name appears in it.

Categories:

In either case, messages can be sorted based on categories if you choose to use them. For IMAP accounts, this option may not be available.

Flag:

If you have flagged items with a due date, it is helpful to sort messages by start date or due date. When sorting in ascending order, flagged items appear last, and when sorting in descending order, they appear first.

Just like Categories, the option might not be available for an IMAP account.

Dimensions:

Certain system administrators care about size, and it isn't always a personal issue. Your company's email servers can get clogged up with images, music, and other heavyweight files that are attached to email messages.

If your system administrator asks you to thin out your inbox, take advantage of this feature: Identify overweight messages and delete or archive them.

Microsoft Outlook categorizes messages as "Tiny," "Small, "Medium, "Huge, or "Enormous," and perhaps it will add "Ginormous" in its next version.

Subject:

While this arrangement is similar to conversations, it doesn't follow the thread of a conversation; it merely groups together messages with the same subject matter. Messages with the same subject do not necessarily belong to the same conversation.

Type:

The items that arrive in your inbox are not always simple messages; you can also receive meeting requests, task requests, etc.

Use the type arrangement so the most interesting messages rise to the top of the list when you want to separate messages from meeting requests and so on.

Attachments:

You may not be looking for a message when you go to your inbox; you may be looking for an attachment. When you arrange your messages by attachment, you have the chance to look for the likely targets first.

Account:

Outlook can collect emails from several different email addresses simultaneously, and each email address gets its own inbox.

Having said that, if you move messages from different email addresses into the same folder, as you might do when you clean up your inbox periodically, at some point you may want to know which messages came from which address or just want to see those messages sent from that address. Click the arrow next to the name of any account you do not want to see if you want to see only the messages sent to that address.

As a result, Outlook only displays messages from accounts that you are interested in. As long as all your incoming mail is consolidated into a single inbox, the account arrangement won't be useful when

Importance: If you need to see high-priority messages first, this is the arrangement to choose.

While the reading pane is off, use the following instructions to arrange your messages:

On the View tab, select the arrangement you want to use from the Arrangement group, and then click Reverse Sort to reverse the order of sorts.

Viewing conversations

No matter if you simply exchange a few emails back and forth with one person or take part in a large group discussion that lasts for weeks, Outlook's Conversations arrangement keeps all related messages on the same subject grouped together.

At a glance, you can view the most recent message in a conversation thread as well as older messages from the conversation at a glance.

Whenever someone replies to a message, either by clicking Reply or Reply All, a conversation begins. All new messages become part of the conversation regardless of who else responds or contributes.

You can enable conversation viewing by following these steps.

- Click the View tab in the Mail module.
- In the Messages group, select the "Show as Conversations" check box.

In your inbox, a small triangle just to the left of the sender's name indicates that more messages are included in the conversation if the conversation display is enabled.

The most recent message that was received in the conversation appears in the Reading pane when you click on a message that's part of a conversation. To view all the messages, you have sent or received as part of the conversation, click the triangle to the left of the message's mail icon.

It is possible for some of these messages to appear in the conversation list even if they are in the Sent folder or have been moved to another folder. Messages that are moved to the Deleted Items folder, however, will not appear in the conversation list.

All messages in a conversation can be replied to. Replying to a message that isn't the latest displays a warning and gives you the option of opening the latest message.

If you want to customize the display of conversations,

- Click the Conversation Settings button on the Ribbon. It will bring up a menu of options. For instance, you may choose to show or hide messages from other folders.

Ignoring conversations

The conversation often comes to a point where you are no longer interested or relevant to the discussion. Off-topic discussions are not uncommon. A simple discussion about when to have the next team meeting ends up being a seemingly endless series of jokes about how hot it gets in the conference room during the winter.

Whenever you no longer wish to follow a conversation, you can easily ignore it:

On the Home tab, click the Ignore button next to the appropriate message.

To ignore a conversation, right-click on any message and select Ignore from the menu; or select any message and press Ctrl+Delete.

As soon as you ignore a conversation, all messages from that conversation, whether in your inbox or in any other folder, will be moved to the Deleted Items folder.

The messages from the conversation that is in your Sent folder are not moved. Your inbox and other folders are free of traces of the conversation.

Additionally, any new messages from the conversation are automatically sent to the Deleted Items folder.

Cleaning up conversations

Most email programs automatically include the text of the original message in the reply when someone replies to a message. Upon the first installation of Outlook, it does this by default.

You can see this setting by:

- Navigating to the File tab > Select the Options button in the left-hand Navigation pane
- In the Outlook Options dialog box, click the Mail tab, then scroll down to the Replies and Forwards section.

As long as everyone replies to a conversation with the text of the previous message, each subsequent message becomes a snapshot of the entire conversation thread up to that point.

As a result, there is a lot of redundancy in the messages because the same information is repeated. Is that needed? Messages that contain information already included in another message can be removed from a conversation if Outlook detects this redundancy. The process is known as "cleaning up" in Outlook.

If you want to clean up a conversation quickly,

- Select the message you want to delete > press Alt+Delete.

To clean up more thoroughly, follow these steps:

- Select the conversation you want to clean up in the Mail module.
  You'll see the message that was most recently received in the conversation. Any folder containing a message from the conversation can be used to do this.
- Click the Clean Up button in the Delete group on the Home tab of the Ribbon.
  A drop-down box appears with three options.
- Select Clean Up Conversation.

When Outlook detects that the messages in a conversation are redundant, it removes them from all the folders they're in and moves them to the Deleted Items folder. However, messages in the "sent" folder remain in the folder. As another option, you can select "Clean Up Folder" to clear out all conversations within the selected folder. It goes a step further by cleaning up all the conversations in the selected folders as well as all subfolders.

As an example, if you have a personal folder inside your inbox folder, selecting the inbox folder and then choosing Clean Up Folder & Subfolder will automatically clear out all conversations in both folders.

Outlook won't remove many messages by itself, so don't be surprised. Outlook takes a conservative approach when it comes to eliminating redundant messages, and it also doesn't move replies that modify a previous message or that are marked as a follow-up.

The following settings will allow Outlook more flexibility in what it can move when cleaning up a conversation:

- Click on the File tab and then Options. A dialog box with Outlook options opens.
- Select the "Mail" button. A window opens with the mail settings.
- The conversation cleanup section can be found at the bottom. Outlook offers several options that control when it moves messages from conversations and when it does not. Choose the settings that work for you.

  Make sure you pay attention to the Cleaned-Up Items Will Go to This Folder text box. Cleaning up items in Outlook places them in the Deleted Items folder by default. However, if you wish to send cleaned-up items elsewhere, such as a folder in an archive file, this is the place where you make the change. Select a destination location from the Browse button.

Simplifying Tasks Using Quick-Steps

It is no doubt any longer a secret that Outlook actions can require multiple clicks of the mouse to complete, like replying to a message and then deleting it. When you perform a particular action in Outlook only occasionally, it's not a big deal, but if you do it regularly, it can quickly get tedious.

Outlook Quick-Steps can come in handy if you perform certain tasks on a regular basis. Multi-step tasks can be reduced to a single click with Quick-Steps.

On the Ribbon's Home tab, under the Mail module, you can find Quick-Steps. Despite Quick-Steps only appearing when you are working with the mail module, this tool can help you speed up most actions in Outlook, such as those involving the calendar and tasks.

Move To:

If you often move messages into a specific folder, use this Quick-Step. The Quick-Step displays Move To: If you have yet to move a message to a folder in Outlook, This Quick-Step replaces the To which folder you moved the message to, such as "Move To: Personal." If you have already moved messages to folders,

To the supervisor:

In this Quick-Step, the selection of a message is automatically forwarded to a particular recipient, but the message is not deleted. Generally, if you are using Outlook in a corporate setting, the message you are reading is forwarded to your manager.

Team Email:

By using this Quick-Step, you are able to open the New Message form and fill in the To field with a specific group of recipients. It's usually set up in a corporate environment by your administrators to open a message already addressed to all members of your team, including your manager and everyone directly reporting to your manager. There's no need to limit your emails to your boss and your colleagues; you can include whomever you choose in a group. Groups are created in the People module.

Delete & Reply:

When this Quick-Step is selected, Outlook automatically opens a message form for replying to the sender and moves the selected message to the Deleted Items folder.

Create A New:

The Edit Quick-Step Wizard, which can be customized to suit your own needs, is technically not a Quick-Step; it merely opens the Edit Quick-Step Wizard.

Every Quick-Step, except reply and delete, requires you to make a decision the first time you use it. This is because Outlook doesn't know yet who your manager is, where your messages should be moved to, etc. As a result, you'll have to tell Outlook what to do when you select a specific Quick-Step. Once you've done this, Outlook will remember what you want it to do whenever you select the Quick-Step.

Here are the Quick-Steps as an example, but each is a bit different:

- Select an inbox message from the Mail module. Any message will do. Move the message without worrying about its actual location. If you're using Move To Quick-Step for the first time, Outlook will not move the message you select; it only needs to know the type of Outlook element you're creating the Quick-Step for.

- The icon in the upper-left corner of the Quick-Steps box appears when you click the Home tab. There might be a "Move To:" text underneath the icon. A dialog box appears for the first time. You might not see "Move To:" Instead, you might see a folder name, such as Personal. If the First Time Setup dialog box has already been pre-populated with a folder name, Outlook is just helpfully suggesting the last folder you moved a message to; the dialog box will still appear.
- Choose where Quick-Step will move the messages. The Move to Folder box will appear. Click the arrow near the end of the box to select a folder. You can also select "Other Folder" if you don't see the folder you want. This will open a window displaying all the folders available. The "Select Folder" option even allows you to create new folders.
- Ensure that the Move to Folder check box is selected. Mark as Read must also be selected if you want each message to be marked as read as soon as the Quick-Step moves it. In order to access Quick-Step's settings, click the Options button. From there you can change the icon, add actions, and create keyboard shortcuts.
- Fill out the Name text box with the name of your Quick-Step. The moment you clicked the Move to Folder box, Outlook also added the folder's name to the Name text box. Skip ahead to Step 2 if you like the name Outlook chose for the Quick-Step. Simply type a different name into the Name text box if you want to name the Quick-Step something else. The Quick-Step should have a name that will make it easy for you to remember what it does. Do you agree that the "Move to Personal Folder" is an appropriate name for a Quick-Step that moves a message to a folder called Personal? Personal also works for those with a preference for brevity.
- To save, click the Save button. This closes the First Time Setup dialog box.

Creating and managing Quick-Steps

Outlook includes a number of Quick-Step templates in addition to the ones shown in the Quick-Step box upon installation. To access these additional Quick-Step templates, follow these steps:

- Click on the Home tab and locate the Quick-Step box's scroll bar (located on the right side of the Quick-Step box; it has one up and two down arrows). Click the arrow at the bottom (the one with a line above it).

  Upon hovering over "New Quick-Step", two options appear at the bottom: Manage Quick-Steps and New Quick-Step.

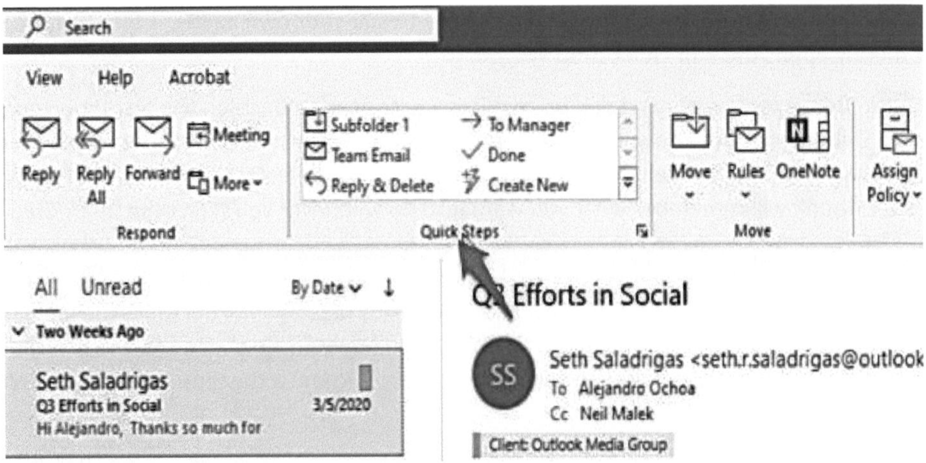

- Select Create New Quick-Step.
  You will see more Quick-Step templates. Any of these templates open the First Time Setup dialog box, which offers choices based on the type of task you selected. If, for instance, you select Move to Folder, the First Time Setup dialog box asks where the messages will be moved, as well as whether you want to mark them as read. Other Quick-Step templates include:
  - Move to Folder: Simply put, this is the same as "Move To."
  - Categorize & Move: The selected message is moved to a specific folder, marked as read, and assigned a color and name for the category. You might not see this if you're using an IMAP account.
  - Flag & Move: The selected message will be moved to the appropriate folder, marked as read, and given a flag.
  - New Email To: Here you can open a form already filled out with a particular recipient's name in the To field.
  - To: Essentially the same as Manager.
  - New Meeting: In cases where you frequently invite the same people to meetings, this Quick-Step will automatically fill in the To field with the invitees.
  - Custom: Clicking this will open the Edit Quick-Step dialog box, which allows you to customize your own Quick-Step.

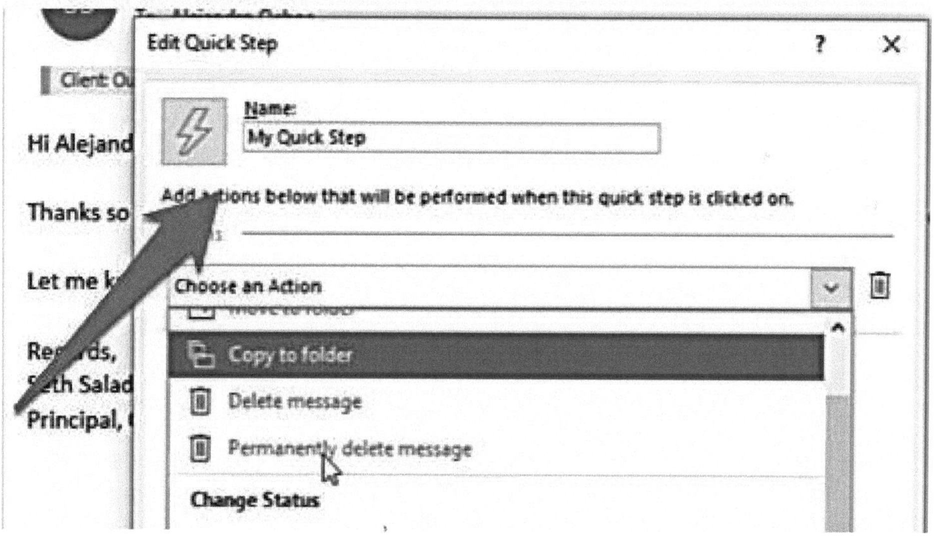

- Choose a template from the Quick-Step menu.
- A First Time Setup dialog box opens as you see when you click Move To Quick-Step.
- Decide on a template name and make your selections. Depending on the task, you'll have to instruct Outlook where a message should go, how to categorize a message, what flags to set, who to send or forward a message to, or whom to invite to a meeting. Be sure to also name this Quick-Step so that you will remember what it does when you input this information into the First Time Setup dialog box.
- Click on the Finish button. You will see the First Time Setup dialog box close.

Creating Quick Parts to save keystrokes

Frequently used text can be saved as a Quick Part to help you save time and effort when typing it into email messages over and over. It's not the same thing as Quick-Steps, but they're ridiculously simple.

The steps for creating a quick part are as follows:

- You can open a reply message in a new window after clicking "Pop Out" while composing it in the Reading pane. This is because you need to access the Insert tab, and you don't get this when composing or replying in the Reading pane.
- The text in the message can be selected by dragging your mouse over it. The selected text will be highlighted.
- In the Text group on the Ribbon, click the Insert tab >select Quick Parts. An option will appear.
- Select "Save Selection to Quick Part Gallery." The dialog box for creating building blocks appears.
- If you don't like the name of your quick part, type a new one. You may like the current name, but you might want to change it. A Quick Part can also be assigned a category and a description, but these options don't make much difference in how you use it, so you can leave them out.
- To proceed, click OK.

You can quickly add a Quick Part to the body of a new email or reply by clicking the Insert tab on the Ribbon, followed by Quick Parts in the Text group, and then clicking the Quick Part you want. Once you try it, you won't want to send an original email again.

CHAPTER 7

MANAGING CONTACTS, DATES, TASKS AND MORE

Learn how to create and manage your Contacts list, including changing how you view your contacts, attaching photos to them, and sending them to other people, as well as how to sort your contacts and use grouped views.

Discover how to create and modify appointments, print your calendar, and work with multiple calendars using the Calendar.

In addition to allowing you to keep your entire list of names and addresses in one place, Outlook also lets you sort, view, find, and print it in many different ways, depending on the type of work you're doing. Furthermore, you can store lists of family and friends alongside your business contacts in Outlook, so you can easily distinguish between them when necessary.

Putting in Your Contacts' Names, Numbers, and Other Stuff

Keeping a large number of names, numbers, and addresses is not difficult, but finding them again in the future requires magic unless you have an application like Outlook. Several programs can save names and related numbers, but Outlook is the most popular choice for working with names, addresses, and phone numbers.

You'll be familiar with the Outlook Contacts feature if you've ever used a pocket address book. You just need to enter the name, address, phone number, and a few tidbits-and you're done!

The quick and dirty way to enter contacts.

A new contact can be added easily to your Contacts list:

- To create a new contact, click the New Contact button in the People module. A contact form will appear.

- Complete the form. It is not necessary to fill out all the fields; only use those that pertain to the contact in question.
- Close the document by clicking Save & Close. It's really that simple. Adding more information to contact later is okay if you don't fill in all the details immediately.

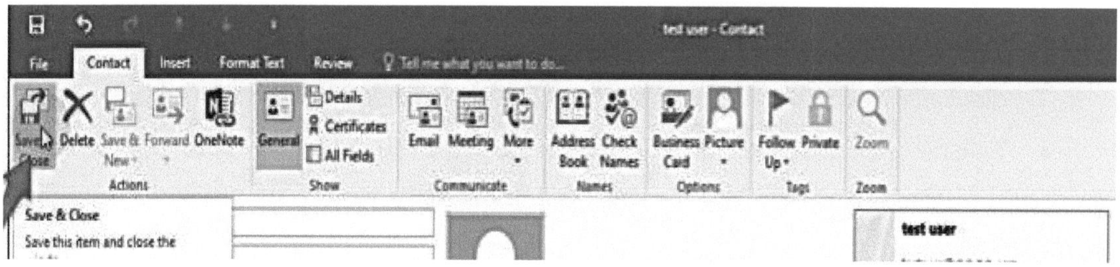

The slow, complete way to enter contact

Depending on your preferences, you can enter literally dozens of details about every person on your contacts list. However, if you just want to include the essentials, that's fine. Here's what you'll need to do to enter every tiny detail for every contact record.

Select "New Contact" from the People module.

You will see the new contact form.

- Choose your full name from the drop-down menu. A window titled "Check Full Name" appears.
- Take one or more of the following actions:
    - To scroll down, click the triangle (called the "scroll-down button") on the right edge of the text box named "Title." You can either type a title (such as Reverend, Guru, or Swami) or select one from the drop-down menu.
    - Enter the contact's first name in the first text box.
    - If necessary, type the contact's middle initial (if any). Leave this field blank if there is no middle initial. The middle name can also be written here if you wish.
    - Enter the contact's last name in the "Last" text box.
    - Select the suffix from the drop-down menu. Select one of the options (like Jr., III, or Ph.D.), or type one in the box (such as D.D.S., B.P.O.E.).
    - To proceed, click the OK button.

  Your Full Name and File NumberAs text boxes are now filled with the name you entered in the Check Full Name dialog box,

- Complete the form by clicking on the appropriate box and entering the requested information. Please leave the box blank if the information is unavailable, for example, if a contact does not have a title. There are more options when a triangle appears after the box. If your choice isn't listed, enter it in the box.
    1. When you enter a name in the Full Name box, you'll see that name appear in the File As box.
    2. You can file a person under a different name by clicking File As and entering the designation you prefer. Your dentist may, for example, be filed under the term "dentist" rather than by name. In the alphabetical listing, the dentist's name appears under "dentist" rather than under the actual name. In your contacts list, you will find the full name and the file as designations. By doing so, you can find your dentist either by name or by the word "dentist."
- Enter the email address of your contact in the "Email" text box. To enter a second email address, click Email 2, pick the second address from the list, and then enter it in the text box.
- Type the business phone number of the contact in the text box beside "Business Phone."
- You will have the option to type in the contact's home phone number if you click on the Home Phone text box. In order to enter numbers other than home and business telephones, click the

triangle beside the number option, choose the type of number you are entering, and then enter the number.

There are four blocks for phone numbers in the New Contact form. The drop-down menu contains 19 different types of phone numbers that you can use according to the types of numbers your contacts have.

- Choose the type of address you want to enter in the Addresses section by clicking the triangle. Choose either a business address, a home address, or some other address.
- Go to the Addresses section and click the button.
  An address check dialog box appears.
- Fill in the appropriate fields with the following information:
  - The street
  - The city, state,
  - The province
  - Region of zipping or postal code
  - Country
- Click OK to close the Check Address dialog box.
- If the address you just entered will be used for mailing, select the "This Is the Mailing Address" box on the New Contact form.
- If you want the address card to link directly to a page, click in the Web Page Address text box and type the page's address. A contact's webpage can be viewed by opening a contact record, clicking on the More button on the Ribbon, and clicking Web Page (or pressing Ctrl+Shift+X). An internet browser will display the webpage.

  The URL of the page can be viewed in your web browser by entering it into the Address box. Within the Web Page Address text box, you can enter the URL for any webpage an Outlook contact has.

Fill in the bottom right notes box with whatever information you want.

There is no limit to what you can enter (preferably something that will help in your dealings with the contact).

- Click the Save & Close button on the Ribbon after you are finished. When you've entered any information you need (or may need) to know about the people you deal with at work, you're ready to start interacting.

Viewing the contacts

The information you enter in Outlook can be viewed in many different and useful ways, called views. By viewing your contact information and sorting the views, you can easily see the big picture of the data you've entered. Within each module, Outlook provides several predefined views. You can edit any predefined view, name it, save it, and then use it exactly as you would the predefined views.

You can change how your contacts list appears by following these steps:

- Select the Home tab from the Ribbon of the People module.
- Choose the view you want from the Current View group. That view will appear on the display. Besides the card view, you can also select the phone view, the list view, or any other view you like.

View Sorting

A few views, such as the phone view of the People module, are organized as simple lists.

You can sort by a column's title once you've clicked on its title if you can't find a contact in a view with columns. Suppose you want to know the names of the IBM employees in your contacts list. If you sort the company column, you can see all the names at the same time.

Here are the steps you need to follow to sort by column name:

- Select the Phone view on the Ribbon under the Current View section of the People module. Contacts are displayed in the phone view.
- Go to the COMPANY column and click the heading.
  Contacts are listed alphabetically from A to Z (or ascending order) according to the "company" column. Now you can find someone by scrolling to that letter. By sorting by company, all contacts appear in alphabetical order of company name.

Upon clicking the title a second time, you will see your contacts sorted in reverse alphabetical order (i.e., descending order).

Rearranging Views

Simply drag the title of a column and drop it where you want it to appear to rearrange the appearance of a view. Below is an example of how to move

You can display the contacts list in the Phone view by clicking the Phone button in the Current View group on the Home tab.

The contacts list may already be displayed in the phone view.

- To the left of the FILE AS column, click on the title of the preferred column and drag it over. Red arrows appear to the left of the clicked column, pointing to the border between two columns. By releasing the mouse button, Outlook will drop the column where the red arrows indicate.
- Let go of the mouse button. You have now dragged the column from the right to the left. The procedure is the same for moving any column in Outlook.

Using grouped views

Sorting doesn't always suffice. After a while, you can easily accumulate several thousand contacts; it only takes a few years. In a long list, if you're looking for something with the letter M, for instance, you'll find it about three feet below the bottom of the screen, regardless of how you sort it.

There is no need for Outlook, Anonymous groups are the solution. You already have several predefined lists that rely on grouping in Outlook.

Lists are displayed in Outlook in several different ways: Sorted lists are like playing cards laid out in numerical order, starting with twos and proceeding to threes, fours, and so on, up to picture cards. In a group view, all the hearts, the spades, the diamonds, and then the clubs are arranged together in a single row.

Here are the steps to viewing the company grouping in list view:

- Select the Home tab under the People module.
- Select the List option from the Current View group on the Home tab.
  Companies are represented by headings with the prefix "company". The number of items under each heading is shown under that heading.
  Expanding or collapsing the contacts under that heading is done by clicking the triangle symbol to the left of the heading.

You can group items according to just about anything you want, provided that the data you enter is accurate.

Below are the steps for grouping by another field:

- Select View Settings from the View tab of the People module.
  A dialog box with advanced view settings for the list appears. You will see that Group By is set to "company."
- group by clicking "group by."
  An appropriate dialog box opens. You will see that Group Items By is set to "Company."
- Select a different field from the company list.
- Specify the sort order by clicking "Ascending" or "Declining."
  Ascending means going from A to Z, and descending means going from Z to A.
- For the dialog box to close, click OK.
- Close the Advanced View Settings: List window by clicking OK.

Identifying Your Friends

A string tied around your finger doesn't do much to remind you to do something that involves another person, and it looks ridiculous as well. Fortunately, Outlook provides a better solution. You can easily remember the name of someone you have promised to call next week by flagging their name in the Contacts list. This will trigger a reminder in your calendar. Flagging isn't limited to contacts. The same effect can be achieved by adding reminders to tasks, emails, and appointments.

The following steps will guide you through the process of adding flags to contacts:

- You can flag a contact by right-clicking on it in the People module. You'll see a shortcut menu appear.
- Select Follow Up. "
  You will see the "Follow Up" menu. Choose the day for your follow-up.
  You have the option of choosing today, tomorrow, this week, or next week. When you flag a contact for a specific day, that contact's name appears in your Outlook Calendar on that day.
- Select the contact, click "Follow Up," then add a reminder.
  A dialog box appears.
  If you need to avoid that person for some reason, you can choose to have the reminder open

and a sound played at the time you specify. Reminders are Outlook's way of telling you to do something.

- The Custom dialog box allows you to select a reminder date from the Date drop-down menu in the Reminder section.
  A calendar is displayed when the arrow next to the date is clicked. Click on the date you want.
- From the drop-down menu, select a time for a reminder.
- Set the reminder by clicking OK.

Using Contact Information

I bet you intend to use all that contact information you entered. Let me show you a few ways to dig through your contacts list for valuable nuggets.

Searching for contacts

It's crucial to enter people's contact information so that you can find them again in the future. Otherwise, what's the point of this whole process?

It's easy to find names in the People module. The module allows you to get names by browsing or searching anywhere. Just use the scroll bar to rapidly move through the contacts. You can search for contacts by entering a name, email address, or any other relevant text in the Search Contacts field and selecting what you want from the results.

Here's how you can do it from card view, which you have not yet used in an official capacity. It is similar to the Business Cards view, except the cards vary in size depending on what information they contain and they are arranged in columns.

In Card view, you can search for a contact by the last name by using the following steps:

- Select the Card view on the Home tab of the Ribbon in the People module.
  You will see the card view.
- You can enter a contact's last name by clicking on the white space below any card.
  This displays the first instance of that name and highlights its card heading.

The company where a contact works may be a factor in your search for their name. Alternatively, you might want to include all the people on your list who live in a certain state or people who make you feel a certain way (now there's something to include in their records). A search tool would then direct you to a specific contact in such a case.

Here are the steps you need to follow to search for a contact using the search tool:

- Enter the desired text in the Search Contacts box of the People module, using your favorite view.
  In your Contacts list, you will find the search box just below the ribbon. Only contacts containing the information you typed appear in your contacts list. If you don't find any contacts matching your search, double-check that you spelled the search text correctly.
- In the search box, click the X button at the far-right end to remove the search filter.

Conversely, if you have only a fragment of the name you're seeking, Outlook can find that fragment wherever it is. When you search for "Geo," you will find George Washington and any other Georges on your contacts list, including Boy George and George of the Jungle (assuming they are all close friends of yours and they are on your contacts list).

Finding a contact from any Outlook module

You'd like to search for a person, but you're currently using another module. That's fine. Any Outlook module can be searched using the Search People box on the Home tab on the Ribbon.

The steps are as follows:

- Find people by clicking on the "Seek People" box. You will find it on the far right of the Home tab of any Outlook module.
- Enter the contact's name.
- To open that contact's record in Outlook, press Enter. When you enter only a few letters of a name, Outlook lists names that contain those letters, so you can choose the contact you had in mind. With the word "Wash," you can search for George Washington, Sam Washburn, and any other people on your list that include "Wash."
- To view a contact record, double-click its name.

Sending a business card

Any Outlook user (or any other application that is capable of displaying digital business cards) is able to forward an electronic business card to another Outlook user. There's no better way to share any contact record in your list with someone else.

The most obvious thing you may want to send this way is your own contact information.

- Create your own contact record in the People module. This record should contain all the information you need to send to someone.
- To send information to a contact record, click twice on the record. Clicking twice opens the contact record.
- From the Ribbon, click the Forward button under the Contact tab. Three options are provided: as a business card, as an internet card (vCard), or as a contact in Outlook.
- Select the format that suits you best.
  If you're unsure, select "Business Card." Both Outlook and Internet cards can be sent using that program. A new message is generated containing the contact information.
- Enter the recipient's address in the "To" text box.
  You can also choose a name from your address book by clicking the "To" button.
- Press Alt+S (or click the Send button).
  A message and vCard will be sent to the recipient.

Double-clicking the icon representing the business card in the message will add the business card to your Contacts list. An additional contact record will be created. You can now add the new name along with all the information on the business card to your contacts list by clicking the Save and Close button.

Alternatively, your business card can be forwarded by clicking the contact record and then selecting the Forward button in the Ribbon. The process is shorter, but you can only forward it as a business card or as an Outlook contact.

Gathering people into groups

A contact group that contains more than one person can be created in your People module. It allows multiple recipients to receive the same message. Contact groups can also be categorized (just as you can with individual contacts), and you can send a contact group as a message attachment to make it available to other Outlook users.

Creating a Contact Group

The process of creating a contact group is simply a matter of naming your list and selecting from the names you've stored on your system. Email addresses are the only pieces of information that a contact group keeps track of—not phone numbers or mailing addresses.

The following steps will guide you through creating a contact group in your People module:

- Press Ctrl+Shift+L on the keyboard to activate the New Contact Group button (or click the People module's Home tab on the Ribbon).
  A contact group list will be displayed.
- Fill out the Name box with the name you wish to assign to the Contact group.
- Select the "From Outlook Contacts" option under "Add Members."
  In the Select Members: Contacts dialog box, you see the names of available members on the left and the Members box at the bottom.
  You can include the email addresses of people not in your contacts list or any of your other Outlook Address Books by following these steps:
    - Pick a new email contact (not from Outlook Contacts).
    - Enter the name and email address of the person you wish to add in the Add New Member dialog box.
    - You can now click OK.
    - Continue with the remaining steps.
- To add someone to your contact group, double-click their name. A box called "Members" appears at the bottom of the dialog box when you double-click on a name.
- If you've picked all the names you want, click OK. That closes the Select Members: Contacts dialog box.
- Using the Save & Close button (or pressing Alt + S) will save and close your document.
  Your contact group will appear in your contacts list once the contact group dialog box has been closed.

Editing a contact group

Like everywhere else, people come and go in contact groups. Fortunately, you can edit the contacts:

- To view the names of your contact groups, double-click the group's name in the People module.
  An identical screen appears from when the group was created.
- Use one or more of the following methods to edit the list:

  - Remove a member from the list: Click the name and select "Remove Member."
  - Select a new member from the names already in your contacts list: Click on the Add Members button and follow the same steps as when you created the list.
  - Add a person whose email address isn't listed in your contacts list: Select the Add Members button, click on New Email Contact, enter the person's name and email address, and then select the OK button.
- When you are finished editing the contact group, click Save & Close.

Adding pictures to contacts

A picture can be included with the contact information you gather-and not just for decoration. Due to the synchronization of many cell phones and other mobile devices with Outlook Contacts lists, you can view someone's picture when he or she calls or texts you. These pictures also appear when you view your Outlook contacts in Business Card view. When you forget names but never forget a face, you can save both names and faces.

You can add a picture to a contact record by following these steps:

- In the People module, double-click the contact whose picture will be displayed.
  The contact's record will appear.
- Click twice on the picture icon in the center of the contact record.
  A dialog box appears.
- Double-click the image you want to add to the location where it is located.
  A picture of the contact will appear right in the contact's record.
- Save and close the document.

CHAPTER 8

UNLEASHING THE CALENDAR'S POWER

How long do people work during the day? Many people spend their days attending meetings. You can't stop meetings from happening, but the Outlook Calendar can simplify the scheduling process and help you schedule time for more meetings.

Getting Around the Outlook Calendar

Since you've been using calendars your whole life, you'll have no trouble understanding the Outlook Calendar. It looks like a calendar, with a row of dates for every day of the week plus weekends. You don't have to think like a computer to understand your schedule.

A mouse click on your calendar is usually all it takes to see more information about something. Click twice if you need more details. And if that is still not enough, there is a feature that can give you just what you need.

This feature actually goes by the name "Date Navigator," but don't confuse it with Casanova's chauffeur. It is a trick in Outlook that lets you change the part of the calendar you are viewing or the period of time you wish to see.

You can navigate through your calendar by following these steps:

- On the Navigation pane, click on Calendar (or press Ctrl+2). The Information Viewer displays your calendar while the Date Navigator appears in the top part of the Navigation pane. The Date Navigator also appears in the To-Do bar. Select Day, Work Week, Week, or Month from the drop-down menus. You will see the button you clicked highlighted.
    - You can see the details of a single date by clicking on it wherever it appears. This displays the events and appointments scheduled for the day you clicked.
    - The Date Navigator can be advanced one month at a time by clicking the triangles next to the month name.
    - With time (so to speak), you'll find the calendar view that suits you best. Most of the time, you can leave Outlook running in the background to keep access to the information you need.

You can find an open date quickly by following these steps:

- Hold down the Ctrl and G keys.
  "G" is short for "go to." The "Go To Date" dialog box opens with a selected date.
- The Date box is currently set to January 15, 2019, but you can change it as you normally would by typing the date you want in the box as usual.
  Typing something like "45 days ago" or "93 days from now" is a great way to change dates. Give it a try. You can change dates in simple English in Outlook. If you want to go to today's date, just click the Today button on the Home tab on the Ribbon.

Meetings Galore: Scheduling appointments

Datebooks are vital to many people. Many people still use paper datebooks. I admit I like these datebooks best for putting stuff in them. However, once it's in, it's hard to find. Electronic gadgets like iPhones, Androids, and BlackBerrys (yes, there are still BlackBerrys!) can also act as datebooks, although sometimes it's hard to add appointments since they're small and don't use a mouse. With many digital gadgets synchronized with Outlook, you can have it all!

With Outlook, adding appointments is surprisingly easy-and finding items you've entered is even easier. You are also alerted when you schedule two appointments at once.

The quick and dirty way to enter an appointment

There are some appointments that don't require much explanation. No need to be a showoff about your appointment with Mom on Friday:

- Make sure the view showing the hours of the day is selected in the Calendar module.
  Examples include a day, a week, and a workweek.
- To view the appointment time on the desired day, click the text box.
- Describe the appointment in detail.
  Mom's lunch, for example.
- Click the Enter key.
  In no time at all, you will have your official appointment added to your schedule.

The complete way to enter an appointment

For lunch dates with Mom, you wouldn't need as much information as you would for your work appointments.

You may want to include:

- Meeting location details
- Agenda Notes for the meeting
- You can show your boss how long you spend with your clients by creating a category.

The following method must be used for giving an appointment a full treatment:

- On the Ribbon, click New Appointment in the Calendar module. The appointment form will appear, or you can press Ctrl+N for the screen to let you add a new appointment to your calendar.
- Enter a brief summary of what the appointment is about in the Subject box.
  Put in a dental appointment, lottery winnings, or whatever you like. This will be displayed on your calendar.
- Enter the address in the Location box.
  If you include this step in your appointment, it is optional, but it might be useful information.
- Remember to include any other details about your appointment.
  You can save driving directions, meeting agendas, or anything else you might need to keep in mind when the appointment time arrives in the large, empty box on the Appointment form.
- Click on the Save & Close button.
  You will now see this appointment on your calendar. If you can't see your new appointment, you may need to click the Date Navigator on the date it occurs.

Changing the dates

Outlook allows you to be as flexible as you like. If you want to change a scheduled item's time, just drag it to the new location.

Follow these steps to change an appointment by dragging it:

- Click on the appointment you wish to change in the Calendar module. The appointment appears with a dark border around it.
- Ensure you're looking at Work Week, Week, or Month-one of the views that display times.
- To change the appointment time or date, just drag the slider.

Changing an appointment's date and time is also possible.

- Double-click the appointment in the Calendar module. An appointment window will appear.
- The selected month's calendar can be viewed by clicking the Calendar icon on the leftmost Start Time box. The calendar is displayed in a drop-down window.
- Select a month by clicking one of the triangles beside its name. You move one month earlier if you click the left triangle, and one month later if you click the right triangle.
- Click on the date you want to view.
- Enter a new time in the Start Time text box, if necessary. A new start time can also be selected by using the scroll-down button.
- If you need to make any other changes to the appointment, do so.
- You can change the information by clicking it and typing over it.
- Click on Save & Close.

Follow these steps to adjust the appointment length:

- Choose a Day, Work Week, or Week view of the appointment in the Calendar module.
  Due to the fact that you cannot see the time without opening the appointment, this process does not work on a monthly basis.
- By moving the mouse pointer, click on the top or bottom handles of the appointment.
  The pointer becomes a two-headed arrow when it is in the right place.
- Slide down the bottom line to extend the appointment time; slide up the bottom line to shorten it.

You can also make an appointment shorter than 30 minutes. The steps are as follows:

- Select the appointment by double-clicking in the Calendar module.
- Select the "End Time" box.
- Enter the finish time.
- Click on Save & Close.

Dates are being broken.

There are times when things just do not work out. It is unfortunate. Despite how hard it is for you to remember, Outlook erases dates that you otherwise cherish with a simple click. Actually, it takes two clicks. You can delete an appointment (after you've called to break it) by following these steps:

- Select the appointment from the Calendar module by right-clicking.
- Click Delete.
  Outlook will cancel your appointment.

We've got to keep seeing each other: Recurring

After completing one thing, you're ready to tackle another. In Outlook, you can easily set up recurring appointments.

Follow these steps to schedule a recurring appointment:

- Select the appointment you'd like to repeat from the Calendar module.
  In the Ribbon, the Appointment tab becomes visible when an appointment is highlighted.
- On the Ribbon, click Recurrence on the Appointment tab.
  A dialog box appears.
  You would have the same appointment every week if you simply click OK in the Appointment Recurrence dialog box. It is possible that you won't be prepared for meeting scheduling indefinitely. It might be a good idea to fill out the rest of the Appointment Recurrence dialog box to be on the safe side
- Enter the starting time in the Start text box. The default time span for Outlook appointments is 30 minutes unless you specify a longer time span by entering an ending time. You can enter an ending time if you wish in the box labeled "End."
- You can choose the recurrence pattern for an appointment by clicking on the Daily, Weekly, Monthly, or Yearly option button under the Recurrence Pattern section.
- In the next section of the recurrence pattern section, specify how often the appointment occurs.
- Put the first occurrence into the start box of the Range of Recurrence section.
- Schedule the end of the appointments.
  Here are the options:
    - Endless (infinity)
    - Finally, after (a specific number of occurrences), It ends
    - on (a certain date)
- Click OK to confirm.
  A dialog box with Appointment Recurrences closes and the Appointments form appears.
- Enter the subject in the Subject field.
- Place your cursor in the Location box and enter the location.
- Click Save & Close.

Occasionally, a recurring appointment is changed as well.

You can edit recurring appointments as follows:

- Double-click on the appointment you wish to edit in the Calendar module.
  A dialog box entitled "Open Recurring Item" appears.
- Choose the Entire Series.
- Press the OK button.
  A recurring pattern appears under the location of the appointment.
- Edit the appointment details.
  Change the recurrence and click the OK button to change the pattern.
- Please click the Save and Close button.

Getting a Good View of Your Calendar

It is possible to organize the information in every section of Outlook virtually any way you can imagine—all by using varied views. The different ways to view a calendar could fill a cookbook, but I'll stick to the standard ways that most people see it. Outlook lets you make calendar arrangements nobody has ever thought of before. In the event that you accidentally create a calendar view, you don't like, you can delete it.

A calendar view consists of four basic views: the Day view, the Work Week View, the Week View, and the Month View. The Schedule view is useful for determining when things were done or when they will be done (such as when you did something and when you won't do something).

All of Outlook's calendar views are accessible from the Ribbon. To see a different view of the calendar, click on the name of the view you want. You can change the view at any time if the one you selected isn't suitable for you. You can arrange your calendar in a more streamlined vertical column by selecting the Schedule view. Other calendars can optionally be added (by clicking Add a Calendar) and their appointments compared hour-by-hour or day-by-day.

Printing Your Appointments

Plain old paper is still everyone's favorite reading material. You will probably need old-fashioned paper and ink to make your computer organizer truly useful, no matter how slick it is. Frankly, I've never been a fan of Outlook's calendar-printing feature. You are probably not to blame if you are unable to print your calendar correctly.

Any module in Outlook can be printed using the same basic steps. The steps are as follows:

- Click on one of the dates within the range of dates you want to print from the Calendar module.
  You can also print a single day by clicking on one date. You can pick a range of dates to print by clicking the first date, holding down Shift, and then clicking the last date. This highlights the entire range of dates you've selected.
- Select File, then Print (Ctrl + P).
  Printing controls are displayed.
- Choose a style from the Settings section.
  There are five basic options: daily, weekly, monthly, tri-fold, and calendar details. Outlook also lets you define your own print styles, so you may eventually have quite a few options

appearing here. The preview feature of Outlook lets you know what a page will look like before you print it.

- Press the Print button. You will get your dates printed.

Adding Holidays

What is the most important day of the week for workers? working days off! Calendar entries for major holidays can be automatically added to Outlook so you don't forget to take time off. You'd never know! Actually, Outlook can add holidays for over 70 countries as well as several major religions automatically.

You can add holidays to your calendar by following these steps:

- Navigate to the File tab.
- Select options.
  A dialog box appears with Outlook Options.
- Go to the calendar.
- You can add holidays to your calendar by clicking the "Add Holidays" button (under Calendar Options).
  A list of countries and religions will appear.
- You can add holidays by clicking on them.
- Once you have saved your changes, click OK.
- Please click OK once you have added the holidays.
- Close the Outlook Options window by clicking OK. Look up your holidays on the calendar.

Handling Multiple Calendars

It used to be regarded as thrilling and dangerous to lead a double life. Nowadays, they're considered underachievers. Get busy with your life, as you only have a small amount of time. It's easy to manage all your calendars with Outlook. Even if you're a mild-mannered person who is just trying to keep things calm, you might want to keep your personal and business calendars separate by setting up two calendars in Outlook.

Creating multiple calendars

The steps for creating another Outlook calendar are as follows:

- Click on the Manage Calendars tab on the Ribbon in the Calendar module to add a calendar.
- Create a new blank calendar by selecting Create New. A dialog box appears. It's like creating a new mail folder, but you're creating a new kind of folder that displays as a calendar.

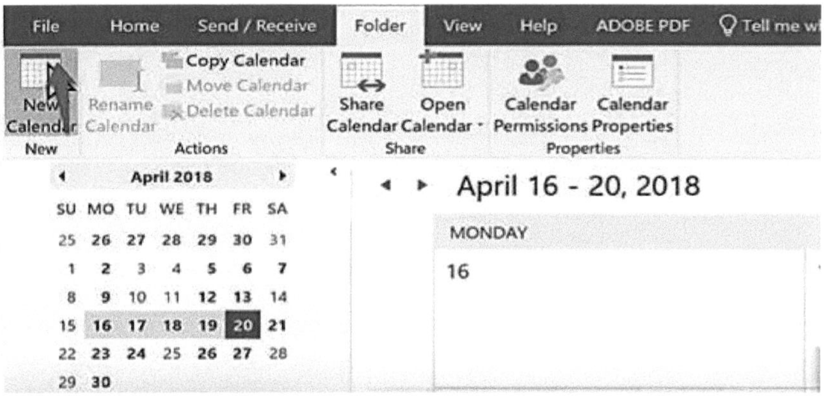

- Name your new calendar by clicking the Name box.
- Your new calendar's name appears next to a blank check box in the Folder pane as you click OK. Checking this box will allow you to view your new calendar alongside your original calendar in the Information Viewer, using the same day, week, or month view. To remove the checkmark, deselect the checkbox. The calendar you selected disappears.

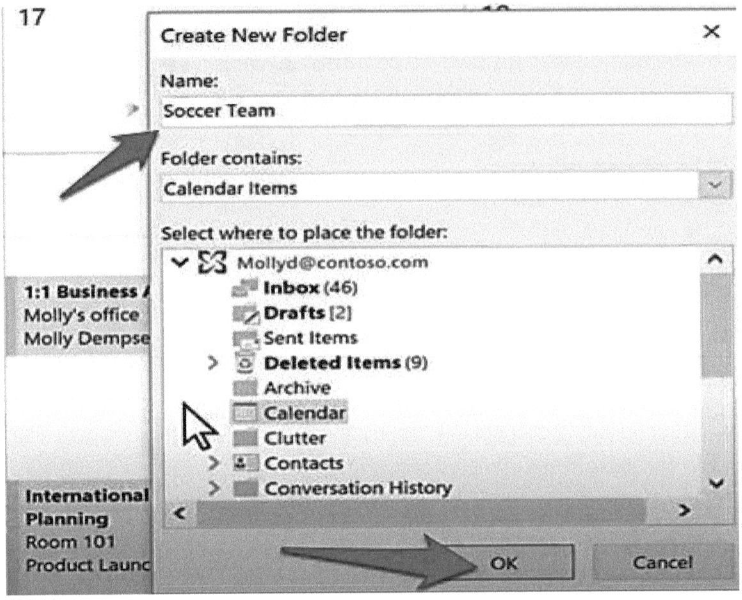

Managing multiple calendars

You can't be in two places at the same time. So you will appreciate the fact that Outlook calendars can be superimposed to make scheduling easier. The top tab of the name of one of the two calendars appears with an arrow when the two calendars are opened side by side. Using that arrow, you can superimpose the two calendars and check if any of your appointments clash.

CHAPTER 9

TASK MASTERY

In Outlook, you can store and manage more information about your daily activities than you might have wanted, but you'll find that it makes it easy to remember and keep track of your daily tasks. Organizing your tasks doesn't have to be a challenge in and of itself.

Entering New Tasks in the Tasks Module

You already have a full schedule; I don't intend to add more work to it. It's not so hard to add a task in Outlook. Outlook offers you lots of information about your tasks, but you can also enter a new task quickly and really quickly.

The quick-and-dirty way to enter a task

You can add a new task to your task list by clicking on the "Add New Task" button (or typing a new task, depending on the view you're using). Just follow the on-screen instructions. Using the regular, much slower way of entering a task is described in the next section if you can't see the box.

Here's how to get a task entered using the quick-and-dirty method:

- Click the button that says "Click Here to Add a New Task" (or type it in manually).
  It's written this way in the Detailed, Simple List, Prioritized, Today, and Server Tasks views; it's written this way in the To-Do List views. These steps cannot be displayed in any other predefined view.
- Put your task's subject in the subject field.
  This will display your task's subject.
- You can now press the Enter key.
  That task will now appear as part of your task list.

The regular way to enter a task

Tasks are entered through the form, which takes a little more time to fill out but allows you to provide much more detail. You don't need to work your fingers to the bone to complete the task—if you enter a subject, you've completed it all. To add due dates and have Outlook remind you to finish what you've entered, you just need to fill out a few more boxes with your information.

To add a task to your task list, follow these steps:

- Click New Task on the ribbon (or press Ctrl+N) in the Tasks module. You will be presented with the task form.
- In the Subject field, type the title of your task.
  Make the subject one that makes it easy to remember what the task is about. Creating a task is mostly for the purpose of reminding yourself to perform the task.
- Click the Start Date box and enter a starting date to assign a start date to the task.

You can skip this step if you have not yet started the task. There is no need for a start date. It's only for your own use.

- For your task, you can enter a due date by clicking the Due Date box and choosing a start date.

As a default, it is just like the Start Date:

- If you want to choose the status of the task, click the Status box. Keep the status set to "Not Started" if you have not yet begun. Alternatively, you can select In Progress, Completed, Waiting on Someone Else, or Deferred.
- Choosing a priority can be done by clicking the Priority box. If nothing is changed, the priority remains normal. Other options include high and low.
- Optionally, check the box next to "Reminder" if you'd like to be reminded before the task is due.
  Leave the remainder blank if you want to forget the task. However, why do the task?
- If you wish, you may click the date box next to the Reminder checkbox and enter the date when you want to be reminded. In the Reminder box, Outlook automatically adds the date you have entered. It is up to you which date you enter. When you choose a date in the past, Outlook informs you that no reminder will be set. There is a calendar on the right of the date box when you click the icon. You can choose the date you want.
- The time that you want to activate the reminder can be entered in the time field. Set the time by typing the numbers for the time. Colon characters and other special characters are not necessary. Type 235 if you want to finish by 2:35 p.m. If you don't tell Outlook otherwise, it will schedule your appointments and tasks during daylight hours.
- Please provide any additional information or notes about this task in the text box.
  You can store directions, supply lists, or anything else you need for your appointment here.
- Finish by clicking "Save & Close."
  There is now a new task listed in your task list, awaiting completion.

Adding an Internet link to a task

In the Task form, typing the name of a website, such as www.facebook.com, sets the color to blue and makes the address a hyperlink you can click to open the website. You can easily save information about a great site by typing or copying the address into your task. Your web browser will open and display the page when you click on the text.

Editing Your Tasks

After you begin a new task, you are faced with the need to modify it immediately. Occasionally, I enter a task quickly and change some of the details later: A due date, a reminder, an extra step, etc. The task can be easily edited.

The quick-and-dirty way to change a task

Outlook offers an easy way to modify tasks in the same way it has an easy way to add tasks. Changing details is limited, but the process is quick.

Here are the steps you need to take if you can see the subject of a task and you want to make changes to the task you can see. For tasks or parts you cannot see, use the regular method, which I will describe in the next section.

You can change a task the quick-and-dirty way by following these steps:

- To edit a part of a task, highlight it and then click on it.
  This is especially useful if you want to edit the task's subject or due date. A blinking line appears at the end of the text, a triangle appears at the right end of the box, or a menu appears.
- Go to the old information. The selected item is highlighted.
- Fill out the new information.
  The old information will be replaced by the new information.
- Then press Enter.

The regular way to change a task

It is a little bit more difficult to change information about a task if it's not on the list you're viewing or if you don't want to be quick and dirty.

Following these steps will enable you to make changes the clean-and-long way (also known as the regular way):

- On the Ribbon, click the Simple List view if it is not already selected in the Tasks module.
  There are other views you can use if you know that they include the task you want, but Simple Listview is the simplest way to view your tasks; it's certain to include the task you're searching for.
- You can change a task by double-clicking it.
  Double-click any component of the task; it does not need to be its subject. You will be taken to the task form. From here, you can modify anything in the box.
- (Optional) Change the task's subject.
  Choosing a subject is up to you. Be sure to name it something that makes it easier to remember. The last thing you want is a computer telling you to do something you don't understand.
- Enter the new start or due dates in the boxes under Start Date and Due Date, respectively (optional).
  You can choose from a variety of date styles: 7/4/19, the first Friday in July, six weeks from now, whatever you like. I am afraid the 12th of Never is not available.
- The status can be changed by clicking the Status box, then selecting a different status.
- You can change the priority by clicking the Priority box and selecting a new priority.
- Optionally, you can turn the reminder on or off by selecting or deselecting the Reminder check box.
  They're simple and harmless, so why not take advantage of them? Make sure you get one now if you didn't ask for one the first time.
- Enter or change the date for your reminder by clicking the calendar icon next to the Reminder checkbox.
  Specify any date you like. There is no need to wait until the due date; you can submit it earlier,

encouraging you to get started. Be sure to set it before the deadline. In general, the reminder will appear on the due date of the task.

- Enter the time when you wish to activate the reminder in the time box.
  Make your entries as simple as possible. You can use entry 230 to enter at 2:30 p.m. You can add or edit the instructions and notes in the text box.
  Detailed information can be added here that isn't really relevant elsewhere on the task form. It's only by opening the Task form again that you can see these details; normally, they cannot be seen in your Task list.
- Finish by clicking "Save & Close."
  That's it. Your task has been changed.

Deleting a task

Getting rid of tasks is the most satisfying part—ideally by completing the tasks you created. You can also delete tasks you no longer want. This version of Outlook isn't able to tell if you have finished any tasks, so you can delete anything you don't want to bother with.

Here are the steps for deleting a task:

- Select the task from the Tasks module.
- Select the "Delete" button from the Ribbon.

Alternately, you can press the Ctrl+D or Delete key on the keyboard. Everything is gone.

Managing Recurring Tasks

Tasks pop up frequently. When you have to do the same thing on different days, To save time and energy on entering tasks that you frequently do, you can set them up as recurring tasks. Outlook has the ability to bring it up again when it is time for the task to be done.

To create a recurring task, follow these steps:

- To open a task from the Tasks module, double-click it. Afterward, a form is opened.
- Press the Ctrl+G shortcut key to recur (or click the Recurrence button on the Task Form toolbar).
  You will be shown the recurring task dialog box.
- To specify how often the task occurs, select "daily, weekly, monthly, or yearly."
  You can set a recurring monthly task to take place at certain times of the month, such as on the 15th of every month or on the second Friday of every month.
- Recur Every lets you specify whether the task will repeat every month or every third day.
  In the case of creating a yearly task, you can set the task to recur each year on the same calendar day (such as the first Friday of June) or on a different day each month.
- Put the first occurrence into the start box of the Range of Recurrence section.
- There are several options to specify the end time of the task (No End Date, End After a certain number of occurrences, or End By a certain date).
- Press the OK button.
  There is a banner at the top of the task form describing how the task is repeated.

- Close the window by clicking on "Save & Close."

Once you've added your task to the list of tasks, you'll see a different type of icon than nonrecurring tasks, so you'll recognize that it's recurring. The icon for regular tasks looks like a tiny clipboard, but for recurring tasks, the icon appears as a tiny circular arrow.

Creating a regenerating task

Regenerating tasks behave like recurring tasks, except they reappear after a specified amount of time has passed since the last time they were completed.

If you use Outlook to schedule your lawn mowings, you can use Regenerating Tasks to generate your schedule. As far as I know, Outlook can't give you a rain forecast without a weather forecaster. Regardless, Outlook will help you keep track of whether you have actually mowed the lawn and make adjustments accordingly.

The steps for creating a regenerating task are as follows:

- Double-click the task to open it.
  A form appears.
- Press Ctrl+G to activate the recurrence option on the Ribbon. A dialog box will appear.
- Activate the option to regenerate new tasks.
- You can specify how long it should take for each task to be generated.
- Click OK to confirm.
  When you select a regeneration pattern for a task, a banner appears on the task form to describe it.
- Please click the Save and Close button.

Regenerating tasks have a different icon than non-recurring ones, so you are able to tell at a glance that they are recurring.

A regenerating task icon looks exactly like a recurring task icon, complete with that tiny circular arrow.

Skipping a recurring task once

In cases where you want to bypass just a single occurrence of a recurring task, there is no need to change the pattern of occurrence permanently; you may skip that occurrence and leave the rest alone.

Follow these steps to skip recurring tasks:

- To change a recurring task, double-click it in the Tasks module. A form is displayed.
- On the ribbon, click Skip Occurrence.
  It will change to the next scheduled occurrence date. There won't be a "Skip Occurrence" button if it's not recurring.
- Then click the Save and Close button.
  Regardless of the new occurrence date, the task will remain on your list.

Marking Tasks as Finished

There's nothing more satisfying than marking off completed tasks—and it's easier too. Right-click on the item you wish to mark as complete from either the To-Do bar or Task list and select Mark Complete. That's all there is to it.

Marking it off

Here are the steps for marking a task as complete:

- Activate the Simple List view in the Tasks module by clicking theSimple List icon on the Ribbon. It doesn't matter which view you choose, as long as the task you want to find appears there. In the event that the task you want to mark as complete isn't in the view you selected, you might try the Simple List, which shows every task you've entered.
- You can mark a task as completed by checking the box next to the subject. The box to the left of the second column is the one you want to choose. When the checkbox is selected, a line is drawn through the task subject. The task is complete.

By switching to the Completed view, you can see a list of the tasks you've marked as complete. It lists all the projects you've completed neatly and nicely.

There are several places in Outlook where tasks can be marked as complete. There is a task list as well as certain views of your calendar, as well as the task list within Outlook Today.

Picking a color for completed or overdue tasks

Upon completion of a task or overdue of a task, Outlook turns the text gray for completed tasks and red for overdue tasks, so you can see at a glance which tasks have been completed and which must be completed. In Outlook, you can choose different colors if you don't like the default colors.

You can change the color of completed and overdue tasks by following these steps:

- Click the Options button on the File tab.
  A dialog box appears with Outlook Options.
- Select Tasks
  A page with task options opens.
- Select the Overdue Task Color.
  A color palette appears.
- You can choose a color for tasks that are overdue.
- Select the "Completed Task Color" button.
  Colors will be displayed.
- Assign a color to completed tasks.
- Then click OK.

The tasks you complete and those that are overdue will appear in your list in the colors you select.

View Your Tasks

Task lists in Outlook can be viewed in several different ways, and a variety of custom views can be added and saved. If you know how to use Outlook's views, you can accomplish a lot.

You can select one of these views from the Current View group in the Ribbon to view your tasks:

- The detail view provides more information than the simple list view. The information is really the same, as is what the status of your tasks are, what percentage is complete, and what categories you assigned to your tasks.
- A Simple Listview shows only the facts. Each task's name and due date (if you assigned one) Add new tasks easily and mark old ones as complete using the simple listview. If more details are desired, no extra information will be displayed.
- Your To-Do List shows all the tasks you've entered along with any flagged emails. Only the items you have directly added to the task list are shown in the other views of the task list.
- Tasks are displayed in priority view according to the priority you have assigned to each task. Thus, you can see what's important and what's urgent.
- In Active view, you can only see tasks you haven't yet completed. Once a task is marked as completed, Active tasks vanish from the active view as they are completed, so you can focus on the tasks that remain.
- You'll see your completed tasks in the "Completed" view. Even though you don't have to deal with completed tasks anymore, you sure do feel a sense of satisfaction when you look at them.
- On the Today view is today's due and overdue tasks, which are essentially today's tasks! Getting a reminder of your workload is a great way to begin the day.
- Compared to Active, the Next 7 Days view appears to be more focused. The Next 7 Days view displays only tasks due to be completed within the next seven days. You'll like it if you like living in the moment.
- Your overdue view is a sign of naughtiness. These tasks should have been done yesterday, but they are still lingering today.
- Using the Assigned view, you can sort your tasks alphabetically by the name of the person you assigned each task to.
- If your company has a task server, you can view the tasks assigned through it. Among its fields are Assigned To, Custom Status, and Custom Priority, all of which are useful in collaborative situations.

Frequenting the To-Do Bar

The To-Do bar in Outlook makes it easy for you to keep track of the things you need to do by pulling them together in one place. Rather than forcing you to check your calendar, check your email inbox, and then check your task list, the To-Do bar makes it easy to see what you need to do at a glance.

Your to-do bar will usually include the following items:

- A list of tasks you've entered
- Several appointments have come up.
- Your flagged email messages are

To-do lists can be confusing at first, since things may appear there that you didn't put there directly. By applying the flag labeled "This Week" to a Monday email message, it will appear for action two Fridays later, even if you forgot that you sent it. To-Do bars keep you from forgetting that's what they're there for.

Adding a new item to the To-Do bar

On the Ribbon,

Click the To-Do Bar button and pick Tasks from the drop-down menu to display the To-Do bar. Besides tasks, you can also choose a calendar or people if you want this information displayed on the To-Do bar.

Right on the To-Do bar is a box saying "Type a New Task." Follow the instructions. When you're done, hit Enter.

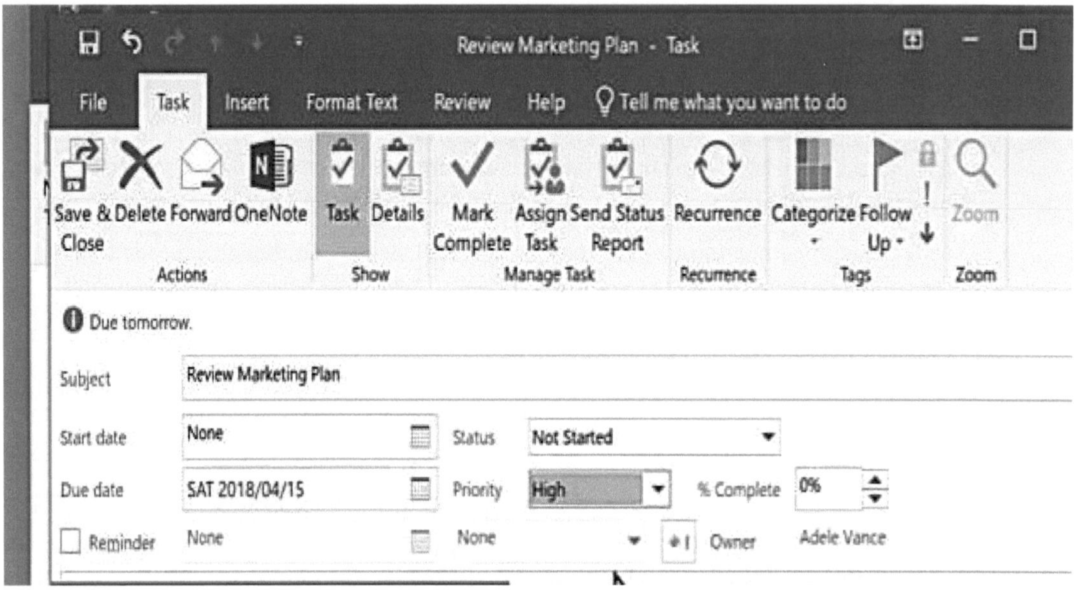

Tasks in the Calendar

Task lists and To-Do bars allow you to keep track of what needs to get done. When you have a free moment, you need to make sure you can do everything that needs to be done. To help you stay on top of upcoming tasks, Outlook displays a daily task list in your calendar. It can be accessed by clicking the View tab, clicking the Daily Task List button, and selecting Normal from the drop-down menu.

Across the bottom of the screen, you see a strip of icons that represent tasks whose due dates fall on the days displayed. When you have too many things on your to-do list for one day, just drag the task to a day when it can be completed. Tasks can even be dragged to a specific time in the calendar to reserve a specific time for completion.

CHAPTER 10

CUSTOMIZING OUTLOOK

"User interface" refers to the display of screens, menus, and other devices on your computer. It takes lots of effort and money for the people who write computer programs to figure out the best way to arrange things on the screen so that a program like Outlook is easy to use.

You can arrange and view the information in an endless number of ways in Outlook. An Organize button shows you the different ways you can slice and dice your Outlook data. Following the "Organize" button, this chapter contains many of the best steps you can take.

Customizing the Quick Access Toolbar

By keeping a few icons for your favorite functions at the top of the screen, the Quick Access Toolbar makes it convenient for you to use them anytime you want. When Outlook is freshly installed, only three Quick Access Toolbar icons are displayed: Send/Receive, Undo, and Customize Quick Access Toolbar. If you use Print, Delete, and so on frequently, you can customize the Quick Access Toolbar to include these functions.

Here are the steps to customizing the Quick Access Toolbar:

- Customize the Quick Access Toolbar by clicking the icon. This is located at the right end of the toolbar. The most popular Outlook functions are displayed in a drop-down menu.
- You can add a function by clicking on its name.
  A Quick Access Toolbar icon appears for the function you select.

Customizing the Ribbon

Several tabs make up the Ribbon, and each Outlook module has its own Ribbon, with its own set of buttons and tabs. It is not helpful to put a button in the wrong place on the ribbon.

Customizing the ribbon is as simple as these steps:

- Click on any part of the ribbon. An options menu appears.
- Select "Customize the Ribbon." A new window opens with Outlook Options.
- Choose the location where you want the command to appear on or off the ribbon. The order of the buttons on the ribbon can be customized by dragging. When adding new commands to the Ribbon, click on the New Group button first and add the commands to the new group.
- Click the OK button.

The right-hand side of the screen has the commands that are already on the Ribbon. They can be removed if you wish. In the left-hand column, you can see the commands you might be able to add to the Ribbon. However, no command will be allowed to be added to the ribbon. In order to add a command to a ribbon, it must be appropriate for the module in which it appears. For example, the Mark Complete command can't be added to the Calendar Ribbon because it isn't useful there. This means that even if you try to add the command to this location, the Add button will be grayed out.

Taking in the Views

The Outlook modules each have their own selection of views and ribbon tabs. Among other things, the calendar has a view that looks like a calendar. Likewise, the Contacts module has a view similar to an address card. The modules all offer the ability to view your data in the traditional row-and-column format.

It's important to understand that each view is designed to make some aspect of how you collect information obvious at first glance. A view can be sorted, filtered, or grouped so that information appears differently. There are innumerable ways to organize your information in Outlook, which you can then view. Information is viewed differently depending on the kind you have and how you intend to use it. With views, there is no way to go wrong because if the old ones fail to work, you can easily make new ones. Experiment as much as you like.

Views usually have a name, which is usually found under the Current View section on the Ribbon. The Current View section may not appear under the Home tab. Click the View tab instead. There is a reason why Microsoft does not always place the Current View section in the same place on every module's Ribbon.

Table/List view

Each module has a table view—a rectangle with rows and columns. It is called the list view by some Outlook commands. As soon as you add a new item (such as adding a task to your task list), a new row appears in Table View. For example, the Tasks module's detailed view is a table view.

Card view

Card View is designed for the People module. Each contact item gets its own little block of information. Each little block shows a little or a lot of information about the item, depending on what kind of card it is.

Calendar views

Viewing dates and setting appointments on the calendar is a fantastic use of the calendar.

Using this module, you can easily switch between Day, Work Week, Week, Month, and Schedule Views from the Ribbon. Each of these views also shows a calendar for the month. A click on any of the dates will display the information for that particular date.

Playing with Columns in Table and List Views

Views based on tables (also known as lists) provide the most detail about the items you've created; you can also use these views to group information in the most efficient way. Despite their dull appearance, table views get the job done.

Information about each item is displayed in the columns of the table view. There is a great deal more data that can be stored in Outlook modules than you can display in a row-and-column format on your screen. In your contact list, for instance, you can access 90 different pieces of information about every individual. Each person would require more than 90 columns to display all their details if they were represented by just one row.

Adding a column

As a new user of Outlook, you only have a few columns in the Contacts list's Phone view. But if you wish, you can easily add more. There is no limit to the number of columns you can display, but the information may appear blurry as you scroll across the screen.

In any table view, you can add a column by following these steps:

- You can right-click on any column title.
  An option will appear.
- From the shortcut menu, choose "Field Chooser."
  A box with the Field Chooser appears.
- Place the desired field in the table.
  If the field you are looking for does not appear in the Field Chooser pane, select All Contact Fields from the drop-down list at the top.
- Close the Field Chooser pane by clicking Close (X) when you're finished, or leave it open in case you want to add more fields later.

Moving a column

It is easier to move columns than to add them. To move columns, just drag their headings. When dragging, you will see two little red arrows to indicate where the heading will end up when you release the mouse button.

Widening or narrowing a column

Columns can be narrowed or widened even more easily than they can be moved. To do so, follow these steps:

- The mouse pointer should turn into a pair of arrows when you continue to move it to the right edge of the column you want to widen or narrow.
  Making that mouse pointer look like a two-headed arrow requires skill. After a little practice, you'll realize how easy it is.
- By dragging the column edge, you can adjust its size. A thin line appears where the two-headed arrow is positioned, which can be dragged to resize the column.

Removing a column

If you do not wish to see certain columns, you can remove them. The steps are as follows:

- You can remove a column by right-clicking on its heading. An option will appear.
- Select "Remove This Column."

If you need to delete columns, don't worry about it too much. Once you remove the columns from the item, the fields remain. If you want it back, you can use the column-adding procedure I described earlier in this chapter.

Sorting Items

Sorting is just an orderly way of arranging your list. There is always an order to list. Changing the order is all that sorting does.

Your list can be sorted as follows:

- If a heading has a triangle, it means the list is sorted by the information contained in that column.
- The list goes down from the largest to the smallest number if the number is in the column and the high side of the triangle is at the top.
- The columns with text are sorted alphabetically. The letter A is the smallest, while the letter Z is the largest.

Sorting in the Table View

In my opinion, this is the easiest way. To sort a column in a table view, click the column heading. All the columns are sorted by what you clicked on; the date, the name, etc.

Sorting from the Sort dialog box

While selecting a column is the easiest way to sort, you can only select one column at a time. It is possible to sort by more than one column.

Here are the steps for sorting two or more columns:

- On the Ribbon, click the View Settings button and then select the View tab.
  Select Advanced View Settings from the dialogue box that opens.
- Select Sort from the menu. A sort dialog box opens.
- Pick the first field you want to sort by from the Sort Items By list box.
- Select carefully; there is a much longer list of fields in the list than is typically available in the view. The list may be confusing.
- You can sort either by ascending or descending.
  If you are sorting ascendingly, select smallest to largest. If you are sorting descendingly, select smallest to largest.
- Sort each field by repeating steps 3 and 4.
- Your list has been sorted. Click OK.

Grouping Items

There is a similarity between sorting and grouping. In both cases, items are arranged by column. In contrast to sorting, grouping creates a collection of related items that you can open or close. You can ignore all the other bunches and only examine those that interest you.

Viewing grouped items

A grouped view identifies the columns that were used to create that view. When choosing People and then List view (which groups your contacts by an organization), triangular icons appear on the left side of the screen. Because the view is grouped by company, it appears next to each icon. Next to the word "company" is the name of the company; the grouped view separates each company in the list.

A small icon appears at the left end of each group heading, either pointing directly to the right or downwards and to the right.

- When an icon points directly to the right, there's more to see; click it to view these items.

- With a triangle tilted down and to the right, there is nothing more to look at; what you see is what you get.

Choosing the company name without the icon selects the group as a whole.

By selecting the company name and clicking the "Delete" button, you can delete the group.

In order to distinguish a selected group bar from the others, it is highlighted in blue.

Viewing headings only

Alternatively, you can open or close all groups at once by clicking on each triangle individually.

Follow these steps for opening or closing groups:

- On the View tab, click Expand/Collapse.
- Either collapse or expand the group.
- By choosing to Expand All Groups or Collapse All Groups, you can expand or collapse all the groups.

Saving Custom Views

If you're used to saving Word documents, you're familiar with the idea of saving views. When you make any of the changes to the view I described earlier in this chapter, you can save the changes as a new view or save the changes to the current view. If you plan to use a certain view repeatedly, it's worth saving.

- Select View from the menu.
- Click on the "Change View" button.
- Select the Save Current View option.
- Give your view a name.
- By simply changing the views you already have, you can do almost anything you want.

Using Categories

Information collection is a valuable asset. If you can't quickly figure out which items are important and which aren't, then you can't get the most value out of a list of contacts or tasks. Outlook's "Categories" feature helps you quickly distinguish between urgent and non-urgent items.

Assigning a category

On the Mail module's Home tab,

Click the "Categorize" button in order to see what categories are available. The categorize button looks more like a small, colorful tic-tac-toe square.

When you click on the "Categorize" button, you'll see a list of color-themed categories. The process of changing the default color of items is fairly simple.

You can assign a category to an item by following these steps:

- If you would like to categorize a mail item, click on it. A highlight appears.

- Choose a category using the "Categorize" button. The item will appear colored according to the category you selected.

It is possible to assign multiple categories to each item, although choosing too many may lead to more confusion than choosing none at all.

Renaming a category

You can get an idea of what Outlook category colors mean to you, but I would prefer to know what each color means so I know why an item is in a particular category.

Follow these steps to rename a category:

- Select all the categories under "Category".
  The dialog box for color categories appears.
- You can rename a category by clicking on it.
  Selecting a category highlights it.
- Rename the file.
  There is a box surrounding the category you selected to indicate that it may be edited.
- You can now rename the category.
  It is renamed after you type it.
- The dialog box for Color Categories closes after you click OK.

A category name you had previously assigned to some Outlook items will automatically change if the category name is changed.

Changing a category color

A category's color and name can be modified. Your ability to identify memorable colors can give you a good indication of how your work has gone or how well you've kept up with a project.

Changing a category's color is as simple as following these steps:

- Choose the "All Categories" option from the "Categorize" button.
  A dialog box appears.
- To change the color of a category, click on the category name.
  Your selected category will be highlighted.
- Select a color from the menu.
  You will see a drop-down menu with the colors you can choose from.
- Pick a color you'd like to assign.
  That color replaces the old one.
- The dialog box for Color Categories closes once you click OK.

Assigning a category shortcut key

Providing each category with a shortcut key allows you to assign a category without touching the mouse. Those tools are very helpful when you'd like a quick way to sort through a screenful of email messages.

Follow these steps to assign a shortcut key to a category:

- Choose the "All Categories" option from the "Categorize" button.
  A dialog box for color categories opens.
    - A shortcut key can be assigned to a specific category by clicking on it.
      The highlighted category clearly shows you have selected it.
    - The Shortcut Key drop-down menu box will appear.
      Shortcut keys will be listed there.
    - To assign a shortcut key, click the desired key.
      Your shortcut key choice appears to the right of the category.
    - The shortcuts can't be assigned to more than one category; it would be confusing. Individual items, however, can have more than one category assigned to them.

CHAPTER 11

SOCIAL MEDIA MAGIC WITH OUTLOOK RSS

Almost everyone uses social media these days. All these web services seem to have attracted a lot of attention, including Twitter, LinkedIn, and other popular ones. It is hard for most people to keep up with what is happening on social media; it changes far too frequently for them to keep up. For example, MySpace was the most popular internet destination in 2006. Since MySpace had become passé by 2009, I was assured by every tech expert that Facebook was the rage. It took only 30 months for MySpace to become yesterday's news. The only constant thing about social media is change. Young people are more and more transferring to Instagram and Snapchat as we speak.

However, social media isn't a passing trend. Business, culture, and public policy are being affected by social media trends. In the world of social media, no matter which particular service is dominant at a given time, you should keep at least a passing knowledge of what's happening since the developments could impact your career and business.

Brushing Up on Social Media Basics

With new social networks popping up and disappearing daily, it can be easy to get confused. In addition to keeping track of your social media subscriptions, Outlook can sync your email, contacts, appointments, and everything else you need to stay organized.

Sending an SOS to RSS

In Outlook, RSS stands for Really Simple Syndication, or Rich Site Summary, which allows you to keep track of all this changing information. You can organize RSS feeds in Outlook in a way that is useful to you by using Outlook's RSS folder. It isn't necessary to understand how RSS works, but knowing that it's available when you need it is nice.

A feed is a delivery method for RSS information. RSS feeds allow for the automatic updating of information over time. You can keep up with new entries and episodes by subscribing to RSS feeds offered by blogs and podcasts.

It might be necessary to make a few settings changes in Outlook in order for RSS to work.

The steps are as follows:

- Navigate to the File tab in Outlook.
- Select options.
  A dialog box appears with Outlook Options.
- Select Advanced.
- The Common Feed List (CFL) should be synchronized with RSS feeds using the Synchronize RSS Feeds box in the Windows Control Panel.
  Go to the RSS section.
- Then click OK.

If you're not a fan of Internet Explorer, Outlook is another way for you to subscribe to RSS feeds. (Windows includes it by default, so it is almost impossible for you to claim it isn't present. However, some people choose not to use it.)

Feeling like a social butterfly

We must not overlook the importance of blogs, podcasts, and news organizations as a part of the social media world. Although you may ignore any of them, it's highly likely that you're already reading or watching many of them.

Podcasts

Digital, downloadable editions of many radio stations' programs, including news, talk, and information stations like NPR, are available. Your favorite radio personalities often refer to these editions as "podcasts." These programs are typically scheduled regularly. Podcasts can be downloaded one at a time or automatically if you subscribe to them.

It is misleading to use the term "podcast." Podcasts are often thought to be compatible only with digital music players. Many people don't listen to podcasts as often as they would like because finding, downloading, and playing podcasts can be hard. In spite of the fact that podcasts were initially designed to play on portable devices, most people today listen to them on their computers. If you have Outlook installed on your computer, podcasts can also be played.

With Outlook, you can access any podcast that you've subscribed to and organize it using the same tools you use to manage email.

Blogs

Several years ago, everyone talked about blogs as if they were a brand new, magical technology, but they're not. If you surf the Internet for news and information, you might come across blogs without even realizing it. The majority of major news services feature a blog section where reporters and commentators discuss breaking news and current events. Blogs are nothing more than web pages that permit frequent updates and display the newest updates first.

Many people today prefer to get their news from bloggers. You can also waste a great deal of time if you surf one blog after another. Microsoft Outlook allows you to keep up with all your blogs in one place and stay on top of the latest gossip. It also keeps you up to date on important news.

Outlook can be used to read blogs, as well as feeds from news providers and more official-looking RSS sources.

Subscribing to an RSS Feed via Internet Explorer

Subscribing to blogs or other feeds is much easier with Internet Explorer than with Outlook. If you have a choice, Internet Explorer is the best place to go. Feeds you set up in either Internet Explorer or Outlook are accessible in both places for reading. In other words, you can set up RSS feeds in Internet Explorer and then have them read in Outlook.

RSS isn't supported by Microsoft's Edge browser.

These steps will allow you to subscribe to a blog or another RSS channel in Internet Explorer:

- Using Internet Explorer, navigate to the RSS feed page for the blog or news source you wish to subscribe to.
- You can subscribe to this feed by clicking the Subscribe to This Feed hyperlink on the page. Alternatively, you can follow these steps:

  - To access favorites feeds, and history, click the start button (top-right corner of the browser window).
  - A task pane appears.
  - In the task pane, select the Feeds tab.
  - Select "Add to Favorites" from the drop-down menu.
- The dialog box for subscribing to this feed opens.
- Press the Subscribe button.
  Click OK to close the dialog box.

Internet Explorer and Outlook can now display the feed.

When you subscribe to a new blog, an RSS Feed folder shows the blog's name in the Outlook Folders list in the Navigation Bar. All you need to do is check the folder. The RSS Feeds folder might need to be expanded to see your newly subscribed content.

Setting Up an RSS Feed in Outlook

As I mentioned earlier, if you use Internet Explorer or Outlook, you can set up any RSS feed and then access it from either. However, it is probably best to use the Outlook method, however, if you are subscribing to a podcast. That way, you can configure the feed's settings while you are there.

Setting up a podcast, or an RSS feed of any type, in Outlook can be done in the following manner:

- Choose Account Settings from the Account Settings menu when you click the File tab.
  A dialog box appears.
- Select "RSS Feeds" from the menu.
  Subscribed feeds appear on the RSS signup page.
- Press the New button.
  A dialog box opens to let you create a new RSS feed.
- Your RSS feed URL should be entered here.
  URLs like this usually look unusually long. http://www.cinemasolo.com/atom.xml Incorrectly entering the address will result in an error.

Here are the steps you should follow:

  - Find the feed you want on the site where it is hosted.
  - Choose XML, RSS, or Feed from the right-click menu. Depending on the site, this could be an orange button or a button that looks like the wireless network icon in the notification area.
  - Select "Copy Shortcut".
  - As soon as you have pasted the address, you can follow the preceding steps and add a new RSS feed.
- Click on the Add button.
  You can make the following changes to your RSS feed subscription using the RSS Feed Options dialog box:

1. Feed Name: The name shown by Outlook can be changed. Feed names can be complicated.
2. Deliveries: Some feeds generate large amounts of information, so you may want to send that information to a special folder or to a completely separate file. Podcasts are a good example. If your company has a limit on the amount of email storage you can have, you might want to save your RSS subscriptions to a separate Outlook account so that you don't run out of space.
3. Using Outlook's automatic download feature, Outlook only downloads a brief summary of each item. This saves disk space, but you must manually download the full text of each item if need be.
4. By selecting the Download Enclosures checkbox, the podcast file and the description will be downloaded.
5. The RSS feed publisher may have a limit on how often your information can be updated. Subscriptions that are updated too often may be canceled. There is an automatic checkbox marked "if there is a limit assigned to the feed you have selected."

- You can now click OK.
- Close this window.

Reading Feeds

The RSS feeds aren't visible from the navigation pane, so you have to open the RSS folder. Reading RSS feeds takes a few extra steps, but once you find your way to the RSS folder, it's pretty simple.

In Outlook, you can read RSS feeds by following these steps:

- On the navigation bar, click the Mail button.
  You will see a list of your mail folders.
- Select the RSS Feeds folder by clicking the arrow.
  RSS feeds are displayed in the folders. There is one feed per folder. A triangle next to RSS Feeds indicates you have not yet set up any RSS feeds.
- You can now read the feed by clicking on the folder that contains it.

Upon subscribing to one or more podcasts and choosing automatic downloads, double-click the attachment in the message file. You will be able to listen to your podcast after your computer launches a player.

CHAPTER 12

MANAGING MULTIPLE EMAIL ACCOUNTS

One email address is common among many people. Once they install the program or someone else does it for them, they set it up in Outlook. Nevertheless, you are not that many people, otherwise, you wouldn't be reading this chapter. You crave more flexibility, more privacy, more small business sidelines, maybe even the ability to use a secret alter-ego like Freaky4U@FantasiesComeTrue.biz on the down-low. The purpose of this chapter is to demonstrate how to set up additional email accounts in Outlook and to provide some advice on how to get those email addresses from various services.

Choosing an Email Provider

The majority of people who have computers at home have an Internet service provider (ISP). Your Internet provider is the one you pay a monthly fee to. ISP options may include your cable provider, telephone provider, or satellite provider, depending on where you live.

Oftentimes, when people get their first Internet connection, they receive an email account from their ISP. It's free and easy for them, and they don't know any better.

Most ISPs provide free email accounts with their ISP services. The setup for their email is usually very straightforward. Every family member can have a personal email address. In fact, they want to be both your email provider and your Internet service provider for free! Don't they sound too good to be true?

You may not be able to shop around for better ISP deals later if you take advantage of this offer. If you have an address with an ISP and have been sending emails from that address for a year or more, changing to a different ISP becomes more difficult since you must notify everyone you know of your new email address. Trying to remember all the merchants and services you used that email address for can be overwhelming because you've probably forgotten all the accounts you created with them. It's so much trouble to have to change email addresses that you'll be stuck with that ISP no matter what their prices are.

Many different services offer free email addresses as an alternative to getting your own address. You can use Gmail from Google, for example, and Microsoft's Outlook.com is free. Microsoft's Office 365 plans also provide a range of options, from basic plans for private users to sophisticated plans for large multinational corporations.

Another popular option is email service provider Mail.com (www.mail.com). Mail.com lets you register for a free email address and check your messages online. In order to use the robust mail-management capabilities of Outlook, you'll have to upgrade to a POP3 or an IMAP account on Mail.com, which will cost you an extra fee per year.

Buying Your Own Domain Name

With a domain name, you can have as many email addresses as you want. You can get one for about $30 a year (maybe a little less). You could, for example, buy the domain roblevsky.com and set up the address Mordecai@Roblevsky.com if your name is Mordecai Roblevsky. Using your first name as an email address might not be possible if you have a more common name, such as John Smith. If need be, you can either change your name or use a variation of your name.

The process of purchasing your own domain consists of three steps:

- Register the domain name.

It must be an available name, of course. Payments can be made just for one year, or several years ahead of time. The cost ranges between $5 and $20 for this service. A domain registrar handles this for you. The most popular is GoDaddy, but there are others as well. A domain registrar manages your contact information and your ownership of the domain name. Well, you are essentially renting it, so long as you keep paying for it. It is yours to use however you wish.

- Sign me up for hosting.

  Hosting is defined as the service that manages your domain's web and email servers. When people enter your domain name in their browser, your hosting company will direct their web traffic to the website associated with your domain. Additionally, your domain can also be set up to send and receive emails through the host's email servers. Hosting can be provided by the same company as the registrar (and admins find this method easier), or a separate company can be used.

- Create the email accounts.

  You can create email accounts and assign passwords to them through the control panel of your hosting company. The accounts can be whatever you wish. You can type anything before the @ symbol. Your domain name follows the @ symbol. Taking this example, Mordecai could set up email accounts for everyone in the Roblevsky family: levi@roblevsky.com, toby@roblevsky.com, and so on.

Setting Up Email Accounts in Outlook

Using Outlook, you can send and receive emails from your new email account once you've signed up for one. The new account you set up will coexist with your existing Outlook email account.

Understanding POP3 vs. IMAP

POP3 and IMAP are two protocols for accessing email accounts, including those you create with your own domain and those you get from your ISP. An email account using a web interface must support POP3 or IMAP in order to integrate with Outlook. In some cases, you can choose between the two.

IMAP uses a server to store all your messages, and your local email client updates your mail folders as soon as you connect to the server. Deleted messages in Outlook are also deleted from the server copy of your inbox. If you use this system on multiple devices, each will show you the most up-to-date version of your email. Each device will display the same version of your inbox. If you use multiple devices to check your email, including a desktop, tablet, or smartphone, then this is the account for you.

When you use POP, however, the focus is on your local PC and email client (for example, Outlook). Sending and receiving messages over POP3 mail servers transfers new messages directly to the local PC without being accessed by the server. Outlook can be configured to delete or save transferred messages as a backup immediately, but the server doesn't check whether a message has been read or responded to.

A desktop computer was probably the only computing device that most people had back then. POP3 dates from the 1990s and was designed for a time when most people had just one device. It does, however, have some advantages. It is possible, for example, to access your old messages from your computer at any time, even if the Internet is not available. Due to your old emails being stored locally rather than on the server, you won't need to worry about the email server storage limit ever being reached. It is also possible to combine the incoming email from two or more POP3 accounts into a single Outlook data file, allowing for easier backups and file transfers.

Almost all email systems that support one of these also support the other, so you can specify which you want. When setting up an account in Outlook (or whatever email client you are using, such as the mail app on your smartphone), you are free to make a choice. You can choose in accordance with your needs.

Collecting the needed information for setup

Setting up an email account using default settings in Outlook isn't difficult; you only need an email address and password. The rest can usually be handled by Outlook. It's easy to set up an email account in Outlook 2019 compared to previous versions, so if you haven't done so for a while, you are in for a pleasant surprise. You can start there if you aren't a techie.

However, if you think you may need to change your email account settings, or if you just want to be proactive, contact your email provider in advance:

- email address and password. You don't have to think about it. They are required.
- incoming mail server address. This address is formatted as text.text.text. Probably in the form of imap.secureserver.com or pop.secureserver.com. Depending on whether you have your own mail server or not, the second and third parts of that might be the same as your domain name.
- Outgoing mail server address It is also formatted as text.text.text. You might see it as SMTP-out.secureserver.com. In the first slot, you will likely see the letters SMTP.
- SMTP stands for Simple Mail Transfer Protocol. Email is sent using SMTP. (POP3 and IMAP provide mail receiving capabilities.)
- Port numbers are used to receive and send mail.
- What encryption method does the server require for encrypted connections?
- For either sending or receiving, the server must use Secure Password Authentication.

Most providers have a support section on their website where you can find this information. To get the correct spellings of the server names and passwords, contact the tech support department of your online service or ISP. (Be sure to ask if they're case-sensitive, which means capitalization is important!)

Setting up an account using automatic settings

As I mentioned above, Microsoft Outlook 2019 makes it easy to set up accounts without having to provide too much technical information. This method works well for people who don't need to mess with settings when setting up their accounts.

These steps will allow Outlook to detect and configure the correct settings for a new Internet email account:

- Navigate to the File tab.
  A backstage view appears.
- Select Account Settings from the drop-down menu by clicking on the Account Settings button.
  A dialog box opens.
- If it isn't already selected, click the Email tab.
  A page for setting up email accounts appears.

- Select "New" from the menu.
  A new account dialog box appears.
- Click Connect after entering the email address for the account you wish to set up.
  Upon entering your email address, Outlook will attempt to automate the setup.
- In the Password box, enter the account's password and click Connect. Outlook then attempts to configure your email account automatically. When it succeeds, a box appears titled "Account Successfully Added." If it doesn't work, close all the dialog boxes and try again by following the instructions in the next section.
- Complete the process by clicking the Done button.

Setting up an account using manual settings

In most cases, Outlook hides from you the complicated mail settings you can specify for your account, as you saw previously. As such, if you are able to set up your email account on Outlook and ignore all those settings, that's wonderful.

Should the automatic method fail, here's what you should do next:

- Navigate to the File tab.
  You will see the backstage view.
- You can change your account settings by clicking the Account Settings button and selecting Account Settings.
  A dialog box opens with account settings.
- If you haven't already selected the Email tab, click it now.
  A page for configuring email accounts appears.
- Select the "New" button.
  The Add Account dialog box opens.
- Enter the email address for the account you wish to create.
- Select "Advanced Options."
  You will see a box that says "Let Me Set Up My Account Manually."
- Click Connect after checking the box.
  The advanced setup appears with icons for various email accounts, including Office 365, Outlook.com, Exchange, Google, POP, and IMAP.
- Set up a new account by clicking on the appropriate type.
- When prompted for a password, type it and click Connect. A screen will appear stating that the account was added successfully.
- Try any of these:
  1. Enter another email address in the Add Another Email Address box, click Next, and follow the prompts to set up the new account.
  2. Setting up Outlook for my mobile phone, too, will send you an email with the link to set up this email address there. A page with instructions will be displayed.

3. Click Done to complete the process.
4. Click "Close" when the Account Settings dialog box appears.

Modifying Mail Account Settings

You can edit the settings of any of the mail accounts that are set up in Outlook. Changing the password for an account on the server (such as if someone got access to it and you need to change it for security reasons) might require changing the stored password.

You can modify an account's settings by following these steps:

- Navigate to the File tab.
  You will see the backstage view.
- Select Account Settings from the drop-down menu.
- If you haven't already selected it, click the Email tab. The Email Accounts setup page appears.
- A page for setting up email accounts appears.
- Click on Change when you have selected an account to modify.

A dialog box for changing accounts opens:

- Edit the account details if necessary.
- Then click Next.
- Click the Done button.
- Click Close to close the Account Settings window.

Changing the Mail Server

When you struggle to access an email account, a technician at your ISP may suggest that you change the port or address assigned to your mail server.

Previously, Outlook made this fairly simple, but not anymore! In order to prevent people like us from messing around with those settings, they dumbed them down.

Even so, there is still a way forward. Instead of the change command, use the repair command.

This is what you need to do:

- Navigate to the File tab. A backstage view appears.
- Click the Account Settings button and choose Account Settings from the drop-down menu.
- A dialog box opens with account settings.
- If it isn't already selected, click the Email tab.
- A page for setting up email accounts appears.
- Click "Repair" after selecting the account to modify.
- The Advanced Options page will appear.
- If you want to manually repair your account, select this checkbox.
- Click the repair button.
- A dialog box with account settings appears.
- The settings for incoming mail are displayed first.
- If necessary, edit the incoming mail settings.

- You might change the mail server or port number, the password, or the encryption algorithm, for example.
- Click the Outgoing Mail tab.
- You will see the settings for outgoing mail.
- Change the outgoing mail settings if necessary.
- You might, for example, change the outgoing mail server or port number, the server timeout delay, and whether authentication is required for SMTP.
- Click the repair button.
- Click the Done button.

Sending Messages from Different Accounts

When you have multiple accounts set up, Outlook adds a new button to the message composition screen. When you click "On Form," you will be able to select your account from which to send the message.

Email messages are always sent from the default account. It may not always be obvious which account should be used. It is sometimes selected as the one that is most frequently used.

Follow these steps to change the default account:

- Navigate to the File tab.
- A backstage view will be displayed.
- Select Account Settings from the drop-down menu by clicking on the Account Settings button.

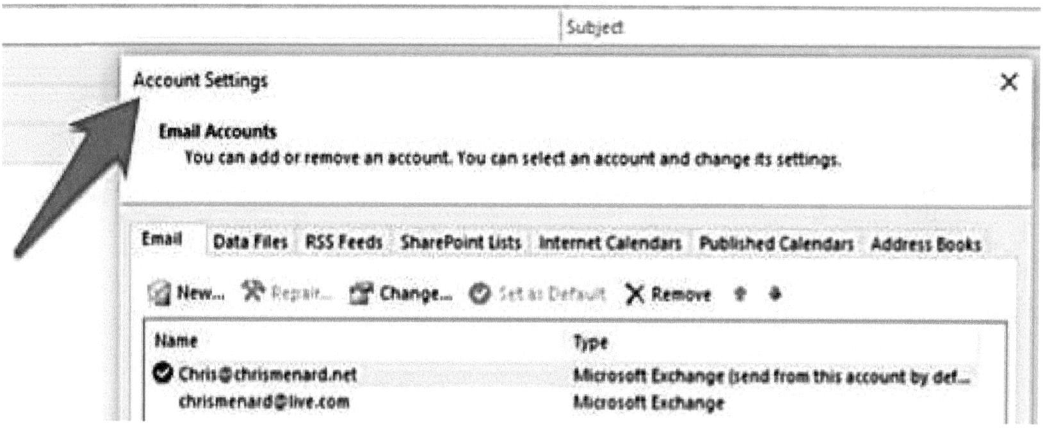

- A dialog box opens with account settings.
- Ensure that the Email tab is selected if it is not already selected.
- A page for setting up email accounts appears.
- Select the default account.
- Set it as the default by clicking on it.
- Click Close to close the Account Settings window.

CHAPTER 13

MERGING MAIL FROM OUTLOOK TO MICROSOFT WORD

In case you're not familiar with the concept of mail merge, it's a way of creating a letter on a computer and then printing different copies of it to different people. It's more than likely you get many merged letters every day. It's called "mail merging" when you send a mass mailing. Junk mail is when you receive a mass mailing.

Making Mailing Label Magic

There is a possibility that you may want to send a message to all of your friends to inform them about a party or a meeting. Hence, mailing labels can be created instantly for everyone on your contact list. Word's Mail Merge feature allows you to integrate the list into itself, so you won't have to worry about exporting files or determining where they went.

Urging merger

Verify that your printer is set up with the correct labels. Next, create your mailing labels by following these steps:

- Go to the navigation bar and click "People."
  A list of your contacts appears.
- In the Actions group of the Ribbon, click the Mail Merge button.
  Hence, the Mail Merge Contact dialog box opens.
- Select Mailing Labels from the list of document types in the Merge Options section.
- From the Merge To list, select New Document.
  This should already be selected but ensure you verify it.
- Click OK to confirm.
  The Mail Merge Helper dialog box opens in Microsoft Word. It tells you that Outlook has created a mail merge document and that the document needs to be set up by clicking the setup button.
- Then click the OK button.
  A dialog box will open.
- Click on the Setup button.
  The Label Options dialog box opens.
- Choose a label brand from the Label Vendors drop-down menu.
  Avery is one of the most popular label brands; generic labels often list an Avery product number equivalent on their packaging.
- Choose the label product number in the Product Number drop-down menu. Check the stock number on your label to make sure it's the same as the one you're choosing. If the stock number isn't available, you can look at the label dimensions in the Label Information section of the Label Options dialog box.
- Click OK.
  The Label Options dialog box closes.
- Click the Close button in the Mail Merge Helper dialog box.
  The Mail Merge Helper dialog box closes.
- From the Mailings tab of the Ribbon, click the Address Block button.
  A dialog box appears.

- The dialog box will close once you click OK to accept your settings. A funny-looking code appears in the first table cell and further down in the table cells: **AddressBlock > > and **Next Record > >. This is known as a "merge code," and it tells Microsoft Word which details to include in your document.
- From the Ribbon, select Update Labels.
  As a result, Word now displays the **AddressBlock > code along with **Next Record > » in every cell of the table, showing that it knows how to fill your label page with addresses.
- You can preview the results by clicking the "Preview Results" button on the ribbon.
  Word displays how the document will appear when printed. If the preview looks good, you can proceed.
- In the Ribbon, under the Mailings tab, click the Finish & Merge button and then select Edit Individual Documents. The Merge to New Document dialog box opens, which allows you to print part of the addresses or all of them in your document. If you choose "all," then the entire range will be printed.
- To proceed, click the OK button.
  The labels have been created.
- To print your labels, select the File tab and then Print.

Making and using a merge template

The blank label document can be saved and used repeatedly if you often print labels.

Here is what you need to do after you've created your labels:

- Go back to the document that contains the merge fields. A document begins with the name of the file, whereas a label begins with the name of the label's output file. Until Preview Results is disabled, its contents look much like the Labels file; select Preview Results to disable Preview Results, and you will once again see the merge codes "Next Record" and "AddressBlock."
- You will need to click File, then Save As.
  A Save As screen will appear.
- Go to the Browse tab.
  You can enter the file's name and location in the Save As dialog box.
- Find the location where you want to save the file.
  You will need to do this only if you want to change the default location for saving the file.
- Put the name of the document in the File Name text box.
  Choose a name that will stay in your memory, like the clever Blank Labels.
- Save the file.
  When the Save dialog box closes, the file is saved.
- Save any remaining open files, and then close Word.

Try this abbreviated procedure the next time you want to do a mail merge:

- Navigate to the People tab.
  You will see a list of your contacts.
- From the Ribbon (under the Home tab, click the Actions group), select Mail Merge.
  A dialog box appears, titled "Mail Merge Contacts."

- Select Mailing Labels from the list of document types in the Merge Options section.
- Choose "Existing Document" as an option in the Mail Merge dialog box.
- Select Browse from the menu.
- Click on "Blank Labels" twice.

Streamlining the mail merge process eliminates a lot of steps, so you can move on to more interesting tasks, such as stuffing envelopes.

Understanding Formal Letter Formalities

It's a letter with standardized text that's repeatedly printed, but with a different address and name on every copy. Regardless of whether you are holding a sweepstake, form letters are still appropriate. One form letter you could write is an annual newsletter to your family and friends.

The following steps will guide you through the process of creating a form letter in Outlook:

- Select people from the navigation bar.
  A list of your contacts appears.
- Go to the Home tab of the Ribbon and click Mail Merge.
  A new dialog box opens.
- Select "Form Letters" under the Document Types menu.
- In the Merge To list, select New Document.
  The option should already be selected, but it is worth checking.
- Click OK to open a blank document in Microsoft Word.
- Fill in the blanks on the form letter.
  It's up to you whether to type the generic parts of the letter first.
- Place the mouse pointer where the merge field should appear.
  You can, for example, type "Dear", followed by the first name of the person.
- In the Mailings tab of the Ribbon, click the Insert Merge Field button and then select the field you want to merge. Field lists consist of every field Outlook can store for each contact. Of course, not every contact will have the same information in every field, so focus on the core fields for consistent results.
- You can check your work before saving the letter and printing it, just as you did with the labels.

With this, you can send a personal, annoying form letter to hundreds of people instead of sending impersonal, annoying form letters to dozens of people.

Merging Contacts from Selected Lists

A letter to everyone on your contact list is probably not necessary. The postage alone could cost a fortune if you end up with thousands of names on your list. If you only intend to send letters to a few contacts, limit your list to just a handful.

You can add people to Outlook by opening the People module and clicking on the first person you would like to add. Press and hold the Ctrl key and select each person you want to include. Use the Mail Merge button and create the merge as you learned in earlier sections of this chapter.

You'll Still Have to Lick It: Printing Envelopes

If you are sending a mass mailing, it is not necessary to print labels; just print on the envelopes. You may even have an envelope feeder on your printer. It's tedious to feed envelopes one by one.

Following the steps I described in the earlier section for creating mailing labels, you can print addresses directly on your envelopes. The main significant difference is that in the 8th step, a dialog box titled "Envelope Options" appears that offers several options for envelope sizes. You'll need to decide which type of envelope you're using (generally, the standard business envelope number 10) and then follow the remaining steps.

Email Merging

Merging emails is another appealing feature of Mail Merge. If you want to send a bulk email message to a bunch of people and customize each message, you can use mail merge email. In most cases, you don't need to use merge email because you can send a single message to as many people as you want. So, Paul, George, and Ringo will not get your "Dear John" message.

In the section "Mastering Formal Letter Formalities", follow steps 1–3. At the fourth step, select Email (rather than New Document) from the Merge To list in the Mail Merge Contacts dialogue box.

With Outlook on an Exchange server, the document is sent directly to your recipient after you click Finish & Merge. It's impossible to fix something you've done wrong. To make sure that your email merge works,

I recommend sending yourself an email first:

- Click on your name in the Contacts list to create your merge message. After you're sure what you want to say has been expressed.
- Select all the people you wish to contact and merge the list.

As a user of Outlook at home, you can disconnect from the Internet temporarily before merging, after which you can switch to your Outbox and approve the collection of messages.

CHAPTER 14

BIG-TIME COLLABORATION WITH OUTLOOK

There is no question that Microsoft writes big programs for big companies that have a lot of money. As you'd expect, Outlook used to be geared more towards employees at big companies. These days, small businesses also require tools to enhance collaboration and teamwork. In an age of remote virtual teams whose members mainly communicate via phone and email, this is especially relevant.

Most companies that use Outlook usually have a server running in the background called Microsoft Exchange Server. By combining Outlook and Exchange, you can do things that Outlook cannot do on its own. With Exchange, users of Outlook can view a colleague's calendar, assign someone else the ability to respond to emails on that person's behalf, and do a whole host of other cool things right from their desktop.

It's hard for most Exchange users to realize that Microsoft Exchange Server is anything more than an extension of Outlook. In the end, it doesn't matter what the technical differences between Outlook and Exchange are; what matters is that Outlook and Exchange can handle a lot of tasks together that Outlook cannot.

Collaborating with Outlook's Help

Your company is probably like many others, where you spend a lot of time in meetings and even more time figuring out when to schedule meetings and agreeing on what to do when they are not scheduled. Meetings and decisions can be planned using Outlook. Even though some of these features are available to all Outlook users, Exchange users will get the most out of them.

Organizing a meeting

Suppose you want to meet with three colleagues. A meeting time is suggested by the first person. You then call the second person—only to discover that the second person isn't available at the time the first person wishes to meet. Then you and the second person agree on a time, only to find that the third person cannot attend.

With Outlook, you can view schedules, choose a date, and suggest a meeting time that everyone can agree upon with just one message.

You can invite several people to a meeting by following these steps:

- The Calendar can be accessed by clicking the Navigation bar (or by pressing Ctrl+2).
- The Meetings tab of the Ribbon can be accessed by pressing Ctrl+Shift+Q (or using the Home tab).
  A new meeting form will open.
- Select the Scheduling Assistant button in the Show group of the Meeting tab on the Ribbon. This displays the attendee availability page. Outlook won't display the Scheduling Assistant button if it isn't connected to an Exchange server. In this case, the button will be labeled "Scheduling."
- To add participants, click the "Add Attendees" button at the bottom of the form.
  A dialog box will appear where you can choose attendees and resources.

- To invite someone to the meeting, simply click on their name.
  When you click on their name, it will be highlighted.
- Depending on how vital a person's participation is to the meeting, click the "Required" or "Optional" button.
  Based on the button you click, the name you select appears either in the Required or Optional box.
- Repetition of steps 5 and 6 will help you choose whom to invite. Select attendees and resources: The Global Address List displays your chosen names.
- You will see your selections on the Attendee Availability page once you click OK. In addition, if Outlook is connected to an Exchange server, the Attendee Availability page shows your team members' schedules so you can see when they have free time. Based on how it connects to the Exchange server, Outlook may not be able to get everyone's schedule right away. Outlook indicates it does not have information about participants' schedules if they do not connect with an Exchange server.
- You can choose the time that works best for you on the timeline at the top of the Attendee Availability page. Your chosen time appears in the Attendee Availability page's Start Time box. Instead of clicking the timeline, you can type the start time and end time of the meeting into the boxes at the bottom of the Attendee Availability page. The Suggested Times window provides a list of available times for everyone you're inviting to your meeting. If you don't see a time that works for everyone, you can pick a time from the list.
- Go to the Meeting tab and click the Appointment button. In the To section on the Appointment page, your invitees are listed.
- The "Subject" box should include the meeting's subject, while the "Location" box should include the location details.
- You can enter meeting-related information in the message box.
- Click the "Send" button.
  Upon submitting your meeting request, the people you invited will be notified, and you will be able to view the meeting in your calendar.

Responding to a meeting request

It's a possibility that you will be invited to a meeting, even if you don't organize them and send invitations, so it's a good idea to know how to respond to such an invitation.

The Meeting tab of the Ribbon displays these buttons when you've been invited to a meeting by email. You can access the drop-down menus for each of these buttons.

- Accept:

As soon as you decide to attend the meeting, Outlook adds it to your schedule and creates an email message to the person who organized it, telling them what you decided.

- Tentative

Your schedule is automatically updated with the meeting. The organizer receives an email message.

- Make a New Time:

You can suggest a different time if the meeting organizer picked an inconvenient one by clicking "Propose New Time."

You have two options:

- In the event that the original time cannot be accommodated, click Decline and then click Propose New Time.
- If you are unsure whether the original time will work and you want to suggest an alternate time, select Tentative and then click Propose New Time.
- Respond

Meeting invitations are sent to you via email, so you can click "Respond" to send a response to the email message without committing to the meeting in any way.

- Calendar

A separate Outlook window displays your complete calendar so you can see how your schedule looks.

If you want to add an explanation to the message, you can select Edit the Response Before Sending in the Accept, Tentative, or Decline buttons' menus, or choose Send the Response Now if you want to deliver the message right away.

In a meeting invitation, if the recipient is working with Exchange, a preview of their calendar is shown for the date and time of the meeting, giving them an overview of their availability. The preview represents only a small portion of the meeting's schedule and displays information about an hour or so before and after the meeting.

Checking responses to your meeting request

Every time you schedule a meeting in Outlook, you send a few emails inviting people to attend, and they respond immediately by either accepting or declining your invitation. Often, I need assistance in remembering who said "yes" and "no." It may be that you have a good memory. However, there is a feature in Outlook that keeps track of who says what.

Here are the steps you need to follow to check on the status of your meeting request:

- Go to the navigation bar and click on Calendar.
  You will see your calendar.
- Check the item by clicking twice on it.
  It opens.
- Choose the "Tracking" button option.
  Click the button itself, not the down arrow below it. To get where you need to be if you accidentally hit the down arrow, click View Tracking Status.
  You'll see the list of the people you invited and their responses.

Sadly, only the organizer of the meeting will know who has agreed to attend. To know whether you will meet that special someone you met in the elevator at a particular meeting, you will have to attend the meeting. When you get an email invitation to attend a meeting, you can check the participants' names via the mail.

Taking a vote

Teamwork and decision-making are constantly emphasized by management gurus. When most of the team members cannot be located most of the time, how can you get the team to make a decision? Take advantage of the voting buttons in Outlook to use it as a decision-making tool.

The voting feature of Outlook email lets you add buttons to emails sent to a group of people. Upon receiving the message, and if they are also using Outlook, recipients can respond by clicking a button. The responses in Outlook are automatically tallied, so you can track the office's momentum.

Following these steps while creating your email message will allow you to add voting buttons:

- You can create a new message by clicking "New Email" on the Home tab of the Mail module. This opens a message form.
- To use voting buttons, click the Options tab on the Ribbon. Select the buttons you want to use. Here are some suggestions:
    1. Approve; Reject Yes; No Yes; No; Maybe
    2. Custom
- The Properties dialog box appears if you choose Custom. Click the "Use Voting Buttons" and type your choices there. Just separate your choices with a semicolon according to the pattern suggested by the suggestions. For example, in order to allow people to vote on lunch menus, for example, list a range of options, such as pizza, hamburgers, and salads. Do not include spaces after the semicolons.
- Click on the set of voting buttons you want to use. A message indicating that you have added voting buttons appears at the top of your message. If you are adding your own custom choices, however, you will need to return to your message by clicking the Close button at the bottom of the Properties dialog box.
- Click on the "Send" button.

Tallying votes

Looking at the subject of a reply will show you who selected what when the replies arrive. A message from someone who chose to approve starts with the word "approve." A rejection message starts with the word "reject."

Check the Tracking tab on the message copy in your Sent Items folder for a full tally of your vote as well. Steps to do so are:

- Select Sent Items from the list of Folders.
  Your sent messages are displayed.
- Click twice on the message you sent for votes.
  It will open.
- Tracking can be accessed by clicking the tracking button. You see who voted and how they voted.
  An indicator shows the total number of votes cast.

Assigning tasks

It is always worthwhile to have someone else do something for you. A task can be assigned to another person, and then the progress of that person can be tracked.

You can assign a task to someone else by following these steps:

- Navigate to Tasks from the Navigation Bar.
  A list of tasks appears.
- Right-click a task in the list of tasks.
  You'll see a shortcut menu appear.
- The Assign Task option is available.
  A form is displayed.
- Just as with an email message, type the name of the person you are assigning the task to in the To field.
  This will appear in the To field.
- Click on the "Send" button.
  The task is now assigned to the individual you selected.

Your task recipient receives a message with two special buttons marked "Accept" and "Decline," similar to the meeting invitation I discussed earlier. As soon as the person accepts, the task is automatically added to their Outlook Tasklist, and if the person declines, the task will not be added to their task list.

Sending a status report

Assign Task is a feature that is extremely popular with people who give out tasks. It is less popular with people who have to do it. If you're more of a frequent task getter than a frequent task giver, look on the bright side, Outlook on an Exchange network can also keep your boss updated on your level of productivity.

On the task form, there are two boxes: status and percent complete. You can remind the top leadership of the value of your contributions by sending status reports if you keep the information in these boxes up-to-date.

Sending a status report is as simple as these steps:

- Go to the navigation bar and click Tasks.
  The task list appears.
- Click on any task twice.
  A form for the task will appear.
- On the ribbon, click Send Status Report.
  There is a message window that appears, and the "To" box contains the name of whoever assigned the task.
- The text box at the bottom of the form is for any explanation you want to send about the task.
  Your text will appear on the form.
- Click the "Send" button.

Depending on your preference, you can send status reports on a weekly, daily, or hourly basis. Hopefully, leaving enough time between status reports will allow some tasks to be completed.

Collaborating with Outlook and Exchange

The features I describe in the rest of this chapter are not available to users of Outlook at home or without an Exchange server. It's not all bad news, however: Microsoft is working to make Exchange-only features available to all Outlook users, so you can use this section as a preview of what's to come.

Giving the delegate permission

Delegating authority is a skill that good managers possess. An assistant can sometimes be tasked with managing the boss's calendar, schedule, and even email if the boss is extremely busy. In this way, the boss is able to focus on what matters, while the assistant takes care of the details.

To name a delegate, follow these steps:

- In the left navigation pane, click the Info button, then click the Account Settings button.
  The Account Settings menu appears.
- Choose Delegate Access from the menu.
  A dialog box appears.
- Select "Add" from the menu.
  The Add Users dialog box appears.
- Identify each delegate by double-clicking on his or her name.
  You will see each name displayed in the Add Users window.
- To continue, click OK.
  In the Delegate Permissions dialog box, you can specify which permissions to give your delegate(s).
- The Delegate Permissions dialog box lets you make any changes you want.
- Once you click OK, the dialog box for Delegate Permissions closes. A list of the names you selected appears in Delegates.
- Click OK to close the Delegates dialog box.

Opening someone else's folder

Teams with close working relationships will often share calendars or task lists; not only can they see what each other is doing, but they can also enter appointments for team members—for example, if sales and service people work in the same office. In your role as a service person, you may find it useful if your sales partner is allowed to schedule appointments while you are out handling other clients.

Your partner will have to access your calendar in order to do this:

- In the left navigation pane, click the Open & Export tab on the File tab.
- Select another user's folder from the menu.
  A dialog box opens.
- Click on the "Name" button.
  A dialog box appears. (This is really the address book.)

- Open a person's folder by clicking twice on his or her name.
  Upon closing the Select Name dialog box, the name you double-clicked appears in the Open Other User's Folder dialog box.
- Click the triangle next to Folder Type.
  A list of folder types appears.
- You can view a folder by clicking on its name. A folder type box appears when you click on a folder name.
- When you click OK, you will be able to access the folder you selected, but where it is might not be obvious. By clicking the Calendar button, you can open the Folder pane so that you can see the other person's calendar. The calendar of the other person appears as a shared calendar in the Folder pane.

Viewing Two Calendars Side by Side

Executives often grant assistants access to their calendars. While the executive is doing something else, the assistant can keep the executive's schedule in order. Having both your boss's calendar and your own calendar displayed simultaneously can be useful when working as someone's assistant. You can view both calendars simultaneously in Outlook if you have the appropriate permissions (rights).

When you open the Folder pane after clicking the Calendar button and opening the calendar of someone else, you will see a section labeled "Shared Calendars." Once you have opened their calendar, the list of people you have Checking a name with the checkbox causes that person's calendar to appear directly beside yours. The date can be changed by clicking on the date you want to see, just as you would when viewing only one calendar.

It may be difficult to read your screen when you have two busy schedules side by side, so you might want to switch to a one-day view to keep it readable. Click the box next to the name of the other person in the Folder pane to return to viewing one schedule.

Setting access permissions

Sometimes, busy executives give their assistants full access to their Outlook accounts and even allow them to edit them right from their desks. With this strategy, the assistant organizes what the executive does, and the executive simply goes out and gets the job done. In this chapter, "giving delegate permissions" is described as granting access permissions, which is much like appointing a delegate.

To allow someone access to your account, you must grant them permission by following these steps:

- Right-click on the account name located in the Folders list in the Mail module.
  You will see your account name on top of the Inbox icon. Right-clicking an account name displays a shortcut menu.
- Permissions for a folder are selected.
  A Properties dialog box opens with the Permissions tab.
- Click on the Add button.
  A dialog box appears. (This is actually the Global Address list.)
- You can grant access to someone by double-clicking their name.
  A box appears at the bottom of the Add Users dialog box with the name that you double-click.

- Once you click OK, the Add User dialog box closes, and the name you selected appears in the Permissions dialog box's Name box.
- In the Properties dialog box, click the name you just added to the Name list.
  It will be highlighted to indicate that it has been selected.
- To change permission levels, click the triangle on the permission level box.
- Select a permission level from the list.

  Permission levels confer a specific set of rights on the person who is assigned them. A reviewer, for instance, can only read items within your Outlook folders and can't add, edit, or remove them. The checkboxes next to the permission level boxes will tell you which rights you're granting. Those rights associated with each permission level will be marked with checkmarks.
- Having been granted permission to view your account as a whole, now you have to give permission to view each folder in the account individually. Almost every folder in Outlook can be made available to another person-even your Deleted Items and Junk Email folders if you like, but not your Contacts folder.
- You can let someone see a folder by right-clicking on it.
  An option appears.
- Select Permissions from the Properties menu.
- Steps 3–8 should be followed.

  If you are granting access to another person, you can either follow these steps for each icon in the Folders list or read through the "Giving delegation permissions" section.

It is impossible to tell whether someone has given you permission to view their data unless you open one of their folders (or unless they tell you), which prevents nasty hackers from stealing several people's passwords at once.

Looking at two accounts

You can set up Outlook so that your own folders and your boss's folders appear together in your Outlook Folders list if your boss grants you permission to view his or her entire Outlook account.

You can add a second person's account to your Outlook view by following these steps:

- Select your account name from the Folders list by right-clicking.
  A menu appears when you right-click on your account name, located above the Inbox icon.
- Select the Data File Properties option.
  A dialog box showing the General tab appears.
- Select the Advanced button.
  A Microsoft Exchange dialog box appears.
- Select the Advanced tab.
  A dialog box for Microsoft Exchange opens.
- Select the "Add" button.
  A dialog box appears.

- The username of the person whose account you want to add should be typed in.

  The username must be entered. There is no list of users to choose from in the dialog box. You will receive an error message if you do not type the username correctly or if the username you type does not exist. In this case, make sure your username is spelled correctly.

- To continue, click OK.
  In the Mailboxes list on the Advanced tab of the Exchange dialog box, the user's name appears in the Add Mailbox dialog box.

- Click OK to close the Microsoft Exchange dialog box.

- Upon clicking OK, the dialog box for Data File Properties closes.

When you add someone else's account to Outlook, the folder pane shows the new person's items. In your Folders list, you'll notice a new section called "Mailbox," followed by the username of the new person; it's where the person's mail-related items are located, including the Inbox. In the Folder pane, select Calendar. The new calendar entry will appear in the My Calendar section, followed by the new user's username; this is where the new user's calendar can be found. So it goes for each module you are permitted to view by the person.

Managing Your Out of Office Message

If going on a vacation is one of the best parts of your job, constantly juggling work emails rather than drinking tropical drinks, you may find it difficult to enjoy your tropical vacation. By turning on the Out of Office message, you can notify all your valued colleagues of your absence. This way, Outlook will automatically reply to every email sent to you based on the information you provide.

Follow these steps to enable automatic replies:

- Select the Automatic Replies (Out of Office) option under the File tab.
  A new window will appear.

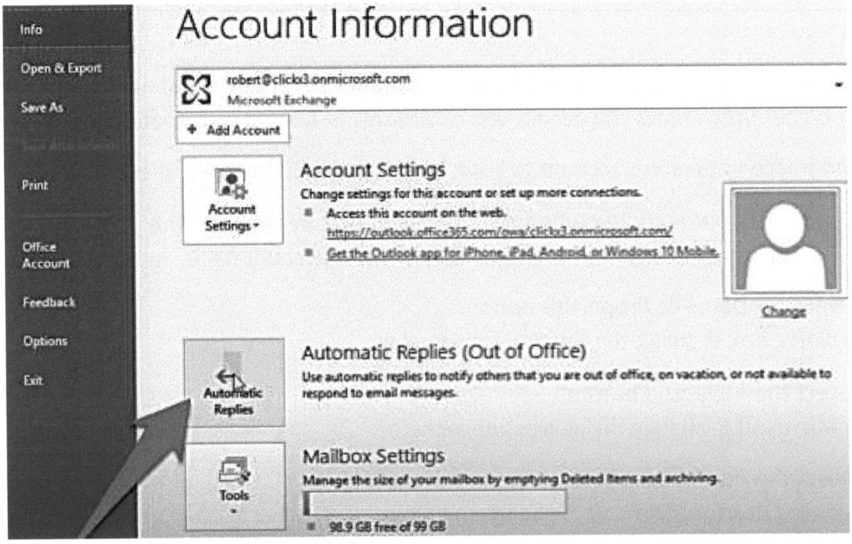

- Select "Send Automatic Replies."
  In the Automatic Replies dialog box, the gray text box at the bottom becomes white, as does the check box labeled "Only Send During This Time Period."

Basically, that tells you to enter a message that will be sent to all the people who email you while you're away. You may also specify the dates and times of your absence so that Outlook stops sending automatic replies after your scheduled absence has ended.

But this is optional. The purpose of an automatic reply is to send you a message, but leaving it blank defeats that purpose. You should fill in the blank with what you want to say.

- Then click OK.

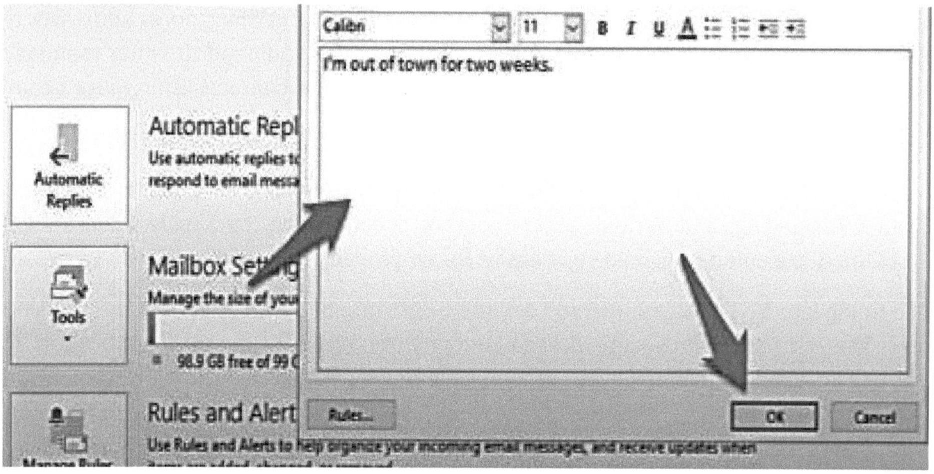

Managing Your Address Books

Besides your regular contact list in the People module, Outlook has several different address books that are actually part of the Microsoft Exchange Server. Names and email addresses are listed separately in the Address Books, so it can be a little confusing.

Contacts (what appear when you click on the People button in the Navigation bar) contain all sorts of personal data, while Address Books (what appear when you click the To button in a new message) contain only email addresses. You can also use an Address Book to send your messages to people on your corporate email system, especially if it is an Exchange Server system.

Let's take a look at your countless address books:

- The Global Address List

Outlook on corporate networks usually includes the names and email addresses of everyone in the company in the Global Address List that your system administrator maintains. With the Global Address List, you can send an email to anyone in your company without having to search for their email address.

- The Address Book of Contacts:

Email addresses in the Contacts Address Book are gathered from the Contacts list. Whenever you click the "To" button in Outlook, your contacts' address book will automatically be populated so you can quickly add people to an email.

- Additional address books:

When you create folders, you can also create separate address books for Outlook contacts. Additional address books can also be created by your system administrator.

Hopefully, you won't see your address book. In a global address list that someone else maintains, such as on a corporate network, all of the addresses of all the people you ever send email to are stored. In such a case, Outlook is perfect. Usually, you don't need to know what an address book is; you just enter the recipient's name in the To field of the message. The recipient's name will be checked for spelling, and Outlook will handle the rest for you. Your computer seems to have a tiny psychic inside who knows just what you need.

Occasionally, Outlook complains that it cannot send a message or can't determine to whom you're sending it. In this case, you have to enter the recipient's address. Outlook ignores addresses that are not in one of its address books or in a format Outlook understands. You need to either manually enter the complete address or add the recipient's name and address to your contacts list in these situations.

Scheduling a Skype Meeting

Some years ago, Microsoft purchased Skype, an online service that offers tools for conferencing, video chat, instant messaging, and various other types of collaboration. By slowly integrating Skype into Microsoft Office, the company is making it easier for employees who work remotely to collaborate.

You need to log into Skype first before you can use Skype meetings or set up Skype for Business on your computer. The Microsoft Office suite already contains the Skype for Business application. Most of the time, you will be able to sign in to Skype with your Windows credentials. However, if you cannot, contact your system administrator.

Setting up a Skype meeting

When you schedule a meeting through Outlook, you can use your Skype software as a virtual conference room. A Skype button is even available on the Meetings page. The invitation appears with a link, "Join Skype Meeting," if you click the button when setting up a meeting. Creating a Skype meeting is as easy as that.

Joining a Skype meeting

Skype meetings are easier to join than to create. Simply double-click on the appointment on your calendar when the meeting time arrives. You'll be automatically taken to the Skype program, and an audio conference will be established as soon as you click "Join Skype Meeting." Skype also allows you to conduct video conferences via the internet. Once you hear people speaking, you can join the conversation. Once your conference is over, click the X in the upper-right corner of Skype.

CHAPTER 15

OUTLOOK FOR THE IPAD AND ANDROID PHONES

Microsoft's new slogan is "mobile-first". CEO Satya Nadella has pledged to make mobile computing a priority for the company. Sales of mobile devices have surged, while sales of traditional PCs are decreasing. Millions of people rely solely on their phones for all their computing needs.

For this reason, Microsoft Outlook is a welcome addition to mobile platforms. Although the products that are now called Outlook for Android and iPad originated from another company, they don't work the same as those on desktop Outlook. These are excellent products. Their differences from desktop Outlook are in many ways an improvement. Additionally, desktop Outlook's features and functions could never be crammed into a tiny smartphone screen, so these products reduce all features down to what's really important.

Understanding the Mobile Difference

According to a Pew Research Center study published in 2011, 35% of Americans owned smartphones. Since then, the figure has risen by nearly 77 percent. With that in mind, I think it's fair to say you have some experience with using a smartphone and how it differs from a laptop or desktop computer. A smartphone or tablet doesn't come with a keyboard; you work through everything by tapping the screen with your fingers. The process is similar to finger painting but without all the mess.

Using Mobile Email

When you're on the go and only have a phone to work with, it's important that you be able to perform useful tasks on your incoming email when you're away from your desk. Messages can be sorted, filed, and marked for later action with the help of mobile Outlook versions, which allows you to process your email more quickly and efficiently. Responding quickly to emails and letting people know you are available is easy with Outlook Mobile. Or perhaps so you look like you're working at your desk while you're actually away.

Reading email

If you use Outlook on your phone or tablet, you'll most likely read your email first. You can see a similar layout to the desktop version of Outlook on the iPad. The left pane shows a list of messages, and the right pane displays the contents of a message. The content of a message is displayed when you tap the message on the left of the list.

A mobile phone running Outlook will display only the message list. The message body can be seen by tapping the message in the message list. Simply swipe your finger across the message body from left to right to return to the message list. The back button on Android devices does the same thing as well.

Replying to email

You can respond to an email message almost as easily as you can read one. An icon with the appearance of a bent arrow appears at the bottom of the message. Click on it to see a menu of reply options, and then select your preferred option.

A message reply form opens. Choosing Reply All or Reply will automatically address the message to your intended recipients. Selecting Edit recipients allows you to add or remove recipients.

Composing an email

How convenient would it be if you could send an email while sitting in a park or in a cab? It is definitely more convenient than sitting at work.

Here are the steps to composing an email message in mobile Outlook:

- You can access the new email icon from the message list pane's top-right corner.

  That icon appears on an iPad as a square with a pencil in it. A circle with a plus sign appears in the lower-right corner of the Android screen. You see the New Message form, the on-screen keyboard, and a blinking cursor in the To field of the message form.

- Put the first few letters of the address or name of the person you are emailing.

  The moment Outlook sees a name or address beginning with the letters you type, it suggests matching names and email addresses. Chances are, the recipient is already listed. Simply tap on his or her name, and the name and address of that person will be automatically added to the To box. If the name doesn't appear, type in the email address completely.

- Enter a subject in the Subject field.
  Although a subject line isn't necessary, it is recommended.

- Enter your message in the main part of the message box.

- Click the "Send" icon. The message pane's upper-right corner displays the "Send" icon. It resembles an arrow pointing to the right. Touch it to send your message.

The steps for forwarding a message look very similar to those for composing a new message, except you have to enter an email address. Otherwise, there is no difference.

Archiving, scheduling, and deleting email messages

One thing that mobile Outlook offers you that is not available on desktop Outlook is the ability to swipe left or right on a message to quickly process it. The name of the thing you're about to do appears in a colored background as you slide messages right or left.

As soon as you install mobile Outlook, you are presented with two options: schedule and archive. There are several other options you can choose from, but you can only have two active at once: one for right-swiping and one for left-swiping.

Other options include:

- Schedule:

If you swipe right to schedule a message, it will be hidden until you choose to show it again. You can choose from a range of options when scheduling a message, from a few hours out to tomorrow. It is also possible to specify a time.

- Delete:

As it says, it's over with a tap. Your mail server might allow you to retrieve it from the trash.

- Archive:

You can choose this option to direct the message to a particular folder that you've chosen beforehand. You must have selected an archive folder for this to work, or else it will not work. Thankfully, it offers to set up an archive folder if you have not already done so. From then on, all your archiving goes into that folder.

- Move:

With this option, the message is moved to a folder, but you are asked to choose which one every time.

- Mark Flagged:

This will flag the message so you'll remember to go back to it just like you do on your desktop.

- Mark Read:

Each message in Outlook is marked as read once it has been read. There may be times when you want to mark things as read without actually reading them. This is a convenient way to do it.

- Mark Read and the Archive:

This combines the two previous choices under the same name.

- No Action

Is it worth it? It doesn't make any difference.

The swipe feature will still be useful even when these settings are left in their original state-Schedule and Archive. Consider, however, which options may work best for you based on how often you use email.

You can change the swiping options by following these steps:

- Tap on the action menu button (three horizontal lines) in the upper-left corner to launch a menu.
  It is sometimes called the "hamburger button" because the three horizontal lines are similar to those on a hamburger.
- Go to Settings (the gear icon).
  You will see the Settings menu.
- Swipe through the available options.
  A screen will appear with swipe options. Sliding your finger down the menu a bit may be necessary.
- Swipe Right or Swipe Left anywhere in the Swipe Right or Swipe Left areas.
  This brings up the menu of options previously listed.

Select your preferred option.

- Return to the settings page by tapping the Back arrow at the top of the screen.
- The Settings screen can be closed by tapping the Close (X) button in the upper-left corner.

Deleting messages

The easiest thing you can do in mobile Outlook is deleting a message. A trash can-shaped icon appears at the top of the message screen. Simply tap it and your message will be sent there. The trash can will not appear if there is no message selected. According to your mail server and Outlook settings, the trash can may be automatically emptied at a certain interval, or you may have to empty the trash yourself.

Managing groups of messages

You can process and read emails much more easily with Mobile Outlook than you can send them. It is still better to compose an email on a desktop or laptop. Although it offers a clever way to handle multiple messages at once, Open the message list and hold your finger on the message you want to check. Check the box next to each message you want to process, and the box will be marked. You can delete all the messages you checked by tapping the trash can icon now.

The flag icon flags them all. You can also select other options such as moving or marking unread messages from the three-dot menu in the top-right corner of the message list on a phone.

Using Your Mobile Calendar

There is also a slick version of the calendar that you see in the desktop version of Outlook on mobile devices. various features, but just those that are most likely to be used and work well on a phone or tablet. Access the calendar by tapping the calendar button below. If you see the message "No Connected Calendars" on this device, you can't use the Outlook Calendar feature for the email account you are using. It is possible to create an additional account that does include a calendar, or you can simply use the calendar feature on iOS or Android.

Navigating the mobile calendar

On a computer screen, calendars occupy a huge amount of space; think about a month's worth of appointments. Mobile devices have such small screens that this can be a problem. Depending on the size of the device, mobile Outlook displays your calendar differently: a tablet displays your calendar one way, a phone another.

The two calendars must therefore be managed differently. If you are using an iPad, you can get to the calendar by tapping on the calendar icon at the bottom of the screen. If you are using a phone, you need to tap the hamburger button, also known as the three-line menu, in the top-left corner, and then tap Calendar.

Similarly, a tablet can show you a full, hour-by-hour view of the entire workweek, whereas a phone can only show you an hour-by-hour view of a single day. Regardless, you can swipe the calendar to see days that come before or after the day you have selected.

Creating a new appointment

A plus sign appears in the upper-right corner of the screen on both versions of the mobile Outlook Calendar. To open a new event, tap the plus sign. The settings for each line of text on the appointment form can be accessed by tapping on the line. In the form, you can adjust the date, time, location, and more.

You should choose desktop Outlook over mobile Outlook if you create a lot of appointments. The mobile Outlook app offers you a convenient way to view your calendar while out and about, but it also makes it hard to handle details on a mobile device, where you can only use taps and swipes. You'll see that entering an appointment in desktop Outlook is much faster than entering one in mobile Outlook. Nevertheless, mobile apps are updated on an almost monthly basis, so chances are it will be much simpler by the time you read this.

CHAPTER 16

TELECOMMUTING WITH OUTLOOK.COM AND THE OUTLOOK WEB APP

We're going virtual! According to experts, the number of people working from virtual offices for virtual companies is expected to increase in the near future.

Our brave new world would not be possible without the Internet. Even though Outlook works great on the web, sometimes it might not be feasible to carry around a full-blown version of it wherever you go, and you may need a little more power than a mobile version provides. However, you can access all of your emails in Outlook.com or Hotmail.com, as well as your contacts and calendar, from any web-connected computer by using Outlook.com. You can improve your results when you use Outlook.com.

Signing In to Outlook.com

The Outlook web-based email service from Microsoft is completely free. The service is somewhat similar to Google's Gmail, but it has a twist: it connects directly to desktop Outlook data. As part of the merge, Microsoft has combined Hotmail and Windows Live into a single email service. The service also now supports Facebook, Twitter, and LinkedIn contacts, as well as your calendar.

A Microsoft account is required to access Outlook.com. It's likely that you have an account that you use to sign in to Windows, for example, OneDrive or Xbox Live. You can use other email addresses at Outlook.com once you've signed in to your Microsoft account.

When you first visit Outlook.com, you'll be taken to a sign-in screen. Create a free account or sign in using your Microsoft account.

Signing into Outlook.com for the first time may display several setup screens. In addition to your language and time zone, you might also have to enter a theme color, a default signature, and your phone number. Click through the setup process to the end, entering any of those details as you wish.

Exploring the Outlook.com Interface

In terms of function, Outlook.com is identical to the desktop version of Outlook, so you won't need to learn any new tricks or techniques, but its visual appearance may differ from the desktop version.

The Outlook.com interface is available in two versions. You can enjoy ad-free, enhanced Outlook.com with an Office 365 subscription. You get a similar experience if you have a free Microsoft account, but with an advertisement on the right side of the screen.

These are some of the notable features.

There are six folders by default on the left side of the screen: Inbox, Junk Email, Drafts, Sent Items, Deleted Items, and Archive.

- To access the other modules in Outlook.com, you use the navigation buttons along the far left side of the screen. Additionally, there are four other modules: Calendar, People, Photos, and Tasks. (Note that Outlook.com does not have a notes module.)
- At the top of the screen, there is a toolbar in place of the ribbon. The commands on the toolbar vary depending on what you're selecting.
- You can choose from the following icons in the upper-right corner (from left to right):
    - Skype: Begin a new Skype conversation

- Manage Premium: This feature is only available if you have an Office 365 subscription, which opens a window that lets you work with premium-only tools and view your storage usage.
- Settings: This allows you to customize some common settings, including themes, conversation views, sorting, grouping, and the reading pane.
- What's New: It displays information about how to use the latest features.
- Help: Displays a help screen for Outlook.com.
- Account: Makes it easy to edit your profile, check your account status, or log out.

Getting Caught Up on Web Email Basics

Outlook.com lets you access your inbox from any browser, so you can read office gossip or delete spam from Nigerian oil magnates.

Reading messages

You can access Outlook.com from any computer with an internet connection since it is webmail. It is possible to access your Hotmail, Messenger, and Outlook.com mail. Most people use the Inbox as a to-do list, and Outlook.com enables this from any computer with an Internet connection.

Here are the steps for reading your messages:

- Navigate to the Inbox folder.
  You will see your messages there.
- Click on the message you'd like to read.
  The message text is displayed in the reading pane on the right side of the screen. When you click on the message name in the message list, the content appears in the reading pane.

Sending a message

Whenever you want to shoot off a quick email from your favorite Internet cafe, you can do so with Outlook.com in a flash. Your message will be ready before you know it.

The next step is to do the following:

- On the Folders page, click Inbox.
  The messages you have in your inbox will appear.
- Navigate to the Ribbon and click New Message.
  A window opens.
- Complete the New Message screen.
  Your recipient's address goes in the "To" box, your subject goes in the "Subject" box, and your message goes in the "main field."
- On the Ribbon, click the Send button.
  Now your message will be sent out.

Choosing the message's importance

When composing a message using Outlook.com, you can select a priority between "High," "Normal, and "Low." Select the Set Importance option from the drop-down menu by clicking the More button above the message composition pane.

Flagging messages

Messages can be flagged to give them extra attention. In your inbox, flagged messages will show at the top. There are no levels, and there is no dated reminder feature (like in the desktop version). If you would like to flag a message, select it and then, on the toolbar above the message, click the More button. The Flag menu will appear.

Organizing Contacts

Having access to your collection of information from anywhere is the whole point of Outlook.com, and keeping track of your contact list is essential.

Viewing the contacts

You can change the appearance of contacts by clicking the Filter icon in the upper-right corner of the Contacts list pane. Then you can choose your preferences.

Adding contacts

The Contacts list in Outlook.com can be updated from anywhere. For instance, if you attend a conference or convention and exchange business cards with many people, you should add their names right away to your contacts list.

It doesn't matter what your device is (whether it's a laptop, tablet, or smartphone), you can access your account remotely to add those new addresses before you leave where you are.

Using Outlook.com, you can add a new contact by following these steps:

- To create a new contact, click the New Contact button in the upper-left corner of the People module.
  A form will appear.
- Complete the New Contact form.
- Then click Save.
  As soon as the form closes, the name you entered in the New Contact box appears in your Contacts list.

You can edit a contact record by opening the record, clicking the Edit button on the Ribbon, and following the same steps.

Using Your Calendar

Outlook.com's calendar design is constantly improving thanks to Microsoft. In other words, what you see today might not be what you see in a few weeks, although the features might not change. For this reason, Web-based applications are both wonderful and annoying.

Viewing your calendar

Experts suggest managing your schedule for the long term, medium term, and short term. In Outlook.com, you can view your appointments in several ways, depending on what you want (or don't want) to see.

The View drop-down menu will be visible in the upper-right corner of the screen. This could be referred to as a day, a week, etc., depending on the current view.

Select the view you wish to use:

- Day shows the day's appointments.
- A week shows a week.
- The workweek corresponds to Monday through Friday.
- A month shows a month.

Outlook.com does not show your schedule details, but you can add and edit items so that you can see the big picture and then deal with the details at your desk.

Entering an appointment

Those who travel extensively probably keep their calendar on a smartphone, but the majority of people use Outlook calendars to manage their meetings and appointments. Since Outlook.com is linked to your Outlook appointments and meetings, you can check where you need to be and with whom from any Internet-capable device. From now on, you'll be able to plan meetings and lunches.

To enter an appointment, follow these steps:

- In the bottom-left corner of the screen, click the Calendar button to switch to the Calendar module.
  Your appointments will appear on the calendar.
- In the left-hand corner of the screen, click the New Event button. You are prompted to create an appointment.
- Add a title by clicking on the "Add a Title" placeholder. Put something about your appointment, like meeting Bambi and Godzilla.
- Enter the location of your appointment in the Location box. Maybe central Tokyo? Or maybe the pediatrician's office?
- A calendar name appears in the top bar of the pop-up window (such as My Calendar). Select the right calendar from the drop-down menu if you have more than one. Outlook.com supports multiple calendars.
- Click on the Start date and choose the start date. The date might not be displayed in the pop-up calendar, so click the arrows next to the month name to see the date.
- Select a start time for your appointment by clicking the down arrow on the first-time box. Select the end time in the second time box by clicking the down arrow.
- The recurrence (repeat) and reminder options can be selected as desired.
- Then click Save.
  Your Outlook.com Calendar will notify you by email when you have upcoming appointments each day. Simply click on the link to view the details.

Moving to an appointment

You can move an appointment to a new time and date by simply dragging it.

Changes other than the date and time of your appointment can be made by following these steps:

- Click twice on the appointment.
- Choose the information you want to change.
- Update your information.
- Then click Save.
  You can delete an appointment by selecting it and clicking Delete on the toolbar.

Exploring Your Options

The options in Outlook.com can be adjusted to a limited extent. Click the Settings button (it looks like a gear) in the upper-right corner of your screen to access the options.

Options can be set at two levels. When you click on the Settings button, a list of the most common and basic ones appears. The options vary according to the module.

To see more options Click View Full Settings at the bottom of the task pane, which opens the Settings dialog box. On the left side of the navigation pane, you can switch between the options for different modules the same way you can do it in the Outlook Settings dialog box.

Automated vacation replies or out of office messages

In the desktop version of Outlook, you can set an "Out of Office" message to notify coworkers when you're out of the office (or simply unavailable). Outlook.com has a similar tool called the Automated Vacation Response. It's an easy and fun way to tell everyone you're on vacation.

All you need to do is:

- At the top of the Outlook.com Mail screen, click the Settings icon.
- At the bottom of the pane, click the "View Full Settings" button. A dialog box appears.
- Go to the middle pane and click "Automatic Replies."
  An automatic reply settings window will appear.
- Enable automatic replies by clicking the "Automatic Replies On" control.
- If you choose the "Send Replies Only During a Time Period" checkbox, please enter the start and end dates and times.
- Fill out the Out-of-the-Office Message placeholder by clicking the "Add a Message Here" button.
- Save your changes.
- Click the Close (X) button.

Don't feel guilty about disregarding those emails anymore. When you get back to the office, remember to turn off auto-replies. If not, everyone will think you are still having fun.

Creating a signature

Outlook.com allows you to create only one signature, which you can include when you want. You might use a very grand and formal signature for business-not only to impress lackeys and sycophants, but also to intimidate enemies. Unless, of course, you only have lackeys and sycophants as friends, in which case you should leave it off your messages. You certainly should go heavy on the praise, Your Royal Highness!

Following these steps will allow you to add a signature to Outlook.com:

- At the top of Outlook.com Mail, click the Settings icon.
- Click View Full Settings at the bottom of the pane.
  A dialog box appears.
- Select Compose and Reply from the middle pane.
  A box appears for the signature.
- Fill in the signature field with your text.
  Format your text as desired using the tools above the box.
- You can now click "Save."
  The Settings dialog box is closed.

Understanding the Outlook Web App

As part of Microsoft Exchange, the Outlook Web App facilitates advanced features of Outlook, such as public folders, shared calendars, and assigned tasks. Many big and small companies run exchanges for this purpose. Many companies that operate on Microsoft Exchange do not offer Outlook Web Apps. If your company does, however, you can access Outlook from anywhere you have a computer: a public library, an Internet cafe, or even the local photocopy shop. A web app is not difficult; the Outlook Web App is a special website that looks and behaves quite a bit like the desktop version of Outlook. It is possible that the Outlook Web App will look different if your organization is using an older version of Microsoft Exchange. However, the fundamental features should remain the same.

Knowing when it's handy

Despite the fact that the Outlook Web App is much more powerful than the desktop version of Outlook, you may find it incredibly convenient to access your Outlook data if you find yourself in the following circumstances:

- Taking a laptop with you on a very short business trip just to check your email may not be practical.
- In cases where you need to work from home from time to time, but don't want to spend time setting up your home computer to connect to the office network,
- Whenever you want to do some simple planning and collaborate with your coworkers, use someone else's computer.
- If you're getting an email on your smartphone (including an iPhone, Android, or another smart device) and want to send a more detailed response than you can with the tiny keyboard,

Signing in and out

Open a browser, enter the address of the page that your company has set up for accepting logins to the Outlook Web App, and then enter your username and password as you would on any other website. Depending on your organization, the process may differ slightly. Ask your system administrator for details.

Using the Outlook Web App is as easy as surfing the Internet. It does not require any special equipment. All you need is your username and password, as well as the address of your Outlook Web App page.

CHAPTER 17

TEN SHORTCUTS WORTH TAKING

The following tips will help you boost your productivity, including turning a message into a meeting and resending it. You can boost your Outlook productivity with these ten accessories, including Skype, OneDrive, and an online service for backing up your data.

Also learn why you cannot have a unified inbox in Outlook as well as how you cannot create a distribution list from an email.

Using the New Items Tool

In whatever module you are in, click the tool on the far-left side of the ribbon to add a new item.

So you change the name and appearance of this icon when you change modules, so it becomes the New Task icon when you switch to the Task module, the New Contact icon when you switch to the People module, and so forth.

Alternatively, you can click the New Items tool just to the right of it to access the menu.

New Items allows you to create a new item in a module other than the one you're in without switching modules.

You might want to create a task while you're answering an email. Choose Task from the list of new items, create your task, then continue working with your email.

Sending a File to an Email Recipient

Using Outlook email, you can send a file with just a few mouse clicks, regardless of whether Outlook is running.

If you're using File Explorer to view your files, you can mark any file for sending to any recipient. Here's what you need to do:

- Use File Explorer to locate the file.
- To send a file, right-click on it.
- You are presented with a menu.
- Choose the recipient.
- A new menu appears.
- Select the recipient of the mail.
- There is a form for new messages.
- The attached file is represented by an icon in the attached box.
- Include the subject of the file and the email address of the person you're sending it to.
- Adding a comment to your message is as simple as typing it in the message area.
- Click the "Send" button.
- The message is delivered to the receiver.

Sending a File From a Microsoft Office Application

Office documents can be emailed directly from the application itself without using Outlook.

Follow the steps below to achieve that:

- Click the File tab while an Office document is open in the application that created it.

- Select Shares
- You will see the share dialog box.
- Under Attach a Copy Instead, choose Word Document or PDF.
- Depending on the share, choose whichever format is most appropriate.
- Choose Word if you want the recipient to be able to edit the document easily, and PDF if that is not what you want.
- Outlook shows a new message form.
- Send the file to the recipient's email address. The subject of the file should be the person's name and email address.
- Under the icon for the file, there is a text box where comments can be entered.
- Click on the "Send" button.
- The Outbox receives your message.

Turning a Message Into a Meeting

Occasionally, after exchanging dozens of email messages about a topic, it would be faster to talk. Creating a meeting from an email message is as easy as clicking the Meeting button on the Home tab (in the Inbox with the appropriate

You can then create a meeting based on the contents of the email by clicking on the New Meeting button.

Finding Something

You can accumulate a lot of items in Outlook in no time, which can then take a while to search through when you are looking for one specific item.

Outlook can help you find items quickly if you type the name of the item in the search field at the top. That launches a quick search, so you can find what you are looking for in no time.

Undoing Your Mistakes

It's time you learned about the Undo command if you didn't already. The Ctrl+Z shortcut key can be used to undo accidentally entered text, as can the Undo button in the upper left corner of the screen in the Quick Access Toolbar.

You can experiment without worrying about the consequences; the worst thing you can do is undo it! It is best if you go back and fix your mistake right away before doing so many things.

Using the "Go to Date" Dialog Box

Any calendar view can be accessed using the "Go to Date" dialog box. You'll find it under the "Go-To" group on the Home tab, under Properties. You can also use Ctrl+G as a shortcut.

Adding Items to List Views

You can add an item to a list at the top of most Outlook lists by typing something into the blank field. Simply click the "Add a New Task" button to begin.

Your new item will be entered into the field once you click on it.

Sending Repeat Messages

Since you might send out one or two messages repeatedly, store them as Quick Parts to save time.

The steps you should follow when finding an Outlook accessory vendor online are:

- Your email message should be addressed.
- The email address of the company appears in my browser.
- Select the Insert tab.
- Click the Quick Parts button in the message body.
- Your saved AutoText item will appear.
- Make certain changes to reflect the name of the product.
- Click on the "Send" button.

Your request can be sent in less than 30 seconds and you can then move on to the next task.

Using this feature requires you to first store text blocks in Quick Parts:

- Choose the text you want to repeat in an email message, appointment, contact record, meeting, or task.
- Select the Insert tab.
- To access Quick Parts, click the Text group button.
- Select "Save Selection to Quick Part Gallery."

You can organize Quick Part text into groups according to their purpose. You could, for instance, generate the text for introductory messages and closing messages for different types of messages and then store them.

Resending a Message

When someone forgets to do something you asked them to do, sometimes you need to remind them.

A new message could be written, telling that person how often you've reminded him or her already.

However, this is quicker and easier:

- Open your Sent Items folder.
- Go back to the message you sent last time and click it twice.
- Decide what action to take.
- You can resend this message by choosing this option.
- Additionally, you could add the following: "Here is another copy in case you didn't receive the first."

CHAPTER 18

TEN ACCESSORIES FOR OUTLOOK

Although Outlook can do a lot by itself, a few thoughtful accessories can make things even easier. Many of my favorite accessories compensate for many of the features Outlook is missing. I also use some of my favorite accessories to access my Outlook data wherever I am.

Smartphones

Smartphones are ubiquitous today, and they are probably Outlook's most powerful accessory.

A smartphone is a cell phone with an integrated personal organizer. If you haven't looked for a new cell phone lately, you should.

Smartphones based on Android and the iPhone are right now the top choices.

Despite the ease with which I can enter and manage data with Outlook, I always have my most important Outlook information with me on my smartphone.

My smartphone even allows me to read my email on the subway (an experience I wouldn't have with a laptop).

A Tablet Computer

The tablet is rapidly becoming an important part of many people's lives. While the Apple iPad is the most well-known and most popular tablet, there are several Android tablets on the market as well. I have an older iPad 2, and it is perfectly sufficient to run Outlook on the go.

The larger screen of a tablet makes reading email more comfortable, which is an advantage for people who receive dozens of emails each day.

Meanwhile, a tablet's lightweight and convenient size allow you to easily scan your email anywhere you are, from coffee shops to restaurants to limos.

E-Learning

With computers, everything changes so rapidly, it's almost impossible to stay current.

It is imperative to keep learning constantly, but how can one find instruction? Search online for resources such as this book you are reading. You can also visit YouTube to watch video clips.

Whether you want to become proficient in Microsoft Office, social media, or even such impressive subjects as calculus, you can learn at your own pace. By spending enough time on the internet, you can become incredibly smart.

Microsoft Office

The first version of Outlook was included in the Microsoft Office 97 suite.

Microsoft offers Outlook as a stand-alone product in some cases (or in conjunction with Internet Explorer), so you may not always enjoy all the benefits of using Microsoft Office and Outlook in concert.

It's easy to send outgoing emails and graphics with Office, while Outlook makes it simple to share your work via the web and email. I would advise using both if need be.

A Business-Card Scanner

With business card scanners, the information on business cards you collect at meetings, trade shows, and conferences can be copied into Outlook with ease. While you can enter all the information manually, using a business card scanner can save you lots of time if you collect more than a few dozen cards per week.

Online Backup

You can choose from several good online backup services at an affordable price, including Mozy and Carbonite. You can restore your information if your computer crashes or if you suffer a fire, flood, or another disaster that destroys your computer. In order to use any of these services, you will need a fast Internet connection. Their monthly fee is worth every penny, and the peace of mind is priceless.

Skype

Skype can be used to host the online meetings you're likely to plan in your Outlook Calendar using an easy-to-maintain virtual conference service.

There's even a button on the ribbon that you can tap to launch a Skype meeting (if you're using Office 365 for business). If you often meet with colleagues who work from home or from different locations, Skype can make your life a whole lot easier.

Microsoft SharePoint

Microsoft SharePoint was most commonly used in large organizations that needed ways of sharing information and collaborating smoothly. Private users and home businesses found the program too complicated and expensive.

In today's world, anyone can subscribe to Office 365 and thereby obtain SharePoint. The price depends on how much service you want and what level of service you want.

You may want to investigate SharePoint as a tool for sharing documents and other information if you regularly collaborate on business projects with a team.

Microsoft Exchange

There are many business-oriented features in Outlook that require a program called Microsoft Exchange to function, such as shared calendaring. The Exchange server allows you to share Outlook data and office tasks with others in your organization.

In Office 365, you can rent Microsoft Exchange accounts. Fees vary depending on the number of optional features you select and the size of the workforce in your company.

OneDrive

File sharing is made easier with OneDrive, a service from Microsoft.

As long as you have a Microsoft account, you have a OneDrive account; both accounts are connected. Sign in to onedrive.com using your Microsoft ID to access the service. The OneDrive icon in the navigation pane of Windows 10 also gives you access to OneDrive.

If you have OneDrive's settings checked, you can determine which files and folders are cached locally, so you can use them even when you are not online. (To access OneDrive's settings, you must right-click on the icon on the taskbar and choose Settings.)

CHAPTER 19

TEN THINGS YOU CAN'T DO WITH OUTLOOK

When you first get Outlook, it won't be able to do any of these ten things.

The geeky programmers tell me you can reprogram Outlook to accomplish many of these tasks using Visual Basic shortcut macros.

In addition to being outside the scope of this book, it's also something that normal, sensible people would not do.

Create a Unified Inbox

There is nothing unusual about having multiple email accounts. People often separate their business and personal accounts.

Most people today use email on two or more devices at once, typically a personal computer and a mobile phone. If you use IMAP, the kind of email made for sending across multiple devices, you can't create a unified inbox on your desktop.

The iPad and Android versions of Outlook offer unified inboxes but not the large desktop client. It would be very useful if it happened, but it doesn't in this version.

Adding a Phone Number to Your Calendar

If Outlook could be configured to automatically look up the phone number of the person you're meeting and insert that number into an appointment record, that would be perfect.

Outlook does not have an address lookup feature like many smartphones. Perhaps another time.

Open a Message From the Reading Pane

Many people use Outlook to keep a record of everything they do and everything they send and receive.

From time to time, you may scroll through your messages again and again to keep track of who you've spoken to and when.

In a very long list, if you select one message to be displayed in the Reading pane and then scroll through the list, you cannot right-click on the Reading pane to open the message you're viewing.

Right-clicking the message in the Reading pane doesn't seem to be too difficult for Microsoft; however, it isn't included.

Performing Two-sided Printing

Printing out your schedule and keeping it in a binder makes it look exactly like a timetable in an old-fashioned planner.

Outlook does not understand how to reorganize printed pages in accordance with whether you view the page on the left side or the right side of the book when you print it.

If this is an extremely minor quibble, but if it's important to you, you'll have to accept one-sided printing.

Searching and Replacing Area Codes

Outlook cannot automatically change all 312s to 708s; you must do that individually. There is a Microsoft utility for changing Russian area codes, but not for changing area codes in the United States!

Printing a List of Meeting Attendees

While preparing for a large meeting organized using Outlook, especially a conference call, it's handy to maintain a list of attendees.

It is possible to keep the meeting item open on your calendar, but you cannot do that if you are presenting at the same time.

Enlarging the Type in the Calendar Location Box

Today, most conference calls are organized in Outlook, and participants can browse the location box on the calendar form to enter their dial-in numbers.

In a hurry after finishing your last conference call, those teeny-weeny numbers are tough to discern, especially when your eyes aren't as sharp as hawks. My recommendation is to use a magnifying glass.

Creating Contact Records for All Email Recipients

If an email is sent to a large number of people, you can create a distribution list from it by copying them all to a group. Dragging a message from one person to the People icon will also create a new contact record for that person.

To create contact records for a group of people, you must create a record for each individual—no dragging and dropping or copying and pasting is allowed.

Tracking Meeting Time Zones

In many cases, Outlook is used to coordinate Skype meetings and conference calls among people in different time zones. In the past, I've often had to schedule phone calls between London, Sydney, and my home in New York.

Using the scheduling tool in Outlook Calendar, users are able to see each other's working hours-if they've set them up but it does not display what time of day it is where they are located.

When arranging a meeting whose timing happens to be awful for someone, it's helpful to know just how awful it will be.

Then, you'll be able to make it a bit less awful.

While there are websites that help you understand the time in various time zones, they do not provide availability information like the one you get in Outlook.

It is therefore imperative to make one guess at a time and to apologize when you make the wrong one.

Easily Backup Outlook Data

Although Outlook has been in the market for over 20 years, it has never been given an adequate means of keeping its data safe.

It's well known that you should always back up all the data on your computer, and you can make copies of your critical Outlook data (some of those tiny memory keys can do the trick, and you can download Outlook data to a handheld computer if necessary), but it's odd that no such feature has ever been added to Outlook itself.

You can stay safe if you get your email service from Microsoft Office 365, which stores all Outlook data safely in the cloud.

CHAPTER 20

TEN THINGS YOU CAN DO AFTER YOU'RE COMFORTABLE

Take Notes

You can easily type a quick note in Outlook's Notes module, and then go on with what you were doing. Ctrl+Shift+N opens a note, you type some text, and then you press Esc to close it.

In the Notes module,

- Click the Notes tab to open the full list of notes.
- To read a specific note, double-click the notes tab button.

As a feature, it dates from Outlook's early days, when it emerged from the primordial ooze.

Customizing the Quick Access Toolbar

With Office 2019/Office 365, there are no menus and instead a ribbon, tabs, and button arrangement of controls.

Though it is an excellent way to get a variety of commands into a small space, I sometimes have trouble understanding how to do many of the tasks I would like to accomplish.

When I find the command I need, I'm often unable to remember how I got there when I need to use it again. It's not all doom and gloom. After you've located the tool you need, you can right-click it and choose to add it to the quick access toolbar.

In that way, a tiny icon is added to that thin strip of icons that sits just above the Ribbon. The Quick Access Toolbar is kind of like a bookmark bar for the commands you want to keep track of, just as you would bookmark websites in your web browser to return to them later.

Every form in Outlook has its own Quick Access Toolbar, which is customized separately from the others. This is useful if you perform some tasks frequently using certain forms only. If you like to print individual email messages from time to time, for example, you can add the Quick Print command when you're reading or composing a message.

Wising Up Your Messages with Smart Art

Designs have the power to make any person look smart if they know what they're doing.

Smart Art is another intriguing feature on the Ribbon's Insert tab that you can use if you don't know what you're doing. With Smart Art, you can add colorful, annotated infographics to your emails.

Follow these steps to learn more about Smart Art:

- Open a new message composition window and click the Insert tab.
- Select the SmartArt button.
- Get a feel for a few designs.

Your Email Messages Should Be Translated

Perhaps your incoming emails are written in a foreign language if they seem so unclear.

You can translate incoming email messages in Outlook this way:

- Select some foreign-language text in a message.
- Simply right-click the selection.
- An options menu appears.
- Select "Translate."
- If you prefer, you can click the Translate button on the Review tab on the Ribbon of the message window.
- The research task pane will be displayed.
- Choose the From and To languages from the menu that is displayed in the task pane.
- Go through the translation in the lower part of the task pane.

If you cannot make sense of what has been translated, you will then know it is not your fault as you have tried your best.

Including Impact Charts

A chart tool is located just beneath the SmartArt button on the Insert tab of the Ribbon.

Your email will appear positively ordered once you use this tool (regardless of how disordered your mind may be).

Chart it up with these steps:

- Choose the chart tool under the Insert tab of your new email message.
- The chart gallery has two parts: a list of general chart types on the left and examples of each type on the right.
- The left side of the page offers a list of general types.
- On the right, you can select a specific type.
- Select OK.

In Email, Use Symbols

To add symbols, such as the euro currency symbol, to your email messages,

- Simply click the Symbol button on the Insert tab while writing an email message
- Select the symbol you want.

If you choose more symbols, you can also insert such clever things as fractions, arrows, and strange hieroglyphics to baffle your recipients into complying with your wishes.

Opening a number of calendars

You can create more than one calendar in Outlook. You might want to do so to track the activities of more than one person or to keep your business life separate from your personal life (which is always a good idea).

The tricky part about keeping multiple calendars is dealing with conflicts between the two. To see two calendars at a time, click the check box next to each calendar name in the navigation pane.

Superimpose Calendars

In Microsoft Outlook, viewing multiple calendars together is quite easy.

- Click the calendar in the navigation pane to open the calendar view.
- Select all the calendars that will appear together in the Navigation Pane.

Microsoft Outlook will then automatically display all checked calendars side by side.

Selecting Dates as a Group

When viewing a range of dates, it's not necessary to limit your view to specific days, weeks, or months. Let's say you would like to analyze a range of dates from September 25 through October 5.

- Click September 25 in the To-Do bar, then (while holding down the Shift key)
- Click October 5, The entire period is selected and appears in the information viewer.
- The Date Navigator will appear once you click on the Folder pane.
- To test it, open the calendar that appears in the left corner.

Pin a Contact Card

The easiest way to view someone's contact information with the message you're reading (in the reading pane or its separate window) is to

- Right-click on a person's email address
- Choose "Open Contact Card." An image of a pushpin is located in the upper-right corner of the contact card.
- Click the pushpin, the contact card will float on the screen
- Click it again to make it disappear.

Have you ever pinned a contact and it's no longer there? No worries. It's probably because Outlook is in its default full-screen mode. A full-screen Outlook screen hides the pinned contact when you click it off. Right-click on the Outlook screen and select Restore Down or Minimize. The contact should now appear.

CONCLUSION

With Outlook you'll get a feature-packed email client with a higher learning curve, but a lot more options to customize your email. Despite the fact that you may use Outlook every day, you might not know all of the cool things you can do with it.

Microsoft Outlook lets you send and receive an email, manage your calendar, store your contact information, and track your tasks. Most people think of Microsoft Outlook when it comes to email, but it has many other features that may be useful to you.

Download and install Outlook today and begin making the most of its features.

BOOK 4

MICROSOFT ACCESS

INTRODUCTION

This book is centered on how you can use Microsoft Access. You might be wondering what Microsoft Access is. It is a relational database program and as you go through this book, you'll find out more about what that even means; but at a very high level you can use Microsoft Access to track customers, orders, or assets, and the list goes on and on of the types of things that you can use or track with a database.

Before utilizing this article, you need to have a basic working knowledge of Microsoft Windows for ease of comprehension. You should be able to operate a computer keyboard and mouse, start applications, maximize and minimize windows, use scroll bars, and understand the backspace and delete keys, among other things. Knowing how to use Microsoft Word and Excel before learning Access is beneficial, as it uses many of the basic concepts taught in Word and Excel. When it comes to studying Access, having a fundamental understanding of the ribbon, editing and formatting text, utilizing the clipboard, and working with rows and columns in a spreadsheet will help.

To get the most out of this book, I recommend relaxing and reading it without trying to do anything on your computer at first, then going back to the beginning and following along with the examples. Don't try to apply what you're learning to other projects until you've mastered how to design a simple database, and if you get stuck or don't understand something, go back over that section or chapter and practice as much as you want until you do.

Most importantly, keep an open mind. Access may appear scary at first, but once you get the feel of it, you'll find that it's simple to use, especially if you follow the steps in this book.

CHAPTER ONE

GETTING STARTED

Welcome to Microsoft Access. In this chapter, we'll go over the basics of Microsoft Access, making sure you know what it is, what it is used for, how it works, and how you can use it. We'll also go over the various components of the Microsoft Access interface and other information needed to get you started.

A Quick Overview

It's not about generating data anymore; every business is generating data. Instead, you need to take control of the data you have, you need to organize it and display it so powerful insights can emerge. That's why there's Microsoft Access.

Access is designed to work with **Office 365** and the new world of cloud-based data so you can see it anywhere. Use templates to build databases faster and create web apps in minutes; everything you need is there, such as **Tables**, **Queries**, **Forms**, **Reports**, and more to help you add data.

Drop-down menus and autofill recommendations make data entry faster and more accurate, and by creating a **web app** you can share your data securely and on any device. In a web app, your information is automatically stored in a sequel database so it's more secure, reliable, and scalable than ever. New database templates have an attractive easy-to-use interface and offer a consistent user experience so you can stop worrying about your users getting lost in the numbers. With multiple user access and permission settings, you can manage and monitor your Access web apps through a SharePoint site.

What Is Microsoft Access?

Microsoft Access is a **database management software** that allows you to **store** an incredible amount of information and then you can **create relationships** between that information. So, this makes it easier to **find** or **sort** what you're looking for and it makes it simpler to use.

Why Choose Access Over Excel?

Now one of the first questions you might have been: why would I use something like Access? I could just enter information into an Excel spreadsheet; why would I ever need to use a database? What you might find after you start to learn about Microsoft Access is that some of the things, you're doing in Microsoft Excel might be best suited in Microsoft Access.

In Microsoft Excel, you can track information here as well. Let's say you want to track orders for a company, you have customer information on the left and the order information over on the right-hand side; so, this works too. Well, it's not that efficient at tracking this information.

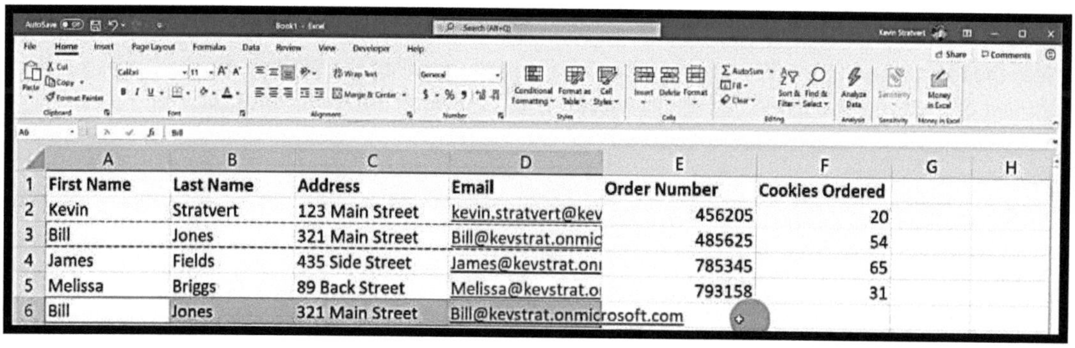

Let's say that a return customer places another order and you want to enter this information. You'll have to enter another row with his new order, and that means you'll take all of the customer information, paste it down in the new row, and then enter in the additional order details. This is what's referred to as a **Flat File**; there are no relationships between this data and it's inefficient, as you've had to copy the customer's data from up there and you've had to paste it in again. So right now, you're carrying this data twice, but with a database, you don't have to do this. You can set up different tables for the customer information and a different table for the order information. Also, with an Excel spreadsheet you're limited to just 1 million rows of data so if you have more data than that, well unfortunately Excel won't be able to handle that.

Also, if you go back to all of the order information, and you want to extract interesting insights from this data, like writing a query, you can do a little bit of that with Excel. You could filter the different columns, insert a pivot table, try to manipulate the data to get an interesting view, but sometimes you just need to be able to write a query and then pull together a report and unfortunately, that's not that easy with Excel; so that's where databases win out.

Microsoft Access is a **fantastic tool for individuals or for small businesses that need to track things**. However, if you're a midsize company or even a large company, you'll probably start to realize some of the limitations of Microsoft Access. So, individuals and smaller businesses can use Access just as a quick and simple database, but if you're a larger company you'll probably look at other options like **Microsoft SQL Server** and other solutions that are more scalable. However, Microsoft Access is a fantastic tool to learn the fundamentals of database design.

How It Works

To learn the fundamentals of database design, we will begin by going through the Access Workspace. Here, you'll learn how to build a database for yourself or your business.

You start by **creating tables**, then you'll go ahead and **create an order entry form**, this way other people can come in and add data to your database, then you'll see how to **write a query** so you can extract interesting insights from your data and then at the very end, you'll **create a report** so you can share some of those insights with others.

With all that said, by the end of this book you should have a pretty good working knowledge of how Microsoft Access works.

Getting Familiar with Access Workspace

Before you jump into Access, it's important to familiarize yourself with the interface to know where everything is, and we're going to take a quick tour of the different areas of the screen that you'll be working with, and if you're familiar with Microsoft you'll understand how these tabs and ribbons work.

Near the top of the window is a large toolbar area, and this is called the **Ribbon**. The Ribbon is a collection of most of the tools you'll need to use in Access, and it's divided into different tabs.

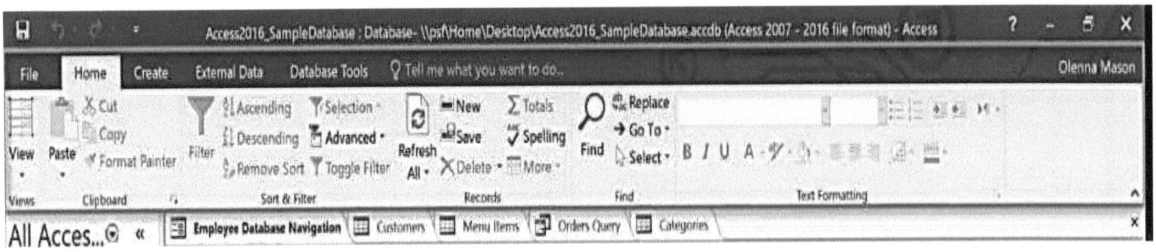

Just above the Ribbon is another toolbar, called the **Quick Access Toolbar**. Here, you can **Save** the current object, and you also have the **Undo** and **Redo** commands. And you can click the drop-down arrow to add more commands to it if you.

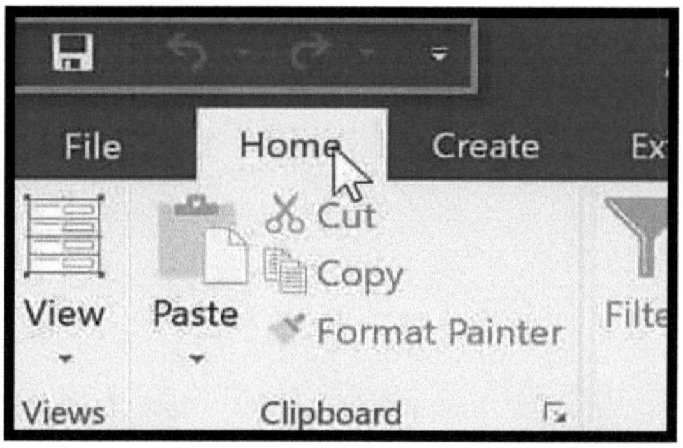

On the left side of the screen, there is an area called the **Navigation pane**. This contains a list of all of the objects contained in your database. All the objects that are currently open will appear on the **Document Tabs bar**, and you'll use this to select which object you want to view or edit.

You can resize the Navigation pane by dragging the right border, and you can minimize it by clicking the double-arrow in the upper-right-hand corner. You can also hide the groups that you're not working on by clicking the bar at the top of each group. For example, you can hide Categories, Products Table, and Sales Unit, so you can easily focus on other tables such as Customers and Menu Items.

At the bottom of the screen, you can use the **Record** Navigation bar to navigate through the different records one at a time. And to the right of that is the Record **search box**, which you can use to search for a specific record.

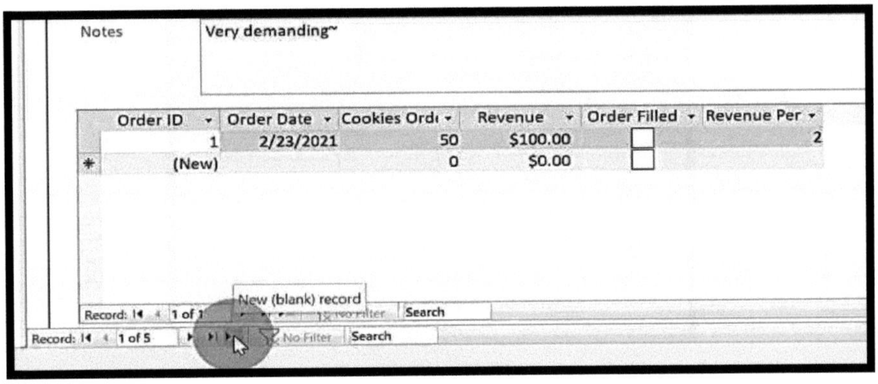

Now let's go back to the Ribbon and look at it in a little more detail. The Ribbon is designed to make things easy to find, so each tab is divided into **groups**. For example, here in the **Forms group**, we have several different commands for working with forms.

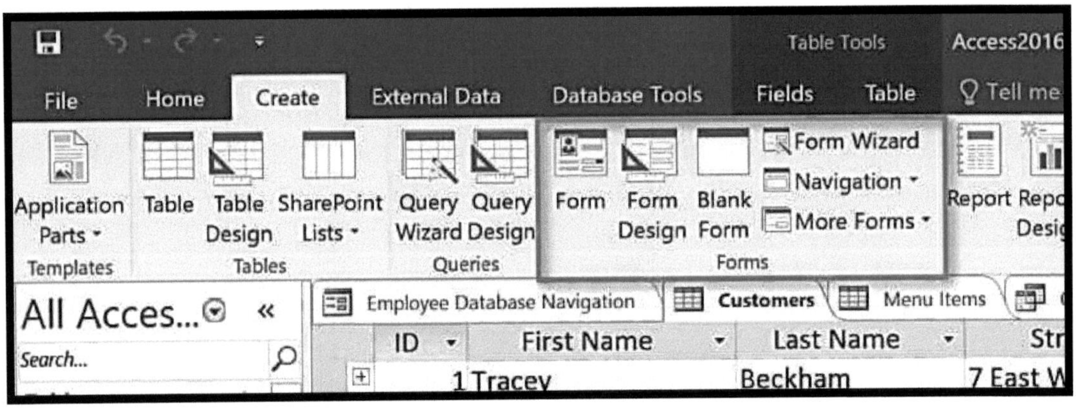

While you're working on a database, you may notice that some extra tabs appear on the Ribbon with a label above them. These will appear automatically depending on what you're working on. When you try to create your first blank database, you have a table that is opened with it. This table hasn't been saved yet but you would notice that at the top it will open up a couple of tabs such as the "**Table Fields**" and "**Table**" tabs, to help you work with tables. As you open up different objects, you're going to notice that this change.

Near the upper-left corner, you'll see the **File** tab. If you click on it, you'll go to the **Backstage view**. This is where you'll find the **Save** options and you can **Open** your databases from here. You can also **Print reports** or **view a Print Preview** and do all the different things that are very similar to Microsoft Word or Excel. To get out of the Backstage view, you can just click the arrow and it takes you back to the workspace.

With the **Home** tab, you can see that there are different tools. If you click on **Views**, you have the **Datasheet** and **Design** Views. If you go to Design View, you can see it's asking you to save your table and you can name your table here also. After that, you will see the view has changed.

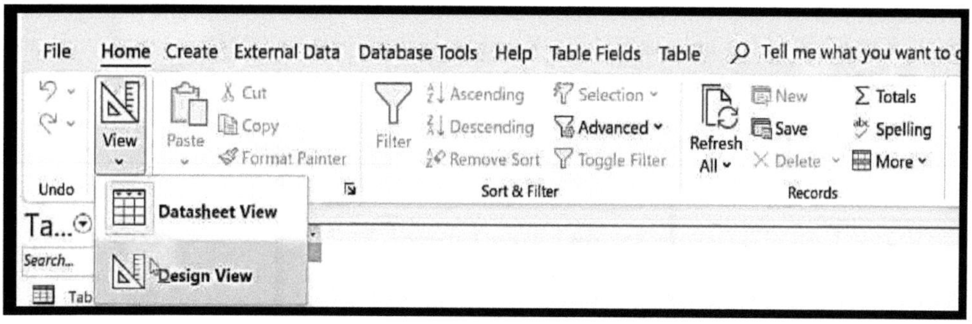

You can switch your views and go back and forth between these views and you'll get to see the differences to those later.

Database designs will often set up a special kind of form called a **Navigation form**. This will appear automatically when you open the database. Instead of using the Navigation pane and the Document tabs bar to navigate through different objects, you'll use tabs within the navigation form. This gives you a much more user-friendly way to work with the database. For example, the employees at a firm only need to work with the Orders, Customers, and Menu Items. They can easily get to these by using the tabs, and some of the tabs will also have sub-tabs on the right. By using the navigation form, you can easily focus on the objects you need without worrying about all of the other objects in the database. It's a great feature, but again, it's something that needs to be set up before you can use it. So those are the basics of how to get around in Access, and in subsequent chapters, we're going to talk about how to open and save databases and objects.

If you take a look through the home tab and check out the ribbons you get to see everything from "**Sort and Filter**" to "**Records**", you have "**Find**" and then "**Text Formatting**", so if you are familiar with Microsoft products some of these looks similar. It can be overwhelming when you see a new program like this but we are going to break it down for easy understanding.

We also do have the **Create** tab. If you take a look at the ribbon this is where you can create tables; you can create more than one table and have those connected through relationships. We'll get to that later. You can also see **Queries**, **Forms** and different parts of them. This tab is focused on all the different things that you can create. You also have **Reports**, **Marcos** and **Codes**. As we proceed, you're going to see how to create Tables, Reports, Forms, and Queries.

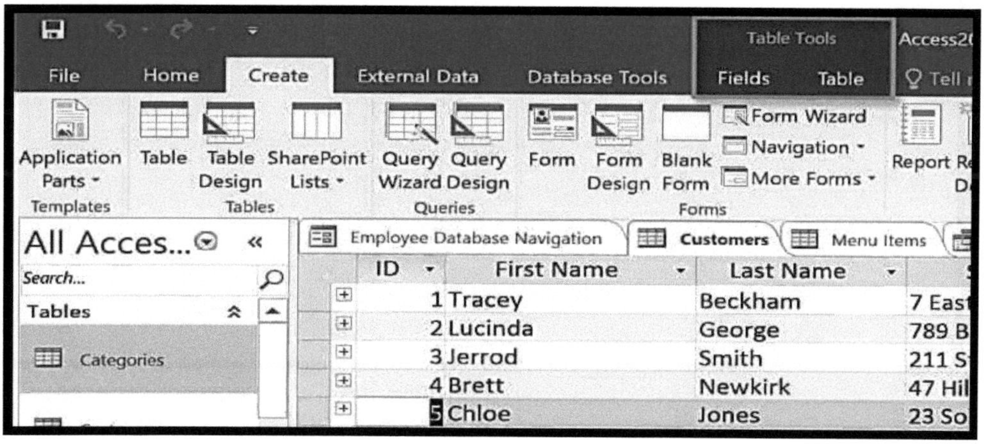

When you start working with databases, you'll use the Navigation pane to open and manage your objects. By default, it **sorts** all of the objects by type. If you want to sort them a different way, you can **click the bar** at the top of the Navigation pane and then choose which sort method you want.

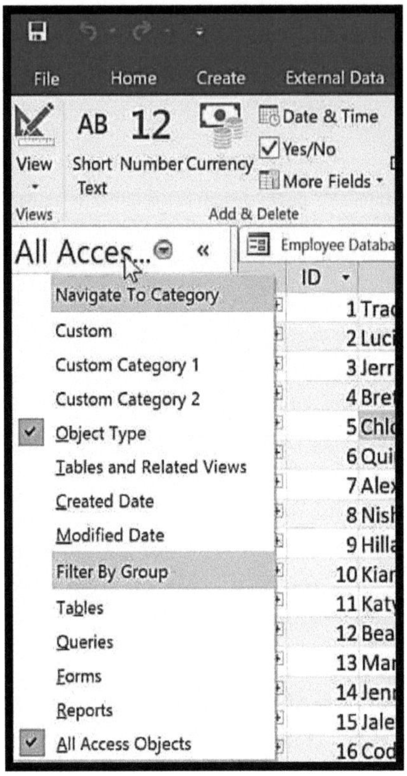

The next tab is the **External Data** tab. With External Data you can **import** something like Microsoft Excel and if you have a contact list with all their information, you can bring that into Microsoft Access. As we proceed, you'll see how to do that as well.

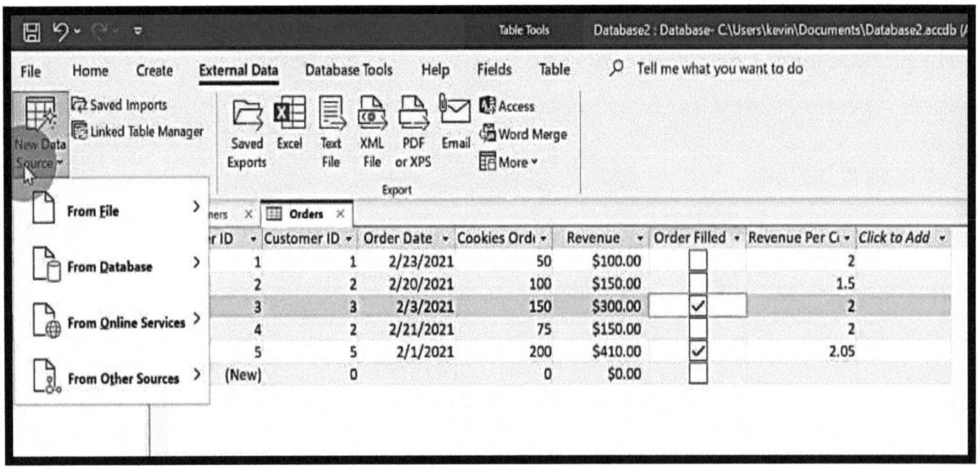

We also have the **Database tools** and this is where you can build your relationships, object dependencies and perform order functions such as **Analyzing** and **Moving data**.

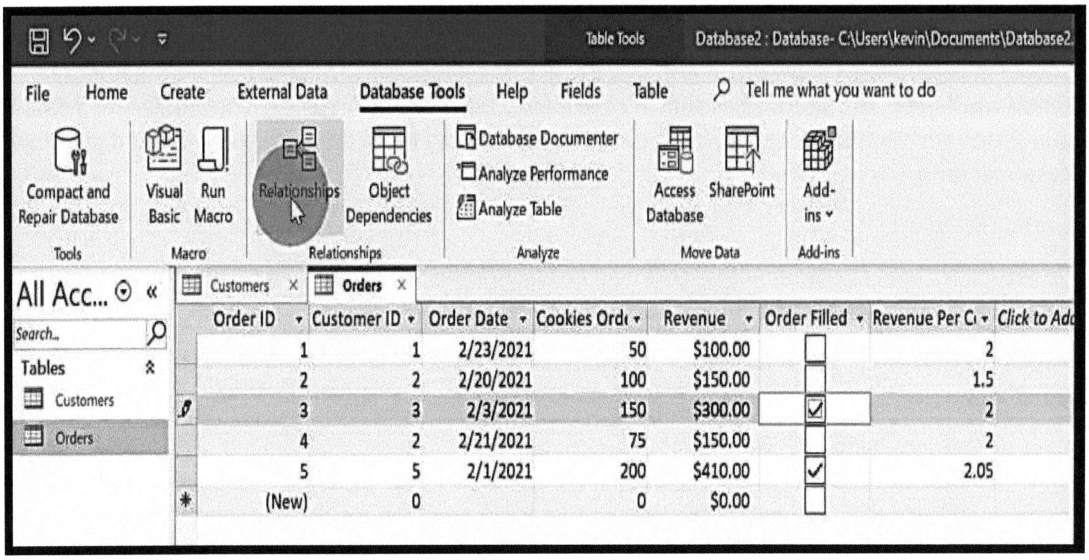

CHAPTER TWO

DATABASE AND TABLES

You're going to learn about the database terminology in this chapter. You'll discover what a database is and what components make up an Access database. We'll talk about how to plan your database, which tables you'll need, and which fields should go in each one. You're going to learn how to create a database and create your first table. You'll also know what primary key fields are and how to enter data into your customer table.

Getting Started

To get started with Microsoft Access go ahead and launch the application and that'll drop you on the start page or right here, what's also referred to as the Home View.

Right here at the top, you can create a new database. You can start with a blank database or work with templates.

Over on the right-hand side, you also see a whole bunch of different templates. If there's a template that matches what you're trying to do, this could help you save a little bit of time and you can go in and just tweak it. There's also a massive collection of templates that you can look at by clicking on the "**more templates**" button.

Below that, you can search for different databases you might have worked on in the past and here you can see all of your recent databases. One nice feature is that when you hover over right here, you'll see a pin icon and when you click on it, it will be added to the Pinned view. This is just a quick way to get back to the files or the databases that you use most often.

If you're just starting, then you go to **create a blank database**. Click on that option and before you save it, give it a name.

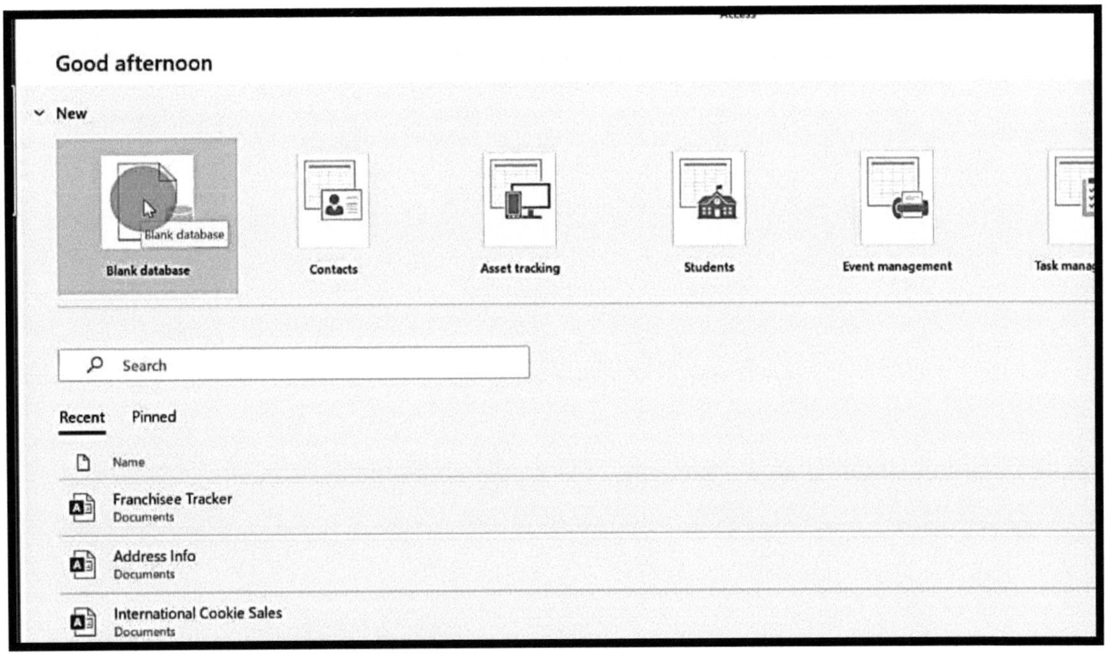

The next thing is for you to select a location quickly for where you want to save it by clicking on the Browse option. When that pops up, you can go then choose whichever place you want to save it. You can also leave it in the default location right there in Documents and all you have to do next is click on **Create**.

Now you have your first database open and it will automatically try to ask you to make a table. Now, we're going to look at tables.

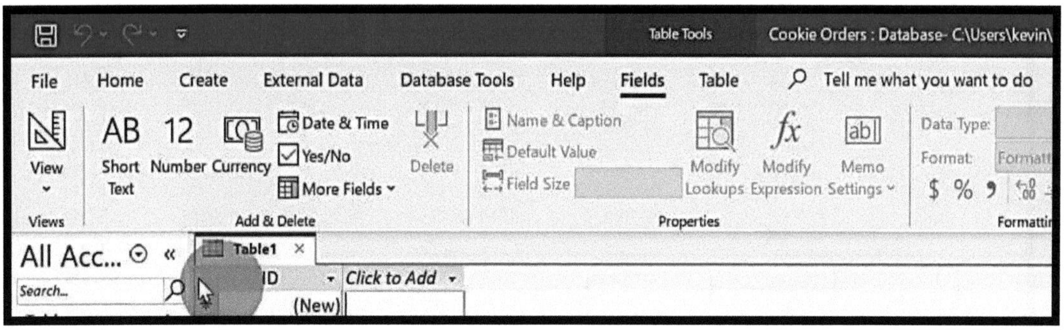

This is an important part of a database, so you must know what a table looks like and its components to help you in building yours easily.

Tables

Tables are the heart of any database because they are where the actual information is stored. In the Navigation pane, double-click on the table that you'd like to open, and tables will be marked with a blue icon next to the name. If you've used Excel, or another spreadsheet program, then you'll probably find tables to be pretty easy to use, and they are very similar, but there are some different terms that we use when talking about an Access table. Each row is called a **Record**. In a table, every record contains a customer's name, address, and other information. Each record will have a **unique ID number**, and this number cannot be changed.

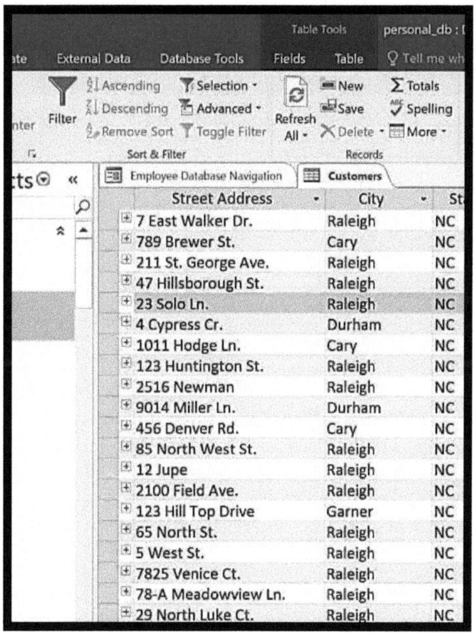

Each column is called a **Field**. And each field contains a different type of information. For example, in the table below we have fields for the Street Address, City, State, and Zip Code.

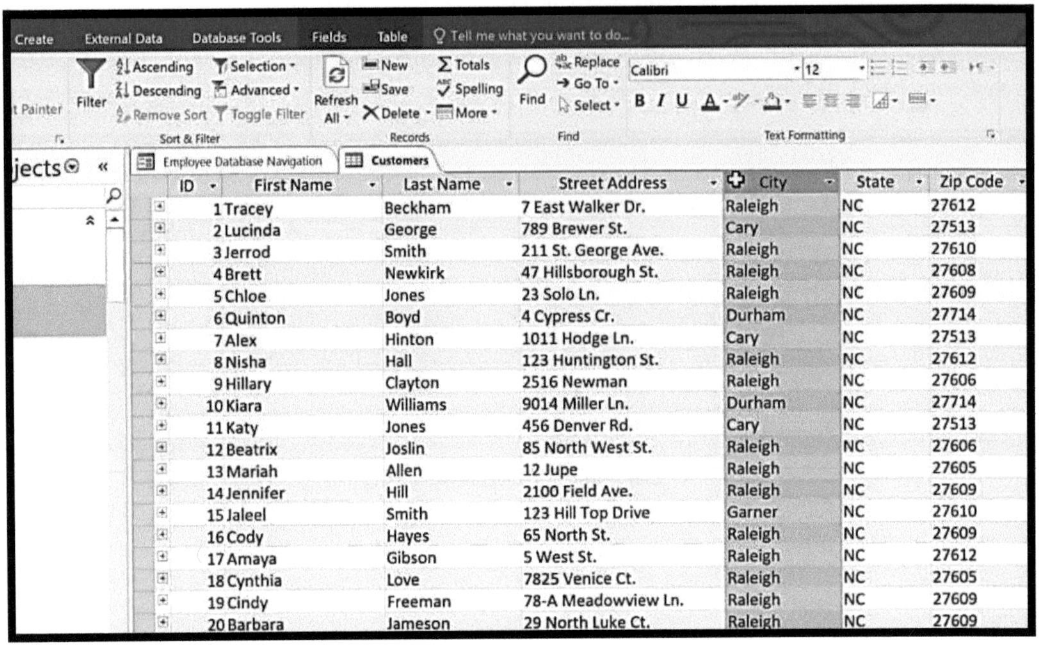

Field Data Types

We have different fields and different types of data that are stored in those fields. The most common type of data that you'll store in them is **Short Text**. So, when you've got anything that's limited to 255 characters such as a surname, a job title, a phone number, now some of you think your phone number that's like a number but we'll come back to the phone number and why we store it as a text; but basically, for anything that's limited to 255 characters or less you're mostly going to use short text.

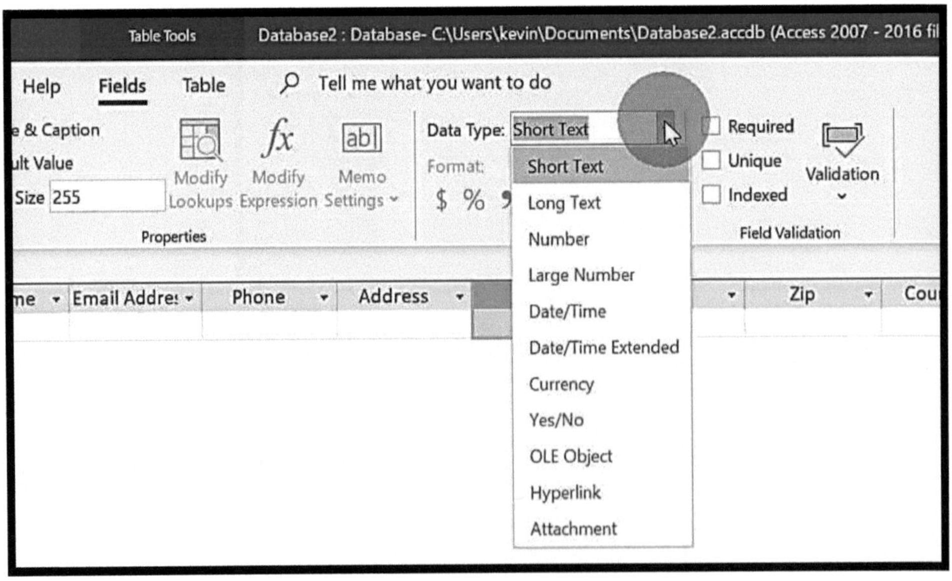

If you need something a bit longer than short text then there's the **Long Text** option and those are for entries that are longer than 255 characters. If you have some notes or some detailed explanations or like a review that's going to be quite a few paragraphs then you would want to use the long text, as it allows for more options there.

Then we have the **Number** data type, and that's any number that is used for a mathematical calculation. Now the reason why you don't use a phone number as a number is that a lot of the time you would want a particular format for the phone number, as the international format, that's better to use as a text. Also, you're never going to do mathematical calculations on a phone number; you're never going to take all the phone numbers and add them or find the average phone number. You just need it as a set of numbers so that's why you should keep that as a text. Also, sometimes your phone number has a zero in front and if you make it a number then it's going to take that 0 away or whatever is in the front, so if 072 will become 72 but if you want to keep that 0 then you need to keep it as short text. So, you use numbers for anything that you want to keep such as recording how many times someone bought or ordered something or some sort of detail like that.

We have the **Date/Time** and that's used to store a date and time or both or one or the other. You can use this option for a birthday or keep track of the day that you started at a company.

Currency is similar to Number but this is normally for monetary value, so for your salary, payments, or something that has to do with money, you can use currency format.

We have **Auto number** and this automatically assigns a unique number to each record. So, it starts with one and then the next trick will be a two and three, giving each one a unique number. That's important for later on when we talk about **Primary keys** and we'll get to that at another point but you can set it to an Auto number and it will automatically do that part for you.

There are other options available. The **Yes/No** option means it's either true or false or yes or no; so, if you've got a question that requires a yes/no, you use this option.

The **OLE Object** is useful if you've got a photo or a music file. If you want a profile picture as part of your database you can use the OLE Object for that.

The **Hyperlink** data type can be used to add a link. It shows you the address web address for a particular web page or if you're storing a document online.

The **Attachment** data type is for any other file that's not supported by the OLE Object such as a word document, if you've got multiple files, you can also use this too.

The "**Calculated**" data type helps you to run a calculation based on particular formulas, equations, or other fields. It's used mainly in queries but you can do a calculated field in a table.

These are all the data field types that you can have in your database and as you create a table, you'll work with a couple of them.

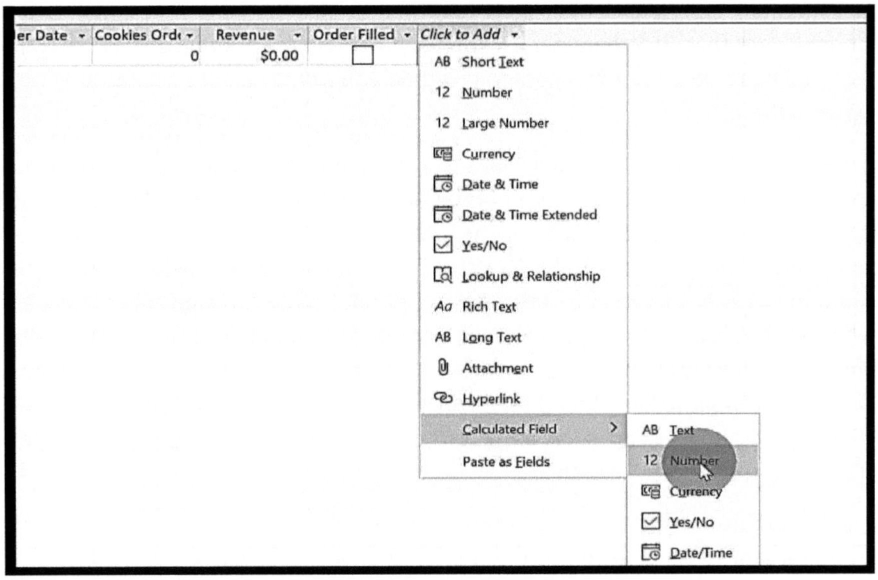

Now you know how the table works, let's proceed with how you can create your database.

Creating A Database

With your blank database open, there's not much here yet but don't worry we're going to go step by step on how to build this out.

Over on the left-hand side, you can see that Access already created a table that you can start building on, and by default, it'll ask you to do it via the datasheet view but if you want, you can use the Design view.

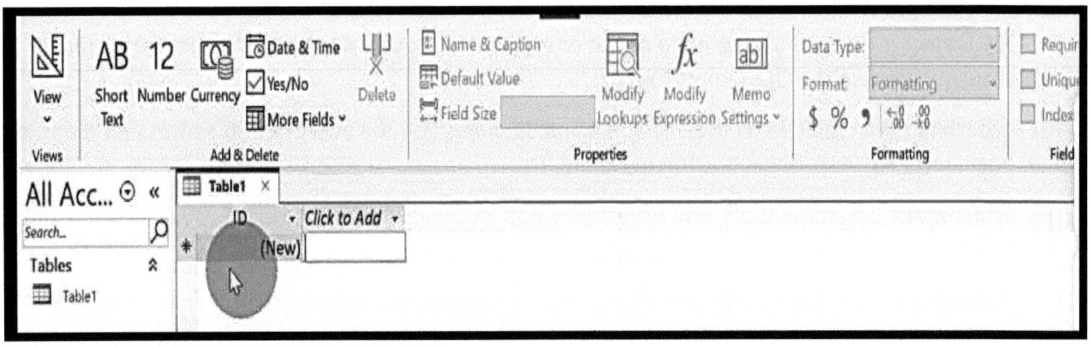

It's going to ask you to save it first, so you're going to give it a name, after which it tries to give you the field names and what type must go into that field. This helps you to set up these columns at the top and what type of data is going to be in them.

Let's get started now by adding some fields to this table. Up here, the first field is referred to as "**ID**" and you can use this table to track all of your customer information. ID on its own isn't that descriptive, so you can update it to be customer ID. To update the name, you can simply double click on the field and then type in your value. Once you're done typing that in, hit the tab key and that will bring you over to the next field.

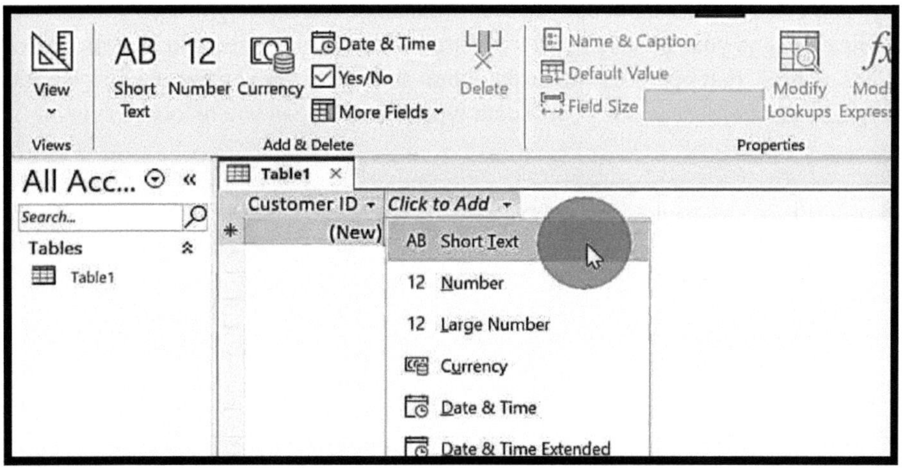

Next, you need to choose the data type. The data type defines what type of data you're going to insert in. If you want to track customer information, this is going to be things like the customer's name (the first name and the last name), the email address, the phone number, and all of those pieces of information or all of those fields will simply be the **Short Text** data type, however, later on when you get to the order sheet and you're going to add numbers and so you'll select a different data type. As you're entering data feel free to look through all the data types to see if one of these best matches the data that you want to enter.

As earlier stated, the Short Text data type allows up to 255 characters which is a pretty good length; so, it'll capture all of the different fields that you want to insert into this table. If you want **Long Text**, that allows it to go much longer but once again short text should be sufficient.

Once you select that data type, now you can type in the name for the field, and when you finish typing that in, hit tab, and you can proceed to select the data type for your next column. You'll go through and add all of the different fields for the customer tracking table.

If you want to add another field that's going to be in the form of notes, let's say you want to include other information like the customer's birthday, this might be a little longer than short text. So, when you click to add a new field, instead of going with a short text, you'll go with a long text and this should now include all of the different fields that you want to include as part of the customer table.

After you've entered all of the different fields, let's see how you can go back and make edits to these.

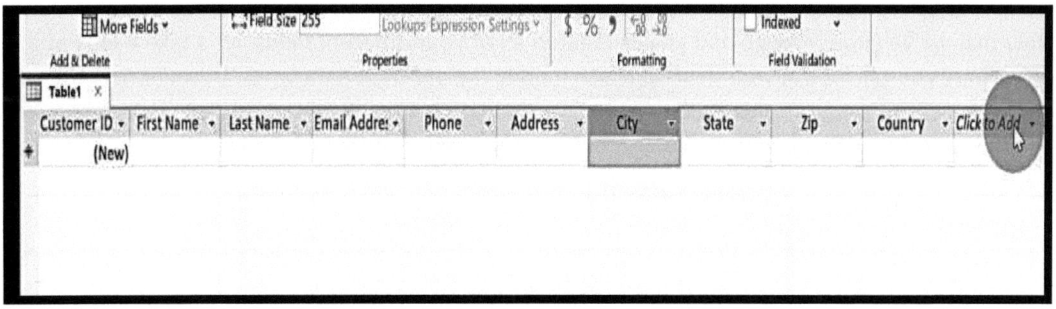

So, let's say you enter something in but you want to change the text, right here you can click on one of these field headers and you can very quickly update the text. Also, when you click on one of these field headers right up here, that opens up the fields ribbon and right here you can always go back and modify the data type, so just because you choose a data type doesn't mean you're locked in on it.

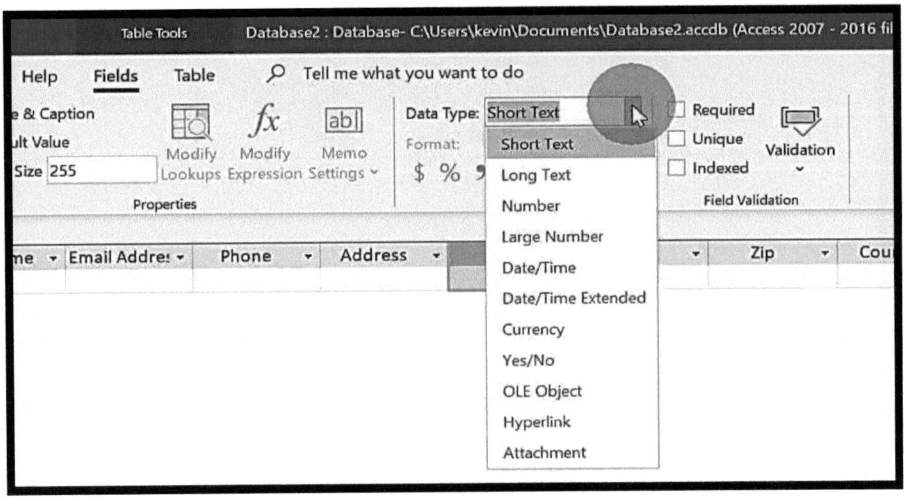

You can also adjust the width of these different fields and here if you **right-click** on one of these field headers it opens up a context menu where you can do all sorts of things. For example, you can **hide** and also unhide fields, you can **freeze** fields if you've ever done that in Excel before, it's the same concept and right down here you can insert a field, so if you want to insert something between an email address and phone, you can do that. You can also **sort**, **rename** and **delete** fields. So, you can take different actions to make sure that the table has the structure that you want it to have.

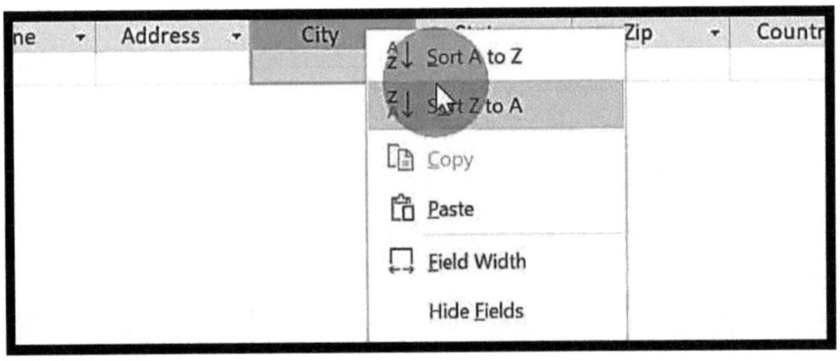

Now that we've gone through and you've created all of your different fields, let's take a look at a quick way where you can look at your fields and also modify the different data types. Down here in the bottom right-hand corner, you can enter what's called the **Design view** and when you click on that, first, you have to name your table. Right now, it's Table1, but you can make it more descriptive.

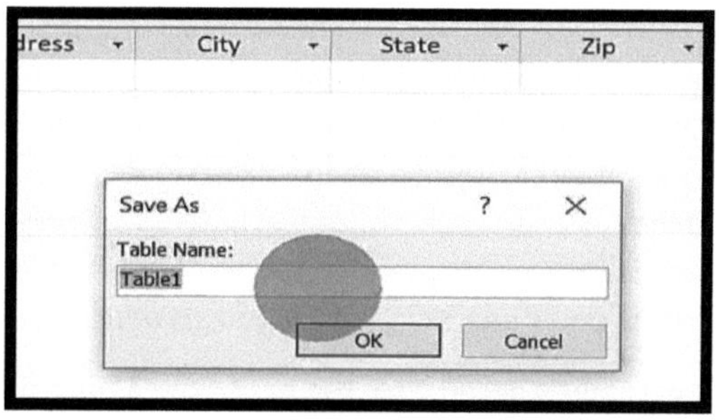

Once you finish typing in the name click on "**OK**". This now drops you into the design view, and here again, you can see all the different field names. You can also add field names here; this is a quick way to add additional fields. Also, on the right-hand side, you can quickly verify what the data type is for each one of your fields, and here too you can click on it and you can change the data type very easily. On the right-hand side, if you want to make your field more descriptive, you can also add a description with some additional details.

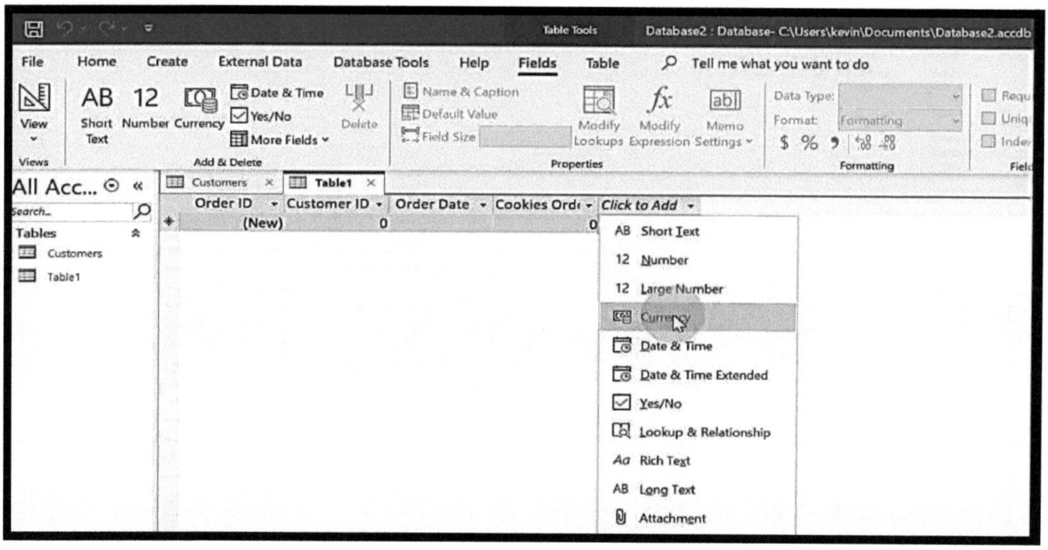

Let's jump back into the **Datasheet view**. To do that right down in the bottom right-hand corner again, you can toggle back and forth between these different views. If you click into the Datasheet view. This drops you back into the datasheet view and you're ready to start adding some records.

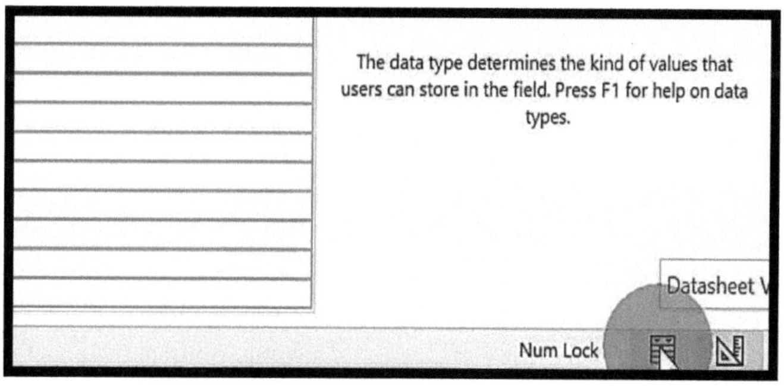

Now you have all of the fields in place, you can start adding your customers. So right here, you'll go in and fill out their details. As you start typing in information here, you'll see that it automatically assigned an ID, so every time you go through and add a new record it'll automatically get an ID assigned to it. As a quick note on this, **ID is also what's referred to as the Primary Key** and this is a **unique identifier**. So, every single customer record that you add to this table will have its unique identifier, so no two customers will have the same ID. You can also use this primary key to connect to other tables later on and don't worry if that sounds complicated, you'll see exactly how you go about doing that.

Once you enter in the first customer's details, hit the enter key and that'll bring you down to the next record and now you can go through and add some additional customers.

After creating a table for your customers' information, you may want to add another table for order or purchase information.

To add another table, go back up to the top ribbon and click on "**Create**". Right over here, you can create a new table. If you click on the one that says "**Table**", this drops you into a new table, and just like before you need to build this out.

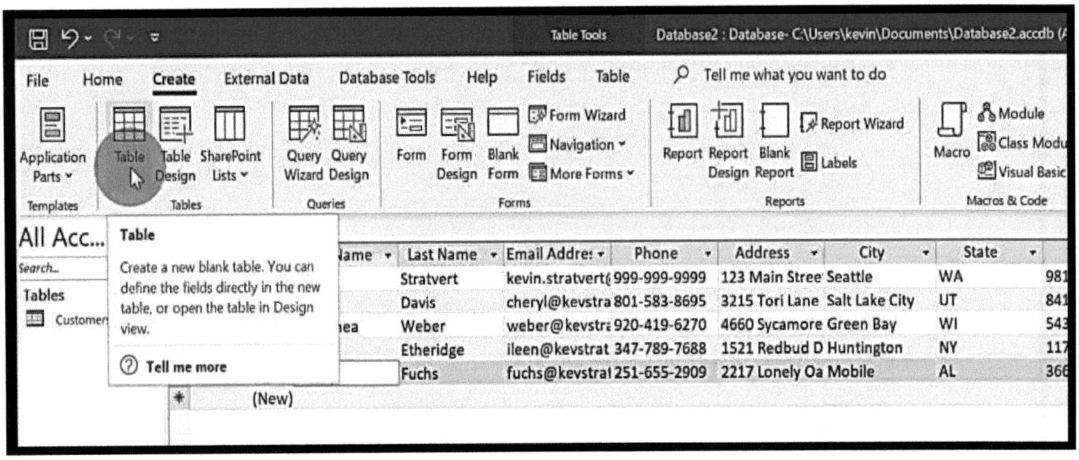

Right up there, the first field is called **ID** and you would want this to be more descriptive. So, you'll double click on this just like you did before and this time you'll call it Order or Purchase ID. For the next field, you may want to tie this back to the customers' table. Once again this is one of the big benefits of databases, you can relate information so you could connect one table to another table. So here you'll click on the tab and for this field, if you want it to be a number data type, you'll click on the number and now you can type in a field name, and for this one, you can name it "**Customer ID**" since you're going to connect based on that primary key from the previous table. As you go, you can add a few more fields. If you want to include the order/purchase date, select the "**Date and time**" as the data type and rename it as "**Order/Purchase Date**".

Next, you can add another field for items ordered or sold and this is going to be a **number data type**, so you'll give it the appropriate name. Now, you need to add another field for the revenue, this is going to be a currency type and you can name this "**Revenue**". You can also add one more field called order fulfilled, this is so your other staff can go in and indicate whether the order has been fulfilled or if it's unfulfilled and they still have to package it and ship it. Here you'll click on the field and right down there, there's the option for "**Yes/No**". So, if they say yes, it's already been shipped, if not, it hasn't, and you'll call this field "**Order fulfilled**".

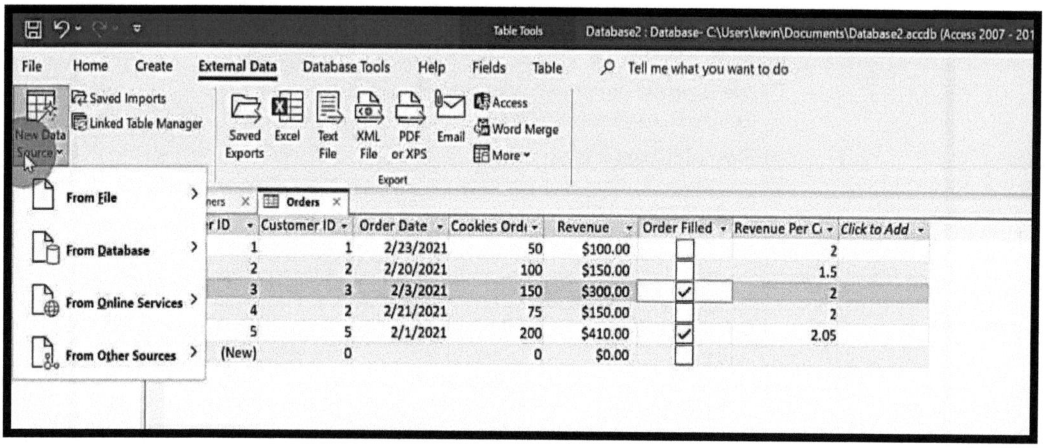

The last field you may want to add is going to be a little bit different. This is going to contain the revenue per item and you have all of the information stored within here to be able to calculate that. Here you already have the revenue and you also have the number of items ordered or sold. So, click on the drop-down where you can select the data type and this is going to be a calculated field; you'll see in a moment what this does. Right down here, go down to "**Calculated Field**" and since the revenue per item is going to be a number, select the number data type.

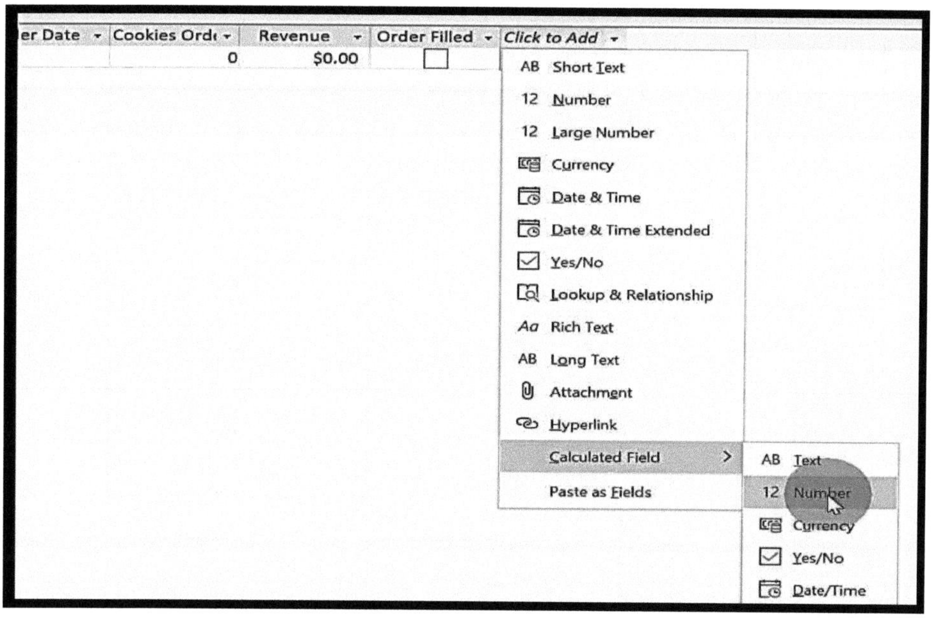

This opens up an expression builder in Access and you can build all types of different expressions. One way to think of it is like entering in a formula. So down here, you can see some of the expression categories; you'll take the revenue and when you double click on that you see it inserts revenue here and since you want to divide by the items ordered or purchased, you'll enter the divide sign and then right down here you'll click on "**Items ordered**". So that's going to be the Revenue divided by the Items ordered, click on "**OK**" and this drops you back into the table view and now you can type in a name for the field. For this one, you'll type in Revenue per Item.

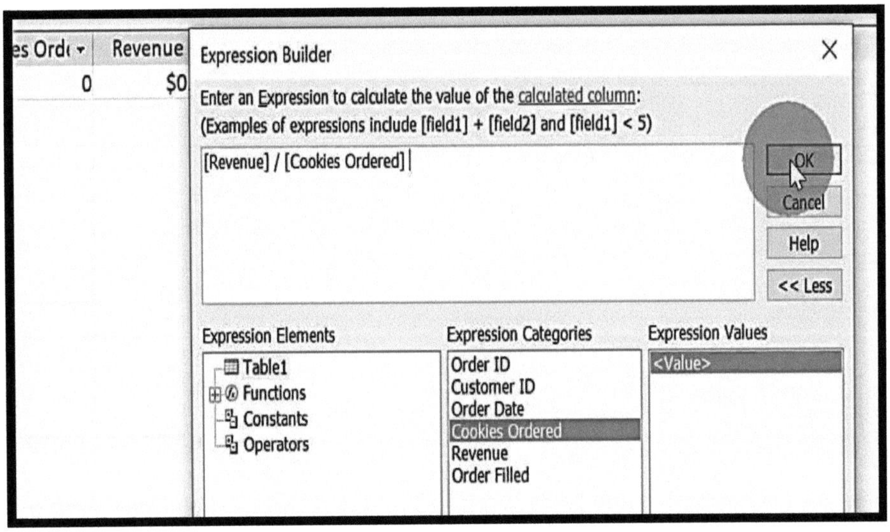

You can now fill in some order details. So, for the first customer, if you remember back on the customers' table, Customer ID 1 is for a particular customer, so right back here in your order table you'll type in Customer ID 1. Next, you'll enter in the order date you entered in an order date and then type in items ordered, and enter your revenue. Now that you have that entered in, one thing you'll see is it automatically calculates the revenue per item. The order hasn't been fulfilled yet, so you'll leave that checkbox unchecked.

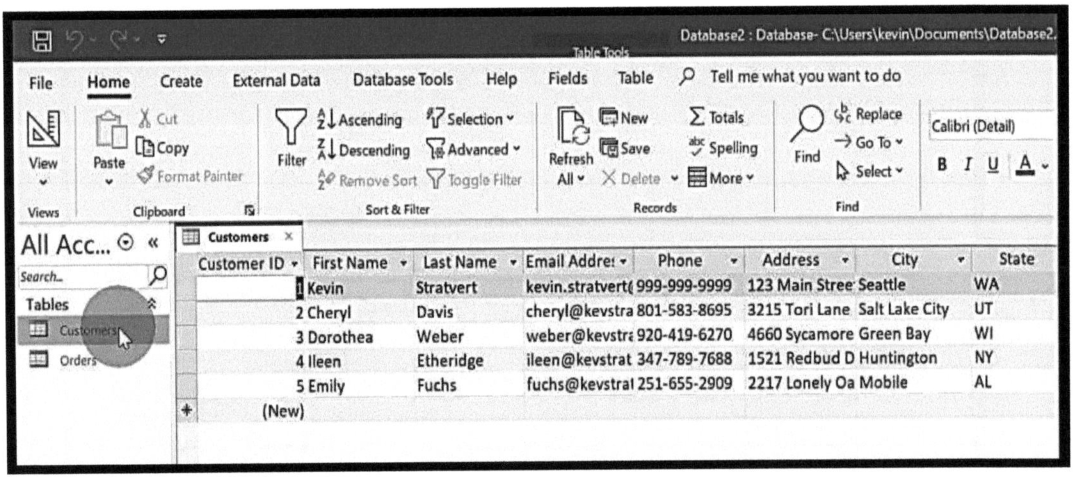

After you've finished filling out the basic structure of your second table, once again if you want to rename it you can go down and click into the design view, and then you'll be able to rename this. Alternatively, you can also close this table and it'll ask you if you want to save it, if you click on "**Yes**" you can now give it a name.

After saving your Orders table, over on the left-hand side you'll see that you have two tables now. You could go up here and close the Customers' table and you'll see that all your tables are gone but don't worry you can easily get them back. On the left-hand side, if you click on the tables, that brings those tables back into view.

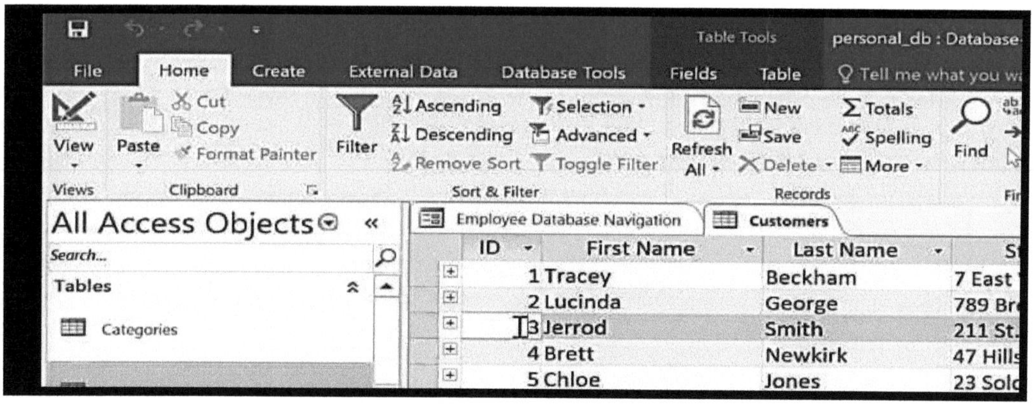

You can also go back to a table and add some more information. Now one of the great things about creating a database here is that with the Customer ID you don't have to repeat the customer information. You simply have to reference that ID and then for the order you can fetch all the customer information, however, you have to **connect this customer ID to the customer table** over there.

How do you do that? Well, at the top go up to database tools on the Ribbon. Once you click on that there's an option for **Relationships**, click on it. Within Relationships, you can define how different tables relate to one another, and over on the right-hand side, you'll see all of your tables. Click on the first table and pull that into the view. Next, click on the second table and pull that over as well. Now you can see your two tables and you can see all of the fields in those two different tables. At the top of each table, you can see a key icon that indicates what the primary key is in that table.

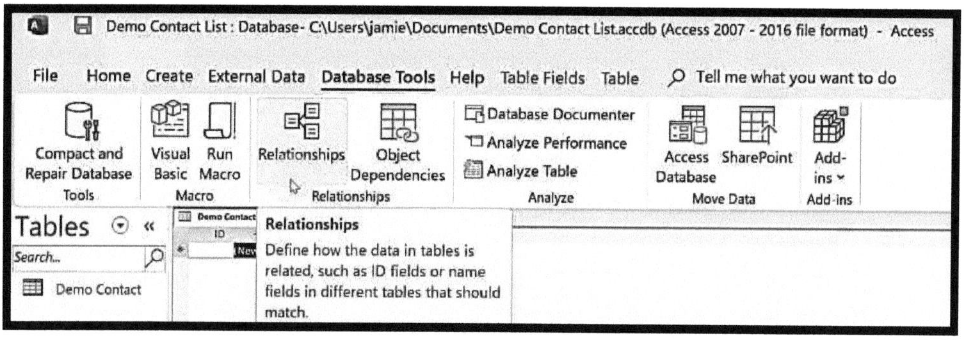

Remember you have the customer ID in your first table and you also have the customer ID in your second table. Now you can simply click on the customer ID in the first table and drag that over to the customer ID in the second table; that opens up a prompt where you can edit the relationship. What this means is that this value or this field in the first table is the same as this field in the second table, so you're saying that these two values or fields are the same.

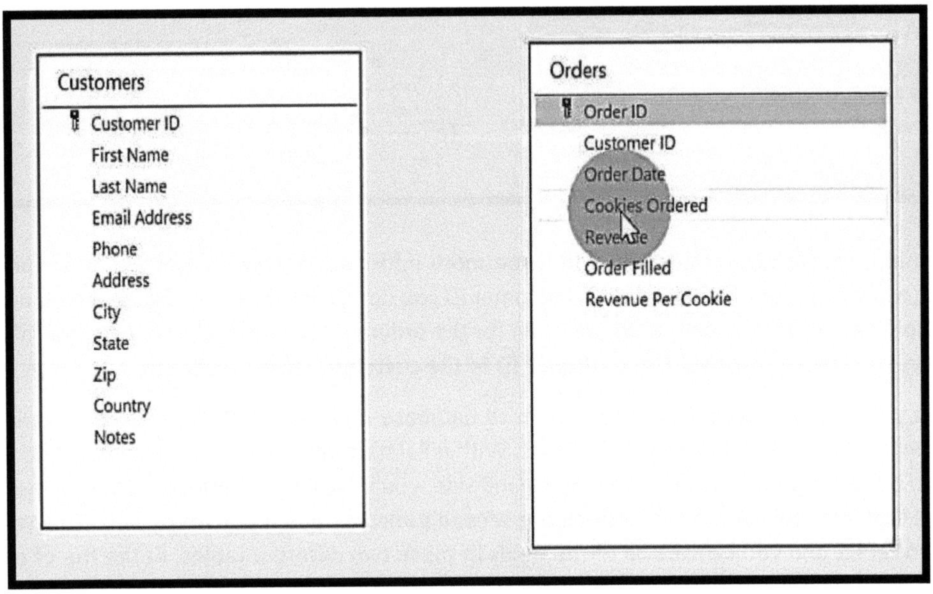

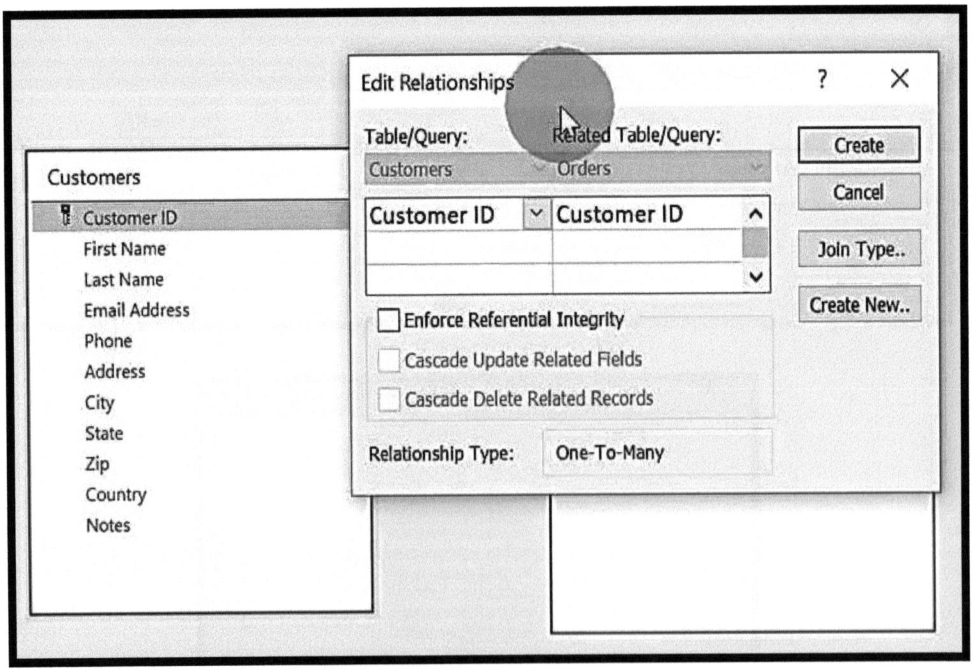

Next, you'll click on create and you'll see now that there's a connection between these two. So, this is how Access knows how these two tables relate to one another.

Once you're all done with this, go to the top and click on "**Close**". This opens up a prompt to save the relationships, you'll click on "**Yes**".

Importing Data

Along with manually entering data into your database, you can also import it from other sources. Right up there on the ribbon, there's the option for External Data. When you click on this, over on the left-hand side you can see all the different places that you can bring data in from. For example, you can bring it in from a File, from a Database from different Online Services; so, there are lots of different ways that you can get data into Microsoft Access.

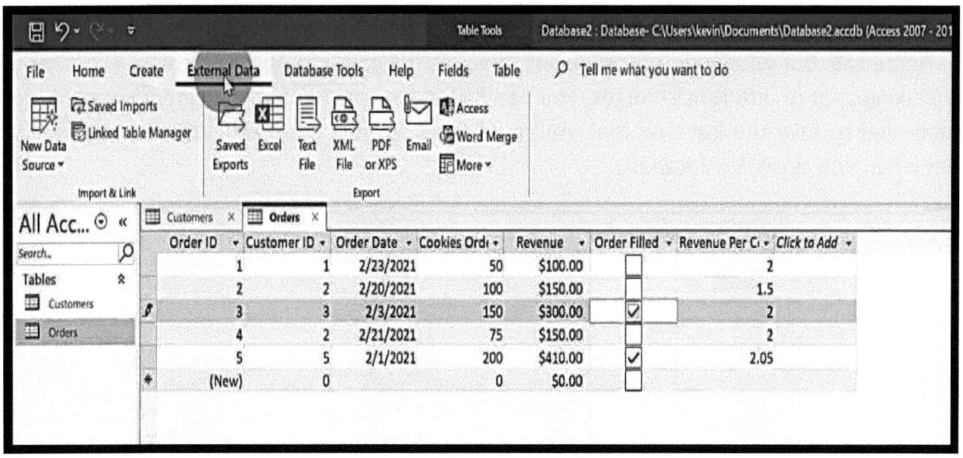

CHAPTER THREE

MORE ON TABLES

In this chapter, you'll learn how to interact with your customer table. We're going to talk about how to add, edit, and delete information from a table. We will also look at some validation rules, working with input masks, you will understand what the Lookup Wizard does and how to create relationships between your tables.

Editing Your Table

Just like in Excel, each box is called a **cell**. If you want to change the information in a cell, all you have to do is **select it**, and then **type in the new text**. After you edit the record, you can save it by clicking **Save** on the Home tab, but whenever you click out of a cell, it's going to automatically save it for you. So, if you're making a lot of different changes, you don't have to worry about saving each record, however, it is a good idea to save the last one that you're working on, just to make sure that you don't lose any changes when you close the database.

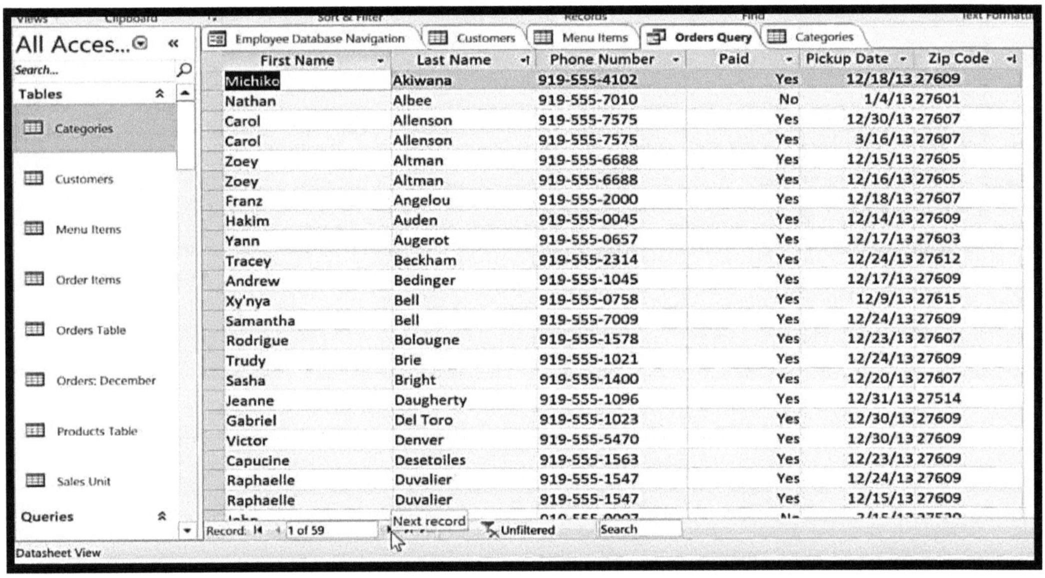

If you are editing more than one record, you can use the arrow keys on your keyboard to move between records; if you prefer, you can use the arrows on the Record Navigation bar. You can also use the Record Navigation bar to **create a new record** and then type in the person's information. If you want to **delete a record**, click the margin on the left side of the record, and then on the home tab, click the Delete command.

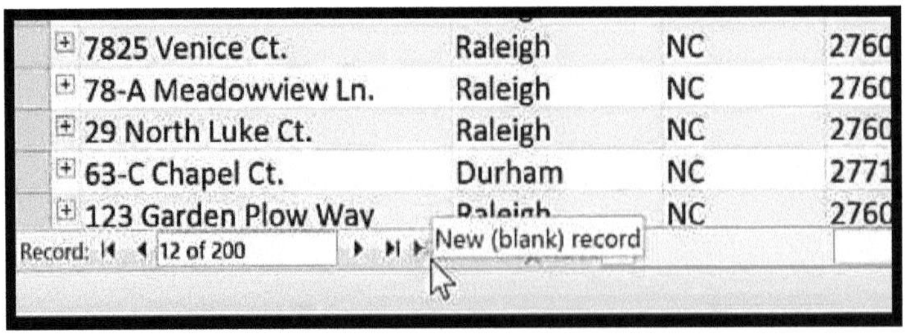

Now, normally you should avoid deleting records because it can negatively impact the database if other objects refer to the record. For example, if you deleted a customer's record, it could cause the information in the Orders Table to be incomplete. But if you're just deleting a brand-new record, then it's generally okay.

For some edits, you may want to use the **Find** and **Replace** options to make a lot of changes at once. For illustration, we're going to open up the Products Table and on the home tab, click the **Replace command** and let's say in some of our seasonal products, we want to change the word "Fall" to "Autumn". Type in the word that you're looking for and the word that you want it replaced with.

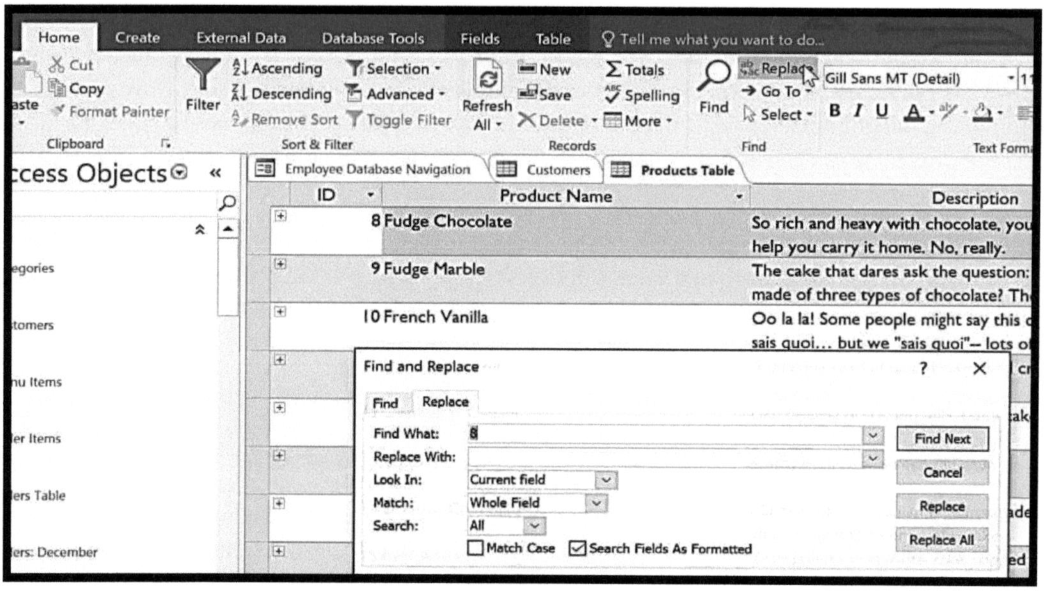

Next, we'll select the Current document, so it will search throughout the entire table and we want to look for matches that appear in Any Part of the field. Click "**Find Next**" to look for the word. When you get to one that you want to change, click "**Replace**" and just go through each one and decide whether you want to replace it.

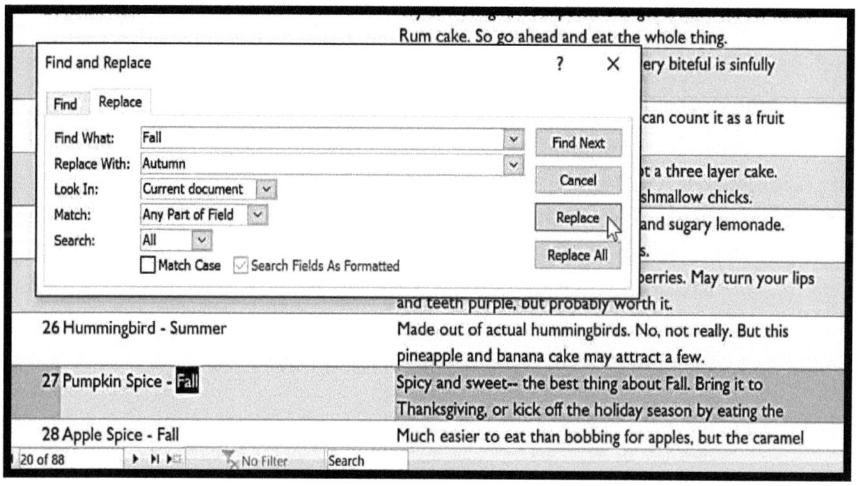

Generally, you don't want to use "**Replace All**" because that doesn't give you the option of skipping any of them. So, you should only use this option if you're sure that it won't replace anything you don't want. As you practice editing tables, keep in mind that most of your changes are saved automatically. So, to be safe, you may want to create new records to practice with.

Validation Rules: AND, OR, BETWEEN, <, >, <=, >=

We're going to be looking at **Validation Rules** in tables. Validation means we want to make sure that people type in the correct information into a table to prevent errors, and a way to do that is to have a **validation rule** which means that the user can't put incorrect data into a field. When talking about Validation Rules, there are two aspects to take note of. One is the rule which is where we specify what values are acceptable for this field so we give it a range, stating what we want in this field, and then there's the validation text and that is what must be displayed if you violate the rule. Now you don't have to put in a validation text, it'll just put up a default error message but if you want to specify a nice little message where you can give the user some help about what they should be entering, you can create your text for your error message.

So, what are the ways that we can make a validation rule? Well, we use operators like if things are equal to a value (=), or if it's greater than a value (**>**), or less than (**<**), or greater than equal to (**>=**), or less than equal to (**<=**). Also, this symbol **<>** is used if we don't want a particular value, so we use a combination of this.

We can also use the operators like "**AND**" if you want two things to be true. If you want one of two things to be true, either one or the other, you could use the "**OR**". If you want to write something and say 'I don't want this then you could "**NOT**" for that case, or you can say "**BETWEEN**" a range of numbers.

There are other operators and formulas you can use but these are the most common ones that we could use. So let's say for example we've got a high school and we only want the senior grades such as grades 10, 11, and 12. So you could say a validation rule could be **10 or 11 or 12**, as the possible values; but maybe you want a high school including juniors and seniors from grades 8 to 12, you could say 8 or 9 or 10 or 11 to 12 (that's one option), or you could say **>= 8 AND <= 12**, or you could say **BETWEEN 8 AND 12**. So those are all options available to you for your validation rule.

So here we've got our table for illustration, and we've got some values into it, so what we're going to do is to do some validation rules. Taking a scenario, for example, let's say **Division** and it can only be **A, B, or C**. So, in the table below, we can go to our validation rule and say it can only be an **A or B or C**. You'll notice that if you click away, whenever you refer to text it'll put double quotes around the text automatically for you,

When you go back to the actual contents of the table if you go to the bottom and enter a new record and state that the division must be division "**F**", if you click away, it will give you an error message stating that you are violating the validation rule and it will keep on giving this message no matter what you type in until you put in one that is suitable like A, B or C before it's accepted.

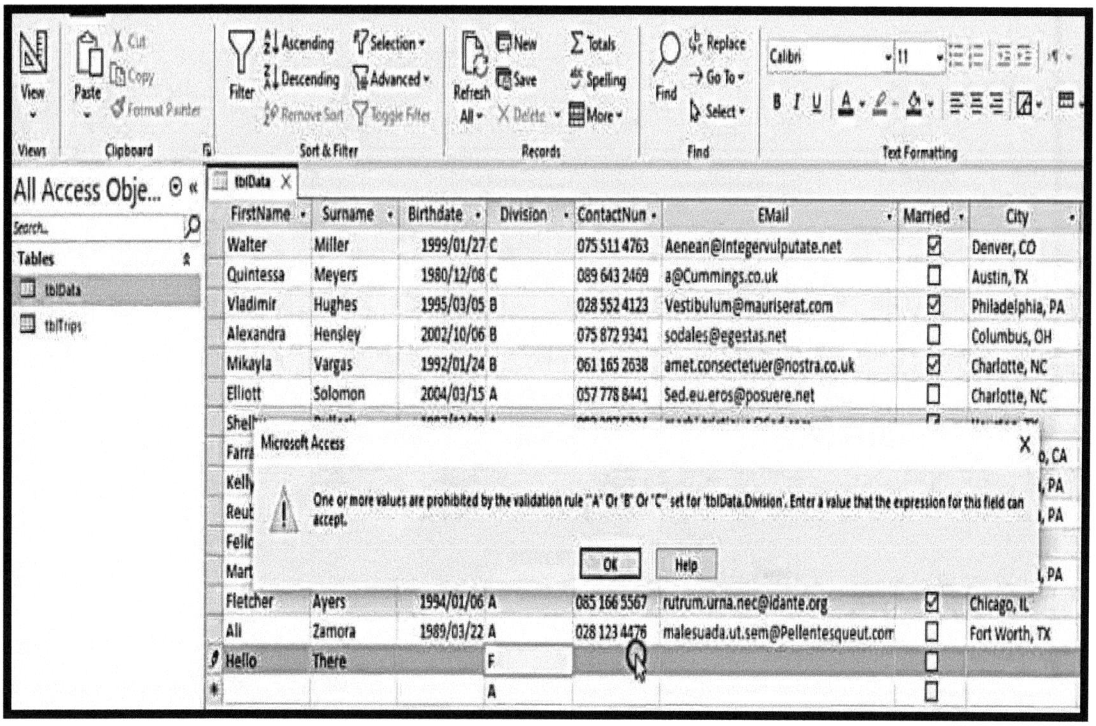

You saw the message that popped up was just a random message, but if you want to customize that message in the table below that contains the validation text, you could say something like "**please enter a division of A, B, or C**". So, you can write whatever you want in the text and that's what will be the error message that pops up. So, when you go to the record and you type in a random letter like "S", it gives you the error message that you put into the validation text. So, the text can be whatever you want but you should give it a type of value that will be intuitive to the user, so that will help them know what they must do to correct that particular mistake.

Another rule that we could have for example is that payments can only be made in a value between 0 and 20000, and if that's the case you can say it must be **>= 0 AND <= 20000**, as the validation that you want. You'll see that with numbers we don't need to put double quotes around the numbers and the same with payments. So, you could do all these types of validation rules and you have text as well.

When it comes to dates you know you're not going to limit the dates but if you want to say you don't want anyone born after a particular date like 1960 for example, then the date must be bigger than 1960 but remember you must specify it as a date, so you need to say **>=1960/01/01**. You'll notice when you click away it puts hashes around the date, so that's a valid date and that's how dates work when it comes to validation.

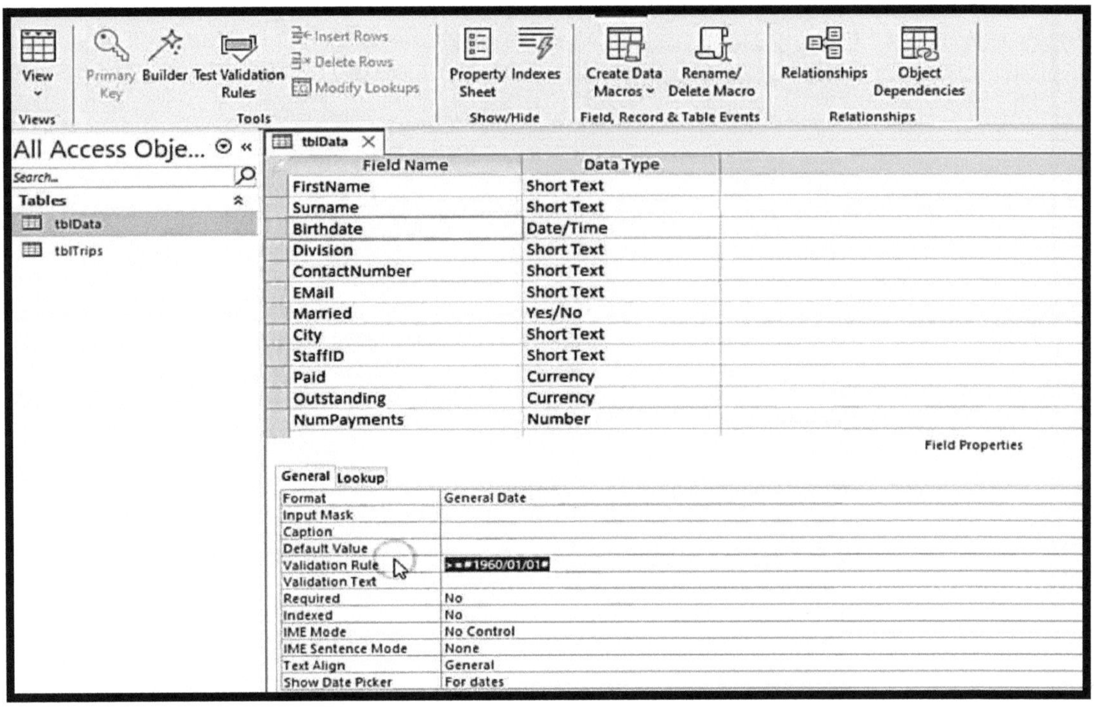

So, there we go. We did some text and some numbers and so what that will do is it will now limit us whenever we enter in those fields. So, when you run it, it should say that there were some violation rules. So just to recap: **change the rule to what you want it to be** and **the text is what must be displayed if that rule is violated**.

Working With Input Masks

Here, we're going to be looking at Input Masks. This is a group of codes that indicate the format of all the values that you want in a field that are valid. So, for example if there's a specific format for what must go into a particular field and you want to specify that, you can use a special code to say this is what it must look like.

When we talk about these fields, there are different options available to us in the different codes that you need to know. Now there are two scenarios with most of the codes, for example, there's a scenario when they must have that particular value and there's one where it could have that value but it could be blank; so, we have the **Compulsory option** and the **Optional option**.

When you want to use a	Compulsory	Optional
Digit (0..9)	0	9
Letter (A..Z)	L	?
Letter or digit	A	a
Character or space	&	C

Furthermore, let's take the scenario of a **digit**. If you've got a number, any number from **0 to 9**, it takes one digit then we will put a **0 as the compulsory code** and a **9 as an optional one**. So, if it's got to be there, you put a 0 and if it could be there but it could be blank, you put in a 9. If you want a two-digit number, it must either be 10 up until 99; you'll just put two zeros there and that will be a compulsory two-digit number.

If you want a **letter** to go in a particular place and any letter from **A - Z**, then if it's supposed to be there you would use the capital **L as the compulsory letter** and a **question mark** (**?**) **if it's optional**, which means it could be a letter or it could be blank.

What happens if it could be a letter or a digit in that particular place? This means in that particular slot it could be a number or a digit; so, it **could be letter A** or **could be 3**. If it's **compulsory,** then you can use the **capital A** but if it could be blank as well or left out then you're going to leave a **small** "**a**" as the **optional** letter or digit.

What happens if it's any character or space? It literally could be anything. It could be a question mark, a letter, or a digit. If it's compulsory then you're going to want to use the "**&**" or what you might refer to as the "and symbol", but if it's possible to be left blank then we can use the **capital C**.

There are lots of other codes but that's where you can start. There are other options available as well. So, for example, if you want **all of the characters to be an uppercase** then you just put a **greater than** (**>**) **symbol** in front of the letters and they will make anything that's after it will then all letters will be in capitals, and for **lowercase**, you can just use **smaller than** (**<**) **symbol** and that will make all the letters into smaller than letters. If you want to display a bunch of text literally like it is and especially if it's part of the code, let's say you wanted to say A, B, C, or M, at the end of a time for example and you didn't want that letter to be confused with the input mask code letter, then you could use quotation marks (" ") around the letter and it will **display the letter as is** but if you want to display one character you could just use the slash (****) for that, so whatever's after that slash will be **displayed as it is**. So, if you put slash L, it won't be the code L, it'll just be the actual character L.

So, let's have a look at some input masks. Let's say you've got Car Registration on your table and you want to put an input mask on that; when you click on Car Registration, you'll see there's an input mask area. You can click on the little ellipse there and it'll take you through a wizard where you can add different input masks, so you can have some default ones that are already set for you or you can set your own.

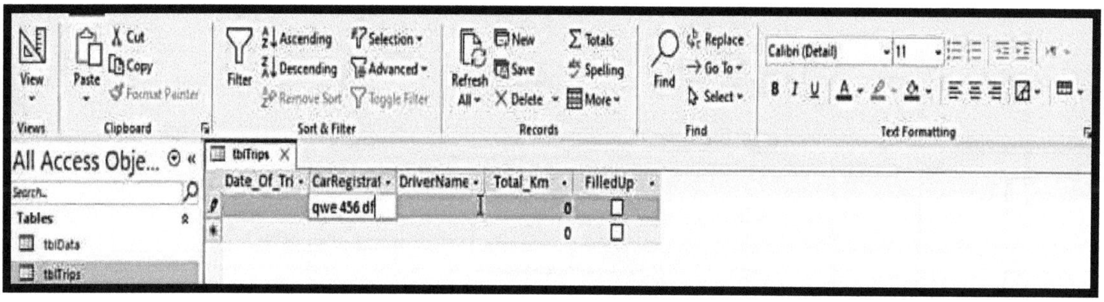

For illustration, if you have **LLL 000 LL** as the original code, what happens if you come to the actual datasheet view once you've saved it, you'll see when you click on Car Registration over there and type in numbers nothing happens and this is because the first three characters have to be letters. After typing in the first three letters, if you keep on typing letters nothing's going to happen because the next has to be in numbers. So, after adding some numbers, if you keep on typing in some numbers nothing's happening nothing's going to happen because you've got to be at the place where you need to add letters.

When you are done, it wouldn't allow you to put anything else in here, just that format. That's a nice little feature to make sure that people put the values in the correct format. Now back to the table where you have your Input Mask, if you put a greater than symbol in front of it and you go back to your table, you'll see that it's all in the capital. So even though your caps lock is off and you're not pressing shift, if you type in letters it's going to automatically put them in capitals. That's a good example of the input mask for Car Registration.

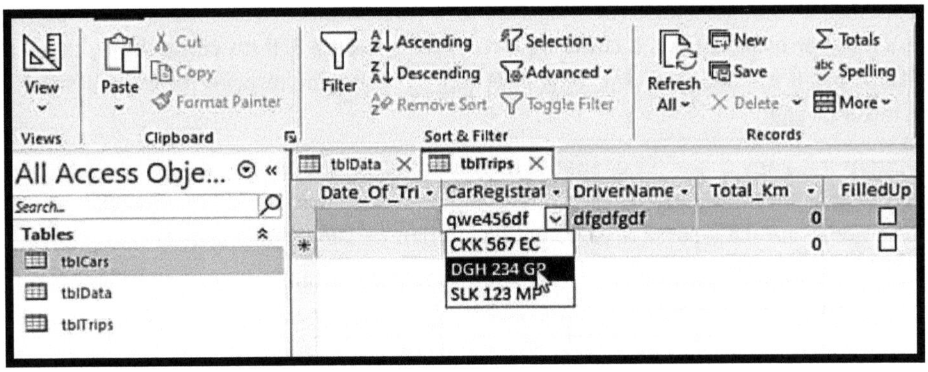

Let's look at another example and that is if you have Contact numbers in your table and you always want the contact numbers to be ten digits. Some countries have ten numbers in their cell phone numbers and then they added another one because they ran out so the last digit could be optional, in that case, you could have 10 numbers or 11. With this in mind, if you go to Contact Number, type in ten numbers and you leave the last one blank, you'll see it accepts it, and if you type in the eleventh number, it accepts that as another option. So that's how you get the optional digits.

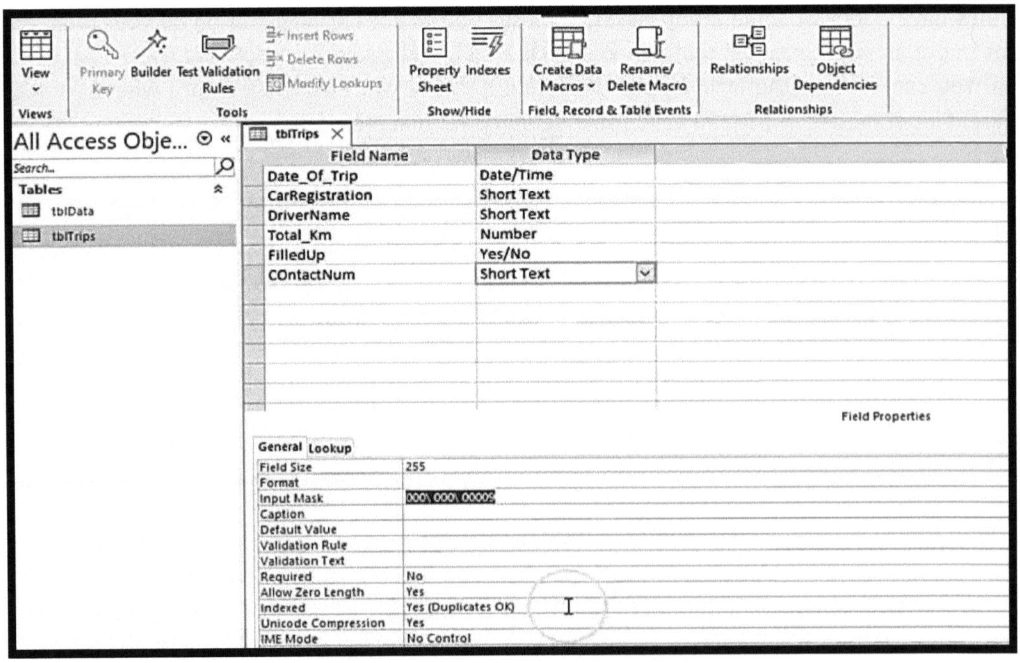

You could have something like a special code, let's say there's a code that you need, and the code contains one letter followed by another letter, followed by one number, followed by another number, or something else. In that case, you've got a compulsory letter, followed by an optional letter, followed by a compulsory number, followed by an optional number, so you could have that particular input mask over there.

For that, if you save it and you go over to Code in the second table, you can type in a letter, followed by a number and that's a valid option. You can also put in two letters and two numbers and two letters and that's also a valid option.

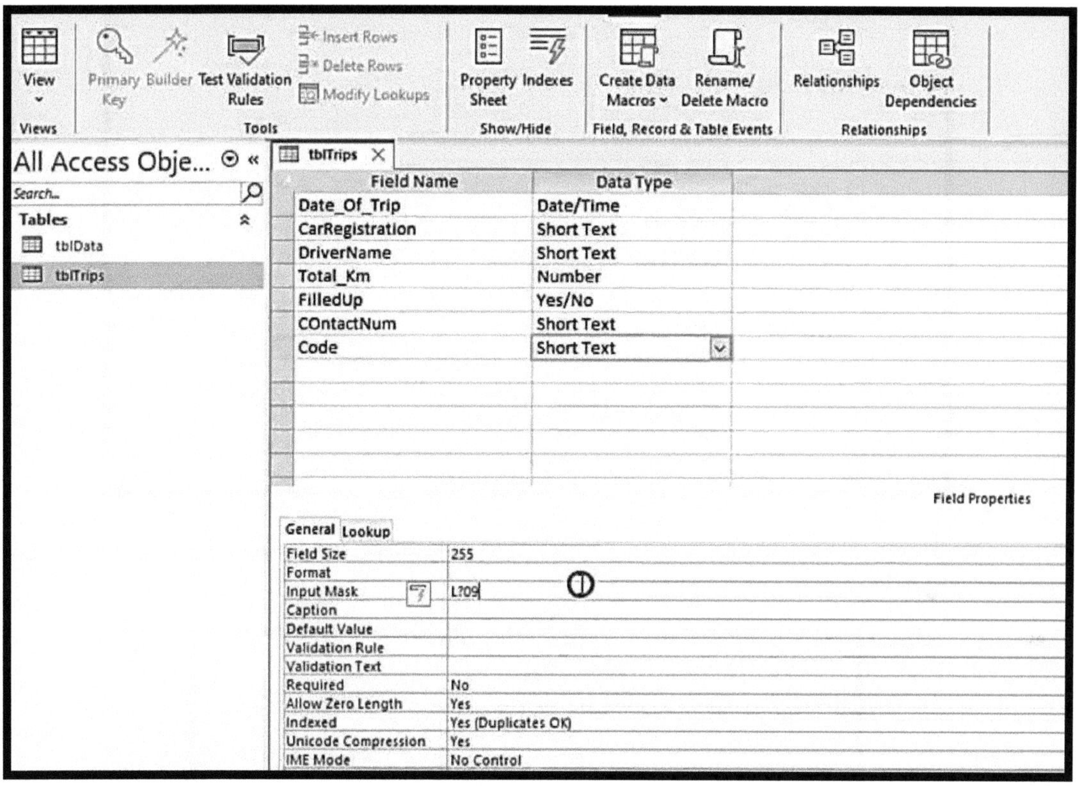

Those are the different types of input masks that you could get and just so you know, while we are on input masks you saw that the code for capital letters is > and that for small letters is <; you can do that on the format as well. If you put a > on the format, you'll notice that it doesn't matter what you type in, it will always be capital letters. If you go to it and change it to a <, it doesn't matter what you type, it'll make it smaller.

So those are options for the format that you use the input mask codes for that as well, for just those two though.

Lookup Wizard

A lookup wizard is used when you want to **limit the user to a certain list of options for a particular field**. If there is a list of options that you want to select from, you would click from a list box or from a combo box to select, and these display the available options for that particular field. So, let's take a look at the table or what it looks like without a lookup wizard.

For example, in the table below you can see Divisions. When you click on it, it allows you to type in the value.

If you don't want that, and you would prefer to have a little box or a little arrow there where you click and you can select one of these options, you're going to come to the Division and change its data type to a **Lookup Wizard**.

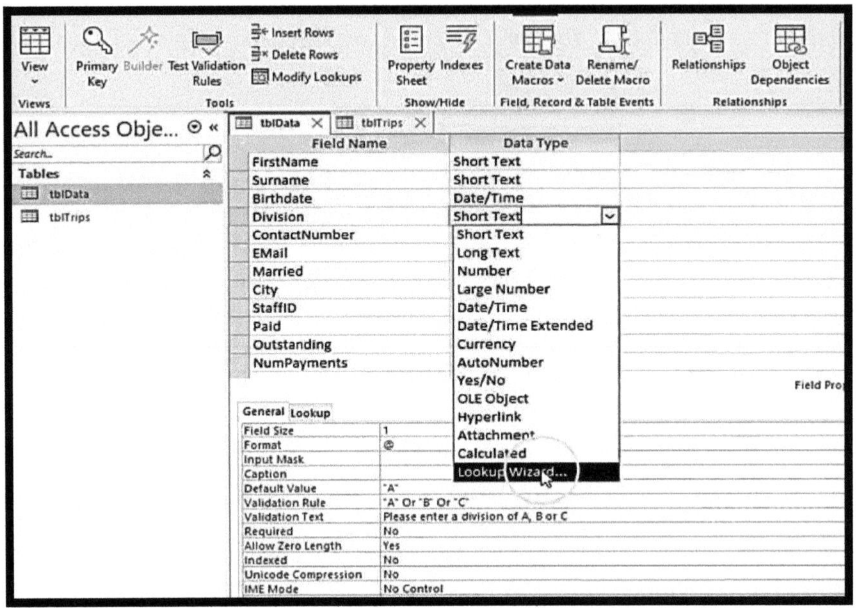

When you go to the lookup wizard, a box will pop up that will guide you through the steps that you want to take.

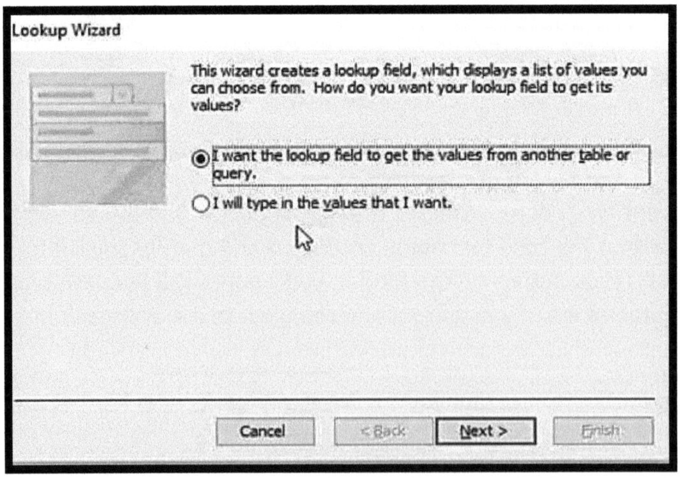

Sometimes you have a list of all the available options in another query or table, if that's the case you would just select the first option and you would go select which table or query it is and select which field from that table where you want to be the list of options. If that is not the case, then select the second option that indicates that you will type in the values you want, and click on "**Next**".

Next, you can specify what your options are. So, you type in the options in the order you want it to be and click on Next.

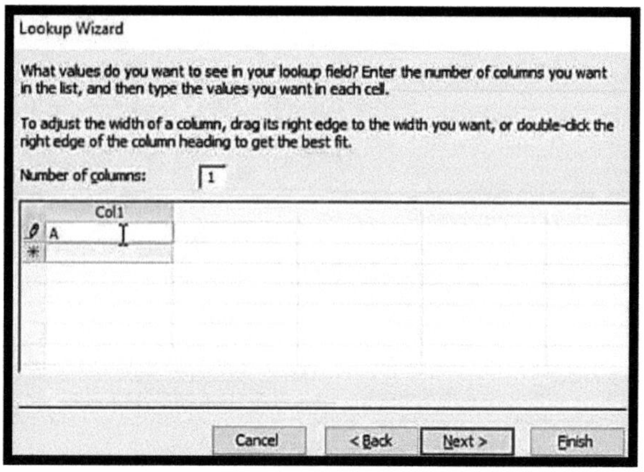

Sometimes you could allow them to add other options that aren't one of those you mentioned, however, you can limit the options to the list you entered, and then you click on **Finish**.

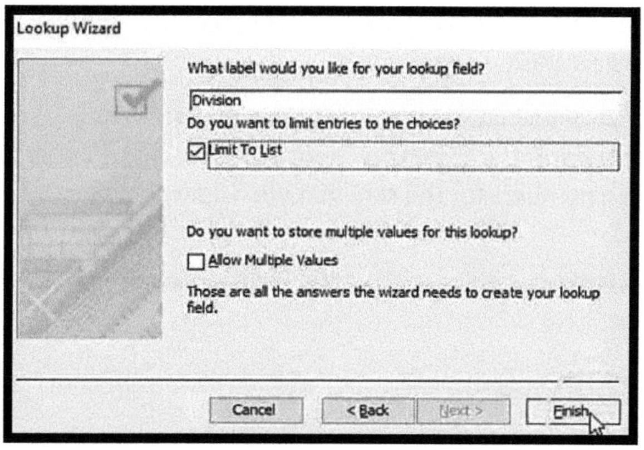

What you'll notice is that the data type changed back to short text. If you go over to "**Lookup**" Under the Field Properties, you'll see a combo box option, a value list options, and the options in your combo box.

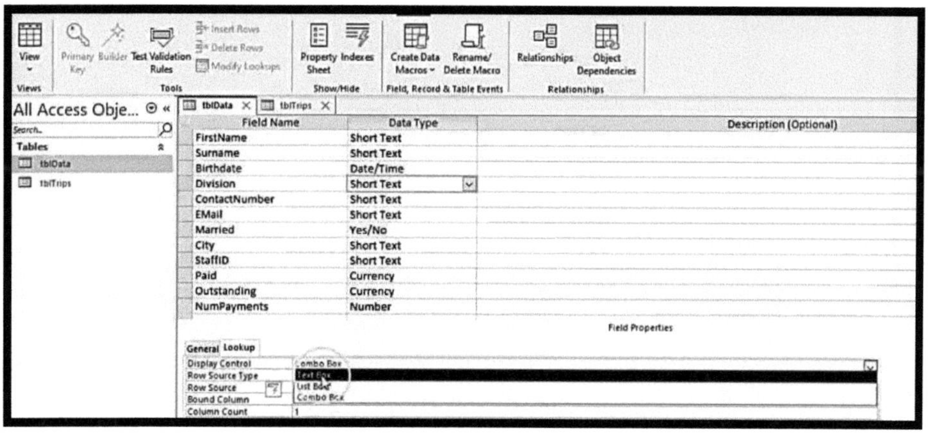

So, what does that look like? When you go to the other table, if you click on division, you'll see there's a little arrow there and if you click on it, it gives you all the options available.

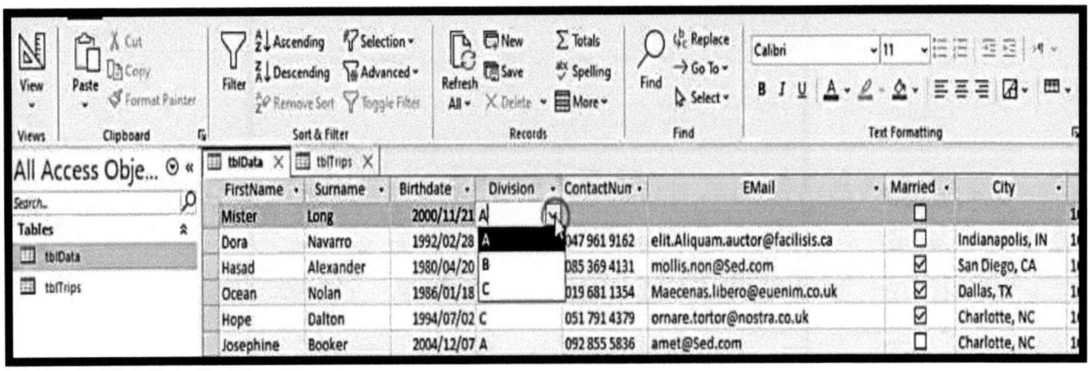

If you want to edit those options after you've done the lookup wizard, then you can just go back to Design in the first table, go to the lookup option and you can change it.

You can also change it to be a list box or a text box if you wanted to, so you can also specify those options.

If you initially said from a query or a table, then the details of that query or table will be here under the value list so you can change it there as well.

Looking at another example, if you have Car Registration and you want a list of all the available cars, then you could use the lookup wizard. Here you can choose to type in the values, go to "**Next**" and then you can specify the registration numbers for the cars that you want; so, you can have a list of all the cars that you've got on your record.

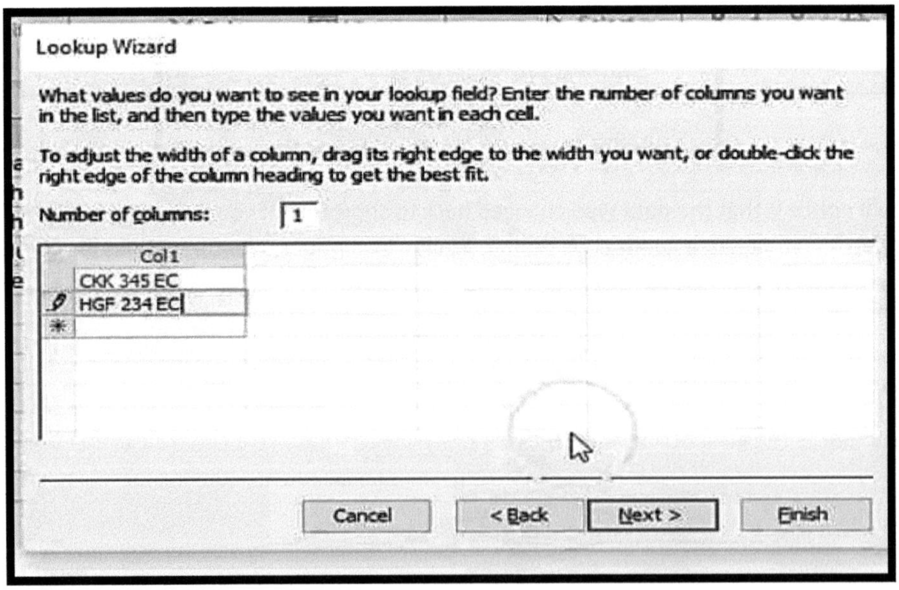

So now when you select a car from the list in the Car Registration field, it'll be one of the ones that you specified.

Also, if you have a list, instead of going to the values you can get it from a table or query. Next, you'll choose which table you want,

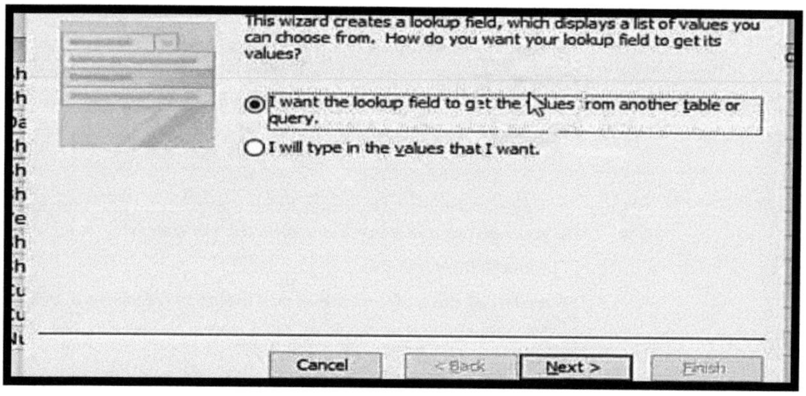

Next, you'll choose the field that you want to get values from.

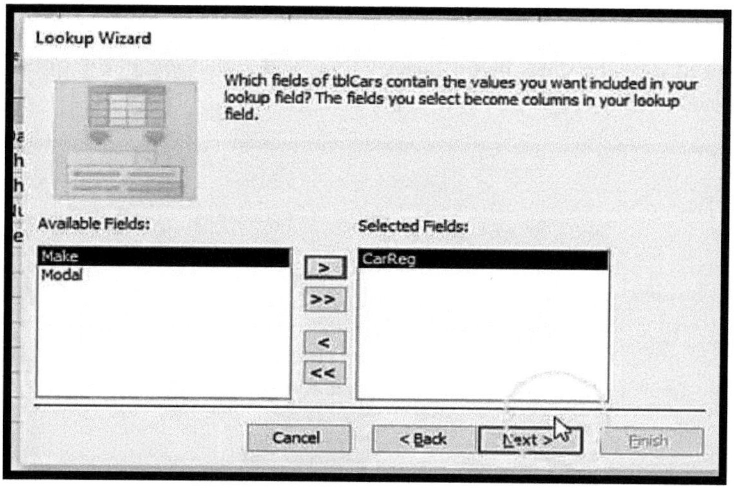

You can sort them if you want to.

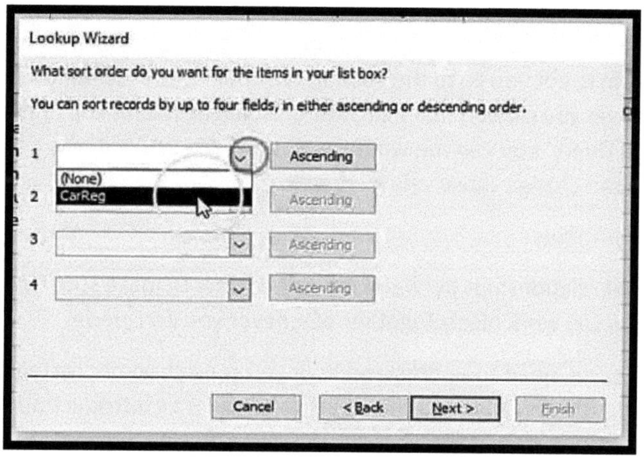

So go next, and click Finish.

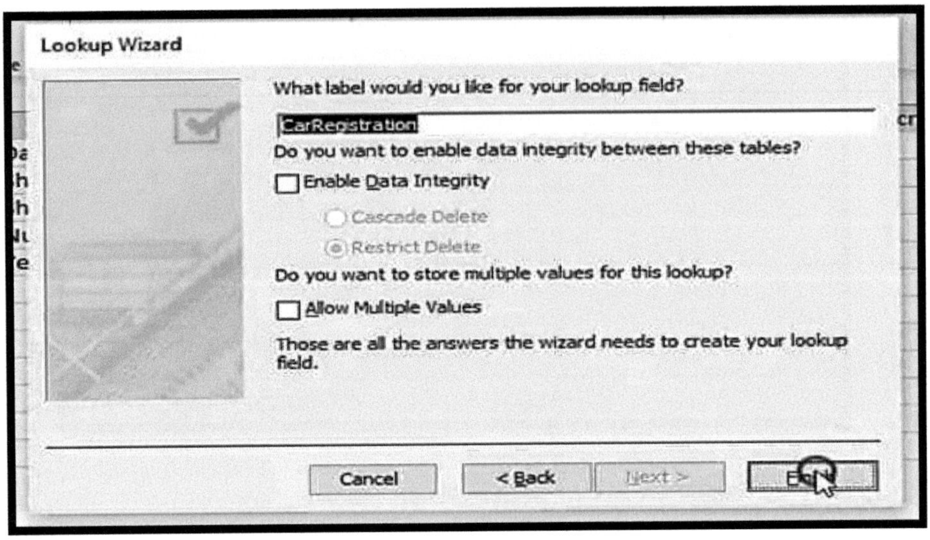

Now, you can go and view the data. If you click on that field, you'll see all the options that were in that table now listed over here for you.

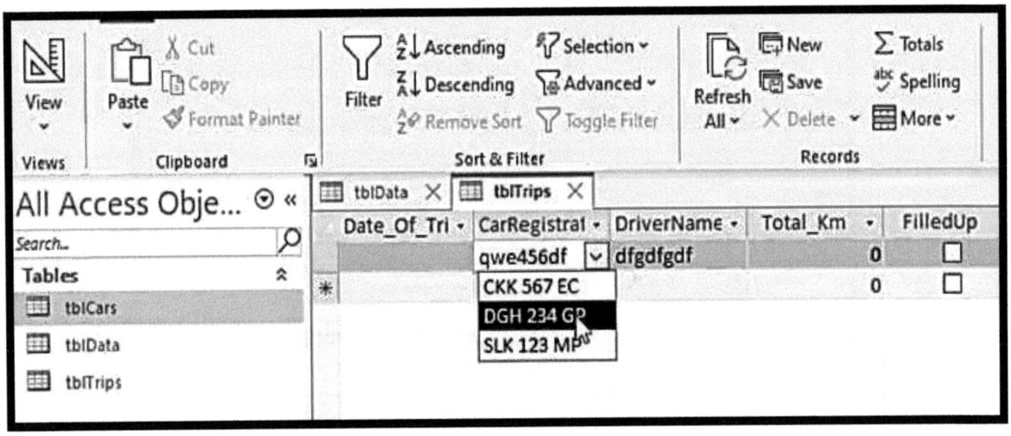

To recap: if you want to limit users to a list of options for a particular field, you just click on the field and it might say Short Text in it, but you go to the Lookup Wizard and you just follow the instructions. If you've already done it, whenever you click on that field, there should be the lookup option where you can change the fields. If it's from a Query, you can see what the options are, and if it was a value list, you can see the values there, and you can change those values as well.

Relationships Between Tables

Here, we will talk about relationships between tables and how to make sure that the data that you store in your database tables can work nicely together whenever you do Queries.

Before we go into that, you need to understand how you can set up or create your tables in such a way that they can be used in relationships and this is just to give you an introduction to what you should think about when you are creating your tables.

When we talk about **Relational Databases**, what we mean is that the **data in your tables are linked to each other** so that the data is common to both tables. This applies if you don't want a table with information on multiple topics and you want each table to store data on a particular topic or aspect. For any aspects that are linked together, you can create some sort of connection to minimize duplication of data or redundancy. So, there are lots of good reasons to have a relational database.

Let's look at the example below. Here, we have a table about trips and then you can see that this includes details of the date of the trip, the car registration and the drivers' details which includes the surname, first name, the contact number, the kilometers for the trip, and then if the person filled up.

tblTrips

Date_Of_Trip	CarRegistration	Surname	FirstName	ContactNum	Total_Km	FilledUp
2021/01/07	SNH 582 GP	Miller	Walter	075 511 4763	456	☑
2021/01/08	HDS 761 EC	Francis	Adena	054 378 1367	256	☐
2021/01/08	BCD 123 MP	Knight	Russell	099 951 5753	129	☐
2021/01/11	FWZ 492 NW	Navarro	Dora	047 961 9162	789	☑
2021/01/11	HDJ 392 EC	Fry	Vladimir	098 447 3645	147	☐
2021/01/11	SNH 582 GP	Baldwin	Wilma	083 952 8772	327	☑
2021/01/11	HDS 761 EC	Vasquez	Randall	038 584 4284	189	☐
2021/01/11	BCD 123 MP	Marshall	Mufutau	026 737 9769	145	☑
2021/01/12	BCD 123 MP	Knight	Russell	099 951 5753	657	☑
2021/01/12	FWZ 492 NW	Palmer	Theodore	069 147 3195	166	☐
2021/01/12	SNH 582 GP	Miller	Walter	075 511 4763	557	☑

That's great, but the problem with this table is that the section about the drivers has a lot of redundant information, and you've got a few people here that appear multiple times. So, there are lots of data that are duplicated, and there are lots of things that can go wrong here. What happens if you fill in the wrong phone number in one record and it is different in the other record? Or if you want to change a phone number? You have to change it in multiple places and if you delete all of those particular trips, there'll no longer be any details about the driver. These are all anomalies that you can have when you've got database tables like this. There's just too much data and you need to find a way to minimize this. You should have another table that is just about the drivers, where you give it a primary key and you can see that each driver now has a unique ID that you can refer to them individually. Now for all that bulky information about the drivers, you don't need all that. What you need now is a connection; you need to find a way that you can attach that first record to the second one. So instead of storing all that data, you store the Driver ID in the Trips table and a driver can be identified in this table by his unique number in the second record where you have all his details. With this in mind, what you want is to take that driver's ID and form some sort of relationship between the driver ID of the drivers and the driver ID of the trips.

They don't necessarily need to have the same field name, they could have different field names, but for convenience, you can use similar field names. Some terminology that we are going to refer to is that the Drivers table is the **Parent Table** that contains the **parent information** of the drivers and the driver ID is what we call a **Primary Key** that uniquely identifies every driver.

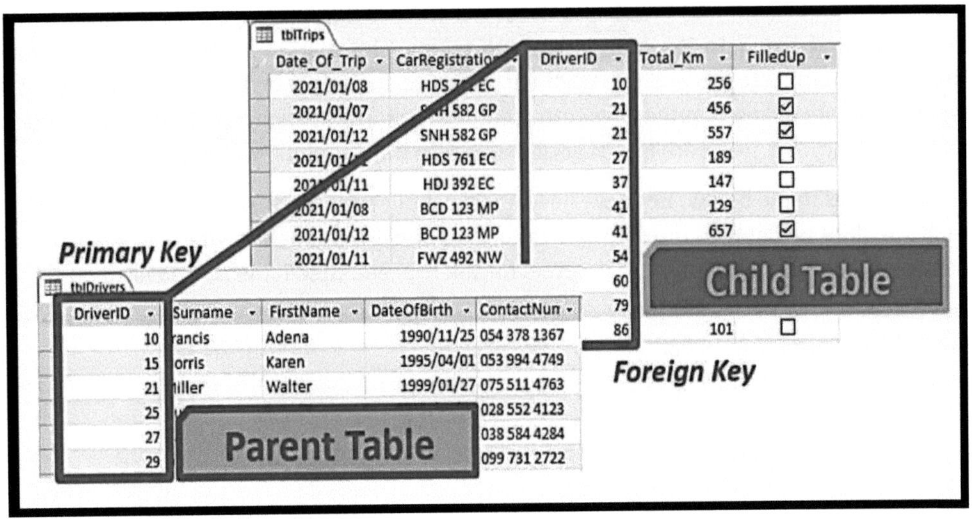

In the trips table, this is the **Child Table** that is dependent on the information in the parent table and although the driver ID is not the primary key, it is a primary key from another table; in other words, we call it a **foreign key**, which means it's a field from another table that makes the trip table link to the driver's table. The drivers' IDs have nothing to do with the details of the trip except that it links to who the drivers are, which is stored in another table. So that's the relationship that you want to create. So, the goal is to create a link between the driver ID of the drivers and the driver ID of the trips.

When you do this a little box will pop up asking how you want to edit this relationship. There's also a section over here that says "**Enforce Referential Integrity**". If you click on that, that is going to tell Access to ensure that your data is accurate and that people don't put anything in there, in other words, orphan records. Now, what's an **Orphan Record**? Well, an orphan is somebody that doesn't have a parent. So, if you think about that for example, you know that the child table is the details of the trips; imagine someone puts in details of a trip and they give the driver an ID that isn't part of the records. So now you've got this random driver that's been allocated to a trip but you don't have any details about them; that's an example of an orphan record.

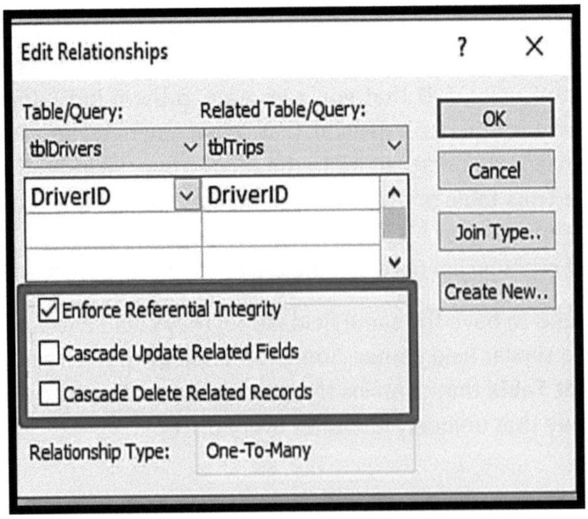

If you want to change a driver's ID in the driver table it's not going to allow you because it's going to affect the fields in the other table. What you could do though is to do the "**Cascade Update**" and the "**Cascade Delete**". What the Cascade Update would do is that if you change driver ID 20 in the driver table to 21, it would automatically go to the trips table and change all the 20s to 21 or whatever you're going to change it to. So, you just change the driver ID and it will automatically change the corresponding IDs in the trip table as well. The Cascade Delete means if you delete driver number 10, it will go to the trips table and delete all the records that contain driver 10 so that your data has integrity.

There's another thing, the **Relationship Type**: **one-to-many**. That's the type of relationship in this scenario, so if you think about the driver table, it can only have one person who has driver ID 10 because that's the primary key but it's possible that the driver ID 10 could occur multiple times in the trips because they could do multiple trips; so that's what we mean by one-to-many Relationship.

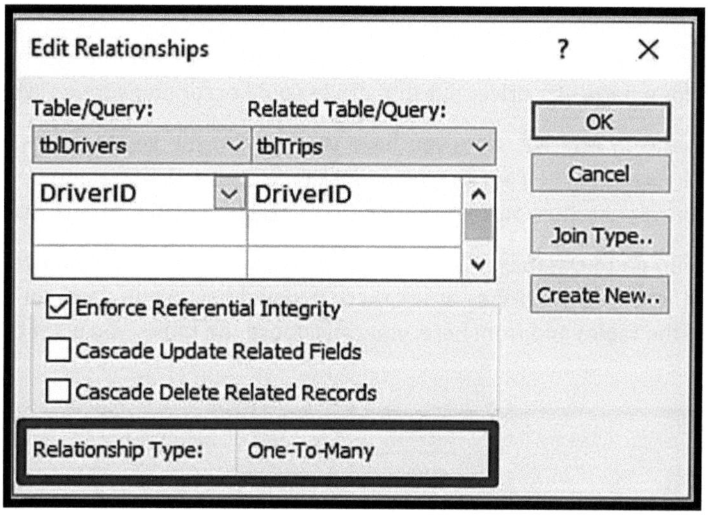

If a driver can only ever do one trip, in that case, it would be a One-to-one Relationship because one driver 10 would occur in the driver table and one driver 10 would occur in the trips table.

There are different types of relationships and there are diagrams that show the relationships with their terminologies.

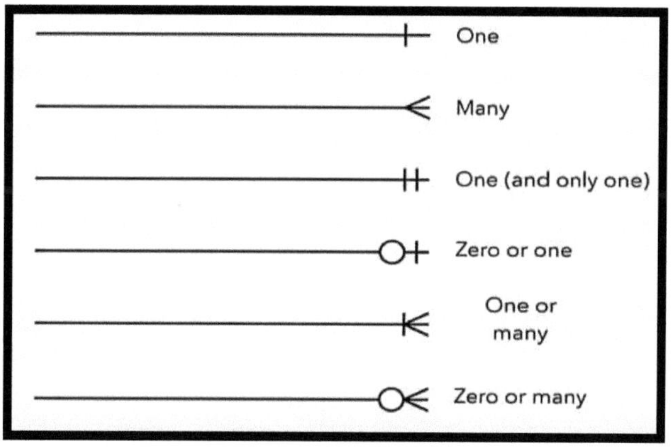

You can see the symbols they use whenever they do an ERD diagram that indicates the relation between tables. When you enforce interference or integrity for a one-to-many relationship, when you click "**OK**" on that, you'll see that one and that infinity sign which means many.

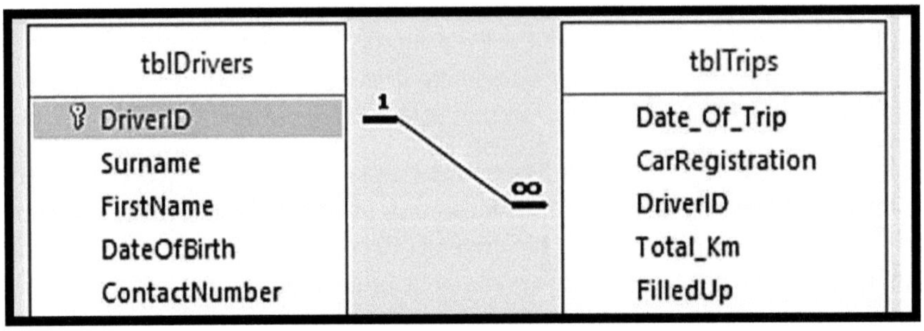

In this case, you can see it's a one-to-many relationship, in other words, there can only be one driver ID in the driver table for a particular driver but that driver could occur many times in the trips table.

So how do you do this in Access? Now you have your tables, the key thing you must do before you proceed is to make sure that they are the same data type and that is the number field. When you are going to create your relationships make sure that all your tables involved are closed.

Now you are going to go to Database Tools and go to Relationships. When you go there, you'll see the tables are already there now but if they aren't there it would be a blank slate, something would pop up where you can add the tables and from here, you can choose the tables you want to add and it'll add the tables.

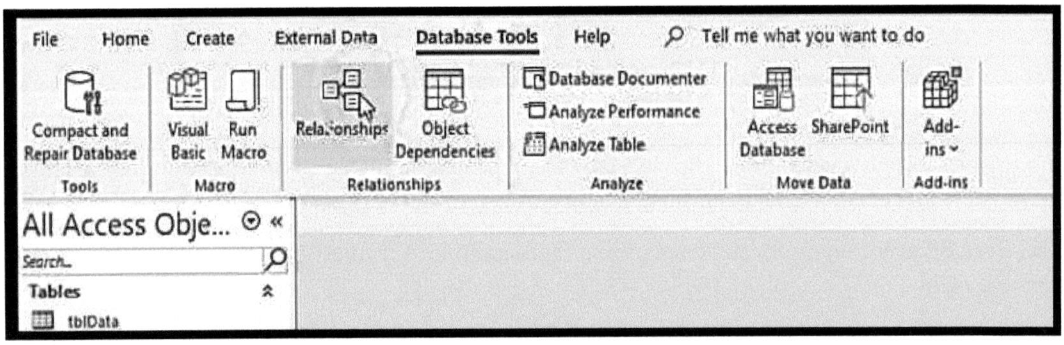

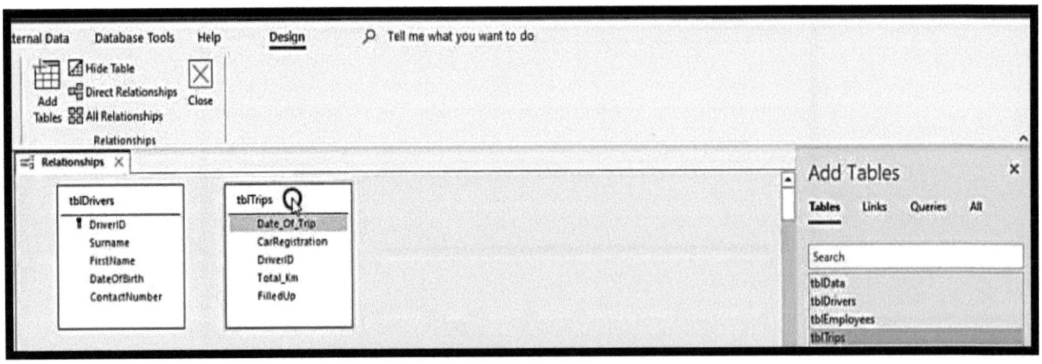

Now if you want to link that driver ID of the first table with the driver ID of the second table, all you have to do is to **drag** it over to the other driver ID and confirm that the drivers' IDs are linked together. Next, in the Relationship type, choose the "**one-to-many**" option, click "**Create**" and it makes that line for you.

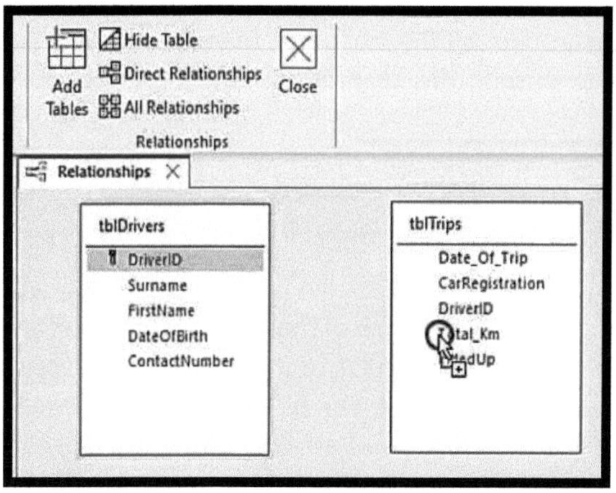

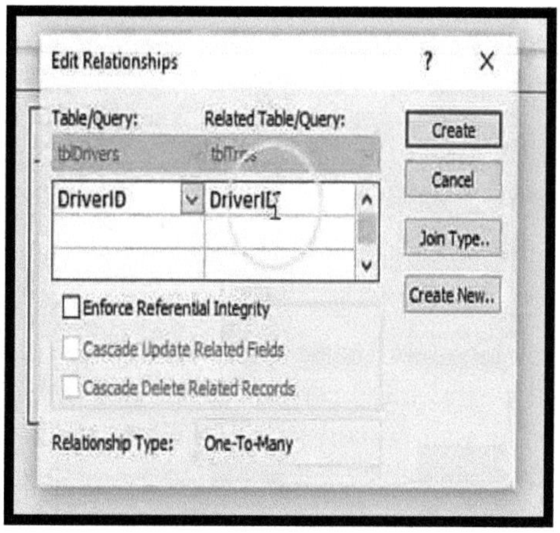

If you want to get rid of it, you can right-click and you can **delete** the relationship.

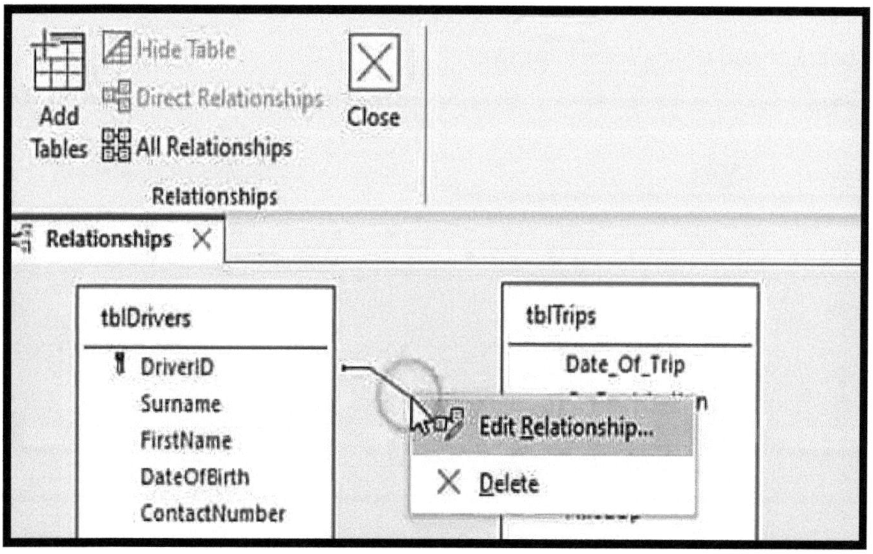

Sometimes if you've made a mistake in a field, if you want to change a primary key or the details of the primary key, it won't allow you because you've got a relationship existing there. So, you need to **undo** the relationship, go make the changes then **redo** the relationship, or you can edit the relationship and you can set the "**Enforce Referential Integrity**", "**Cascade Update**" or "**Cascade Delete**" if you want to do that.

So, that's how you create a relationship; you simply take the one field, drag it to the field that you want, and then when that box pops up make sure that all the details are correct, and make sure that you save it, after which you can close and now you can work with your tables again.

CHAPTER FOUR

WORKING WITH FORMS

So far, you've been entering data directly into the table view, but let's say you want to have other people in your organization come in and add data. This might not be the most user-friendly view, instead, you might want to create a form that makes that easier, so in this chapter, we're going to be looking at Microsoft Access Forms.

What Is a Form?

A Form is a way to **view** and **edit** the data, particularly in a table, sometimes in a query, in a more aesthetic way.

If you remember when we did tables, we worked with a whole bunch of rows and columns and you saw all the data, now when you are dealing with other people, they're going to use the database. They might not like your current layout; it might not be very user-friendly to them. It's also nice just to view each record individually and be able to see all that information and that's whether you want to change it, view it, or even insert a new record. So, what you would do is to convert that table into a form; it doesn't change the table, it just means you create this other thing called a form that connects to the table. Whatever you do to the form will happen to the table, so if you change something in the form, it'll change it in the original table.

The image below shows what a Form looks like.

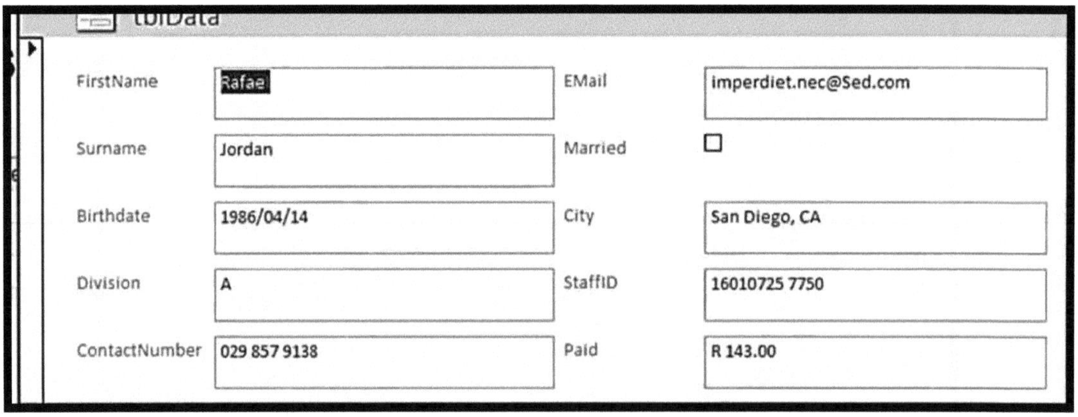

As you can see it's a lot nicer to look at, a lot more laid out, you can see one record in full, you can adjust how the data is presented, you can make the columns a bit smaller, and that makes it look more appealing to the person that's going to use it.

Getting Familiar with Forms

Before we get into forms, let's just get some terminology right so you understand. Over here on the left side, there are little blocks on the side and those are called **Labels**. They will display whatever the names of your fields are, so we've got the first name with no space and you can see that's how it is displayed there.

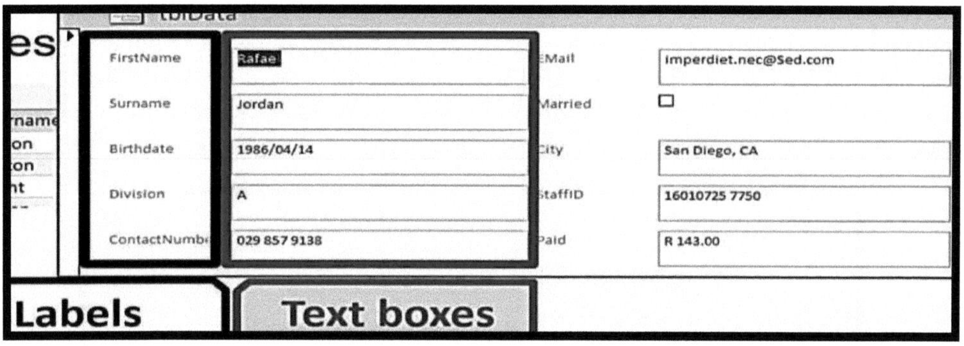

The boxes next to them are called **Text boxes** and that's where the data for those fields will be put; so, for each record, when you move through the different records those will change depending on which record you are in.

Now, let's go into Access so you can develop your first form.

Creating Your Form

To create a form, go to the Create option and navigate to Forms, and right there in the middle, there is a section for Forms.

There's a form wizard that will help you through the process, you can also start from blank. Click on the **Form Wizard** and it's going to take you through a step-by-step process of trying to find out what you want in your form.

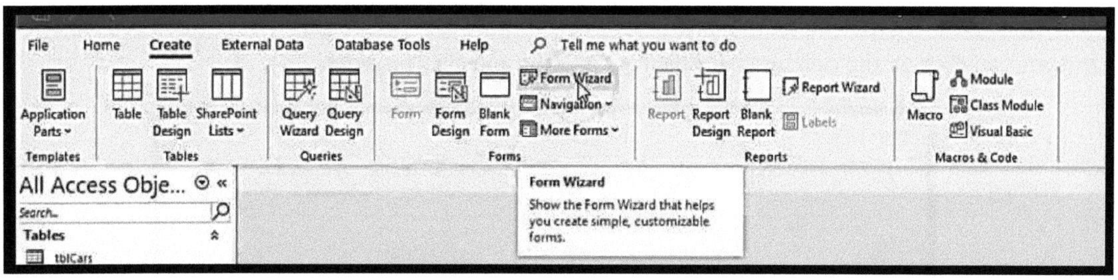

After clicking on the Form Wizard, it's going to pop up a little window.

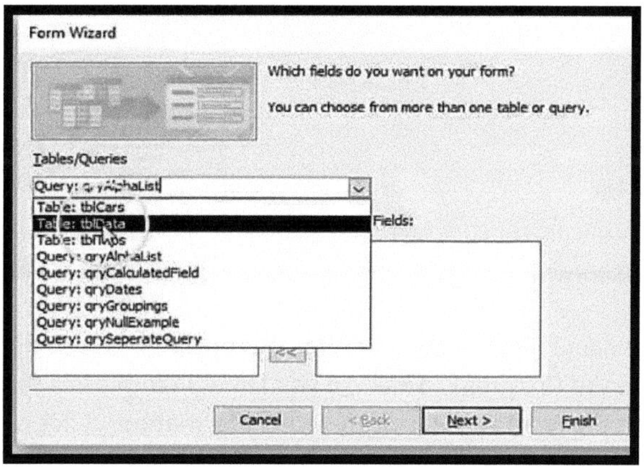

You can base a form off of a query or off of a table, but for this illustration, go to a table, and then you'll see all the fields available to you and here you can say which fields you want to put into your form and the ones that stay here are the ones you don't want on your form. So, if you want all of them, you can just click on the two arrows there and it puts all of them there or you can just add a couple of them in if you want. After selecting the fields, you want, click on **Next**.

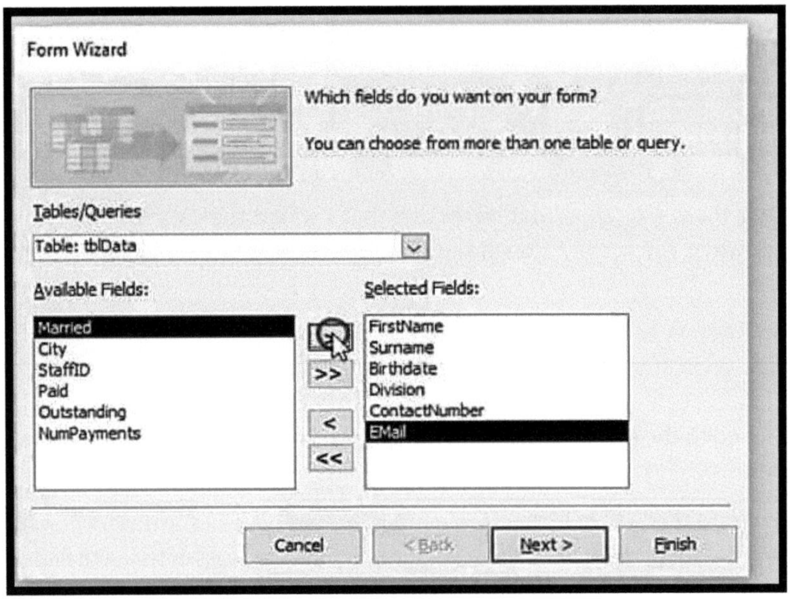

Now, you get to choose the type of layout you want. There are lots of different options available to you and you can click on the default one which is Columns and click on **Next**.

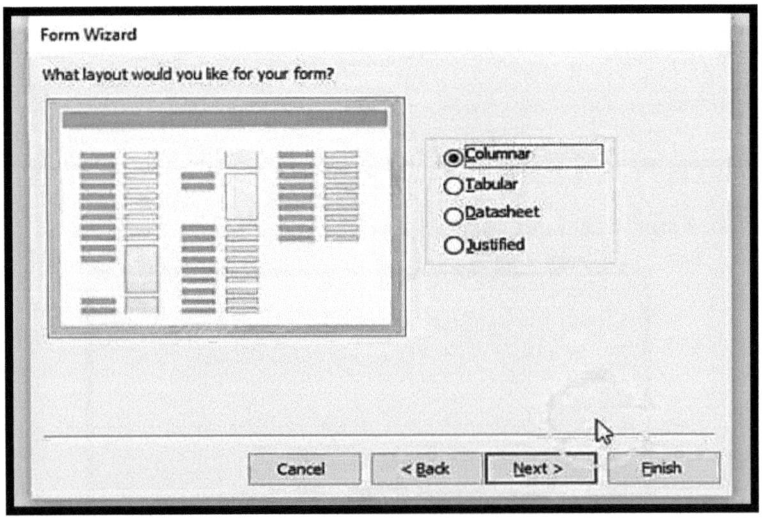

Now, it's time to give it a name. Remember your naming conventions, you don't want to give it the same name as a table, you can use the prefix "**FRM**" so you know that this is a form and you can click on the option that says to "**open the form so you can view the information**". Click on "**Finish**" and there you go; you've got a nice little layout all done for you.

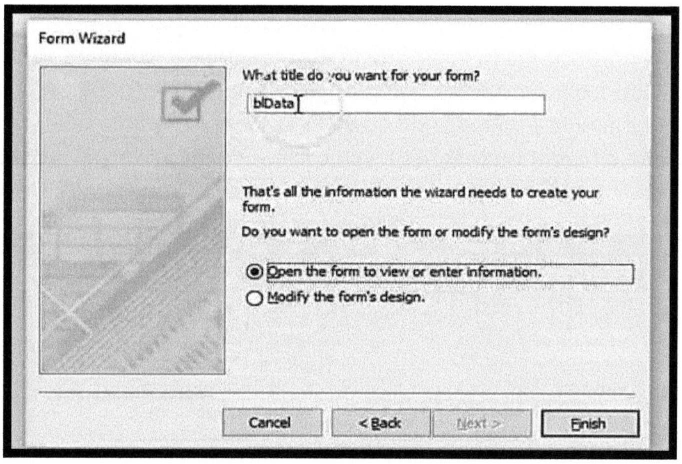

Your first record doesn't have all the data entered so you can start adding some information and when you are done, click **Save**.

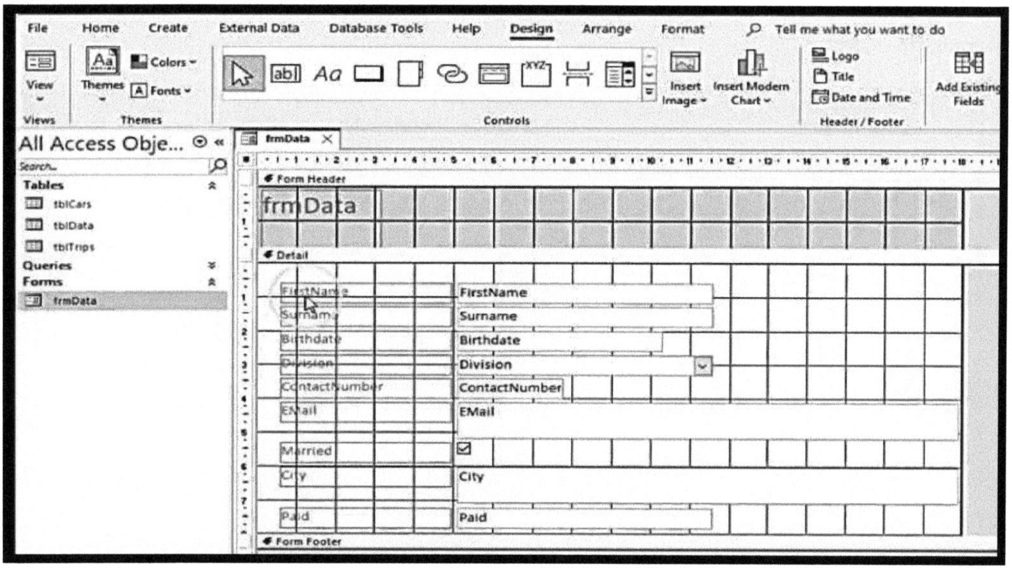

If you had to go back to the table, there's a link now between the fields here and the fields in the form. At the bottom of the form, you can move through the different records and you can see all the individual records. That's how easy it is to make a form.

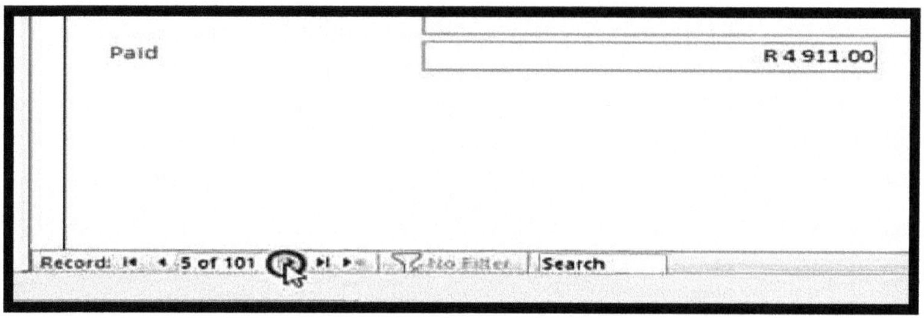

If you prefer to use the **Form Design** instead, click on that instead. This now drops you into a new form, so this will be a lot easier for others to come into and start adding information and they can also use it to review information. Now because you connected the customer table to the orders table, you'll see here all of the customer information shows up and down below it has all of the related order information. So, if you jump through the different records here, you can see your customers and the orders or purchases associated with them.

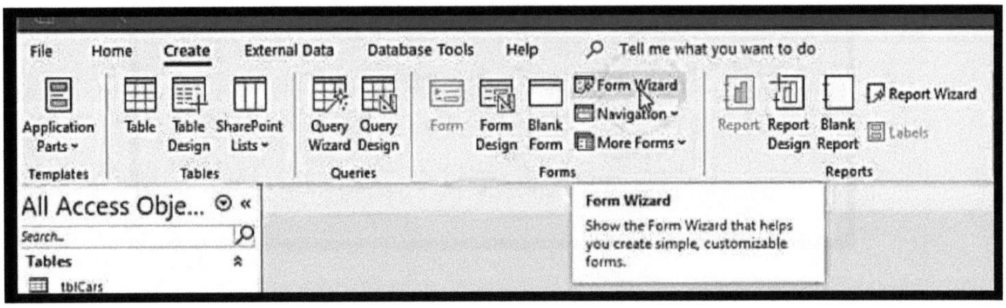

Right at the top, you can design what your form looks like, so you can choose different themes and colors, you can add different controls to your Form, you can even add a logo if you want to personalize it with your company logo.

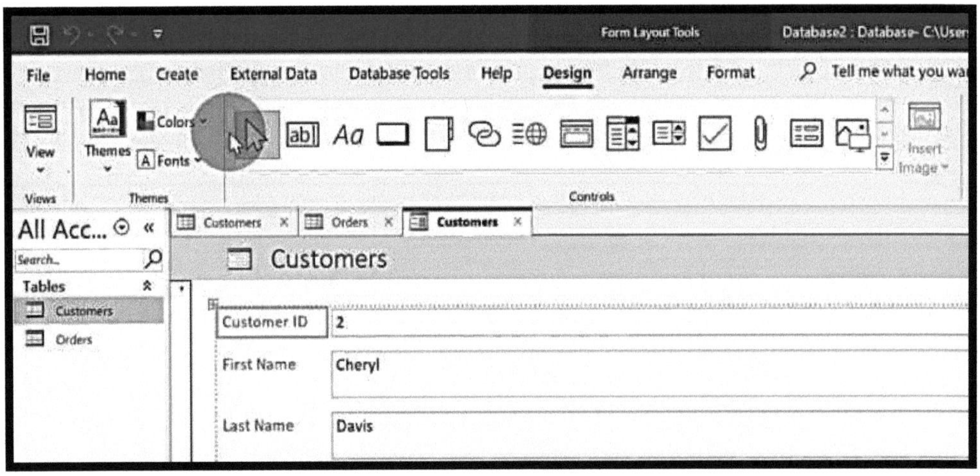

Down in the bottom right-hand corner just like we could do in the table view, you can launch the design view and within the design view, you can modify what this form looks like. Right up here there's a Form header, there are also details and you can see a footer. You can take these different elements and move them around; you can design the form how you want it to look.

Over on the left-hand side right down here, you can also launch the **Form view**.

If you're going to have people in your organization going through and filling out forms, this will likely be the view that they see. Here, they'll see all the customer information but they won't be able to modify the form. To add a new customer record, you simply go down to the bottom and you can click on the "New record" icon to add a new blank record. When you click on that you can then go through and you can fill out the form and right down here someone could go in and add some additional IDs. This makes it easy to get new data into your database without having to see the tables and all of the details of the database.

Once you're all done customizing your form you can close out this form by clicking on the "**X**" icon at the top and clicking on Yes to save it, then you could give it a name and once you're done typing in the name click on "**OK**".

Over on the left-hand side, you'll see that it added a new category so you have your tables and now you have a new form. You can double click on the Form and this will open up that Form again.

Customizing Your Form

If you've closed your form and you want to change a few things about how it looks, you can right-click on the form on the left-hand side of your workspace and then go to the design view.

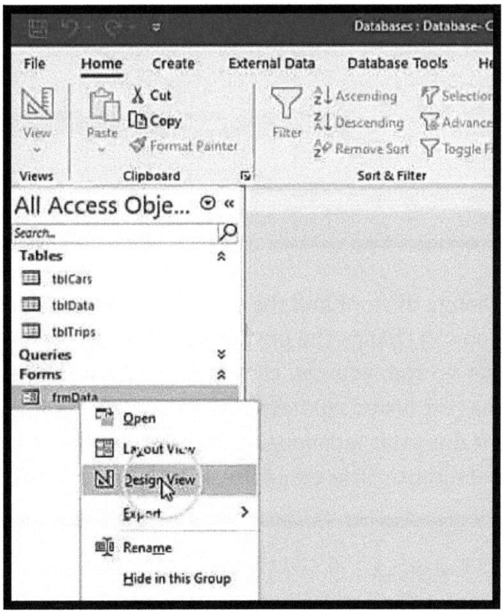

In this view, you can change the labels into whatever you want, make it just a little bit better to view and it's not going to change the data because the data is going to go into those text boxes.

At the top, you have the header which is what is always displayed at the top, and at the bottom, you have the footer. You can **right-click** on those areas and you get some options like colors, grids, and other options.

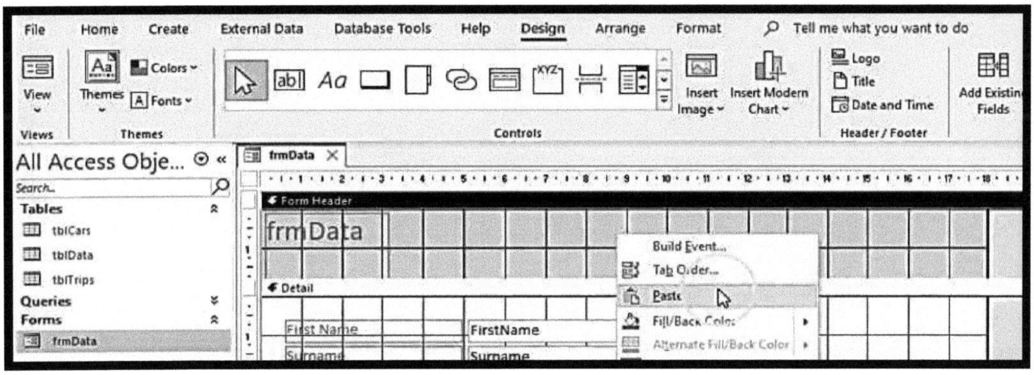

There are lots of other things you can add. Over the top there you can see you can add a logo, a title, or the date and the time. If you want this in the footer you can just click on that, cut it and then come down to the bottom and then just paste it.

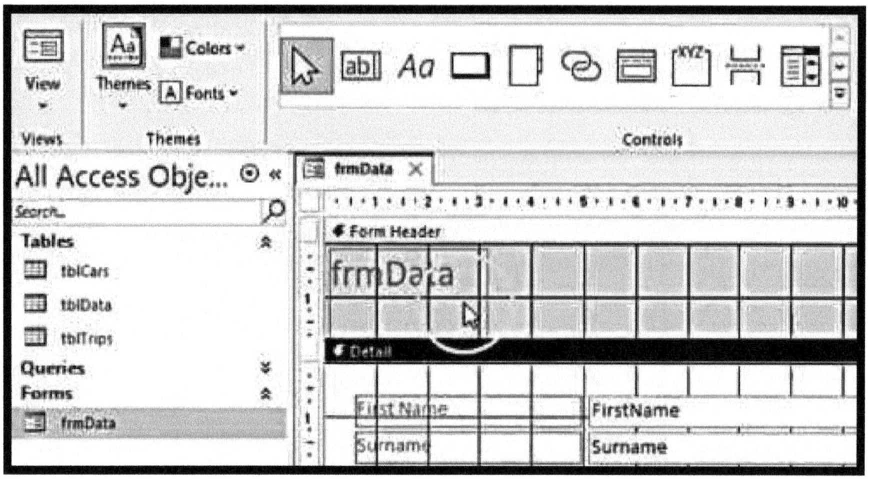

If you go to Home, you can change the font and the color of all those different things there. Another thing that you can do in forms is you can change the positioning of these values. So, let's say you want the first name and surname to be like on top, you can click and move them to the position you want to place them. if you want to place the text boxes separately, then click on the little gray box close to the text box and when you move them, it does this individually. So, if you don't click on those boxes, it'll move the whole combo but if you click on those little gray boxes, you can move them individually.

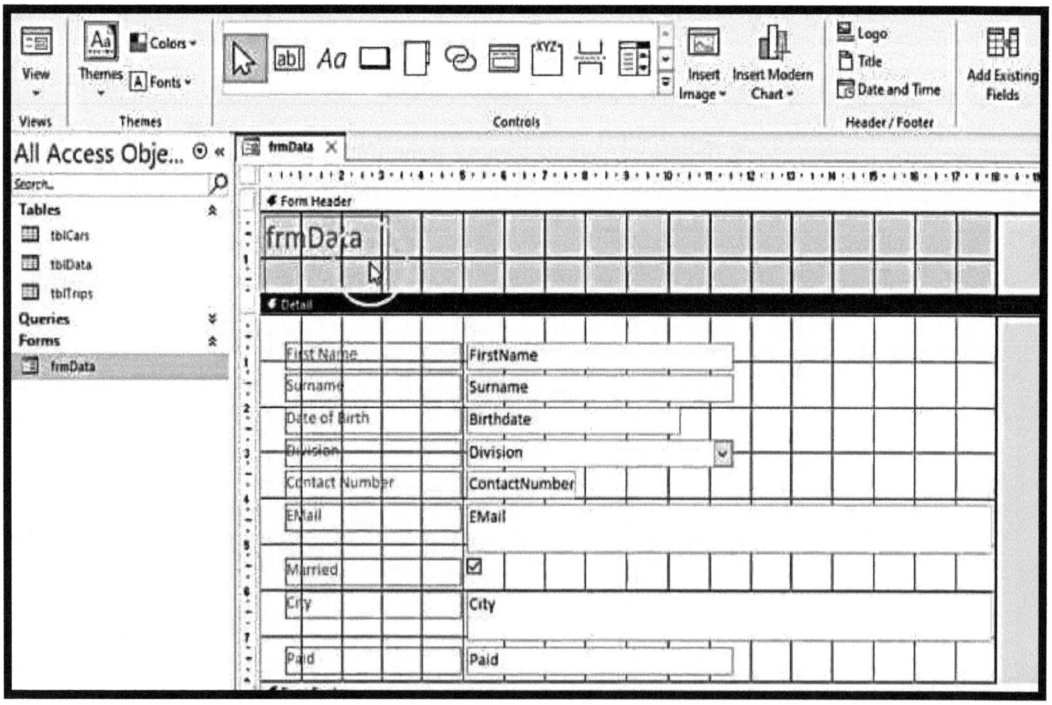

Also, if you **right-click** on a particular form, you have a lot of options to help you customize your form.

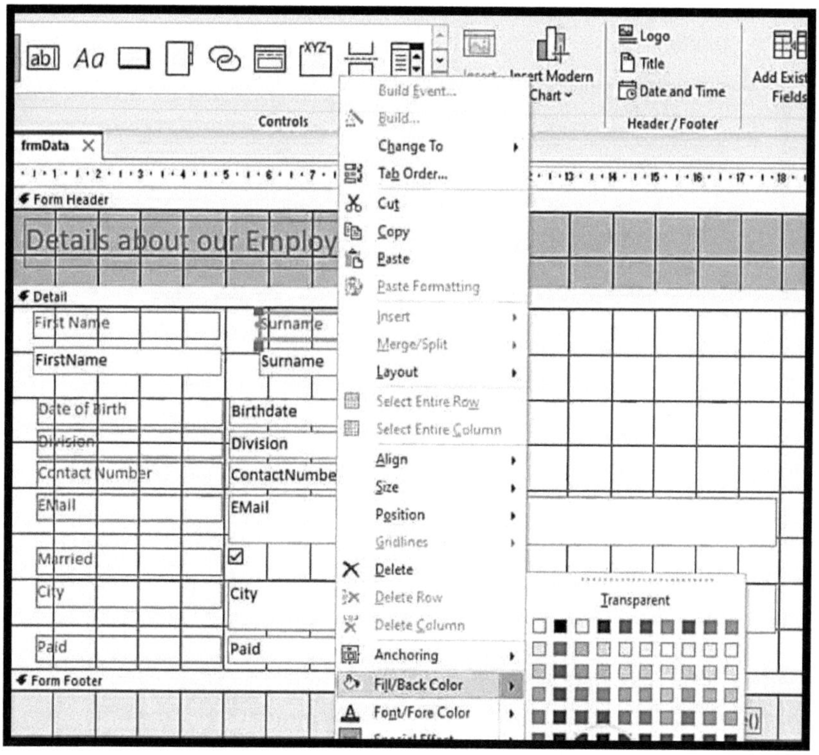

You can decide to go for **Conditional Formatting**. If you're familiar with Excel, this is the same as Excel, so you can put in a new rule.

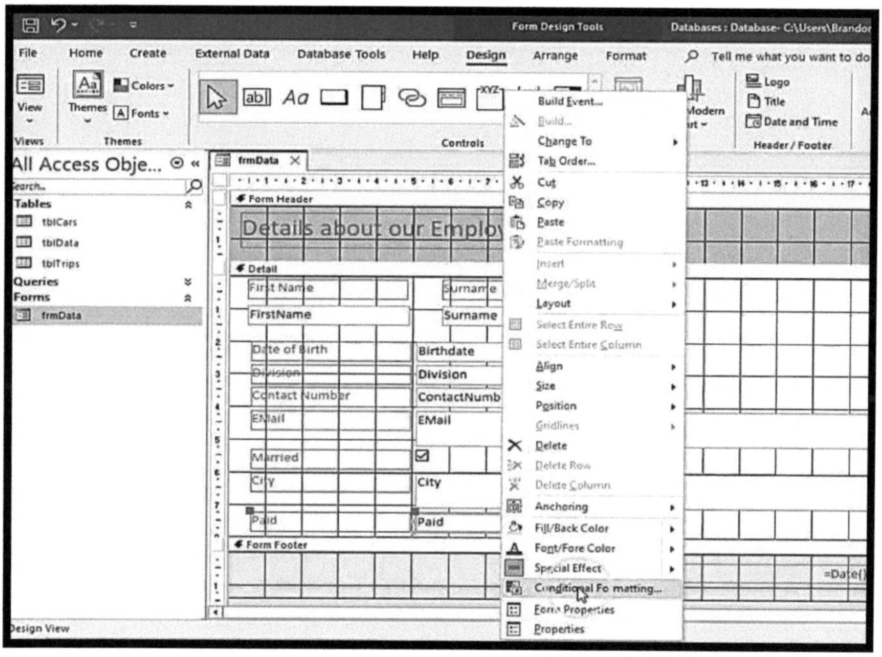

You can also change the background of the whole form, as you saw earlier with the background full color, but you can also click on different aspects as well. You just have to right-click and change the fill color of these particular components.

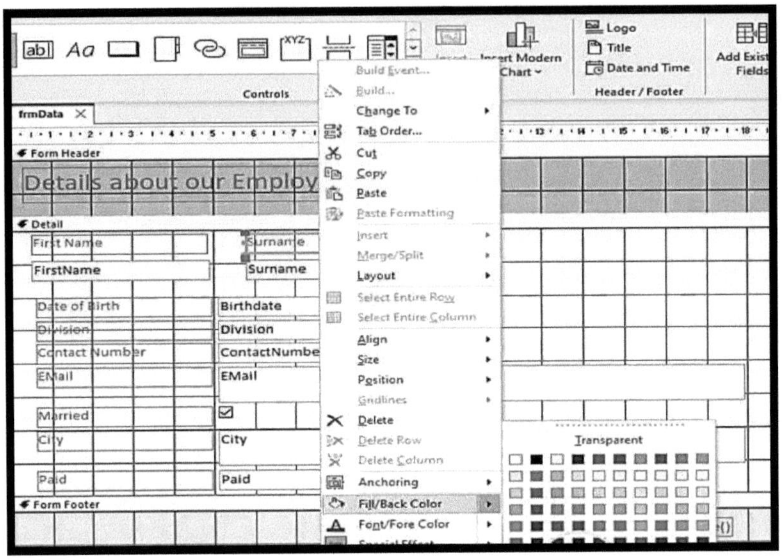

If you want to do it to a whole bunch of these components or all the labels, you can just click on one label, press and hold the **CTRL** key, then you can click on multiple ones at the same time, so that means they're all selected. So, when you change their fill color, it'll change all of them.

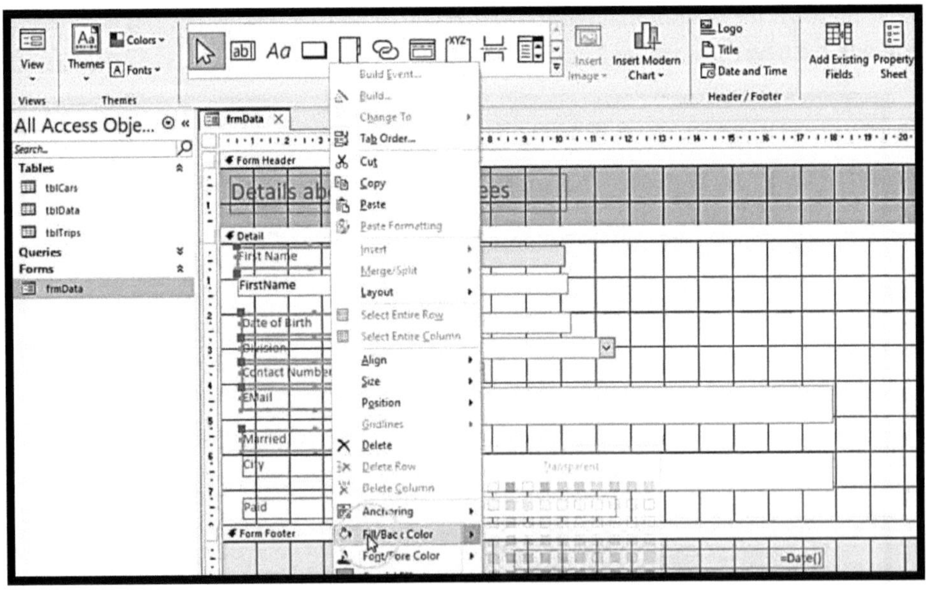

So, there are a lot of options to help you customize your forms and if you're not too sure you can always just right-click over here and see what options are available to you. In summary, you can create forms by going into the **Create** tab, **Form wizard**, and once you've got a form, you can just go to the **Forms' Design view** and change the things that you want to make it look a little bit better.

CHAPTER FIVE

QUERIES

Much of the power of databases comes from being able to run queries and in a moment, you'll see how you can run and also write your query. In this chapter, we will cover the basics of what a query is and how you can make a basic query.

How It Works

In the previous chapters, you learned that tables are the places where we store all the information in a database, we store information about particular topics in different tables and store massive amounts of data. When you've got massive amounts of data, it's very difficult to find the particular data that you want and so that's why we have an aspect of databases called Queries, and what Queries allow us to do is display just the relevant data that you want to be based on some criteria or conditions; so, you can get just the information that you want and then you can save these results to view them later or edit them or to use them in reports inside.

So, the benefit of queries is that you can take all these massive amounts of data and get just the information that you want. Let's use an example, if you go to the bank, the bank has a database with everyone's records and transactions and all the details of what things that they've done but when you go onto an atm or go into the bank they will bring up just your record, so they will run a query which runs just to get your specific information with your account details and your transactions, and that way they can only deal with yours; they don't just sift through massive amounts of data to find just you're in particular records they can just get your specific data.

When we deal with queries there are some things that we need to be able to use in comparing something to another when we talk about the criteria or the conditions. Remember the **operators** that we discussed in the previous chapters, so, you can check if something's equal to another by using the equal sign (=). These other symbols will be familiar to you, for example, if you want to check if something's bigger you must use the "greater than" (<) symbol, if you want to find "smaller than" fields, for example, if they paid less than 200, you use the less than (<) symbol, if you want "Larger than and equal to" you use the symbol "**>=**" or if you want "Less than and equal to", you use the symbol "**<=**".

Also, if you can compare two things to say that they are equal, you can compare things to see that they are not equal to and if you remember this symbol is usually written as the less than symbol and the greatest than symbol together, making it look like a diamond (**<>**).

Remember also that different data types need to be referenced in different ways; how do you know what symbol to use for the different data types? Well, let's just review that quickly.

If you've got numbers, you can just leave the numbers as is, you can say it's **= 200** or **< 200**, you don't have to change anything to the number.

That's easy but when you get to text then you must remember to put double quotes around the text. So, if you want a certain field to be "Smith", then you must put "Smith" around double-quotes. When it comes to dates, if you want to find all the dates before a particular year, let's say 2020, you can't say less than 2020 because 2020 isn't a date, that's a number. So, you must refer to dates when you're referring to the criteria. So, for example, if you want before 2020, then it must be **<#2020/01/01#**. You'll notice there that when we refer to dates you see that we put hashes around the dates; so, make sure that you mention the day of the month when you're referring to dates and then the dates will be in hashtags.

If you've got a Yes/no field, that's very easy. It might seem like it's text but it's not and you can just say that the field is equal to yes or equal to no with no double quotes around, or you could use the words true or false. There are no double quotes or hashtags around it, you just leave it as.

Now that you know this, let's go try a couple of examples.

Creating A Query

Before you create a query, it's very important to know what the data looks like in the main table because if you're going to do queries you need to refer to it correctly based on the data type and the type of information that you're looking for.

If you want to create your first query, it's very easy. You just click on the "**Create**" tab, go to the Query options and click on Query design. A box will pop up and it'll ask for the table you want to use, so you're going to double click on that and it adds it to the ones that you are using and you can now close the tables option. If you didn't do it correctly and you want to add more tables you can just go click on the "**Add Tables**" button at the top and it will get you back to the options to add tables.

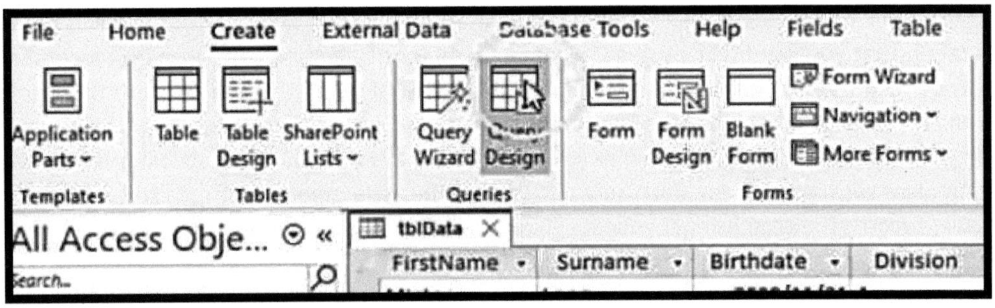

Sometimes you may not want to display all the fields in a table, if you just want a list of all the names and the surnames then in the table below, you can select which fields you want and then you can run the query by clicking on the "**Run**" button at the top. You can either run it or you can go and look at the **datasheet view**; both of them do the same thing technically.

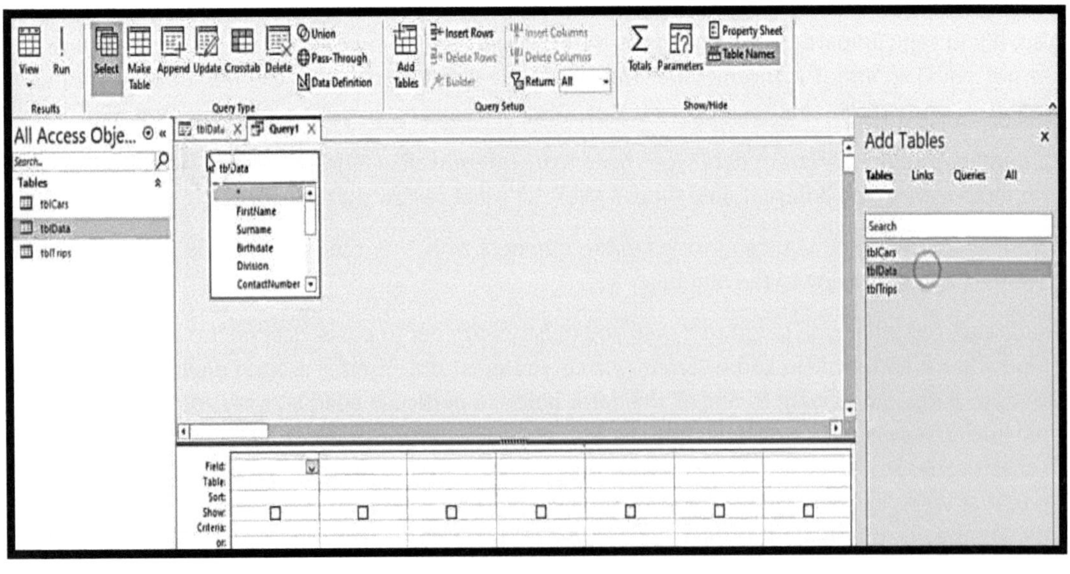

If you want to sort the information, go back to the design view and you can edit this table, click on **Sort**, and now if you **run** it, you can see it's been sorted how you want it. When you are sorting, sometimes if you've got multiple sort criteria, it must read from left to right. So just remember when you are sorting, make sure that you get your order correct.

If you want to save this query so you can access it at a later time, you can just click on the Save button and it will ask you to name the query. Like we did with forms, you can add a prefix such as "**QRY**" to your data name so this way you know the difference between queries and tables, and once you've done that, you've got your query that you can access at a later time.

If you close this and you want to go open up the query you can just double click on it and there you have all the data that you saved.

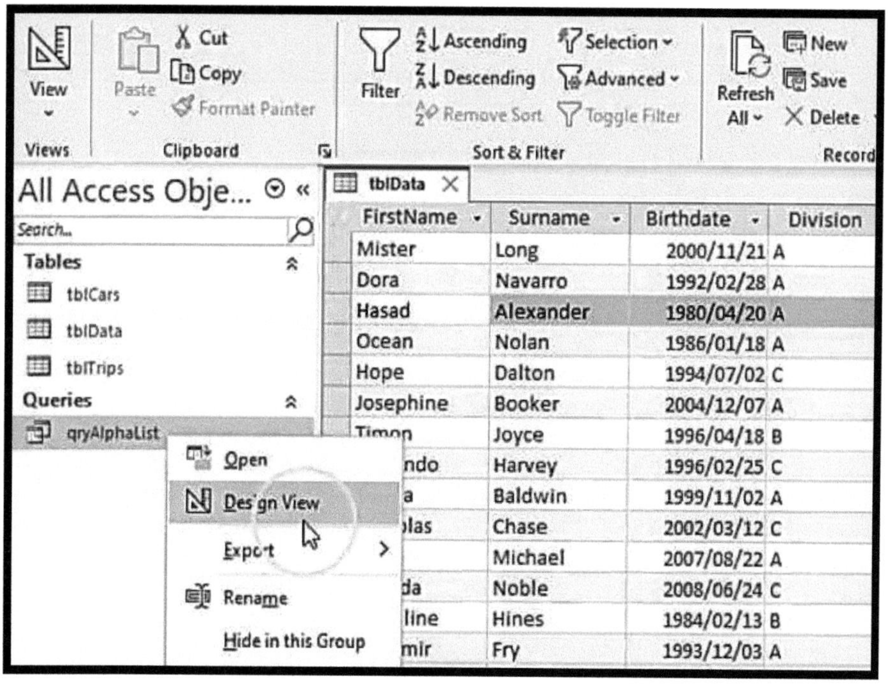

If you close it and you want to make changes to it you can also right-click on it, go to design view, and that way you can change the properties here or you can just go through those options available to you for your list and any data that changes in the original table will then also be reflected in the query; so, you don't have to keep updating it.

Working It Out

Using the table below, we want to find all the people that made three payments. So, if we go back to our query, Under the Fields section, select the Surname, First name, and the number of payments.

In the criteria section, here is where you put the criteria so you can say = **3**; because it says equal, you can just put the 3 and that's fine as well. If we go view this query, you can see we just have a list of those that paid 3 times.

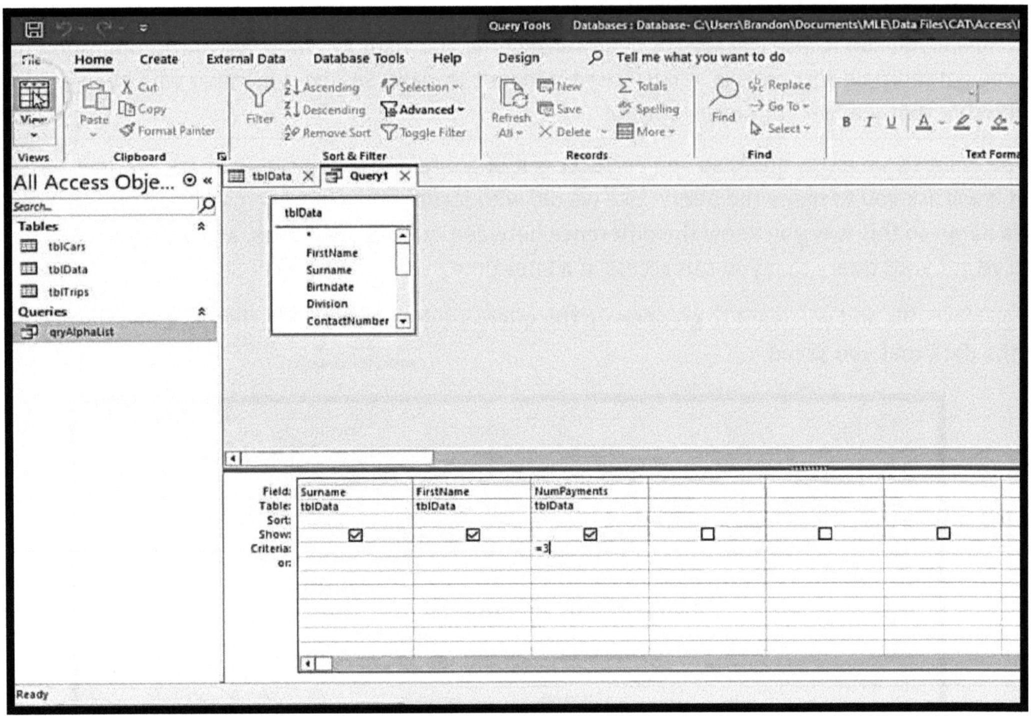

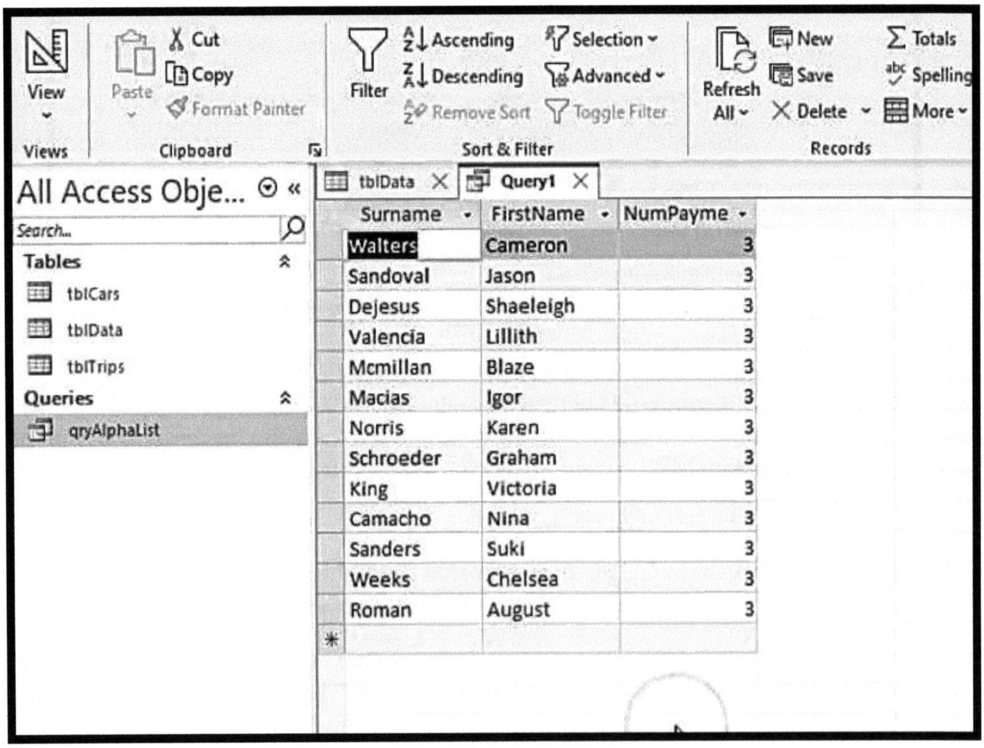

If you don't want to see the Number of payment field but you just want the names and surnames and you want to use it as a criterion, then you can click on the box under the "**Show**" section, if you go to

"**Run**', you can see the same data with that field hidden. So, note that if you hide a field and there are no criteria in it, when you go back it'll probably take it away when you save it.

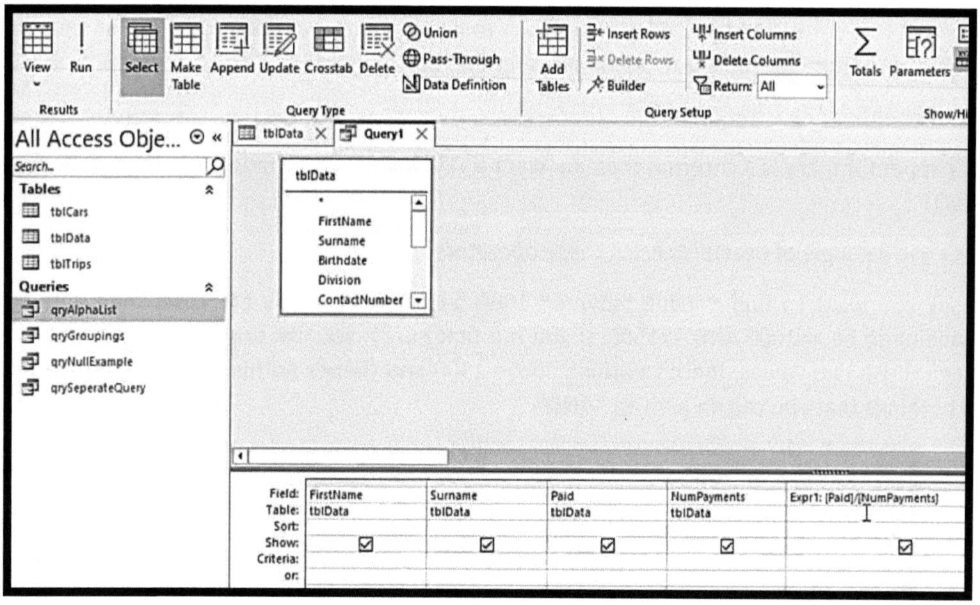

Also, if you want all the people that made one, two, or three payments, that means it must be less than and equal to 3 (**<=3**), so when you run it, you can see it displays all the people that made one to three payments.

Another example is that if we want to see all those in the outstanding column that was above 1000, we use the criteria of **>=1000** and we can see all the outstanding amounts that are above a thousand.

If we also want all those that have more than 1000 outstanding but they have only made one payment, this is to be **>=** 1000 but the number of payments is =1.

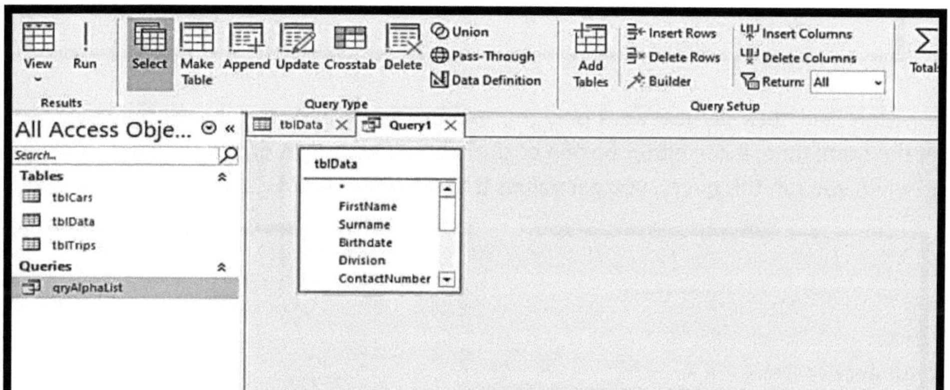

AND, OR and NOT

What happens if you have multiple criteria in the same field?

If we've got two criteria and you want them both to be true, then we're going to use an operator called the "**AND**" operator. So, you'll put "**AND**" between the two criteria.

If you want either one criterion or the other criteria, then we would use the "**OR**" criteria.

What happens if there is a criterion that you want but you want the opposite? In that case, you can use the "**NOT**".

So, let's use a couple of examples using these operators.

Let's say you want to find a value between 1000 and 1500. If that's the case, then you want your outstanding to be **>=**1000 AND **<=**1500. If you run that query, you can see all the outstanding amounts between those two values, there's nothing above 1500 and there's nothing below 1000. Those are the types of things that you can do with an "**AND**".

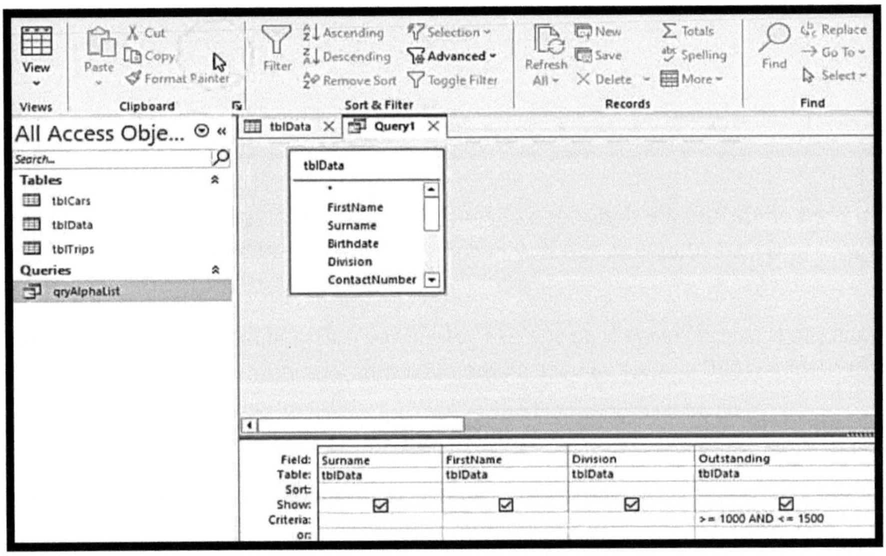

In the Division field, let's say you want only those in division A or C, this means they can't be in both A and C at the same time; it can either be one or the other. If you say A or C, it puts it in double-quotes for you and when you run this query, you get values that fall under A or C.

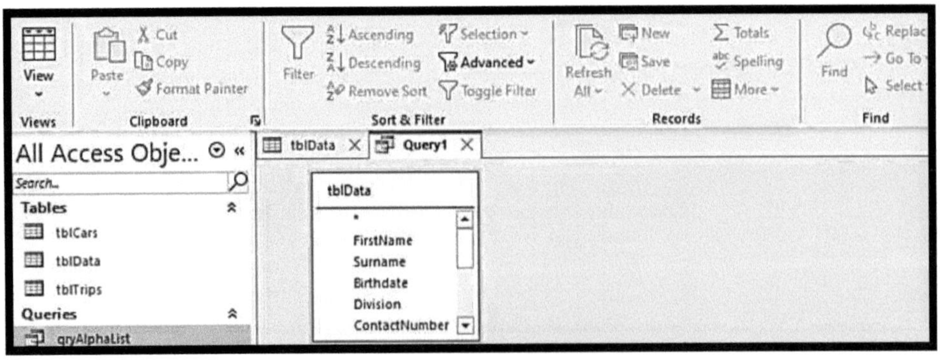

For the number of payments, let's say you only want those who made 1, 3, or 7 payments, you can say it's a 1 or 3 or 7. When you run it, only the 1s, 3s or 7s will be displayed.

Now let's say you want all those that did not make 3 payments so you could either say **<=** 3 or **not** =3 and if you run that you'll get the same results.

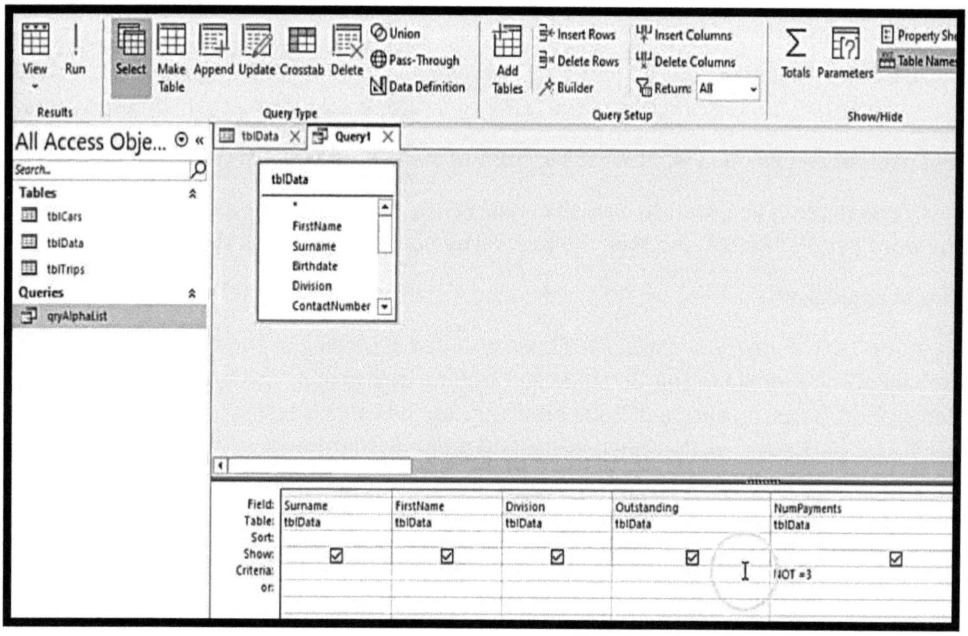

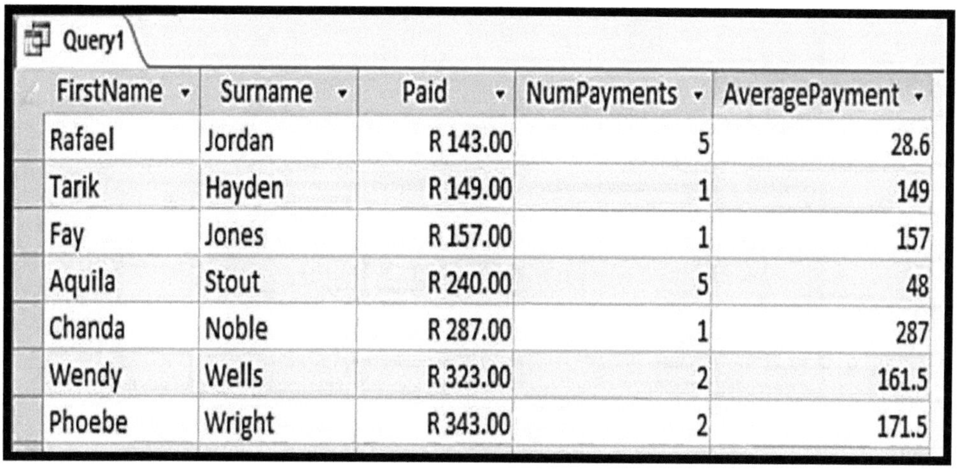

Query1

FirstName	Surname	Paid	NumPayments	AveragePayment
Rafael	Jordan	R 143.00	5	28.6
Tarik	Hayden	R 149.00	1	149
Fay	Jones	R 157.00	1	157
Aquila	Stout	R 240.00	5	48
Chanda	Noble	R 287.00	1	287
Wendy	Wells	R 323.00	2	161.5
Phoebe	Wright	R 343.00	2	171.5

LIKE and NULL

Let's take this scenario where you've got a field that has different types of car models. You can see it's a text field and it's got a lot of information including the year the car was made, as well as the type of car it is and what type of fuel it uses. This is a whole lot of information in one field but the problem is if you want to find just the cars in 2019 (for example), that's going to be difficult because you can't say where the car model equals 2019. In this case, we're going to have to use the "**LIKE**" operator with wildcards.

First of all, let's learn what a wild card is. The first wildcard is the Star symbol (*). This is the symbol for nothing or anything. So, when you put this symbol in a query, you are saying that it can be replaced by anything or nothing. So, if you are looking for all the cars in 2019, you know that the value starts with 2019. So, this can be written as **2019*** and this would be displayed as all the fields that have 2019 at the beginning and something or nothing after it, that is: **LIKE "2019*"**

Now if you type that into criteria or condition for query, it will put in automatically the "**LIKE**" for you and the double quotes around it, you can type it in yourself but that's what the "**LIKE**" does; it doesn't equal 2019*, it's **like 2019***. So, the function of the "**LIKE**" operator is to indicate that you used the wild card.

What happens when you don't want what's in front of a word but you just want the back of the word?

In the above example, if you want to have the word Petrol, you'll just have a star symbol for anything and then the word Petrol; ***Petrol**, and then the like will be put in for you with the quotes around it.

Grouping in Queries

Looking at the table below, you can see it's been grouped according to the married field and in the next field, we've got the average of the Outstanding Field for each group. This query gives a summary of the outstanding field, based on each particular grouping. So, the value is telling us that's the average amount outstanding for all the groups that are married and then the bottom value is the amount or the average amount outstanding by the people that are not married. So that's how we can create statistics and we can group data and get statistics on those different groupings.

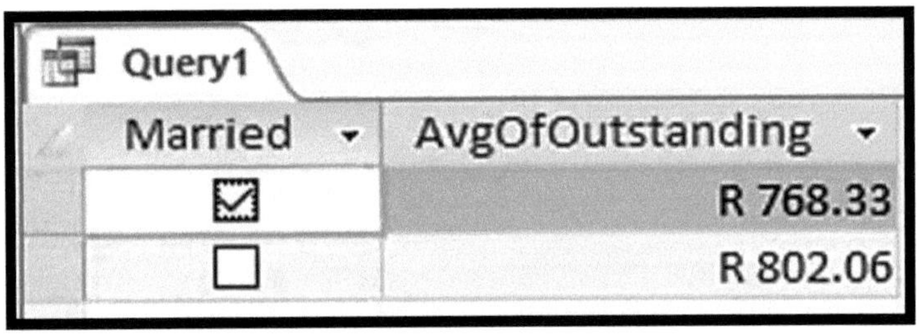

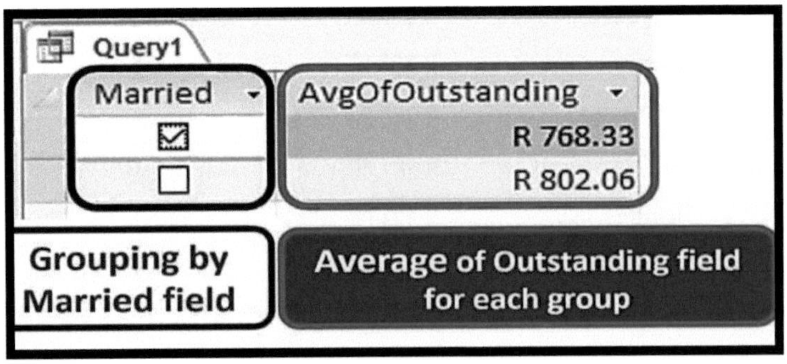

The table below shows another grouping that we can look at and, in this case, you can see that there is a **Division** (A, B, and C), and then we've also got "**Married**" as part of the grouping. In this case, we've grouped it first by division and then by marriage, so we've got two groupings here. Over there we've got the Paid field in an aggregate form, one scenario is to find the minimum of Paid and the other one is to find the maximum, so we have the lowest amount paid and how much is the most paid, but for each group. What this means in this table is that we have all the Division As that are married and all the Division

As that are not married, and you can also see the minimum and the maximum amount paid by the as that are married and the as that are not married. The same applies to Divisions B and C.

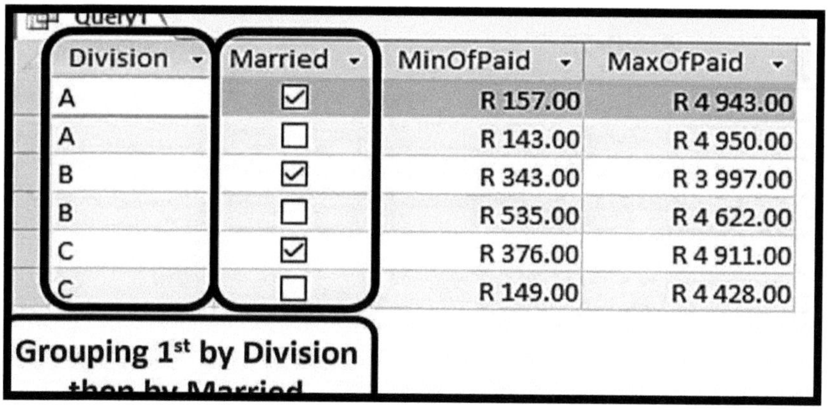

Division	Married	MinOfPaid	MaxOfPaid
A	☑	R 157.00	R 4 943.00
A	☐	R 143.00	R 4 950.00
B	☑	R 343.00	R 3 997.00
B	☐	R 535.00	R 4 622.00
C	☑	R 376.00	R 4 911.00
C	☐	R 149.00	R 4 428.00

This has been grouped by multiple groupings here and we found an aggregate field on the same field but different stats on that same field. So how do we do this in a query?

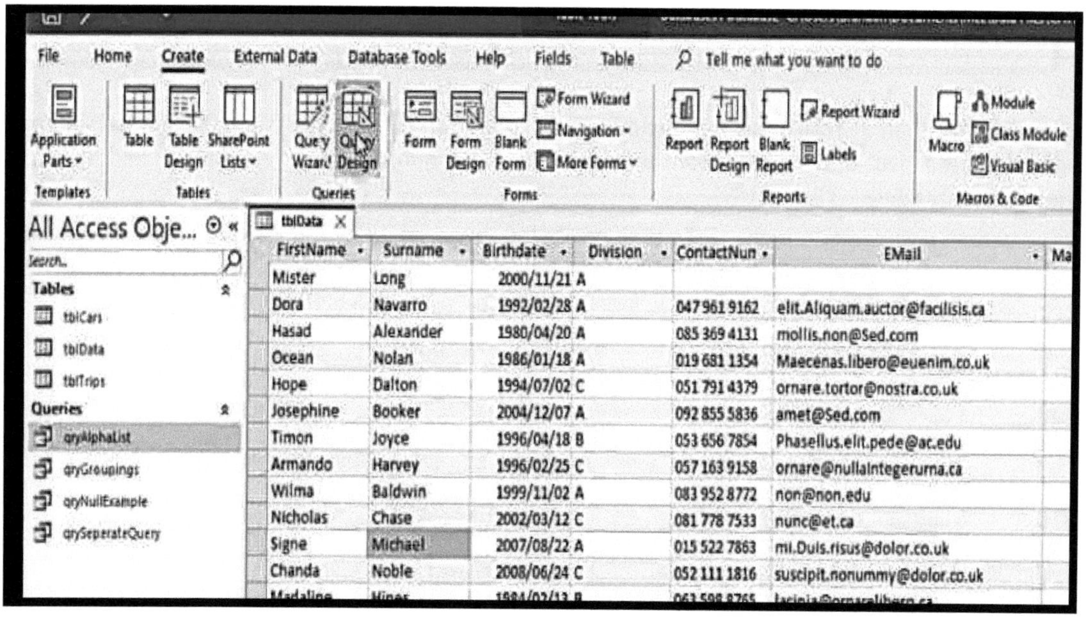

FirstName	Surname	Birthdate	Division	ContactNun	EMail
Mister	Long	2000/11/21	A		
Dora	Navarro	1992/02/28	A	047 961 9162	elit.Aliquam.auctor@facilisis.ca
Hasad	Alexander	1980/04/20	A	085 369 4131	mollis.non@Sed.com
Ocean	Nolan	1986/01/18	A	019 681 1354	Maecenas.libero@euenim.co.uk
Hope	Dalton	1994/07/02	C	051 791 4379	ornare.tortor@nostra.co.uk
Josephine	Booker	2004/12/07	A	092 855 5836	amet@Sed.com
Timon	Joyce	1996/04/18	B	053 656 7854	Phasellus.elit.pede@ac.edu
Armando	Harvey	1996/02/25	C	057 163 9158	ornare@nullaIntegerurna.ca
Wilma	Baldwin	1999/11/02	A	083 952 8772	non@non.edu
Nicholas	Chase	2002/03/12	C	081 778 7533	nunc@et.ca
Signe	Michael	2007/08/22	A	015 522 7863	mi.Duis.risus@dolor.co.uk
Chanda	Noble	2008/06/24	C	052 111 1816	suscipit.nonummy@dolor.co.uk

Using the table above, we are going to create a query, select the data table, and get all the fields. Under the fields, we select the data that indicates if they were married or not and we also select the value for outstanding. You can put all the fields that you want in the order that you want.

For this table, we want to group it by "**Married**" and we want to find the average of outstanding. To do that, we're going to come up here to the tool called **Totals**.

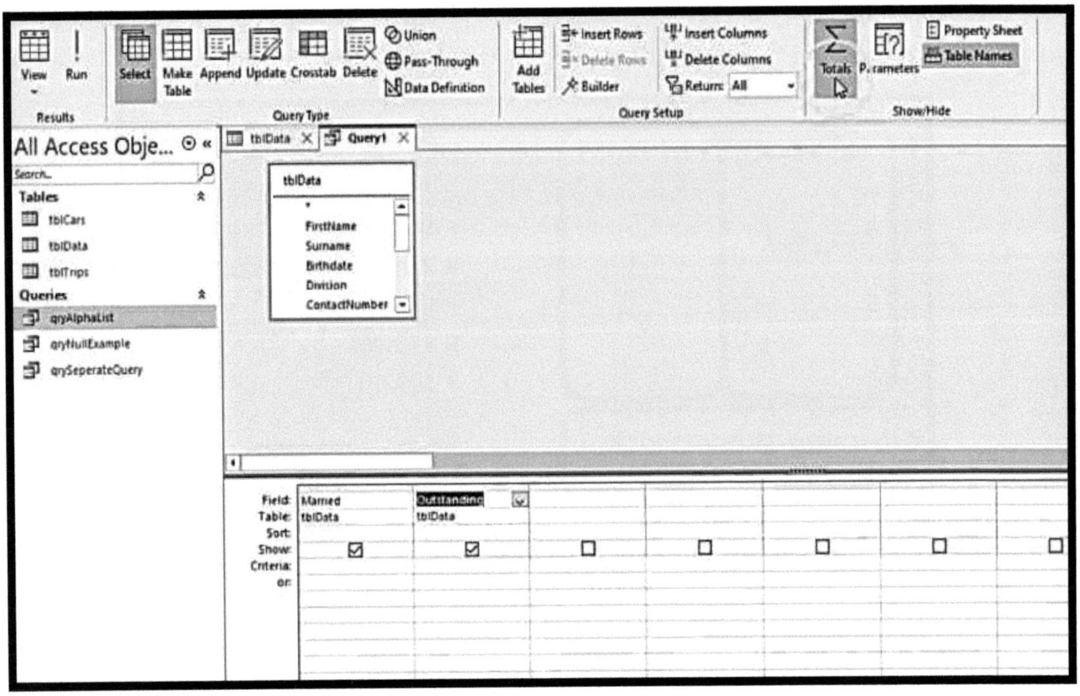

If we click on that, a "**Total**" **row** will be added and it allows us to group by things. Recall that we want to group by "Married" but we don't want to group it by outstanding, instead, we want to find the average for the outstanding.

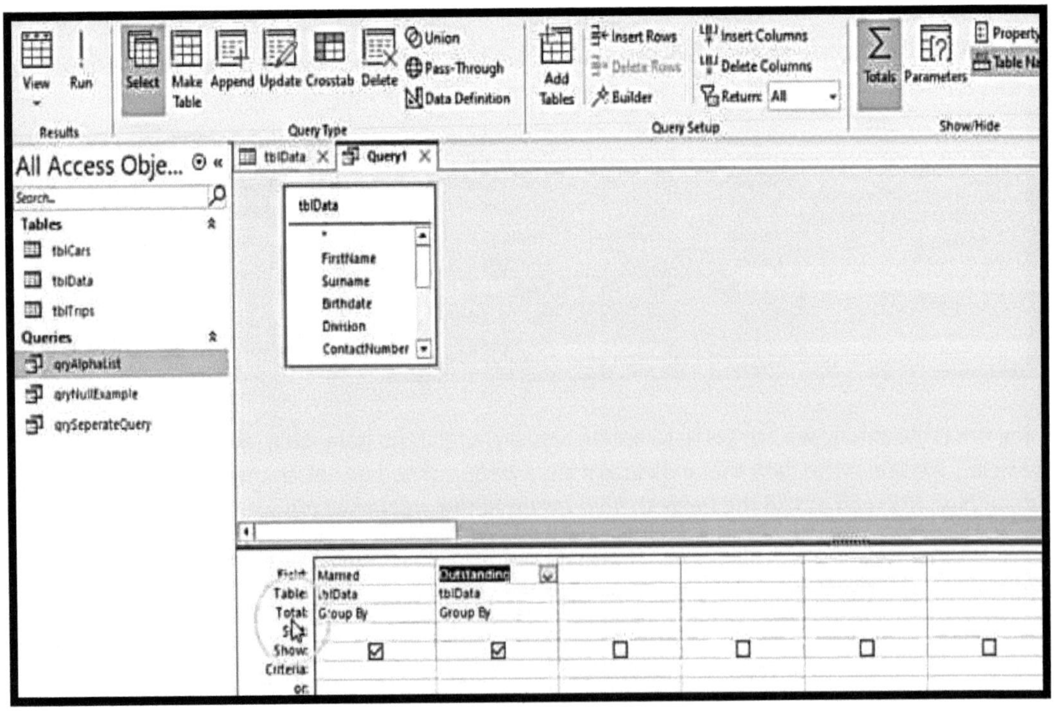

When we click on it, there are a lot of available options and because we want to find the average of the outstanding based on this grouping, we click on "**Avg**".

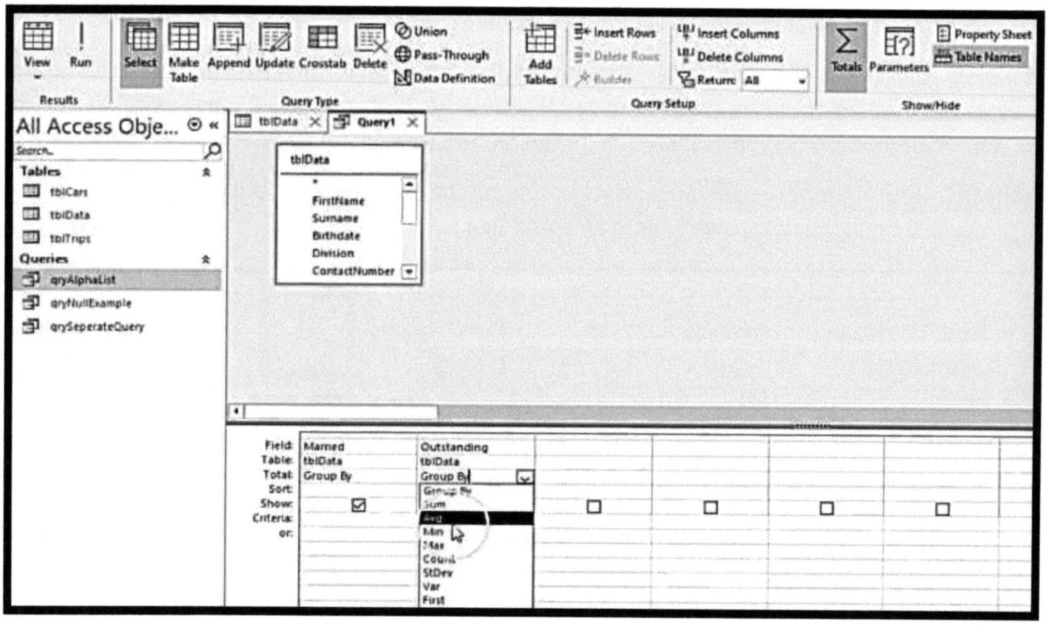

If we run the query, we've only got two options and you can see average outstanding for the different marriage groups.

If we want to find the average outstanding for the different divisions, then all we would do is go back to our design view and change that field to "**Division**". In this case, if we run the query, we can see the average outstanding for the different divisions.

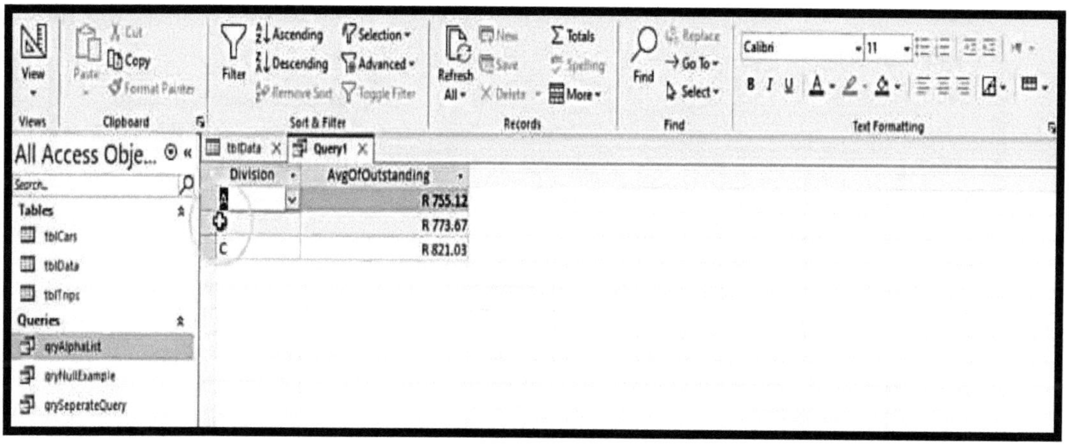

So that's how we can group data that is similar and find statistics on each of those groupings.

To recap: if you struggle to see the option there you must click on "**Totals**", and that's where you get this "**Total**" **row**. You can then specify what you are grouping by and the fields you want to find statistics on. You change the group bar to the statistic that you want. Remember to save the query when you are done.

Calculations In Queries

First of all, recall the little tip on tables that stated that you shouldn't store any data in your database tables that can be calculated. Now, how do you calculate?

In queries, you can calculate the values, so you can have a special column that has the calculated value or calculated field and there are three steps to calculated fields. If you don't know how to do a calculator field it is recommended that you follow this three-step program to help you do your calculated fields.

For the first step, in the unused column, that is a column that's not used in your query design, right at the top of the column in the field row, enter in an equal sign.

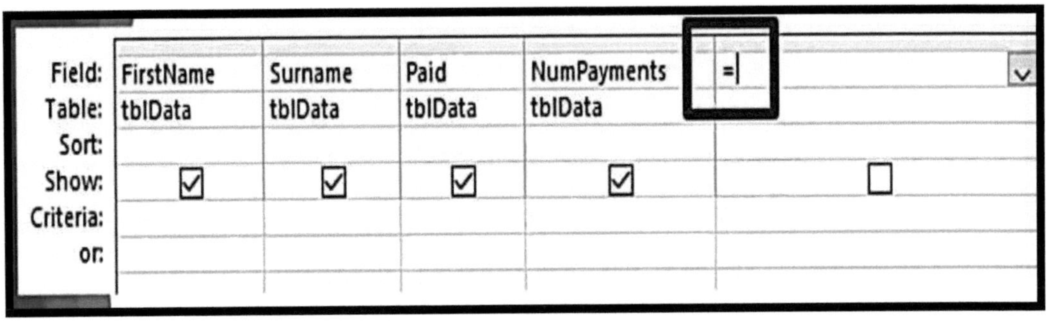

Next, you are going to enter in a formula and if you need to use values from other fields you can use those field names but remember you must use them as they are spelled in the database. Let's say in this scenario you want to work out the average number based on the number of payments, so if you've paid 100 dollars and you made 5 payments, the average payment would be 100 divided by 5; so, you want to take that paid field and divide it by the Number of payments Field. Now, you'll type in the formula. You can type in normal numbers but if you are referring to field names in those calculations, use them as they are spelled.

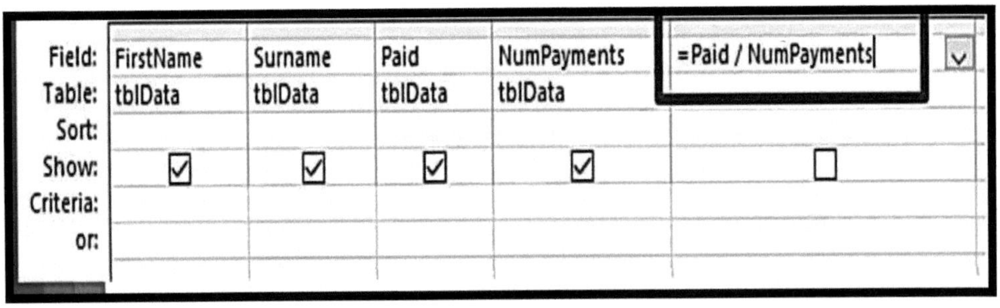

The next step is you press enter, and once you've done that some text will be added automatically to your calculated field in the front. There'll be a colon and then you'll have the actual calculation that you just entered. You'll notice there that the formula has got square brackets around it now, it is recommended that you do not use spaces in your field names. If you do have a database where there are spaces in the field names, then in your calculation make sure that you put square brackets around the names otherwise it's going to think that they are two different fields so just put square brackets around your fields if there are spaces. If there are no spaces just type them as it is and then Access will do the rest for you.

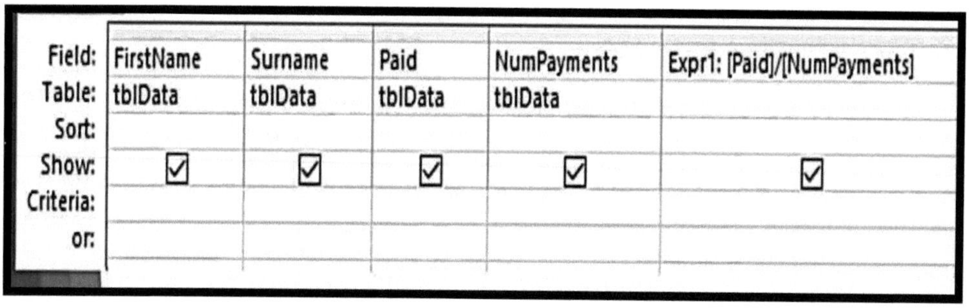

After entering your formula, you will notice that there is "**Expr1**" attached before the colon. You can replace this text and give it your value.

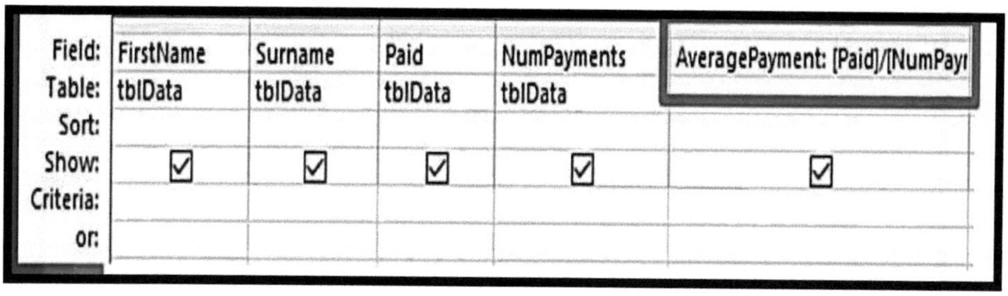

Now, you've got a calculated field called Average Payment and the calculation for it is taking the paid field and dividing by the Number of payments Field.

What happens if you make a mistake and you spell something wrong? Well, a box will appear when you go to view the query. When a box pops up and it asks you for a value for a field, whereas it should get it from the field, that means you've spelled something wrong, that's how you know you must go and correct it.

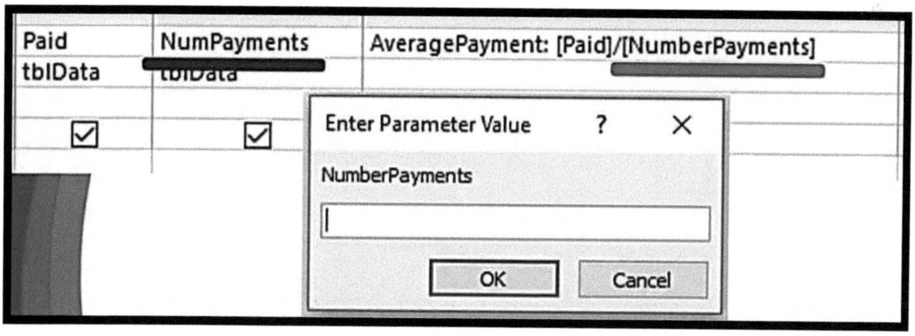

So just to recap our three-step program on how to do a calculated field:

- In the unused column at the top (in the field row) enter an equal sign (=).
- Enter a formula and if you're using fields in your formula make sure that you spell them correctly. If they have spaces, remember to put them in square brackets.
- Press enters. A text will appear with a colon in front of it, just change that little text that just appeared to whatever name you want to put at the top.

If you know the format of the calculated field you can just type in as is but if you're not too sure, or if you forget, just follow this three-step program.

So, let's do some calculations with the database below and for this data, we want to find the average number of payments.

First, we are going to create a query. So, in the field row, we put the first name and surname, the payment field, and the number of payments.

In the next column, we're going to put an equal sign (=), type in our formula, so we take the paid field divided by the number of payment fields (but remember it must be spelled the same), we press enter and it will put in the Expr1 and the fields in square brackets. At this point, you can replace that word with yours, then we run the query and we can see the average payment for each name.

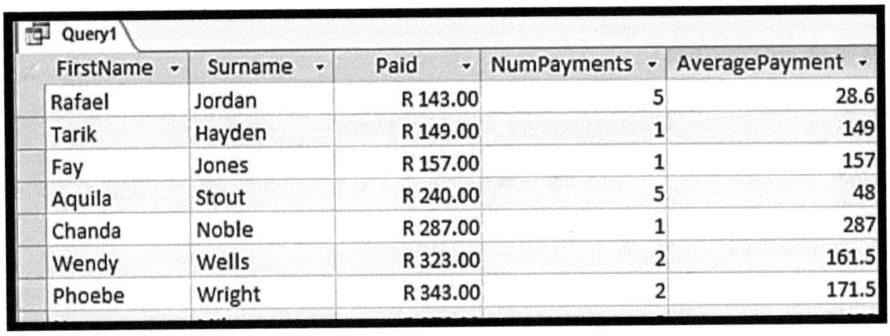

Query1

FirstName	Surname	Paid	NumPayments	AveragePayment
Rafael	Jordan	R 143.00	5	28.6
Tarik	Hayden	R 149.00	1	149
Fay	Jones	R 157.00	1	157
Aquila	Stout	R 240.00	5	48
Chanda	Noble	R 287.00	1	287
Wendy	Wells	R 323.00	2	161.5
Phoebe	Wright	R 343.00	2	171.5

If you want to change that value, go to the column, right-click on it, go to properties and you can change that format to some sort of decimal number or currency.

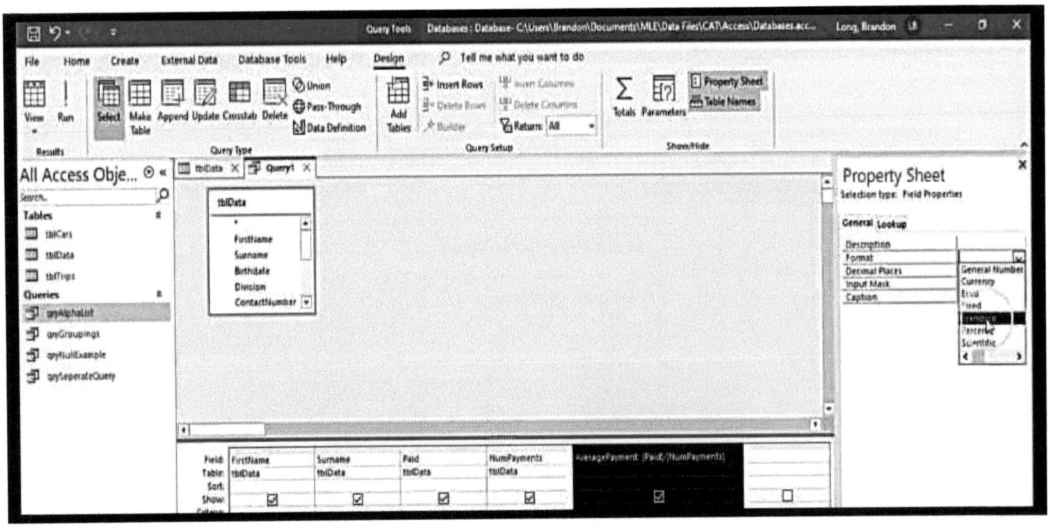

If you want to add in a VAT column, let's say you want to find the VAT amount on how much they've paid which is 15 percent of the payments, so what you'll have in the next column is **=15/100*Paid**.

You can press enter or click away, change that to the amount, and if you run it, you can see that there's a display of the VAT amount which you can change to a currency as well. So, these are the steps involved in doing calculations in queries.

CHAPTER SIX

REPORTS

In this chapter, we're going to look at how you can create a basic report and add a couple of features to it. You'll see how to do some calculations in your reports and how you can do a specific report based on a query.

What is a Report?

A report is a way to organize or summarize the data so that it can be printed.

If you have a table with a lot of data, if you have to print it just like that it would be very difficult to see what all the data is and it's quite complicated seeing it in that table format, especially if you're going to be presenting that printed document to someone else. Instead, you can take that table and the data that's in that table or a query and base a report on it to make it look a little bit better.

The report is very similar to Forms except for you're just doing it in a printed format instead of a form.

Some Terminologies

At the top of a report that's where we put the **Labels** for the fields so that you know what each column is about, and then the section below that is the **Textboxes** that will contain the data from the table or the query that you are referring to.

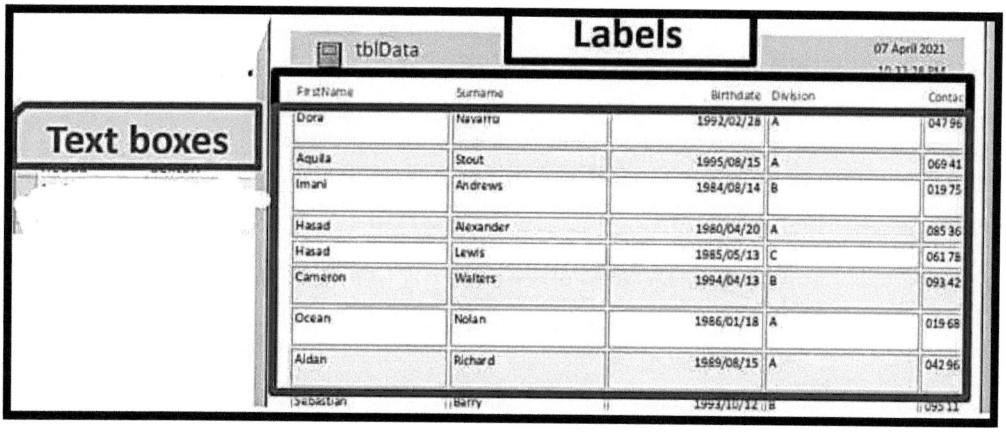

The image below shows what it would look like in the Design view.

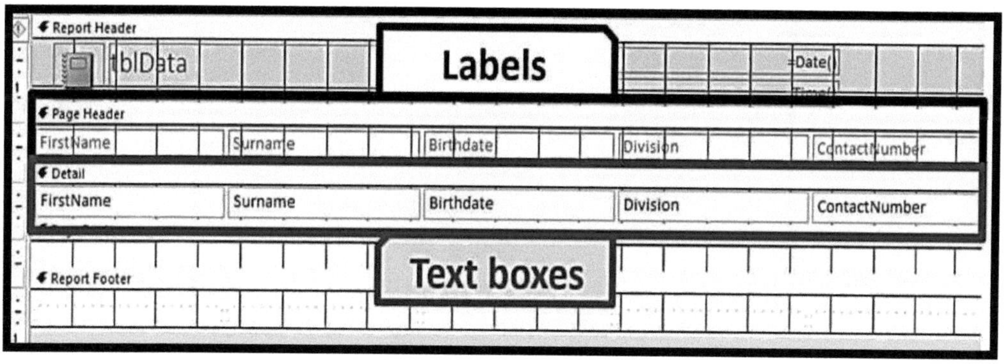

You'll see that there's a page header and in the page header you have the labels that'll display as it is, so you can change those and it won't affect anything. However, in the detail section, it has the text boxes which have the actual link to the fields in the table, so whatever is in the table will be displayed there. You can't change these because it's going to show the value for the fields.

You'll notice in your report there are a lot of blocks for each row in the design view. It only shows one and it will just repeat this detail section again and again until it gets to the bottom of the page and then it will recreate the page with a brand-new page header and then continue with those details.

Controls in Reports

There are some controls that you can add to reports which are very similar to what we did in forms.

You can add a label and if you want to specify some other information on the report you can use a label. If you want to connect to an image or show a little picture on it you can use an image component. If you just want to draw a line or separate something with a line you can use the line component. You can use all of these options and we're going to try them now in the actual report.

Creating A Report

Using our database from the previous examples we're going to create a new report. To do that, click on **Create** and go to the **Report** options. You can design a report, you can have a blank report and just add values to it or you can use the **Report Wizard**. For this illustration, we are going with the last option.

After clicking on that option, you can specify the values you want in your report and the fields you want in your report.

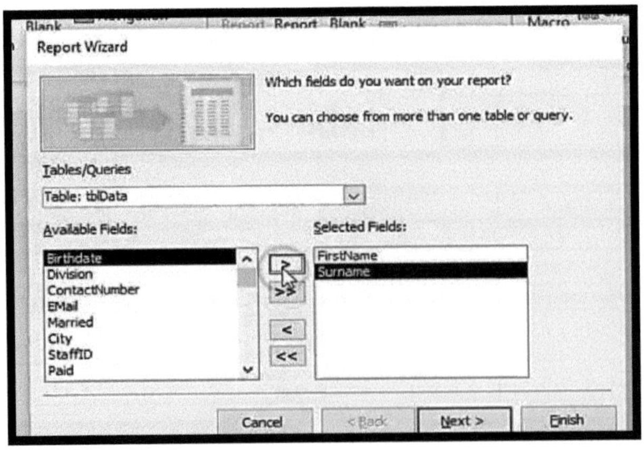

When you click on "**Next**", it'll ask you if you want to do a grouping. We'll talk about grouping as we proceed.

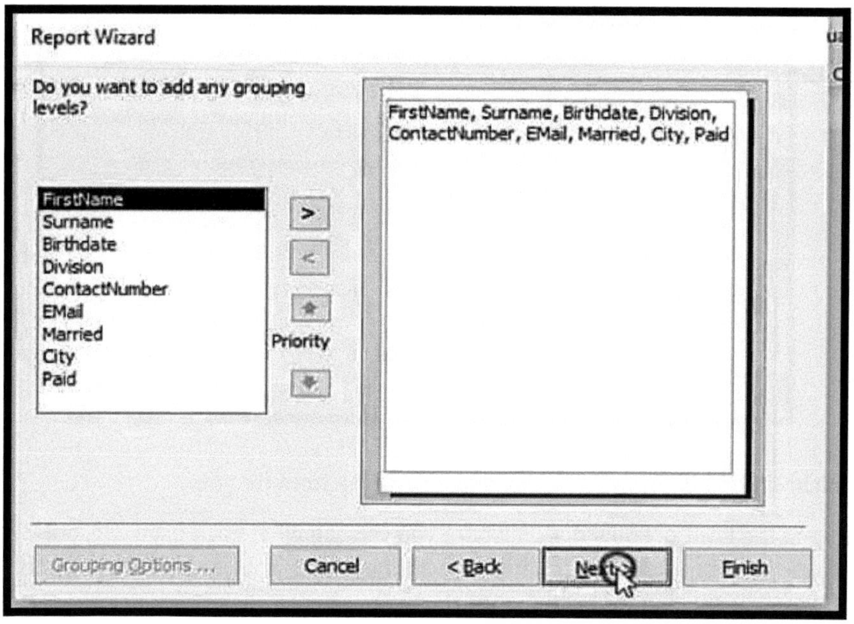

Next, you can specify if you want to sort data by a particular field.

You can also select the layout for your report.

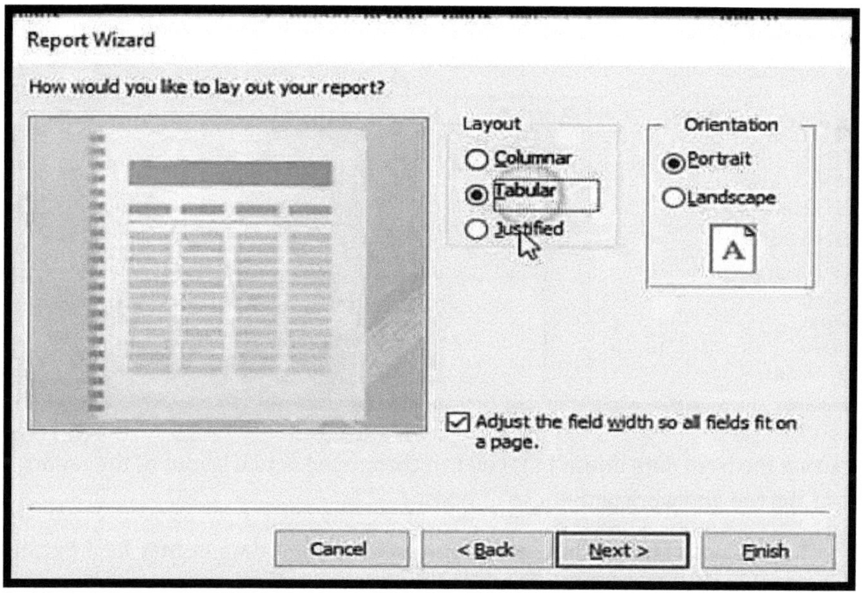

Lastly, you will be asked to give your report a name or title. Remember the naming conventions, and since this is a report, you can use "Rep" as your prefix so that you know the difference between the table data and the report data.

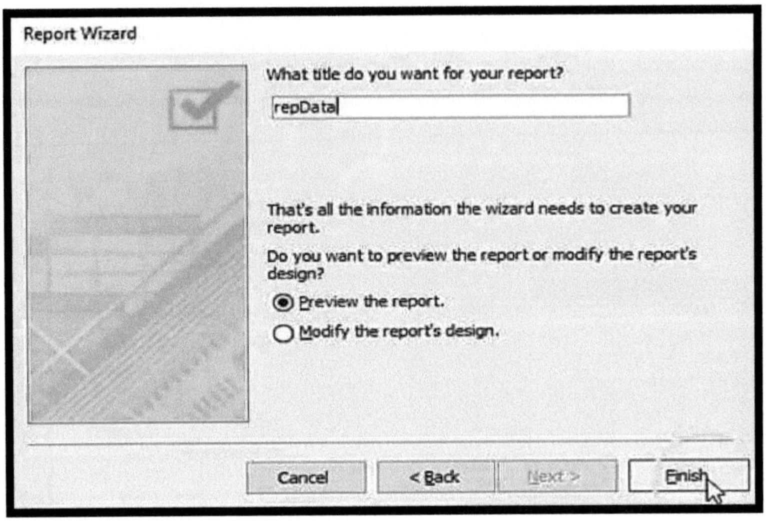

Click on "**Finish**" and it will show the design and display the form for you.

This opens up in the Print preview option where you can see what each page will look like and you can go straight to printing, but if you're not satisfied with the report, you can edit it to look better. To do this, go to the design view and now you can change things over here.

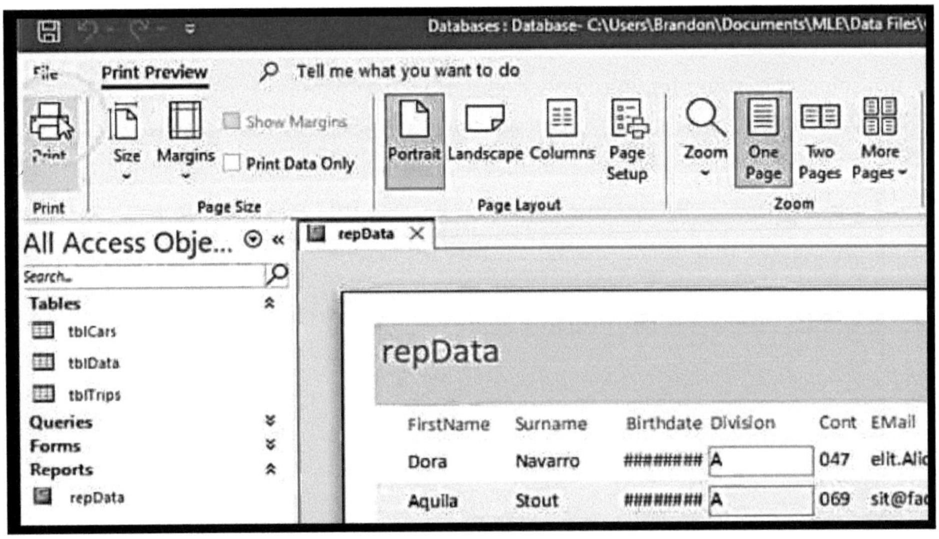

If you notice that the birth date doesn't fit, you can change the actual layout of the report, so you look at the options at the top and work with those options.

If you see hashes on your report, that means you can't see the data in that field because it's not big enough. To avoid that you can just rearrange some of these fields to make them a little bit spaced out and make them a little bit bigger.

If you want to make it justified or you want to make it left-aligned or right-aligned, you can do all those types of things by selecting Home and seeing all the options over here.

You can change the labels and it won't affect the data but for the text boxes, leave as is because that's going to fetch the data from the table and repeat this detail section.

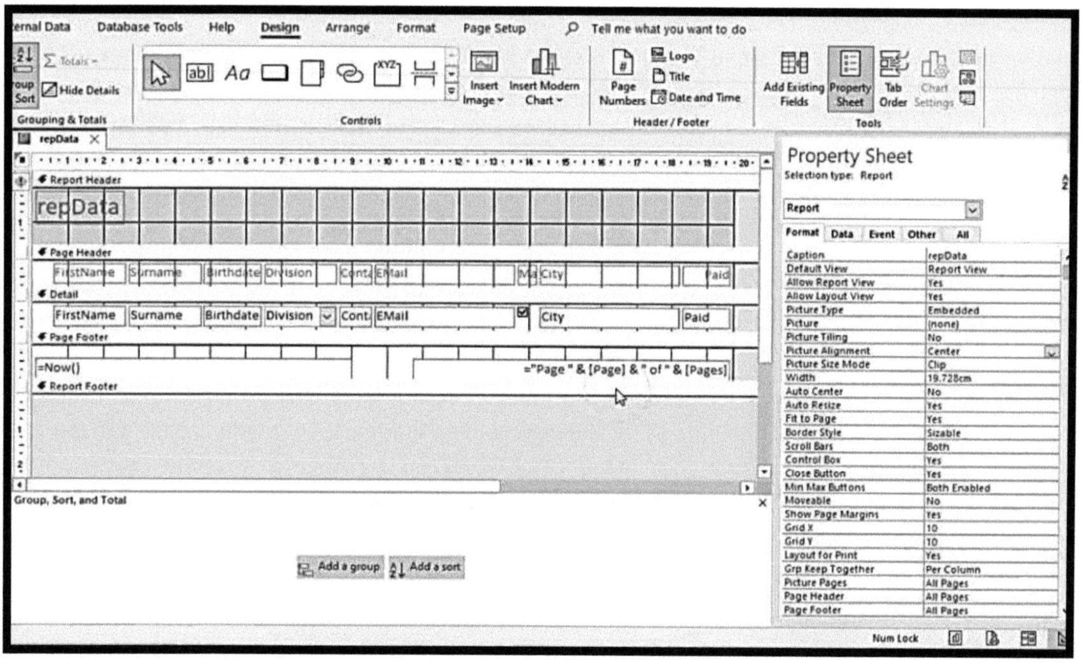

You can also do other features to your report such as adding another label at the top or editing a particular label. Let's say you want to put a line between each record. In that case, under your Design tab at the top, click on the Line icon and it will put a nice line across. You can right-click on that line, go to Properties and you will see other options specific to that particular line. Please note that anything you do in detail will be repeated for each record that is displayed.

Just like we had in the Forms, you can make your report have a little footer, so you drag that down and customize the footer. You can add your little details and anything that you want to change you can just right-click on it and go to the properties and you've got all those options over there to change.

Calculations In Reports

Before we proceed, you need to know what control you can put those formulas in, and the control you need to use is the **Text box**. You can add a text box to your report when you design and then write the formula inside the text box. You can't write standard text in a text box; if you want to write standard text then you will use the label as we showed you in the previous chapters.

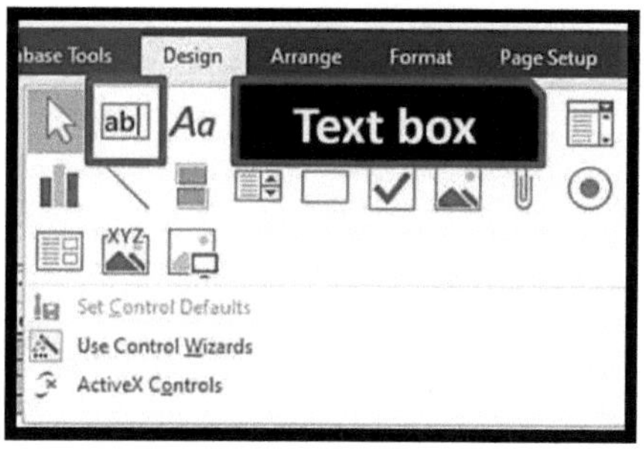

So, what are the types of calculations that you can use?

If you want to add up all the values in a particular field, then you will use Equals sum (**=SUM**), and then in brackets, you will write down the name of the field that you want to do the summing on. This is very similar to Excel except that instead of cell references like A1 to A10 you'll just write the name of the field you are referring to. You'll notice that the field name is in square brackets, that's the format is. If you do not have spaces in your field names then you can write them just as is and they will put the square brackets in for you but if you have spaces then just use a square bracket. Another one is the "**MAX**" which finds the biggest value out of that particular field, there's the "**MIN**" which is the smallest value and then there's the "**AVG**" and that's if you want to find the average amount paid.

There's also a "**COUNT**" where you can count how many records there are, so you can count all the emails for example but because you're counting all the records you don't need to specify a particular field you could just use a *.

The first four functions can only be used with numerical fields or fields that have numbers. You can find the average amount paid or the sum of all the outstanding amounts, so anything that has to do with currency or numerical value, you can use those first. **COUNT**, on the other hand, can be used in any field, so when you use the **COUNT**, you can use any field in it or you can use a *.

Please note that you can't use these functions in the page header or page footer, this is because you can't find the sum of all the values on a particular page or some or the max of all the values on a particular page. These will not work in the page header and footer so don't put them there.

If you do want to put a formula in the page header or footer, you can put the current page number. To do this, just say = **[PAGE]** keyword and that'll give you the current page number. If you want the total number of pages, that'll be = **[PAGES]**.

If you want to use the NOW field, which is **=NOW ()**, that will give you the current date and the time, and then you can format that to display just the time or the date.

So, let's go to the report we designed earlier. Right-click on it, go to the design view so you can change the details of your report. You'll notice that by default it's already added a couple of options like the "**Now**" and that of the pages.

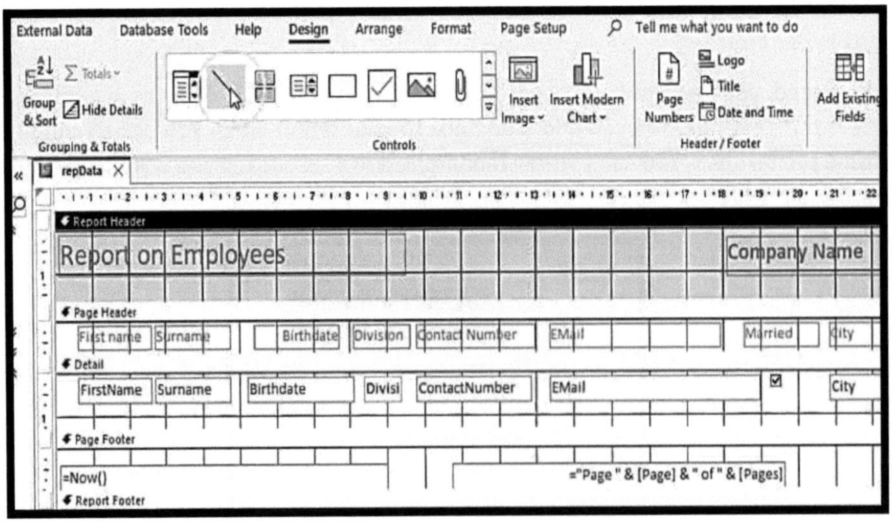

Let's say you want to find the average of everything in the report, it's ideal that you put that in the report footer. First, click on the text box and it'll give you a label and a text box. To find the sum of all the "Paid", that's going to be written as **=SUM(Paid)**.

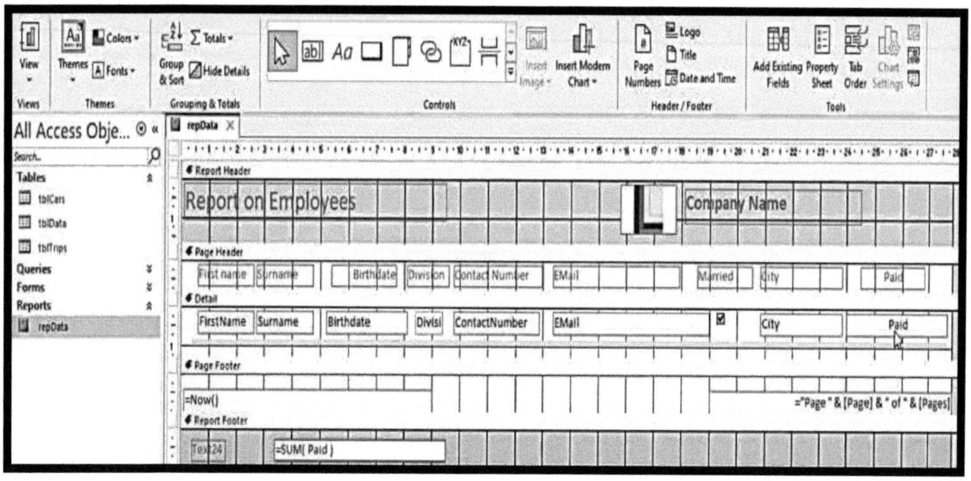

Notice that there are no square brackets because you typed just as it is, however, it will put the square brackets in for you. Remember to give it a label as well.

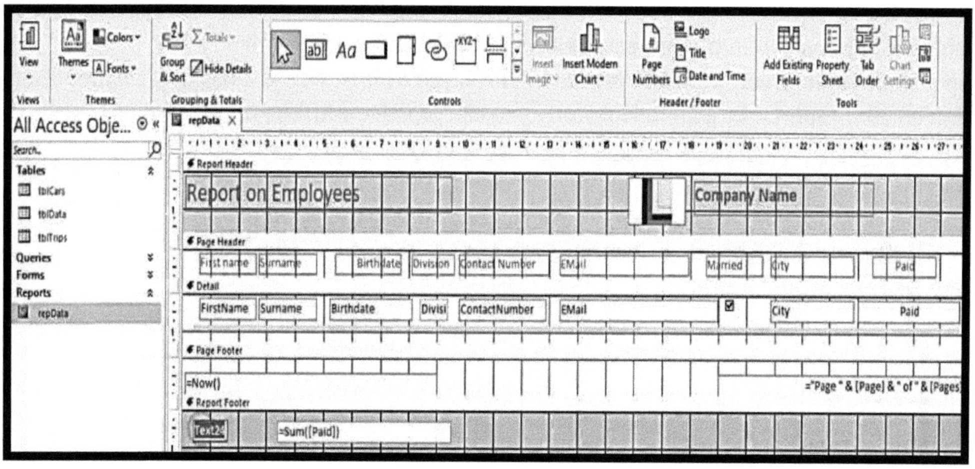

You could also put it in the report header, but that's not a great place to put it but keep in mind that we put all our formulas in the report footer.

Next, go to view and in this case, you can use the Report view, and when you scroll down to the bottom where you have the report footer you can see that the total amount Paid is now included.

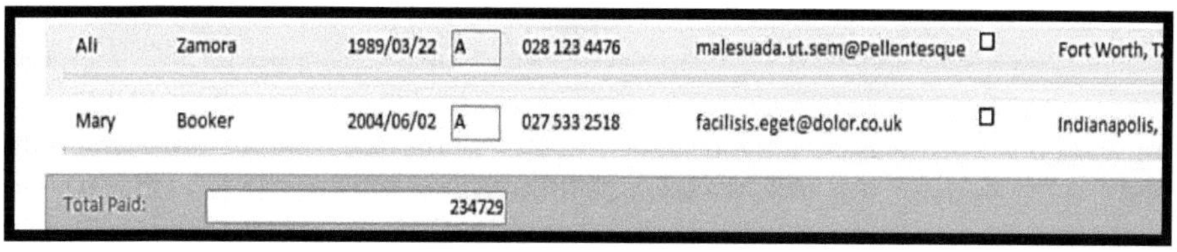

If you want to make changes to it, go back to your design view, **right-click** on it and go to Properties. Here, you can either change it to currency or add decimals to make it look better.

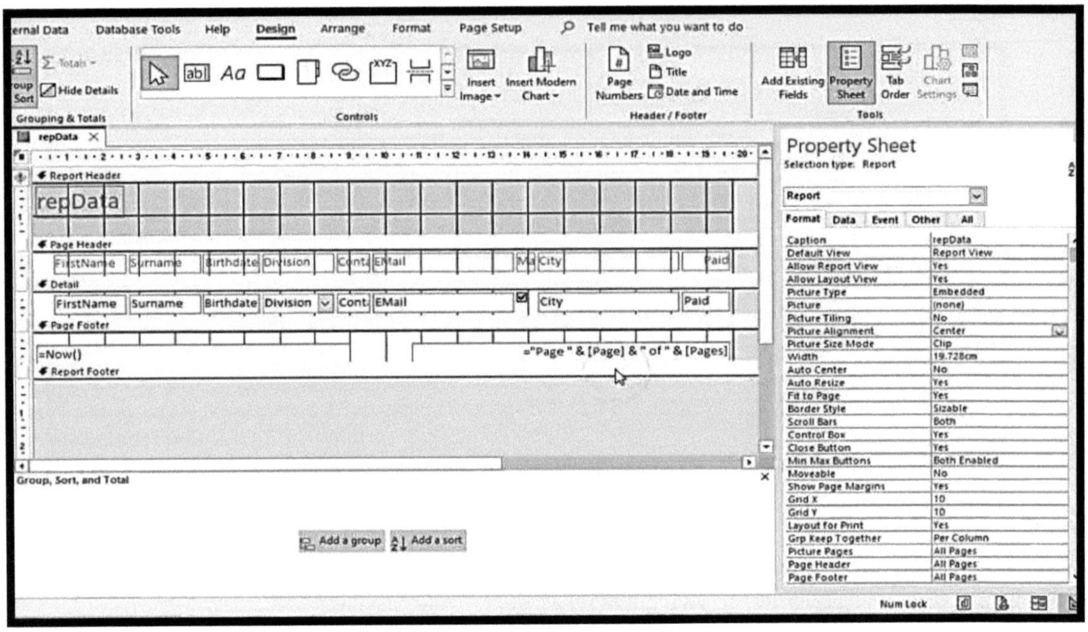

Remember to always put in the correct spelling and you can't put the title in a text box, you can only do that in a label because you want the text to be displayed as it is. This is a basic calculation that you can perform on your reports. To gain full mastery, you can practice with the other formulas.

Reports Based on Queries

Let's take a look at this example: let's say we've got a report based on all the records in the data table below and then on the right-hand side there you can see the city. The last two letters of the city indicate its state, so Texas would be **TX**.

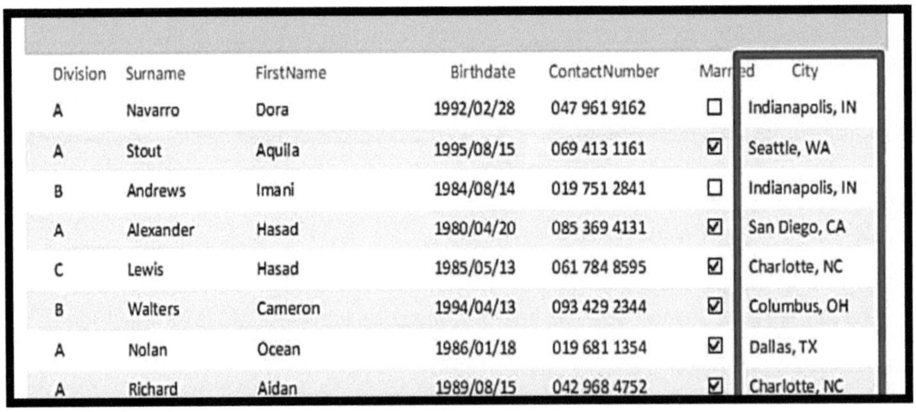

Division	Surname	FirstName	Birthdate	ContactNumber	Married	City
A	Navarro	Dora	1992/02/28	047 961 9162	☐	Indianapolis, IN
A	Stout	Aquila	1995/08/15	069 413 1161	☑	Seattle, WA
B	Andrews	Imani	1984/08/14	019 751 2841	☐	Indianapolis, IN
A	Alexander	Hasad	1980/04/20	085 369 4131	☑	San Diego, CA
C	Lewis	Hasad	1985/05/13	061 784 8595	☑	Charlotte, NC
B	Walters	Cameron	1994/04/13	093 429 2344	☑	Columbus, OH
A	Nolan	Ocean	1986/01/18	019 681 1354	☑	Dallas, TX
A	Richard	Aidan	1989/08/15	042 968 4752	☑	Charlotte, NC

What if we want only the fields that are in Texas, in other words, we want all those where the state (last two letters) is a TX? And not only that, we want all those where the division is a C. So, we want all those in Division C of Texas.

This is a very specific report that we want but how can we do that? Well, we could do groupings and that could be quite complicated but then we've got everything in that report and we want to report with just that data. If that's the case, then we're going to have to create a query first and then once we've done the query, we will base the report off of the query.

Let's go look at the query. If you're not yet familiar with queries, you can go back to the previous chapter where we covered that.

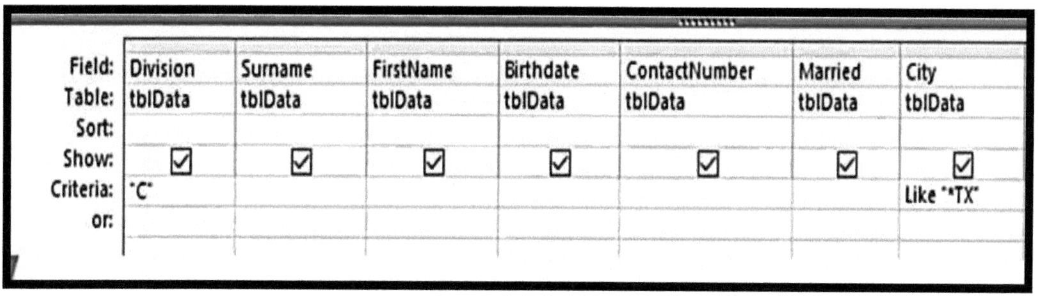

Looking at the query above, on the right-hand side, you can see the city is **LIKE *TX**, in other words, all those fields that end in the letter **TX**. Again, if you're not sure what that means go back to the section where we talked about the **LIKE** operator and **wildcards** in our query chapter. You would also notice that the division is Division C.

After running the query, we get a specific set of results and after saving the query we can go and **create the report** using the **Report Wizard**. When you start, make sure that you select the query that you just created, and in that way, you can create a report which has only specific results.

Next, we'll follow the instructions, and after we finish, we'll end up with a report for this query and there you can see it's a very small report, there's not much information and as we've done in previous lessons, we'll go to the design view and just play around with the fields so that they can all fit in.

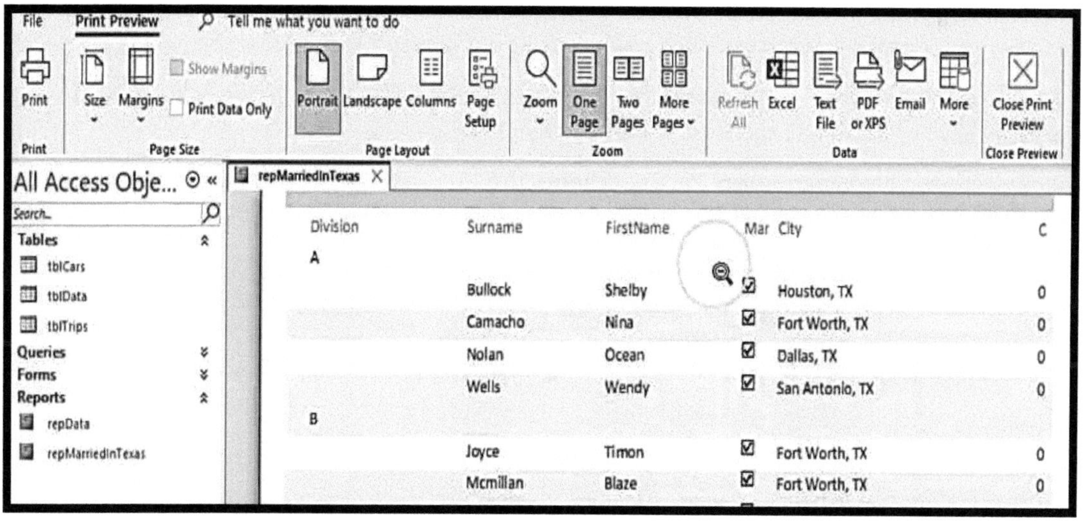

This is how you create a report that doesn't have all the information on the table but only the specific information that you need at that moment.

Grouping In Reports

If you have data with different divisions and you want to report on that but you don't want the divisions split up but a more organized report with the Divisions properly arranged, this is where grouping comes in.

So how do we group the data?

If you were creating the report from the beginning using the **Report Wizard** you would eventually come to this step where you would have been asked what you want to group it by.

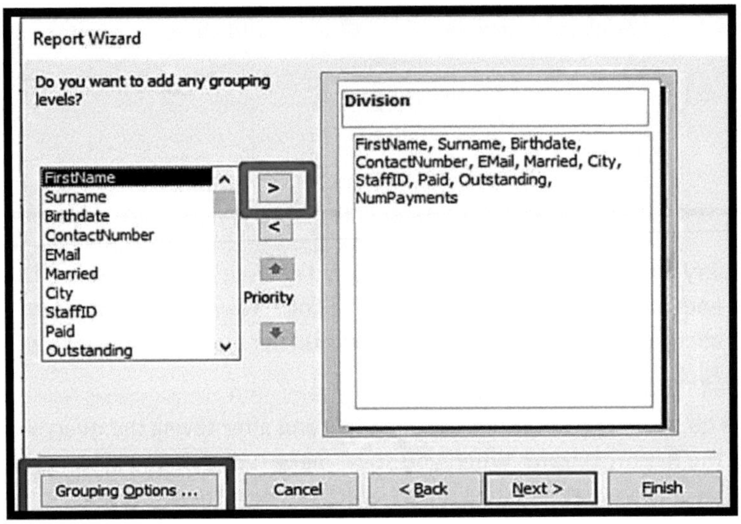

You can click on that button to specify which field you want it to be grouped according to and then you have the grouping options at the bottom where you can specify the details of that grouping.

You can also group by multiple fields but one field will be first and another field will be second; so, you can have a secondary or third value for your grouping but it does it in the order that you specify.

What happens if you are in your report and you haven't done this when you created it? How can you get a grouping when you didn't specify it in the beginning?

Well, at the top bar under Design you'll see there's the "**Group & Sort**" button and if you click on that a box will appear at the bottom and you'll see there's a little place here where you can add a group.

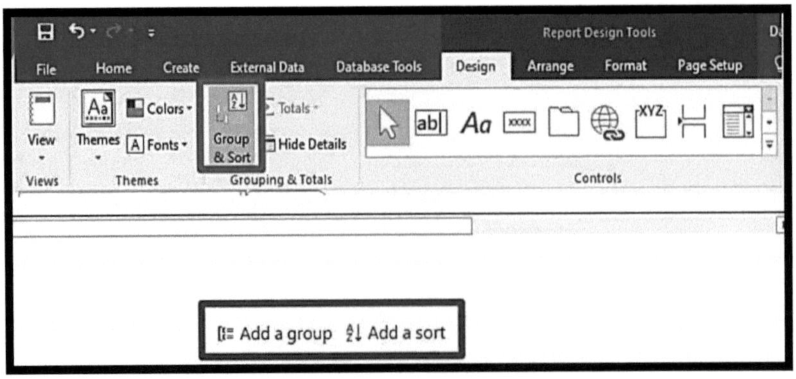

This can also be done for sorting, so if you want to sort by a particular record and you forgot to do it in the Report Wizard you can specify those ideas here.

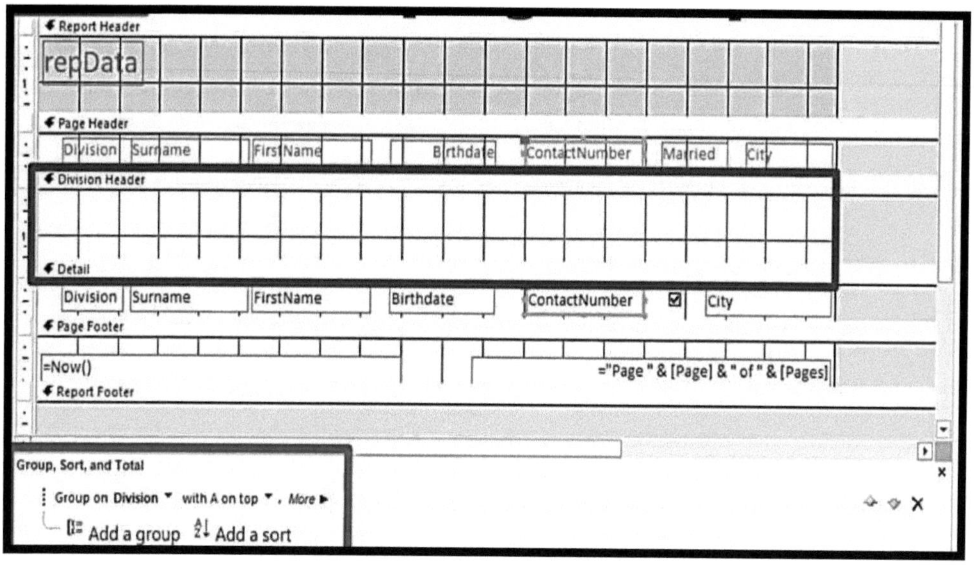

If you click on the "**Add a group**", it'll ask you which field you want to group by and if you select the field at the bottom, you'll see it's added to the grouping. You can still add more groups and add more sorts but obviously, it'll be the next level and then a header will also be added at the top. So, you can have a header for each division.

The key thing to remember for this is that if you are wanting to add a grouping to your report after you've created it, just remember to click on that **Group & Sort** button. You'll get the option to add groups and change the group details.

CONCLUSION

This book is about the newly released Microsoft Access 2025. There are many things you can do with this software application that will be so easy to miss out on some. However, this can be averted if you go through this well written piece.

Throughout this book, the importance and uses of Microsoft Access is discussed, starting with the description of what a particular function is and a step-by-step explanation of how to carry out each of the functions.

The book starts with an explanation of what Access is all about, databases and what makes up a database. Subsequent sections discuss the creations and uses of tables, queries, forms, and other things that are associated with Microsoft Access 2025. The conclusive chapter talks about uncommon tips that will be useful for you when working with this newly developed software.

The complete book is a success, and I am sure it will be useful for anyone who wants to use it as a guide manual for the newly released Microsoft Access 2025, whether a novice, an average user or a professional user. Read it thoroughly, faithfully follow the outlined steps to carry out a procedure and operate Access like a wizard.

Enjoy!

BOOK 5:

MICROSOFT POWERPOINT

INTRODUCTION

Microsoft PowerPoint is a program that is used mostly for presentations. It was created by the duo of Robert Gaskins and Denis Austin at a software company known as Forethought Incorporations. This software was released on April 20th, 1987, and was initially made for Macintosh computers alone.

Three months after the creation of this software, it was acquired by Microsoft for a fee in the region of about $14 million. This was the first very important acquisition by Microsoft, and they set up a new business unit specifically for PowerPoint in Silicon Valley. This was where Forethought had been located.

Afterward, PowerPoint became a part of Microsoft Office. At first, it was offered for the Macintosh alone in the years 1989 and 1990, along with several other Microsoft apps. Starting with PowerPoint 4.0, which was created in 1994, PowerPoint was integrated into Microsoft Office development and incorporated shared common parts and a unified user interface.

PowerPoint's market share was significantly lower initially, and this was before it was introduced as a version for Microsoft Windows. It then grew very fast with the growth of Windows and Office. Since the late 1990s, PowerPoint's market share of presentation software all over the world has been estimated at 95 percent. Originally, PowerPoint was made to provide visuals for group presentations, mostly within business organizations, but over the years, this has changed as it has now been widely used in many other communication situations, including business and beyond.

The impact of this widespread global use of PowerPoint has been felt like a powerful change throughout society, with very strong reactions that include advice on it being used less, used better, or used differently.

The very first PowerPoint version, which was created in 1987, was used to produce overhead transparencies. The second and third, in 1988 and 1990, were also used to produce color on about 35mm slides. The third version, produced in 1992, brought the video output of virtual slideshows to digital projects, which over the years could completely replace the physical transparencies and slides. About a dozen important versions have since then infused so many additional features and modes of operation and have also made PowerPoint available way beyond the Apple Macintosh and Microsoft Windows, with the inclusion of versions for IOS, Android, and Web access.

VERSIONS OF POWERPOINT

PowerPoint 1.0

The innovations included in this version are: various slides in a single file, arranging slides with a slide sorter view and also a title view, which has the precursor of an outline view, speakers' notes pages fixed to each slide, printing of audience handouts with various slides on a single page, text with different outlining styles and all-encompassing word-processor formatting, and graphic shapes with fixed text that can be used to draw diagrams and tables. It also has its manual and is shipped with a hardbound manual.

The 1.0 version created overhead transparencies on a black and white Macintosh specifically for laser printing. Presenters were now able to control their overheads directly and would no longer need to work through the person who had the typewriter. PowerPoint took care of the creation of overheads such that they all look alike and just a single change can reformat them all. Typographic fonts were considered way better than Orator typeballs, and charts and diagrams could be brought in from MacPaint, Excel, or MacDraw, all thanks to the new Mac Clipboard.

The system requirements for this version were Mac OS X original Macintosh or better, System 1.0 or higher, and 512 RAM.

PowerPoint 2.0

The innovations for this version included the color, added word processing tools, the find and replace option, checking of spellings, color schemes for presentations, a guide to the selection of color, the ability to change the color scheme later, and the option of shaded coloring for fills.

This version also added color 35mm slides, enabling the transmission of the resulting file over the modem to Genigraphics for imaging on Genigraphics. As of then, Genigraphics was the foremost leader of the professional service bureau, having developed its own Digital Equipment Corp. After a significantly short time, Geingraphics also switched to PowerPoint.

This version's system requirements include a Mac original Macintosh or better, a system 4.1 or higher, and 1 MB of RAM. Windows 286 PC or higher, Windows 3.0, and 1 MB of RAM are required.

PowerPoint 3.0

The innovations for this version included the first application, which was designed majorly for the Windows 3.1 platform with full support for TrueType fonts, templates for presentation, outline view editing, new and modern drawing, the inclusion of freeform tool, auto shapes, flip, rotate, align, scale, and transform inserted pictures into drawing basics to make them editable, transitions between different slides in the slide show, progressive builds, and the incorporation of sound and audio.

The animations that were included were the "flying bullets," where bullet points flew into the side one after the other, and some degree of Pen Computing support was also included.

This version also added the video-out to feed the new video projectors with effects that were capable of changing a collection of synced slide projects. It also added fades and dissolves, as well as other transitions and animations of text and pictures, and could also infuse video clips with synced audio.

The basic system requirements for this version included Windows 286 or higher, Windows 3.1, and 2 MB of RAM. A Macintosh Plus or higher, System 7 or a much higher version, and 4 MB of RAM are required.

PowerPoint 4.0

The innovations included in this version include the auto layouts, word tables, hidden slides, the AutoContent Wizard, and the rehearsal mode.

The standard Microsoft office look that is shared with Excel and Word was also introduced in this version with tooltips, a status bar, and Full OLE 2.0 with in-place activation.

The system requirements for this version include Windows 386 or higher versions, Windows 3.1, 8 MB RAM, Mac 68020 or better, System 7 or higher, and 8 MB RAM.

PowerPoint 7.0

The innovations for this version include new animation effects, autocorrect, black and white view, insert a symbol, real curves, and meeting support features, which include "Meeting Minder."

There is also a total rewrite of the product from scratch in C++, with a full object model and internal VBA programmability.

The basic system requirements for this version include Windows 386 DX PC or a higher version, Windows 95, and 6 MB of RAM.

PowerPoint 8.0

The innovations in this version include the office assistant, file compression, the save to HTML option, "Pack and Go," AutoClipArt, and transparent GIFs.

The system requirements include a Windows 486 PC or higher, 8 MB of RAM, a PowerPC Mac or better, and 16 MB of RAM.

PowerPoint 9.0

In this version, the innovations include the three-pane browser view (which has a list of selectable slide miniatures or titles, a large single slide, and notes), autofit text, save to web option, presentation conferencing, picture bullets, aliased fonts, and animated GIFs.

The system requirements for this version include a Windows Pentium 75MHz+, Windows 95 or higher version, and 20 MB of RAM. Mac 120MHz+ or better, MacOS 8.5 or higher, and a minimum of 48 MB of RAM.

PowerPoint 10.0

The innovations in this version include the "install from web option," the usage of clipart on the web, and the use of exchange and SharePoint for storage and collaboration.

The system requirements include Windows Pentium III, Windows 98 or higher, 40 MB of RAM, and Mac OS X 10.1 (puma) or later, but this won't run under OS 9.

PowerPoint 11.0

This version features innovations such as tools being visible to the presenter during slide shows (notes, thumbnails, time clock, re-ordering and editing of slides), "package for CD" to write the presentation and viewer app to CD.

The Microsoft Producer for PowerPoint 2003 was a free plug-in from Microsoft that used a video camera to create Web page presentations with a talking head narration perfectly coordinated and timed to the existing PowerPoint presentation for over the web delivery. The Genigraphics software that sends a presentation for imaging 35mm slides was also removed from this version.

The system requirements include a Pentium 233 MHz+, Windows 2000 with SP3 or later, and 128 MB of RAM. 256 MB RAM, OS X 10.2.8 or later.

PowerPoint 12.0

This version features innovations such as a new and distinctive user interface, employing the use of a changeable ribbon of tools

throughout the top to have the menu and toolbars replaced, SmartArt graphics, a lot of additions to the graphics of text and drawings, and an improved presenter view. In this version, the auto-content wizard was taken away.

Another very important change in this version was from a binary file format that was in operation from 1997 up until 2003, to a new XML file format that has been predominantly used in recent versions.

The system requirements include Windows 500MHZ processor or higher, Windows XP with SP2 or more recent, 256MB RAM, and Mac OS X 10.4.9 or newer than 512MB RAM.

PowerPoint 14.0

This version features innovations such as the single document interface (SDI), sections within presentations, a unique reading view, the redesigning of backstage functions, the save as video option, the insert video from web feature, embed video and audio, an upgraded edition for video and pictures, and broadcast slideshow.

The system requirements include a Windows 500 MHz processor or one that is much higher, Windows XP with SP3 or a more recent version, 256 MB of RAM, with 512 MB of RAM recommended for video. Intel processor, Mac OS X 10.5.8 or later, and 1 GB of RAM are required.

PowerPoint 15.0

This version features the change in the default slide shape to the 16:9 aspect ratio, an online collaboration by various authors, a unique user interface designed specifically for different touch screens, enhanced audio, video, animations, and transitions, and an additional change to the presenter view. Clipart collections, which include insertion tools, were taken off but made available online.

The system requirements include a 1 GHz processor or one that is quite faster, an x86-or x64 bit processor, Windows 7 or a more recent edition, 1 GB of RAM (32-bits), 2GB of RAM (64 bit), a Mac Intel processor, Mac OS X 10.10 or a more recent version, and 4GB of RAM.

PowerPoint 16.0

This version includes the "Tell me" feature to search for program controls, the PowerPoint designer pane, morph transition, real-time collaboration, the zoom to slides or section to slides option, and the unique presentation translator for a real-time translation of the spoken word of the presenter to on-screen captions in just about any of the 60+ languages, with the system analyzing the text of the PowerPoint presentation as context to improve the precision and impact of the transitions.

This version's system requirements include a 1 GHz or faster processor, an x86 or 64-bit processor, Windows 7 with SP1 or a much later version, and 2 GB of RAM.

CHAPTER 1

INTRODUCING POWERPOINT PRESENTATIONS

What is a slide?

A slide can be defined as a single page of a presentation. As a collection, a group of slides can be known as a "single deck." A slide show is also said to be an exhibition of several slides or images on an e-device or a projection screen.

A presentation slide used to be a 53mm slide, typically viewed with a slide projector or transparent overhead projector before the advent of the personal computer.

In the digital age, a slide is most commonly known as a single page developed with the use of a presentation program, which includes MS PowerPoint, Apple Keynote, Google Slides, Apache OpenOffice, or LibreOffice. It is equally possible to make them with a document markup language, for example with the use of the LaTex class Beamer.

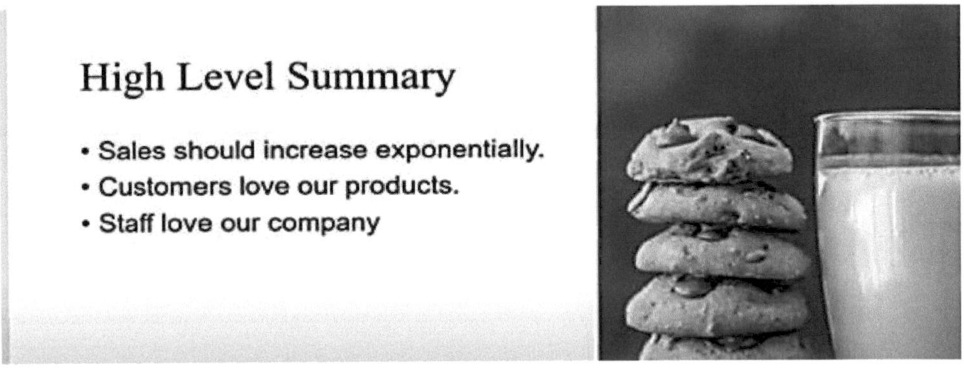

What's Embedded in a Slide?

A slide can have so many different types of content inserted into it. By clicking on the appropriate icon on one of the different slide layouts or choosing the object of choice in the images or illustrations group on the Insert Ribbon, you can insert images, clipart, shapes, screenshots, and much more.

There are various options for including graphics on the slide. To have an image added, navigate to the insert tab icon, then **choose pictures,** and then select **the picture from the file**. In the window that appears, you should be able to select the file to be added to the slide. There may need to be some adjustments, but the picture will fit well. By adjusting the size of the image using the handle in one of the corners, you can make the image bigger or smaller. This way, it will not begin to look short and squat or tall and skinny.

Additionally, if a PowerPoint picture is reduced in size, it will usually appear fine and nice. However, some photographs will become pixelated only if they are enlarged. An additional option will appear the moment an object is inserted, allowing users to choose from a variety of available designs.

To create a shape, select **the preferred shape from the shapes icon option**. If you click on a**ny shape and release any of the shapes on the slide**, a standard size shape can be obtained and the size can also be reduced to match the chosen or desired option. If you choose the alternative, hold and drag the shape on the slide. This will require that you draw out the shape. In addition, you can resize, move, and color your shapes as necessary.

Starting PowerPoint.

When you open PowerPoint, it will take you to the start screen first.

Next an interface with various options such as a blank presentation or a presentation with themes will be displayed. Its best to choose a blank presentation to begin from the scratch.

Here is what a blank presentation looks like.

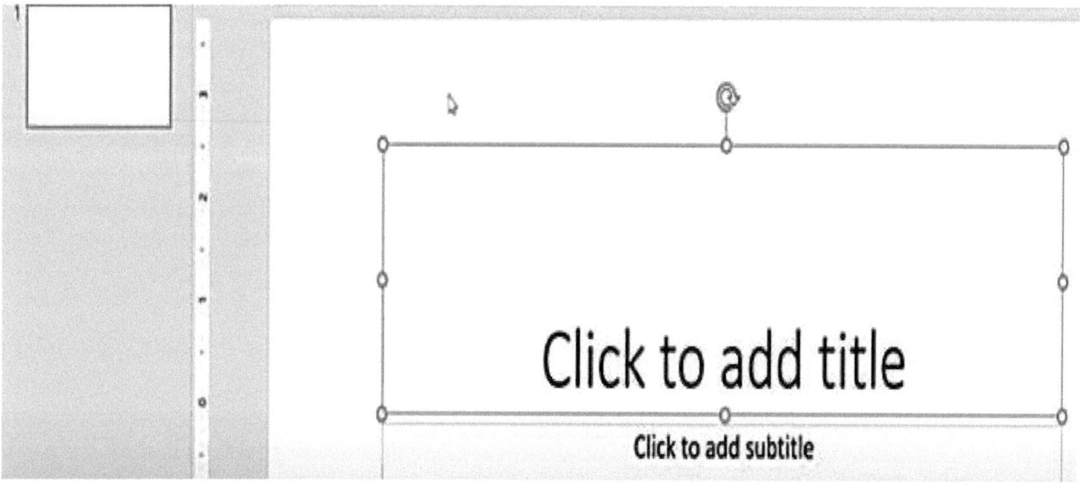

The User Interface of PowerPoint

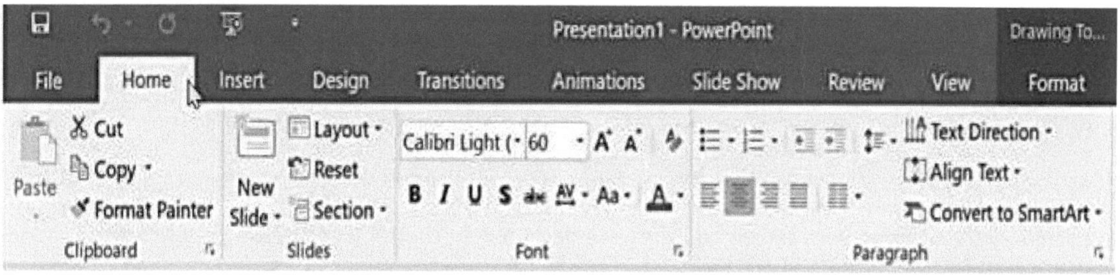

The user interface of PowerPoint 2025 for Windows starts up with the backstage view. With this view, you can design multiple ways to make the next presentation using a template or theme. There is also the option to open a previous presentation. When new changes are made in this view, the actual PowerPoint presentation can be viewed in this interface.

File Menu

When the file menu is selected, **the backstage** menu that contains all the save, share, create, and print options for the presentation can be seen.

Quick Access Toolbar

Usually, the customizable toolbar appears at the top of the ribbon. You can add icons to the toolbar that represent commands that you use frequently. You can also position the quick access toolbar at the bottom of the ribbon.

Ribbon

Several groups of buttons can be found on the Ribbon, including tabs and groups of buttons. A number of these groups also include galleries, including galleries for themes and themes colors.

Slides Pane

The slides pane can be found on the left-hand side of the interface. The pane displays the thumbnails of all the slides that are in the open presentation.

Task Pane

In the task pane, there are more options, which show when you select an option in one of the ribbon taps. Selecting the format background button within the design tab of the ribbon, for instance, opens up the format background task pane.

Status Bar

The status bar is simply a horizontal strip with information about the presentation, such as the slide number, the theme applied, and so on. The status bar also includes zoom and view options.

Notes Pane

The notes pane can be found right below the slide currently in use, and it is also where the speaker's notes are written for the current slide.

It is worthwhile to note that none of the content in this pane is visible on the actual slide when the presentation is ongoing, although it can be made visible in both the **Notes Page** view and also the **Presenter** view.

View Buttons

Most importantly, there are three buttons on the status bar shown towards the left-hand side of the zoom-in and zoom-out options:

- Normal: If in some other view, like the slide sorter view, select the normal button located on the status bar to change to the normal view. Shift-clicking this button enables the slide master view to be scrolled.
- When this button is toggled, it switches from any other view to the **slide sorter view.** This view helps to show the zoom-able thumbnails of every slide in the currently open presentation. When this same button is shift-clicked, it helps get to the Handout Master view.
- When this button is toggled, it can help switch from any other view to the reading view.

Unraveling the Ribbon

As the name implies, the ribbon consists of a long strip of tabs and buttons running across the top of the main window within the PowerPoint interface. Ribbon commands almost all the functions that are necessary for working with slides, and they are designed so that one can easily locate the commands a user needs to complete a specific task.

Nevertheless, PowerPoint also has a single menu, which is called the **File Menu.** Additionally, it includes a single, customizable toolbar called Quick Access Toolbar. Additionally, all other commands can be located within the different tabs on the ribbon within the user interface.

Collapsing Ribbon

If more screen is needed while creating a presentation, the whole ribbon can be hidden while still displaying the tabs by **pressing Ctrl + F1** on the keyboard. The same combination of keys, i.e., **Ctrl + F1,** needs to be pressed again to restore the ribbon. Alternatively, you can double-click any of the visible tabs.

The ribbon has a lot of interface elements, some of which are highlighted below.

Tabs

The ribbon has fixed tabs such as Home, Insert, Design, etc. Each of these tabs has a set of tools used to make and edit other slides. Much more than these visible slides, there is a hidden tab that is called the **Developer tab,** that can easily be enabled.

Contextual Tabs

These tabs are very special tabs that are located right within the ribbon but are not always visible. These unique tabs only make an appearance when work is being done on a particular slide object, which can be adjusted with the use of special options. An example of this kind of tab is the **shape format tab,** which is always highlighted in green when a shape or another drawing object is being chosen on the slide.

Gallery

A gallery can be defined as a collection of styles or properties. In most cases, galleries can be seen as drop-down galleries. To have access to the drop-down gallery, simply click the **more button**. For example, all the available styles can be seen as little thumbnails in the shape styles gallery.

More Button

In the ribbon tab, clicking "More" will expand a gallery. All options will become visible this way. If the gallery has many options, you can use the two arrows above the More button to scroll inside the gallery without expanding it or right after expanding it.

Down Arrow

There is always a small triangle pointing downwards next to many buttons, which indicates the downward arrow. By clicking this arrow, you will be shown a gallery of additional options or submenus pertinent to the selected button.

Dialog Launcher

Dialog launchers are small squares with arrows that are found in the lower right-hand corner of several ribbon groups. By their very name, dialog launchers display dialog boxes with more improved options when selected.

Editing Text

Although most slideshows contain lots of animations, charts, and sounds, it is not uncommon for them to have text as well. A list of highlighted items or a chart with information can be inferred by text. In fact, even the simplest of presentations should have some text included, so it makes sense.

Text is just how facts and figures are passed to the audience, although this doesn't mean the text should be plain and unattractive to the slideshows. PowerPoint 2025 helps to provide a way to make text with custom styles and include it in the slides.

When a new PowerPoint slide is created, two objects are also created automatically. A first object is a text object, and the second object is underneath the text object in an open space where the selection of an object can be made easily. PowerPoint 2025 works with rectangular objects, wherein they can be adjusted and moved to a particular point on a slide.

Text box by default

The default text box created in a new slice is just at the top of the main object rectangle. As the text indicates, simply click the text box, and then there will be a prompt to enter the new text that you wish to be displayed.

Any text can be typed into the object, but if too much text has been typed or the chosen font is too large, there might be a need for the text box to be resized. Resizing a text box is usually the same as resizing any object in PowerPoint. Circles that are placed on each side of the rectangular area and in the corners can be used to make the text smaller or bigger.

In the "insert tab" option, a section named "Text" contains the tools to help build text objects in the presentations.

Section of the Text Menu

PowerPoint gives a lot of text box objects. The "Text box" button helps to add a basic object where just about any text can be included on a slide. The "WordArt" button on its own helps to add clipart and text in unique and specialized styles offered by Microsoft. A date and time text object can also be added if need be. It helps to provide a unique date and time on the slideshows in such a way that they can be time-stamped as the slides are being run.

In slides with slide numbers, a small text box object appears in the corner where it is easy to identify the slide number. The number of slides added to the presentation will appear as each slide is scrolled past, making this feature very useful if the number of slides being added to the presentation needs to be kept track of. Having this information handy can help a presenter know what slide they have reached in a presentation.

Headers and footers are commonly used with word documents, but they can still be used in PowerPoint presentations. A header shows text at the very top of a presentation. This header can also be repeated with each new slide.

The primary difference between a header and a footer is that a footer appears at the bottom of a slide instead of at the top like the header. In PowerPoint presentations, headers and footers are useful for showing information without having to type it manually on each slide.

A particular title can be added to a footer object, for example, if one is needed for a presentation. There are also page numbers in footers, also known as slide numbers. When it comes to copyright notices, footers are extremely beneficial, as they can be positioned at the bottom of the slide away from the other critical content.

You can add any of the text objects to the slide by selecting them. To add a text object, you need to select the slide first. A red outline surrounds the slide that is very active in the left panel. The slide that is active can be viewed in the center window, as well.

Adding a Text Object

It is just as easy to add a text object as any other type. There's no need to worry about how the process works for each object; it all depends on the text object to be added.

The most common text object that is always added to a slide is the basic text. The object will be displayed just about anywhere on the slide. It can be displayed as a bullet or a number list. To add the text, simply select the **text box** in the Text menu section. A new text box with the default fonts and colors is shown, then new content can be added.

Adding A New Slide

When a blank presentation is created in PowerPoint, a default slide that has a title slide layout appears. Simply click into **placeholders** in the title slide and type the preferred text that needs to appear as the title and subtitle of the presentation.

Inserting a new slide into the presentation and specifying the exact placeholder that should be displayed on the slide is how another slide can be added to the presentation. The slide layout selected for the slide inserted determines where the placeholders are displayed. After the slide has been inserted, you can change the placeholder layout that will be applied to the slide.

To fix a new slide in PowerPoint, which will include a title and content slide, the layout, select the **Home tab in the Ribbon**. Right after that, select the **new slide button** in the slides button group.

A new slide will then be added. You can always repeat this process to continuously add new slides.

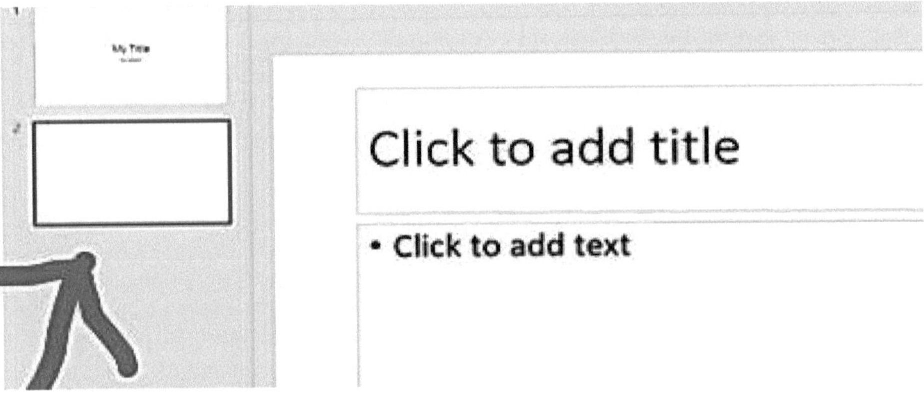

In the alternative, to fix a new slide with a much different slide layout, select the **Home tab** in the Ribbon. Then select the **drop-down aspect of the New Slide button** in the slide button group. The drop-down menu that will be displayed will show the other slide layouts that are provided. Choose the name of the preferred slide layout in the drop-down menu to fix a new slide with the chosen layout.

Adding Some Style

The PowerPoint designer automatically generates design ideas that users can select from in order to improve slides created by users. As you add content to a slide, our designer automatically matches it with a professionally designed layout in the background.

Request for design ideas at any point in time by selecting the design, then designing ideas on the ribbon. The first time using the design option on PowerPoint, it might request permission to get design ideas. If there is a need to use the designer option, simply click on **the Turn on button**. Immediately, this became popular. Whenever a slide is created, PowerPoint automatically opens designs. With time, PowerPoint learns what designs the user likes and brings back the appropriate design at the right time.

After completing all of the steps above, scroll through the design suggestions that should have been presented in the Designs Ideas pane on the right-hand side of the window, and select the design that best matches the content that is being created, or close the window if you do not need a design at this time. A slide will change according to the design chosen. Another idea can also be chosen from the pane, or you can choose to return to the design of the original slide. To undo a particular design you have just chosen, either by mistake or because you simply want to choose another option, all you need to do is **press CTRL + Z.**

Displaying Presentation

The presenter view in PowerPoint 2025 is a purpose-built view that follows the slide show view. This view is of immense benefit in an environment that has at least two displays, which include a laptop and a projector, or at least two monitors attached to the desktop system. The projector or the secondary display then shows the very slides in what is commonly called the "slide show view." However, the presenter view can be seen on the laptop or the primary display.

It's worthwhile to note that much more than just the slide will be displayed. Be rest assured that the audience that is viewing the project or secondary output can see only the slide.

Although the presenter view in older versions of PowerPoint had to be turned on manually, the newer versions of PowerPoint detect if two displays are connected and then turn on the presenter view. However, it's necessary to check and make sure that this view is displayed on only the primary view and not on the secondary view also.

If it is just the single display that is being used, it can still be changed to presenter view from within the slide show view. Move the cursor over the bottom left area of the slide to locate the six watermarked navigation icons. Select the last icon and, from the menu that pops up, choose the Show Presenter View option. This will bring up the PowerPoint presenter view. Note that PowerPoint 2025's presenter view has an enhanced view interface that includes the ability to zoom into a specific part of the slide and then move across the slide area.

Furthermore, the pause, resume, and restart buttons give more control over the timing of slides. Also, there is easy access to the Laer Point tool, which gives direction to the viewer's eye with a bright-colored point of light.

The presenter's view also shows the following

Toolbar

Show taskbar: This is a toggle that alternates between showing the Window taskbar and hiding it. Select **once** to ensure the taskbar is made available. It can be a very useful option to have full access to the taskbar if at any point there is a need to access any open application. Once this has been done, get back to the slide show and select **the Show Taskbar option** once again to hide the taskbar.

Display Settings

This feature, when selected, opens a drop-down menu wherein there are options to swap presenter view and slide show. There is also an option to duplicate the slide show. If in a single display environment, these options will be grayed out. In a dual-display environment, the Swap Presenter View and Slide Show can be used to swap displays between the primary and secondary displays. The Duplicate Slide Show option will display the Slide Show view on both displays.

Timer

This can be found just above the slide preview area. The timer helps estimate the time passed since the slide show must have commenced. **In addition,** PowerPoint offers two buttons for pause/resume and restart.

The preview area displays the active slide. The very same slide is simultaneously shown full screen on the projected screen, which is the secondary display.

- **Next Slide:** This is located on the top right side. A preview of the next slide is also located there.
- **Notes Pane:** This section helps to show the notes for the active slide. In this section are two buttons to make the text larger and also make the text smaller.
- **Pen:** This is useful for drawing or making annotations.
- This is also similar to the pen, but it is virtual.
- **Ink color**: This changes the pen color.
- **Pen or Highlighter Eraser:** This helps to erase any pen or highlighter marked areas (it is always grayed out if there are no areas that are marked on the slide.
- **Erase all ink from the slide:** This ensures that all ink markings are removed. This option is also always grayed out if there are no marked areas on the active slide.
- **In this feature,** the user is given the option to show (visible) or hide (hidden) the cursor while the presentation is being presented. The automatic option is always turned on as it is the default setting. As a result, the cursor appears when the mouse or trackpad is being moved for a while, and it is then hidden when no movement is being made.
- **See all the slides:** Bring up all the slide stacks and choose any slide you prefer by clicking on this option. With this option, it is easy to switch slides in or out of sequence from a grid that you alone can view. The audience will only see the slides that have been selected.
- **Custom Show:** This feature opens a sub-menu that displays all the custom shows that are available in the opened presentation. Note that this option will also be grayed out if there are no custom shows created in the presentation that is being viewed.
- **Hide Presenter View:** This helps to return to the slide show view when in a single display environment.
- **Screen:** Gives room to show a black or white screen to the viewer's

Saving Presentation

To ensure all of the hard work doing the content will not be lost in the future, it is of utmost importance to safeguard presentations. There is also an option to print the presentation if the hard copy is needed to be shared in a meeting, for example.

To save a presentation, **select a file,**

and then click **on the save as option** then choose PC to save to the computer in use and then type a preferred name for the presentation in the file name box and then click **on the save button.**

It's best to have the content saved while still working on it. This can give a guarantee that the files are well saved. Simply click on the **Ctrl + S button** often to have this done.

Closing presentation

After you have concluded and saved the presentation, you need to close it. In this case, it seems more like gathering papers and placing them neatly in a file or folder and then returning the file to its proper location.

To close a file, click on the **close button** that is displayed at the top right corner of the PowerPoint window. As an alternative, select the **File tab** and then choose Close, or simply use the keyboard shortcut **Ctrl+W**.

Have you noticed the presentation has disappeared from the computer screen? Never mind it is tucked safely somewhere on the hard drive where it can be gotten back much later when it is needed.

CHAPTER 2

EXPLORING 'BEHIND THE SCENES'

Welcome to Backstage View

The backstage view can be accessed from the File tab that is located at the extreme left of the ribbon. Though it has the look of a tab, when the file tab is clicked open, the header brings up a menu on the left-hand side of the backstage view. It is worthwhile to note that there are no other menu options in the PowerPoint 2025 window interface.

The backstage view provides easy access to frequently used features for managing PowerPoint presentation files. As well as organizing files, this view enables users to save and create new files, search for hidden metadata or personal information, print slides, and set file options.

When the backstage view is opened up, there are about fourteen options that can be found on the left-hand side of the pane. Extra options can also be added by any third-party PowerPoint add-in. Each of these fourteen options opens an individual tab.

When PowerPoint is launched, and there are no presentations, the home tab will be made visible within the backstage view. If the backstage view is accessed with a presentation open already, the home tab will still be made visible. The fourteen tabs that can be seen include Home, New, Open, Info, Save, Save as, History, Print, Share, Export, Close, Account, Feedback, Options.

To return to the backstage view from the normal view, simply select the back arrow button that can be found at the top of the left interface.

Taking into account the themes and templates

Microsoft PowerPoint templates give the widest range of choices when it comes to designs, which makes them perfect for modern and stylish presentations and pitches. PowerPoint themes, on the other hand, give room for lots of presentation topics, allowing you the freedom to select the very best presentation templates designed for any project.

Microsoft provides different PowerPoint templates for free and premium PowerPoint for subscribers of Microsoft 365. Each template for PowerPoint enables the use of stylized charts and graphs while ensuring that a professional tone is kept. Whatever you choose to do, be it hosting a healthcare seminar, teaching a high school course, having to design a fashion magazine and lots more, any choice made will be supported by PowerPoint.

Despite the fact that Microsoft's PowerPoint presentation templates are designed professionally and require very little customization, they can also be heavily tweaked. A presentation can be customized easily if more hands-on approach is needed. Each PowerPoint template can also be customized to meet specific needs.

Changing a Theme

To change a theme, simply open your presentation in PowerPoint. On the Design tab, in the Themes group, choose the theme that best suits your needs from the list. If you hover over any of them, a preview option can be seen. When you're satisfied and ready to continue, just click **the theme**. It will be applied to the whole presentation.

Unless the previous theme is a PowerPoint default theme, which will appear under the list of default themes, when a new theme is applied to PowerPoint since it typically affects all slides, the previous theme will be completely removed.

How to Apply a New Theme to Selected Slides Alone

Simply open the presentation in PowerPoint and locate the design tab. Choose the **slide of the theme** that needs to be changed. If different slides are to be selected, simply hold the Ctrl/Cmd button and click on the slides simultaneously.

Select the theme you want to apply. You can preview a theme by hovering the mouse over the list of themes before selecting one. Note that should the overall theme of a presentation change much later, these slides will be affected as well.

Importing a Theme

To bring in a theme from another source, start by opening the presentation in PowerPoint, and on the Design tab, in the Themes group, select the **drop-down arrow**. Click on the **Browse for Themes** and a new window will be opened where the presentation that contains the theme to be imported will be selected.

Do note that the selected theme will be applied to all of the slides, and any previously used theme will be removed totally.

Editing the colors of Themes

Based on the selected theme, it can have different colors or different palette presets that enable a user to select and also change the color of the items in the presentation. The colors of the themes can be edited by opening the presentation in PowerPoint, navigating to the design tab option and clicking on the variants group if the theme has variations.

The color palette can be changed by selecting the variant drop-down arrow, choosing colors, and then selecting any preset. The customize color option can be used if a specific palette should be applied to the presentation. A new window will appear with the colors that will work best for the presentation. To the right of the window, you will see a preview.

To change the background color, select the **variants drop-down arrow** and select the **background styles**. You can choose any of the presets or select Format Background to create a background that can be solid colors, gradients, images, or even patterns.

If you choose to do this via the Format Background options, you will see a panel on the right side of the screen. To apply a customized background, simply select **apply to all**. If the background is changed with the use of the Format Background option, all the background pictures will be taken off.

Keeping a Presentation Safe

It is essential to have your presentation saved so that when the need to use it arises, it can be called back at any point in time. PowerPoint presentations can be saved either on the computer or on the OneDrive platform.

Computer

To save on the computer, simply click on the **File tab** and select the **save option**, then select **computer** and then the location where it should be saved on the computer, either the desktop or a particular folder for easy access.

OneDrive

OneDrive is a Microsoft cloud-based service that helps a user get connected to other files. OneDrive helps a user to keep and protect their files and also share them with those they want to share them with. A file saved on the OneDrive platform can also be accessed anywhere on all devices.

To save a PowerPoint presentation on the OneDrive, simply select the **save as option** and then choose the OneDrive option. This will ensure that you can make use of your presentation from anywhere around the world.

Introducing the Presentation

Pinning a Presentation for Easy Access

It is always necessary to 'pin' a PowerPoint presentation; this way a user can get to it easily when the need arises rather than search through the entire presentation that must have been created.

To pin a particular presentation for easy use, simply click on the File tab and then go to the backstage view. Select "open." This will give way to recent presentations. When you move the mouse over the presentation that needs to be pinned, a pushpin icon will emerge next to it. An already pinned presentation appears with this icon. All other presentations will be subordinated to the pinned presentation, so it can easily be accessed. Whenever the need to unpin a presentation arises, click on the pushpin icon again to unpin it.

CHAPTER 3

SLIDE EDITING

Moving from slide to slide

As earlier discussed, slides are single pages of a presentation. Collectively, a group of slides may be known as a slide deck.

You can switch from one slide to the other, especially during presentations, by clicking the **left mouse button.** That is all that is needed to know at the primary level. A user can also move around the slide by using the shortcut buttons to move to specific parts of the slide. A user can also right-click and use the shortcut menu that moves around.

To move slides automatically, have the presentation open and on the slide show tab, click on the **from the beginning button.**

The first slide will then appear in the slide show view. When done with the first slide and need to move, click to **move to slide 2**, then tap the spacebar to proceed to slide 3, click on the **right arrow key** to proceed to slide 4, then press the enter button to proceed to slide 5. Above, I have listed various ways a user can choose to proceed to the next slide.

To go back to the previous slide, simply select the backspace key or the left arrow key. These two keys are dedicated to going back to the previous slides.

To jump to any slide at any point in time, simply right-click on **anywhere in the slide** and then choose **the see all slides** option. This will show a list of thumbnails where all the slides can be seen. Then select any slide you choose to move to. This method makes it easier and faster to get to the desired slide.

Working with objects

There would be no point in PowerPoint slides without objects. Text, images, and charts ensure that the presentation has meaning and content to ensure that blank slides are meaningful to the audience. The use of objects in a presentation should be kept to a minimum when it comes to adding objects. Slides should not be crowded with too many objects that the main point of the slide is lost.

Most of the objects embedded in slides are text objects, which enable a user to type text into the slide. Every slide has its layout, which has one or more placeholders. A placeholder is simply known as an area on a slide that is kept basically for text, clip art, a graph, or some other type of object. As an example, a slide that makes use of the title layout has two placeholders for text objects: one is for the title and the other is used for the subtitle. The slide layout task pane is used to choose the layout when a new slide is created. This can be changed much later, and more objects can be added to the slide. Objects can also be moved, deleted, or resized to suit one's taste and meet specified needs.

There are many types of objects that can be added to a presentation, including clip art, graphs, shapes, etc. Content layouts enable you to create slides using a variety of tools that are displayed on the drawing toolbar that is located at the bottom of the screen. You can also add additional objects to a slide using the icons that are displayed in the center of the slide. All objects have rectangular sections on their sides. Depending on the contents of the object, the rectangular area might or might not be filled visually. Regardless, the outline of the object is visible when selected.

In addition to overlapping objects, they can also be grouped together. Although they shouldn't usually overlap, sometimes their overlap creates a distracting effect. Some clipart can also have text overlaid on it.

Selecting Objects

Before any edit can be made on the slide, an object that contains what needs to be edited has to be selected first. As an example, a user can begin typing away to make an edit on the screen. Instead, the object that contains the text that needs to be edited has to be selected first. Furthermore, other types of objects must also be selected first before their contents can be edited.

It is worthwhile to note that the normal view must be activated first before individual objects can be selected on the slide. With the use of the slide sorter view, the whole slides can be selected but not the individual elements they have on them.

Rules to keep in mind when selecting objects

To choose text objects so that their text can be edited, have the point of insertion moved over the text that needs to be edited, then click the **OK button.** You can double-tap the text by using a mouse or touchpad. The text insertion point and a rectangular box will be visible just around the object. A user can follow the text insertion point as they type.

The functions of other kinds of objects are slightly different. As soon as the object is selected, the rectangular box will appear around it in order to alert users that the object has been hooked. You must then drag the object around the screen or alter its size after it has been hooked. You cannot edit its content after it has been hooked.

The Ctrl Key: Much more than just one object can be selected by clicking on the **first object** and then holding down the CTRL key while selecting additional objects.

Click and drag: This option is another unique way to select an object or more than just one object. Simply make use of the insertion point to drag a rectangle all around the object that needs to be selected. Point to a specific location above and to the left-hand side of the object that needs to be selected, then drag the mouse in a downward manner to the right-hand side until the object is surrounded by the rectangle. When the button was released, all the objects right within the rectangle would have been selected.

Pressing the Tab key is another amazing option to select objects. Click on the **Tab button** once to choose the first object on the slide. Press the Tab button again to choose the next object. Continue to press the Tab button until all the objects that need to be selected have been selected. Pressing the Tab button to choose objects comes in handy when it isn't so easy to point to the objects that need to be selected. This can be a problem and it can happen if the object that needs to be selected is under another object, or if the object is empty or invisible and the user is unsure of the location.

Resizing or moving objects.

A box will appear around selected objects when they are selected. Closer inspection of the box reveals it has love handles on both edges, one in each corner. They are used for adjusting the size of an object. You can also move objects on the slide by taking the edge of the box between the handles.

The object will have every tendency to position itself with close objects if it is moved or its size is changed. The alignment lines will appear when the object is moved into alignment with other objects on the slide. The object will adjust to the alignment shown if the mouse button is free when the alignment mark appears. Additionally, many types of objects will also have a rotating handle over them, which is represented as a circular arrow. The object can be rotated effortlessly by grabbing the handle and moving it around in a circle.

To change the size of an object, choose the object to be selected and then grab one of its love handles simply by clicking. Ensure you hold down the mouse button and then move the mouse to change the size of the object.

The different handles on an object give different ways to change the object's size.

- The handles at the corners ensure that both the height and the width of the object can be changed.
- The handles at the top and the lower edges make it possible to have just the height of the object changed.
- The handles on the right and left edges can change only the breadth of the object.

If the Ctrl key is held down while one of the love handles is dragged, the object will remain centered at its current position on the slide with its size adjusted. Furthermore, try to hold down the shift key while moving an object by making use of the corner love handles. The combination of both helps to maintain the proportion of the object when resizing it.

Adjusting the size of a text object's text does not mean the size of the text within the object will change; it will only change the size of the frame in which the text is contained. The change in the width of a text object is equal to the change in margins in a word processor. This helps to make the lines of the text wider or narrower. If the text within a text object has to be changed, the font size must also be changed.

You can move an object by selecting just anywhere in the outline box except on a love handle, then dragging it to its new location. Users are not required to click precisely on the outline box for shapes and other graphic objects to move them- they can click and drag anywhere within the object to move it. For text-containing objects, it is necessary to click the outline box itself in order to move the object.

Selecting Text

Text can be highlighted in PowerPoint in various ways with the use of the text highlight color that can be found on the Home tab in the ribbon. It is worth noting that highlighting uses color to help users focus their attention. However, the color might not be made visible to some members of the audience while the presentation is ongoing. As such, combining highlighting with another effect like bold can make the text more visible. Bear in mind also that highlighting is just one of the various ways to help the audience focus attention. Different strategies can be used in PowerPoint to help keep attention focused. Strategies include creating hierarchy, making spotlight effects, inserting images, and adding animations.

The tabs for buttons and ribbons might show in a variety of ways depending on the size of the screen and the settings of the control panel. The ribbon tab can show various names. The picture tools format tab, for example, may display picture format, whereas the drawing tools format tab may display drawing format or shape format.

Using Cut, Copy, and Paste

When it comes to human-to-computer relations or user interface design, cut, copy, and paste are related commands that give us an inter-process communication technique for moving data directly through the interface of a computer. Copying ensures a duplicate is created of selected data, while cutting removes it from its original location.

In both cases, the selected data is saved in temporary storage, known as the clipboard. In the following step, the data is copied from the clipboard and pasted where the paste command is used. Any application that supports the feature will be able to access the data, allowing for easy data transfer between applications. With the cut, copy, and paste feature, users can recall particular data and have it transferred from one place to another.

Duplicating an Object

The cut, copy, and paste commands can be used to duplicate an object in PowerPoint. Simply right-click on **the object** then click on the **copy option** or left-click down and drag the mouse over the object that needs to be copied navigate over to the location where the object is needed, right click again, and select **the paste option**. The object would have successfully been duplicated. Note that objects can be duplicated in different locations as long as they are needed.

Using the clipboard task pane

The clipboard task pane helps users gather up to about 24 items of text or graphics from any office program and then successfully paste them into the presentation. To launch the Clipboard task pane, simply select the **dialog box launcher** in the Home Tab on the Ribbon at the side of the Clipboard group.

Once the task pane has been opened to paste any item from it, just click **the item** and it will be inserted automatically.

Using the undo command when a mistake has been made

Undo is a command in different computer programs. It helps to take off the last change that has been made to a particular document and revert it to an older state. In some more developed programs, such as graphic processing, the undo command negates the last command done to the file that is being edited. In most applications, which also include PowerPoint, the undo command can be activated by simply pressing the Ctrl+z or the Alt+Backspace keybindings. The undo command helps to quickly make changes to any mistake that must have been made rather than having to press backspace continuously.

Working with slide layouts

Slide layouts are designed to contain formatting, positioning, and placeholder boxes for all the content that can be seen on a slide. Placeholders can be described as the dotted line containers located on the slide layouts that help to hold contents such as titles, body text, SmartArt graphics, pictures, clipart, videos, and sounds. Slide layouts can also contain the fonts, colors, effects, and the background (which collectively is known as the theme) of a slide.

PowerPoint presentations have inbuilt slide layouts that can be adjusted to meet the user's basic needs and can also be shared with other people who also make use of PowerPoint presentations.

The slide master view of a PowerPoint presentation allows you to change the standard slide layouts that are custom created. As you view slide masters in the thumbnail panel on the left-hand side, a slide master will be displayed at the top, followed by layout masters that define different layouts that can be used, such as the title slide and the title and content slide.

To add a defined slide layout to a given slide, choose the slide first and then, on the toolbar ribbon, choose Home, then Layout, and select a layout from the vast options that are being displayed

If there is a need to customize the definition of a slide layout that can be later applied to individual slides, on the toolbar ribbon, choose view and then slide master. The layout master will be displayed as a thumbnail in the thumbnail pane under the slide master. Simply select a **layout master** in the thumbnail pane and start customizing.

Deleting a slide

After a presentation has been created, the user then needs to go over and view it to make changes if need be. Deleting a slide can come in handy at this point in time. To have a particular slide deleted, select **the slide** (note that one or more slides can be deleted at once), single slide, right click, and then choose the delete slide option.

If multiple slides are to be deleted, hold the Ctrl button and choose the slides that need to be deleted, right-click and then choose the delete slide option. Lastly, to delete a sequence of slides, press and hold the shift button and then select the sequence that needs to be deleted, right-click and then choose the delete slide option.

Duplicating a slide

By copying a slide either from the presentation that is being worked on or from another, you can save a lot of time and reduce your energy usage. When a slide or several slides are copied from a presentation to a new location, either within the same presentation or to another presentation, the theme of the new slides can be easily specified.

To have a slide duplicated, select the view option and then clock normally in the presentation views groups. In the slide thumbnails located on the left-hand side, select the **desired slide** that needs to be copied. To select multiple or sequential slides, select the **first slide**, then press the shift button and click on the last slide that needs to be added. In the alternative, select the first slide, press the shift button, and then press the cursor down button until it reaches the last slide. To select multiple non-sequential slides, press the Ctrl button and then choose each slide as needed.

After selecting the slide(s) to be deleted, right-click the **selected slides** and select the **copy option**. Then move the cursor to the desired location where the slides should be and then right-click again and, under the paste options, do one of the following:

- To use the theme of the location the slide is going to, select the "Use destination theme option." This will ensure that the copied slide matches the destination of the theme.
- To keep the theme of the presentation that is being copied, choose the "keep source formatting" option. This means that the slide will save its original theme format and not the presentation theme.

When the slide is pasted to its new location, the paste options button will automatically be displayed near the slide that is pasted in the normal view. As an alternative, the Ctrl+V button can be used to paste the slide in the desired location.

Finding and replacing text

With the use of the find and replace text option, editing can be easier and done as a whole rather than having to look for the word to be replaced line after line.

To do this, navigate to the Home tab, in the Editing group, select **replace.** In the box named "find what," enter the word that needs to be found and replaced, and then in the "replace" box, enter the word that should be the replacement.

To complete the find and replace at once do one of the following:

To replace the currently chosen occurrence of the word, simply choose to replace, and to replace all the occurrences of the word, simply select the replace all option.

Rearranging slides in slide sorter view

The Slide sorter view offers users a view of their slides in thumbnail form.

With this view, it is very easy for sequences of slides to be sorted and organized in preparation for printing the slides. The slide sorter view can be reached from the taskbar below the slides window or the view tab option located on the ribbon.

It is possible to rearrange slides in the slide sorter view before printing them, either due to a change of opinion about how the arrangement should look or to correct an error before the slide is printed.

Below is the very simple way slides can be rearranged in a slide sorter view.

Change to Slide's Sorter view and all of the presentation slides will be displayed as thumbnails. Click and move the Slide in the desired location once you have located the slides you need to arrange. All slides in the presentation will be renumbered after the slide moves to its new location. After you have arranged everything that needs to be arranged, simply return to the normal view to continue editing the presentation.

CHAPTER 4

WORKING IN THE OUTLINE VIEW

Calling up the Outline

The outline view in PowerPoint helps to show the presentation as an outline made of the titles and the main text from each of the slides. Each of the tiles will show up on the left-hand side of the pane that contains the outline tab, which includes a slide icon and a slide number.

The main text will be centered right under the slide title. Graphic objects will be displayed only as little notations on the slide icon in the Outline view. Working in the outline view is especially useful when global edits need to be made. All that needs to be done is to grab an overview of the presentation, then adjust the sequence of bullets or the slides, or simply apply formatting changes.

The outline can be opened or recalled from the tab located on the navigation panel, which appears next to the slides by default. If by accident, the navigation panel is closed, the slide show outline tab will not be displayed until the panel's display is restored.

To do this, simply click on the **view tab** in the Microsoft PowerPoint menu, then double click on the "Normal" option that can be found far to the left-hand side of the menu. If you are already in the normal view, the navigation panel will be reopened on the left-hand side of the PowerPoint slides.

Finally, click on the **Outline tab** from the navigation panel to see and edit the slide show outline. If the outline tab still cannot be seen, simply hover the mouse around the vertical that can be found on the left-hand side of the screen, click the **left mouse button,** and move the line to the right-hand side until the outline tab appears on the left pane and is now visible.

Selecting and editing an entire slide

In PowerPoint, there are three options when a group of slides is to be selected to add formatting, such as adding an animation effect or transitioning to all of them at once. To choose a group, make use of either the slide sorter or, better still, the slide panes located on the left-hand side of the screen. Alternate between these two views with the use of the icons located on the status bar at the lower part of the screen.

Selecting all the slides

All the slides being chosen differ slightly based on whether the slide sorter is being used or the slide pane.

If using the slide sorter, select view, then slide sorter. The very first slide in the deck will be selected. To select all of the slides in the presentation at once, press the shift key and select **through to the last slide**.

If using the normal view, select **view,** then normal. In the slides pane, choose the first slide, then press the shift button and select **through to the last slide**, to select all the slides in the presentation at once.

Select a group of consecutive slides

This process is very easy and very fast as well. Simply choose the first slide in the group of slides that needs to be selected. Hold down the shift key and then select the last slide that needs to be added to the group. This includes the first slide as well as the last slide, and also all the slides in between the first slide and the last slide. Consecutive slides can also be selected by moving across the slides that need to be selected.

Select "No consecutive slides

This process is quite straightforward. Choose the first slide in the group that needs to be selected. It doesn't necessarily have to be the first slide of the presentation. Hold down the Ctrl key while selecting the specific slide that needs to be added. Note that slides can be chosen in no particular order.

It is possible to edit the slide master in much the same way that you can edit a slide in the normal view. This way, the changes will be visible throughout the presentation. Changes will be made to the slide master tab and they can be moved to all other tabs on the ribbon; **however,** the slide master view will remain unchanged.

If you would like to change the theme font on all slides, select **the view tab** option and then select the slide master view command under the presentation view groups. After selecting the slide master view command, the active tab will appear. If they haven't been selected yet, select the slide master for all slides.

Click on **the fonts command** in the edit theme group located on the slide master tab. Hover over each option with the mouse to show a live preview of the font option on the master slide. Select the **menu option** to choose it and then click on the **closer master view** command to return it to the normal view.

To make changes to text color or alignment on all slides at once,

Click on the **View tab**, then choose the Slide master view command, which is usually located in the presentation views group. The slide master tab will then be displayed as active. Choose the **slide master** for all the slides if that option is not chosen already, and then choose the text that needs to be adjusted and make the changes as it suits you.

Click on the **home tab**. Whenever the slide master view is in use, the supporting layouts are shown in the slide task pane on the left. To create the font cascading effect, select the command in the font group. A live preview will appear on the master slide as you move the mouse gently over the font options. Click **on the color option** to select a particular font. In the task pane on the left side of the screen, the font color change will be reflected on all slides.

Next, select an **alignment command** to make changes to the alignment of the text that has been selected. Finally, choose the slide show view or the normal view command that is located at the lower part of the window to leave the slide master view and be able to view the changes already made in the presentation slides.

To insert pictures on all slides also, click on the view tab option and choose the slide master view command in the presentation views groups. This will ensure the slide master tab becomes active and will be displayed. If it hasn't already been done, select the Insert tab option.

Next, select the **Picture command** in the Illustrations group. The Insert Picture dialog box will be displayed. Find the picture that beads should be inserted into, then select the **picture** and click on the **insert button**. The picture will then be displayed on the slide master. Ensure the picture is in the right location by selecting the picture and then holding it down with the mouse. Then drag it to the desired location on the slide and release the mouse button.

On a final note, select the **slide show view or the normal view command**, which can be found at the bottom of the window, to leave the slide master view and view the changes that have been made in the presentation slides.

Selecting and Editing One Paragraph

To ensure the text on the slides is much easier to read, there might be a need to change the line spacing between text and paragraphs in the presentation slides. It is worthwhile to note that the alignment and indentation of the lines of the etext can also be adjusted.

Line spacing, which includes single space, double space, and so on; vertical alignment, which also includes the top, bottom, or middle, can be seen in the menu buttons on the Home tab of the ribbon located in the paragraph group;

In the paragraph dialog box, there are more detailed spacing options available.

- On the presentation slide, choose the text that needs to be adjusted

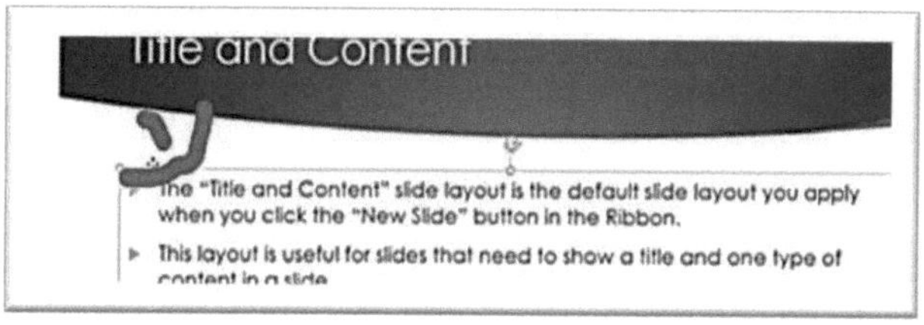

- Select the **Home button** and, in the paragraph group, choose the dialog box launcher. This will open up the paragraph dialog box.

The following options will be made available in the paragraph dialog box:

Alignment

To change the horizontal placement of the text in the alignment box, simply select **either the left, right, center, justified, or distributed option**. The justified option helps to add spacing between words such that the lines of text will touch both the left and the right margins, leaving out just the last line of the paragraph, which uses only the normal word spacing.

The distributed option is similar to that of the justified, but the last line also touches both the left and the right margins, with space added in between words and letters as deemed fit.

Indentation

To include an indentation or make changes to the amount of indentation before a text, choose or type a particular number in the "before text box". Special options can be used to indent the first line only or to add an ident that hangs. Note that indentation is measured only in inches and it can be any whole number or decimal, such as 1.2 or 3.4 inches.

Spacing

To make changes to the line spacing above or below a paragraph, type or choose the arrows located next to before or after. This number can either be a whole number or a decimal, such as 6.5 or 8.

To have the spacing above and within a paragraph changed, simply make use of the line spacing options: single, 1.5 lines, or double. Alternatively, select the **exact paragraphs** and then add a point value between 0 and 1584 in the **box.** Note that the bigger the point value, the wider the spacing. Multiple options can also be selected and a value will be added to the **box**. Make use of any number that is less than or equal to 9.99; the value of 1 word equals a single space, while a value of 3 would be equal to triple spacing.

Promoting and Demoting Paragraphs

"To promote a paragraph" simply means moving it up one level in the outline. This can be done by either using the promote button, dragging the paragraph with the mouse in an upward

manner, pressing the shift+tab button, or making use of the shortcut key (Alt + Shift + left arrow). Note that a paragraph can be promoted by placing the cursor just anywhere in the paragraph and then pressing the shift+tab button or by pressing the promote button that can be found on the outlining toolbar. Choosing to promote helps to move the chosen paragraph to the next higher heading level (just one up) on the left-hand side. You may have trouble doing this with the mouse, as the arrow changes from a single arrow to a four-cornered arrow.

Demoting Paragraphs

In contrast to promoting, to demote a paragraph means to move it downwards to just one level. This can be done by either using the demote button, pressing the Tab button, dragging the paragraph with the mouse in a downward manner, or making use of the shortcut key (Alt + Shift + Right arrow). The demote option helps to move the chosen paragraph to the next lower heading level, i.e., down one level to the right-hand side.

Adding a new paragraph

To add a new paragraph, simply click on **the Tab key.** This will help change the new slide to a paragraph on the former slide. Position the insertion point at the beginning of the paragraph and then click on the enter button. A new word will be inserted just above the cursor position.

Adding a new slide

In its initial form, a presentation has just one slide, which can be used for very brief presentations. To make the presentation last quite a while, you can add new slides. You have two options:

- Right from the Home tab, choose the new slide button in the slides group.
- Use the shortcut button to get a faster result: I secretly pressed Ctrl+M.
- Right-click on the slides pane located on the left-hand side, then select the **Choose New Slide Option**.

Make use of any of the three options and a new slide will be added. Note that you can add as many slides as possible depending on the content of the presentation.

Text boxes, word art, or shapes that move

When pictures, shapes, or other objects are added, they are stacked automatically into individual layers just as they are added. Individual shapes can be moved along with other objects or groups of objects in a stack. Objects within a stack can be moved up or down at a time, or they can be moved to the bottom of the stack in a single move. Objects can also be overlapped when they are drawn to create different effects. To move a shape, object, or text box, simply click on the border of the word art, textbox, or shape that needs to be moved. If more than one box needs to be moved, then press the CTRL button down and click on the different borders. Once the cursor changes to the four-headed arrow, move it to the desired location. To move in just small increments, press and hold down the CTRL button while pressing an arrow key. To move the object in a horizontal or vertical motion only, press and hold the shift key while it is being dragged.

To move a text box, word art, or shape forward or backward in a stack, simply click the **WordArt, shape, or textbox** that needs to be moved up or down in the stack and on the drawing tools format tab, select **either the "bring forward" or "send backward" option**.

Note that there will always be an option to keep moving the object one layer forward (the "bring forward option") or to the top of the stack (the "bring to font option). The send backward has similar options: down one layer (the send backward option) or to the bottom of the pile (the send to back option).

Collapsing and expanding the outline

There are times when a presentation might have many slides and grasping its overall structure might seem very difficult, even when looking at the outline. Collapsing an entire outline doesn't have to mean that the whole-body text will be deleted; it will only hide the body text so that there will be a focus on the order of the slides that are in the presentation.

A presentation's outline is restored when the collapsed body text is restored, allowing it to focus on the details once again. Users can collapse and expand an entire presentation, or they can collapse and expand only specific slides during a presentation.

To collapse the entire presentation, right-click anywhere in the outline and then select the collapse option. Collapse all or use the keyboard shortcut (Alt + shift +1). To expand the presentation, right-click and then choose the expand option—expand all or press Alt + shift +9.

To collapse just a single slide, right-click **anywhere on the slide** and then select the collapse option—collapse from the menu that appears. To expand a single slide, right-click the **slide that is already collapsed and needs to be expanded**, and then choose the expand option.

CHAPTER 5

PROOFREADING YOUR PRESENTATIONS

Checking for spelling as you go

Are you worried about making mistakes while typing? You don't have to be as PowerPoint helps to provide different proofing features, which include the spelling and grammar tools, which ultimately help to produce professional and error-free documents.

To run a spell check, right from the review tab, simply click on the **Spelling command**. This will bring up the spelling pane and make it visible.

For each error in the presentation, PowerPoint will seek to offer one or more suggestions. The suggestions can be selected and then select the change option to correct the error that must have been created.

PowerPoint will scan through each error that must have been committed until all have been reviewed. After the last error has been reviewed, a dialog box will be displayed showing that the spelling check is complete. Click on the **OK button** to complete this.

Ignoring spelling errors

The spell check is not always the correct option. Sometimes it will indicate that a particular word is spelled incorrectly when actually it is not. This often happens with people's names or nicknames, which, of course, cannot be found in the dictionary. If PowerPoint indicates that there is an error, there is a way to ignore this by following the options below:

- **Ignore:** This will help to skip the word without making any adjustments to it.
- **Ignore all:** This will skip the word along with changing it, and in addition, it will skip all other instances of the word that might appear again in the presentation.
- **Add:** This adds the word to the dictionary, so it will never show up as an error again. Always ensure that the word is well spelt before choosing this option.

Automatic Spell Check

By default, PowerPoint will automatically check the presentation for possible errors. There might even be no need to run a separate check with the use of spelling commands. These errors are always shown with the red wavy lines.

To make use of the automatic spell check feature, A menu will be displayed by simply right-clicking on the **underlined word.** Choose the correct spelling of the word from the list of possible suggestions. The corrected word will then be shown in the presentation.

Modifying proofing options

PowerPoint provides users with options to customize the way text is checked, allowing them to have more control over the checking process. You can customize how PowerPoint detects spelling errors by changing the automatic spell check options. In addition, you can turn on the grammar check option to help detect contextual spelling errors such as their vs. they're. Other types of grammar errors will not be detected by PowerPoint. Therefore, a proofread will still be necessary to catch any grammatical errors in the presentation.

To turn on the grammar check options, select the **file tab** to gain access to the backstage view.

Choose options and this will open up a dialog box and make it visible and then choose the proofing option, and then check the box located next to the check grammar with spelling and then click **OK**.

To have a spelling and grammar check run, move over to the review tab and choose the spelling command. All the spelling and grammar errors will be displayed with a red wavy line.

To make adjustments to other options, select the file tab to gain access to the backstage view. Choose options and a dialog box will be displayed. Click on **proofing** and the PowerPoint options dialog box will provide a variety of options that users can choose from:

- Check your spelling as you type. Uncheck the box if there is no need for PowerPoint to check for spelling errors automatically. This automatically checks and also helps to hide different spelling errors.
- Hide spelling and grammar errors: Check this box if there is a need to hide the red wavy line that marks spelling and grammar errors. Note that this option is always disabled automatically if "check spelling as you type" is unchecked.

N.B: If the automatic spell check has been turned off, there is still an option to run a check by simply going to the review tab and clicking on the spelling command. Also note that making adjustments to any settings in PowerPoint options, including proofing, affects all of PowerPoint, not just the current presentation.

Spell checking after the fact

If it is best to ignore the nagging by PowerPoint about spelling, it is OK to check spellings the conventional way by simply running the spell checker after typing all of the document. The spell checker has a way of working through the whole presentation, chasing up every word in its list of correctly spelled words and bringing up any misspelled words to the user's notice. The spell checker performs this task without giggling or having to snicker. As an included bonus, the spell checker also provides users the opportunity to tell it that the user is right and it is wrong, and also that it should check how to spell words just the way the user does. The following steps will give an insight into how spellings should be checked for an entire presentation:

- If the presentation that needs to be spell-checked is not open yet, ensure it is open.
- Open the review tab option on the ribbon and then select the spelling button located in the proofing group.
- Have your fingers tapping on the desk. A little patience should be exercised here as PowerPoint will be checking for spelling and grammar errors that might seem embarrassing.
- Keep calm and coordinated if PowerPoint finds a spelling error. If it does, it will switch over to the slide that contains the error, highlight the word that seems offensive, and also display the words that must have been misspelled along with a suggested word as a correction.
- Choose the correct spelling and select the change button. As an alternative, you can also choose to ignore it. If you agree that a particular word is misspelled, go through the list of corrections that PowerPoint has to offer and choose the one that is preferred, then click on the change button. If you also like the way the word is spelled the first time (maybe it is an unusual word that cannot be found in the PowerPoint spelling dictionary), simply click the ignore button. If there is also a need for PowerPoint to ignore all occurrences of a particular misspelling in the presentation currently in use, click on the ignore all button. Similarly, if there is a need for PowerPoint to correct all occurrences of a particular word that has been misspelled, simply select the "change all" button.
- When the message "spell check complete" pops up, then all is set and good to go.

Note that PowerPoint always checks to spell in the whole presentation, starting with the first slide, unless a single word to be checked is specified or a group of words by highlighting them. PowerPoint also checks how titles are spelled, body text, notes, and text objects that are added to slides. It does not check the spelling of objects such as graphs or charts.

Using the Thesaurus

A thesaurus, which can also be known as a dictionary, is a reference work used to search for synonyms and also antonyms of words. They are most often used by writers when they need to find the best word to express a particular idea.

PowerPoint contains a built-in thesaurus that can be used to quickly show users synonyms for a word they have typed while preparing slides. Making use of the thesaurus can be quite easy:

- Right-click on a particular word that has been typed and a menu with a list of synonyms for the word will be displayed (PowerPoint sometimes throws one or two antonyms into the list just to contradict it) and then choose the synonyms for the word from the menu that pops up.
- Choose the particular word that should replace the word that has been selected. PowerPoint will then replace the original word with the word that has been chosen.

If the Thesaurus is chosen from the synonyms menu or the Thesaurus button is clicked upon in the ribbon that shows in the margin, the Thesaurus section of the research task pane will be displayed with the synonyms that are listed. The thesaurus provides users with the option of looking up words to search for more synonyms. For example, if the word "falloffall offhosen from a list of synonyms, another set of words can be obtained. You can continue to click words to search for other synonyms for as long as needed until you are ready to get back to the main work.

Capitalizing correctly

The PowerPoint change case command helps users capitalize their text properly when preparing slides. The following steps below show how it can be used:

- Choose the text that should be capitalized.
- Select the Home tab on the Ribbon and then click on the Change Case button in the Font Section. Doing this helps to show a menu of change case choices.
- Carefully study the available options for a moment and then choose the most preferred one.

The case options include:

Sentence Case: In this case, only the first letter of the first word in each of the sentences is capitalized. All the other words are changed to lowercase.

Lower Case: In this option, all the words are changed to lowercase.

Uppercase: In this option, all the letters are changed to capital letters.

Capitalize each word: In this case, only the first letter of each word is capitalized.

Toggle Case: This option changes the uppercase to lowercase and turns the lowercase into uppercase to create a type of ransom-note look.

Using the autocorrect feature

PowerPoint has an autocorrect feature that can correct spelling errors, grammar errors, and style errors automatically as they are being made. For example, if "teh" is typed by accident, PowerPoint will help change the text automatically to "the." Furthermore, if the first word was not capitalized while typing, PowerPoint will capitalize it automatically. The autocorrect feature can even go as far as detecting errors in multi-word phrases. For example, if "there are" is typed, autocorrect will change it to "there are".

CHAPTER 6

POWERPOINT NOTES

Understanding Notes

Printing out PowerPoint presentation notes is basically printing slides with notes for yourself. They are mainly intended to help the user recall some of the slideshow's finer details in a very fast and accessible manner. Notes appear on the presenter's screen during the presentation, but not on the main screen, so that anyone viewing the slideshow will not be able to see them. The slide show and notes should be printed together.

Adding notes to a slide

The Notes pane at the bottom of the normal view can be used to add speaker notes to PowerPoint decks. Presenters can see speaker notes during a slide show and they can also be printed. When PowerPoint slides are created, notes can be added, edited, and formatted.

Speakers' notes should include the following:

- Key talking points
- Key reminders to tell an anecdote or a story
- Optional or supplementary information
- Key references or sources
- Links to documents related to websites or websites
- Reminders to begin a video or an animation
- Prompts to start-up or gestures to another speaker or someone in the audience
- Reminders as regards taking lunch or a break
- Reminders for the audience to follow up or a call to action

If there is a need to check out the notes much later during the presentation, there will be a need to keep them short since there is quite a limited amount of space to check them in the presenter's view. Notes can also be formatted, and they can also be shared as a standalone document or as an alternative to the traditional PowerPoint handouts.

Adding an extra note page for a slide

The Notes pane is a box that is displayed below each slide. An empty Notes pane will pop up with text that says, "Click to add notes." Type the speaker's note in there. If the notes pane is not made visible or if it has been minimized to the barest minimum, click on notes on the taskbar located across the lower part of the PowerPoint window.

A vertical scroll bar will be displayed on the side of the notes pane if notes exceed the allotted length of the notes pane. The notes pane can also be enlarged by pointing the mouse directly at the top line of the pane and then moving upwards after the pointer changes into a double-headed arrow.

Displaying notes on a separate monitor

Presentations can be made with the use of two monitors. The presenter view, when used, is a great way to view your presentation with speaker notes on a particular monitor, e.g., your laptop, while the audience sees the notes-free presentation on another monitor, probably a larger screen that has been projected to. To execute this procedure and split the view between the projectors in this manner, a connection must be made to the second screen.

On the slideshow tab, in the monitors' group, choose the use presenter view option. This should leave the window display settings open. In the display settings dialog box right on the monitor tab, choose the monitor icon that would be used to show the speaker's note, then choose the "this is my main monitor" check box. Once this box has been selected as unavailable, the monitor will automatically be designated as the main monitor.

Choose the monitor icon for the second monitor. This is the one that will be viewed by the audience. Then choose the "extend my Windows desktop onto this monitor" check box.

CHAPTER 7

PRINTING YOUR PRESENTATIONS

The quick way to print

In PowerPoint, slides, speaker notes, and handouts can be created and printed to serve the audience in case they get lost while the presentation is ongoing.

To print, simply choose **the file to be printed** and then select the **print option**, or use Ctrl + P for the very fast print option.

For the printer, choose the printer that should be used to print the document. Navigate to the settings option and then select the desired options;

Slides: From the drop-down menu, choose either to print all slides, selected slides, or just the current slide. Alternatively, in the slide box, type in the slide number or numbers that should be printed and separate the numbers with a comma.

Print Layout: Choose to either print the slides, speaker notes, an outline, or handout. The outline prints just the text embedded in the slides without the images. The notes of a presentation display the slide and similar speaker notes that can be found below it. If handouts are preferred to be printed, they can be printed on several slides on one page, making use of different layouts, some with space for taking notes.

Collated or uncollated: Select whether the sheets should be collated or uncollated.

color: Choose if you prefer the document pointed, grayscale, or pure black and white.

Edit Header and Footer: Choose to adjust the header and footer before printing.

Finally, choose the number of copies that should be printed, then click on **the print button**.

Printing from the backstage view

To print from the backstage view, click on **the file menu**. This will leave the backstage view open. Choose the **print option** in the sidebar to show the options available. As an alternative, select the **Ctrl+P** shortcut on the keyboard.

Options that can be found within the Print tab option of the backstage view include:

Print

Click on the **print button** much later, after all, other options have been chosen.

Copies

Choose the specific number of copies that should be printed.

Printer

Choose the particular printer from the drop-down menu list that is preferred to print the document. This can either be a physical or even a virtual printer that makes PDF files.

Print what

Select from the **drop-down menu list t**o check out more options. Choose whether **to print all the slides, sections, or just a selection of slides**. It is also possible to enter the current slide or a custom range of slides.

If a custom range of slides is chosen to be printed, there will be an option to type in each slide number as well as the slide range. There is also an option to print hidden slides. If the hidden slide option is grayed out, this means there are no hidden slides in the presentation currently open.

Print Layouts

Click this option to show a drop-down list. There is an option to print regular slides or slides that are on the pages of the notes. Another option is to print just the text outline of the presentation. Furthermore, there is also an option to print different handout layouts that also include slide thumbnails.

- **Frame Slides:** This adds a black frame around the slides. This feature can be a very useful one when printing white or light-colored slides on white paper.
- **Scale to Fit Paper:** This feature helps to adjust the size of the printed slides to match the given size of the paper.
- **High Quality:** This feature helps to produce a much better quality output. It is worthy of note that when there are too many slides, this feature can increase the amount of time spent printing.
- **Print Comments and Ink Markup:** this feature is made available if the presentation has any comments or ink annotations.

Collated

This option opens up a drop-down list. Collated permits the user to print in sequence even if many copies are to be printed. If, for example, four copies of a ten-slide presentation are to be printed, making use of the collate option will print all of the ten slides first and then print extra copies of the ten slides as well. If the uncollated option is chosen, four copies of the first slide will be printed, and then four copies of the second slide will also be printed.

Color

This is a default option that makes all documents printed in color form, provided the printer selected has the capacity to print in color.

Print more than one copy.

PowerPoint provides users with the option of printing more than one copy of either a page or the whole document. All that needs to be done is to enter the numbers of each copy that should be printed into the box that pops up after the user has selected the print option. For example, if a user enters four, the printer will print four different copies of that particular page.

Changing printers

If there is access to more than one printer, a user can choose to change a printer for any of the following reasons:

If the printer connected to the laptop or computer initially is not colored and the printer has to print a colored document, the user can change the printer by connecting another printer to the laptop or computer. When the print option is selected from the PowerPoint, the user will also choose the preferred laptop (the new one that has just been connected) and then print out the document.

Faded Ink: A user can also choose to use a different printer if the ink of the printer currently in use has dried. The documents printed from a printer with drained ink are often not always clear, hence the need to change printers.

Printing part of a document

There are times when just a portion of a presentation is needed in hard copy. All the user needs to do is to go to the very page that is needed and then use the shortcut command (CTRL + P). This will immediately send that very page to the printer and print out the document, leaving out other parts of the presentation.

Using Print Preview

The print preview is a feature that enables users to view the pages that are about to be printed, giving users the opportunity to see these pages the exact way they will look when they have been printed.

By having a preview of what the layout will look like when it's printed without actually printing it just yet, users can check for possible errors and ensure these errors are fixed before they continue with the printing option. This feature is often used by every user and is useful for making sure that the layout is exactly the way the user expects it to be and is void of any form of error.

Microsoft PowerPoint also has this feature, and it helps users check the pages of each slide well before they are printed.

Setting up a slideshow

A slide show can be defined as the presentation of a series of still images known as slides on a projection screen or allowing users to use an electronic device most often in an arranged sequence. The changes can be controlled by the person presenting the slide or be automatic at regular intervals.

A slideshow can be a presentation of images mainly for their artistic value, and they are sometimes accompanied by text for description and to enhance better understanding. It can also be used to make clarify, reinforce certain information, ideas, comments, or solutions suggested. Slideshows can be conducted with the use of an apparatus known as the carousel slide projector or an overhead projector, but nowadays, the use of an electronic video device and a computer running presentation software is most common.

To set up a PowerPoint presentation so that it runs automatically, simply follow the following steps below:

- On the slide show tab, select the **setup slide show option.**

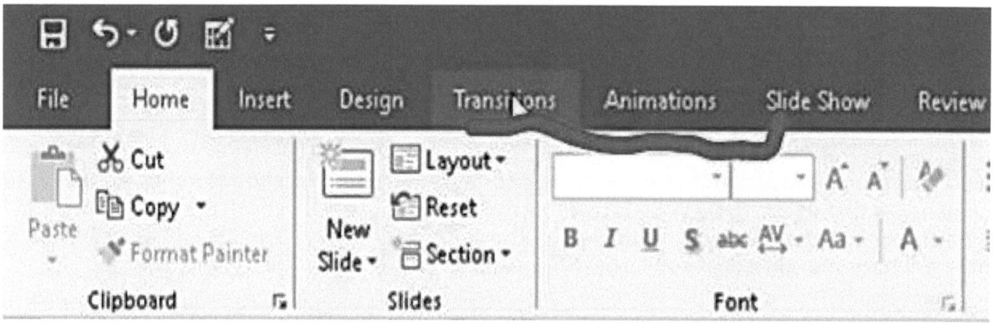

- Under the show type, choose one of the following options: to allow the people viewing the slide show to have control over when they can advance the slides, choose the Presented by a speaker (full screen) option; to present the slide show in a window where viewers would not have control over the advancing of the slides, choose the browsed by an individual (window) option; to continue to loop the slideshow until the viewers press the Esc key.

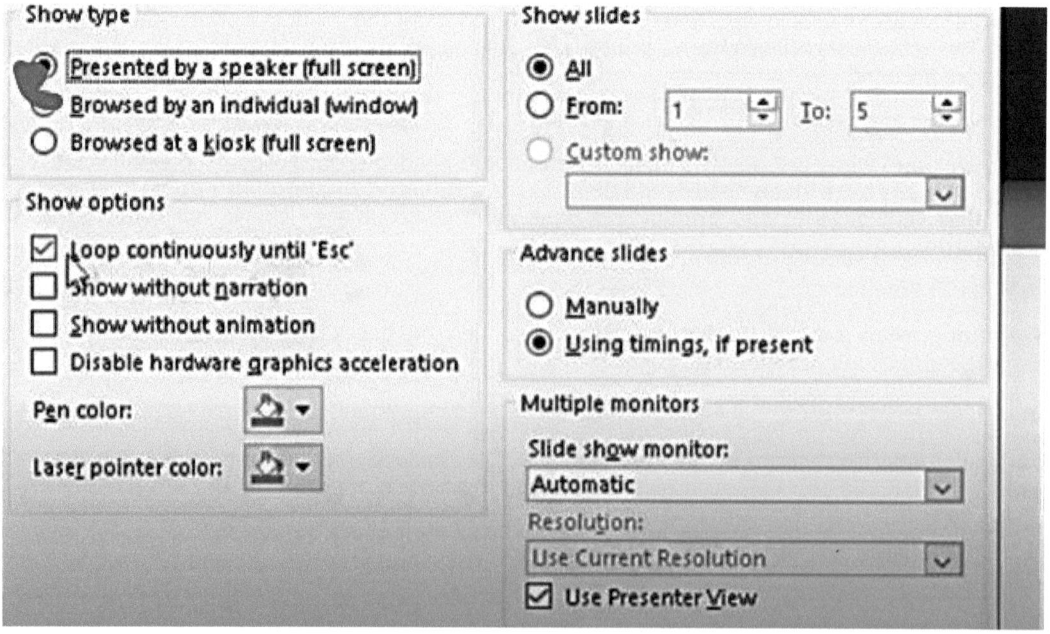

Starting a slideshow

To gain access to the slideshow, choose from any of the following options:

- Click on the **slide show button** displayed on the status bar to play the presentation from the active slide. To leave the slide show, press the Esc key. Furthermore, there is an option to hold the Alt key while pressing the same button to view the slides in a quarter-screen view. Note that the quarter-screen view covers just the top left part of the screen, and it gives the opportunity to see PowerPoint's editing view in the areas left.
- The slideshow tab can also be accessed from the ribbon, after which the user will choose from one of the following options:
    - **From Beginning:** This option ensures that the slide show starts from the first slide of the presentation. The F5 function key can also be used for this.
    - **From Current Slide:** This option permits the slideshow to become active from the current slide in view. The shift + F5 shortcut key can also be used for this.

To leave the slide show view, select any of the following options:

Press the Esc. (Escape) key or right click **on the slide** and then select the **end show option in the resultant menu.**

It's worthy to note that the slide show is just a display view and no form of editing can be done on the slides at this time. However, the slide view contains some sort of navigation and annotations. If the mouse is moved over the lower part of the slide, six watermarked icons will pop up.

The icons include the following:

Previous slide

Click on this icon to navigate to the previous slide. The following keys can also be used: P left arrow, or backspace.

Next Slide

Click on this icon to move to the next slide, or press N, the right arrow, or the spacebar keys.

Laser Pointer Devices

A menu with the following will pop up when you click on this button.

This helps to direct the audience's eyes with a brightly colored point of light. The laser point color can be set up within the dialog box. To gain access to the dialog box, select the slide show tab from the ribbon and then choose the Set Up Slide Show button.

Pen: This helps the user create drawings or annotations.

A Virtual Highlighter: This also performs almost the same function as a virtual highlighter pen.

Eraser: This helps to take off any pen or highlighter-marked areas. If there are no highlighted areas, this feature will be grayed out.

Erase all Ink on Slide: This helps to take off all ink markings. It is also grayed out if there are no marked areas.

Ink color: This changes the color of the pen.

View all of the slides

Select this **button** and this will bring up all the slide decks and a choice can be made from any of the slides. With this feature, there is an option to easily switch in or out of the sequence from a grid that can be seen only by the user.

Zoom into the slide

With the use of this tool, a part of the slide can be zoomed into and pan around. Zooming works more swiftly if a tablet such as the Microsoft Surface is being used.

More Options

This icon, when chosen, opens up a menu with some options close to the menu and includes the "show presenter view" option to gain access to the presenter view.

Working in presenter view

The presenter view in PowerPoint 2025 can be defined as a purpose-built view that complements the slide show view. The presenter view is more beneficial in an environment that contains at least two displays, like a laptop and a project, or better yet, two monitors attached to the desktop system.

The projector or the secondary display will then show the actual slide in what is popularly known as the "slide show view." However, the presenter view can also be seen on the laptop or the primary display.

Though the presenter view in older versions of Microsoft PowerPoint is always turned on manually, recent versions detect automatically if there are two displays and then the presenter view will be turned on. However, it is necessary to check and make sure that the presenter view shows up on the primary and not the secondary display. If a single display is being used, there is still an option to change to presenter view form in the slide show view. Move the mouse over to the lower part of the slide to find the watermarked navigation icons as explained above. If the last icon is selected from the resultant menu, choose the "show presenter view" option and this will bring up the presenter view.

It is also worthy of note that PowerPoint 2025's presenter view has an enhanced web 2.0 interface style and also has many other features in the presenter view's console, like zooming to a particular slide and then panning all over the area of the slide. The pause, and restart buttons have also been dedicated to giving more control over the timing of slides. There is also easy access to the laser pointer tool, which helps to direct the audience's eyes with a light of very bright color.

The presenter view has the following options:

Toolbar

There are basically three options here.

You will find a toggle that alternates between showing the window taskbar and hiding it. Select just once to allow the taskbar to be seen, after which the option will then be seen as highlighted. It can be very useful to gain access to the taskbar if there is a need to gain access to any application that is open. Once at the slide show, click the **show taskbar** option again to taskbar hidden. The option will then no longer be highlighted.

When this option is clicked, it displays a drop-down menu where there will be options to swap between the presenter view and the slide show view, and also one other option to have the slide show duplicated. In a single display environment, these options will be grayed out, but in a dual-display environment, the swap presenter and slide show view can be used to swap displays between the primary display and secondary display. The duplicate slide show option will then show the slide show view on both displays.

End Slide Show: This takes the user away from the presentation. It can also be duplicated with the use of the Esc keyboard shortcut.

Timer

The timer is shown above the slide preview area. It helps to predict the time elapsed since the start of the slide show. Furthermore, PowerPoint offers two buttons for pause/resume and also for restart.

Preview Area

The preview area always shows the active slide. The same slide is simultaneously shown full screen on the screen that is projected (the secondary display).

Next Slide

On the top right-hand side, the preview of the next slide can be found.

Notes Pane

This option shows the notes for the slide in view. There will be two buttons located here, one to make the text larger and the other to make the text smaller.

Controlling your presentation with the keyboard and mouse

With the use of the keyboard shortcut commands, it is very easy to control the presentation with the keyboard.

Below are shortcuts that are used to control PowerPoint presentations.

- **F5** is used to start a presentation from the beginning.
- **Shift + F5** starts a presentation from the active slide
- **Alt + F5** helps to show the presenter's view.
- **Alt + S** helps to switch to the slide view.
- The left and right arrow keys help to change the slide to next or previous.
- Help to take the slide to the specific number as preferred by the user.
- The escape key helps to end a presentation.
- **The H key** helps navigate to the next slide if the next slide is hidden.
- **The T key** helps to set new timings when rehearsal is ongoing.
- The keyboard shortcuts work well with videos that are imported from the computer, but they don't work with online videos.
- The mouse can also control the presentation, as it can be used to click on the slide that should be viewed next.

Using the laser pointer feature

The laser pointer feature helps to show a bright red dot on the screen which can be moved around with the use of the mouse. It is not quite as nice as making use of a real laser pointer, but in cases where there happens to be none, the PowerPoint laser pointer can fill the gap. The following figures show the laser pointer in action.

The laser pointer can be activated in one of the following ways:

The first is to select or tap the **pen and laser pointer buttons** that are in the presenter view, then select the laser pointer. Next, make use of the mouse to move the laser pointer all around the slide.

The second option to ensure the laser pointer is activated is to click on the **Ctrl key** and hold it down, then hold the left button on the mouse pad and move the mouse. The laser pointer will then be displayed on the screen and it will move just like the mouse is moving.

The third option applies only if the device in use is a tablet computer or a computer with a touch screen monitor; simply press down very firmly on the slide and the laser printer will pop up. Drag it around with the use of the finger and lift the finger off the screen to hide the pointer. When the mouse button is released, the laser pointer disappears automatically.

Scribblings on your slides

Users can choose to scribble on the slides to attract the audience's attention to a specific part of the slide. The pen tool can be used to make drawings on a slide. Choose the pen and laser button in the presenter view and then choose the pen tool.

With the use of the mouse, draw on the slide, hold the left button of the mouse down, and move the mouse around as best as possible to ensure a mark is left.

Here are some more thoughts worthy of remembering:

Instead of making use of a solid pen, a transparent highlighter can also be used. Simply choose the highlighter pen tool and then make use of the mouse to establish a mark on the slide. Take note that the highlighter tools work optimally when there is a light background on the slide. If need be, the color of the marks left by the pen tool on the slide can be changed. Select the pen and laser pointer buttons, then select the **ink color** and choose the preferred color to be used. When all has been completed with the presentation, there is always an option to keep the scribbled annotations in the presentation.

This way, there won't be a need to scribble all over again the next time. On a touchscreen device, there are additional buttons in the lower-left corner of the presentation used to call up the laser pointer, highlighter, or pen.

Rehearsing your slide timings

In PowerPoint, the duration of Transition and the timing of transition are not the same. The time of transition is actually the slide stays active during a slide show before moving onto another slide. Basically, during a slide show, users can choose to move to the next slide by pressing the enter key on the keyboard or by tapping on the next slide with the use of the mouse.

On the other hand, making use of timing, on the other hand, helps set slides to advance on their own and remain active only for a specified time.

To check out the slide timing before actually presenting the slide itself, choose the slide show, then select rehearse timings.

Choose the next option and click the **mouse** or press the right arrow key to move to the next slide. The time for the slide that is active is always shown on the right-hand side of the pause icon. Choose the pause icon for a temporary stop on the recording and then click on the resume button to continue the recording. When all of this has been done and all seems perfect, click on the "yes" button to have the slide timings saved or "no" to discard them and repeat the process. There is also an option to press the Esc key to stop the recording process and exit the presentation.

Using custom shows

The custom show option in Microsoft PowerPoint 2025 offers the opportunity to make several slideshows that look the same and are stored in a single presentation file. For example, if there is an opportunity to make a presentation about the benefits of a company to the management and non-management staff, a presentation that has the slides for all the benefits of the company can be created, and then a custom show that contains just the slides listing the benefits of the company to the non-management staff can be made available. The entire presentation can then be given to management, while the custom slide is given to non-management staff.

A presentation can have as many custom slideshows as deemed fit by the user. Each of the custom shows is just a subset of the main and complete presentation, composed of chosen slides from the complete presentation.

To make the custom show, follow the steps below:

On the slide show tab located on the ribbon, select the custom slide show that can be found on the start slide of the show group, then select the **Choose Custom Shows from the menu** that is displayed. This then shows the custom show dialog box. Next, click the new button and then define the custom show dialog box that is being displayed. Set a name for the custom show in the slide show name field and add the slides that need to be shown in the custom slide show.

Note that all the slides made available in the presentation are listed in the list box that can be found on the left-hand side of the custom show dialog box. To include a slide in the custom show, choose the slide that should be added and then select the **"Add" button**. The slide will be displayed in the list box on the right-hand side of the Define Custom Show dialog box.

If by mistake, a slide has been added, it can actually be taken off. Simply deselect the **slide that needs to be removed** in the list box that is on the right-hand side of the **Define Custom Show dialog box,** and then click on the remove button.

The up and down arrows close to the right edge of the Define Custom Show dialog box can be used to change the arrangement of the slides in the custom show. Clock on the OK button and then go back to the custom show dialog box and finally click close to leave the Custom Show dialog box.

Displaying a custom show

To display a custom show, start by opening the presentation that has the custom show. Then open the slide show tab located on the ribbon, select the **Custom Slide Show button** and then select the **custom shows from the menu** that pops up. This shows the Custom Show dialog box which enlists any custom show that is in the presentation. The preferred custom show can be selected and begin the show simply by tapping the show button.

Hiding slides

If going through all the hassle of making a custom show can be avoided but you still have a need to exclude some slides from a presentation, there is no need to delete the slides. All that needs to be done is to hide them. To hide a slide, choose the slide that should be hidden and then click on the **Hide Slide option** in the Slide Show Ribbon tab. To make the slide visible (unhide), choose the slide and then click on the same button again. (There is an option to know the slides that are hidden by simply checking through the slides in the slides tab located on the left-hand side. If there is a slash through the slide number, then the slide is hidden.

Showing your presentation online

Microsoft PowerPoint 2025 has an online presentation option that enables it to share presentations with other people in a remote setting over the internet easily. All you need is a Microsoft account, which you should already have if you have Office 365. If you don't have a Microsoft account yet, simply sign up for one. To present a PowerPoint presentation online, follow the steps below:

- Select the **present online button** in the Slide Show Ribbon tab.

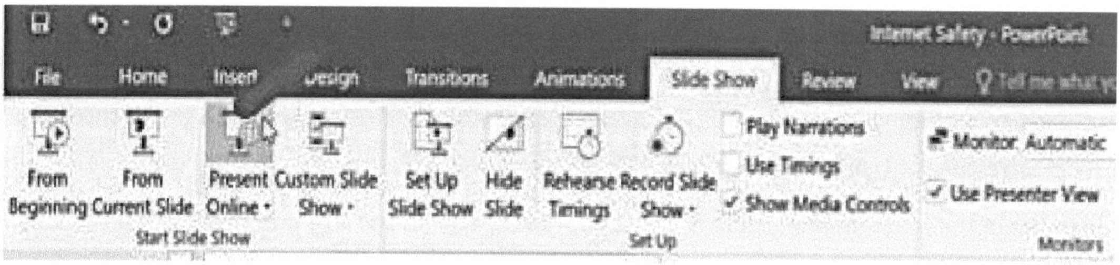

Having this done helps to bring up the present online dialog box option.

- Select the **connect button** so as to be connected to the internet.

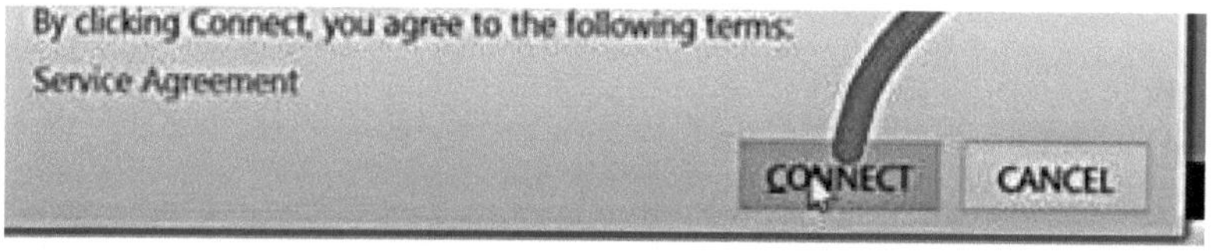

If the box to enter pops up, enter your Microsoft account username and password.

- To have an email sent to the participants that should be in the meeting, click the **send in email option** and then complete the email by including recipients and any other text that should be added to send the email. Participants can click on the link sent to open the presentation in their web browser. To copy the presentation link to the clipboard, simply click the copy link option and then paste

  the link directly into an email to share with participants at the meeting. Alternatively, there is an option to paste the link directly into the address box of any web browser to see the presentation.
- When all the participants are set to see the presentation, click on the **Start Presentation button**. This starts up the presentation, and the participants can also follow the presentation from their own web browsers.

CHAPTER 8

FONTS AND TEXT FORMATTING

Changing the look of text

The look of a text can be changed at any time if the need for it arises. Simply highlight the text that should be changed, right-click on it, and choose the font option from the text menu.

When the font menu option has been selected, a window will open up with different text options that can be chosen. In this window, an alternative font from the Latin text font drop-down menu or any other special style displayed in the preference window can be chosen. Effects such as the strike-through or the display text in all capital letters can be chosen. Though the theme sets a font color, alternative font color can also be added, including color for the underline should the need for it arise.

All text can be adjusted, styled, moved, and designed when added to the slides. Slides can have themes configured, but the text has options that seem very flexible in PowerPoint.

Changing the size of the character

Changing the size of the font for all the slides in the presentation can be done easily on the slide master. All PowerPoint presentations have a lid master with related slide layouts.

Locate the view tab icon and choose "slide master." In the slide master view, the slide master is usually located at the very top of the thumbnail pane with the layouts that are similar right under it. Select the slide master to have the font on all the slides in the presentation changed, or choose any of the layout options to change the font size on that particular layout.

On the slide master layout, choose the level of placeholder text to which the font size should be changed to. Finally, go to the home tab and enter your preferred font size in the font size box. There is also an option to type in the preferred size in multiples of **1**, such as **10.3** or **106.4.** The bullet will be increased or decreased automatically, as the case may be, so as to match the changes in the size of the font.

To change the font size of selected text only, select the text that should be changed. The shortcut command **Ctrl+A** can be applied to select all the text at once. On the Home tab, choose the preferred font size in the font size box. There is also an option to type in the size preferred within the limits of **1** and **1638**, in multiples of **5**, such as **10.5** or **125.5.**

Choosing text fonts

As an alternative to changing fonts in a slide one after the other, the entire font in a presentation can be changed at once. Bear in mind that the font pairing header (header font and body font) is a very important design decision in PowerPoint.

Click on the **view tab icon**, then choose the **slide master**. On the slide master tab, select the **fonts drop-down menu**. Choose the preferred font that should be on all the slides in the presentation. It is not compulsory to choose from the predefined font pairs on the menu; simply select the **customized fonts** at the lower part of the menu to choose a preferred font.

On a final note, close the master view. The text throughout the presentation will then be updated automatically to the newly chosen font.

Adding color to text fonts

To change the color of a text font on a slide, select the **text** whose color should be changed. Navigate to the Home tab then choose the preferred font color, and then choose the color that the text should be changed to.

If a presentation is being built with many slides, it's best to have the slide master customized first. This way, the text color, and other design styles will be set ahead of time, and each new slide that is created will come with the preferred text color.

When the color of the text is changed on the slide master, the changes are also applied to the text on the multiple slides at the same time.

On the View tab, click on the **Slide Master**. On the left-hand side of the thumbnail pane, choose a layout that has the text that needs to be changed to a different color. On the mini toolbar that is displayed, select the **preferred font color** and then choose the color that the text should be changed to. After all of this has been done, select the **close master view option** and then go back to the presentation.

Adding shadows

Adding shadows to texts, images, shapes, or other objects can be a unique way to add flair to a presentation. However, the presence of too many visual effects might take the audience's attention away from the main message in the slide. There should be a perfect balance between text and shadows added to objects in the presentation.

Open the PowerPoint presentation and navigate over to the slide that has the objects that will be worked upon, then choose the image. On the picture tools and format tab, select the **picture effects button.** If a shadow is to be added to a text, the button will be named "text effects". In the drop-down menu, choose the "shadows" sub-menu. The submenu has a wide selection of shadow effects. When each option has hovered over, it opens up a live preview. Play around to pick the one that is most preferred. Choosing this option will have the settings added to the object with immediate effect. The same thing occurs when a shadow is applied to a text object. The very same also goes for objects like shapes.

Indent the bulleted list

A bulleted list enables a writer to make a list that stands out from the main text without the implication of chronology or order of importance that a numbered list might show. There is no specific symbol that must be used for the bullets, though. Small circles or little squares are often used. It is also important to note that in a bulleted list, the first word should be in capitals and every sentence should always end with a full stop.

Having to indent a line of a list makes a multilevel list. This means the line indented is a subtopic of the line above it. Indenting a list is quite useful if there is a need to make an outline or a hierarchy.

Creating a numbered or bulleted list

Bulleted or numbered lists help to arrange the text and also display a sequential process in the PowerPoint presentation, making it look well composed and organized.

Navigate to the view tab option that can be found in the presentation views group, then select normal. Right on the left-hand side of the PowerPoint window, select a slide thumbnail that you prefer to add bulleted or numbered text to. When on the slide, choose the lines of text in a text placeholder or table that the bullets or numbers should be added to. Navigate to the Home tab, within the paragraph group, then choose the bullets or numbering icon.

It is worthwhile to note that the color, size, or style of the bullets or numbers in the PowerPoint presentation can be changed. To change a bullet or number, fix the cursor at the start of the line that should be changed. To change more than one bullet or number, choose the text in all of the bullets or numbers that should be changed. This can be done using the Ctrl + A command.

Navigate to the home tab, select the arrow next to the bullets or the numbering button, and then choose the bullets and numbering option. When in the bullets and numbering box or the bullet tab or the numbered tab (based on the kind of list that works best for you), choose the style changes as it suits you. The styles include;

- Bullet or number format
- color
- Size (to change the size of a bullet or number so that it becomes a particular size that fits the text, select the size, and enter a particular percentage).
- Starting number (from the numbered tab, enter the specific number that is preferred in the start box)
- Pictures (to make use of a picture as a bullet, on the bullet tab, choose the picture, then scroll to look for the preferred picture).

- Symbols (to add a character from the list of symbols to the bullet tab temporarily, click on the customize option from the bullet tab and then select the OK button. Symbols can be added to the slides from the style lists.
- Select Home, then Convert to SmartArt from the drop-down menu.

The Setting tab stops

If there is a need to specifically have control over where the text is placed within the text box or a particular shape, or a need to have the text aligned, add the tab stops with the use of the ruler.

Move to the view tab, right in the presentation views group, and ensure normal is chosen. If the ruler cannot be seen, on the View tab, in the show group option, select the ruler check box. Click on the paragraph to be edited, then click on t**he ruler** at the very point where a tab stop should be added. When the tab stops with the shape looking like a capital L, it will then be displayed on the ruler as it is clicked.

The cursor can now be moved with the tab key on the keyboard and typed into the tabs. Note that the default tab stops (the square markers colored grey below the ruler) can also be changed simply by clicking **and dragging them.**

Spacing Out

PowerPoint offers the opportunity to make changes to the amount of space between each of the lines in a list or a paragraph. The line spacing can be reduced minimally to fit more lines on a slide, or it can be increased to make it more legible and improve readability.

In some cases, PowerPoint can change the font size automatically when the line spacing is being adjusted. This means increasing the line spacing too much might make the text smaller.

Choose the text that should be formatted, navigate to the home tab and get to the paragraph group, choose the line spacing command, and choose the preferred line spacing option from the menu. The line spacing will automatically be adjusted. If there is a need to make changes to the line spacing with more precision, choose the line spacing options from the drop-down menu. The Paragraph dialog box will be displayed, giving room for the line spacing to be fine-tuned and also allowing changes to the paragraph spacing, which is simply the amount of space that is included before and after each of the paragraphs.

Lining things up

Objects can be aligned in PowerPoint in different ways. Objects that can be aligned include images, shapes, placeholders, or text boxes. However, charts, tables, and SmartArt objects can also be aligned. The most popular way to have objects aligned is by using the align command that is displayed on the multiple tabs in the ribbon.

Note that buttons and ribbon tabs may show in a different way as it is dependent on the size of the button and the settings in the control panel.

To align objects with the use of the align command in the ribbon, choose the objects by selecting the first object, then pressing the shift or control button while clicking on other objects. If selecting a textbox or placeholder, locate the edge and press the shift or the control button. Multiple objects can also be selected by selecting the **blank area a**nd moving them around.

Select the **format tab option** in the ribbon. This tab might be displayed as picture tools format, picture format, drawing tools format, drawing format, or shape format, all of which are dependent on the objects that have been chosen. As an alternative, you can also choose to click on the **Home tab**.

Click on the **align button** in the arrange group and a drop-down menu will be displayed, after which the align selected objects option should be chosen. If it is not chosen, click on the **aligned selected objects again** and then select **the align button again**. Choose one of the alignment options: align left, align center, align right, align top, align middle, or align bottom.

The selected objects are then aligned as follows:

- Align left aligns the object along the left edges.
- The align center helps to align objects in a vertical manner through the centers.

- Align right helps to align objects along the right edges.
- Aligning the top ensures objects are aligned along the top edges.
- Align the bottom aligns objects in a horizontal manner along the middle.
- Align Bottom aligns objects along their bottom edges.

It is worthwhile to note that the option to align is not always available when the right-click button is selected and the context or the drop-down menu is displayed.

Right-click on the align option in the ribbon and click on Add to Quick Access Toolbar from the drop-down menu so as to add it to the quick access toolbar which is located either above or below the ribbon.

Aligning an object or objects with respect to the slide

An object or objects can also be aligned with the slide. It's usually best to align an object (which can be grouped at times) to a slide. To align an object or objects to a slide with the use of the align command in the ribbon

- Choose the **object or objects** by selecting the **first object** and then pressing the shift or control key while choosing the other objects. If choosing a text box or a placeholder, press the shift or control button at the edge. Multiple objects can also be chosen by clicking on the blank area and moving them around.
- Choose the **format tab on the ribbon**. This tab might be displayed either as picture tools format, drawing tools format, drawing format, or graphic format depending on the objects that have been chosen.
- Select the **align option** in the arrange group and a drop-down menu will be displayed.
- Choose the align to slide option if it is not already selected.

Aligning objects with the use of guides

Static guides can also be shown and used to align objects. Smart guides might also be shown when these objects are moved. To use guides when aligning objects;

- Right-click on any **blank part of the slide**.
- Choose grids and guides from the drop-down menu.
- Choose guides from the submenu so as to make the guides visible. Note that these guides are not the same as gridlines or smart guides. PowerPoint shows a vertical and horizontal guide by default.
- Move the guides to the preferred location. A guide can also be right-clicked on so as to add another guide.
- Select each of the objects and move them to the guide in order to align them.
- Do the same for other objects.
- Right-click on any blank part of the slide.
- Choose the grids and guides from the drop-down menu.
- Choose guides from the submenu to keep the guides hidden.

Copying and aligning objects

When lots of objects are being copied, they can also be aligned so they do not appear to be disorganizing the slide. To copy an object and help it align with the first object, simply do the following:

- Choose an object to start with. If the object happens to be a textbox or a placeholder, click on its edge.
- Press the Ctrl and Shift buttons together and then move the object to another location in order to make a copy. If the control button is tapped while dragging the object, PowerPoint also

creates a copy of the object. If both the Shift and Ctrl buttons are tapped, the object will remain aligned with the former object.

- Press the Ctrl + Y buttons so as to create more than one copy of the object. If more copies need to be made, continue pressing those buttons.

Make use of the alignment tool to ensure PowerPoint objects are well aligned and also make slides that are well designed. The bad side to this is that there are no built-in keyboard shortcuts to help align objects. However, you can consider adding align to the Quick Access Toolbar.

Making Columns

Text can be arranged into columns in a text box or shape. The spacing can also be adequately specified between the columns. This is how to go about it:

- Place the mouse on the text box, placeholder, or shape border and right-click, then click on the format shape option.
- Locate the right-hand side of the window and choose the text options, then choose the text box.
- Select columns and then enter the number of columns needed in the number box and the amount of space each column should have in inches in the spacing box.
- If the column button is grayed out, that might be because the work is being done on a table.
- Note that any text that is entered will be displayed in columns. If the object that is right-clicked is contained already in the text, the text will be The balance of the text can be adjusted in the columns by adding or removing line breaks: Select the enter button to add the text to the column that follows, or tap the backspace or delete button to take off the blank lines and move the text to the column that follows. Note that the size of the box that contains the columns can also be changed so as to increase the columns if need be.

Creating Fancy Text with WordArt

Microsoft has fused WordArt into PowerPoint so its formatting can be applied to any bit of text in the presentation simply by highlighting the text and adding the WordArt formats. To do this,

- Choose the **text** that should be added to WordArt formatting. The text can be just about anywhere in the presentation. For example, WordArt formatting can be added to any part of the slide.
- Choose the **drawing tools format tab option**. This option has a WordArt style group. The ribbon group also has a lot of pre-conjured WordArt styles and buttons that enable the user to have control over text fill, outline style, and text effects like shadows and glowing.
- Select the **more button** located at the lower part of the scroll bar to the right-hand side of the predefined WordArt styles. This will help display the WordArt styles gallery quickly.
- Choose the WordArt style that most closely resembles the formatting that should be applied. You don't have to worry if none of the choices in the gallery fits the effect needed. There is an option to tweak the text's appearance much later.
- Play around with other WordArt controls in the WordArt styles group of the drawing tools' format tab.

CHAPTER 9

DESIGNING YOUR SLIDES

Getting design ideas

When creating a PowerPoint presentation, there is definitely a need to make the presentation stand out and also attract the attention of the audience with unique design ideas. For this to be done, a captivating presentation design to show the content in a very intriguing manner will be needed. Having a professional take a look at things sometimes goes a very long way. With the use of the PowerPoint deck, they can provide a revamp to any of the already created designs. Lovely designs that would stand out won't just be displayed in the air.

Remarkable designs can be gotten out of PowerPoint with the use of templates. If you are new to PowerPoint or just getting used to it and are not a PowerPoint designer, it's best to start by having a review of templates that have already existed. Without a doubt, they are the fastest and easiest way of getting lovely designs into PowerPoint.

These templates are definitely not the most exceptional designs out there, but they can be useful as suggestions that can in turn bring out the best in a designer such that he or she gets more design ideas.

In PowerPoint, you don't just start as a designer; templates that are already existing can help get a better perspective that is needed to become a full-fledged PowerPoint designer. Before going through PowerPoint templates, it's necessary to note that there are basically two main types of templates: slide layouts and the master slide layout. A "slide layout" is a term that is mostly used in PowerPoint for a template that has just one slide. A template for the whole of a presentation is known as a "master slide."

Working with Themes

The themes group in the designs tab provides users with the option of choosing a theme to add to their Microsoft PowerPoint presentation. It's worthy to note that PowerPoint 2025 comes with loads of crafted themes that have the ability to give your presentation that professional look. If you are into arts and crafts, designing themes by yourself shouldn't be a big deal.

A theme can be defined as a set of design elements that are added to one more slide in a presentation. Each theme has different basic components:

- A set of unique colors that work hand in hand. Each theme has about four colors that are capable of being used as text or backgrounds and six different colors that can be used as accents.
- A set of unique fonts that look lovely when put together. Each theme should have a font that is used just for headings and another font used for regular text.
- A couple of background styles, which are a mixture of background colors and effects that include patterns or gradient fills,
- A couple of design effects include line and fill styles and also some line styles.

Microsoft Office has about 40 different predefined themes, although the exact count is always reviewed from time to time as designers from Microsoft bring up new innovations and discontinue the older ones. Each theme can be seen in four different colors, which means that there are over 100 predefined themes that can be applied to a slide.

Applying themes

To add a theme to the whole of a presentation, simply choose the preferred theme in the themes group that can be found on the Design tab. If the preferred theme is not available and can not be seen, make use of the scroll buttons on the right-hand side of the themes group to show the themes that are inclusive.

When a theme has been chosen in the theme group, variations on the theme will then be displayed in the variants group, after which there will be an option to choose the variants that should be used.

Making use of the theme colors

All PowerPoint themes have a built-in color scheme that has sets of colors selected by color professionals. Microsoft paid this group of people a lot of money to argue the benefits of using mauve text on a teal background. created color schemes can be used, or there is an option to make one if there is an assurance that it will be better than the one designed by Microsoft.

Without the use of color schemes, everyone is at liberty to choose freely from the about 16 million colors that PowerPoint has offered to be included in slides. Each color theme has a total of 12 colors, with each color designated for particular use as shown in the list below:

- **Four text/background colors:** These four colors were created mainly to be the primary colors for the PowerPoint presentation. One color from each pair is used solely for text, while the other is used for the background. Both colors can be used for the same purpose, though it might make the slide very difficult to read.
- **Six accent colors:** These sets of colors are used for different parts of the slide that add to the basic text and background colors.
- **Two hyperlink colors** are used only when the presentation has a hyperlink.

When a theme is added, the color scheme for that particular theme is also added alongside the other elements of the theme. However, PowerPoint allows the color scheme to be changed from the scheme that is associated with the theme. As an example, a theme like Opulent can be added, and then the color scheme changed to the scheme from the Verve theme.

Include a color Variant

To have the basic color scheme used for slides changed, choose a theme first. Then select the **more button** that can be found at the lower right corner of the variant's gallery in the design tabs and click the **colors** option. This opens up the color variants that are available.

Customize a color Scheme

If, unfortunately, you wouldn't like to make use of any of the color schemes that are included with the built-in themes, there is an option to design a personal color scheme by following the steps below;

- Choose a color scheme that is similar to the intended one.
- Choose the "custom colors" option from the lower part of the color variants list. This will open up the dialog box to create a new theme color scheme.
- Select the **button for the color** that should be changed. For example, if the first color should be changed, choose the Accent 1 button. A gallery of choices of colors will be displayed.
- Choose your most preferred color since a lot of colors will be made available.
- If you can't find your choice in any of the colors displayed, select the "more colors" button option. This will help bring up the color dialog box. If the custom tab was last used when the dialog box was used, the custom tab will be chosen also.
- Click **on the color of your choice** and then select the "OK" button. Repeat these steps for any color that should be changed and then click on save.

The use of theme fonts

Theme fonts are close to theme colors, but there are fewer choices to choose from when it comes to theme fonts. Though there are about 12 colors in one theme, there are just about two fonts; one for the headings and the other for the body of the text.

The theme fonts can be changed in the design tab located in the ribbon if the fonts used with the theme that has been chosen for the presentation will not be changed. The chosen fonts are added throughout the presentation.

It is worthwhile to note that having the theme font changed is not the same as making changes to the font through the font controls that can be found in the font group of the Home tab.

When the font controls on the home tab are used, direct formatting is applied. Direct formatting helps to override the font setting defined by the theme momentarily. As a general rule, theme fonts should be used, so ensconce the fonts that would be used all through the presentation. Direct formatting shouldn't be used all the time. It can be used when a word or two needs to be created in a font that is quite different from that of the other presentation.

The font used in a theme can be changed by choosing the More button located at the lower right-hand side of the variants gallery, choosing fonts, and then choosing the Custom fonts option. Having this done brings up the create new theme fonts dialog box. This way, the font used for both the headings and the body of the text can be changed.

Applying the effects

Another very important part of the PowerPoint themes is the theme effects, which add subtle variations to the look of the presentations graphically. Theme effects are added automatically whenever a theme is applied. However, theme effects can be added from a totally different theme by choosing the **More button** on the lower right-hand side of the art gallery and selecting the theme. This will open up the Theme effects gallery and then any of the theme effects listed can be chosen.

Changing the slide size

It is always a good idea to have knowledge of changing the slide size in PowerPoint, as there might be a need to adjust the presentation to more than one device. It is thus not a waste of time to understand how to properly scale a presentation in Microsoft PowerPoint.

To change the PowerPoint slide size in a presentation, choose the design tab of the toolbar ribbon, choose the slide size located close to the far-right end of the toolbar, then choose the standard **(4:3 aspect ratio),** the widescreen option **(16:9),** or a custom slide size.

Note that whenever PowerPoint is unable to scale content automatically, it brings up two different options:

Make use of this option to add to the size of the slide content when scaling to a much larger slide size. Selecting this option can lead to the content not being a perfect fit on the slide.

Ensure fit: Choose this option to reduce the size of the content when scaling to a much smaller size. This can lead to the content being displayed as smaller, but all the content in the slide will be seen.

Applying background styles

PowerPoint 2025 most of the time makes presentations with a plain white background. Though it might be very fine in some cases, there might be a need to spice things up at other times so the background looks even finer. There are four major ways the background can be changed with PowerPoint.

Using a solid fill

Solid colors can offer contrast to slides, but you must ensure that any text or graphics that are displayed on the slides can still be seen. For example, if a red background is chosen, any text or image on the slide should be displayed in light colors. To ensure that solid background colors do not overshadow the text and graphics, the transparency in the background can also be changed to ensure the color shows darker or lighter.

To have the background changed to a solid color, simply

- Right click **on the slide**.
- Switch to the customized group and then choose the format background icon. The format background pane will then be displayed
- Choose the **solid fill option**
- Choose the **preferred color icon** and a pull-down menu will be displayed with a list of a palette of colors.
- Choose **a color.** PowerPoint helps to fill the slide background with the color that is chosen.

- Though it is optional, to have the transparency color changed, move the transparency slider either to the left or to the right. Note that the higher the value of transparency, the lighter the background color will show.
- To adjust the background of all the slides in the presentation, select the **apply to all** button. If the "apply to all" button is not selected, PowerPoint will make changes to the background of the active slide only.
- Finally, in the format background pane, select **the close icon**.

Using a gradient fill

A gradient shows just one that fades slowly into a different color, for example, from green to orange. To define a gradient in simple terms, define it as one or more stops, transparency, and gradient direction. Stops simply mean where the colors in the gradient begin and end. Transparency is simply how much one can see through when a color is being displayed. The gradient direction is simply how the gradient is shown, which includes vertically or horizontally.

To define a gradient background, follow the steps below:

- Right-click on **the slide**.
- In the customize group that appears, choose the format background icon. The format background pane will then be displayed.
- Choose the gradient fill option. The format background pane will show more options for defining a gradient.
- Open the type list box and choose an option, either linear or rectangular.
- Open the direction list box and choose a direction gradient.
- In the angle box, select the up or down arrow to add to or reduce the angle, respectively.
- Move the gradient stop sliders either left or right. The far left and far-right parts of the stop position slider help to a state where the gradients start and end on the slide.

Using other background effects

Choosing a picture background

A picture, like clip art or a photograph that has been taken with a digital camera, can be displayed in the background. After a picture has been used as a background, its transparency can be adjusted in such a way that the title or subtitle on each slide can be read easily.

To add a picture background to any slide, simply follow the steps below:

- Select the **design tab**.
- Locate the custom group and then select the format background icon. The format background pane will be displayed.
- Choose the picture or texture fill option. The format background pane will show options for including a picture in the background.
- Choose one of the buttons below:
    - **File:** helps to bring back an image saved on the computer, like a digital photo. When the insert picture dialog box is displayed, choose the preferred picture and click "open."
    - **Clipboard:** This pastes a cut or copied image that has been previously used in another program like Photoshop.
    - **Online:** This shows a library of clip art images that can be chosen from. Simply type a description of the preferred picture. When the clip-art images are displayed, choose the preferred one and then select the insert button.

CHAPTER 10

ANIMATING YOUR SLIDES

Using the transition tabs

A slide transition is simply how a particular slide is taken away from the screen and the next slide is shown during a presentation. PowerPoint provides lots of entertaining and different slide transition schemes. The secret is to be very careful so as not to make use of too many different schemes in just one presentation. Be choosy and have in mind the way the scheme should be before adding them to the slides. Test the schemes out by opening the slide show and checking out their effectiveness. ***Ensure it's quality rather than just quantity.***

To apply transitions to a presentation, select the **slide sorter view**, which can be found at the lower part of the screen. Thumbnails of all the slides in the presentation will then be displayed. Select the transition tab. The transition tab has the transitions for this slide group. From this group, select a special effect that should be applied between the previous slide and the next slide when the transition is ongoing.

Creating a slide transition

A slide transition can be defined as the visible effect that takes place when a slide is moved from one to the next when a presentation is ongoing. It is worthy of note that in a transition effect, the speed can be controlled, sound can be added, and the look can also be tweaked so as to meet the needs of the user.

To add transitions to slides, simply do the following:

- Choose the slide that needs a transition to be added.
- Click on **the transition tab** and then select a transition.
- Choose a transition so as to see the preview.
- Choose the **effect options** to select the direction and nature of the transition.
- Choose **a preview** so as to check what the transition looks like.
- Click on the "apply to all" button so as to add a transition to the whole presentation. Select transitions and choose none to have a transition removed.

Using the Morph transition effect

PowerPoint has so many interesting animations and transitions. The morph transition is just one of the very recent additions to the library.

The morph transition allows the creation of a smooth object animation from one slide to another. This particular transition provides the illusion of the growth of an object or objects between two different slides. When used the right way, the morph transition can add to the creation of a great slide show.

To make use of the morph transition, the slide will need to be duplicated to make use of the transition, after which the presentation can be opened. Right-click on the slide that should be duplicated on the left-hand side of the pane. In the menu that pops up, choose the duplicate slide option. This will display an exact copy of the slide on the left-hand side of the pane. Choose it.

Navigate to the slide that has been duplicated and then select or move the object to the place that will be the endpoint of the transition. Move to the "Transitions" tab and choose the "Morph" transition. If there is a need for a preview, choose the preview option in the transition tab. The transition will then be played.

Using advanced animations

An animation effect can be triggered to start playing at the beginning of an audio or video clip, or even while an audio or video is playing. An animation effect can also be triggered to start when the shape or object that will be animated has been selected. To trigger an animation effect to start when it is selected,

- Choose the shape of the object that needs an animation.
- Navigate over to animations, then advance animation, and then add animation. Choose the preferred animation.
- Next, go on to animations, and then advanced animations, and finally the animation pane.

- In the animation pane, choose the preferred animated shape or the other object that should be triggered to play when it is selected.
- In the advanced animation group, select trigger, then select On Click and then choose the object.

Trigger an animation to start during a video or an audio clip.

To trigger an animation to start during a video or audio clip, a bookmark should be infused first at the preferred point in the clip.

- Add a bookmark to the video or audio clip at the very point where the trigger effect should start.
- Move to animations, then advanced animations, and then the animations pane.
- In the animation pane, choose the animation effect that should start at a particular point during the audio or video clip.
- In the advanced animation group, choose the trigger option and then select the preferred that should trigger the animation to start.

Timing your animations

In PowerPoint, when an animation effect starts, the speed at which the animations occur can be controlled. The effect can also be repeated or rewound so as to get the right look and feel for the presentation.

Set the start time.

Click on the **mouse** to start the animation effect in coordination with the other animation effects on the slide. On the slide, choose the text or object that has the animation effect that the timing needs to be set for. On the animations tab, choose the start down arrow and select one of the following:

- To begin the animation when the slide has been chosen, select "On Click."
- To begin the animation at the very same time as the former effect in the list, choose the previous option.
- To begin the animation effect right after the previous effect in the list is over, choose the "after precious option."

Fix the speed.

To run an animation effect at a very fast or much slower pace, all that needs to be done is to change the settings of the time.

On the slide, choose the **text or object** that has the preferred animation effect that should be set for the speed. On the animations tab, in the Duration box, enter the number of seconds that the effect should run.

CHAPTER 11

THE ART OF INKING SLIDES

Drawing on your slides

When showing a presentation, drawings can be created onscreen with the use of a digital pen so as to put emphasis on a specific point or show possible connections between two or more points.

To make drawings during the presentation, simply do the following:

- Turn the pen on in the slide show.
- On the slide show tab, make sure the use presenter view option is turned on.
- On the slide show tab, select **either "from the beginning" or "from the current slide options"** to begin the slide show.
- When you go back to the presenter view, some help buttons will pop up at the bottom left-hand side of the slide. The option that looks much like a pen shows how the mouse pointer will be displayed to the audience. Choose the pen button and then select Pen from the menu that pops up.
- Select and move the mouse to draw on the slide.
- To turn off the pen, simply right-click the active slide, move over to pointer options, select arrow options, and make the preferred choice from the options listed.

Change the color of the pen.

By default, the color of the pen is usually red, although there are lots of other colors available for use. To change the color of the pen, move over to the presenter view and choose the pen icon at the bottom of the slide. From the pop-up menus that are displayed, choose the ink color and then select the color of your choice.

Keep or discard the ink annotations.

When the slide show has gotten to the end and you are about to return to the normal view, PowerPoint provides users with the opportunity to save their ink annotations. Choose the discard option to get rid of the annotations. Choose the "keep" option to help save the ink-pen annotations on the slides.

Erase Ink on slides

Oftentimes, ink might have been left on the slide by mistake, or some of the drawings needs to be taken off before the presentation. To take it off, simply right click on the slide that has the ink that needs to be taken off, move to the pointer options, and then do one of the following:

- Select the **eraser** and then hold down the left button of the mouse and move the eraser over what needs to be erased.
- Choose the "**erase all ink on slide**" option so as to take everything that has been written or drawn off the slide.

Making use of the ruler

Make use of the ruler on the Draw tab of the ribbon to draw lines that are straight or that will align with a set of given objects. The ruler adjusts to any location that is wanted, be it vertical, horizontal, or any angle in between. It has a degree setting such that it can be adjusted at whatever angle is needed by the user if deemed fit.

- Open the file and then click the options button.
- Click on the **customize** ribbon tab located in the options dialog box.
- In the box on the right-hand side of the dialog box, click the **checkbox called "draw."**
- Select **OK** to close up the options dialog box.
- Click on the d**raw tab button** and the ruler on the ribbon will then be displayed.

Draw a line or align objects.

Choose the slide where a ruler is needed. Click the **ruler** on the draw tab option so it can be displayed on the drawing surface of the slide. Fix the ruler at the preferred angle (use a finger to drag the ruler either up/down or left/right, use two fingers to rotate the ruler to the preferred angle, use three fingers to rotate the ruler by five-degree increments). Draw a line and click on a pen highlighter on the Draw tab option and start drawing.

To align single items, choose each of the items and move them until their selection handle is snapped to the ruler.

To align more than one item at once, select different items by pressing the Ctrl button while tapping each of the items. Move the set of objects until they are snapped to the ruler.

A shape will be aligned to the ruler at the edge, and then an object like an icon, picture, or textbox will be aligned to the ruler on its bounding box.

Control the ruler with the use of the mouse.

Move the ruler by choosing and moving with the use of the mouse. To stop dragging the ruler, set the mouse button free. Rotate the ruler in a one-degree increment by moving the scroll wheel on the mouse. The ruler moves where the mouse pointer is pointed. Note that rotation has to be done with the use of the mouse scroll wheel; it cannot be done with laptop trackpads.

Control the ruler with the keyboard.

If the computer in use isn't touch screen activated, or the preferred mode of control is the keyboard, use this combination of keys to manipulate the ruler after the ruler button to activate it on the slide has been chosen. To control the ruler with the use of the keyboard, click on the ruler on the draw tab option to make the ruler display on the drawing area of the slide. Choose the ruler with the aid of the mouse. Then press Shift + F6 to enter the ruler manipulation mode and use the keyboard shortcuts to manipulate the ruler.

The arrow buttons help to move the ruler up, down, left, or right. Windows can use the Alt + Ctrl + arrow key combination to help rotate the display of the monitor. The monitor-rotation effect is always controlled by the graphics card of the computer. If this effect is not activated for the computer in use, it will take precedence over the ruler shortcut key and this will mean that the monitor display will now be rotated 90 degrees when the Alt + Ctrl + right or left arrow key is selected. The monitor rotation can be reset by pressing the Alt + Ctrl + arrow up key.

If the ruler manipulation combination key is to be used, turn off the monitor rotation feature by right-clicking on the computer in use and choosing a command like Graphics Properties or Graphics Options. Look for the Hot Keys command and then disable it. After the monitor rotation feature has been turned off, the Alt + Ctrl + arrow key combinations can both be used on the ruler and also on shapes that must have been fused into slides in PowerPoint.

Hide the ruler

Choose the **ruler** on the draw tab menu to ensure it disappears from the drawing part of the slide.

Converting Windows ink to text

To change an ink drawn somewhere on the slide to text, simply draw or write with the use of the digital ink, choosing the ink by tapping on it or making use of the Lasso selection tool. Move over to the draw tab and choose the ink to **text option**, or just choose the convert your ink option. If the desired result is not obtained, choose the "more options" button located on the right-hand side of the converted shape so as to see other options that can be adopted.

Converting window ink to shapes

To change an ink drawn somewhere on the slide to shapes, simply draw or write with the use of the digital ink, choosing the ink by tapping on it or making use of the Lasso selection tool. Move over to the draw tab and choose the ink to shape option, or just choose the convert your ink option. If the desired result is not obtained, choose the "more options" button located on the right-hand side of the converted shape so as to see other options that can be adopted.

Converting window ink to a math formula

To change an ink drawn somewhere on the slide to a math formula, simply draw or write with the use of the digital ink, choosing the ink by tapping on it or making use of the Lasso selection tool. Move over to the draw tab and choose the ink to math formula option, or just choose the convert your ink option. If the desired result is not obtained, choose the "more options" button located on the right-hand side of the converted shape so as to see other options that can be adopted.

Animating Windows Ink

- On the draw tab option, choose Lasso Select and then choose the ink that should be added to the animation.
- Click on the **animations tab** and then choose the replay or rewind option from the animation effects gallery. The replay will start with the ink not displayed and then it will play the ink stroke in the very same sequence in which they were drawn. The rewind option begins with the ink displaying and then erases it from the screen in a reverse manner.
- By default, the animation effects start on click. This setting can be changed on the right-hand side of the animation tab.
- By default, the animation effect lasts only two seconds. This time can be changed on the right-hand side of the animations tab, and a delay can be added before it starts if need be.
- On the slide show tab, choose the **From Current Slide** option in order to view the animation in the slide show view.

The Make Ink replay seen on the Ribbon

- Open the file and then select the **options**.
- Choose the customize ribbon tab in the options dialog box.
- Check to see if the customized ribbon box is set to Main Tabs or All Tabs.
- In the long box below the main tab or all tabs option, choose the entry for the draw tab.
- In the lower part of the dialog box, open the list box that is named "Reset" and choose "Rest Only" as the Ribbon Tab.
- Finally, select the **OK button** to close the dialog box. The ink replay button has now returned to the draw tab.

CHAPTER 12
SLIDE MASTER

Working with Masters

Modifying the slide master

The slide master has a default placeholder that is basically for the title slide, footer, date, subheadings, and more. Any formatting changes that will be made should be in the plain area. It is also possible to choose from PowerPoint's themes in order to design the slide master. To select the elements in the slide master, it is also possible to select the master layout and then uncheck the placeholders that are not needed.

Modifying the handout master

To make some adjustments to the handout master,

- Click on **view the presentation views** and then handout master, or simply hold down the shift key while tapping on the slide sorter view button. The handout master will then be displayed, and it usually includes a handout Master tab on the ribbon.
- Play around with it. The handout master shows the arrangement of handouts for slides of different pages. It can include handouts for slides printed on two, three, four, six, or even nine pages, and also the manner of arrangement for printing the outlines. Using the slides per page control on the handout master tab, it is possible to transition from one handout layer to another. The downside is that the slides and outline placeholders that appear in the handout master cannot be moved, resized, or deleted. It is possible to add or modify elements that should appear on each page of the handout, such as the name and phone number, as well as a page number.
- Select the **close master view button** that is located on the handout master tab on the ribbon. PowerPoint will then go back to normal view.
- To confirm the changes, print a copy of the handout. Handout Master elements cannot be seen until they are printed, so it's okay to print at least a page to check if the changes were effective.

Note that when the handout pages are printed, the slides are adjusted based on the slide master. The way the slide master is displayed cannot be changed from the handout master.

Modifying the Note master

Notes pages have a reduced image of the slide, and any note that is typed goes along with the slide. When printed, note pages are adjusted to suit the Notes Master. To make changes to the Note Master, follow the steps below:

Click on **view**, then **presentation view**, and then **Notes Master**. The Notes Master will then be displayed.

- The notes master has two main placeholders: one for the notes and the other for the slide. These objects can easily be moved or adjusted, and the format of text can also be changed in the notes placeholder. Elements that should be displayed on each handout page can also be changed or adjusted. It is also essential to note the convenient placement of the footer, date, header and page number.
- Select the **close master view button** and PowerPoint will go back to its normal view. So as to verify if changes have been made, print notes.

Using masters

Master slides are very special and powerful slides. Typically, they deliver the same commands slightly adjusted for different slides. This depends on how well-formatted the slide master is and whether the changes affect all slides in the presentation or just a few.

Slide Master

The slide master is the topmost slide that is found in the slide master view. It is the very first slide in the slide pane in the slide master's view. It is usually a little bigger than the master slides.

Formatting: The changes made to the slide master will always affect all the other slides in the presentation. When a particular theme is chosen for presentation, what is really being done is assigning a theme to the slide master. Due to the fact that formatting the commands that have been given to the slide master apply all through the presentation, the theme design, and the colors are also applied to all the slides. If there is a need for a logo to appear, simply place the logo directly on the slide master.

Layouts

Start by choosing a slide layout and then adding the title and content so as to create a new slide. In the slide master view, PowerPoint offers a single layout for each type of slide layout in the presentation. By choosing and formatting the layout in the slide master view all over again, all the slides in the presentation that were created with the very same slide layout can be reformatted. For example, to have the fonts, alignments, and other formats on all the slides changed, choose the title layout in the slide master and then change the master styles on the title layout.

Each layout has control over its own little fiefdom in the presentation. A fiefdom also includes slides that are created with the very same slide layout.

Master Styles

Each master slide, i.e., the master slide and layout, provides the opportunity to choose and make changes to the master styles. The master style controls how the text will be formatted on slides. By changing the master style on a master slide, the look of the slides all through the presentation can also be changed. For example, when the master style font changes, the fonts in all the slide titles in the presentation will also change.

Slide master layouts and PowerPoint slides are designed according to the "trickle-down" principle. As soon as a master style is formatted on the slide master, the formats trickle directly down to the layouts, followed by the slides. The master style that is formatted on the layout trickles down to the slides that were created using the same layout. This chain-of-command relationship is created to work right from the top down, with the master slide and layout giving orders to the slides below. Slides receive orders from layouts, and layouts, in turn, receive orders from the Slide Master, in order to maintain consistency of design.

Overriding the master text style

Other than making adjustments to fonts in the slides one after the other, it is possible to change the default fonts for the whole presentation. The font pairing (header font and body font) is a very important design decision when it comes to PowerPoint.

- Choose the **view,** then slide master.
- Navigate to the slide master tab and then choose the fonts drop-down menu. Click on the preferred font that should be used for all the slides in the presentation. It is not compulsory to select from the predefined font pairs on the menu; select the "customize fonts" option located at the lower part of the menu to choose the preferred font.
- Select the **Close Master View** option to have the master view closed. The text throughout the presentation will then be automatically updated to the new font.

Hide background objects

Microsoft PowerPoint provides various sets of pre-made themes that can be used to make presentations compact and very easy to follow. They are all well-designed and will also save lots of time that would have been spent on making them themselves. On the other hand, the fixed formatting can sometimes be mixed up with just how certain slides have been imagined. Luckily, these background graphics can be removed from slides in PowerPoint without letting go of the particular theme.

If there is a need to hide background graphics, simply follow the steps below;

- Open PowerPoint
- Open the file that needs to be adjusted.
- Click on the **design tab** located at the top of the screen.

- Choose the format background menu option located on the far right-hand side of the toolbar. A new menu will then be displayed on the same right-hand side.
- Finally, select the **"hide background graphics box."**

On the very next slide, the chosen theme will disappear. Other slides that must be chosen will be different from the rest of the presentation.

There is also an option to hide background graphics on all the slides in a presentation. This is much easier if the format of the theme is preferred but the graphics are a turnoff. The background graphics can be hidden easily by following the steps below.

- Choose any slide from the presentation.
- Click on **the Design tab** and locate the format background menu.
- Tick the option to hide background graphics.
- Click the **apply to all** button on the lower part of the menu. This will have the background graphics hidden on each slide in the presentation.

Removing the background from slides

There are times when the situation is just the very opposite, and removing all other things on the slide but leaving the graphics will be the preferred option. If you come into contact with a theme with the desired graphics but the background image is sticking out, it can be easily removed while the graphics remain intact.

- Navigate to **the design tab**.
- Choose the format background option under the customize tab section.
- Choose a solid fill from the menu that appears
- Click on **the paint bucket icon** that can be seen.
- Choose the white color under the palette of the theme color.
- Finally, click on **the apply to all options**.

The background or watermarks will then be removed, and the theme format and graphics will be left unchanged.

Using headers and footers

A header or footer is the text that is displayed at the very top or in the lower part of each page of the slide in the presentation. It is usually a standard text or line that might have the name of the company, trademark symbols, time, or date information embedded in it. Here are a few things to know about headers and footers.

- Headers and footers can be tweaked with font color, type, and size.
- It is not compulsory to use a header or footer from the title slide.
- Only footers are added to slides; headers and footers can be added to handouts and note pages.
- Lastly, slide header or footer information can be separate for handout and note pages.

Headers and footers allow information like the page number that recurs to be displayed. To add headers and footers to slides, simply open the presentation and then tap the insert tab option. Locate the ribbon and navigate the text grouping commands to select headers and footers. This will open up the header and footer dialog box with various options. Choose the preferred options and select **apply to just this slide or apply to all slides.**

Note that there is no option for the header simply because the header cannot be displayed on slides. Only footers can be displayed on slides.

Adding a header and footer to notes or handout pages

To add a header or footer to notes or handout pages,

- Locate the ribbon and click on **the Insert tab**.
- Move over to the text section and select the header and footer.
- When in the header and footer window, right on the slide tab, check the boxes for the date and time, the slide number, and also the footer so as to add those elements to the footer of each of the slides. The footer option gives the user the opportunity to enter custom text to display in the footer.
- In order to add elements to the header and footer of note pages and the printed handouts of slides, right within the header and footer, select the notes and handouts tab.
- Take a look at the boxes for the date and time, page number, header, and footer to add those elements. The header and footer options provide the option to enter custom text to show in the header and footer.
- To add the chosen header and footer elements to all the slides, note pages, and printed handouts, select the **apply to all option.**
- To make more header and footer adjustments, choose the Insert tab in the ribbon. If it has not been chosen, select the header and footer option in order to open the header and footer window all over again.

Editing the header and footer placeholders directly

A handout master can be used to create and make adjustments to presentation handouts. Changes will be displayed on all the pages of the printed handout. The number of slides per page, page orientation, and the size of the slide can all be changed. Page numbers, headers and footers, dates and time can be applied or removed. There is also an option to include pictures in the background.

Headers and footers in the placeholder group can be changed as well. By default, the placeholders are displayed at the top or bottom corners of the handout master page.

Select the text in a placeholder (header, footer, date, or page number) to make adjustments to it. Follow the steps below;

- Turn headers and footers on or off: In the placeholders group located on the ribbon, clear the checkbox of the item (header, date, footer, or page number) that needs to be turned off.
- Edit or add content in the text placeholders. In addition to text changes, commands can also be used in the insert tab option to include graphics or other types of content.
- Format text shape and display: Choose the text and make use of the options on the format and home tabs so as to make the necessary adjustments.
- Change text placeholders: Move a text placeholder to change its position. Use the text box size handles to have the size changed.

Yes, you can serve two masters.

Yes, in PowerPoint, two masters can be served at the very same time. When there is a need to make use of different themes in one presentation, there will be a need to make use of more than one slide master, with each slide master representing a theme.

Creating another slide master

- On the View tab, select the slide master.
- Do one of the following in the slide master: In the edit master group, choose the insert slide master option. When a new slide master is inserted following this procedure, the new slide master will then be displayed under any previously existing slide master without any theme color or theme effect. Custom themes can, however, be applied with the use of any option in the background group.

- On the slide master tab, within the edit theme group, choose themes, and then choose a new theme. If the presentation has just one slide master, choosing a new theme with the use of this procedure will change the theme of the other slide master with the new theme being put in place.

Restoring lost placeholders

While playing around with the masters in PowerPoint, there is every possibility that a placeholder layout might have been deleted. To retrieve this,

Navigate to the slide master view

It is worthy of note that this procedure can work only for the actual master and not for each layout. If a placeholder has been deleted from a layout, it needs to be recreated by making use of the insert placeholder option.

Call up the Master with the placeholder that is missing

Right on the ribbon, click the master layout button in the master layout group. A dialog box displaying the master layout will appear. One of the most confusing dialog boxes a user can see is the master layout dialog box. When it is called up for a master that already has all its placeholders, the checkboxes on the master layout dialog box will be grayed out. In the event that a placeholder or multiple placeholders are deleted, the checkboxes for the missing placeholders will still be displayed.

Lastly, choose the text boxes for the different placeholders that should be restored and click on the OK button.

Working with templates

Templates are actually the start of creating very good-looking presentations. If a presentation is created from scratch, starting with a slide that is blank, this might be quite difficult to do. Creating a presentation isn't so much of a tough task, but creating one that looks very good can be a completely different story.

But amazingly, PowerPoint offers loads of templates that have been designed by professionals who have a clear-cut understanding of color combinations, balance, and all other things that have to do with art.

Pptx Templates make use of a very special extension .pptx, but ordinary PowerPoint presentation files (.ppt) can also be used as themes or templates. This means that any of the custom-created presentations can be used as a template. If lots of changes are made to the presentation masters, the very same presentation can be used as a template for other presentations that are created. Or rather, the presentation can be saved as a template by making use of the.pptx file extension.

Creating a good-looking presentation can be very difficult, even for professional artists. For those who are not artistically inclined and are not willing to learn, it is almost impossible to create one.

Creating a new template

If there happens to be no template that comes by default with PowerPoint that is appealing, it is best to create one for yourself. All that needs to be done is to create a presentation with the masters and the color scheme will be set up just as you want it to be. You can then have it saved as a template.

Here are a few things to remember about templates:

- If very few changes need to be made to one of the default templates, open the template by making use of the open command, then make the adjustments and use the save as command to ensure the template is saved under a new name.
- Personal presentation templates can also be created. Simply create the template as a normal presentation and then include the number of slides that should be included.
- Select a particular place to have all the templates stored. The path to this storage location should be known so as to create a new presentation based on the templates.

Creating a presentation based on templates

To create another presentation based on a template that has just been created, the PowerPoint must be configured to search for templates that are personal. To do so, select File, then Options, and then click on the Save button. Enter the path to the templates folder in the default personal template's location box, and then when all of this is done, click on the OK button.

Once the template location has been configured, a new presentation can be created on the basis of one of the templates that was created by selecting File, New, and Personal to show the list of all the personal templates.

Working with presentation sections

When creating a very large PowerPoint presentation, sections that can be expanded and also collapsed can be created. This can help ensure the slides are well organized and also make editing easier. To have this done, follow the steps below:

- Locate the slide list, right-click on the very slide where the new section should begin, and choose the Add section option from the drop-down menu.
- Right-click on the option labeled "Untitled Section" and choose the **"rename section" option.** Type in the name of the preferred section and, if need be, repeat the process from the start to have other sections created and named also.
- To move a section, right-click on **the name of the section** and make use of the move section up and move section down options.
- To close or open up a particular section, select the collapse icon on the left-hand side of the section name. All sections can also be minimized or maximized at once simply by right-clicking on **the name of the section** and then choosing either the collapse all or the expand all option.

Furthermore, these settings can be accessed by selecting the slide sorter under the view tab option. Note that this kind of segmentation is a very good way to have an overview of the logical flow of all the slides at once and check if there are any needed changes. For instance, there might be a reason to break a slide into two or three other slides or to have some slides merged together.

CHAPTER 13

INSERTING PICTURES

Exploring the many types of pictures

A way to improve the way PowerPoint 2025 looks is to add pictures to one or more slides. The graphics can be informative, like having a chart showing the result of sales, or decoration such as a cartoon smiley face that symbolizes good news or something to cheer the audience up. Display clip art of images or images that have been stored on the computer's hard drive, which includes pictures from digital cameras, can be added to the presentation.

Bitmap pictures

In bitmap graphics, an image is shown on the screen as a collection of very small squares called pixels, which come together to form a pattern. Bitmap images, also sometimes called raster images, are the most commonly used type of graphics. Almost all photographs and images on the internet are bitmaps. Below are some features of bitmap graphics:

- The size of the file is usually dependent on the data and resolution (pixel x dimensions) x Bit Depth (the number of colors per pixel) and also the compression technique that is used to merge the data together.
- All bitmaps are rectangular and act in almost the same way.
- Bitmap file formats include TIF, BMP, and GIF. JPG. The only difference between the formats can be seen in the sizes of the files.
- Most bitmap programs include Photoshop, Corel Paint, Photo Impact, Windows Paint, and also a painter.

Victor, give me a vector

Vector graphics are a way of representing pictures by designating positions and drawing lines or geometric shapes corresponding to them. When it comes to vector graphics, the image is usually saved as a file that has instructions for drawing it, as opposed to bitmap graphics, where the image is saved as a collection of pixels. One major plus with vector graphics over the use of bitmaps is that a picture in vector can be increased or reduced without any loss in quality. Vector graphics are usually seen in the form of clipart, fonts, and logos. Adobe Flash is a program that makes use of the vector format so well, especially on websites. Features of vector graphics include small file sizes, files that are independent of resolution and bit-depth, vector graphics can be any shape as opposed to just the rectangular dimension, and most vector programs contain illustrator and freehand Corel Draw.

Inserting pictures into the presentation

To liven up a presentation, pictures can be added, either from those saved on the computer or from the web.

From the Web

- Choose the preferred location where the picture should be inserted on the slide.
- Locate the Insert tab option in the images group and select pictures, then choose online pictures.
- In the search box that appears, type in the preferred image and then tap the enter key.

From the computer

To insert a picture from the computer, do the following:

- Click on the **preferred location** where the picture should be inserted into the slide.
- Locate the **Insert tab icon** in the images group and choose pictures, then select **this device**.
- In the corresponding dialog box that opens up, browse through the pictures and then choose the preferred picture that should be inserted and then click on **the insert button**.

From the stock library

- Choose the very location where the picture should be inserted on the slide.
- Locate the insert tab option and in the images group, choose pictures by clicking on stock images.
- For much easier access, simply type in the image that is preferred and click on the insert option.

Moving, sizing, and stretching pictures

When a picture is added to a slide, there might be a need to adjust the size or move the picture to a different location entirely. To adjust the size of an image, simply follow the steps below.

- Choose the picture that needs to be adjusted. PowerPoint will automatically display handles on the picture that has been chosen.
- Drag the mouse arrow just over the handle. The mouse arrow will then point in two directions.
- Hold down the left button of the mouse and move the mouse. PowerPoint will then adjust the size of the chosen picture.
- Set free the left mouse button when you feel satisfied with the new size of the picture.

To move an image, follow the steps below.

- Drag the mouse pointer right over the edge of the image that should be moved. The mouse will change to a four-way pointing arrow.
- Hold down the left button of the mouse and move the mouse. This way, PowerPoint will move the image.
- Set free the left button of the mouse when satisfied with the new position of the image.

When an image must be added to a slide, there might be a need to delete it. To delete the image, follow the following steps below:

- Choose the image that should be deleted. PowerPoint will show the handles around the chosen image.
- Click on **the delete button**. PowerPoint will then delete the chosen image.

Adding style to pictures

With PowerPoint 2025, you can draw lots of attention to pictures by adding styles and features that include borders, shadows, and reflections.

To add a style effect to a particular picture, choose the picture and open the picture tools tab on the ribbon. Then choose the preferred picture style.

It is worthwhile to note that PowerPoint has about 28 different predefined picture styles that are displayed in the gallery picture. Each of these styles is a simple combination of three different types of formatting that can be applied to pictures: shape, border, and effects. If needed, these formats can be applied individually.

It is worthwhile to note that if any one of these predefined picture styles is used, the picture will be updated automatically if the theme of the presentation is changed much later. This means that one of the predefined styles should be used whenever possible.

Applying a picture border

Picture borders can be applied to a picture simply by opening the picture tools and then clicking on the format tab and clicking on the picture styles group. This will display the picture border menu, which provides the option of choosing the border color, weight (the width of the borderlines), and also the pattern of dashes that should be used. Note that if a shape has been added to a picture, the border will also be added to the shape.

Applying picture effects

The picture effects that can be found in the picture styles group (located on the format tab option) provide the option of applying various interesting types of effects to the pictures. When the button is tapped, a menu showing the following effect options will be shown:

- **Shadow:** Aids in the addition of a shadow to the image. One of several predefined shadow effects can be chosen, or there is also an option to call up a dialog box that helps to adjust the shadow as desired.
- **Reflection:** helps to develop a mirrored image of the picture under the original picture.
- **Glow:** applies a glowing effect all around the perimeter of the picture.
- Soft edges help to make the perimeter of the pictures soft.
- **Bevel:** develops a 3-D beveled display.
- 3-D Rotation: This turns the picture in such a way that it creates a three-dimensional effect.

Applying artistic effects

- Choose the **desired picture** where an artistic effect should be added.
- Click on **picture tools**, then format, and then choose artistic effects.
- Move over to the options to have the picture previewed and then choose the preferred one.

Removing picture backgrounds

Microsoft PowerPoint offers a bit of picture wizardry in its ability to remove a background from a picture. To accomplish this part of the photo editing magic, follow the steps below:

- Choose **the picture whose background** needs to be taken off.
- Locate the picture tools option and then click on **the format tab** and click on **the remove background button icon**, which can be found in the adjust group option.

When this is done, PowerPoint will try to check which part of the picture is the subject of the picture and which portion is the background. PowerPoint will then develop a bounding rectangle that has what it perceives to be the subject in the picture. It will then check the colors in the picture to ascertain what it thinks are the background portions of the picture. The background will then be shown in purple.

After which the Background Removal tab will show on the Ribbon and when selected the background goes off

- If need be, change the size of the bounding rectangle so as to enclose the subject well enough.
- If need be, make use of the Mark Areas to keep and mark areas to take off buttons, and also change the location of the background of the picture. For instance, if an area that is a part of the subject is displayed as a background, choose the Mark Areas option to keep the button. Then choose the option of either clicking on the area that should be added or clicking and moving a line across a very large portion of the area to be added. PowerPoint will make an attempt to know the part of the picture that has been marked and add that area to the subject of the picture.

Bear in mind that the part to be included doesn't need to be circled, nor is there a need to be very accurate. PowerPoint will try its best to decipher the parts of the image to add based on the mark. Furthermore, if PowerPoint has by mistake taken a portion of the background as a subject, select the Mark Areas to take off the button and then click or draw a line around the area that should be taken off. If PowerPoint also misinterprets the mark, press the Ctrl + z button to undo the action or select the delete mark button icon and then choose the mark that should be deleted.

- Repeat the 4th step until the picture's background has successfully been taken off.
- Select the "keep changes" button icon. The slide will then go back to normal with the background of the picture now taken off.

Correcting sharpness, brightness, contrast and color

- Choose the preferred picture.

- Click on **picture tools**, then click on **format**, and then choose **Select Corrections**.
- Move over to the options menu to have the picture previewed and then choose the preferred one from the options displayed.

Compressing pictures

- Choose the **preferred picture**.
- Click on the **picture tools**, then choose the **format option**, and then click on **Compress Pictures.**
- Make a choice from the options displayed and click on the "OK" button.

Working with 3D Models

Of all the Microsoft Office suite applications that now have the capacity to make use of 3D objects, PowerPoint happens to be the best of them all. This use of 3D means the addition of improved visual elements that don't necessarily have to be self-created before making use of them.

To use 3D objects, open PowerPoint and then open a blank presentation or a previously existing one. Click on **the insert button** so as to switch tabs. Under illustrations, select 3D models to open up the drop-down menu. Choose from an online source, or there is also an option to bring in your preferred artwork by selecting from a file. Choose from a category or make use of the search bar at the top to locate the preferred item. Once located, double-click on the item. The size can be adjusted by clicking on the handles of the object and dragging it.

CHAPTER 14

DRAWING ON THE SLIDES

Some general drawing tips

Zooming In

While presenting a slide with the use of the presenter view, any preferred part of the slide can be magnified on the screen. To do this, simply follow the steps below.

- Locate **the bottom right-hand corner** of the slide and choose the **Slide Show view option**.
- In the bottom left corner of the presenter view, choose the magnifying glass icon. A very bright rectangle will appear on the slide in the presenter's view. The mouse pointer will change to the image of a hand.
- Direct the mouse to the part of the slide that should be enlarged and then click on the zoom in on-screen option. The mouse pointer will then change to a hand to show that the zoom effect is now active in the presenter's view and can be moved to another part of the slide.
- The zoom effect can be turned off by pressing the Esc key or clicking on the magnifying glass icon. The onscreen slide zoom will then return to standard view.

Displaying the gridlines and guides

PowerPoint gridlines help position objects on the slide more accurately. Gridline alongside rulers and guides helps to position and snap slide objects such that they can easily line up a set of slide objects accordingly.

Gridlines are parallel and vertically dotted lines that are shown on the slide, looking almost like a dotted graph. These gridlines cannot be printed or be visible in the slide show view. Basically, by default, these gridlines are not even visible on the slides.

Below is how to make use of gridlines to make working with PowerPoint much easier.

- Open a blank presentation in PowerPoint. Note that gridlines are visible by default.
- Click on **the View tab** of the ribbon to make gridlines enabled and also ensure that the gridlines checkbox is selected. In the same vein, if gridlines are preferred to be hidden, simply deselect the gridlines checkbox.
- The gridlines will then be displayed on the slide.

Guides

As opposed to gridlines, which are always at a fixed point, guides can be moved about to help with the arrangement of objects on a slide.

Click on **the guides checkbox** located on the view tab and turn them on. A horizontal and vertical guide will be displayed on the slide. They can be selected and moved about anywhere on the slide. Select and move the guides to fix them on the slide. Select and move the object with the use of the guides to have the object fixed correctly. Hold the Ctrl key and click and move the guide to create another. Finally, clear the "guides" checkbox to turn the guides off.

Sticking to the color scheme

By default, PowerPoint chooses a particular color that should be used once a presentation is created. If the default color suits what you intend to create with the slide, simply adhere and completely stick to it without having to change the color. This will help save time and ensure the presentation is completed faster.

Saving frequently

Losing a file when it is almost completed can be very painful, as this will mean you have to start all over. Avoid file loss by using the shortcut command button Ctrl + S as often as possible when creating a presentation. This will ensure that your file is saved in the desired location and not lost eventually.

Remember to use Ctrl + Z.

The Ctrl + Z command button is basically used to undo if a mistake was made or there is a need to make certain adjustments while creating a presentation. Save yourself the stress of using the backspace or delete button by applying this command button.

Drawing simple objects

With the use of a digital pen or the keyboard, drawings can be made in PowerPoint 2025. See objects and how to draw them below.

Drawing straight lines

The line button can be used to draw straight lines on the slides. Simply follow the steps below.

- Locate the shapes group on the insert tab and click on the line button.
- Direct the cursor to the location where the line should start from.
- Select and move the cursor to the position where the line should end.
- Set free the button of the mouse when you get to the preferred location. The line should end.

After the line has been drawn, the ribbon will show the drawing tool tab. The controls in the shapes and styles group can then be used to make changes as desired.

Bear in mind that a line can be forced to be perfectly horizontal or vertical by pressing down on the shift key while drawing. If the shift key is pressed down and moved diagonally while drawing, the line will be constrained to a perfect 45-degree angle.

Drawing rectangles, squares, ovals, and circles

To draw any of the above-mentioned shapes, follow the steps below.

- Locate the insert tab, click on the shape button, and then select the preferred shape button.
- Direct the cursor to the location where one corner of the rectangle should be fixed.
- Select and move to where the opposite corner should be located.
- Set the mouse button free.

Note that the steps for drawing an oval shape are just the very same ones for drawing a rectangle. The only difference is that you just have to select the oval button rather than the rectangle. To draw a perfect round circle or a square, choose the rectangle or the oval button, but ensure to press down the shift key while drawing.

Creating other shapes

To create other shapes, just ensure you choose the very shape in the shapes button and if that shape cannot be found in the shapes button, choose another similar to it and press down the shift key while drawing.

Creating a text box

The process of adding a text object is similar for all types. The most common text object that is usually added to a slide is the basic text. To add the text, select the textbox in the text menu section. A new textbox with default fonts will then be displayed where new content can be inserted. A new textbox object is usually displayed as an empty space or a small square. Text can be typed into the square, which is usually set as the default font. The text can also be rotated with the arrow at the top, just like an image can also be rotated. Rotating the text will change the way the text is shown both vertically and horizontally.

Styling the shapes

Choosing a shape style offers the opportunity to add preset colors and effects to rapidly change the appearance of the shape or the textbox.

- Choose **the shape of the text box** that should be changed.
- Locate the format tab and select the more drop-down arrow in the shape styles group.
- A drop-down menu of styles will then appear. Make a choice from the list that is displayed.
- The shape will then be displayed in the style selected.

Choosing a shape fill

To change the shape's fill color;

- Choose **the shape or textbox** that needs to be changed.
- Locate the format tab and select the shape fill drop-down arrow. The menu for the shape fill will then be displayed.
- Drag the mouse over the different colors. Select the "more full colors" option to see more color options.
- The shape will be displayed in the selected color.

Creating the outline of the shape

To change the outline shape,

- Choose the **preferred shape** that should be changed.
- Locate **the format tab** and choose the shape outline drop-down arrow. The shape outline will then be displayed.
- Choose **the preferred color** and if there is a preference for the outline to be transparent, choose the no outline option.
- Finally, the shape will be displayed in the chosen outline color.

Applying shape effects

Applying a shadow to a shape can make it look like it is floating over the page, and it can also help to include a contrast between the shape and the background.

- Choose **the shape** that should be changed.
- Locate the format tab and choose the shape effects drop down arrow. In the menu that is displayed, move the mouse over the shape and choose the shadow that should be used.
- The shape will then be displayed in the shadow that is selected.

Flipping and rotating objects

In PowerPoints, objects such as text boxes, shapes, and pictures can be flipped or rotated.

- Choose **the object** that needs to be rotated.
- Click on the rotation handle at the very top of the object, and then begin to move the handle in the preferred direction.
- To flip an object, choose the object, arrange, rotate, and choose between flip vertical and flip horizontal.

For a more accurate movement, choose "arrange," "rotate," and then "more rotation options" so as to see the menu options.

Rotating an object 90 degrees

Objects can be rotated in PowerPoint by moving the rotation handle and entering the very precise rotation amount or by making use of the keyboard shortcuts. Rotating an object means having the object turned around its center. Shapes, images, and other types of objects can be rotated in both the normal and slide master view.

Using the rotating handle

An object can be rotated by moving the rotation button that shows right at the top of the object. To rotate an object with the use of the rotation handle, follow the steps below:

- Choose **the object** by selecting it. If a textbox or placeholder is being selected, click **on its edges**.
- Tap the left button on the mouse and move the rotation handle that shows in the middle of the above to either the left or the right. This is known as free rotation. If the shift button is pressed down, PowerPoint will confine the rotation to just 15-degree increments.
- Once all this has been done, set the mouse button free.

Do note that the rotation handle will always be displayed once an object has been selected.

Drawing a complicated picture

When working on a PowerPoint presentation, there are times when the basic shapes in the program might not meet the specific needs. Below are steps to create a complex shape if need be.

- Start by inserting two different shapes that should be combined in the presentation. To insert a shape, navigate to the "insert shapes" option and then select "shapes." Choose the preferred shape and then draw it on the slide, pressing down on the shift key if need be.
- Move to the format tab options and select the merge shapes drop down menu. Several options will be displayed for merging. Options such as

Union: This helps to merge two shapes together such that they both become one and can be used like any other shape in PowerPoint.

Combine: this looks much like merging. The only difference is that there will be an empty space where the shapes overlap.

Fragment: This helps to combine shapes, but there will be a different area to work with the three components.

Intersect: This option takes off everything, leaving only the intersection of the two shapes. This will make it similar to the very opposite of the combined option.

Subtract: This option subtracts from a shape what the other shape has already covered; this eliminates any form of duplication.

The five options above offer lots of tools for bringing different shapes together to make just one that suits your needs.

Changing layers

You can change the order in which objects, placeholders, and shapes appear on a slide. The slide objects are always stacked in the order in which they are inserted, with the most recent object at the very end. By clicking on the "bring and send" command button on the right menu, it is very easy to reorder the layers.

To do this, simply choose the object and right-click to choose one of the following options.

Bring to Front: Bring to front helps to take the object to the very top of the layer, while bring forward helps to take the object up one layer in the stack.

Send to Back: Send to back takes the object to the bottom of the layer, while send backward takes the object in the layer one step backward.

CHAPTER 15

WORKING WITH CHARTS

Adding a chart to the presentation

Charts can be created in PowerPoint. To add a chart to PowerPoint, simply follow the steps below.

- Click on **the insert button**, then select **chart**, and choose the **preferred chart** option by double-clicking on it.
- Change the placeholder data in the worksheet that is displayed with the information at your disposal.
- When a chart is inserted, some little buttons will be displayed next to its upper right-hand corner. Make use of the chart elements button to display, hide or make adjustments to things such as axis titles or data labels. You can also choose to make use of the chart styles button to swiftly make adjustments to the color or style of the chart.
- After all of this has been done, close the chart or worksheet.

Copying and pasting a chart or data from Excel

There are times when there might be a need to copy a chart from an Excel spreadsheet to a PowerPoint presentation so as to show the summarized data to the viewers or display the visualizations in the dashboards and reports.

To copy a chart from Excel to PowerPoint, simply make use of the copy and paste option from the clipboard. Move to the Excel spreadsheet and choose the range that should be copied. Click on "home" and then select "copy" and then navigate to PowerPoint and click on the "paste option." In PowerPoint, the simple paste option or the paste special option can be used. This is based on whether the chart will be pasted with a special format or pasted as an image. It is also possible to make use of keyboard shortcuts such as **CTRL + C** and **CTRL-P** to copy and paste the chart where necessary.

- If there is a need to insert a range of charts or data from Excel into PowerPoint, name the range in the excel spreadsheet and add the excel spreadsheet object to the PowerPoint presentation.
- Navigate to PowerPoint and select the "paste special option" from the home menu. After this has been done, add a Microsoft Excel worksheet.
- Ensure the paste link is chosen and then add the Microsoft Excel worksheet object.
- After all this has been done, the table will then be added.

Changing the chart type

To add any of the new types of charts (or any other chart) to a presentation, choose the insert button, then select the chart option, then choose the type of chart that should be inserted. There is also an option to choose the chart icon in the area that is displayed when some new types of slides are created. While being in the box, there is also an option to add text, table graphics, and some other contents, after which the preferred type of chat will be inserted. No matter the way the chart is added, it will be displayed in the presentation placeholder data and a window will pop up that looks much like a mini-Excel spreadsheet will then be displayed. Enter the data or make some adjustments to the data. There is also an option to edit in the excel button to ensure it is opened up in excel and then it can also be edited there.

Change the chart data on a slide

If a PowerPoint presentation has a chart, the chart data can be adjusted directly in PowerPoint; it doesn't matter if the chart is infused or linked to the presentation. The data can in turn be updated or refreshed in a linked chart without the need to go to the program in which the chart was created.

- Right on the slide, choose the chart that is preferred. The chart tools contextual tab will be displayed at the very top of the PowerPoint window. If the chart tools tab or the design tab is not displayed, ensure the chart is chosen and selected. It is worthwhile to note that the design tab under the chart tools is just not the same as the default design tab in PowerPoint. The chart tools tab will be displayed only when a chart has been chosen and the design layout or format tab under it brings up different commands in relation to the chosen chart.

- Under the chart tools, on the Design tab in the data group, choose the edit data option.
- Choose one of the following options next:

To make adjustments in PowerPoint, choose the Edit data option. This will open up a window from the spreadsheet.

To make adjustments to the data directly in Excel, choose the "edit data in Excel" option.

- Make the necessary adjustments and it will be shown in the chart in PowerPoint.

Adding rows and columns

When there is a need to show precise content with numbers and text on a PowerPoint slide, like financial information, the best option is a table. There are times when the columns or rows might not be the expected width or height. There might be a need to add or remove some rows or columns so that the table will appear the preferred way. To begin with, we will learn how to create a table based on

Create a table by moving to the insert tab option and selecting the table command. When the table grid has been opened, move the mouse pointer to choose the number of columns and rows as preferred. The table should also consist of cells, which are boxes that will contain information. Data, text, or numbers can be added to the table also by choosing the mouse in a cell, including data, and then making use of the tab key to change from cell to cell.

To add a row to a table, right-click on the table and click on the insert button. Locate the menu option and choose either the insert rows above or the insert rows below. This process can be repeated to insert a second row as well.

To delete a row or a column, *right-click on the row or column* that should be deleted and click on the delete button. The menu that will be displayed will display three different options: Delete rows, Delete columns, or Delete table. Choose the preferred option.

To add a column, right-click on the table and click on the insert button. From the menu displayed, choose between the insert columns to the left or the insert columns to the right. The seller's name and header title can also be added.

Editing the source data

If the PowerPoint presentation has a chart, the chart data can be adjusted directly in the PowerPoint. It doesn't matter if the chart is infused in or linked to the presentation. The data can also be updated or refreshed in a linked chart without necessarily having to go to the program in which the chart was created. Follow the **(changing the chart data on the slide option).**

Changing the chart layout

Right after a chart has been created, the look can be changed immediately. In place of adding or having to change chart elements or making adjustments to the chart manually, a predefined layout or style can be added very swiftly to the chart. There are various useful predefined layouts and styles that can be chosen from. That being said, the layout or style of a chart can be tweaked by changing the layout and style of the chart elements that are made use of in the chart manually. Note that the tweaked layout or format cannot be saved, but if it needs to be used again at a much different time, the chart can be saved as a chart template.

Select a predefined chart layout

- Choose the chart that needs to be formatted. It will show the chart tools, including the design, layout, and format tabs.
- On the Design tab, in the chart layouts group, select the **preferred chart layout.**

Changing the chart style

To change the style of the chart, do the following:

- Choose the chart that should be adjusted and the design tab will show
- Right from the design tab, choose the **more drop-down arrow** located in the chart styles group.
- Choose the preferred style from the menu that is displayed.
- The chart will then be displayed as chosen.

The chart formatting shortcut buttons can also be changed to swiftly add chart elements, have the chart style changed, and also filter the chart data.

Using Treemap and Sunburst charts

A treemap can also be said to be a chart of nesting categories used basically for providing a hierarchical display of data, where each color also shows a different category.

When compared to standard charts, treemaps have a number of advantages.

- They make the presentation display very unique, securing the interest of the audience.
- They can also be used with the backgrounds in the image, giving way to the addition of flags, product images, logos, buildings, and other necessary graphics.
- They make it quite easy to note patterns within groupings by grouping data into color-coded options.
- Branches with top categories are indicated by much bigger rectangles. Smaller rectangles are used mainly for subcategories. This makes a more complex hierarchy very simple to comprehend.
- Furthermore, treemaps can display proportions between the various branches. The rectangle size is proportionate to the size of the data shown. They are very effective for displaying important relationships.

Create a map chart by following the steps below;

- Choose an image background. Locate the insert option and choose pictures, then pick the preferred photo. Choose the image option and click on format under the picture tools. Choose the color > recolor to make changes to the color of the image.
- Insert a chart. Locate the insert option and select the chart icon. Choose Treemap in the left column and click on the OK button. By default, an Excel table will be displayed above the chart by default. All the entries can also be replaced with the preferred data. Place the chart over the image that has been inserted and then make changes to the size of the chart in order to fit the image.
- Finally, move the chart back downwards from the image. Choose the image and press the Ctrl + X button. Right-click on the chart and choose the format data series option. Locate the task pane and choose the bucket icon for fill and line, then image or texture fill, paste image, and clipboard.
- Finally, input each category and that will be all.

It is a chart that symbolizes hierarchical data; the rings or circles represent each level of the hierarchy, with the innermost circle being the top. Unlike a doughnut chart, which shows how the rings on the outside relate to the rings on the inside, a sunburst chart with varying levels of categories demonstrates how the rings on the outside interact with the rings on the inside. There is a color-coded sunburst chart that shows each group and a breakdown of the subcategories.

Follow the steps below to make and insert a sunburst chart in PowerPoint.

- Locate the Insert tab option and click on **the chart in the illustration group**.
- In the insert chart dialog box right on the left pane, select Sunburst and click on **the OK button**.
- The diagram will then be infused into the slide alongside the Excel spreadsheet.
- Enter the data into the spreadsheet and close the spreadsheet afterward.

- The sunburst chat has successfully been created.

When a sunburst chart is created in PowerPoint, the labels are automatically added to the chart. If there are no labels on the chart, simply right-click on the chart and choose the option to add labels from the context menu. Alternatively, there is also an option to click on the chart elements button at the upper right of the chart and check the check box for data labels.

Using map charts

Most people are not aware that PowerPoint already has an integrated map function. The map option is available in PowerPoint 2025. It can be used to visually illustrate data by coloring, highlighting, and formatting individual countries. To have a map added to a slide

- Locate the ribbon in PowerPoint and select the **Insert Tab option**.
- Select a **chart** in order to open the insert chart dialog.
- In the list of all charts, click on the map.
- Select the **OK button** to add a general world map to the PowerPoint presentation.

CHAPTER 16

WORKING WITH SMARTART

Understanding SmartArt

SmartArt is a way to change ordinary text into something that is more appealing to the user and audience. It helps to have attention drawn to very important information or make information much easier to read and comprehend. This can also be called a diagramming tool in PowerPoint and provides the option of creating a visual display of information.

SmartArt graphics can be created to match the look and feel of the presentation and can also be used to make process flow, cycle diagrams, pyramids, and organizational charts.

Creating a SmartArt Diagram

The easiest way to make a SmartArt diagram is to make a new slide and then enter the bullet list just like the list will be displayed as normal text and then convert the text to SmartArt. Follow the steps below:

- Create a new slide with a title and a content layout.
- Type in the bullet list. Make use of one or two bullets, but ensure to keep the list as short and concise as possible.
- Right-click anywhere in the list and choose the convert to SmartArt option. Choose the SmartArt type that should be used. If the SmartArt type preferred is not displayed in the menu, choose more SmartArt graphics to show the choose a SmartArt graphic dialog box.

Click on the **OK** button and the diagram will have been created. The diagram can also be modified at any time.

PowerPoint provides eight different types of SmartArt diagrams.

These diagram types are mentioned below.

- **List:** This shows just a simple list. Some of the list diagrams display information that does not have any specific organization; others help to show information in such a way that depicts a step-by-step progression, like the steps in a task.
- **Process:** This shows a process in which steps flow in a step-by-step manner.
- **Cycle:** This shows a process that occurs in a continuing cycle.
- **Hierarchy:** this helps to show the structure of the relationship, like in the organization chart.
- **Relationship:** this helps to show the items that are basically related to each other. This includes different types of radial and Venn diagrams.
- **Matrix:** this helps to display four items arranged in a quadrant manner.
- **Pyramid:** This shows how elements are built upon one another to form a structure.
- A picture helps to show information in a variety of different formats that add picture objects into the chart design.

Tweaking a SmartArt Diagram

After you create a SmartArt graphic, you may need to modify your PowerPoint presentation. Some examples of modifications include adding or removing shapes, altering their order of appearance, and altering graphic types and layouts.

- On the first slide, select the frame of the SmartArt graphic to choose the whole graphic.
- Choose the SmartArt tools design tab and then select the more button in the layouts group.
- Choose the **more layouts button** and then choose a SmartArt graphic dialog box will be displayed.
- Select the **process category**.
- Click on **the continuous block process layou**t. Note that not all graphics might appear nice when a change to a different graphic type is made.

- A new graphic type will be added when you click on the OK button.
- Locate the SmartArt Tools Design tab and select the right to the left button. This will change the graphic direction. If the right to the left button is selected again, the graphic will go back to its original position.
- Select the **text pane button**. If the text pane does not show instantly, a text pane should be shown on the left-hand side of the graphic.
- In the text pane, change the word "production" to "manufacturing."
- Click the **close button** on the text pane that is close to it.
- On the graphic, choose the quality inspection shape and then select **the delete key** to take it away.
- Finally, save the presentation.

Working with organizational charts

Be it for business or for other reasons, creating an organizational chart with the use of SmartArt in Microsoft PowerPoint is very easy.

Move over to the **insert tab** option and then select SmartArt. In the option that indicates choosing a SmartArt graphic, choose the hierarchy category that is located on the left-hand side. On the right-hand side, choose an organizational chart layout like an organization chart. When all of these have been completed, click on the OK button.

Next, choose a box in the SmartArt graphic and then type in some text. Type in the text that should be in place of the placeholder text. Choose an additional box in each SmartArt graphic and then type text in those boxes as well.

As an alternative, there is also an option to type in the text in a given text pane rather than typing directly into any of the boxes. If the "Type your Text Here" pane is not being displayed, select the control located on the edges of the SmartArt graphic. To have a box inserted, choose the box that is already in place and located closest to where the new box should be added. On the design tab, select the "Add shape" option. Type the new text directly into the new box through the text pane.

Deleting a SmartArt graphic or a shape in a SmartArt graphic

A shape can be removed from the text pane of a SmartArt graphic or from within the SmartArt graphic itself.

Overview of deleting shapes

If a line of level 1 is deleted while there is still an existing line of level 2, the first line of level 2 text will then be moved to level 1 (i.e., the top shape). It is nearly impossible to delete a shape if it is part of a much larger shape. For instance, in the Bending Picture Accent list layout in the list type, it is not possible to delete just the small round shape from any of the three displayed larger shapes. All of the associated shapes must first be selected together. Note that associated shapes are the shapes that come together to make a much larger shape.

It is also impossible to delete background shapes like the arrow shapes or, better still, the divisor bar located in the counterbalance arrows layout.

To delete a shape from a SmartArt graphic, click on the border of the shape that should be deleted and press the delete button. After this must have been done, **it is important to note the following:**

- To delete a shape, it is best to click on its border. If the shape that should be deleted has text embedded in it and the inside of the shape is selected other than its border, there is every possibility that some of the text will be deleted when the delete button is pressed.
- If you make an attempt to delta a shape that is combined with other shapes to make a much larger shape, i.e., an associated shape, the shape won't be deleted in its real sense, only the text that is contained in the shape will be deleted.

- If a top-level shape or its text has been custom created already, any adjustment added to that very shape might be lost when a much lower-level shape is then promoted.
- If a connecting shape is chosen, like a line or an arrow that connects or links shapes one to another, just the text in that shape will also be deleted, as it is impossible to delete the shape in its entirety.
- In the layout of some SmartArt graphics, like the Matrix layout, the number of shapes in the SmartArt graphics is usually fixed, and as such, the shapes cannot be deleted.

Deleting clipart or pictures from a SmartArt graphic

If certain pictures or clipart have been added to the SmartArt graphic, they can be taken off by having the fill settings for the very shape that has the picture or clipart changed.

- Right-click on the border of the SmartArt graphic shape that a picture or clip art should be taken off of. If the picture or the clip art is added as a background to the whole SmartArt graphic, right-click on the border of the SmartArt graphic.
- Locate the shortcut menu and click on the **format shape or the format object option**.
- Get on the fill tab option and click on the no fill, solid fill, or gradient fill option based on what the picture or clip art is being replaced with, then select the **fill options that are preferred**.
- After all these have been done, click on the **close button**.

CHAPTER 17

ADDING SOUND AND VIDEO TO YOUR PRESENTATIONS

Adding sound to a slide

Adding audio clips to PowerPoint presentations can be so much fun. Some might be wondering where sound should be added to a presentation. There might be a couple of reasons why an audio clip should be added, from adding a voice-over to just inserting a little background music so as to make the presentation more fun and less boring.

PowerPoint can insert audio clips from various files. The files include Mp3, WMA, or WAV files. These files are automatically infused into the presentation. Also, note that it is very good practice to put the audio files in the very same folder in which the presentation has been saved before inserting them into the presentation.

Follow the steps below to insert audio clips into PowerPoint.

- Move over to the slide where an audio clip should be added. From the insert tab of the ribbon, select the down arrow found below the audio button so as to drag up the drop-down menu. Right within this menu, select the Audio on My PC option.

After that has been selected, it will bring up the Insert Audio dialog box. Move over to the folder that has the audio clip saved and select the "any audio file option."

- Then, click on the **"insert" button**. Getting this done helps add the audio clip to the active slide and also includes it as part of the presentation. This, of course, has the ability to blow up the size of the file of the PowerPoint presentation. If the file size is a problem, you can simply link the file to the presentation rather than including it in the presentation.
- If linking the audio clip is preferred over including it in the presentation, click on the down arrow located next to the insert button. Getting this done will bring up a menu that has two different options: the Insert option, which is the default option that includes sound right within the presentation, and the Link File option, which, as discussed earlier, does not add the audio file to the presentation, only links it. If this is the preferred option, make sure to always copy the audio clip to the very same folder as the presentation before the file is added to the slide. If, by any chance, the presentation needs to be moved to another computer, the entire folder can be copied rather than just the presentation file.
- Immediately an audio file has either been linked or inserted, PowerPoint will have an audio file placed at the middle of the slide.
- It is worthwhile to note that when the mouse is moved over the audio icon in the normal view, the player controls bar will be displayed. A very similar bar will also be displayed in the slide show view. The play button on the player controls can be used to play the audio file. If the preferred option is for the sound to play automatically in the slide show view without having to click on the play button, click on the audio icon first so that the playback contextual tab will be made active on the ribbon. This tab provides various options to control the playback of the audio that has been chosen. Find the start option. Selecting this option will bring up a drop-down list. From this list, select the Automatically option. This step ensures that the audio clip will play automatically when the slide that has the clip shows up in the slide show mode.

Always note that the default option is the click sequence. If this option is chosen, the sound will be played only when the space bar has been tapped or the cursor is selected in slide show mode to move to the next event. Furthermore, a third option can be displayed when clicked on in the drop-down list. This option shows the player control, and the play button must be selected before the audio file will play.

- When all of this has been done, save the presentation.

Investigating sound files

A presentation should be tested for sound before it is saved and then presented to an audience. This could result in hiccups during the presentation if the sound is not properly linked or inserted. You can avoid this drama by selecting the "from the beginning" option from the slide show tab and testing the sound. Always keep in mind that in order to hear sounds, the computer must have a sound card or sturdy speakers.

Setting audio options

Looping the sound

There are instances when the length of the sound doesn't cover the whole of the presentation, but it can be looped such that it plays over and over again. This feature is mostly used when the sound effect is quite cool and subtle, like the sound of waves crashing, and you deem it fit to continue playing as long as the slide is still in view. To loop an audio clip, just click on the loop button until the stop checkbox is found in the audio options group.

Hide the sound icon.

By default, the icon that shows an audio clip is made to be displayed on the slide while the side show is ongoing. If the sound has been set to play automatically, there is every possibility that the icon being displayed won't be needed. The audio options group has a checkbox with the heading "hide" when it is not being played, but it just hides the icon when the sound is not playing; the icon will be shown when the sound is being played.

Fade in and out.

The fade in and out options help to fade the audio clip in and out slowly. By default, these controls are always set at 0. This way, the audio clips begin and stop at the highest volume. By changing one or both of these controls to a value like 3 or 4 seconds, the sound in or out can be gently faded for a cooler effect.

Audio Clip and Trim

Selecting the trim audio buttons helps bring up the trim audio dialog box. This dialog box helps to choose only the parts of the audio clips that should be played in the presentation by providing an option to choose the start and end times of the audio clip. The start and end times can be selected by dragging the green pointer indicating start and the red pointer indicating stop over the audio file waveform (the stop or endpoint can often be determined by looking at the waveform displayed in the Trim Audio dialog box). The time for the start and stop can also be entered in seconds in the start and end time boxes. Ideally, the start and end trim points should be chosen when the audio file is naturally silent so as to avoid the audio file ending abruptly.

Adding video to your slides

It doesn't matter if all you are doing is selling an idea, a product, or even a strategy; the attention of an audience will be drawn faster with creative videos. It is very important to keep in mind that a PowerPoint presentation is simply as good as the video inserted into it. If you need to make a very engaging, stylish, and informative video, this can be done within minutes. Simply follow the steps below to get this done promptly.

Inserting a video from the PC

- Select the slide that is preferred, then move to the menu, and then click on the inset icon.
- In the top right-hand corner, click on video, then choose the video on my PC option.
- Search for the preferred video and select the insert button.
- Make changes to the settings in the video format tool tab so as to ensure the video plays just the way you want it to. If the video should take the center stage immediately after it is selected, ensure the slide has been set to Play Full Screen and Start Automatically.
- Take a preview of the presentation to ensure the video plays just the way you prefer it, and that will be all.

Insert a video from YouTube

To insert a video from YouTube, simply choose the insert video option and then choose the online video option. Insert the web address for any video from YouTube, Vimeo, etc. Note that an internet connection will be needed for this to work during the presentation.

- Open the YouTube website, search for the preferred video, and click on the share button that can be found below the video frame to choose the embed option.
- Copy the embed code. Ensure that the code is the right one so that the whole process goes on smoothly. Do not copy the URL on the first "share" pop-up that is displayed on the screen. After selecting the "embed" icon, wait a little for the pop-up screen named "embed video". After this, select the **copy button** in the bottom right-hand corner. The correct code should begin with "iframe width."
- Navigate to PowerPoint and choose the slide that the video should be added to.
- Select the **insert button.**
- Click on **video** and choose the video from the website.
- Choose the **video and select the playback option.**
- Select the start menu so as to choose the way the video should be played.
- Ensure you are online when the play button is selected. Remember once again that an internet connection is needed for this.

Setting video options

The playback option in PowerPoint can be used to control how and when a video should be displayed in the presentation. The video can be played in Fullscreen mode; alternatively, it can also be resized to preferred dimensions as specified. The volume can also be adjusted, the video played again and again (loop) and the controls of the media can also be shown.

Play a video in the click sequence, either automatically or when clicked.

- Get back to the normal view and select the video frame on the slide.
- Right under the video tools, on the playback tab in the video options group, in the start list displayed, choose one of the options below;

Automatically, the video plays automatically when the slide show is displayed.

When clicked, the video will only play when it is clicked upon.

In the click sequence, the video plays in sequence with the rest of the actions that must have been programmed on the slide, like animation effects. This option does not need a click. The video can be triggered to start with any mechanism that brings about the start of the next action on the slide, like pressing the right arrow key.

Play a full video screen

A video can be played such that it fills up the entire slide when a presentation is being delivered. Based on the resolution of the main video file, it might appear distorted when enlarged. Ensure to always preview the video before adding it to the presentation. This way, if the video is shown as distorted or blurry, adjustments can be made with the use of the undo button (Ctrl + Z) to change the full-screen option.

If the video is set to be shown as full screen and it is also set to start automatically, the video frame can be moved off the slide into the grey area. This way, it will not be shown on the slide or shown briefly just before the video goes full screen.

- Locate the normal view, choose the preferred video frame on the preferred display that should play full screen.
- Right underneath the video tools, on the playback tab, in the video options group, choose the "play full screen" checkbox.

Resize a video

If there is no need for a video to be played full screen, there is an option to resize the video to the preferred dimension.

- Right in the normal view, choose the video that should be adjusted.
- Do one of the following:

To adjust the formatting option, click on any corner sizing handle until a two-headed arrow is displayed, then move the arrow until it fits the frame.

To keep the middle of the video right in the same place, press down on the Ctrl button while moving. By default, this should keep the proportions (have the aspect ratios locked) of the video as it is being moved.

Basically, most videos have a resolution of 680 x 480 and a 4:3 aspect ratio. The default aspect ratio for PowerPoint 2025 presentations is 16:9. If the aspect ratios of the presentations and the video are not the same, the video will appear distorted when it is played.

- To be specific about the height-width proportions of the video,

Right under the video tools option, which is on the format tab in the size group, select the dialog box launcher.

Select size and then, right under the scale, choose the lock aspect ratio check box.

Under the size and rotate options, fill in the sizes in the height and width boxes or under scale in the scale height and scale width boxes that are displayed, and fill in the preferred percentage of size, in tandem with the video's original size.

Preview the video

- In the normal view, choose the video frame.
- Select the play button. Find the play option in the preview group that can be found both on the format and playback tabs under the video tools.

Set the volume of the video

- Right underneath the video tools, on the playback tab in the video options group, select the video option and select one of the following: low, medium, high, or mute.

Make a Video Loop

Make use of the loop feature to play a video in a repeated manner and continuously when a presentation is ongoing. Right under the video tools, on the playback tab in the video options group, click on the loop icon until a stop option is displayed in the checkbox.

After watching a video, you can rewind it.

To rewind the video after it has been played during a presentation, go right under the video tools option on the playback tab in the video options group, then choose the rewind option after playing in the checkbox.

Compressing Media

Compressing images in a Microsoft PowerPoint presentation can help to reduce the overall file size of the presentation while also saving disk space on the device it is saved on.

Open the PowerPoint presentation that has the images that should be compressed and then click on a photo. After this has been done, it will be in the picture format tab automatically. After this, click on the "Compress picture" icon in the adjust group. The compressed pictures window will then be displayed. When in the compressed picture options group, there will be an option to choose if the compression should apply to the selected picture only or not. If this option is unchecked, PowerPoint will compress all the images found in the presentation, and this option overrides any other option that has been set on the images.

CHAPTER 18

MORE THINGS TO INSERT ON YOUR SLIDES

Inserting Tables

The feature of creating tables in PowerPoint is similar to that found in Excel. Even though it doesn't have all the features that Excel has, its tables can also be used for basic functions like creating small sets of data within a slide. In addition, PowerPoint uses a spreadsheet format just like that of Excel to make the data an intuitive part of the slideshows. Although Excel is able to display a lot of data in its own interface, PowerPoint tables are typically very small and can only hold a few records that would fit on just one or two slides.

Standard templates are used for the layout of objects on a new slide. A blank template is one such object, where you can add what you want to the slide. The black template area displays six icon buttons. On the first row you see the insert table button. Simply because it displays an image that resembles that of an Excel spreadsheet, it is very easy to identify. There is an icon for each object that can be inserted. On the right side of the table, for example, is an icon depicting a graph sheet. Upon selecting this icon, the slide will display a graph. In the case of merely inserting a table, click the icon with the table image.

Tables are usually made up of rows and columns. Rows depict the horizontal sections of the table and also contain data for just one record. Columns are basically the fields that makeup one item in a record. If more records are required, use Excel, or take just a few records, use them in the slide show, and have all the other records in the Excel spreadsheet.

Note that the table will be the very same color as that of the presentation theme. The first row is always slightly darker so as to give an indication that it is the first row in the header row. In spreadsheet software such as Excel, the very first row is most often used as the header cell. The header cell shows the column names. These column names are also made to state the type of data that will be displayed in each of the fields.

Customizing a Table

After a table has been created, its look can be tweaked to suit the taste of the user. Once a table has been created, the design tab will be created, and this has lots of menu options where colors and styles can be changed.

The first section, labeled "Table Style Options," defines the header rows and columns. By default, the header row check box is always selected, which then shows a darker color in the first row so as to show that it is the header row. This color can also be added to the first column to give the rows a header section too.

There is another light and dark color in banded rows, making it easier to tell each row apart. In a table with a list of numbers, the Total Row option will place a darker color in the lower portion. Columns can be done in a similar manner. By doing so, a more distinct table can be created.

The style color for the table is determined in the table styles section. Depending on the theme, PowerPoint 2025 selects a color style, but you can change the coloration with any of the available styles. A flat color, a darker color, and a lighter color are some of the color combinations available for placing headers and banded rows.

Creating a table in a content placeholder

The placeholder on a slide is used to represent the containers on the slide, which are shown with a dotted external border.

Placeholders can hold text, pictures, tables, charts, SmartArt graphics, and media clips. Placeholders will help to ensure that the design across an entire presentation is consistent and that the design across the whole presentation is the same to ensure that positioning and formatting can be made the same.

- Open the Home tab located on the Ribbon and then select the New Slide button in the Slides Group to add a slide with a title and content layout.
- Click on **the table icon** in the middle of the content placeholder. This will bring up the insert table dialog box.

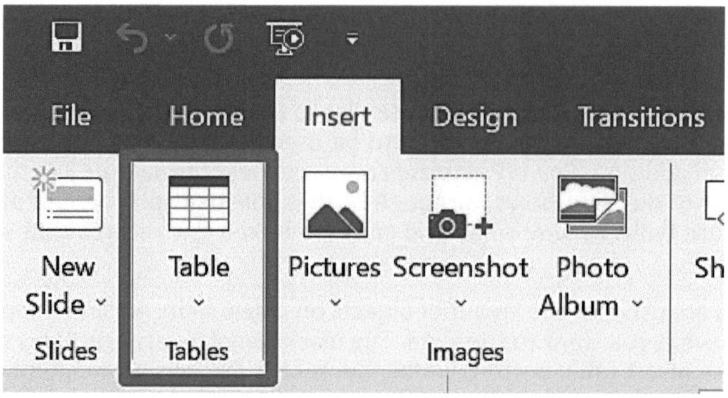

Place the number of columns and rows.

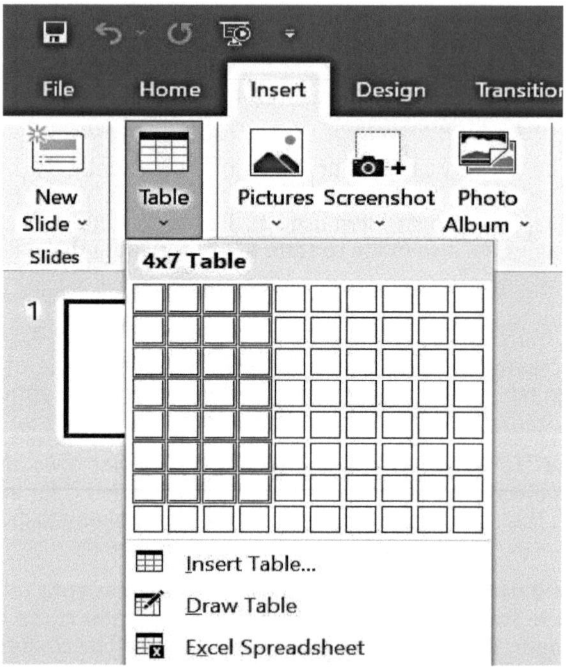

- Type in the needed information into the table cells. This can be done by clicking anywhere on the table and then starting typing. There is also an option to move from cell to cell by clicking on the tab key or the arrow keys.

Drawing a table

The draw table command gives users the option to draw tables that are quite complicated on screen with the use of a very simple set of drawing tools, as the insert option won't fit. This command is necessary for creating tables that are not just a simple grid of rows and columns, but complex conglomerations in which some of the cells span more than one row, and others span more than a column.

To draw a table:

- Locate the Insert tab on the ribbon and then select **the table button** in the tables group and select t**he draw table** from the menu that is displayed. PowerPoint will then change the cursor to a pencil.
- Draw the entire shape of the table by moving the mouse to create a rectangular boundary for the table. When the mouse is set free, a table with a single cell will be created.

- Click on the **"draw table"** button that is located in the table tools design tab. The mouse pointer will change into a much smaller pencil when this button is selected.
- Cut the table into smaller cells by moving the lines across the table. For instance, divide the table into two rows, focus the cursor somewhere around the left edge of the table, and then select and move a line across the table towards the right edge. When the button on the mouse is released, the table will split into two rows. Keep moving the line to cut the table into smaller cells. For each slice, focus the cursor at the edge where the new cell should start and select, then move to the other edge.
- If the size of the line or the style that has been drawn needs to be changed for a particular part, make use of the pen style and pen weight drop down controls in the draw borders group on the table tools design tab. The style of an already drawn line can be changed by tracing over the same line with another style.
- If, for instance, a mistake is made while drawing the table cells, select the eraser (the icon that has the eraser image displayed) button in the draw borders group and then erase the line segment drawn in error.

Adding style to a table

By using the design tab menu under the table tools menu, you can format the table styles in PowerPoint. This menu controls all the styles used in the table, including the cell format, border color, shadow effect, and table background. By using this menu option, you can easily customize a table by using the quick style options, but if you need to use the advanced styles, you can choose from the other available options. If a quick style is preferred, any modern style can be easily applied to a chosen table.

However, if more style options are needed, draw the border section that has the pen color. Note that line properties can be very useful also. To have the border in a PowerPoint table changed, choose the pen color first, and then choose the cells where they should be added, moving over to the borders option. After all of this has been done, select the border style to be used.

Working with the layout tab

This tab contains all the options that can be used to set up the layout of the document pages to your liking. There is a possibility of setting margins, applying themes, as well as controlling page orientation and size. The document can be divided into sections, lines can be broken, line numbers can be displayed, and the paragraph indentation and lines can also be fixed.

To change the layout, arrange the slide content with the different PowerPoint slide layouts as will suit your taste or to help improve the clarity and readability of the content in the slide.

- Choose the slide that the layout should be changed to.
- Choose the Home option, followed by the Layout option.
- Choose the layout that is preferred.

In the layout, there are placeholders for text, pictures, videos, charts, and shapes. There's plenty more. The layouts also have formatting for objects such as theme colors, effects and fonts.

Inserting WordArt

PowerPoint 2025: WordArt formatting can be applied to any portion of text. WordArt is also shown on the insert tab on the Ribbon, which offers a very convenient way to insert a text box with text that has already been formatted with WordArt formatting.

To insert WordArt, follow the steps below:

- Locate the slide that the WordArt should be inserted into, select the **Insert tab** on the ribbon, and then choose the **WordArt button** in the text group. A gallery of WordArt styles will then be displayed.
- Choose the **preferred WordArt style** that should be used. The WordArt text box is then inserted on the slide.

- Select the **WordArt text box** and then enter the preferred text. There is also an option to add any other WordArt formatting as it suits the user.

Using Hyperlinks

A hyperlink, or simply a link, is a reference to data that a user can follow by clicking or tapping. A hyperlink points to an entire document or to a particular element within a document. The text that is linked is called anchor text.

To automatically insert a hyperlink when typing in PowerPoint, type a web address URL or email address into the text that has the object in a slide. Press the spacebar or the Enter key on the keyboard. To take off the inserted hyperlink automatically, press the backspace key once just immediately after pressing the spacebar or the enter key.

As an alternative, hyperlinks can also be added to a slide object in a text that has a slide object. Choose the slide object or text right within the text that has the slide object to attach the link. Then choose the link button in the links button group on the Insert tab of the ribbon. When this has been done, the "Insert Hyperlink" dialog box will be displayed. The right-click selection can also be used. After that, the link command can be tapped in the pop-up menu that is displayed to open the same insert hyperlink dialog box.

The chosen text will be displayed in the text-to-display field of text links. This field displays the selection in the document in the case of linked objects. Choose the ScreenTip button to add a screen tip to the hyperlink that is displayed when the mouse is moved over it. Then you can enter the text to show in the field and then select the OK button to save the screen tip, then return to the insert hyperlink dialog box.

Creating a hyperlink to another slide

As an alternative, to have a hyperlink inserted into a presentation slide, choose the "place in this document" option from the link to list on the left-hand side of the Insert Hyperlink dialog box. Next, select **the slide** that should be linked from the "select a place in this document list", then click on the **OK button** to put in the hyperlink to the slide that has been chosen.

Creating a hyperlink to another presentation or to a website

To insert a hyperlink to a file that is already existing or a webpage, click on the existing file or the web page option in the link to the list located on the left-hand side of the insert hyperlink dialog box.

To have a web page linked to another web page, type the URL of the web page into the address field. Use the Look-in feature or the drop-down menu to find a file or a page that you can link to for searching, or use the current folder, browsed pages, or recent files buttons. After locating the file that should be linked, click on it to have it selected. After creating the hyperlink, select the **OK button** to put the hyperlink into the document.

Adding action buttons

Action buttons can be added to PowerPoint in the slides. These actions ensure users perform the action that is associated with a slide object by selecting or dragging the mouse pointer right over the object.

Most basic actions are linked to slides. Web pages and email addresses, and also run programs. Actions can be added to any slide object also, but they are often associated with action buttons in PowerPoint. PowerPoint has an "action buttons" shape category that helps to insert buttons for basic slide navigation actions.

To insert action buttons into PowerPoint, select **the insert tab** option in the ribbon, then choose the shapes option from the drop-down button in the illustrations button group. Select the action button face to insert from the action buttons category. Finally, select and move the mouse over the part of the slide where the action button should be added. The action button will then be displayed in the exact location when the mouse button is released.

After the action button has been added, the action settings dialog box will then be displayed. This dialog box will be used to set the action that the button will perform. However, for most of the action buttons in PowerPoint, the action has already been programmed. There are basically only two events for which the actions can be programmed: "Mouse click," which is basically used when the user clicks the button, and "Mouse over," which is when the user rolls their mouse over the button.

Choosing button shapes

To choose an action button, follow the steps below:

- Select **the Insert tab**.
- Choose the shapes command in the illustrations group. A drop-down menu will then be displayed with the action buttons placed at the very bottom.
- Choose the preferred action button.
- For Fx insert the button into the slide by choosing the preferred location. The action settings dialog box will then be displayed.
- Choose the mouse click or the mouse over the tab. Choosing this mouse click tab shows that the action button will perform that very action only when it is clicked. Choosing the mouse over the tab means the action button will perform its own action when the mouse is moved over it.
- In the part that has the action on click, choose the Hyperlink to option, and then click on the drop-down arrow and select an option from the menu.
- Finally, check the "play soundbox" if a sound should be played when the action button is selected. Choose a sound from the drop-down menu, or choose the Other sound option to make use of a sound on the computer. When all of this has been done, click on the OK button.

Editing a button:

- Choose the action button.
- Select the Insert tab option.
- In the links group, choose the action command. This will ensure the action settings dialog box is displayed.
- Finally, click on the edit action or hyperlink and then click on the OK button.

Inserting equations

There are times when there might be a need to have a math equation inserted into PowerPoint. If the built-in equation doesn't meet the needs required, the equations can be edited by hand, most importantly those that have very complex symbols and structures so as to meet the needs of the user.

- Locate the **Insert tab** and choose the equation symbols in the group.
- In the drop-down list, a variety of equations will be displayed; select the preferred one.
- The chosen equation will be shown automatically on the active slide.
- The drawing tools and equation tools will be displayed. Choose the design and the symbols and structures can be edited in the formula by using the tools in the structures group.

To input the formula manually:

- Locate the Insert tab and select the **equation in the symbols group.**
- Select the **ink equation** at the lower part of the list.
- The handwriting board will show; just move the mouse so as to enter the formula. If there are continuous errors in inputting data, select the erase button and choose the error part to erase it all. If an identification error also comes up, choose the select and correct button and then choose the symbol that should be adjusted, then choose the correct one in the list that pops up in order to replace it. If there is a need to have the formula entered again, select the "clear" button and everything on the handwriting board will be erased.
- Finally, ensure that all the symbols or structures are right by tapping the insert button at the bottom right-hand corner. The formula will then be displayed on the active slide.

CHAPTER 19

COLLABORATING ON PRESENTATIONS

Understanding Collaborations

Working together with a team to complete a project, there might be a need to pass files back and forth, making use of email addresses or file sharing, and then it becomes pretty hard to stay in sync on the particular project.

The solution to all of this is to make use of PowerPoint in order to stay in sync and work together as a team. Microsoft has included this feature in PowerPoint so as to help people solve the problem of teamwork so that different users can work at the same time.

When collaborating, the first thing that will be done is to have the PowerPoint presentation saved on OneDrive; this way it will be stored in the cloud for collaboration. To save the file to OneDrive, locate the save as option and choose OneDrive. This will have the presentation saved to the cloud, so others can work together with you.

To invite those that will work with you, locate the share button close to the upper right-hand side and tap it to invite others or team members to the presentation.

A new window where others can be added will then be displayed in the PowerPoint presentation. Enter an email address to invite anyone to collaborate on the presentation. Also, you can make a choice on how the people that are invited to collaborate with you will work by selecting between the "can edit" and "can view" options. This way, the control they have over the files can be monitored.

Working with OneDrive

Most of the features in Microsoft Office are tailored towards saving and sharing files online. OneDrive is Microsoft's online storage space that can be used to save, edit, and share presentations and other files. OneDrive can be accessed from a computer, smartphone, or any other device that has access to the internet. To get started with using OneDrive, all that is needed is a Microsoft account. If you don't have one, do sign up for one.

Features of Onedrive

- **Access to files anywhere:** When files are saved on OneDrive, they can be accessed from any computer, tablet, or just about any device that has access to the internet. New presentations can also be created from Onedrive.
- **Backing up files:** Having files saved to Onedrive guarantees an extra level of protection. If, for instance, the computer in use develops a fault, OneDrive will still keep the files safe and they can still be accessed by any other computer.
- **Share files:** It is quite easy for files to be saved on OneDrive with friends and those you work together with. There is also an option to choose their roles, if they can only read the files or if they also have access to edit the files. This option is quite amazing for the sole purpose of collaboration, as lots of people will be able to edit the presentation at the same time, which is also known as co-authoring.

Sharing a OneDrive presentation

When a PowerPoint presentation is saved to OneDrive, your team can collaborate on the presentation at the same time. For this to happen, everyone needs to work with PowerPoint 2010 or a more recent version.

To share a presentation

- Open the PowerPoint presentation and choose the share option in the top right-hand corner of the ribbon when all is set to collaborate with your team. If the presentation has not been saved on OneDrive yet, there will be a prompt that will allow that to be done.
- Locate the send link box and enter the email address of those that will be working on the presentation with you so you can share it with them. If the person's contact info is already stored on your device, all you have to do is enter the person's name. A drop-down list will be displayed and this will help determine if the invitees can edit the file or can only read the file.

- Finally, when all of this has been done, click on the share button. If the file has already been saved to the cloud (OneDrive), an email invitation will be sent to the invitees.

Whenever anyone is viewing or editing the presentation, the person's thumbnail picture will be displayed in the top right-hand corner of the ribbon. PowerPoints will display a prompt when anyone enters or leaves the presentation.

Alternatively, in the share pane section, the name of anyone viewing or editing the document at that point in time will also be displayed. There will also be an indicator on the thumbnail of the slide and on the current slide that will show the exact place where the person is working on the presentation. You can move the mouse over the indicator or select it to find out who is editing the presentation.

For accessibility's sake, PowerPoint also provides the option for the automatic display of the editor's name to be turned on. To turn it on, select file, then options, then click on advanced, choose display, then click on the option to "Show presence flags for selected items."

After all of this has been done, save the file.

Working with Teams

If you are making use of Microsoft Teams to meet with other people that are working together with you but remotely, there are about 7 different options you can use to share PowerPoint slides with them.

- **Share the whole screen or desktop.**

This is the most common method that most people use because it is the closest to what is done in a meeting room when the participants are seated around the table. In teams, select the sharing option known as "Desktop." This will enable the audience to see everything that is shown on the desktop, and if the PowerPoint is not in full-screen mode, they will only see the wallpaper and any other application that is open with just the confidential information being displayed. They will also be able to see the small team window, except it has been minimized.

Once the slide show is started in PowerPoint, all the features will work, including animations and transitions. A presentation remote can equally be used to move through slides, or there is also an option to use the arrow keys. Since the slide takes up the whole screen, the teams' controls won't be visible; this means chats and questions from the audience might be missed. If Alt + Tab is used to switch to the Teams window in order to see the chat discussion, the audience will be able to see this also because they can see everything on the screen.

- **Share the slide show window.**

Due to the fact that Teams provides the option of sharing any window that is open on the computer, another option is to share the window that has the slideshow in it.

Before sharing is started in teams, start the slideshow in PowerPoint first. This will show the slides on the whole screen. Make use of Alt + Tab to move back to the Teams window. Locate the teams' sharing options and choose the window that shows the slideshow (make sure the slideshow is chosen and not the PowerPoint regular window).

Once again, since PowerPoints will be used in full, all features should work, and this is close to presenting in a meeting room before an audience. Since it's just a window that is shared, the audience will not see anything else on the screen. Due to the fact that the slides take up the entire screen, the teams window with chats won't be displayed. However, there is an option to switch to the Teams window with the use of the Alt + Tab buttons so as to check the chat conversation without the audience being able to see what is being done.

- **Share the editing window with a clean look.**

The first two options basically won't provide the option to see other documents or notes that may be on the screen due to the fact that the slides cover the whole screen. If there is no need for animation or transition features and it's okay for the audience to see just the edit view of the PowerPoint, there is an option to make use of a view that reduces the PowerPoint interface. This way, the slide will be the main focus.

In the normal editing view, have the slide thumbnails minimized by moving the vertical divider over to the left until the text showing that thumbnails are now available is displayed. Move the horizontal divider at the lower part of the slide downwards to Options notes that can be found underneath the slides. Click on the collapse indicator and this will collapse the ribbon in the bottom right-hand corner of any ribbon.

Finally, have the slide in the editing window maximized, and PowerPoint will have a much cleaner look than the normal editing view.

Move to the team sharing option and choose the window for this presentation. When moving through the slides, the animations and transitions won't be seen since the slideshow mode is not in use. Any media insert will not run automatically and must be played manually. Make use of the arrow keys or the page down and page up buttons to move through the slides.

- **Run the slideshow in a window and share that window.**

PowerPoint now allows you to run a slide in the window it is located in without taking up the whole screen. Many presenters are not aware of this option. This option is referred to as the "browsed by an individual mode" or the "reading view." To make use of this mode, locate the slideshow on the ribbon and select the setup right-hand slide show button. In the dialog box that is displayed in the show type section towards the left-hand corner, change the option by selecting the radio button for "Browsed by an individual." Choose the OK button to make sure the changes are saved and then leave the dialog box.

The slides will always run in this PowerPoint window instead of the full screen if you are in slideshow mode. The default value of "presented by a speaker" can, however, be changed after a meeting has concluded.

In the team sharing options, select the window for this PowerPoint presentation. Begin the slideshow mode in PowerPoint. This will show the slide show with some other added controls at the upper and lower parts of the window and, if possible, black bars on the upper/lower part of the left/right based on the size of the window and concerning the size of the slides. This can be a slightly different look for the viewers if they are used to seeing the full-screen version of the slideshow mode.

Because this is a slideshow mode, all animations and transitions work as expected. The only difference, apart from the Fullscreen show mode, is that the laser pointer, inking, and some other features will not be available. Since the sharing window is all that is in use, the audience will not see any other documents that may be open on the screen. To see the full teams window, you can select the small teams' window located on the lower right-hand side of the screen, and the audience will not have to see this either. This ensures that you remain in touch with every chat discussion while the presentation is ongoing. Since this option is sharing just one window, it uses less bandwidth than if it were sharing a full screen, and this will help users on lower speed connections have a much better experience with little or no lag or distortions of any form. This option might offer the very best combination of PowerPoints features and presenter control options.

- **Using the PowerPoint sharing option in Teams**

Microsoft has created both Teams and PowerPoint, creating a presentation method that is simply unparalleled among other meeting tools. You can load a PowerPoint file from the SharePoint library or the computer by using the PowerPoint sharing option. The slideshow will be displayed via PowerPoint on the web inside the teams. This option provides some advantages, which include the use of less bandwidth as compared to screen sharing and does not require that PowerPoint be installed on the computer or the use of PowerPoint on the web features, the availability of co-presenting options, and also less usage of battery life as compared to screen sharing.

Though all of these advantages are lovely, the greatest disadvantage is the fact that this method makes use of PowerPoint on the web, which does not offer support for all features of PowerPoint yet.

Due to the fact that this method is infused into Teams, the slideshow occurs in the Teams window, and the Teams control bar at the top of the slideshow can be seen while the presentation is ongoing. You also have access to all the team options in the window. Since the slideshow is in the Teams window, the rest of the screen is also available for other documents that could be referred to.

- **Using the presenter view to show the audience the slides while you see the presenter view**

If the preferred option is to make use of the presenter view to see the notes while the audience sees just the slides.

- **Along with the PowerPoint slides, show a video.**

Microsoft Teams allows you to make presentations with video beside the PowerPoint slides so that facial expressions can be seen easily with the slides. In this case, the whole screen would be shared, with slides and video organized on the desktop. You can control the size of both the slides and the video in this way. Aside from making use of built-in features within PowerPoint and the operating system, this does not necessarily require any software to be installed.

Using Comments

Comments are used mainly to give feedback to colleagues after a presentation. This can help the person involved know what to do as regards the presentation, or it can also be in the form of an appraisal.

CHAPTER 20

EXPORTING YOUR PRESENTATION TO OTHER FORMATS

Creating a PDF file

PowerPoint provides the option of saving presentations as PDF files. Saving your PowerPoint document as a PDF file is a very fast way to produce a PowerPoint file that is ready for reviewing, printing, and also emailing. The PDF will have all the formatting that has been applied retained, whether or not the recipient has the styles, fonts, or themes installed on the computer. Also, PDF files are a secure way to transfer presentations to anyone because they cannot be edited or tweaked. To create a PDF file, open the PowerPoint presentation, choose the file option, and select the Save As option. Find the file format downwards and choose the PDF option, then click on the Save button to have the PDF file created.

Always note that animations and sounds are not always activated in a PDF document. PDF files can only be edited with the use of special software.

Packaging your presentation on a CD

When a PowerPoint presentation is created, the plan is to present it in person from a place different from your office or where it was created. This would require the presentation to be copied on a storage device such as a USB drive or having it burned on a CD. **Note** that there is no need for a CD for each presentation; different presentations can be on a CD.

To package files on CD, follow the steps below:

- Locate the file menu in the ribbon
- Choose export from the options menu.
- Select the package for the CD option and then select the package for the CD command (this will open up the options dialog box).
- The space to enter the name of the CD will pop up. Enter a preferred name for the presentation.
- Keep in mind that the active presentation file has already been added to the list.
- To include more PowerPoint files, click on the "Add" button. This will open a new window where other slides can be selected. Double click on the preferred presentation and then repeat the same process to add other slides.
- Finally, click on the copy to CD option to start the burning process.

Note that PowerPoint will add a special program in the package known as the PowerPoint Viewer, which can show the presentation on a computer that does not have PowerPoint installed on it.

Making a Presentation Available on Microsoft Stream

To export a video option from a PowerPoint presentation. Select a file, then choose the export button, and then choose the "Create a video option." As an alternative, locate the recording tab on the ribbon and select the export to video option.

You can select either the HD (720p) 1280 x 720, medium file size, or the Standard (480p) 852 x 480, the smallest file size. Then, after creating the MP4 file, upload it to the stream.

It should be noted that recorded PowerPoint files are frequently very large, and as such, they will take some time to upload to the stream.

CHAPTER 21

THE TEN DO'S AND DON'TS OF POWERPOINT

Thou wilt frequently save thy work

Always ensure you save your work as you are entering the text so as not to lose information and have to begin again. While working on the presentation, use the Ctrl + S button to save on the fly.

Thou shalt store each presentation in its proper folder

Store all presentations properly and in their appropriate folder. This is so that you will be able to locate the file easily when you need it. Make use of the save as an option whenever there is a need to save any presentation.

Thou shalt not abuse thy programs' formatting features.

Do not abuse any program's formatting feature. Features that cannot be tweaked should be left the same way they are and used that way.

Thou shalt not stealth copyrighted materials

Avoid the use of copyrighted materials in your presentation, either as text or as graphics. Ensure your presentation is as natural as possible.

Thou shalt abideth by thine color scheme, auto layout, and template

Try to make use of the same color scheme and template. Make sure the presentation is focused on passing information rather than being very complicated.

Thou shalt not abuse thine audience with an endless array of cute animations or funny sounds

Keep your audience glued to the screen with the use of relevant information. Avoid the use of animations and funny sounds that might make them distracted.

Keep thy computer gurus happy.

Keep your presentation highly informed; this will make tech professionals really happy.

Thou shalt backeth up thy files day by day

Make it a priority to always back up your files each day, especially on the cloud (Onedrive). This way, you are sure your files are saved.

Thou shalt fear no evil for Ctrl + Z is always with thee

While entering text or graphics, do not be alarmed if a mistake arises and you mistakenly delete some vital information. Make use of the Ctrl + z command to call back any information that might have been lost.

Thou shalt not panic

Once you save your presentation frequently using the Ctrl + S command, you have no need to be frightened or panic.

CHAPTER 22

TIPS FOR CREATING READABLE SLIDES

Try reading the slide from the back of the room.

This way, you'll be able to easily identify errors and make corrections.

Avoid small text.

Be sure to disengage from the use of small text and make use of text with very legible fonts. This way, the audience will be able to view the presentation easily and not have to squint their eyes.

Please limit yourself to five bullets.

For the sake of being precise and to avoid the presentation being overloaded, ensure you only have 5 bullet points at most on each of the slides in the presentation.

Avoid excessive verbiage, leading to excessively lengthy text that is not only redundant but also repetitive and reiterative.

Keep the text in the presentation as simple as possible.

Use consistent wording.

Make use of simple words consistently.

Watch the line endings.

Keep a tab on the line endings. Make sure the words end well.

Avoid unsightly color combinations.

Make sure you combine the colors well. Do not use colors that do not match. This can make the whole presentation outcome rather unbearable.

Use only two levels of bullets.

Avoid too much inconsistency. Do not use more than two levels of bullets; this will make the entire presentation look simple and unique.

Avoid bullets altogether if you can.

If possible, totally avoid the use of bullets. You can choose to use number points rather than bullets.

Keep charts and diagrams simple.

Avoid the use of complex diagrams. Make sure the charts and diagrams employed are simple enough.

CHAPTER 23

TEN WAYS TO KEEP YOUR AUDIENCE AWAKE.

Don't forget your purpose.

Always keep in mind the reason the presentation is being created. Make sure not to deviate from the purpose of creation. All text, graphics, and sound attached must be tailored to achieve this purpose.

Don't become a slave to your slides.

Don't let the appearance of the slide overpower your consideration of the purpose for which the slide is being created. Be the master! Be in charge!

Don't overwhelm your audience with unnecessary details.

Be detailed but specific; do not bore your audience with details that are quite irrelevant.

Don't neglect your opening

Make sure the opening of your presentation is catchy and enterprising; this will keep the audience tuned in and focused.

Be relevant

Ensure your choice of information is relevant to the topic and purpose for which the slide is being created.

Don't forget the call to action.

Always remember to add the call to action where necessary and also at the end of the presentation.

Practice makes perfect.

Keep practicing; you will only become a professional with constant practice.

Relax

Trust yourself and don't panic.

Expect the unexpected.

Keep your mind open; anything can creep up. Be sure and confident that you have everything under control.

Don't be boring.

You don't want your audience to fall asleep while the presentation is ongoing. As much as relevancy is important, ensure your slide isn't boring. Animations in line with the topic and purpose of creating the slide are highly needed.

BOOK 6:
MICROSOFT PUBLISHERS

INTRODUCTION

Getting the perfect design for a particular purpose could be time and money-consuming. Imagine spending hours trying to develop a captivating strategy for your publication. It is energy-draining, but it is also a sure way to spend a lot of money.

But then, imagine if you did not have to struggle to spend so much on that design; imagine if you need not spend so much money on getting the job done; how incredible would that be?

If you did imagine for a second what that would look or feel like, congratulations! You have successfully pictured the Microsoft Publisher.

Publisher allows you to create practically any form of publishing project you want. If you wish, you can build something from the ground up. If you want simplicity and speed, you can choose from hundreds of predefined free themes available within the application. Microsoft Publisher is a desktop publishing program that lets you create professional calendars, flyers, postcards, posters, newsletters, business cards, certifications, and even websites.

Microsoft Publisher is intriguing because it enables you to generate professional-looking content without learning a program designed specifically for designers. It has a wide variety of pre-installed templates and free templates that can be downloaded from within Microsoft Publisher. You can even utilize mail-merge features to send content to a list of customers.

This guide aims to walk the user through the entire process of creating a publication with the Publisher app. You can accomplish a lot with the Publisher app. Everyone may develop and publish their digital content in today's world of do-it-yourselfers, and computer software companies like Microsoft offer applications that make it simple to do so.

You may have basic programs for text documents (Microsoft Word), numerical data (Microsoft Excel), and presentation content (Microsoft PowerPoint) (Microsoft PowerPoint).

Microsoft Publisher is a program for creating professional print publications. Anyone can utilize the application's desktop publishing tools, both economic and straightforward. As a result, it's an excellent choice for small enterprises, ordinary users, educators, parents, or freelancers who need to develop publications but lack graphic design skills. They can use the tool to produce professional-looking branding and sales materials.

This guide will give you a greater understanding of the program, what it does, how to use it, and some resources to help you get started.

What is the purpose of Microsoft Publisher?

You can operate as simply with Microsoft Publisher as you can with Microsoft Word, but instead of text editing tools, you'll be modifying page layouts and graphic material.

You might be wondering what Microsoft Publisher can accomplish for you.

The answer is practically everything you'll need to design and publish your materials correctly.

- Personalized birthday cards
- Event posters
- Newsletters for small businesses and organizations
- Professional business cards are just a few examples.
- Programs and flyers
- Brochures and postcards

That is, of course, merely the tip of the iceberg. MS Publisher shines when it comes to editing and personalizing that content. You can start from scratch, use templates, or use your already existing content.

You will discover after this study that you have gained a new skill simply by depending on and learning from the jewels provided in this guide. Even if this is your first time designing and publishing a document, I believe you are in the correct spot and have nothing to fear. You don't have to be an expert. This tutorial will provide you with enough ammunition to accomplish any design or publishing in the future. With that said, sit back and relax as we get down to business.

CHAPTER ONE

WHAT MICROSOFT PUBLISHER IS ALL ABOUT

One may be wondering what Microsoft publisher means, considering we already have Microsoft Word, which enables you to get your writing done. The difference lies in their functions, as we will see in the latter parts of this study.

Microsoft publisher is a publishing application on a desktop. Microsoft publisher is focused on the design and art of the written content. Just like the name implies, it is a tool used by small businesses or private individuals who might want to get handy and make designs themselves, as they may not have professionals at hand

From its initial release in 1991 and its subsequent series of upgrades through the years, Microsoft publisher, which is included in higher-end editions of Microsoft Office, was developed to create visually rich and professionally looking publications without having to expend money and time in a complicated desktop publishing application.

It is fantastic to know that you can make simple things like greeting cards, labels, flyers and programs, personalized birthday cards, professional business cards, and even more complex stuff like professional email newsletters, postcards, brochures, yearbooks, and catalogs.

Microsoft publisher is, however, only included in some versions of Microsoft.

Difference Between Publisher and Word

Both Microsoft Publisher and Microsoft word are quite capable software programs in and of themselves. While there are numerous similarities between the two applications, they each serve a distinct purpose. The distinctions between Microsoft word and Microsoft publisher will be discussed in this section.

Each place's importance on content and design distinguishes Microsoft word from Microsoft's publisher. Microsoft word puts content first, while Microsoft publisher puts design and layout first. Both software programs provide an immense number of feature-rich possibilities.

Both Microsoft publishers and Microsoft word have an enormous influence on today's computer users. In this essay, the purpose of each program and what it excels at will be discussed. Then with the help of this article, you will find out the distinction between the two to assist you in deciding which one to use on your next job.

In-depth Look into What Microsoft publisher is

Microsoft Publisher is a desktop publishing tool that allows you to make professional calendars, flyers, postcards, posters, newsletters, cards, certifications, and even websites.

What makes Microsoft Publisher so appealing is that it allows you to create professional-looking content without learning a program designed exclusively for designers. It comes with a large number of pre-installed templates, as well as additional templates that can be downloaded for free from within Microsoft Publisher. You may even use it to deliver articles to a list of consumers using mail-merge features.

You can develop almost any type of publishing project you wish. You can construct something from the ground up if you so desire. You can also choose one of the hundreds of predefined free themes accessible within the program if you prefer simplicity and quickness.

What purpose did Microsoft Publisher serve? It, too, made the playing field more level. Design firms no longer enslaved businesses. People didn't have to pay a lot to receive high-quality brochures, flyers, or other marketing materials.

History of Windows Publisher

On September 16, 1991, Microsoft Publisher was initially announced. It was created with the novice computer user in mind. Microsoft Publisher was built primarily for folks who did not want to spend a lot of time messing around with their computers, and Windows 3.0 had just been released.

Microsoft Publisher is an excellent low-cost option for anyone who wants to create something that appears professional while still saving money.

The Uniqueness of Windows Publisher

Microsoft Publisher's primary goal is to make the content seem good.

If Microsoft Word's goal is to focus on content, Microsoft Publisher's goal is to make that content appear attractive, and it succeeds.

"Content Sells," you may have heard, and although that is true, there is something to be said about how that content is delivered.

Microsoft Publisher's most exciting feature has to be this. You can make what you say look beautiful with minimal effort... literally. People are drawn to high-quality eye candy, and they will read what is offered if presented appealingly.

Consider the last time you received a flyer. If you took it from their hands, the only thing that would entice you to read what was written on it would be its attractiveness. If it were merely printed on white paper with no pictures, images, text placement, it would be a bore to read. However, if the text was strategically put, with color and visuals in just the right places, there's a good chance you would read it.

The concept of "Page Wizards" was initially presented by Microsoft Publisher. Page Wizards were created after Microsoft noticed that many users lacked the requisite design abilities to create a visually appealing page. In reality, software engineers felt dissatisfied when they made tools to perform the job, but customers didn't know how to utilize them. The software engineers came up with the idea of employing Wizards to help users jump-start their project, essentially doing a lot of the design work.

Professional design firms still scoff at using Microsoft Publisher to develop any project, but it's not about the tool; it's about making quality design accessible to the average user who has no idea what good design is.

Today, Microsoft makes software available to everybody, regardless of whether or not they are professionals in that sector. This is why you'll find what amounts to "Page Wizards" in every Microsoft Office program now available. This is the philosophy driving every Microsoft Office product, which has allowed individuals to accomplish more than they previously imagined possible.

Varieties of Publications One Can Create Using Microsoft Publisher

One of the things I like about Microsoft Publisher is that it can be used at home and in enterprises. The Real Estate industry, out of all the industries that utilize Microsoft Publisher, is the one that has benefited from it the most.

One can use Microsoft Publisher to make business cards and flyers for their open houses. They can even design high-end, professionally designed brochures without paying a professional design agency a dollar.

Real estate agents are not computer nerds, and they don't have to be to get the most out of Publisher. They merely need to turn on their computer, open Microsoft Publisher, choose a layout that they believe will work well for their potential client, and click PRINT. Microsoft Publisher is cost-effective and facilitates providing high-quality projects by any standard.

Another feature of Microsoft Publisher that I appreciate is that it is not limited to business use. You can use Microsoft Publisher to make flyers to advertise a local yard sale or to put images of your lost pet on telephone poles with your phone number that can be quickly removed by tearing the flyer off.

Students can use Microsoft Publisher to create school projects that look fantastic. The list goes on and on, and your only limitation is your creativity. Anything you can envision, Microsoft Publisher probably has a template for it, and if it doesn't, you can create it from scratch with Microsoft Publisher.

Microsoft Publisher and Microsoft Word have a lot of similarities. The built-in templates available to you are perhaps the most evident. The built-in assistance system is another crucial aspect that is sometimes ignored when comparing the two.

If you're unsure how to perform something, write your question into the search box, and each product will respond with suggestions. Each product's help system is fantastic, and you should check it out if you haven't already.

The most straightforward approach to distinguish Microsoft Publisher in Microsoft Word is to demonstrate which software you would use in a specific situation. I've listed various projects that I'd use in both Microsoft Publisher and Microsoft Word below.

For instance, I'd use Microsoft Publisher to complete the following tasks:

- A printed calendar
- A flyer
- A poster
- Banners
- Business cards
- Brochures
- Greeting cards

Each of the topics described above is focused on the visual and design aspects.

There are several things you can do in both Microsoft Publisher and Microsoft Word, but that does not imply you should.

You could, for example, make a business card in Microsoft Word, but it would be wiser to do so in Microsoft Publisher because the focus of this project is on design. And, as I previously indicated, when it comes to design, Microsoft Publisher should be your go-to software solution.

Similarly, you could write a newsletter in Microsoft Publisher, but it would be quicker to do so in Microsoft Word because the project's focus is on content. And, as I previously indicated, when it comes to content, Microsoft Word should be your go-to software solution.

On Your Next Project, Which Application Should You Use?

How do you know which software solution is best for the project you're working on? When choosing the tool to employ on your next project, start by asking yourself, "What is the primary purpose for this project?" If it's all about the content, Microsoft Word is the way to go. If the style and layout are more critical, Microsoft Publisher is the way to go.

CHAPTER TWO

THE ESSENTIALS OF MICROSOFT PUBLISHER

This section will show you how to use Microsoft Publisher to assist build a wide variety of desktop publishing tasks. This section evaluates the layout, tool, and how to include content that can be employed to prepare newsletters, brochures, and more.

The Publisher Ribbon

After creating a new publication, you need to become more familiar with the rest of the publisher interface.

Microsoft introduced Ribbon as a user interface feature in Microsoft Office 2007. It's next to the Quick Access Toolbar and below the Title Bar. Home, Insert, Page Layout, References, Mailing, Review, and View are the seven tabs. Each tab contains a subset of related instructions. It provides easy access to the most regularly used commands you'll need to finish a task.

We are therefore going to start from the top of the screen and work our way down to understand the components of the interface better. Firstly, let us head towards the left-hand corner of the screen, where you will see the quick access toolbar.

The quick access toolbar is also present in other Microsoft Office products, so it might not be a stranger to you. It contains icons for frequently performed commands within the application, that is, commands that you frequently perform. It is customizable, but by default, the Save, Undo, Redo, and Print preview icons are available.

If you want to add additional icons to this toolbar to save time in the fitting, you can click the down arrow next to the Print preview button.

You will see a drop-down list of other popular commands; next to each, there will be checkboxes.

If you want to remove any of those, you will select them from this list. It will remove the checkmark, and it will also remove it from the quick access toolbar.

If you are looking for a command not displayed on the list, select the **More Commands** option.

Selecting that will open up the publisher options dialogue box. Within this area, you can search through the commands listed on the left-hand side. By default, the publisher will display a long list of popular controls. However, if you cannot find the command or commands you are looking for, you can select **all commands** from the choose **commands** drop-down menu.

From this alphabetical list, find the command you would like to add, select it and click the **Add** button. This moves the command icon to the list on the right-hand side.

Reordering The Commands

You can reorder the commands by selecting the order and using the **up, down** arrows located at the right-hand side of the dialogue box.

You can reposition these commands to be wherever you like them to be. It is all a matter of your personal preference. Then, you will repeat these steps until you have all the desired icons for all of the frequently performed commands that you have been moved over to this list on the right-hand side.

When you are finished, you can come down to the bottom of the dialogue box and click **OK.**

Just below the quick access, you will see the ribbon, which is made up of tabs such as:

- Home
- Insert
- Page design
- Mailings
- Review
- View

Within each tab, you will notice that your command icons are grouped, and those groups are labeled such as

- Clipboard
- Font
- Paragraph
- Styles
- Objects
- Arrange
- Editing

If you check the tabs on the ribbon, you can see that the available commands have changed. For example, the Home tab has various formatting choices for text and the ability to insert objects such as text boxes, images, tables, and shapes. You also can organize these choices within your publication by using some of the commands displayed in the arrange group. Editing groups help you find specific text or formats in your document. The replace feature allows you to recognize a particular text or format and replace it with another.

The commands for adding elements to a document can be found on the Insert tab. You can add pages and catalog pages to your collection. You can change the orientation and size of the page by clicking the Page Design tab. This page also contains commands for customizing your website's color scheme, font style, and background. On the mailing page, you can run a form letter to create a magazine with an address or an online catalog with product information.

Spell checking, Internet dictionary lookups, thesaurus usage, and text translations into other languages are available on the **Checks tab.** In this view, you can show the master page and show or hide page features such as scratch areas and baselines. Even so, the View tab offers a variety of choices for customizing your document's zoom level and handling multiple open publication windows.

The Pages Pane

The page's pane is likewise seen underneath the ribbon at the left-hand aspect of the screen. Selecting the left arrow will disguise this pane.

Or you may pick to show it once more via way of means of choosing the proper arrow, that is withinside the equal role because the previously gift left arrow.

If you need to resize the pane, you may hover your mouse over the proper part till you spot a double left and proper arrow.

Afterward, you could drag your mouse left or right and notice that it resizes the pages panel in reaction to your movement.

Page thumbnails will seem withinside the Pages window, permitting you to hastily flow among all the pages withinside the e-book through clicking on them.

The Work Space

In the center of the screen, you`ll observe a workspace wherein your clean web page is shown. The scratch location refers to the extra area surrounding the clean web page. Objects like photographs may be saved right here till you are equipped to encompass them into the very last product. It's due to this that once you turn at the scratch location, this stuff can be displayed on this clean area across the pages, irrespective of what web page you are on to your book.

This white area will continue to be seen even after you switch off the scratch location. However, you may now not be capable of seeing your objects in this location.

To navigate round to your e-book, use the proper-hand aspect of the screen, right away underneath the e-book location, wherein you may locate scroll bars.

The Status Bar

Then test the status bar at the bottom of the screen. On the left side, you will see where the 1/1 web page is written.

Adding pages to your publication and navigating each web page will help you remember where you are in your publication. Click Web Page 1/1 to cover the aspect window. Click again to see it again.

this is a seamless way to open and bring this place closer as needed. Note that when you move the cursor in the document, the exact function of the cursor is displayed in the reputation bar, the position of the item function is displayed in the status bar.

Click a small XY image or item length to display a dimension box.

The measurement field helps to display the measured value at the exact position of the selected item in the publication.

On the right-hand side of the status bar, you will notice two icons for viewing the document.

The first one on the left is the single page view which is the default. And the one on the right is the two-page spread view.

In addition to these icons, you may also see a zoom slider. Use the slider to move the slider to the left, or use the to zoom in or out on the eBook publication.

The percentage of your zoom is displayed to the right.

Let us also discuss the **show whole page** icon. This allows you to display the entire page within the screen automatically.

Whenever you need to remove a selection from the reputation bar, remember to click the reputation bar appropriately and select the selection you need to remove.

In this case, you may see a checkmark in the entire lot and on the left side of each selection.

CHAPTER THREE

FAMILIARISING WITH THE RIBBON TABS IN MICROSOFT PUBLISHER

The File Tab

The File tab is a component of the Office Ribbon in Microsoft Publisher and other Microsoft Office programs that offers you access to file functionality. The Open, Save, Close, Properties, and Recent file options, for example, are all accessible from the File tab.

The Home Tab

The Home tab is the default tab in Microsoft Office products such as Microsoft Publisher, Microsoft Word, Microsoft PowerPoint, and others. The Home tab in Microsoft Word is seen in this illustration.

Characteristics of the Home tab

Users can access a variety of functions using the Microsoft Publisher Home page, including the ones listed below.

- Change the font's size, color, and type.
- Make text vivid or underlined.
- Change the color of the text highlighting.
- Aligns the text on the page to be left-aligned, centered, or right-aligned.
- Contains bulleted or numbered lists.
- You can increase or decrease the space between lines and paragraphs.
- Indentation can be scaled up or down.
- Frames can be added, edited, or deleted via text, text fields, and tables.
- You can add or change the title type.
- Draw a horizontal line in the center of the page.
- Replace or search for text.

The Page Design tab

The web page layout tab is the primary vicinity to move in deciding on a background in your publication.

On clicking at the web page layout tab, it opens up the web page layout ribbon, and a number of alternatives open as much as assist you to adjust what the layout web page appears like.

The web page layout tab lets in you extrude the residences of a web page such as; its margins, its orientation, and its size. In addition to this option, this tab consists of instructions associated with color schemes, font schemes, and additionally converting the background of your page.

Changing The Background in Publisher

- In changing or applying a new background, all you need to do is click on the **background** option at the right-hand side of the ribbon.
- This opens up a variety of different options which you will pick from. If you choose to, select a **solid background** or a **gradient background,** better still, you could select the **more background** option if you do not want a plain solid background.
- After selecting the **more background** option, this opens up the **format background** dialogue with a few more options.
- Within the **Picture or texture fill** option, there you should see a button giving you the chance to insert a picture.
- Go ahead and click on that.
- After selecting that, the **insert pictures** dialogue opens up
- You could select an option that fits your preference. In this case, we will go with stock images.
- This opens up the **stock images** dialogue
- Select an image that best fits your product
- Click on **insert**

- This calls you backwards to the **format background** dialogue. Since we already have the image selected, click **OK**.
- You now have a beautiful background on your flyer

The Insert Tab

To insert a picture into the publication you are making, the insert tab is the way to go.

The **insert tab** gives options of commands for inserting objects within the publication you are developing. Asides from inserting objects, you also have the freedom to insert pages or catalog pages too.

On selecting the **insert tab,** you will find within it, all sorts of different items that you could bring into your publication.

You can insert:

- Tables
- Pictures
- Shapes
- Calendars
- Borders & Accents
- Advertisements

Inserting and Adjusting Text

- To add text, you can go to the **home tab.** Within the home tab, you will find an option to draw a text box.
- Click on that. And from there, you will notice that you can now place your text box wherever you want the text to appear on your design. It is also resizable.
- Type in your text into the text box. You can choose to increase the text, change the font, adjust the color.

Mailings Tab

The **mailings Tab** embodies the **mail merge** tool. The Microsoft Publisher mail merge tool allows you to send bulk email messages to many subscribers in a contact list, such as e-newsletters. You must first construct an address list in Publisher before using the mail merge wizard to mail merge your publication.

What you can do with **mail merge** is that you can use it to send out a holiday card or greeting card. You can use mail merge to customize every single card. For example, you have a hundred different friends that all have different names, you can customize the cards with their names on them.

It greatly simplifies creating customized messages for a large number of people.

The Review Tab

It is Ribbon's sixth tab. This page provides you with several useful commands for modifying your document. It assists you in proofreading your text, adding or removing comments, tracking changes, and so on. Proofing, Comments, Tracking, Changes, Compare, and Protect are the six groupings of linked commands on the Review tab.

The review tab includes commands for spell checking, accessing the web for dictionaries, using the thesaurus, and translating material into other languages.

The View Tab

The View tab is next to the Review tab on the toolbar. You can toggle between Single Page and Two Page displays using this tab. You can also use it to manipulate layout tools like limits, guides, and rulers. Its main goal is to provide you with a variety of methods to view your content. Document Views, Show/Hide, Zoom, Window, and Macros are the five groupings of related commands on the View tab.

When we select the view tab, we can examine master pages, as well as reveal or hide items on the page such as guidelines, rulers, the scratch area, and baselines. You can adjust how you view your document with the zoom feature, as well as how to handle numerous publication windows that are open at the same time, all from the View tab.

CHAPTER FOUR

PUBLICATION TYPES

The publisher has a large number of templates for practically any sort of publishing, including brochures, newsletters, and greeting cards. If you can't find a design that you like, you can always change one to fit your needs or start from scratch.

After launching the publisher application, you will land on the **start page**. Here, you will find the **Home, New,** and **Open Views.**

Within the **home view,** you will find the most popular actions that you will take when you first start using Publisher.

- The first option gives you the chance to launch into a blank document
- The second gives you the opportunity to launch into a blank Publisher file.
- By selecting the next option, you can select more blank sizes.
- If you move towards the right-hand of the screen, there, you would discover the **more templates** option. This option will let you into a window of a broader set of templates.

Select the **New view option** at the bottom of the **home view** by the left side corner of the screen will give you an even more comprehensive view of all the different types of new files that you can create.

It shows you a more comprehensive set of templates. Using templates is a very easy way to start with Publisher. It could turn out that you find the exact template for each task.

Also, you have the liberty to look for different templates or still, use suggested searches to find templates.

The last action icon on the start page is the Open View. When you click on Open, this will give you the freedom to navigate to OneDrive, through your PC to locate the Publisher files you might have created in the past.

The Pages Pane

In chapter two, we previously discussed the pages pane which is located at the bottom of the ribbon on the left-hand side of the screen. By clicking on the left arrow, you can hide this pane.

The Page Navigation window lets you to view and scroll through your publication's pages. It also has options for adding, moving, and deleting pages. To access the Page Navigation pane, go to the Ribbon's View tab and then to the Show group. The Page Navigation check box should be selected.

Using Text in Publisher

You'll have to change the text you type into Publisher in order to make it fit your publication. While most of Publisher's text capabilities are similar to those found in other Office programs, a handful is tailored to Publisher's specific publishing needs.

Using Text Boxes

In Publisher, text is included in text boxes, blocks of text that you might choose to put anywhere in the publication. The Text Box Tools tab appears on the Ribbon when you create or select a text box. This tab offers commands for adjusting and formatting your text box and its contents.

To Insert a Text Box:

- Select the Insert tab on the Ribbon, then the Text group.
- Use the Draw Text Box command to create a text box.
- The cursor will become a pair of crosshairs. To construct the text box, click anywhere on your publication and drag your mouse.
- Now you can begin entering into the text area.

Adding Images from Your Computer into the Publication

Images are a terrific way to spice up your publication's aesthetic appeal. Adding images to your publication may be necessary depending on the type of publication you're making. A product brochure, for example, would not be complete without at least one image of the product. Publisher's photo features make it simple to insert and alter images in your papers.

You'll discover how to insert, fit, and format photographs, as well as clip art, in this session. You'll also learn how to prepare your photographs for publishing and how to make advanced modifications to your photographs using other Office products.

- Select the **Insert tab** then find the **illustrations** group.
- Select the **picture** option
- Locate the picture you want to use.
- Then, select the **insert** button

Adding Images to Your Publication from The Web

- You could again use the built-in function called Insert Online Pictures under the Insert Illustration group in Publisher to insert online images, pictures, or Clip Arts.
- Locate the **Online Pictures** button on the **Insert tab.**
- To find an image or Clip Art, start typing a description of what you are searching for, such as "cookie."
- Select the image you would love to use, then click the insert button.
- You can now alter the image after inserting it.
  Color Schemes
  Colors that complement each other, alongside those that give contrast, will most likely make up your color scheme. Multiple schemes can be attained and kept for usage in a wide range of publications.
  Creating A Color Scheme
- Create a publication or open an already created one
- On the Publisher task pane, locate **color schemes**
- Below the task pane, select **create a new color scheme**
- In the scheme color section, from the new pull-down lists, select the desired color.
- Select/type a name of your choice for your color scheme and input it into the **color scheme name box.**
- Select **save**

Applying A Color Scheme

The publisher usually brings with it, several color schemes pre-installed. These color schemes have an impact on your document's text, graphics, and shapes. You can choose to use one of these or better still, design your custom scheme. Custom color schemes are organized alphabetically in their area of the list, according to the name you gave them.

- Create or open a publication
- On the Publisher task pane, locate **colour schemes**
- From the options of color schemes, select your preferred color scheme

Inserting Shapes and Borders

One great way to induce some flavor and interest in your publication is by applying shapes. There are numerous shapes available for your use and you have the freedom of customizing them using your colors.

- To get started, locate the **Insert tab** and select it.
- After that, you would have successfully opened it and you will have to select the **shapes** option
- Scroll through the plethora of shapes to find a fitting one.
- Select the shape you intend to use.
- At this point, after selecting, you can begin to move your cursor to fix the size and location of the shape.

Borders

Adding A Border to A Master Page

- Select **Page design** and click on **master pages**

- Select your master page, and locate **Edit Master Pages**
- Click **home,** select **shapes.** Getting to the **basic shapes** section, there you can select a **rectangle**
- Place your cursor on the page and drag to draw the page border
- Right-click on the border, next, click on **format auto shape**
- Locate the **color s and** the **lines** tab then choose a color and line type, next, select **OK.**
- On the master's page tab, select the option titled close master page.

How to Add A Line Border Unto A Single Page

- With a page selected, click **home.** Locate **settings** and click.
- In the **basic shapes** category, select the **rectangle.** Next, drag to bring the page border. Take the rectangle to the rear if there are subsisting elements on the page, such as pictures or text boxes, by choosing the rectangle and clicking Send Backward > Send to Back. Everything will be piled up on top of the rectangle as a consequence of this.
- On the border, do a right-click next, click on **format auto shape.**
- Select a color and line type from the **Color s** and **Lines tab**, next, select **OK.**

How to Add A Pre-designed Pattern Border To A Page

- Select the page and then go to **Home > Shapes**.
- Click the Rectangle Button pictures located within the Basic Shapes area, next, pull to bring out the page border.
- Right-click the periphery and select **Format Auto-shape** from the menu.
- Then select BorderArt from the **Color s, Lines tab.**
- Select the desired border from the ready to use Borders list, then select **OK.**

How to Create A Custom Border

Include a custom border made from an image file, an image that was scanned, or a drawing software picture for more flair. Publisher border art will be utilized to store your custom border.

- Click **BorderArt**, then Create Custom in the dialogue box that displays.
- Select a picture by selecting it with the cursor.
- Locate the image you would like to use in the **Insert Pictures** dialogue box, whether it be on your system or using Bing's Images search.
- Next, click on **Insert** after selecting the image.
- Locate the **Name Custom Border** dialogue box and name your custom border, then click **OK.**

Creating Catalogs In Publisher

Beginning with a catalog publication that has already been designed.

Microsoft Publisher allows you to create catalogs in two ways:

If you only want to list a few things in your catalog, you may choose to begin by choosing one of Publisher's pre-existing catalog publications. (Select Print Publication then Catalogs, then select the design you wish to use in the New Publication task pane.)

You could include your text and portrayals to the placeholder content and graphics in the pre-existing publishing catalog. If you like, you may change the page layout, font scheme, and color palette of your catalog to make it look different.

Build integrated pages for a catalog with the Mail and Catalog Merge Wizard.

You can construct catalog pages by utilizing catalog merge to unify the documents to a version if you wish to develop a huge catalog that displays numerous elements and you keep records of these items in a spreadsheet, table, database, or other data file.

You may choose to mix several records from a data source into one publication using catalog merge. Catalog merge can be used to make pages befitting any project with one or more items per page, such as a catalog, directory, or photo album. Text and images can be combined in your publication.

Start, using a fresh empty one-page publication to develop unified pages utilizing catalog merge. This document will be used to create your catalog merge template. You arrange the setup for the merged data in your catalog merge template.

You structure the pattern for the merged data in your last publication pages in your catalog merge template. In the catalog merge region of your catalog merge template, you may put merge fields as placeholders for the input in your data source. When you finish designing a merge, the data from your data source populates these merge fields, and the catalog merge area appears again and again in the merged publication pages until all of the records from your data source are displayed.

When you've finished a merge, you can choose whether to add the merged pages to the end of an existing publication or keep them as a new one. After that, you can update these pages just like several similar Publisher publications. When you edit the details in your data, you can store your catalog merge template and utilize it in the production of subsequent sets of merged pages.

Using the Mail and Catalog Merge Wizard With a pre-designed catalog publication,

You can choose to use the cover design from a previously designed catalog publication if you intend to use the Mail and Catalog Merge Wizard to produce substance for your catalog.

To start, select a pre-designed catalog publication and construct a beginning and ending cover befitting the catalog. Save the publication and close it after deleting all but the first and last pages of the catalog.

Make sure your catalog merge template has an identical page size and similar page view (which is usually a two-page spread) as the cover of the catalog when you build it. Select the option- **Booklet** under Publication type in the Page Setup dialogue box to accomplish this (File menu, Page Setup command).

When you're finished with your merge, choose the option which states: **Add to an existing publication** from the drop-down menu and choose your saved catalog cover publication. The combined pages are added to the top of the cover publication by the publisher. Select one page on the page sorter and pull it to the desired spot in order for you to be able to restructure the pages, transfer the back cover page to the stop of your project.

CHAPTER FIVE

FORMATTING IN PUBLISHER

Page Layout

In Publisher, you can fix the page size, paper size, and page orientation. The region of your publication is referred to as page size. The paper size used for printing is the paper size or sheet size. The portrait (vertical) or landscape (horizontal) layout is oriented.

The **Page size,** size of the **paper,** and the orientation, whether it be **landscape/portrait,** are **all factors to consider.**

Any publication style you choose in Publisher contains the most common orientation and size of the page for that type of project. In the menu where you find Page Setup of the Page Design tab, you may adjust the size of the page and orientation and preview your changes. In the Backstage View's Print tab, you can adjust the paper size and direction, as well as preview your changes and the relationship between the page size and the paper size. See Print for further information on the Print tab.

You may manage the page's location within the paper and evade mismatched prints by altering the magnitude or structure of your page and document individually. You have the freedom too to print on a document that is bigger in comparison to your publication and trim it to the completed size to generate a bleed off the border of the page.

Sizing and Orientation of Individual pages, papers

Single pages in a multi-page publication cannot change the size and orientation of pages and papers. To make a multi-page publication with varied singular pages, you initially have to produce various publications for every distinct paper size, page size, or orientation, after which you manually compile the printed publication.

Choosing a Page Size

The magnitude of your publication is discerned with this technique. For example, irrespective of whether your printer can deal with a sole enormous paper or whether you will publish the poster as a series of sheets of paper existing side-by-side, you can use this method to select the size of the printed publication to print a poster that measures 22 inches by 34 inches when printed (called tiles).

- Select the **page design** tab
- Select Size from the Page Setup group and click the icon that indicates the page size you want.

For example, select Letter (Portrait), which measures 8.5 x 11" from the drop-down menu. If you don't find the size you're looking for, select More Preset Page Sizes or build New Page Size to make your own.

Changing the Paper Size

- The paper sizes you can publish on are discerned by the printer you're using. Consult your printer's manual or look at the dimensions of paper that are presently designated for your printer in the Print Setup dialogue box to see what sizes your printer can publish on.
- Ensure that the sizes of the page and paper are identical if you want to publish your edition on sheets of paper that fit the page size. Alter simply the paper size if you need to print your publication on a different dimension of writing — for example, to build a bleed or to print several pages on a single sheet.
- Select **Print Setup** from the File menu.
- Select the desired paper size from the Size list in the **Print Setup** dialogue box's Paper section.

Change the page's orientation.

Regarding the page orientation, you can switch between Portrait and Landscape.

- Select the Page Design tab from the drop-down menu.
- Select Portrait or Landscape from the Orientation drop-down box in the Page Setup group.

Formatting Master Pages

- On the ribbon menu, click on **page design**
- Next, go towards the top right of the mesh and select **master pages**
- Type one character in the Page ID (1 character) text field to represent this master page. A-Z, 1-9, AS AN EXAMPLE
- Type a more descriptive name for this master page in the Description text box.
- ***OPTIONAL***: Select a double-page master to make this master a double-page master.

NOTE: It is sometimes necessary to have different backgrounds for the left and right pages. When making a book, for example, all left pages might have the edition's name, and all right pages might have the label of the chapter.

- Select **OK**.
- A new master page is displayed.
- Fill in the blanks with every text and graphic element displayed on this master page.
- Select **Master View** OR **Deselect Master View** from the View menu.

Guide Lines and Grid Lines

In this section, we will be looking at the layout guides. As seen on your screen, the margin guidelines are blue.

From the **view tab,** you can choose how you view them. By locating the **boundary** option, you can select/deselect the **option of the guide** beneath it to get rid of the margin guides or place them.

You can also decide to select/deselect their **rulers** if you need them to emerge or not.

There is also the **baseline** option within the **view** tab.

Nevertheless, if you wish to adjust the positions of the above properties, you have to do that in the **page design tab**.

- After selecting the **page design tab,** choose the **option of the guide**
- This opens up a set of options that you can select from
- You can decide on getting rid of ruler guides if you set that up, and you have the liberty to choose built-in ones.
- Nonetheless, if you wish to alter or put it in manually, you can select the **page design tab,** find **guides,** and on the dialogue box which occurs, scroll down to click on-grid **and baseline guides**
- This will reveal a set of boxes that will enable you to set up the grid lines manually
- Adjust your margin guide by setting it on a two-page master or not, according to your preference.
- Locate **grid guidelines** to set up the columns and rows
- After setting up, click **OK.**

Formatting Text in Publisher

The publisher comes with several typography commands that may enhance your writing. Although Publisher's makers highlight this as a critical feature, it's worth noting that many of these effects are limited to a few fonts, like Calibri, Cambria, and Gabriola. Even so, if you're using these fonts, the typography commands can help your text look better.

The Text Box Tools tab contains typography controls. Choose your text and then click the desired command to apply it. Specific commands, such as Stylistic Sets, have options drop-down menu.

- Select the **Format tab.**

There are six typographic commands in Publisher, as shown below in the illustration:

- The introductory letter of the specified text is enlarged using **Drop Cap.**
- **Number Style** enables you to select from four different number spacing and alignment techniques.
- **Ligatures** are attached letters to make it easier to read specific mixtures of letters.

- **Stylistic Sets** enable you to select from various font decorations, usually in the form of exaggerated serifs or flourishes.
- **Swash** is a decorative treatment for capital letters.
- **Stylistic Alternates** are letter variations that are ready in a mixture of styles.

These text formats play a massive role in beautifying and arranging your text.

CHAPTER SIX

SAVING, SHARING, AND EXPORTING YOUR PUBLICATION

How To Save Your Publication

- To conserve your work, go up to the left-hand of the mesh and select the **File tab.**
- This opens up what is called **backstage.**
- Within that screen, click on **save as.**

This opens up the **save as** menu, and you have the liberty to choose any location you would prefer to insert it in.

- After selecting a location, type in a record title and choose **save**

Your file has been saved.

How To Share Your Publication

- To share your work, go up to the left-hand of the screen and click on the **File tab.**
- This opens up what is called **backstage.**
- Within that screen, select **share**
- This opens up options that let you either share the file as an attachment, a pdf or if you saved it in OneDrive, you could make it as a convenient link.

Exporting Your Publication

- To export your work, go up to the left-hand of the screen and click on the **File tab.**
- This opens up what is known as **backstage.**
- Within that screen, click on **export.**
- Within the **export menu,** you will see different options allowing you to send out as a PDF/XPS, HTML, or a chance to alter the file type.
- When you export the publication in a PDF/XPS format, it will maintain the same look as what you have within your Publisher
- Within the **change file type** option, you can export it into a publisher file, a PNG, a GIF, and many more.

CHAPTER SEVEN

PRINTING YOUR PUBLICATION

After the work is done and the publication is thoroughly designed, you can decide to distribute it electronically or print it.

- To print out your publication, head to the **File tab** at the top left of the screen.
- This will take us to the **backstage view**. It is recommended to run the **design checker** as it will help you locate and correct any errors before printing or distribution.
- Select the button
- If there is an error inside your edition, you will find it indicated in the design checker Window on the right of the screen
- To find out, select the error using the cursor
- Then, you will be equipped to detect the exact error the checker is referring to. From the illustration above, it is on page one.
- After fixing the error, revert to the backstage view and click on **print**
- This will take you to the print preview window
- Input the number of copies you intend printing
- Next, you have several settings that you can alter to fit your requirements. Among those are two settings for advanced printing tasks: Double-sided printing and Collating
- Double-sided printing allows you to publish on both the front and the back of a particular sheet of paper. To enable this setting, click **print one-sided.** Then switch to **click on both sides.** You can flip the switch on the long boundary like a book or the short boundary like a legal pad.
- Collating lets you assemble copies of your edition in the correct order which can save you lots of time and effort. This setting should already be enabled. If not, click to open the menu then choose **collated.**
- Then choose the color settings for the edition by clicking **composite RGB**.
- If you have many pages in your publication, it may be beneficial to be equipped to access numerous pages at once. You can do this by selecting the icon at the bottom right corner of the screen.
- To view the other aspect of the sheet, transition between front and back in the bottom left.
- When you are satisfied with your settings, click the **print** button on the upper left corner

Publisher's Advanced Printer Setup dialogue box

Establish your printer with a wizard so will have the leisure to publish envelopes and double-sided publications, print color plates separately, and include printer's markings using these options in Publisher.

Separations Tab

Output Section

Choose the selections for printing a whole or detachments, which plates to print, and at what resolution.

Print Color As-. To choose the color model, choose one of the following choices:

RGB Composite To publish a combination whose colors are specified by the RGB color model, use this alternative. For desktop computer printing, this proves to be the most reasonable option.

Grayscale Composite To publish a combination whose colors are specified as shades of grey, use this option.

CMYK Composite To publish a composite whose colors are characterized by the CMYK color model (cyan, magenta, yellow, black), use this alternative.

You must have a color PostScript Level 2 or later printer attached to your computer to use the CMYK option.

If you choose this alternative without initially specifying a color printer in the Print dialogue box, a hint emerges prompting you to do so.

Separations-

If you choose this option, a separate plate will be published for each ink used in the edition.

Hint that if you choose Separations, the choices Don't print blank plates and Halftone Screen are both enabled.

These plates-. Select one of these alternatives to distinguish the inks used in the printing process. The **Halftone Screen** settings may differ pivoting on which alternative you select from this list.

Note: This alternative is only ready if Separations are checked.

All Defined Inks-. Choose this option to print a separate plate for each ink you specified for the publication, whether or not it is used.

Used Inks Only-. Choose this alternative to publish individual plates for each ink used in the publication.

Convert Spot to Process-. Choose this alternative to modify any spot colors in the publication to their CMYK equivalents and print them as a fraction of the whole process of color separations.

Don't Print Blank Spaces- To stop publishing plates for ink that is employed in the publication but not on a particular page, check this box. As long as this check box is chosen, Publisher, for instance, publishes only a black ink plate for a page with just black text. The publisher uses cyan, magenta, yellow, and black ink plates to publish pages with color images.

Resolution

The choices that emerge are different relying on your printer. Other resolutions, in addition to the (default), are 525 x 525, 600 x 600, 1200 x 1200, and so on. The resolutions on the list are usually expressed as dots per inch (dpi). Depending on your printer or imagesetter, they may be in dots per centimeter (DPC). The printer driver decides the usable resolutions that you are using. Printer resolutions are predetermined and usually cannot be changed. A handful of printers can merely print at a single solution.

Halftone Screen Section

Select the alternatives for specifying the lines per inch (LPI). The settings may vary depending on which selection you chose from the list of These plates.

Note: these settings are exclusively available if Separations are checked.

Use customs settings below-Set adapted line screen angles and frequencies for process-color and spot-color plates by checking this box. The Print plate, Frequency, and alternatives of the Angle vary relying on whether you chose **All defined inks, Used inks only**, or **Convert spot to process.** The **print plate, Frequency,** and **Angle** options differ if you desire to modify the default choices, talk to your commercial printer.

Print plate-Select the boxes beside the ink plates you wish to print.

Frequency For all screens and halftones on any plate, choose a custom line screen frequency.

Angle For all screens and halftones on each plate, enter a custom line screen angle.

The publisher's default frequency is 133 lines per inch (LPI). The following are the fixed angles in Publisher:

- Forty-five degrees black (K).
- One hundred and five degrees in cyan (C).
- Seventy-five degrees magenta (M).
- Ninety degrees yellow (Y).

Spot colors change depending on the number of spots employed. The first spot color is printed at a 45-degree angle by the publisher. 105, 75, 30, 60, 90, 135, 15, 165, 120, and 0 are the other spot colors printed.

Reset This Tab

When you click this button, all of the parameters on this tab are reset to their default defaults.

Page Settings Tab

Choose which printer's markings you'd like to utilize. These marks emerge on the surface of the page and are only printed if the paper size is larger than the publication's page size.

Print Output Section

Choose your output's orientation and imaging. Under **Sample**, you can see the results pertaining to the selection of these settings.

Note that only PostScript printers have the options below:

Vertically flip - To publish a vertical reflection of your publication, tick this box. When publishing to film on an imagesetter, this alternative is extensively employed to ensure that the image scans correctly when the emulsion side of the film is down (as when burning a press plate).

Horizontally flip - To publish a horizontally portrayed impression of your publication, tick this box. When printing on film on an imagesetter, this alternative is naturally employed.

Negative Image-To publish a negative illustration of your publication, tick this box. When printing to film on an imagesetter, this alternative is most commonly utilized to ensure that the picture scans positive when burned into a press plate.

Sample Area

Sample Area is a representation of how the settings you choose affect the images and text. If you choose **to do a horizontal flip** and **Negative picture**, for instance, the **simple image** shows a white-colored creature sprinting toward the leftover right-aligned text. Black will be the stage paper.

Printer's Mark Section

Select from several various printer marks. These marks are published outside the page and can solely be generated if the paper size is broader than your publication's page size.

Crop Marks- click to tick this check box in order to print marks revealing where a page will be cropped to fit the intended magnitude of the edition.

Job information- Click to tick this check box in order to publish information on each plate regarding the print job. The file title of your edition, the date it was published, and the page number are all contained in the job information. If you're printing separations, you'll also need to know which color ink the plate is for.

Registration marks-If you wish to align the printing of two or more press plates, you will need registration marks on a single page, choose this check box.

Density bars- choose this tickbox to publish a density bar with a scale ranging from 10% screen to 100% fill. This bar would be employed by a commercial printer to ascertain the mandatory plate exposure time.

Color Bars (CMYK and Separation only) choose this tick box to build a color bar that a marketable printer can utilize to discern how a substantial area of ink looks on the page. You can only use this alternative when you choose the Composite or Separations CMYK.

Mark style You must tap the icon in order to alter the mark style.

Bleeds Section

Allow bleeds and show the proportion of spruce using bleed marks if you wish to.

Note that only PostScript printers have the alternatives below:

- **Allow bleeds**- Check this box to allow objects that prolong beyond the page's edge to be published up to 1/8 inch outside the page size. By default, this check box is selected.

- **Bleed Marks-**This check box is only available if **Allow bleeds** is chosen. To print bleed markings that show the size of a document, tick this option.

Graphics and Fonts Tab

Select how fonts, image resolution, cropping, and down-sampling will be dealt with all through the printing process.

Fonts Section

Strictly Use Publication Fonts- Make this pick if you wish your printer utilize only the font version specified in your document.

Authorize Printer To Replace Fonts Make this choice if you wish to use your printer's resident fonts that have the same title as the characters in your edition, whether they're in read-only memory (ROM), random access memory (RAM), or on a hard disk drive. This alternative might make your file print faster because it is smaller, but it may change the appearance of your edition.

Pictures Section

Print All Pictures at Full Resolution-. To publish the linked images as high-resolution graphics, pick this as a choice.

Print Linked Pictures at Low-resolution- To publish each of the connected images, choose this option to use the low-resolution placeholder graphic saved in the publication.

Do not Print Any Pictures- To publish a box instead of a single image, select this alternative. When you merely wish to publish a brief indication of the layout that illustrates the placement of photographs, this alternative may be useful.

Note: This alternative supersedes all input images, including linked and embedded images, as well as clip art. Border art and AutoShapes will invariably be published. Images employed as fills in AutoShapes, text boxes, and WordArt, as well as pictures utilized as page backgrounds, will be published.

Crop and Downsample Pictures Section

Commercial Composite **RGB** does not possess these options.

Note: To pick Composite **CMYK**, you are under an obligation to have a color PostScript Level 2 or later printer chosen in the **Print** dialogue box.

Color and Grayscale Pictures- Choose this tick box for print excellence, next, fill in the dots per inch (dpi) range for the color and monochrome pictures. For every single printing process, the standard least and most will vary. For high-class printing, the standard lowest is 300 dpi, and the normal highest is 450 dpi.

Line Art (1 bit) For higher-quality printing, select this box, then specify the line art's dots per inch (dpi) range. The standard least and most resolutions are 1200 dpi and 1500 dpi, respectively.

Reset this Tab- If you wish to return all of the options on this tab to their customary values, select this control.

BOOK 7:
MICROSOFT ONENOTE

INTRODUCTION

If you are constantly thinking of tasks that you need to do, or you have new ideas for something you are working on, or you need to quickly write down something someone just told you about so you can easily recall later, you may end up collecting these in different places, and none of them in sync. A lot of this information may disappear and chances are that you may forget about them.

In this guide, you will learn about an effective tool that will help you put things in place and how to use it effectively if you want to increase your productivity.

David Allen, the author of "Getting Things Done" says, your mind is for having ideas, not holding them. Therefore, we need to find a way to record and store these ideas externally, so we can concentrate on actually getting these things done. That's where OneNote comes in.

OneNote is a digital notebook, but it's much more than that. I'm sure you're going to agree with me, by the time you get to the end of this book.

To show off what OneNote can do, this guide is divided into sections, with each section covering the different aspects such as introduction, getting started, organizing your notes, adding contents to your notes, and how you share them, among others. All these put together will help you understand how this works.

CHAPTER ONE

GETTING STARTED

This section equips you with all you need to get started. Here, you will learn what OneNote is about, how you can get it and all there is to know about the Interface to become familiar with how it works.

So, let's get started.

What Is OneNote?

OneNote is a free digital note-taking app from Microsoft that you can use to capture ideas and thoughts.

It is a tool that automatically syncs and saves your notes so you can access them anywhere whether you're on your phone, PC, or any type of device that has access to the web. So if you're using it on a computer and accessing it across any of your devices, you could go to your phone, install the app, log into your account and access the same information from there.

You can use it to create notes and add content to them and it stores your notes in the cloud. So you can use them on whatever device you happen to be on. It's very similar to other note-taking tools like Evernote and Google Keep.

This is great if you're probably in a classroom taking notes, or if you are in a business meeting or just your for everyday life, family life, planning a vacation, and you just want to put all these ideas in one spot.

There are different versions of OneNote like OneNote 2013 and OneNote 2016, but the official version of the app now is just called OneNote. It is available in Windows, and also for Mac, plus, you get it for free, and it syncs across all your devices. You can install the OneNote app on IOS and Android devices as well, and there's a web version that works on almost any modern browser. So, you see, there are many ways you can use OneNote to be more organized and productive.

How To Get OneNote

You may be thinking of how to take advantage of OneNote and how to get to OneNote. Well, the good news is OneNote comes with Windows 10, so if you have a Windows 10 device you already have one note.

How do you get to OneNote? What you can do is you simply go down to the search field and just type in "**OneNote**" and what you'll see happen is OneNote will show up as the best match on the list. You can now go ahead and click on it and the OneNote app will open up. If you are a Windows 11 user, unfortunately, this application doesn't come pre-installed. For that, you just need to visit Microsoft Store and there, you can find OneNote.

Another way you could also get OneNote is from your web browser. What you could do is simply open up your web browser and then go to **office.com**. This is how you're going to access OneNote. Click on **sign-in** and if you have a Microsoft account you sign in with your Microsoft account. If you don't have a Microsoft account you could also create one for free by clicking on the create one button.

As earlier said, OneNote stores all your notes in the cloud, and because they're in the cloud it can sync those notes between all the devices you're using. So if you create a note on your phone, you'll immediately see it on your desktop. For the syncing to work you need to log into OneNote and to do this you need a Microsoft account; this could be one that you use at work or school with office 365 or from another Microsoft service like Outlook and so on.

Once you sign in you'll land on office.com and you'll see that you can access OneNote by clicking on the OneNote icon. In addition to OneNote, you could also get Word, Excel, PowerPoint, through **office.com**, and if you have a phone whether an Android device or an iOS device, you can also download the OneNote app from the App Store or the Google Play Store and you can install that on your phone.

Here, I'm going to show you the Windows 10 version of the OneNote app but the look and feel are pretty similar across all the different platforms it's supported on.

The Hierarchy

A great feature in OneNote is that you can create your hierarchy structure. There are three main levels, the highest level Is the **Notebook**, and just like the physical notebook, they can hold many pages. The next level down is **Sections**. Think of this like chapters in a book. And finally, there are **Pages**. These hold your actual notes.

If you need more hierarchy levels, you can cluster multiple sections into a **section group**. After you created a group, you can drag sections into the group. Another level you can create is **Sub-pages**. To be able to use them, you need to have at least two pages in a specific section.

For example, let's say you have a page with your meeting notes from Excel Conferences, you can create sub-pages with the notes for the individual sections. So you have a piece called Session Notes Excel Conference, and sub-pages for each section.

To **create sub-pages**, just click on the page you want to turn into a sub-page and select "**Make subpage**". It's going to indent the title. You can even have another level for sub-page that says sub-page of a sub-page. You don't have to use these as much, but if you have a lot of notes and need a detailed structure, they can be really helpful.

So, to summarize, these are the different hierarchy levels, you can have: Notebook, section group, section, page, sub-page one, sub-page two. Just start with a structure that makes sense to you, you can always change it or add to it later.

The Interface

So, you have got Microsoft OneNote installed, what you have to do next is to click on the application and if you are starting this application for the first time you need to **sign in** with your Microsoft account, if you don't have one then you can click on "**Create one**" and it will take you to the Microsoft website where you can create your brand-new account from Microsoft.

The beauty of this application is that whenever you log into your Microsoft OneNote it will automatically sync and download all the notes that you have created before, but if you are doing it for the first time then you won't see any notebook and you would have to create one.

The first thing that I want to introduce in OneNote is how it looks like. If you go up there in the left-hand corner, you have this thing that's called "**Show Navigation**". Go ahead and click on that, and that way you can organize your notes. The three main elements of OneNote are the **Notebook**, the **Section**, and the **Page**.

Pages are stored in sections, and sections are stored in notebooks. At the top here you can see the notebook account, then along the left-hand side, you can see the sections, which consists of Events, Meetings, Projects, Quick Notes, and Research.

If you click on the top bar, you will see all your notebooks. You could think of this as an actual physical notebook and you can have many different notebooks. If you want to create a new notebook, you come up to the notebook section at the top, and you can see your current notebook if you have any and also the option to look for more existing notebooks. Then you can come to the bottom and click on "**Add new notebook**" and that will give you the chance to give your notebook a title, after which you'll have a second notebook which you will see at the top.

To switch between notebooks, all you need to do is press on it, choose the other one and it will take you back to the other notebook.

Within a notebook, you could set up what are called **Sections**. Under the notebook, you have one section, and next to that you can see the **pages** within the sections.

If you want to create a new section within this notebook, you just go to the "**Add section**" button, give the section a title, and then you can add new pages to it by pressing "**Add page**". If you've already got a page when you press "Add page", it adds another and you just need to type the title in over on the right side to name it.

Now for example you create a chemistry notebook, in the real scenario what will happen is you will have a chemistry book and in that chemistry book, let's say there are five units and inside those five units, there are 20 chapters and then within the 20 chapters you have the pages. This layout helps you to quickly create the whole setup and structure. So let me quickly show you how you can do that. Now, under "**Section**" just make a right-click and there you will find options that say "**New Section Group**", click on that and you can give it a name as your Unit One.

What you will do next is to rename the section as your Chapter One. Let's say this Unit has got two chapters, you can create one more section and give it the name of Chapter Two. What you may likely do next is to place these two chapters inside this Unit One; for that, you will simply drag it and drop it inside, so in this way, you can easily place these chapters inside this unit and if you click on the drop-down arrow under Unit One, then you can see how these chapters are hiding inside this subgroup section.

Not only can you create subgroups sections, but you can also create other groups as well. For that simply right-click and create a "**New Section**". Here you can see that you have created one Subgroup and inside that subgroup section you have added two sections, and then you have created the next Subgroup and inside it, you can have another section. So, whenever you want to move it, you can easily do that by dragging it upwards or downwards.

In each chapter, you can add multiple pages, then define the topic name.

On each page, you can make a note. You can not only type there but you've got several options to draw, write or create handwritten notes and add images, links, and videos. We will look at these in detail in the next section.

Let's say you create a notebook to hold all your recipes. You can either right-click and select New Notebook, or you click on the Add notebook down here. The sections in this notebook could be the different types of dishes. So, you can create a section for appetizers, one for main dishes, one for vegetarian dishes, one for desserts, and so on. Within the sections, you add separate pages for the actual recipe.

Apart from recipes, you can have notebooks for ideas for videos with different sections depending on the progress, like brainstorming, research, or scripting. You can also have a notebook for private things like shopping lists, vacations and so on.

So, you could organize your notebook as much or as little as you want but you have a lot of tools where you could set different hierarchies to your organization for OneNote, so quite a bit of power there.

CHAPTER TWO

ADDING CONTENTS AND FORMATTING

Once you have your page, you can now start formatting it. In this section, we will cover different ways you can format your page to make your note look appealing. Let's go.

Changing The Background of A Page Or Section

The default color of a page in OneNote is white but you can change it to whatever color you want. To do this, go to the "**View**" **tab** and click on the **Page Color** drop-down. You will see some color options from which you can select any color of your choice.

Sometimes in meetings, especially if your handwriting is terrible, if you write on a blank page you tend to go off all over the place by not sticking to a line, and you end up with a messy page. What you can do is use the View menu to add lines to your page. So, you just go to your Insert tab, choose the type of line you want, you can have very wide or narrow lines depending on what you need and once you have your lines you can use them as a guide. You can also go full screen with the button in the top right-hand corner so you haven't got any of the navigation bars on the left-hand side and you can just write all over the page.

To add lines, go to the "**Rule Lines**" option and if you click on that you can select any effect of your choice.

Where you have your Rule Lines options, are that of the Grid Lines as well. So, if you want to change to add that effect, you can select it here as well. You can also change the color of the Rule lines as well.

Below that you can see another option that says "**Always Create Pages with Rule Lines**", so if you pick that, whenever you add a new page, the new page by default would be created with rule lines intact.

If you wish to change the color of your section, all you have to do is right-click on the section, go to "**Section Color**" and you can change the color from here.

Working With Text

On Your page in OneNote, you can work with type text pictures videos hyperlinks and even printed pages, but first, let's start working with texts. You can type freely in the page body and you can format the text, just like you do it in Word.

Once you've typed in a name for your page in the top left corner press enter or click on the page to start typing. As you start typing you might notice that there is a small box around the text that you're working with; this is called a **Content Container** in OneNote.

The content container is there to hold the texts and allow you to move them around on the page. So, once you've typed some information in the page you can drive the content container from the grid bar at the top to move it around and place it somewhere else on your page. You can also resize the content container and make it narrower or wider and adjust the text flow.

If you click outside the content container or somewhere else on the page, you can start typing and OneNote would create another new content container for you. These content containers can move around separately on the page and they can even overlap each other.

To **combine two content containers,** you can either **copy the text** or **cut the text from one container**, then **click into the other content container and paste** it there, or you can **hold the shift key** while you **drag one content container onto another** to combine the two containers.

As you're typing out your notes, you might want to do a little bit of formatting or give your content a little bit of structure. Click on the Home menu, and in the formatting area you can change the font style, color, size, bullets, and indentation, you can also choose the ellipses and that will give me a whole bunch of other options like Cut, Paste, and more. You can also use the drop-down with predefined heading Styles to add some structure to your content.

Now that you know how to work with text you are ready to get productive with OneNote.

Inserting In OneNote

The Insert Tab is a powerful one. With this tab, you can perform functions such as insert tables, printouts, attach files, embed videos, links, audios, and insert symbols and equations, among others.

Inserting A Table

OneNote is designed for quick and easy content capture and that philosophy extends into tables.

To insert a table, you can click on the "**Insert**" tab from the menu, click on the "**Table**" tool and draw out the width and height of your table. Please note that this is not an Excel spreadsheet, so there's no calculation in there.

Oftentimes, you don't know how big the table is going to be until you start to work with it, however, there's a quicker and easier way to create a table in OneNote. Simply type a word and then press the **tab key**; OneNote will automatically turn what you typed into a table. Type something into the second cell, then press the tab again to add another column to your table. Keep doing that until you've defined all of the headings on your table and once you're ready press the "**Enter**" **key** to insert a new row and you can keep adding new rows by pressing enter from the last cell of the last row.

When you click into a table the table toolbar appears. Click on it to add more columns or rows, to delete columns, rows, and even the whole table at once. You can also add a background color to individual cells.

The table tools in OneNote are simple and easy to use but not as advanced as tools like Excel and Word. If you want to share a live excel spreadsheet that can be edited by multiple contributors in real-time, we recommend sharing the excel file via teams SharePoint or one drive.

If you have a table prepared somewhere else, such as Microsoft Excel, for example, you can just copy it from Excel and paste it here and you will have that table created. However, this will appear as a table with no calculation here, even though it came from Excel, but you can use it to store your finished calculations in OneNote just by copy-pasting it.

Please note that OneNote 2019 has the functionality of using excel tables within OneNote, but the OneNote for windows 10 version doesn't.

Attaching Files

You can attach files from your computer to OneNote. Note that the file you attach will be available in OneNote wherever you choose to access your OneNote account. This on the other hand increases the file size of the OneNote itself, so if you add a lot of video or audio files to your OneNote, it will get big faster and as this is synchronized over the internet, the synchronization process will take a long time if you add multiple gigabytes of data to the OneNote. So, the better way to do it is to upload your videos to OneDrive and link it back to your OneNote for easier access.

Inserting Pictures

You can also drop in pictures. If you go to the Insert tab and choose pictures, you can choose from a file, from your camera if it has a camera on the device you're using, or you can go online and choose using Microsoft Bing search, so if you just chuck in a keyword, it will bring back images that you can then just place straight into the page.

Embedding Videos

You can also embed video and audio into your pages, so if you choose "**Online Video**", you'll see it asking for the URL of the video. When you paste in the URL and click "OK", it will drop in the video. What's interesting here is that you can also play the video straight inline, inside of OneNote. So, you don't even have to open up a browser and go to YouTube or any other site to see the video.

You can click the button to see what sources are supported but one of the sources that are supported is YouTube. This is useful in cases where you have a tutorial for something and you want it within a page to work with that.

Inserting Printouts

This is a very interesting and useful feature. So, if you had done a presentation and want to insert it into OneNote, when you click on the Printout icon and open it, it will add all the slides you had in your presentation to your OneNote. The same is possible if you have a pdf on your desktop or in some folder. You can just drag and drop it here and it will ask you what you want to do with it such as "**Upload to OneDrive and insert link**", "**Insert as an attachment**", or "**Insert as a printout**". and if you click on "Insert as a printout", you will have your documents inserted as pages within OneNote. So, what can you do with that?

For example, when you are traveling you typically have multiple things you get via emails such as hotel reservations or BnB, or you have your train tickets and maybe some time plans or brochure you might have downloaded. You can add all of these things into one page and have them handy wherever you are as you can sync your OneNote to your mobile. This way it's easy to have all the information and QR Codes you might need just in your pocket without printing it all out.

Recording audio and Videos

You can also record audio and videos. This is the record audio functionality which works similar to the dictate function on the home page but instead of transcribing what you say, it will just record the audio and place the audio file within OneNote. Note that If you create a lot of audio files within OneNote, it will just blow up at some point.

Adding Links and Hyperlinks

It's easy to create and share links to specific parts of your notes, either to the entire notebook, to particular sections or pages, or a specific paragraph on a page. You can use these links whenever you want others to find related notes quickly or when you want to create an interactive table of contents that will make it easier for others to navigate your notebook. You can also add the link to an official document or an outlook email so that you can share the contents of your notebook.

To link to a Notebook, switch to another notebook then right-click on the current notebook and you can see the option "**Copy Link to Notebook**". Click on that, go back to your previous notebook, right-click and paste. while pasting click on the option that says "**Keep Source Formatting**", and your link would appear. You can follow a similar method for the other links.

To link to a Section, go back to the notebook and select the section you want to create a link for. Right-click on that and click on "**Copy Link to Section**", then go to the area where you want the link and simply right-click to paste it.

To link a page, you have to first go to the page that you want to link, you click on that and you can see an option that says "**Copy Link to Page**". Select that option, go to the area where you want it, and just paste it.

Now to link to a paragraph, all you have to do is to get the four-headed arrow, right-click and select the option that says "**Copy Link to Paragraph**". Now you go back to your notebook or any area you want to have this link and paste it.

There is another feature provided by Microsoft OneNote. If you click on any of these links, let's say you click on a link to a paragraph, the paragraph will open and if you want to go back to the previous notebook quickly, at the top left corner you will see the back option, click on that and you will be navigated back to your original notebook. This shortcut works for all the links.

If you want to link to a document, just go to any section of a page, right-click on the screen, and select the "**Link**" option. You can then select the text you want to display the link. If you copy this link and try to paste it in a word document, then it will get pasted as a link to a document and if you hover your cursor to this link, it says press **CTRL K** to follow the link, and if you do that you will be redirected to your OneNote notebook.

Inserting Meeting Details

One thing that can be very useful on the other hand is the Meeting Details icon. OneNote is a prime example of a tool that can be used for creating Meeting Minutes and Microsoft also considers OneNote to be the meeting minute tool at least for corporate use. So, if you click here on the "**Meeting Details**" icon, it will open a tab with all your meetings. This takes all the information about the event from outlook or your calendar app depending on what you have connected to Windows 10 OneNote and if you click here on "**Expand**", you can see the details of the meeting including the participants who have been invited.

Drawing

The draw tab is most useful if you can work with the pen on your pc. For example, if you have a touchpad or a graphic tablet you can use it in combination with your pc because then you can use OneNote to work with that. If you only have your mouse, you can still work with that, and you have a collection of pens and pencils here that you can use to draw on your canvas. You can also change colors or thickness or whatever you need. You can also always use the eraser to delete lines.

Adding Shapes

You can also add shapes, so for example if you want to show the connection with an arrow you can just use the arrow here. You can also show graphs or basic shapes and there's just a default collection of shapes you can use. You can also drop in shapes. There's a bunch of different shapes you could drop onto the page and even math equations if you're studying math or using OneNote for research.

Ink to Shape

If you choose a pen from the top bar and then start to draw your shapes, you will see that they're a bit rough. The first option you have is the shapes button at the top. You can drop that down, choose a shape, and then use the pen to draw the shape onto the page.

Once you have the shape on the page you can select it and you can resize it and move it around, so if you take the circle in the middle of the object and put your mouse on it you can move that shape around. If you touch the shape you can pull up a menu and then you can do things like rotate and flip. At any time, you can use the Lasso button in the top left to grab all of the objects on the page in one go, and then you can manipulate them all at the same time. So, you can resize, delete, rotate them, and so on.

The next option you have is this “**Ink to Shape**” button at the top this is slightly different in that you can start drawing freestyle the shape and it will complete it for you and straighten it.

It supports several shapes like triangles, circles, rectangles, and some others as well. Once you've used Ink to Shape you can manipulate the object just like you could with any of the other shapes, so you can move it around, rotate it, flip it, delete it, or copy it.

Researching

You also have the "**Researcher**" **tab** where you can use the Bing search engine within OneNote to get information about certain people, theories, locations, and enter them as links. For example, if you search for OneNote, you'll get relevant topics including websites. If you want to quote them you can just add them to the list by clicking the "**Add to Page**" button and you'll get the link with the title already on your page and you can always refer back to them.

Pen And Inking

Switching to an iPad, I would like to show off another feature of OneNote which is Pen support and Inking. With the Pen support and Ink feature in your OneNote Pages, you can draw and you have a selection of pens so you can change the colors and the thickness of the pen. You also have highlighters and other types of pens up there.

You can also change the view to a different paper style, you can have lines, so if you were taking notes in a meeting, you could write them along the lines just like a normal notebook.

You can do the same with the mouse if you don't have a device that supports a pen but it's not as accurate. In your Windows 10 device, you will get the same page in OneNote as well.

There are many more features that can be very helpful depending on your needs. For example, you can hand draw shapes and then apply ink to shape, and OneNote will convert them automatically into shapes with straight lines and clean corners.

Ink to Text

Another cool thing with OneNote is that you can turn ink into text.

This is useful if you're using the pen to write handwritten notes and then you want to turn them into editable text. You just write your text down just as you would normally in OneNote, then you can use the lasso tool to select the text, what you can do next is choose the "**Ink to Text**" button which will then turn it into some normal text.

Once you've done that you can then select it, change the color and edit it just like you would do for a normal piece of text.

OneNote can also convert your handwriting into typed text even if you have horrible handwriting. First, select what you want to be converted with the Lasso Select, and then just click on the "**Ink to Text**" right there. You may not use this feature often if you are faster when typing, but it can be useful if you prefer to hand-write your notes.

Math

What this allows you to do is to write freehand on your OneNote page with your pen including math formulas. Once you've written your formula you can **use the Lasso tool to select** what you've written and then you can use the "**Math**" **button** on the top right-hand corner to identify the formula.

Once this is identified, it will give you a set of actions that you can now carry out. For example, in a case where you have to "Solve for X", it will figure out what X is, based on what the formula is that's been written. You can add the answer back into the page just with a click and then once the answers are there you can manipulate that just like normal text. This is a basic example and you can get much more complex formulas.

With these, you can add symbols and equations and this is powerful if you're into math or want to create a worksheet for your kids. You can do nearly everything.

With the Math tab, you can Convert, Solve and Graph Math equations. This is a nice feature for students who need help with their math homework. You can type in an equation or problem, click on Math, select an action under that such as Evaluate, Solve or create a graph, and you will get all you the steps you need.

Not only does the Math feature help convert written equations to text; when you click "**Select an option**" and "**solve the equation**" if you're interested, it can even show the steps on how to solve it.

Tags

Tags are a great way to categorize and group notes together that aren't in the same notebook or section. So, let's say you have a notebook for something you want to consider doing soon. By using the drop-down tag this note up there, you can select different default tags to add to the notes or parts of the notes.

So, you can tag it with "**Remember for later**", so you don't forget about it. You may have another notebook with a page to check out for something later. You can also tag this note with the same "Remember for later". So now you have marked two notes that are in totally different notebooks with the same tag, meaning although they are in a different hierarchy, they're still grouped and it's easy to find your tag notes later on.

So, let's say you want to check all notes that are tagged with "Remember for later", all you have to do is to click the Search icon there on the left, then either just click in the search bar and select from the tags it will suggest down there, or just start typing what you're looking for, select the tag that you want, and you're going to get a list of all the notes containing this tag.

You can either use the default tags, but I recommend you also create your tag system through your custom tags. You just need to click the "**Create New Tag**" down there, and you can then name it and select an icon for it. Currently, there isn't a way to rearrange the tags, like you could in previous versions of OneNote. New ones will always be added below, so don't go overboard and just keep the number of custom tags limited. So, in short, use tags to categorize notes.

There are many applications for tags, just use them depending on your needs. So, for example, let's say you want to stay on top of tasks you assigned to your team members, or use these to highlight important parts in your learning notes. They make it easy to find what you need when you need it.

Sticky Notes

Sometimes you may scribble down things you don't want to forget on a piece of paper, then they either get lost or you spend a lot of time searching through my bag. Sticky Notes make things a lot easier now. The Sticky Notes are integrated into the OneNote app on mobile, so if you just want to write something down quickly, you'll just open the OneNote app and click on the Sticky Notes button down there, and start writing. You can also find all your previously collected Sticky Notes right here.

Now when you get back to the office, and you want to see those notes, you directly go to the Sticky Notes app. It's a built-in app in Windows 10. To get to it, click on the Start button and type in "sticky", and you're going to see it pop up there.

So on your computer or laptop, you're going to find it as a separate app, whereas on your mobile, it's a part of the OneNote app. These are automatically synced, so you have your notes with you at all times, no matter where you took them.

Wiki Linking

A slightly obscure but interesting feature of OneNote is Wiki Linking. Wiki Linking allows you to quickly generate links to other pages or even to automatically create pages if they're not already there. The best way to understand Wiki Linking is to see it in use. So, on your page, if you enclose a sentence in double square brackets, you'll notice that as you add the last brackets it automatically converts the line of text to a link. Now if the text within that page lines up with the title of another page in this section, OneNote will automatically create a link to that page but if there wasn't a page within the section that has the same name as that text, OneNote will automatically generate the page. You'll notice that it's got a slightly dotted line underneath it, so if you click on that dotted line, it now takes you to that newly generated page, with the title of the page already in place and you can now start filling out the content.

So, you can use Wiki Linking to create a table of contents that builds the pages needed automatically. This is one of those instances where OneNote gives you tools that allow you to generate content very quickly. Try Wiki Linking and you'll be well on your way to getting more productive with OneNote.

Extracting text from images

Another great feature is **Optical Character Recognition** (**OCR**), which means OneNote can re-text inside an image. This has two great benefits. Number one, it lets you copy text from pictures so you don't have to manually type it out. So, for instance, you copied the screenshots in your notes, when you right-click on it, you get the option "**Copy Text from Picture**". You can open a new page and copy the text from the image there.

The second benefit is that you can search for text inside an image. This would be of great help If you usually collect a lot of information from videos where you take notes. Later when you look for something specific, you can use the search feature and with the OCR functionality, it will also search for the text in the picture, and show you all relevant information, images included.

Focusing with Immersive Reader

A great feature you would love to use when you want to focus on a text, or for proofreading is the **Immersive Reader**. It's now built into several applications like Word, Outlook, and also OneNote. If you're on a page, you can c**lick on Immersive Reader from the View tab**. It will show you the page without visual crowding, which makes it easier and faster to read. You can change the settings for text size and color themes to whatever fits your needs.

Here you can also change the font, and an interesting one is Comic Sans, which is a font that may not be visually appealing for a professional environment, but it works great for kids. The way the As and Gs are shaped is how kids learn to write in school. You could also use this to practice reading with your kids. You can also use the play button down here and have the text read out loud for you.

With the voice settings, you can change the voice speed, and select a female or male voice

Translating

You can also translate a sentence by using the "**Translate**" button and you can translate either a Selection or the whole page. So, if you choose "**Selection**", you can choose which language it should be translated from and to, hit "**Insert**" and it will change the text directly on the page for you.

Moving and Zooming

All of the elements that you put down on the page can be moved. So if you hover over the content with your mouse, a border pops up and if you take your mouse to the top of the border on the gray bar you can click on and move that content over to the right, and then you can start using this area again with some more content.

The page itself is infinite so you can zoom out and add lots more or move items that exist already way over on the right-hand side. So, you see you can make the page as big as you like.

Creating Page Templates

A template is a pre-designed layout that you can apply to a new page to provide a level of consistency and safe time. OneNote comes with several written templates including that of page backgrounds and to-do lists. You can create your template or customize an existing one. You can use OneNote templates as they are, change them as you like, or create your own.

To create a OneNote template, first, you have to **add a new page** then go into the **View tab** and click on **Paper Size**. Here you can see different options such as Paper size and Print margins. If you want to print out your OneNote you may want to make a selection because if you have auto it's the continual note that goes on but you can select Letter size or any other size; whatever your print dimensions are you can select from here.

After you select the size of your choice, you have to select the orientation and fill in the rest of the options. You can enter some data into the page and then click on the button that says "**Save current page as a template**", a pop-up will appear and you have to write the name of your template. If you want every page in the Section to use this, tick the box that says to **set this as the default template for new pages in the current section** and save it.

If you want to check out your newly-created templates or insert this template, then you have to go to the "**Insert**" **tab** and you can see the option for **Page Templates**.

Click on the drop-down under that and click on **Page Templates** again and to the right, you will see all available templates. Under the "**My Templates**" option you will see your created templates and if you click on "**My Created Template**" you will see that a new page has been added. So, whenever you create a new template, a new page will get added.

A OneNote template is a page design that you can apply to new pages in your notebook to give them an appealing background, a more uniform appearance, or a consistent layout. To apply a OneNote template to a page, go to the "**Insert**" **tab** then click on Page Template drop-down, click on Page Templates again, and to the right, you will see the different templates available. Now you can select a template of your choice and a new page will be created with the selected template. Similarly, if you select another template then another page will be created with that selected template. You can then proceed to close the window and work on your newly-created template.

If you want to get rid of the page template that you don't want anymore, just go to page templates navigate to the bottom of the window where you have a drop-down, click on it, select "**No Default Template**" and your next page won't have it.

If you just want to get rid of it altogether, you open up "My Templates", right-click, then select "**Delete**" and it's gone. You no longer have that template.

You can play around with it, learning how to customize these page templates the way you want.

CHAPTER THREE

SAVING, SYNCING, AND SHARING

After working on your notebook, several actions can be taken to ensure you keep track of your note. This section covers those actions which you can take to keep your notebooks secure. You can decide to share your notebook, a page, or a section with others, export or just save them.

To know more about how you can achieve these, let's go.

Sharing

OneNote allows you to share note pages and notebooks through emails, permissions, and links, and you can control how much access you give to recipients and decide if they can view or edit that content.

There are a few different ways that you can share notes in OneNote, so let's take a look at them. One important point to know when it comes to sharing OneNote notebooks is that you can only share notebooks that have been stored in OneDrive. So, if you have a locally saved notebook, you're going to need to move it into OneDrive before sharing it. If you have different notebooks for business and personal, you can store your business notebook in OneDrive and your "Personal" Notebook can be stored locally in your downloads folder.

If you want to share your "Personal" Notebook with anybody, you need to move it to OneDrive. To do this, open your "Personal" Notebook, navigate to "**File**" and then go down to your share option and what you'll see in there is it says to share this notebook, so you'll need to put it in OneDrive or SharePoint. You're going to put that in OneDrive by selecting it. You could also browse for it if you don't have it listed here, you can name your notebook and choose "**Move notebook**". You will get a message saying the notebook is now syncing to the new location, click on "OK". Now that you've done that you'll see that you now have all of your share options come up because you've now put it in a location where sharing is available.

Let's go through each of these sharing options one by one.

Sharing With Others

The first option you have here is to share with people. What you can do here is you can invite people via email to share my notebook with them. Type in the email addresses are the people that you want to share with, and of course, if you have any contacts stored in your address book you can select them from there as well.

You can then choose what level of access I want to give them to this particular notebook. So, do you want them to be able to edit or just view this notebook? Type in your message and then click on "**Share**". As soon as you do that, you can see underneath where it says "shared with" it's showing you all of the people that you've shared this notebook with now.

At some point, if you decide you don't want to share this notebook with a contact anymore for some reason, you can essentially revoke access as well. All you need to do here is right-click on the contact's name and you have two options, which are either to "**Remove user**" or "**Change permission**".

If you want to remove the user entirely, select "**Remove user**" and the user's access has now been revoked.

Sharing With Links

The second option you have is to create a sharing link. If you want to share your notebook with your entire team or wider organization, sometimes it's easier to copy and paste a link to the notebook into an email and use a group distribution list that you've got in outlook to send it out, and again you can choose if you want to create a view link or an edit link. So, if you want to create a link to this notebook that only people can view you would click "**Create a view link**".

You would then take this little link that it's created, **copy** (CTRL +C) and then **paste** (CTRL + V) into an outlook email and send it out. You'd do the same for an edit link and send it out to an outlook distribution list.

Sharing With Meetings

The third option you have in here is to share your notebook with a meeting. Notebooks are great for taking notes during online meetings, but sometimes it's quite hard for one person to keep up with all the information being exchanged. So, you can share your notebook with others during an online meeting so that multiple people can help keep the notes up to date; so essentially, you're all collaborating on the same notebook and adding notes into it. Click "**Share with Meeting**", and if you have any online meetings in progress they would be listed in a window and you can choose which meeting you want to share your notebook in.

What will happen is a link to the notebook is shared with everyone in the online meeting and the meeting details are added to the notes page.

Sharing with OneDrive

Finally, you have the move notebook option which you saw how to use before when you moved your notebook into OneDrive.

It's not necessary to always share the entire notebook; if you prefer you can share a single page from the notebook only. So, if you want to share a page from your notebook, make sure you've clicked on the page that you want to share, go up to the Home tab and you'll see you have an option here to "Email page", keyboard shortcut Ctrl + Shift + E. What it does is it will open up an outlook email, it's going to attach the contents of that page, you can then select who you want to share this with and send that email off. The user receives an email that's got the name of the notebook and an open button. If they hit that it'll open up the notebook in the browser and if that user makes a change, it will sync that change and, in your notebook, you'll see the change, with a little marker indicating that it was added by the user.

Syncing And Saving

One important thing to remember when you're sharing notebooks is that there could be multiple people accessing and editing a notebook at any given time. So due to this, your notebook must be synchronized to ensure data is updated immediately. Synchronization is automatically turned on in OneNote as it's a key feature but you do have the option to change this and sync manually if you prefer.

To do this, go to the **File** tab and ensure you are in the "**Info**" area. what you'll see over on the right-hand side is you have a button here to view sync status and this is going to show all of the notebooks that you have and whether they're synchronized and the last time they were updated.

The option automatically selected is to sync automatically whenever there are changes. So, if you've shared this notebook with five other people and they're all in this notebook making changes, any changes they make are automatically synchronized so that you can see them.

However, if you prefer to **sync manually** you have the option as well, and as soon as you do that it puts a cross over each of these notebooks just to let you know that they're not currently synchronizing and you can choose which one you want to sync.

So, if you click "**Sync Now**", the icon is going to change and any updates made since the last sync will be updated. If you have a lot of notebooks in here, to make this easier you have a single button at the top as well, to "**Sync All**".

Password Protection

Another thing that you might want to do when sharing your notebooks is **password-protect sections of the notebook**. This prevents unauthorized access to specific sections of your notebook that you don't want people to see. All you need to do here is select which sections you want to protect. If you want people to be able to see everything that's in one section but not in another section, all you need to do is right-click and select "**Password protect This section**", click "**Set Password**" and now if you share this notebook, people aren't going to be able to see what's on that particular section.

To unprotect, right-click, go back into "**Password protect This Section**", select "**Remove password**" and type it in to remove that protection.

Exporting Notebooks

Many of us use OneNote notebooks to organize our lives. It is important to **know how to backup your OneNote** because technology can fail and when it does you want to make sure your materials are safe and secure and so it is necessary to back up your work. It is highly recommended you do this at some point, especially as you get towards the end of the school year and you leave for the summer or as you're working with other members of a team. Backing up your notebook in OneNote is a great idea.

There are cases where you may not have access to OneDrive, let's say you're getting a new job, you're moving to a different district, for some unforeseeable reason your server goes down or you need access when you're not online, you need to export your OneNote. To do this, go to your "**Files**" tab, and instead of choosing "**Save**" you're going to choose "**Export**". You'll get several options. You can export a page, that's handy if you just want to share a page with somebody; you can share a section, or for your use, you can export your entire OneNote notebook, and you are going to have file types to choose from. You can export them as PDFs but you can't edit those in the future. To export your OneNote as a package so you can upload it into OneNote in the future as its full functioning self, you'll need to select "**OneNote Package**" and then you just click the export button.

You're going to tell it where to save on your device and you have your OneNote notebook saved on your hard drive for loading whatever you want, wherever you want, whether you are online or not.

This can be done for group notebooks as well, so if you're in charge of your group's notebook this is a good idea to do every once in a while, but it's also handy for individual notebooks.

CHAPTER FOUR

OUTSTANDING FEATURES IN THE DIFFERENT VERSIONS OF ONENOTE

This section summarizes the amazing features and benefits you can get with OneNote, giving you the specific versions where these features are available.

The feature of **Copy Text from Pictures**. This is only **available in the version of OneNote that comes with Microsoft office**. Remember that you can get this for free.

To copy text from pictures, well first off bring a picture into your OneNote file and when you click on the image you can extract the text by simply right-clicking on the image, and then there's an option you'll see that says "**Copy Text from Picture**". It's going to use OCR to extract all of the text from this image and you'll see all of the text from the image was extracted and pasted into the OneNote.

Microsoft OneNote makes it easy to consume content using the **Immersive Reader**. It improves the reading experience and then also you can have OneNote read to you. This is only **available on the OneNote for the windows 10 app**, and this is the app that comes pre-installed with Windows 10.

To use the immersive reader, go up to the toolbar on top and select "**View**". Within the View ribbon, click on Immersive Reader and this opens up your text. First off, it removes all of the distractions around the text, increases the font size, and makes the spacings a lot better so it's easier to read the content. You can change my text preferences; highlight different parts of the speech and you can have OneNote help you focus on each line.

Another feature is the **ability of OneNote to read to you**. On the bottom of the Immersive reader, there is a play button, and next to that you'll see some voice settings, here you can set the voice speed and also choose whether it's a female or a male voice.

You can **create Sticky Notes in the OneNote mobile app** and have them automatically synchronize and show up on windows 10. First, you need to get the OneNote app. You can get that through the app store or the play store; it's free to download and install. Along with typing text, you can also insert a photo or change the color of the sticky note. On your Windows 10, go down to our taskbar, and type in sticky notes. This is an app that comes pre-installed and comes with windows 10, this launches the sticky notes app and you will see your note are in synchrony with that of your mobile app. This is a very nice way if you want to add sticky notes on the go and then get them onto your Windows 10 PC and vice versa. You can also add sticky notes on your Windows 10 PC and then they'll show up in your OneNote app.

You can have Microsoft OneNote help you **write meeting notes**. This is handy if you've agreed to take notes for a meeting that's taking place. It is **available in both versions of OneNote** and **you need to be using Microsoft Outlook to be able to take advantage of this**. This inserts all of your meeting details; you can see the subject, date, location, and participants and once the meeting takes place you can start typing in your notes.

You can **easily take tasks from your Microsoft OneNote** and you could add it to your Microsoft to-do task list. Click over on the left of the text and there's your cursor, then go up to the Home tab, and over on the right-hand side there's an option for Outlook Tasks. When you click on that, it opens up a menu and you can now choose when the task is due. Once you pick your desired action, it adds a flag next to the item within your OneNote and this will now be synchronized over to Microsoft To-do and then also tasks within outlook. If you go to your Microsoft To-do task list, you'll see the task that you just added from OneNote. So, it synchronizes in both places and in your Microsoft Outlook, you will see that same task in your task list synchronized over from OneNote. This functionality is **only available in the version of OneNote that comes with Microsoft office**.

You could **tag items in your OneNote** to help you get back to them more easily in the future. This is **available in all versions of OneNote**. To make sure you remember something so could come back to it in the future, a tag can help you with that. Go to the Home tab and in the middle, you can see all the different tag options. You can even add your tags. Once you add the tag over on the left-hand side, you'll see that it added an important marker next to the item. If you want to get back to this tag in the future, right up above in the same area you can click on find tags. This now opens up a pane over on the right-hand side and there you see a summary of all of your important tags. So, this makes it easy to get back to the item. You can also see other tags that were used in the document and a category for tasks or dates those items are due.

There are different ways you can group your tags; you can group them by section of your OneNote, by title, date, and note text. So, there are many ways you can get back to the content that's important to you.

Microsoft OneNote makes it very **easy to collaborate with others**. If you want to share a page with others, you can simply right-click on that and there's an option to copy a link to this page. Not only are you limited to copying pages, but you can also copy sections. Right at the top of a section, right-click, and here too you can copy a link to the entire section. Within outlook now you can simply go in and paste a link to the meeting notes and then other people can click on this and they'll be able to see your OneNote and then also work on your OneNote with you. This is **available in both versions of OneNote**.

You can easily **password-protect a section of OneNote**. This is **available in both versions of OneNote**. If you don't want anyone to be able to see a part of your section, when you right-click on the section you have the option to password-protect this section. While using this feature, please note that if you forget your password, you won't be able to get back your data so be very careful and make sure you remember your password.

You can easily **convert handwriting into text that you can then edit** on your computer. This **works in both versions of the OneNote app**. Doing this is extremely simple. Go up to the top toolbar, click on Draw, and over on the far right-hand side there's an option to turn Ink to Text. Clicking on that converts your hand-drawn text into editable text on your computer.

Did you know that you can set it so your **OneNote window always appears as the topmost window** especially when you're taking notes? You might have multiple windows open and your Note canvas within OneNote falls to the back. It would be nice to keep it always on top. Within OneNote click on the View tab and over on the far right-hand side of the ribbon there's the option to always keep OneNote on top. When you select this, you can click on your browser window or other content and your Note canvas will stay on top. This is **only available in the OneNote app that comes with Microsoft office**.

You can **convert your OneNote into a full-screen mode** to help you with note-taking. This is **available in both versions**. In the top right-hand corner of the canvas, there's the full pageview. When you click on it, this gets rid of the ribbon and all the chrome of OneNote so you only see the canvas. Together with staying on top, this is a nice way to take notes on your PC.

You can **easily translate text** in OneNote and this is **available in both versions of OneNote**. To access "**Translate**", go to the bar on top, go over to Review and in the middle of the Review ribbon there is the option to translate. You could translate the selected text and you could open up a mini translator.

You can now choose what language it starts in and what language you want to translate to. After that, you have two options; you could insert it as is and this will replace the English text that I have over there, alternatively you can just copy it and paste it into your page.

To use the **Mini Translator**, within the Review ribbon click on "Translate" and then activate the Mini Translator. With the Mini Translator activated, when you hover over a word, you'll see a text box appear over there, and when you move your mouse over it, it'll show you the translation for this one specific word.

Did you know you could **access previous versions of a OneNote page** especially if you're storing your OneNote in OneDrive or SharePoint? This is **available in both versions of the app**. When you right-click on the translate option on a page where you have to translate, there at the bottom of the list there's the option to Show Page Versions. When you click on this you see a previous version of this page where you just had the English text and if you go to the current version, this version now has the translated text. As you work on a page over time, you'll see more versions stored.

If you start taking a lot of notes, sometimes it's nice to **have multiple instances of OneNote open**. If you currently have one copy of OneNote open and you want another one, it's easy to do that. Simply navigate down to your taskbar and right-click on the OneNote icon and then you can click on OneNote and this will open up another copy of OneNote. This too **works with both versions of OneNote**.

You can **use OneNote to solve math equations**. This is **only available in OneNote for Windows 10**. First, off you highlight the equation, next, you go up to the top toolbar and click on "**Draw**". Within the drawing ribbon, all the way over on the right-hand side click on the option that says "**Math**". When you click on Math, this shows you your equation over here, and below, you have a drop-down list where you can select an action. when you click on this you can solve the equation. Another nice feature here is that there's another drop-down menu and when you click on that you can view the steps for solving the equation.

You can **dictate to OneNote and it'll type up everything you say**. Click on the Home tab and if you go to the right-hand side there's the option to "**Dictate**". Click on that and you can now start speaking. This is **available in the OneNote for Windows 10 app**.

CONCLUSION

Having gone through this guide, there is no denial that OneNote is an amazing tool to get you more organized and productive in your work or personal life. Take advantage of OneNote today and stay organized.

ONEDRIVE FOR BUSINESS

Here, you are going to learn everything about getting started with OneDrive for business. We will begin with understanding what OneDrive for business is and cloud storage in general, how to access OneDrive, navigate around the sites, organize contents, and how to organize contents stored within OneDrive. We will talk about accessing information from the recycle bin and how to share information with callings. We are going to learn how to sync OneDrive to your computers so that you can easily access information in the cloud.

OneDrive for business is referred to as an integral part of office 365 or SharePoint server and provides a space in the cloud where you can store, share and sync your work files. With OneDrive, you can update and share your files from any device with OneDrive for business. This can also enable you to work on office documents with other people at the same time.

UNDERSTANDING CLOUD STORAGE

Cloud storage is a cloud computing model where data is stored on remote servers which can be accessed through the internet or a cloud. It is maintained, operated, and managed by cloud storage service providers on storage servers that are built on virtualization techniques that are managed by Microsoft 365.

To access OneDrive for business, the easiest way is to navigate to the URL **portal.office.com**

This will bring you to the initial log-in page of office 365. No matter where you work, you can simply place your email address into the log in page, for example, someone@.com, and then your password

Once you select sign in, it will bring you to the office 365 portal where you can access all the different elements of office 365 including OneDrive for business.

Now OneDrive for business is your work "OneDrive" whereas if you have a home license for office 365 you will be using OneDrive. When I select OneDrive, it opens up a new tab and I can see this page.

The first time you come to OneDrive, it gives you some information on what this product does, helping you get started. As it says, it has a place to store files, pictures and they are always at your fingertips no matter what device you are working from. You can also collaborate in OneDrive for business sharing documents and presentations and finally, OneDrive for business and office makes a great product that we all love keeping things easy to find and access.

Now that we have accessed one drive for business let's go ahead and explore the set. On the top left-hand corner of our OneDrive for business, you have the waffle

This allows you to navigate to different applications in office 365 products.

You can also use the office 365 button to navigate back to the portal.

You will notice when you select one of the tiles it opens up a new tab. Just below the waffle and your office 365 home button, you have your search button.

You can search for files and folders. Below that, you can see that you have your navigation around OneDrive for business

You can navigate to the files that you recently created or even files that other people had shared with you.

You also have the recycle bin as shown below.

There is also a section to create a new group.

Although some organizations do disable it, at the centre of the OneDrive for the business screen, you should expect to see all the different files or folders that you uploaded for OneDrive for business. You have got a column heading along the top that you can use to sort or filter files in OneDrive for business.

Up at the top, you can create new files, upload files, and sync OneDrive to your local computer or

At the top right corner, you can sort by name or modify by going into the individual columns within this view.

You can switch between the previews views which can easily show you a preview of this file.

This will open up all recent activities as well as additional information about the file you selected.

Above these options, we have our settings icon, where we can go in and change our settings to OneDrive for business or other applications within the office business account.

We can be notified of the note with notify icon and also the help icon if you need any help.

If you want to sign out from OneDrive for business, you can simply select your name in the right-hand corner.

And here, you have the **about me** section, you can view your account and of course, sign out.

As you can see the interface for OneDrive for business uses the standard plus your side applications. You can access additional information among the top list and eventually you can see your files and folders added to the centre section of your screen.

Now that we have taken a look at the interface for OneDrive for business, We can go ahead to start creating some folders, so we have ways to organize the contents. From the top left corner, select new icons.

It gives you the ability to add several different files, including links, surveys, and work plus. I also can create folders.

Here, create a folder for the personal expenses port. Give it a title and simply click "Create"

And it will add a new folder showing you that it was modified a few seconds ago.

You can also drag and drop folders that are stored locally on your computer. To do this, open the explorer window.

And from your Explorer window, you will see some folders there including documents and other information stored on the computer.

If you want to take a folder from your local computer, do that by selecting, dragging, and dropping this folder.

It uploaded the folder as well as any content that was resident in that folder.

Uploading a folder to OneDrive for business is as simple as selecting the "New", choosing the folder, and naming the folder. In addition to being able to create folders in OneDrive for Business. You can also use OneDrive for businesses to store photos and videos. Create another folder to store photos and videos by going to "New" and choosing a new folder.

To navigate into the folder, select it and it will bring you to this new page. You can still see that this folder is empty

But there are pictures and videos on my local computer that I like to be included in the folder.

So, I am going to navigate into the video section by simply dragging and dropping the video I want to include in my OneDrive for business.

BENEFITS OF MICROSOFT ONEDRIVE FOR BUSINESS

OnLive is one of the simplest cloud storage systems among other cloud storage apps. When the OneDrive for business was introduced by Microsoft as part of their 365 office it increased its use by a lot of organizations.

OneDrive has a reputation for the safe storage system, yet many organizations does not know how important and useful the OneDrive for business is to the safety of their data systems. OneDrive has a lot of rich features that assure the safety of your data and easy access to your files and as a matter of fact, the easiest cloud storage app that is very easy to use and collaborate with.

HOW TO USE ONEDRIVE FOR SMALL SCALE BUSINESS

The Microsoft OneDrive for business is exceptionally good for business, people and organizations. It is a cloud storage system and also a file-sharing collaboration that can compete with any other cloud storage app or system anywhere in the world of the web-based online storage system. It has competed very well with Google Drive Enterprise, Dropbox for business, and a whole lot of other professionally based business storage app systems. It is interesting to note that OneDrive for business starts at $5 per user per month on the annual plan list. It includes one terabyte (TB) storage and it can store up to 15 gigabytes of files. It can share files in 100 secured online and it can sync local files and folders on offline mode.

The Microsoft OneDrive for business is ideal for small or medium-sized businesses that are using Microsoft solutions in the form of Microsoft Exchange, Microsoft Office 365 as well as Microsoft teams. You can use OneDrive for business in so many ways as well as in personal cloud storage and file sharing.

OneDrive because of its adaptability for various users on the different operating systems is an excellent choice for small, medium, and big businesses. It has various interesting and helping features which makes it an excellent collaboration with artificial intelligence which gives it an excellent file and folder organization as well as excellent project execution in conjunction with a 100 % security online storage and governance option.

This system of online web storage (OneDrive) can be used with Microsoft 365 for jobs that require high integrity and collaborations that gives you easy access to your files and sharing with colleagues, co-workers, friends, and family anywhere at any place no matter the type of device you are using.

The OneDrive for business has grown over the years and it can now boast over 270 files types which include Microsoft office files and even more complex challenging Autodesk CAD or AutoCAD FILES.

INSTALLING AND SETTING UP MICROSOFT ONEDRIVE FOR BUSINESSES

In this section, we will be looking at how you can install and set up OneDrive for Business. The steps for installing are written below:

1. Install the windows version of the new OneDrive for Business Sync Client.
2. Click on the new OneDrive for Business for sync client, it will download that file, then you need to wrap this so that you can save it if you want to. When you have done that, it's going to come right up.
3. Then the next thing that comes up after the download of OneDrive for Business will be a setup pop up page.
4. Type the email address of your work account
5. After putting your email address, the next thing is to click on the sign-in and this will put you to account office 365 and the next thing will be a pop of question asking you whether you want a personal or work or school account?
6. Click on work or school, it will take you to the office 365 work or school account log-in page.
7. When you are here, type your password and then click " sign-in" and it will bring you to your OneDrive folder.

8. After signing in with your office 365 information, it's going to show you where your OneDrive for Business sync folder is going to be.

If you want to change that location, you can, the big folder is just under your windows profile, and another pop up tells you that "File already exists in this OneDrive folder"

9. Then on the option, you have to choose "Use this location"
10. It will take you to the next page, where you can select the folders or files that you want to sync to your computer.
11. Choose the folders you have already that you want to download to your co-hard drive.
12. When you click "Next" your OneDrive will be ready for you and will be synced to your PC. You can add files by moving them to your OneDrive.
13. After that, click "Open my OneDrive – Tech impact folder" then you will see that I now have my OneDrive -Tech impact on the left side of my computer as well as my folder to the right

THE RECYCLE BIN

In this section, we will be discussing how to use the OneDrive recycle bin, so let's get started.

Firstly, open OneDrive in File Explorer and you can see here that I have a word document that I have created for this lesson.

I will delete the document that I have prepared by clicking on delete and a pop will appear to explain to me what's going to happen if I click "yes' to the delete, it explained that OneDrive does not use the same recycle bin as the one on the desktop.

The easiest way that you can see a recycle bin in your browser will be to right-click on your OneDrive and click to view it online.

At this point, it will open your OneDrive, click on the menu dropdown and go to the recycle bin.

This will bring you to the screen as above, refresh the page, you will notice that the word document has appeared at the top of your recycle bin.

The next thing to do is restore it. To restore it, click and check the box, then click and restore on the menu bar

The pop will appear to say that it has restored 1 item.

And this now has reappeared at its original location on OneDrive. You will notice a blue cloud icon to show that it is stored only on OneDrive. You must know that personal files in a recycle bin are only saved for thirty days and school or business accounts are saved for 93 days.

You must also know this, if your recycle bin is full, the oldest file existing on your bin will be deleted after 3 days.

HOW TO RESTORE FILES

The following are the steps to restore your file on OneDrive:

1. Go to the OneDrive web page
2. Select "Settings"
3. After that, you will select "Restore your OneDrive"
4. After doing that, select a date
5. You can make it 14 days or more or less.
6. After selecting a date, click "Restore "
7. There will be a pop up asking you "Are you sure you want to Restore your OneDrive" Just click "restore"

Your file will automatically be restored.

INDEX

B

C

D

E

F

G

H

I

J

K

L

M

N

O

P

Q

R

S

T

U

V

W

Z

This book has been produced in line with the EU GPSR guidelines about the safety of products.

The General Product Safety Regulation is the European Union's updated framework for ensuring that all consumer products, including books, are safe for consumers.

This book has been printed by Libri Plureos GmbH. The printer has issued safety certificates for the materials - like ink, paper and glue - being used.

The product identifier is: 9789403876320

The author is responsible for the content of the book and had it produced by Bookmundo.

Should there be any questions in regard to the safety of the product, please contact us.

Bookmundo
Delftsestraat 33
3013AE Rotterdam
The Netherlands
info@bookmundo.com